Evaluation of short-term tests for carcinogens

Evaluation of short-term tests for carcinogens

VOLUME 1

Report of the International Programme on Chemical Safety's Collaborative Study on *in vivo* Assays

Editors
John Ashby
Central Toxicology Laboratory, ICI plc
Cheshire, UK
Frederick J. de Serres
Michael D. Shelby
Barry H. Margolin
National Institute of Environmental Health Sciences, North Carolina, USA
Motoi Ishidate, Jr
National Institute of Hygienic Sciences
Tokyo, Japan
George C. Becking
IPCS, Interregional Research Unit
North Carolina, USA

The right of the University of Cambridge to print and sell all manner of books was granted by Henry VIII in 1534. The University has printed and published continuously since 1584.

Published by
CAMBRIDGE UNIVERSITY PRESS
on behalf of the
WORLD HEALTH ORGANIZATION

Published by the Press Syndicate of the University of Cambridge
The Pitt Building, Trumpington Street, Cambridge CB2 1RP
32 East 57th Street, New York, NY 10022, USA
10 Stamford Road, Oakleigh, Melbourne 3166, Australia

First published 1988

Printed in Great Britain at the University Press, Cambridge

British Library cataloguing in publication data

Evaluation of short-term tests for carcinogens:
report of the International Programme on
Chemical Safety's collaborative study on *in vivo* assays.
1. Carcinogenicity testing
I. Ashby, John II. International Programme
on Chemical Safety
599.08′765 RC268.65

Library of Congress cataloging in publication data

Evaluation of short-term tests for carcinogens.
Includes indexes.
1. Carcinogenicity testing. I. Ashby, John.
II. World Health Organization. [DNLM: 1. Biological
Assays. 2. Carcinogens–analysis. 3. Mutagenicity
Tests. 4. Mutagens–analysis. QZ 202 E9231]
RC268.65.E95 1988 616.99′4071 87–22440

ISBN 0 521 34152 3
(Volume 2 ISBN 0 521 34153 1)

TM

Contents: Volume 1

Contributors (Volumes 1 and 2)

Ilse-Dore Adler, Institut für Säugetiergenetik, Gesellschaft für Strahlenund Umweltforschung, D-8042 Neuherberg, Federal Republic of Germany

Rosario Albanese, Pharmaceuticals Division, Imperial Chemical Industries PLC, Alderley Park, Nr. Macclesfield, Cheshire SK10 4TG, England

James W. Allen, Genetic Toxicology Division, U.S. Environmental Protection Agency, Research Triangle Park, North Carolina, 27711, USA

Jeffrey A. Allen, Department of Mutagenesis and Cell Biology, Huntingdon Research Centre Ltd., Huntingdon, Cambridgeshire, PE18 6ES, England

Ole Andersen, Odense University, Institute of Community Health, Department of Environmental Medicine, J.B. Winslows Vej 19, DK-5000 Odense C, Denmark

Diana Anderson, British Industrial Biological Research Association, Woodmansterne Road, Carshalton, Surrey SM5 4DS, United Kingdom

Janos Arany, Institut d'Hygiene et d'Epidemiologie, 14 rue J. Wytsman, B-1050 Brussels, Belgium

Manicketh Aravindakshan, Bio-Medical Group, Bhabha Atomic Research Centre, Bombay, 400085, India

Michael J. Armstrong, Merck Sharp & Dohme Research Laboratories, West point, Pennsylvania, 19486, USA

John Ashby, Imperial Chemical Industries PLC, Central Toxicology Laboratory, Alderley Park, Nr. Macclesfield, Cheshire SK10 4TJ, United Kingdom

Robert A. Baan, Medical Biological Laboratory TNO, P.O. Box 45, 2280 AA Rijswijk, The Netherlands

Lorraine C. Backer, (National Research Council Fellow), Genetic Toxicology Division, U.S. Environmental Protection Agency, Research Triangle Park, North Carolina, 27711, USA

James P. Bakke, SRI International, 333 Ravenswood Avenue, Menlo Park, California 94025, USA

Peter Bannasch, Abt. für Cytopathologie, Institut für Experimentelle Pathologie, Deutsches Krebsforschungszentrum, Im Neuenheimer Feld 280, D-6900 Heidelberg, Federal Republic of Germany

Helmut Bartsch, Division of Environmental Carcinogenesis, International Agency for Research on Cancer, 150 Cours Albert-Thomas, 69372 Lyon Cedex 08, France

Christian L. Bean, Merck Sharp & Dohme Research Laboratories, West Point, Pennsylvania, 19486, USA

Brita Beije, Department of Genetic and Cellular Toxicology, Wallenberg Laboratory, Stockholm University, S-106 91 Stockholm, Sweden

Ron D.L. Bell, Mutagenesis Section, Environmental Health Centre, Department of National Health & Welfare, Tunney's Pasture, Ottawa, Ontario, K1A OL2, Canada

Edilberto Bermudez, Department of Genetic Toxicology Chemical Industry Institute of Toxicology, P.O. Box 12137, Research Triangle Park, North Carolina, 27709, USA

Hyman C. Birnboim, Department of Experimental Oncology,

Ottawa, Regional Cancer Centre, General Division, 501 Smyth Road, Ottawa, Ontario, K1H 8L6, Canada

Jack B. Bishop, Cellular and Genetic Toxicology Branch, Toxicology Research and Testing Program, National Institute of Environmental Health Sciences, P.O. Box 12233, Research Triangle Park, North Carolina, 27709, USA

David H. Blakey, Mutagenesis Section, Environmental Health Centre, Department of National Health & Welfare, Tunney's Pasture, Ottawa, Ontario, K1A OL2, Canada

Stephen D. Blowers, The British Industrial Biological Research Association, Carshalton, Surrey SM5 4DS, United Kingdom

Peggy W. Braden, Biology Division, Oak Ridge National Laboratory, Oak Ridge, Tennessee, 37831, USA

Rolf Braun, Zentralinstitut für Genetik und Kulturpflanzenforschung der Akademie der Wissenschaften der DDR, 4325 Gatersleben, German Democratic Republic

Giorgio Bronzetti, Consiglio Nazionale delle Ricerche, Instituto Di Mutagenesi e Differenzimento, Via Svezia 10, 56100 Pisa, Italy

Gisèle Brun, International Agency for Research on Cancer, 150 Cours Albert Thomas 69372 Lyon Cedex 08, France

Victor E. Bulyzhenkov, Institute of Medical Genetics, Academy of Medical Sciences, Moskvorechie Street 1, Moscow 115 × 78, USSR

Brian Burlinson, Genetic and Reproductive Toxicology Department, Glaxo Group Research Ltd, Ware, Hertfordshire, 5612 ODJ, United Kindgom

Byron E. Butterworth, Chemical Industry Institute of Toxicology, P.O. Box 12137, Research Triangle Park, North Carolina, 27709, USA

Katherine T. Cain, Biology Division, Oak Ridge National Laboratory, Oak Ridge, Tennessee, 37831, USA

Richard D. Callander, Imperial Chemical Industries PLC, Central Toxicology Laboratory, Alderley Park, Nr. Macclesfield, Cheshire, SK10 4TJ, United Kingdom

James Campbell, Environmental Health Research and Testing, Inc., P.O. Box 12199, Research Triangle Park, North Carolina, 27711

Vera E. Campbell, Bionetics, National Center for Toxicological Research, Jefferson, Arkansas, 72079, USA

Angelo carere, Istituto Superiore di Sanita, Viale Regina Elena 299, 00161 Rome, Italy

Bohuslav Cerny, Czechoslovak Academy of Sciences, Institute of Nuclear Biology and Radiochemistry, Videnska 1083, 14220 Prague, Czechoslovakia

Ramesh C. Chaubey, Bio-Medical Group, Bhabha Atomic Research Centre, Bombay, 400085, India

Pawan S. Chauhan, Bio-Medical Group, Bhabha Atomic Research Centre, Bombay, 400085, India

Frank H. Cheng, Radiation Research Laboratory, Department of Radiology, University of Iowa, Iowa City, Iowa, 52242, USA

Susan Chiu, Moe-Biohazard Laboratory, P.O. Box 213, Resources Road, Rexdale, Ontario, M9W 5LI, Canada

Ivan Chouroulinkov, Unité De Čancérogénèse Expérimental Et De Toxicologie Léńetique (E.R. 304) I.R.S.C.C.N.R.S., BP 8, 94802 Villejuif, France

Nancy C. Chromey, Haskell Laboratory for Toxicology and Industrial Medicine, E.I. du Pont de Nemours & Co., Inc., P.O. Box 50, Elkton Road, Newark, Delaware 19714, USA

M. Gillian Clare, Shell Research Ltd., Sittingbourne Research Centre, Sittingbourne, Kent ME9 8AG, United Kingdom

Dean A. Cole, Radiation Research Laboratory, Department of Radiology, University of Iowa, Iowa City, Iowa, 52242, USA

Robert D. Combes, School of Biological Sciences, Portsmouth Polytechnic, King Henry Building, King Henry I Street, Portsmouth PO1 2DY, United Kingdom

Claudio Corsi, Istituto di Mutagenesi e Differenziamento CNR, Via Svezia 10, 56100 Pisa, Italy

Colette Coton, Mammalian Genetics Laboratory, Department of Biology, C.E.N.S.C.K., B-2400 Mol, Belgium

Renata Cozzi, Centro di Genetica Evoluzionistica del CNR, c/o Dipartimento di Genetica e Biologia Molecolare, Città Universitaria – Piazzale A. Moro, 00185 Rome, Italy

Riccardo Crebelli, Istituto Superiore di Sanità, Viale Regina Elena, 299–00161 Rome, Italy

Roderick H. Dashwood, School of Biological Sciences, Portsmouth Polytechnic, King Henry I Street, Portsmouth, PO1 2DY, United Kingdom

Brian J. Dean, 32 The Poles Upchurch, Sittingbourne, Kent ME9 7EX, United Kingdom

Gary M. Decad, Department of Materials Toxicology, IBM Corporation, 5600 Cottle Road, San Jose, California, 95193 USA

Gilbert Decat, Mammalian Genetics Laboratory, Department of Biology, C.E.N.S.C.K., B-2400 Mol, Belgium

Marcella de Ferrari, Istituto Nazionale per la Ricerca sul Cancro, V.le Benedetto XV, 1ϕ – Genova, Italy

Renata Del Carratore, Istituto di Mutagenesi e Differenziamento CNR, Via Svezia 10, 56100 Pisa, Italy

Loretta Lyon Delehanty, Department of Genetic Toxicology, Chemical Industry Institute of Toxicology, P.I. Box 12137, Research Triangle Park, North Carolina, 27709, USA

Frederick J. de Serres, Office of the Director, National Institute of Environmental Health Sciences, P.O. Box 12233, Research Triangle Park, North Carolina, 27709, USA

Nico de Vogel, Dept. of Radiation Genetics & Chemical Mutagenesis, University of Leiden, Sylvius Laboratories, Wassenaarseweg 72, 2333 AL Leiden, The Netherlands

Dennis Dietz, Research Triangle Institute, P.O. Box 12194, Research Triangle Park, North Carolina, 27709, USA

Rose M. Disher, Baylor College of Medicine, Department of Pharmacology, One Baylor Plaza, Houston, Texas, 77030, USA

David J. Doolittle, Toxicology Research Bowman Gray Technical Center, R.J. Reynolds Co., Winstonsalem North Carolina 27102

George R. Douglas, Mutagenesis Section, Environmental Health Centre, Department of National Health & Welfare, Tunney's Pasture, Ottawa, Ontario, K1A OL2, Canada

Frank W. Edens, Department of Poultry Science, North Carolina State University, Raleigh, North Carolina, 27612, USA

Udo H. Ehling, Institute für Säugeziergenetik, Gesellschaft für Strahlen- und Umweltforschung, D-8042 Neuherberg, Federal Republic of Germany

John J. Eldridge, School of Pathology, Middlesex Hospital Medical School, Riding House Street, London, W1P 7LD, United Kingdom

Barry M. Elliott, Genetic Toxicology Section, ICI, Alderley Park, Nr. Macclesfield, Cheshire SK10 4TJ, United Kingdom

Gregory L. Erexson, Environmental Health Research and Testing, Inc., P.O. Box 12199, Research Triangle Park, North Carolina, 27709, USA

Jenny Erna, Institut d'Hygiène et d'Epidémiologie, 14, rue Juliette Wytsman, 1050 Bruxelles, B-Belgique

Josef Exner, Czechoslovak Academy of Sciences, Institute of Nuclear Biology and Radiochemistry, Videnska 1083, 14220 Prague, Czechoslovakia

Rudolf Fahrig, Fraunhofer-Institut für Toxikologie und Aerosolforschung, Nikolai-Fuchs-Str. 1, D-3000 Hannover 61, Federal Republic of Germany

Emiliana Falcone, Istituto Superiore di Sanità, Viale Regina Elena, 299–00161 Rome, Italy

Elena G. Feldt, A.N. Sysin Institute of General and Environmental Hygiene, 10 Pogodinskaya Str., 119121 Moscow, USSR

Mario Fiore, Centro di Genetica Evoluzionstica del CNR, c/o Dipartimento di Genetica e Biologia Molecolare, Città Universitaria – Piazzale A. Moro, 00185 Rome, Italy

Roy Forster, Life Science Research, Roma Toxicology Centre, via Tito Speri 14–00040 Pomezia, Roma, Italy

Manton R. Frierson, Department of Chemistry, Case Western Reserve University, Cleveland, Ohio, 44016, USA

Kazuo Fujikawa, Drug Safety Evaluation Laboratories, Central Research Division, Takeda Chemical Industries Ltd., 17–18 Jusoumachi 2-chome, Yodogawa-ku, Osaka 532, Japan

Chie Furihata, Department of Molecular Oncology, Institute of Medical Science, University of Tokyo, 4–6–1, Shirokanedai, Minato-ku, Tokyo 108, Japan

Alvargo Galli, Istituto di Mutagenesi e Differenziamento CNR, Via Svezia 10, 56100 Pisa, Italy

Sheila M. Galloway, Merck Sharp & Dohme Research Laboratories, West Point, Pennsylvania, 19486, USA

Walderico M. Generoso, Biology Division, Oak Ridge National Laboratory, P.O. Box Y, Oak Ridge, Tennessee, 37831, USA

Howard P. Glauert, McArdle Laboratory for Cancer Research, University of Wisconsin, 450 N. Randall Avenue, Madison, Wisconsin, 53706, USA

Michael Goddard, Mutagenesis Section, Environmental Health Centre, Department of National Health & Welfare, Tunney's Pasture, Ottawa, Ontario K1A OL2, Canada

G. Jay Graepel, Haskell Laboratory for Toxicology and Industrial Medicine, E.I. du Pont de Nemours & Co., Inc., P.O Box 50, Elkton Road, Newark, Delaware 19714, USA

Ulrich Graf, Institut fur Toxicologie, ETH and University of Zurich, Schorenstrasse 16, CH-8603 Schwerzenbach, Switzerland

Elizabeth Griesiger, Department of Pathology, Medical College of Ohio, 3000 Arlington Avenue, Toledo, Ohio 43699, USA

Alison C. Griffin, Immunology Section, ICI PLC, Central Toxicology Laboratory, Alderley Park, Nr. Macclesfield, Cheshire SK10 4TJ, United Kingdom

Hans Jörg Hacker, Abteilung für Cytopathologie, Institut für Experimentelle Pathologie, Deutsches Krebsforschungszentrum Heidelberg, Federal Republic of Germany

Inger Porsch Hällström, Wallenberg Laboratory, University of Stockholm, S-106 91 Stockholm, Sweden

Carol M. Hamilton, SRI International, 333 Ravenswood Avenue, Menlo Park, California, 94025, USA

Barbara L. Harper, Dept. of Preventive Medicine and Community Health, Division of Environmental Toxicology, University of Texas Medical Branch, Galveston, Texas, 77550, USA

David Hart, Imperial Chemical Industries PLC, Central Toxicology Laboratory. Alderley Park, Nr. Macclesfield, Cheshire SK10 4TJ, United Kingdom

George G. Hatch, Toxicology Division, Northrop Services, Inc., Environmental Sciences, P.O. Box 12313, Research Triangle Park, North Carolina, 27709

Makoto Hayashi, Division of Mutagenesis, Biological Safety Research Center, National Institute of Hygienic Sciences, 1–18–1, Kamiyoga, Setagaya-ku, Tokyo 158, Japan

Philip R. Henika, U.S. Department of Agriculture, ARS, Western Regional Research Center, 800 Buchanan Street, Albany, California, 94710, USA

Sydna L. Herren-Freund, U.S. Environmental Protection Agency, Health Effects Research Laboratory, 26 W. St. Clair Street, Cincinnati, Ohio, 45268, USA

R. Marian Hicks, School of Pathology, Middlesex Hospital Medical School, Riding House Street, London WIP 7LD, United Kingdom

Lori A. Hughes, Biology Division, Oak Ridge National Laboratory, Oak Ridge, Tennessee, 37831, USA

John M. Hunt, Department of Pathology and Laboratory Medicine, University of Texas Medical School, P.O. Box 20708, Houston, Texas, 77225, USA

Edith Hüttner, Zetralinstitut für Genetik und Kulturpflanzenforschung der Akademie der Wissenschaften der DDR, DDR-4325 Gatersleben, German Democratic Republic

Ingrid Ingwersen, Institut für Säugetiergenetik, Gesellschaft für Strahlenund Umweltforschung, D-8042 Neuherberg, Federal Republic of Germany

Naomichi Inui, Biological Research Center, Japan Tobacco Inc., 23 Nakogi, Hatano, Kanagawa 257, Japan

Motoi Ishidate, Jr., Division of Mutagenesis, Biological Safety Research Center, National Institute of Hygienic Sciences, 1–18–1, Kamiyoga, Setagaya-ku, Tokyo 158, Japan

Nobuyuki Ito, First Department of Pathology, Nagoya City University Medical School, 1 Kawasumi, Mizuho-cho, Mizuho-ku, Nagoya 467, Japan

Vladimir I. Ivanov, Institute of Medical Genetics, Academy of Medical Sciences, Moskvorechie Street 1, Moscow 115478, USSR

Jorn C. Jensen, National Food Institute, Institute of Toxicology, Morkhojgaard, Morkhoj Bygade 19, 2860 Soborg, Copenhagen, Denmark

Dag Jenssen, Department of Genetic Toxicology, Wallenberg Laboratory, University of Stockholm, S-106 91 Stockholm, Sweden

Karen Jones, Immunology Section, ICI PLC, Central Toxicology Laboratory, Alderley Park, Nr. Macclesfield, Cheshire SK10 4TJ, United Kingdom

Kevin M. Joy, School of Pathology, Middlesex Hospital Medical School, Riding House Street, London, WIP 7LD, United Kingdom

Jana Kalouskova, Czechoslovak Academy of Sciences, Institute of Nuclear Biology and Radiochemistry Videnska 1083, 14220 Prague, Czechoslovakia

Hiroo Katayama, (Visiting Scientist from Saitama Medical School, Japan, Abteilung für Cytopathologie, Institut für Experimentelle Pathologie, Deutsches Krebsforschungszentrum Heidelberg, Federal Republic of Germany

Brian J. Kilbey, Institute of Animal Genetics, University of Edinburgh, King's Building, West Mains Road, Edinburgh, EH9 3JN, United Kingdom

Julia Kilcullen, Imperial Chemical Industries PLC, Central Toxicology Laboratory, Alderley Park, Nr. Macclesfield, Cheshire, United Kingdom, SK10 4TJ

Ian Kimber, Immunology Section, Imperial Chemical Institute PLC, Central Toxicology Laboratory, Alderley Park, Nr. Macclesfield, Cheshire SK10 4TJ, United Kingdom

Ulf Kliesch, Institut für Säugetiergenetik, Gesellschaft für Strahlen- und Umweltforschung, D-8042 Neuherberg, Federal Republic of Germany

Andrew D. Kligerman, Environmental Health Research and Testing, Inc., P.O. Box 12199, Research Triangle Park, North Carolina, 27709, USA

Gillers Klopman, Department of Chemistry, Case Western Reserve University, Cleveland, Ohio, 44016, USA

Sohei Kondo, Department of Fundamental Radiology, Osaka University School of Medicine, Nakanoshima Kita-ku, Osaka 530, Japan

Ludmila M. Konstantinova, Institute of Medical Genetics, Academy of Medical Sciences, Moskvorechie Street 1, Moscow 115478, USSR

Douglas Kornbrust, Merck Sharp & Dohme Research Laboratories, Department of Safety Assessment, West Point, Pennsylvania, 19486, USA

Yulia V. Korogodina, Institute of Medical Genetics, Academy of Medical Sciences, Moskvorechie Street 1, Moscow 115478, USSR

Ronald A. Krol, Department of Family Medicine, Medical College of Ohio, Roche Building, Detroit Avenue, Toledo, Ohio, 43699, USA

Kim Lainson, Institute of Animal Genetics, University of Edinburgh, King's Building, West Mains Road, Edinburgh, EH9 3JN, United Kingdom

Claude Lasne, Unité de Cancérogénèse Expérimentale et de Toxicologie Génétique (ER-304) – I.R.S.C.-C.N.R.S., B.P. No 8, 94802 Villejuif Cedex, France

Paul A. Lefevre, Imperial Chemical Industries PLC, Central Toxicology Laboratory, Alderley Park, Nr. Macclesfield,

Cheshire, SK10 4TJ, United Kingdom
Marvin S. Legator, Dept. of Preventive Medicine and Community Health, Division of Environmental Toxicology, University of Texas Medical Branch, Galveston, Texas, 77550, USA
Alain Leonard, Mammalian Genetics Laboratory, Department of Biology, CEN/SCK, B-2400 Mol, Belgium
Paul A. Lindholm, Radiation Research Laboratory, Department of Radiology, University of Iowa, Iowa City, Iowa, 52242, USA
David M. Logan, Biology Department, York University, North York, Ontario, M3J 1P3, Canada
Erica K.N. Loh, Sri International, 333 Ravenswood Avenue, Menlo Park, California, 94025, USA
Paul H.M. Lohman, Medical Biological Laboratory TNO, P.O. Box 45, 2280 AA Rijswijk, The Netherlands
Nicola Loprieno, Dipartimento Scienze Deu' Ambiente Der Territorio, 56010 S. Piero a Grado, 00185 Pisa, Italy
Carol A. Luke, Medical Department, Brookhaven National Laboratory, Upton, New York 11973, USA
James T. MacGregor, U.S. Department of Agriculture, ARS, Western Regional Research Center, 800 Buchanan Street, Berkeley, California, 94710, USA
Jan Magnusson, Wallenberg Laboratory, University of Stockholm, S-106 91 Stockholm, Sweden
Alexei M. Malashenko, Scientific Research Laboratory of Experimental Biological Models of the Academy of Medical Sciences of the USSR, Moscow Region, USSR
Christian Malaveille, International Agency for Research on Cancer, 150 Cours Albert Thomas, 69372 Lyon Cedex 08, France
Barry H. Margolin, Biometry and Risk Assessment Program, National Institute of Environmental Health Sciences, P.O. Box 12233, MD B3–02, Research Triangle Park, North Carolina, 27709, USA
Tsuneo Masui, First Department of Pathology, Nagoya City University Medical School, 1 Kawasumi, Mizuho-cho, Mizuho-ku, Nagoya 467, Japan
Taijiro Matsushima, Department of Molecular Oncology, Institute of Medical Science, University of Tokyo, 4–6–1, Shirokanedai, Minato-ku, Tokyo 108, Japan
Douglas McGregor, Developmental Toxicology, Inveresk Research International, Ltd., Musselburgh EH121 7UB, United Kingdom
Antonella Melchiori, Istituto Nazionale per la Ricerca sul Cancro, V.le Benedetto XV, 1ϕ – Genova, Italy
Anthony L. Meyer, Shell Research Ltd, Sittingbourne Research Centre, Sittingbourne Kent, ME9 8AG, United Kingdom
Vladimir A. Mglinetz, Institute of Medical Genetics, Academy of Medical Sciences, Moskvorechie Street 1, Moscow 115478, USSR
Valerie M. Miller, Medical Department, Brookhaven National Laboratory, Upton, New York, 11973
Jon C. Mirsalis, Cellular and Genetic Toxicology Department, SRI International, 333 Ravenswood Avenue, Menlo Park, California 94025, USA
Youichi Miyamae, Laboratory of Genetic Toxicology, Hatano Research Institute, Food and Drug Safety Center, 729–5 Ochiai, Hadano, Kanagawa 257, Japan
Mary Morgan, Department of Pathology, Medical College of Ohio, 3000 Arlington Avenue, Toledo, Ohio, 43699, USA
Pasquale Mosesso, Life Science Research, Roma Toxicology Centre, Pomezia, Italy
Bernard Most, Toxicology Division, Northrop Services, Inc., Environmental Sciences, P.O. Box 12312, Research Triangle Park, North Carolina 27709
Gregory Muller, Rohm and Haas Company, Toxicology Department, Spring House, Pennsylvania, 19477, USA
Tetsuro Murota, Laboratory of Genetic Toxicology, Hatano Research Institute, Food and Drug Safety Center, 729–5 Ochiai, Hadano, Kanagawa 257, Japan
Naguib Nashed, Johann Wolfgang Goethe-Universität, Theodor-Stern-Kai 7, Haus 75A, D-6000 Frankfurt/M, Federal Republic of Germany
Steven B. Neal, Toxicology Division, Lilly Research Laboratories, Division of Eli Lilly and Company, Greenfield, Indiana 46140, USA
Angelika Neuhäuser-Klaus, Institut für Säugetiergenetik, Gesellschaft für Strahlenund Umweltforschung, D-8042 Neuherberg, Federal Republic of Germany
Isaac Neuteboom, Dept. of Radiation Genetics & Chemical Mutagenesis, University of Leiden, Sylvius Laboratories, Wassenaarseweg 72, 2333 AL Leiden, The Netherlands
Riccardo Nieri, Istitut di Mutagenesi e Differenziamento CNR, Via Svezia 10, 56100 Pisa, Italy
Daniel J. Nordlund, Department of Family Medicine, Medical College of Ohio, Roche Building, Detroit Avenue, Toledo, Ohio, 43699, USA
Leonard Oppenheimer, Merck Sharp & Dohme Research Laboratories, West Point, Pennsylvania, 19486, USA
Brian G. Ormiston, Medical Department, Brookhaven National Laboratory, Upton, New York, 11973
Dennis A. Pagano, Cellular and Genetic Toxicology Branch, Toxicology Research and Testing Program, National Institute of Environmental Health Sciences, P.O. Box 12233, Research Triangle Park, North Carolina, 27709, USA
Mauro Pala, Istituto Nazionale per la Ricerca sul Cancro – V.le Benedetto XV, 1ϕ – Genova, Italy
Fabrizio Palitti, Centro di Genetica Evoluzionistica del CNR, c/o Dipartimento di Genetica e Biologia Molecolare, Città Universitaria – Piazzale A. Moro, 00185 Rome, Italy
Moreno Paolini, Istituto di Mutagenesi e Differenziamento CNR, Via Svezia 10, 56100 Pisa, Italy
Silvio Parodi, Lab. Cancerogenesi Chimica – Istituto Nazionale per la Ricerca sul Cancro – V.le Benedetto XV, 10-Genova, Italy
David Paton, Imperial Chemical Industries PLC, Central Toxicology Laboratories, Alderley Park, Nr. Macclesfield, Cheshire, SK10 4TJ, United Kingdom
Ladislav Pavlik, Czechoslovak Academy of Sciences, Institute of Nuclear Biology and Radiochemistry Videnska 1083, 14220 Prague, Czechslovakia
Michael A. Pereira, U.S. Environmental Protection Agency, Health Effects Research Laboratory, 26 W. St. Clair Street, Cincinnati, Ohio 45268, USA
Christine E. Phillips, Imperial Chemical Industries PLC, Central Toxicology Laboratory, Alderley Park, Nr. Macclesfield, Cheshire SK10 4TJ, United Kingdom
Henry C. Pitot, McArdle Laboratory for Cancer Research, University of Wisconsin, 450 N. Randall Avenue, Madison, Wisconsin 53706, USA
Jacques Pot-Deprun, Unité de Cancérogénèse Expérimentale et Toxicologie Génétique (ER-304) – I.R.S.C.-CNRS Laboratories de Recherche Appliquée sur le Cancer, BP 8, 94802 Villejuif, France
Neville Pritchard, Imperial Chemical Industries PLC, Central Toxicology Laboratory, Alderley Park, Nr. Macclesfield, Cheshire, SK10 4TJ, United Kingdom
Gregory S. Probst, Toxicology Division, Lilly Research Laboratories, Division of Eli Lilly and Company, Greenfield, Indiana, 46140, USA
Raymond J. Proudlock, Department of Mutagenesis and Cell Biology, Huntingdon Research Centre Ltd., Huntingdon, Cambridgeshire, PE18 6ES, United Kingdom
Claes Ramel, Wallenberg Laboratory, University of Stockholm, S-106 91 Stockholm, Sweden
Marilyn J. Ramsey, SRI International, 333 Ravenswood Avenue, Menlo Park, California, 94025, USA
Kurt Randerath, Baylor College of Medicine, Department of Pharmacology, One Baylor Plaza, Houston, Texas, 77030, USA
Claudio Ricciardi, Istituto Superiore di Sanità, Viale Regina Elena, 299 – 00161 Rome, Italy
Lauren B. Rickard, Haskell Laboratory for Toxicology and Industrial Medicine, E.I. du Pont de Nemours & Co., Inc., P.O. Box 50, Elkton Road Newark, Delaware 19714, U.S.A.

Ruggero Ricordy, Centro di Genetica Evoluzionistica del CNR, c/o Dipartimento di Genetica e Biologia Molecolare, Città Universitaria – Piazzale A. Moro, 00185 Rome, Italy
Elizabeth Riley, Central Toxicology Laboratory, Imperial Chemical Industries PLC, Alderley Park, Nr. Macclesfield, SK10 4TJ, United Kingdom
Kenneth J. Risko, Biometry Risk Assessment Program, National Institute of Environmental Health Sciences, P.O. Box 12233, MD B3–02, Research Triangle Park, North Carolina, 27709, USA
Leslie C. Rogers, Toxicology Division, Northrop Services, Inc., Environmental Sciences, P.O. Box 12313, Research Triangle Park, North Carolina, 27709
Herbert S. Rosenkranz, Department of Environmental Health Sciences, Case Western Reserve University, Cleveland, Ohio, 44016, USA
Annelies H.M. Rotteveel, Dept. of Radiation Genetics & Chemical Mutagenesis, University of Leiden, Sylvius Laboratories, Wassenaarseweg 72, 2333 AL Leiden, The Netherlands
Alison V. Ruck, Shell Research Ltd., Sittingbourne Research Centre, Sittingbourne, Kent, ME9 8AG, United Kingdom
Patrizia Russo, Istituto Nazionale per la Ricerca sul Cancro, Viale Benedetto XV, 1ϕ – Genova, Italy
Haruko Ryo, Department of Fundamental Radiology, Osaka University Medical School, Nakanoshima, Kita-ku, Osaka 530, Japan
Michael F. Salamone, Moe-Biohazard Laboratory, P.O. Box 213, Resources Road, Rexdale, Ontario, Canada M9W 5LI
Charles B. Salocks, Department of Materials Toxicology, IBM Corporation, 5600 Cottle Road, San Jose, California, 95193, USA
Janice L. Sames, Rohm and Haas Company, Toxicology Department, Spring House, Pennsylvania, 19477, USA
Leonardo Santi, Department of Clinical and Experimental Oncology-University of Genoa, V.le Benedetto XV, 1ϕ I–16132, Genoa – Italy
Timothy Schofield, Merch Sharp & Dohme Research Laboratories, West Point, Pennsylvania, 19486, USA
Jörg Schöneich, Zentralinstitut für Genetik und Kulturpflanzenforschung der Akademie der Wissenschaften der DDR, 4325 Gatersleben, German Democratic Republic
Harvey E. Scribner, Rohm and Haas Company, Toxicology Department, Spring House, PA, 19477, USA
Antony G. Searle, MRC Radiobiology Unit, Harwell, Didcot OX11 ORD, United Kingdom
Gary A. Sega, Biology Division, Oak Ridge National Laboratory, P.O. Box Y, Oak Ridge, Tennessee, 37831, USA
Yousuf Sharief, Environmental Health Research and Testing, Inc., P.O. Box 12199, Research Triangle Park, North Carolina, 27711, USA
Jennifer Shaw, Imperial Chemical Industries PLC, Central Toxicology Laboratory, Alderley Park, Nr. Macclesfield, Cheshire, SK10 4TJ, United Kingdom
Michael D. Shelby, Cellular and Genetic Toxicology Branch, Toxicology Research and Testing Program, National Institute of Environmental Health Sciences, P.O. Box 12233, Research Triangle Park, North Carolina 27709, USA
Hiroko Shibuya, Laboratory of Genetic Toxicology, Hatano Research Institute, Food and Drug Safety Center, 729–5 Ochiai, Hadano, Kanagawa 257, Japan
Tohru Shibuya, Laboratory of Genetics, Hatano Research Institute, Food and Drug Safety Center, 729–5 Ochiai, Hadano, Kanagawa 257, Japan
Tracey Smith-Oliver, Chemical Industry Institute of Toxicology, 6 Davis Drive, P.O. Box 12137, Research Triangle Park, North Carolina, 27709, USA
Toshio Sofuni, Division of Mutagenesis, Biological Safety Research Center, National Institute of Hygienic Sciences, 1–18–1, Kamiyoga, Setagaya-ku, Tokyo 158, Japan
Karen L. Steinmetz, SRI International, 333 Ravenswood Avenue, Menlo Park, California, 94025, USA
Dennis A. Stephenson, MRC Radiobiology Unit, Harwell, Didcot, Oxon OX11 ORD, United Kingdom
Reggie H. Stevens, Radiation Research Laboratory, Department of Radiation, University of Iowa, Iowa City, Iowa, 52242, USA
Gary D. Stoner, Department of Pathology, Medical College of Ohio, 3000 Arlington Avenue, Toledo, Ohio, 43699, USA
Jeremy A. Styles, Imperial Chemical Industries PLC, Central Toxicology Laboratory, Alderley Park, Nr. Macclesfield, Cheshire SK10 4TJ, United Kingdom
Krishna Sundaram, Bio-Medical Group, Bhaba Atomic Research Centre, Bombay, 400085, India
Kurt E. Suter, Preclinical Research, Toxicology Department, Sandoz Limited, CH-4002 Basle, Switzerland
Rumiko Takezawa, Department of Molecular Oncology, Institute of Medical Science, University of Tokyo, 4–6–1, Shirokanedai, Minato-ku, Tokyo 108, Japan
Caterina Tanzarella, Centro di Genetica Evoluzionistica del CNR, c/o Dipartimento di Genetica e Biologia Molecolare, Città Universitaria–Piazzale A. Moro, 00185 Rome, Italy
Ad. D. Tates, Dept. of Radiation Genetics & Chemical Mutagenesis, University of Leiden, Sylvius Laboratories, Wassenaarseweg 72, 2333 Al Leiden, The Netherlands
Christopher M. Teaf, University of Arkansas for Medical Sciences, Department of Pharmacology, Division of Interdisciplinary Toxicology, Little Rock, Arkansas, 72205, USA
Raymond R. Tice, Medical Department, Brookhaven National Laboratory, Upton, New York, 11973, USA
Hiroyuki Tsuda, First Department of Pathology, Nagoya City, University Medical School, 1 Kawasumi, Mizuho-cho, Mizuho-ku, Nagoya 467, Japan
C. Kim Tyson, SRI International, 333 Ravenswood Avenue, Menlo Park, California, 94025, USA
Tatijana B. Uvarova, Scientific Research Laboratory of Experimental Biological Models of the Academy of Medical Sciences of the USSR, Moscow Region, USSR
Ruby Valencia, Dept. of Zoology, University of Wisconsin, Zoology Research Building, 1117 West Johnson Street, Madison, Wisconsin, 53706, USA
Rita Vellosi, Istituto di Mutagenesi e Differenziamento CNR, Via Svezia 10, 56100 Pisa, Italy
Demetra A. Vlachos, Haskell Laboratory for Toxicology and Industrial Medicine, E.I. du Pont de Nemours & Co., Inc., P.O. Box 50, Elkton Road, Newark, Delaware 19714, USA
Ekkehart W. Vogel, Dept. of Radiation Genetics & Chemical Mutagenesis, State University of Leiden, Sylvius Laboratories, Wassenaarseweg 72, 2300 RA Leiden, The Netherlands
Irena I.F. Vyskubenko, A.N. Sysin Research Institute of General and Community Hygiene of the USSR AMS, Moscow, USSR
Paul A. Watkins, Pharmaceuticals Division, Imperial Chemical Industries PLC, Mereside, Alderley Park, Nr. Macclesfield, Cheshire SK10 4TJ, United Kingdom
Carol M. Wehr, U.S. Department of Agriculture, ARS, Western Regional Research Center, 800 Buchanan Street, Albany, California, 94710, USA
Barbara Westbrook-Collins, Environmental Health Research and Testing, Inc., P.O. Box 12199, Research Triangle Park, North Carolina, 27709, USA
G. Ashley Wickramaratne, Central Toxicology Laboratory, Imperial Chemical Industries PLC, Alderley Park, Nr. Macclesfield, Cheshire SK10 4TG, United Kingdom
Dieter Wild, Institut für Pharmakologie und Toxikologie, Universität Würzburg, Versbacher Strasse 9, D-8700 Würzburg, Federal Republic of Germany
Anita C. Wilton, Medical Department, Brookhaven National Laboratory, Upton, New York, 11973
Peter K. Working, Department of Genetic Toxicology, Chemical Industry Institute of Toxicology, P.O. Box 12137,

Research Triangle Park, North Carolina, 27709, USA
Friedrich E. Würgler, Institut fur Toxicologie, Schorenstrasse 16, CH-8603 Schwerzenbach bei Zurich, Switzerland
Errol Zeiger, Cellular and Genetic Toxicology Branch, Toxicology Research and Testing Program, National Institute of Environmental Health Sciences, P.O. Box 12233, Research Triangle Park, North Carolina, 27709, USA
Heide Zerban, Abteilung für Cytopathologie, Institut für Experimentelle Pathologie, Deutsches Krebsforschungszentrum Heidelberg, Federal Republic of Germany
Viacheslav S. Zhurkov, A.N. Sysin Institute of General & Environmental Hygiene, 10 Pogodinskaja str., 119121 Moscow, USSR
Annalisa Zunino, Istituto Nazionale per la Ricerca sul Cancro – V.le Benedetto XV, 1ϕ – Genova, Italy

Preface

The International Programme on Chemical Safety (IPCS), a joint effort of the United Nations Environment Programme (UNEP), the International Labour Organization (ILO), and the World Health Organization (WHO), aims at protecting human health and the environment from the adverse effects of the ever-increasing number of chemicals on the market and in the environment. One of the priorities within the IPCS is the development of methodology which can improve the evaluation of risk presented by such chemicals. This includes the coordination of international collaborative studies designed to validate the usefulness of certain methods for the assessment of risk from chemical exposures and to encourage the development of such methods to produce internationally comparable results.

A multitude of short-term assays, both *in vitro* and *in vivo*, and in organisms as diverse as bacteria, insects, animal and human cells, and whole animals, have been proposed as alternatives or supplementary procedures to the traditional long-term rodent cancer bioassays. The world-wide use of these numerous tests, the concern about the choice of one or more appropriate short-term assays for screening, the sensitivity and specificity of these tests, the need for complementary assays within screening methodologies, and the role to be played in testing schemes by short-term *in vivo* assays, supported the need for inter-laboratory collaborative studies on an international scale.

In 1981, the IPCS initiated a large international collaborative study of short-term assays used to identify chemical mutagens and carcinogens. The objective of the first phase of the study was to identify one or more *in vitro* eukaryotic assays suitable for general use as a complementary test to the *Salmonella* mutation assay, whereas the second study of *in vivo* short-term tests was directed towards the development of an extensive data bank of responses obtained from a wide range of *in vivo* assays to two structurally related carcinogen/non-carcinogen pairs. Taken together, these two phases have shown that international collaborative efforts are an effective forum to resolve important issues in toxicology when objectives are clearly defined and an international body such as the IPCS can be accepted to be in a position to offer an effective planning and coordinating mechanism.

The successful completion of this international collaborative effort has been due to many people, including the enthusiastic response of the scientists throughout the world who volunteered to take part in the study. The Steering Group members deserve special thanks for their unselfish efforts on behalf of the IPCS over the last six years. I am particularly grateful to Dr John Ashby for his outstanding scientific contributions to this study and to Dr Frederick de Serres, Chairman of the Steering Group, for his scientific contributions as well as the many hours spent organizing the study to ensure a successful completion. Lastly, I wish to thank the Rapporteurs, Dr Michael Shelby, and Dr Barry Margolin for their valuable efforts.

Dr M. Mercier
Manager, IPCS

PART I

Background and summaries

1

Introduction

The development and validation of short-term assays to test environmental chemicals for mutagenic and carcinogenic potential is generally believed to show promise for the identification and control of human carcinogens. In the paper of McCann *et al.* (1976) a high percentage of carcinogens was shown to give a positive response in the Salmonella assay and a high percentage of noncarcinogens gave a negative response. Similar data were reported by Sugimura *et al.* (1976) and by Purchase *et al.* (1976). One of the problems encountered in the International Program for the Evaluation of Short-term Tests for Carcinogens (IPESTTC) (de Serres and Ashby, 1981) was the high frequency of positive responses *in vitro* shown by 7 of the 14 chemicals that were believed to be noncarcinogens. An interesting aspect of this same study was that these 7 noncarcinogens were predominantly or totally negative in a series of *in vivo* short-term tests. The number of *in vivo* short-term tests in IPESTTC was limited, but the total data base developed in this study suggested that although some noncarcinogens were positive *in vitro* that they were negative in *in vivo* short-term tests. Two good examples of these differences were provided by the noncarcinogens in the benzo(a)pyrene/pyrene (BP/PYR) and 2-acetylaminofluorene/4-acetylaminofluorene (2AAF/4AAF) pairs. Thus, the origin of the present study was IPESTTC and the purpose was to acquire a comprehensive activity profile from a broad range of *in vivo* short-term tests as a step toward understanding how various tissues and genetic endpoints in intact animals respond to chemicals defined as genotoxic *in vitro*.

It was in this context that a collaborative study on *in vivo* short-term tests for genotoxicity and carcinogenicity was proposed by the International Programme on Chemical Safety (IPCS)[1] and the National Institute of Environmental Health Sciences (NIEHS) of the U.S.A. as a participating institution in that program. The general goals and designs of the study were outlined by an ad hoc Working Group[2] who met by invitation of the IPCS in

[1] The International Programme on Chemical Safety (IPCS) is a collaborative program of the World Health Organization (WHO), the International Labour Organization (ILO) and the United Nations Environmental Programme (UNEP).

[2] Participants: Dr J. Ashby, Prof N.P. Bochkov, Dr M. Draper, Dr B.E. Matter, Prof T. Matsushima, Dr F.J. de Serres, Dr M. Shelby and Prof F.H. Sobels.

Geneva, April 30-May 1, 1981. The plans were consolidated by and IPCS Working Group[3], who met in Geneva, November 13–14, 1981 (Ashby *et al.*, 1983). The subsequent coordination of the *in vivo* study was the responsibility of a Steering Committee derived primarily from the Working Group (F. de Serres, Chairman; J. Ashby, M. Ishidate, Jr., B. Margolin, M. Shelby, M. Draper, and G. Becking, IPCS Secretariat). The *in vivo* study was launched in March 1983 with the shipment of samples of each of the 4 test chemicals to 61 investigators in 15 different countries. With the completion of the *in vitro* study (Ashby *et al.*, 1985) at a meeting of investigators held at St. Simons Island, Georgia, on October 23–25, 1983, full attention was given to the completion of the *in vivo* study. During 1983, additional investigators were added to the study to provide data on new *in vivo* assays. Because the total number of investigators had increased to about 90, 8 Coordinators[4] were recruited from the *in vivo* investigators to oversee the work being conducted by investigators performing identical or similar assays. As whole animal bioassays are often more difficult and time-consuming to perform that *in vitro* assays and because of the need to establish an appropriate series of doses for the test chemicals, the Coordinators were in close communication with the investigators in each group. The Steering Committee met with the Coordinators on January 12–13, 1984, in Brussels to review the progress and problems associated with each group of assays[5]. A plan was developed to enter the data into a computer at NIEHS for statistical analysis and for comparison of data from different laboratories performing the same assay. At that time it was expected that studies of the 4 test chemicals would be completed by October 1984 and a meeting of investigators could be planned for February 1985. By September 1984 it was evident that additional time would be required for many of the investigators to complete their evaluation of the 4 test chemicals. A second meeting of the Steering Committee and Coordinators was planned for November 14–17, 1984; the purpose was to review the status of the current data base, to obtain estimated completion dates for individual investigators, to discuss problems encountered with particular assays, and to set a time for the meeting of investigators in 1985. In preparation for the November 1984 meeting, all of the Coordinators prepared status reports that included the progress of each investigator. These were reviewed in detail and in some cases investigators were eliminated where it was evident that they would not be able to provide the data initially commissioned. Two new Coordinators (W. Generoso and R. Fahrig) were appointed to oversee the work in the rodent dominant lethal test and the mouse coat colour spot test, respectively. During 1984 several new investigators joined the study bringing the total number of primary investigators to 97 from 16 countries. After this review it was decided, in a subsequent meeting of the Steering Committee on November 18, 1984, that there would be a sufficient number of experiments completed during the Spring of 1985, to plan a meeting of investigators in May 1985. A deadline of March 1, 1985, was established for receipt of the final reports from each investigator by the Coordinators with a deadline of March 31, 1985, for Coordinator acceptance of these final reports. Drafts of final manuscripts were requested at the time of the meeting of investigators for purposes of review and interlaboratory comparisons.

The meeting of investigators was held May 15–21, 1985, at Cap d'Adge, France, and was organized by the French government and cosponsored by the French Ministry of Health, IPCS and the Commission of European Communities (CEC). At the meeting of investigators, representatives from each of the laboratories performing an assay met to prepare a series of reports: (1) assay system group reports, (2) summary reports on the germ cell assays and the liver-specific assays, and (3) reports to summarize the total data base on each pair of compounds BP/PYR and 2AAF/4AAF. Draft reports were prepared during the course of the meeting and finalized during the next two months. A meeting of the Editorial Board[5] was held at NIEHS from July 24–August 6, 1985. The purpose was to review the work group and investigator reports, and to develop a schedule for the completion of those reports and for the preparation of the Introduction and Overview for the publication of the *in vivo* study.

The organization of this study was financed primarily by the IPCS, together with some of its participating institutions. Meetings of the Steering Committee and Coordinators in Brussels were financed by the government of Belgium and their National Institute of Hygiene and Epidemiology. The funding of the work performed by the individual investigators was provided, in the majority of cases, by the individual investigators who incorporated the work into their research programmes. This could only occur with the goodwill and belief in the project by the senior management of the 97 laboratories involved from universities, research institutes, and industrial research facilities throughout the world. In addition, a number of governments that support the IPCS provided additional financial assistance for this study. These include the governments of Belgium, Italy, Japan, The Netherlands, the United Kingdom and the United States. Special recognition is due the U.K. Department of Health and Social Security for underwriting the cost of the rat carcinogenicity studies of 2AAF and 4AAF.

Some 97 principal investigators have taken part in the present study. In the main part, these investigators have had other collaborators who are listed as joint authors on the investigator reports in this book. In addition, most of the *in vivo* short-term tests currently available are represented.

Current strategies for the detection of potential human carcinogens and mutagens often rely on the

[3] Participants: Dr J. Ashby, Dr G.R. Douglas, Dr M. Draper, Dr M. Ishidate, Dr A. Leonard, Dr N. Loprieno, Dr B.E. Matter, Prof T. Matsushima, Dr R. Montesano, Dr F.J. de Serres, Dr M. Shelby, Prof F.H. Sobels, Dr M. Stoltz and Dr M. Waters.

[4] Coordinators: Dr E.W. Vogel, Dr W.M. Generoso, Dr. I.-D. Adler, Dr D. Wild, Dr T. Tice, Dr B.E. Butterworth, Dr D. McGregor, and Dr M. Salamone.

[5] Editors: J. Ashby, F.J. de Serres, M.D. Shelby, B.H. Margolin, M. Ishidate, Jr., G.C. Becking.

Table 1. *Assay systems and endpoints employed in the IPCS collaborative study on short-term in vivo tests for mutagens and carcinogens*

I. CYTOGENETIC ASSAYS
- Chromosomal Aberrations
  - Bone marrow – mouse, rat, Chinese hamster
  - Ascites tumor cells – mouse
- Micronuclei
  - Bone marrow – mouse, rat
  - Blood – mouse
  - Foetus – mouse
- Sister-Chromatid Exchange
  - Bone marrow – mouse, rat, Chinese hamster
  - Blood – mouse
  - Intestinal epithelium – Chinese hamster

II. LIVER SPECIFIC ASSAYS
- Initiation/Promotion
  - GGT positive foci – rat
  - Enzyme altered foci – rat
- Unscheduled DNA Synthesis – mouse, rat
- Increased S-phase Cells – mouse, rat
- Cytogenetics
  - Chromosomal aberrations – rat
  - Micronuclei – rat
  - Sister-chromatid exchange – rat
  - Ploidy changes – mouse, rat
- Primary DNA Effects
  - Single strand breaks – rat
  - Protein crosslinks – rat
  - Unwinding – mouse
- DNA Adducts – (see Miscellaneous Assays)

III. MISCELLANEOUS ASSAYS
- Carcinogenesis – mouse, rat, quail
- Sebaceous Gland Suppression – mouse
- Epithelial Hyperplasia – mouse
- Ploidy Changes in Hepatocytes – mouse, rat
- Lung Cell 6-Thioguanine Resistant Mutants – Syrian hamster
- DNA Adducts
  - ELISA – liver, spleen, blood – rat
  - Radiolabelled chemicals – liver – rat
  - Post-labelling – liver, lung, kidney – mouse, rat
- Immunotoxicity
  - Cytotoxic T-cells – rat
  - Natural killer cells – rat
- Transformation
  - Peritoneal macrophages – rat
- Host-Mediated Assay
  - Mouse/Saccharomyces
  - Mouse/Salmonella
- Urine Mutagenicity – Salmonella

IV. SPOT TEST
- Coat Colour Spots – mouse
- Melanocyte Mutants – mouse

V. SPERM ASSAYS
- Dominant Lethal – mouse
- Sperm Morphology – mouse, rat
- Unscheduled DNA Synthesis – mouse, rat

VI. DROSOPHILA TESTS
- Germ Cells
  - Sex-linked recessive lethal
  - Chromosome loss
- Somatic Cells
  - Mutation and recombination – wing spots, eye spots

development of short-term *in vitro* and *in vivo* genotoxicity assays. A recent tabulation of the *in vitro* data from genotoxicity tests on 70 chemicals classified as noncarcinogens in NCI/NTP two-year rodent bioassays (Shelby and Stasiewicz, 1984) shows that substantial numbers of noncarcinogens give a positive response in one or more *in vitro* assays.

That paper alerted us to the more general problem of how to cope with the problem of "false positives" in mass-screening programs that utilize *in vitro* assays. The clue derived from IPESTTC – that noncarcinogens which are active *in vitro* may be inactive *in vivo* – has provided the basis for the present study. Genotoxicity test data generated for new chemicals include data from both *in vitro* and *in vivo* assays and it is hoped that the present study will provide a basis for interpretation of what appears to be a discordant test data base.

1 Assays and endpoints

The studies conducted as a part of this collaborative effort range from bacterial tests carried out to confirm the *in vitro* mutagenicity of the four test chemicals to long-term rodent tests of the carcinogenicity of 2AAF and 4AAF. The vast majority of studies addressed the question of whether or not there is evidence of interaction between a test chemical and the DNA of rodent somatic cells. Numerous endpoints were employed and tissues or cell types ranging from skin to germ cells have been investigated and are summarized in Table 1. Specific information on assays and protocols is available in the investigators' reports.

2 Cytogenetics

The three most commonly employed cytogenetic endpoints – chromosomal aberrations, micronuclei, and sister-chromatid exchanges – were reported on by several laboratories in the study. Ten sets of bone marrow chromosomal aberration data were submitted: seven from mouse, two from Chinese hamster, and one from rat. Chromosomal aberration data from mouse ascites cells and liver epitheloid cells were each reported by one investigator. There were 21 reports on the induction of micronuclei: 15 from mouse bone marrow, three from mouse blood including a report on foetal blood, two on rat bone marrow, and one from rat hepatocytes. Ten SCE studies were carried out with six done in mouse bone marrow. There was one report each on rat bone marrow, mouse lymphocytes, Chinese hamster bone marrow, Chinese hamster intestinal epithelium and liver epitheloid cells. The majority of studies were conducted using gavage administration of the test chemicals, but as is seen in Table 1 of the Overview Chapter, intraperitoneal injections were also used and permit a limited comparison of the influence of different routes of exposure.

3 Liver assays

The liver, a major site of metabolism and frequently the organ in which tumors are induced, was the subject of a wide range of studies. The largest number of laboratories reported data on unscheduled DNA synthesis (UDS) in the mouse (2) or rat (9). Two of the rat studies were unique in that one reported data from weanling rats while a second studied neonates exposed by lactation from treated dams. An endpoint related to UDS that received considerable attention in this study was the frequency of S-phase hepatocytes, for which seven laboratories reported results.

As pointed out in the previous section, cytogenetics data were reported for aberrations and SCE in epitheloid cells and micronuclei in hepatocytes. In addition, the ratio of diploid to tetraploid hepatocytes was reported for both rats and mice.

The induction of DNA strand breaks was studied by both the alkaline elution method and by the fluorimetric analysis of DNA unwinding. The induction of DNA–DNA and DNA–protein crosslinks was studied by one laboratory.

Six laboratories investigated the initiation or promotion activity of the test chemicals by assaying for altered foci in the livers of treated animals. The studies included gamma-glutamyl transpeptidase-positive foci as well as other enzyme or storage properties, a test for promotion of liver tumours, and a novel method involving the colonisation of host's liver by hepatocytes from a treated animal.

4 Miscellaneous assays

There were a number of studies conducted that did not easily fit into the above categories, so they were grouped under this heading.

Three studies were carried out to address directly the carcinogenicity of the four chemicals. Two-years rat cancer studies were initiated on 2AAF and 4AAF. All four chemicals were tested in the mouse lung adenoma assay as well as the quail study where in ovo treatment was followed by histopathology at 12 months.

Effects on mouse skin were studied by sebaceous gland suppression and epithelial hyperplasia. Evidence of DNA damage in the rat forestomach was investigated using unscheduled DNA synthesis as an endpoint. Induction of gene mutation was studied by determination of 6-thioguanine resistant lung cells from Syrian hamsters.

Three methods were used to study the binding of the test chemicals to DNA from the liver, lung, kidney, spleen, or blood. These included the use of radiolabelled test chemicals, ^{32}P-post-labelling, and ELISA (Enzyme-linked Immunosorbent Assay).

Effects of the four chemicals on the immune system were studied by methods employing T-cells or natural killer cells of rats.

Three laboratories conducted studies of determine the effect of test chemicals on the ability of peritoneal macrophages from treated animals to grow in soft agar.

Two general types of studies were carried out using microbes as indicator organisms for mutagenicity – the host-mediated assay and urine mutagenicity. The mouse was used as the host organism and either yeast or Salmonella were recovered from liver, lung, or kidney. In yeast, mutation, gene conversion, or mitotic recombination were the endpoints monitored; and for Salmonella, reversion of the histidine locus. Urine or urine extract from rats and mice was assayed for the ability to revert histidine auxotrophs of Salmonella.

Five laboratories employed the mouse spot test to detect the induction of mutations in somatic cells. Three laboratories used coat color analysis while two assayed microscopically for mutant melanocytes.

Although not intended as assays to detect potential carcinogens, rodent germ cell assays were included in the study to give a more complete profile of the *in vivo* activity of the four test chemicals.

Dominant lethal tests were conducted in male and female mice and in male rats. The induction of abnormally shaped sperm was also investigated in both species, and evidence of DNA damage, as determined by unscheduled DNA synthesis, was studied in rat spermatocytes and in mouse sperm.

Two classical tests for germ cell damage in Drosophila were conducted. Five laboratories reported results from the sex-linked recessive lethal assay while two reported on the chromosome loss test. Three sets of data were submitted for three separate systems that detect the induction of genetic damage in somatic cells. Two systems monitor for eye spots while the third detects sectors on the wing of Drosophila treated in the larval stage.

5 Computerization of the study database

The Steering Committee recognized, early in the course of the collaborative studies, the need for a central repository for the test data generated. Computerization of such a study's data is appealing, despite the substantial cost involved, because: (a) it provides an effective management tool with which to monitor study progress, and more importantly, (b) it yields a compact, highly functional data repository that facilitates statistical analysis. Both the *in vitro* and *in vivo* databases were computerized at the National Institute of Environmental Health Sciences. The computerization was performed by Ms T. Nieman under the supervision of B.H. Margolin.

The possibilities for extensive analyses provided by computerization are clearly discernible in Chapter 5 of this book. There, Margolin and Risko have focused on the rat hepatocyte UDS and mouse bone marrow assays, and:

- (i) assessed sources of assay variability, such as strain and laboratory;
- (ii) devised appropriate statistical analyses, and applied them uniformly to all data from each assay; and
- (iii) began exploration of potency as a meaningful consideration in the intercomparison of results from different laboratories, or with different chemicals.

As experience accumulates with computerized databases, it is increasingly clear that their possibilities for extensive analysis warrant their cost of construction.

6 Evaluation and publication of results

This book contains the complete results of the *in vivo* study of benzo(a)-pyrene, pyrene, 2-acetylaminofluorene, and 4-acetylaminofluorene. Included are chapters submitted by each of the participating investigators and these include methodologies, data, and investigators' interpretations of results. Overview chapters on groups of similar assays present the outcome of discussions held at the meeting of investigators in May 1985. More comprehensive overview chapters were developed following that meeting. Three of these deal indepth with the overall results from specific organs or tissues: bone marrow – Chapter 6; liver – Chapter 48; germ cells – Chapter 100. Total results obtained with each of the chemical pairs are presented and discussed in Chapter 3 (BP/PYR) and Chapter 4 (2AAF/4AAF). Issues involved in the statistical analysis of data obtained from in vivo assays are discussed in Chapter 5 and focus on analysis of bone marrow micronucleus and liver unscheduled DNA synthesis data. Chapter 2 presents an overview of the results of the entire study and conclusions that were reached based on those results.

Finally, four appendices provide important background information that underlies the concept and conduct of this study. The appendices present the source and purity of the test chemicals, a review of the evidence for their *in vitro* genotoxicity a presentation of the evidence for their carcinogenicity, and an attempt to predict their carcinogenicities based on structural characteristics and results of genetic toxicity studies.

This study was conducted with the ultimate purpose of improving our ability to protect human health from the potentially harmful effects of environmental chemicals. To this purpose, the further evaluation and use of the data presented herein is encouraged.

7 Acknowledgements

The editors are deeply indebted to the following individuals for their assistance in the preparation of many of the manuscripts that comprise this book: Wendy Singleton, Edith Akyea-Mensah, Janine Sandoz, Mona Rao, Kathy Foster, Joyce Arnold, and Betty Linthicum. Special thanks are due to Isabelle Naegelen for her efforts in the organization and conduct of the meeting of investigators and to Elizabeth Von Halle for technical editing of the manuscripts. We are particularly grateful to Dr Barry Margolin and Dr Kenneth Risko for the creation and technical realization of the many figures in Chapter 2.

8 References

Ashby, J., F.J. de Serres, M. Draper, M. Ishidate, Jr., B.H. Margolin, B.E. Matter and M.D. Shelby (Eds) (1985) Evaluation of Short-term Tests for Carcinogens: Report on the International Programme on Chemical Safety's Collaborative Study on *in vitro* Assays. *Progress in Mutation Research*, Vol. 5, Elsevier/North Holland, Amsterdam.

de Serres, F.J. and J. Ashby (Eds.) (1981) Evaluation of Short-term Tests for Carcinogens: Report of the International Collaborative Programme. *Progress in Mutation Research*, Vol. 1, Elsevier/North Holland, Amsterdam.

Ashby, J., F.J. de Serres, M. Draper, M. Ishidate, Jr., B.E. Matter, and M.D. Shelby (1983) The two IPCS collaborative studies on short-term tests for genotoxicity and carcinogenicity. *Mutation Res.*, **109**, 123–126.

McCann, J. and B.N. Ames (1976) Detection of Carcinogens as Mutagens in the Salmonella/Microsome Test: Assay of 300 Chemicals: Discussion. *Proc. Natl. Acad. Sci. USA.*, **73**, 950–954.

Purchase, I.F.H., E. Longstaff, J. Ashby, J.A. Styles, D. Anderson, P.A. LeFevre and F.R. Westwood (1976) Evaluation of six short-term tests for potential carcinogenicity and recommendations for their use. *Nature* **264**, 624–627. Also *Br. J. Cancer*, **37**, 873–959.

Shelby, M.D. and S. Stasiewicz (1984) Chemicals Showing No Evidence of Carcinogenicity in Long-term Two Species Rodent Studies: The Need for Short-term Test Data. Environ. Mutagen., **6**, 871–878.

Sugimura, T., S. Sato, M. Nagao, T. Yahagi, T. Matsushima, Y. Seino, M. Takeuchi and T. Kawachi (1976). Overlapping of carcinogens and mutagens. In *Fundamentals in Cancer Prevention*, P.B. Magee, Ed.: 191–215. University of Tokyo. Tokyo/University Park Press, Baltimore, Md.

John Ashby, Michael D. Shelby[1] and
Frederick J. de Serres[1]
Imperial Chemical Industries PLC, Central Toxicology Laboratory, Alderley Park, Cheshire (United Kingdom) and[1] National Institute of Environmental Health Schences, Research Traingle Park, North Carolina, USA

2

The IPCS in vivo collaborative study: overview and conclusions

1 Introduction

This discussion is concentrated on those issues of major importance to the effective use of whole mammal genotoxicity assays in the identification of potential human carcinogens and mutagens. This focusing has been enabled by the separate consideration of allied topics in a series of review chapters and appendices. Thus, the source and purity of benzo(a)pyrene (BP), pyrene (PYR), 2-acetylaminofluorene (2AAF) and 4-acetylaminofluorene (4AAF), the 4 chemicals employed in this study, are described in Appendix 1. Confirmation of their general genotoxicity *in vitro* is outlined in Appendix 2, and a discussion of their carcinogenic status is presented in Appendix 3. The complete database derived from the study is shown in the present Chapter (Table 1). Detailed discussions of the data as they relate to the performance of the several classes of assays are provided in the Summary Reports, and two chapters are devoted to an evaluation of the database as it illuminates the relative genotoxicity *in vivo* of BP and PYR, and 2AAF and 4AAF (Chapters 4 and 5). Chapter 3 contains a discussion of the statistical handling of data and emphasizes, by example, the pressing need to place positive test response criteria on a logical foundation. Chapters 3, 4 and 5 are necessary complements to the present overview – their separation was enforced by the magnitude of the database assembled and the consequent problems posed when attempting an overall evaluation and summary.

Chapters 4 and 5 are unique in collaborative studies. Towards the end of the final meeting of investigators, Dr R Forster was asked to lead a discussion of the total genotoxicity database generated for BP and PYR, and to present the conclusions in this volume (Chapter 4). Similarly, Dr R. Combes undertook to review the database for 2AAF and 4AAF (Chapter 5). These two authors were not involved in the design or planning of the study, although they had both presented data as investigators. The purpose of these two chapters is to enable two independent viewpoints regarding the general significance of this study to be presented. To optimise this initiative, no editorial influence was exercised by the editors over these 2 chapters. This inevitably leads to differences in emphasis, and sometimes to differences in conclusions, among Chapters 2–5. But this adds vitality to the study and emphasises the pressing need for all involved in genetic toxicology to strive towards a

Table 1. *Qualitative results of the study as reported in the individual investigator chapters. These results occasionally differ from those discussed in the several summary chapters because, in some cases, workgroups applied common test criteria to investigators' data resulting in a change in the qualitative conclusion. The symbols used in this Table were:?, representing data insufficient to draw firm conclusions; this is distinct from the weak positive (pink) designation; o, ip, and der represent use of oral, intraperitoneal, and dermal routes of exposure, respectively; and * represents studies which have yet to be reported.*

IPCS CSSTT *IN VIVO* STUDY — SUMMARY OF QUALITATIVE RESULTS

	ASSAY	CHAPTER NUMBERS	INVESTIGATOR	BP	PYR	2AAF	4AAF	ROUTE OF EXPOSURE
1.	CYTOGENETICS							
1.1	Chromosomal aberrations							
1.1.1	Mouse bone marrow	8	I.-D. Adler	Positive	Negative			o
			,,			Positive	?	ip
1.1.2	,,	9	V. Zhurkov	Positive	Negative			o
			,,			Negative	Negative	ip
1.1.3	,,	10	A. Leonard	Positive	Negative	Negative	Negative	o
1.1.4	,,	11	B. Harper	Positive	Negative	Negative	Negative	o
1.1.5	,,	12	A. Malashenko	Positive	Negative	Positive	?	o
1.1.6	,,	13	E. Huttner	Positive		Negative	Negative	o
1.1.7	,,	14	C. Luke	Positive	Negative	?	Negative	ip
1.1.8	Mouse-ascites	13	E. Huttner	Positive		Negative	Negative	o
1.1.9	Rat bone marrow	15	R. Albanese	Positive	Negative			o
			,,			?	Negative	ip
1.1.10	Chinese hamster bone marrow	16	J. Pot-DePrun	Positive	Negative	Positive	Weak positive	o
1.1.11	,,	17	J. Allen	Weak positive	Negative	Weak positive	Negative	o
1.2	Micronuclei							
1.2.1	Mouse bone marrow	19	D. Wild	Positive	Negative	Positive	Negative	o
1.2.2	,,	20	D. Blakey	Positive	Negative	Positive	Negative	o
1.2.3	,,	21	I. Chouroulinkov	Positive	Negative	Positive	Negative	o
1.2.4	,,	22	B. Harper	Positive	Negative	Negative	Negative	o
1.2.5	,,	23	M. Hayashi	Positive	Negative	Positive	Negative	o
1.2.6	,,	24	V. Ivanov	Positive	Weak positive	Positive	Weak positive	o
1.2.7	,,	25	J. Jensen	Positive	Negative			o
1.2.8	,,	26	D. Jenssen	Positive	Negative	Positive	Negative	o
1.2.9	,,	27	U. Kliesch	Positive	Negative			o
			,,			Positive	Negative	ip
1.2.10	,,	28	C. Coton	Positive	Negative	Positive	Negative	o
1.2.11	,,	29	M. Salamone	Positive	Negative	Positive	Negative	ip
1.2.12	,,	29	M. Salamone	Positive	Negative			o
1.2.13	,,	30	J. Styles	Positive	Negative	Positive	Negative	o
			,,	Positive	Negative	Positive	Negative	ip
1.2.14	,,	31	P. Chauhan	Positive	Negative	Positive	Negative	ip
1.2.15	,,	32	K. Suter	Positive	Negative	Positive	Negative	o
1.2.16	,,	33	A. Tates	Positive	Negative	Positive	Negative	o
1.2.17	Mouse blood — weanling	34	J. MacGregor	Positive	Negative	Positive	?	o
1.2.18	,, — fetal	34	,,	Positive	Negative	Negative	Negative	o
1.2.19	,, — maternal	34	,,	Positive	Negative	Negative	Negative	o
1.2.20	,, — fetal	32	K. Suter	Positive	Negative	Positive	Negative	o
1.2.21	Rat bone marrow	35	B. Beije			Positive	Negative	o
1.2.22	,,	36	P. Watkins	Positive	Negative			o
			,,			Positive	Weak positive	ip
1.3	Sister chromatid exchange							
1.3.1	Mouse bone marrow	38	R. Tice	Positive	Negative	Positive	Weak positive	ip
1.3.2	,,	39	J. Allen	Positive	Weak positive	Positive	Weak positive	o

Positive (red circle) Negative (blue circle) Weak positive (pink circle)

Table 1 (*Cont.*)

IPCS CSSTT *IN VIVO* STUDY – SUMMARY OF QUALITATIVE RESULTS

	ASSAY	CHAPTER NUMBERS	INVESTIGATOR	BP	PYR	2AAF	4AAF	ROUTE OF EXPOSURE
1.3.3	,,	40	A. Tates	Positive	Negative	Positive	Weak positive	o
1.3.4	,,	41	S. Galloway	Positive	Weak positive	Positive	Weak positive	ip
1.3.5	,,	42	F. Palitti	Positive	Negative	Positive	Negative	o
1.3.6	,,	43	A. Zunino	Positive	Negative	Weak positive	Negative	o
1.3.7	Rat bone marrow	44	D. Vlachos	Positive	Weak positive	Positive	Weak positive	o
1.3.8	Mouse lymphocytes	45	A. Kligerman	Positive	Negative	Positive	Negative	o
1.3.9	Chinese hamster bone marrow	46	S. Neal	Positive	Negative	Positive	Negative	ip
			,,	Positive	Negative	Positive	Negative	o
1.3.10	Chinese hamster – intest. epithel.	46	S. Neal	Positive	Negative	Positive	Negative	o
2.	LIVER SPECIFIC ASSAYS							
2.1	Initiation/promotion							
2.1.1	Enzyme altered foci	50	P. Bannasch			Negative	Negative	o
2.1.2	GGT positive foci	51	M. Pereira	Positive	Negative	Positive	Negative	o
2.1.3	GGT positive foci	52	H. Glauert	Positive	Negative	Positive	Positive	o
2.1.4	GGT positive foci – promotion	53	N. Ito	Positive	Negative	Positive	Weak positive	o
2.1.5	GGT positive foci	54	J. Hunt	Negative	Negative	Positive	Negative	o
2.1.6	,,	73	M. Hicks			Positive	Negative	o
2.2	Unscheduled DNA synthesis							
2.2.1	UDS in mouse liver	56	B. Butterworth	Negative	Negative	Negative	Negative	o
2.2.2	UDS in rat liver	56	B. Butterworth	Negative	Negative	Positive	Negative	o
2.2.3	,,	57	J. Ashby			Positive	Weak positive	o
2.2.4	,,	58	J. Mirsalis	Negative	Negative	Positive	Weak positive	o
2.2.5	UDS in mouse liver	58	J. Mirsalis			Negative	Negative	o
2.2.6	UDS in rat liver	59	B. Beije			Positive	Negative	o
2.2.7	,,	60	C. Salocks			Positive	Weak positive	o
2.2.8	,,	61	D. Kornbrust			Positive	Negative	o
2.2.9	,,	62	D. Doolittle			Positive	Negative	o
2.2.10	UDS in weanling rat liver	63	B. Burlinson	Negative				o
2.2.11	UDS – neonate rat, lactation exposure	61	D. Kornbrust			Weak positive		o
2.3	S-phase hepatocytes							
2.3.1	Rat	56	B. Butterworth	Weak positive			Positive	o
2.3.2	Mouse	56	B. Butterworth				Weak positive	o
2.3.3	Rat	58	J. Mirsalis	Negative	Negative	Positive	Positive	o
2.3.4	Mouse	58	J. Mirsalis			Weak positive	Positive	o
2.3.5	Rat	59	B. Beije				Positive	o
2.3.6	Rat-weanling	63	B. Burlinson	Positive				o
2.3.7	Rat	60	C. Salocks			Negative	Positive	o
2.3.8	,,	62	D. Doolittle			Negative	Positive	o
2.3.9	,,	61	D. Kornbrust				Positive	o
2.4	Cytogenetic analysis							
2.4.1	Aberrations in liver epitheloid cells	65	M. Clare	Weak positive	Negative	Weak positive	Negative	o
2.4.2	SCE in liver epitheloid cells	65	M. Clare	Weak positive	Negative	?	Negative	o
2.4.3	Micronuclei in hepatocytes	66	A. Tates	Negative	Negative	Positive	Negative	o
2.5	DNA strand breaks							
2.5.1	Alkaline elution	68	E. Bermudez	Negative	Negative	Negative	Negative	o

Positive   Negative   Weak positive

Table 1 (*Cont.*)

IPCS CSSTT *IN VIVO* STUDY — SUMMARY OF QUALITATIVE RESULTS

	ASSAY	CHAPTER NUMBERS	INVESTIGATOR	BP	PYR	2AAF	4AAF	ROUTE OF EXPOSURE
2.5.2	Alkaline elution	69	B. Elliott	Negative	Negative	Negative	Negative	o
2.5.3	,,	70	P. Russo	Negative	Weak positive	Positive	Positive	o
2.5.4	DNA-DNA / DNA protein crosslinks	70	P. Russo	Positive	Negative	Positive		ip
2.5.5	DNA unwinding	71	H. Birnboim	Negative	Negative	Positive	Negative	o
2.5.6	,, (with araA)	71	H. Birnboim	Positive	Negative	Positive	Negative	o
3.	MISCELLANEOUS ASSAYS							
3.1	Carcinogenesis							
3.1.1	2-year rat study	73	M. Hicks			Positive	*	o
3.1.2	Mouse lung adenoma	74	G. Stoner	Positive	Negative	Positive	Negative	ip
3.1.3	Quail egg	75	G. Hatch	*	*	*	*	
3.2	Supplementary assay							
3.2.1	Rat forestomach UDS	76	C. Furihata	?	Negative	Negative	Negative	o
3.2.2	Sebaceous gland suppression	77	C. Lasne	Positive	Negative			der
3.2.3	Epithelial hyperplasia	77	C. Lasne	Positive	Negative			der
3.2.4	Rat hepatocyte ploidy	78	J. Styles	Positive	Negative	Positive	Negative	o
3.2.5	Mouse hepatocyte ploidy	78	J. Styles	Positive	Negative	Positive	Negative	o
3.2.6	Syrian hamster 6-TG resistant lung cells	79	N. Inui	Positive	Negative	?	Negative	ip
3.2.7a	32P-post labelling — rat liver	80	R. Disher	Positive	Negative	Positive	Weak positive	ip
3.2.7b	32P-post labelling — rat lung	80	R. Disher	Positive	Negative	Positive	Weak positive	ip
3.2.7c	32P-post labelling — mouse liver	80	R. Disher	Positive	Negative	Positive		ip
3.2.7d	32P-post labelling — mouse lung	80	R. Disher	Positive	Negative	Positive		ip
3.2.7e	32P-post labelling — mouse kidney	80	R. Disher	Positive	Negative	Positive		ip
3.2.8	Radiolabelled test compounds	81	R. Combes			Positive	Positive	ip
3.2.9	ELISA rat	82	R. Baan			*	*	o
3.3	Immunotoxicity							
3.3.1	T-cells — rat	83	R. Stevens	Positive	Negative	Positive	Weak positive	ip
3.3.2	Natural killer cells — rat	84	I. Kimber			Positive	Negative	o
3.4	Transformation of peritoneal macrophages							
3.4.1	Rat macrophages	85	O. Anderson	Negative	Negative	Negative	Negative	o
3.4.2	,,	85	N. Nashed	Negative	Negative	Negative	Negative	o
3.4.3	,,	85	J. Styles	Negative	Negative	Negative	Negative	o
3.5	Host mediated assays							
3.5.1a	Mouse / yeast — mitotic recomb. — liver	86	G. Bronzetti			Negative	Negative	o
3.5.1b	Mouse / yeast — mitotic recomb. — lung	86	G. Bronzetti			Negative	Negative	o
3.5.1c	Mouse / yeast — mitotic recomb. — kidney	86	G. Bronzetti			Negative	Negative	o
3.5.1d	Mouse / yeast — mutation — liver	86	G. Bronzetti			Weak positive	Negative	o
3.5.1e	Mouse / yeast — mutation — lung	86	G. Bronzetti			Negative	Negative	o
3.5.1f	Mouse / yeast — mutation — kidney	86	G. Bronzetti			Weak positive	Negative	o
3.5.2a	Mouse / yeast — gene conversion — liver	87	R. Forster	Positive	Positive	Positive	Positive	o
3.5.2b	Mouse / yeast — gene conversion — lung	87	R. Forster	Negative	Negative	Negative	Negative	o
3.5.3	Mouse / Salmonella — liver	88	D. Pagano	Negative	Negative	Negative	Negative	o
3.6	Urine mutagenicity							
3.6.1	Rat urine extract	89	R. Combes	Positive	Positive	Positive	Positive	ip
3.6.2	Rat urine extract	90	S. Blowers	Positive	Positive	Positive	Positive	ip
3.6.3	Mouse urine extract	91	B. Harper	Weak positive	Weak positive	Positive	Positive	o

Positive ● Negative ● Weak positive ●

Table 1 (*Cont.*)

IPCS CSSTT *IN VIVO* STUDY – SUMMARY OF QUALITATIVE RESULTS

	ASSAY	CHAPTER NUMBERS	INVESTIGATOR	BP	PYR	2AAF	4AAF	ROUTE OF EXPOSURE
3.6.4a	Rat urine	92	R. Crebelli	Positive	Positive	Positive	Positive	ip
			,,	Positive	Positive	Positive	Positive	o
3.6.4b	Guinea pig	92	R. Crebelli			Positive		ip
3.6.5	Rat urine	93	C. Malaveille	Positive	Positive	Positive	Positive	ip
4.	MOUSE SPOT TEST							
4.1.1	Coat color	95	R. Fahrig			Positive	Negative	ip
4.1.2	,,	96	E. Huttner			Negative	Negative	o
			,,			Negative		ip
4.1.3	,,	97	A. Neuhauser-Klaus	Positive	Negative	Positive	Negative	ip
4.1.4	Melanocytes	98	A. Searle			Positive	?	ip
4.1.5	Melanocytes	99	A. Wickramaratne			Negative	Negative	ip
5.	MAMMALIAN GERM CELL STUDIES							
5.1	Dominant lethal							
5.1.1	Female mice	102	W. Generoso	Positive	Negative	Negative	Negative	ip
5.1.2	Male mice	103	P. Chauhan	Positive	Negative	Negative	Negative	ip
5.1.3	,,	104	U. Ehling	Positive	Negative	Negative	Negative	ip
			,,	Negative	?			o
5.1.4	,,	105	T. Shibuya	Positive	Negative	Negative	Negative	ip
			,,	Negative				o
5.1.5	Male rat	106	J. Bishop			Negative	Negative	o
5.2	Sperm abnormalities							
5.2.1	Mouse	108	M. Salamone	Positive	Negative	Positive	Negative	ip
5.2.2	,,	109	M. Salamone	Weak positive	Negative	Negative	Negative	o
5.2.3	,,	110	J. Arany	Positive	Weak positive	Positive	Negative	ip
5.2.4	,,	111	B. Harper	Positive	Negative	Negative	Negative	o
5.2.5	Rat	112	J. Bishop			Negative	Negative	o
5.3	Unscheduled DNA synthesis							
5.3.1	Rat spermatocytes	113	P. Working	Negative	Negative	Negative	Negative	o
5.3.2	Mouse sperm	114	G. Sega	Negative	Negative	Negative	Negative	ip
			,,	Negative				o
6.	DROSOPHILA ASSAYS							
6.1	Sex-linked recessive lethal							
6.1.1	SLRL assay	116	E. Vogel	Negative	Negative	Positive	Weak positive	o
6.1.2	,,	117	K. Lainson	Negative	Negative	Negative	Negative	o
6.1.3	,,	118	U. Graf	Negative	Negative	Negative	Negative	o
6.1.4	,,	119	R. Valencia	Negative	Negative	Negative	Negative	o
6.1.5	,,	120	V. Ivanov	Negative	Negative	Negative	Negative	o
6.2	Chromosome loss							
6.2.1	Germ cell	121	J. Magnusson	Negative	Negative	Negative	Negative	o
6.2.2	,,	116	E. Vogel	Negative		Negative		o
6.3	Somatic mutation and recombination							
6.3.1	Eye spots	116	E. Vogel	Positive	Negative	Weak positive	Weak positive	o
6.3.2	,,	122	K. Fujikawa	Positive	Negative	Positive	Negative	o
6.3.3	Wing spots	118	U. Graf	Positive	Negative	Weak positive	?	o

Positive (red circle) Negative (blue circle) Weak positive (pale circle) common understanding of the essential goals and the best means for their achievement.

2 Context and aims of the study

The use of *in vitro* assays for the identification of genotoxins is established; what remains to be attained is agreement on the number and types of tests required. Genotoxicity assays using whole mammals have a less clearly defined role, despite the fact that their use is increasing (Figure 1) and that they figure in some legislative guidelines for the conduct of mutagenicity tests. Whole mammal assays are usually more resource consuming than their *in vitro* counterparts and the use of animals for experiments which can be conducted *in vitro* is to be

Figure 1. Number of papers published between 1970 and 1984 on the five classes of *in vivo* rodent genotoxicity assays shown. The mouse micronucleus assay, the chromosomal aberration assays, and the SCE assays were restricted to the bone marrow, and the UDS papers are concerned only with the rat liver. The searches were conducted using the MEDLINE database using appropriate keywords and exclusions. For example, the SCE search was constructed to include papers with the words 'sister chromatid exchange' or 'SCE' within their titles/abstracts/keywords, together with the word 'mouse/rat/mice or rodents'. In that case, papers were excluded which contained any of the following keywords: peripheral blood, *in utero*, transplacental, sperm..., germ cells, gonad, oocyte, zygote, or *in vitro*. The lines through the datapoints are visual estimates of a best fit curve. The implicit errors in these graphs are negligible as each is only designed to emphasize that genotoxicity assays conducted in whole mammals are increasing with time. To set these figures within a broader context, 538 papers were published in 1983 which contained the keywords 'Salmonella' or 'Ames', and 'mutagenicity' within their titles, abstracts, or keywords.

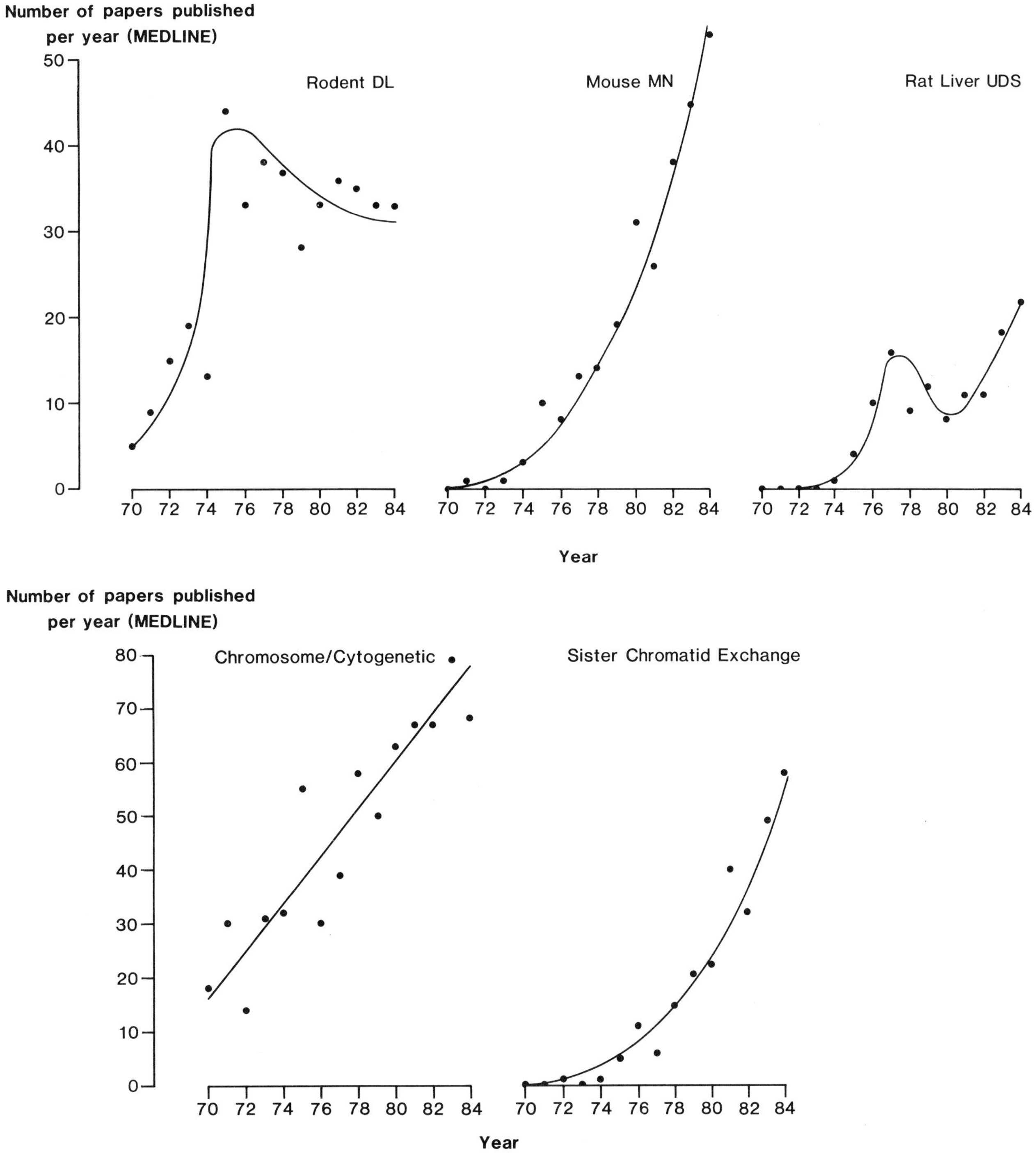

discouraged. These facts combine to indicate that the use of whole animal genotoxicity tests should be reserved for situations where specific questions require an answer. Six such situations were open for consideration or evaluation by the present study. These are listed here and have been developed to provide the framework of this overview:

a) The waning role of whole mammal assays for the primary definition of genotoxic hazards.
b) Rodent host-mediated and urine microbial mutagenicity assays.
c) The growing role of whole mammal genotoxicity assays for hazard assessment studies on *in vitro* genotoxins.
d) Identification of rodent germ cell mutagens.
e) Cancer research, e.g., the study of tumor initiation and progression, and of carcinogen organotropy.
f) The definition of non-genotoxic mechanisms of carcinogenesis.

The chemical structures of the 4 test chemicals are shown in Figure 2, and it is apparent that they form two pairs of structurally related chemicals, BP/PYR and 2AAF/4AAF. The former member of each pair is a reference genotoxic mammalian carcinogen while the second member of each is either non-carcinogenic or of considerably reduced carcinogenic potency (Appendix 3). Knowledge regarding the absolute carcinogenic status of 4AAF and PYR is of little immediate consequence to the present study, as discussed later, but they are both predicted to be non-carcinogenic by reference to structure-activity databases (Appendix 4). Each of these 4 chemicals is genotoxic *in vitro*, as evidenced by the data shown in Figure 2 [derived from an earlier study (de Serres and Ashby, 1981)] and by the repeat *in vitro* data generated as part of the present study (Chapters 123, 124 and 125 and Appendix 2). These chemicals, therefore, present the challenge of discerning ways to distinguish between two potent genotoxic carcinogens (2AAF and BP) and two structurally-related genotoxic non-carcinogens (4AAF and PYR). An indication that whole mammal genotoxicity assays may contribute to this resolution was provided by data generated in the earlier study (de Serres and Ashby, 1981), as illustrated in Figure 2. However, that particular *in vivo* database was too limited to draw firm conclusions in this respect; in particular, it was confined to assays in the mouse bone marrow and testes and this left unmonitored the possible genotoxic activity of 4AAF and PYR in the target organs for carcinogenesis of 2AAF and BP (the liver in the case of 2AAF, and the skin, lung and forestomach in the case of BP).

It was apparent at the start of the study that 4AAF and PYR were weaker genotoxins *in vitro* than were 2AAF and BP, respectively. However, the inadvisability of using such data to anticipate the relative carcinogenic potency of chemicals is now well-documented, and is illustrated further here by the *in vitro* mutagenicity profiles of the rat/mouse carcinogen 4-chloro-*o*-phenylenediamine and the structurally related non-carcinogen 4-nitro-*o*-phenylenediamine (Figure 3). The

Figure 2. Results for the four test chemicals as reported in the International Program for the Evaluation of Short-Term Tests for Carcinogens (de Serres and Ashby, 1981). Data are only shown when both members of each carcinogen/non-carcinogen pair (2AAF/4AAF or BP/PYR) had been evaluated. It was evident that while each of the chemicals were considered genotoxic *in vitro*, only the carcinogens of each pair were genotoxic *in vivo*. The present study, employing a wide range of tissues and assay systems, was initiated to evaluate the general validity of this apparent resolving power of whole mammal assays.

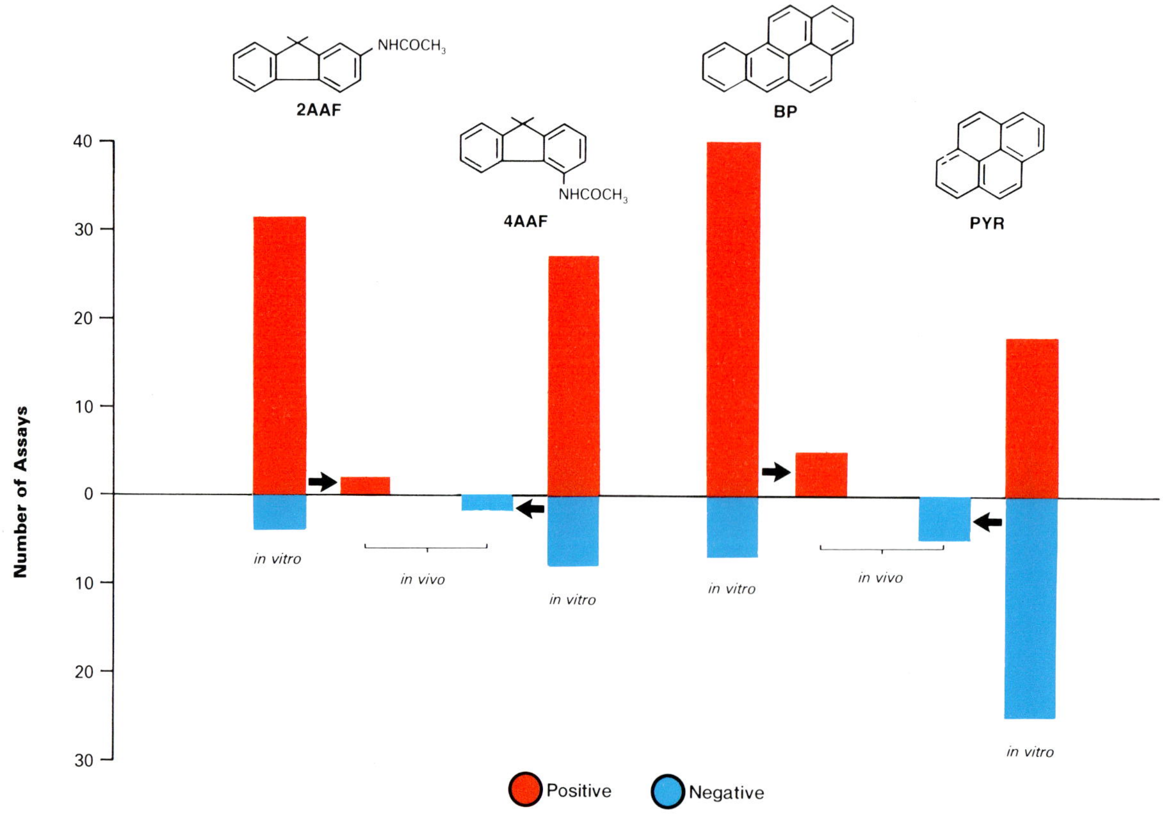

non-carcinogen also serves to complement this study in that it has been defined as non-carcinogenic after adequate experimentation, unlike 4AAF and PYR. Nonetheless, it is genotoxic in a range of *in vitro* genotoxicity assays in addition to the Salmonella test results shown in Figure 3 (Shelby and Stasiewicz 1984). This example, therefore, endorses the pressing current need to devise methods of qualifying *in vitro* genotoxicity data if accurate predictions of rodent carcinogenicity are sought.

The present study, therefore, had the broad aim of acquiring a profile of the genotoxic activity in rodents of these four chemicals using as wide a range of tissues and endpoints as possible. In this way it was hoped to ascertain if the genotoxicity profile of each chemical *in vivo* would more closely approximate its observed rodent carcinogenicity than its *in vitro* genotoxic activity.

3 Discussion

The results of the study are discussed using the six headings listed in the previous section of this chapter.

The waning role of whole mammal assays for the primary definition of genotoxic hazards

Certain rodent mutagenicity assays were in use before the advent of practical *in vitro* genotoxicity tests. These, such as the rodent bone marrow cytogenetic and the germ cell mutagenicity assays, remained in use during the past decade when *in vitro* tests flourished, and they are still employed. But a clear functional separation of *in vitro* and *in vivo* roles was missing during this transition period. Thus, Jenssen and Ramel (1980) have studied the use of the Salmonella assay in combination with the mouse micronucleus test as a screening battery, and Styles and Penman (1985) have similarly considered the Salmonella assay as a complement to the mouse spot test. Likewise, Mirsalis and Butterworth (1980) originally described their *in vivo* rat liver UDS tests as a primary screening test. This lack of clarity regarding the role of *in vivo* assays is illustrated later using the example of the dominant lethal assay, and here, by the fact that the bone-marrow cytogenetic tests are regarded as equivalent to *in vitro* chromosomal aberration assays in some regulatory guidlines. This has the disadvantage that the test protocol employed for screening may be less detailed than that required for hazard assessment. For example, Gocke *et al.* (1981) have described the results of screening 31 chemicals using the mouse bone-marrow micronucleus assay. Only 2 compounds were positive, but a very limited test protocol was employed (3 dose-levels, 1 timepoint, 12 mice in total) which would be inappropriate for determining if an *in vitro* genotoxin was genotoxic *in vivo*, i.e. for hazard assessment purposes.

Figure 3. Relative mutagenicity to Salmonella strain TA98 of the carcinogen, 4-chloro-o-phenylenediamine, and the non-carcinogen, 4-nitro-o-phenylenediamine, when tested in the presence of Aroclor-induced Sprague Dawley rat S9 mix (10%) in a pre-incubation assay (20 min.). These data are from the U.S. National Toxicology Program database and are representative of data from tests conducted under a range of other experimental conditions (E. Zeiger, personal communication). The carcinogenicity test results of these two chemicals is discussed in NTP Technical Report No. 63 (1978) and No. 180 (1979), respectively.

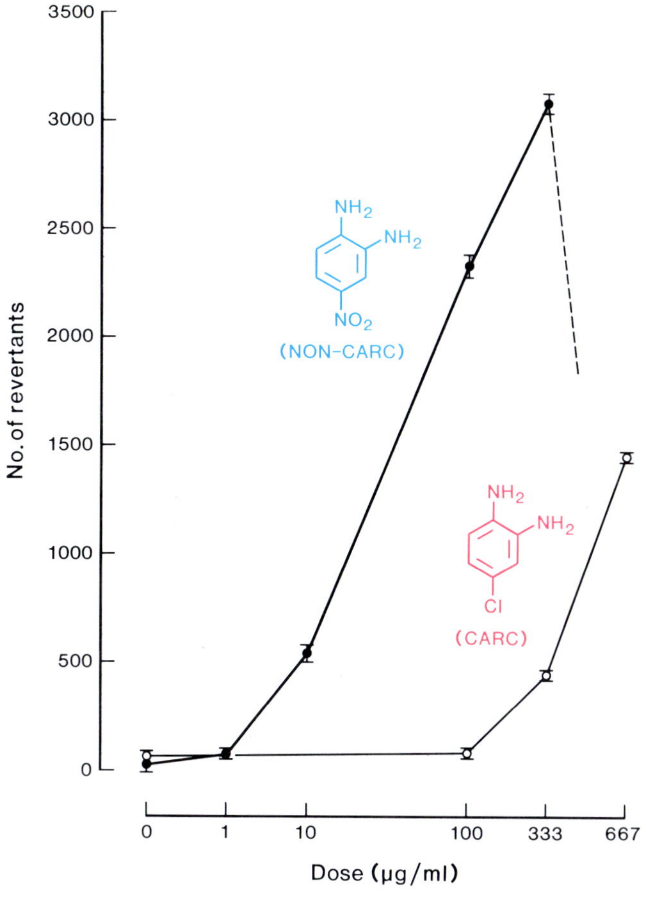

When the two IPCS studies of short-term tests were initiated it was decided to regard *in vitro* assays as distinct from *in vivo* assays; thus, it was implicit in the design of the present study that whole mammal assays have no role to play when screening for genotoxins. Rather, the only uses which can be justified by data are those discussed in the next 5 sections of this chapter. One of the most consistent presentations of the unique role of whole mammal assays has been made, over the years, by Legator. This started with his discussion of the Bridges 3-tier testing strategy (Bridges, 1974, 1976) in 1975 (Legator and Williams, 1975) and has recently been reiterated in the form of a suggestion by Legator and Harper (1986) that a 'reverse tier' approach to testing should be adopted, i.e., that testing should commence in rodents. The present study was designed with the knowledge that *in vivo* genotoxins exhibit genetic toxicity *in vitro*, and on the assumption that *in vitro* assays will be used to define genotoxins such as the present four test chemicals (Appendix 2). *In vivo* assays will then only be employed when more refined or specific predictions are required regarding the potential mutagenic/carcinogenic consequences of exposure of humans to *in vitro* genotoxins. Within the above context it is evident that assays in Drosophila are inappropriate substitutes for whole mammal hazard-assessment studies. Further, the insensitivity of the Drosophila germ-cell assays (6.1 and 6.2; Table 1) to the present 4 genotoxins provides a salutory warning of the possible inadequacy of these systems for primary screening. For the latter purpose, the somatic cell mutation assays seem to possess more appropriate sensitivity (6.3 in Table 1). Overall, it is concluded that assays in Drosophila have no unique role to play in screening programmes, albeit they are particularly suitable for more comprehensive analyses of the genetic activity of known mutagens (see for example, Vogel *et al.*, 1985).

Rodent host-mediated and urine microbial mutagenicity assays

One of the reasons why the genotoxicity of a chemical *in vitro* might differ from that observed *in vivo* is that mammals with absorb, distribute, metabolize and excrete a test chemical in a manner largely unrelated to its interaction with S9 mix and target cells *in vitro*. This being so, then for the aquisition of *in vivo* genotoxicity data, the simplest expedient could be to expose the marker organisms to the chemical *in vivo*, or to its metabolites, followed by culture of the cells *in vitro* to determine induced genetic changes. The two major classes of assay that enable this interaction to occur are the host-mediated mutation assays (3.5; Table 1) and the urine mutagenicity tests (3.6; Table 1). Despite the fact that these two classes of test were conducted in several laboratories, the latter using both mice and rats and the oral and intraperitoneal injection routes of administration, neither provided any data to discriminate the two potent carcinogens from among these 4 *in vitro* genotoxins. Reference to Table 1 shows that the host-mediated assays were generally negative while the urine assays were uniformly positive. A more detailed analysis of the poor performance of these assays is presented in Chapters 4, 5 and 72, but it is concluded here that they have nothing to offer when attempting to qualify *in vitro* genotoxicity data. This conclusion is formally limited to the marker organisms and test protocols adopted for use in the study, but it is considered unlikely that changes in these parameters would affect this conclusion.

The peritoneal macrophage transformation assay (3.4.1–3; Table 1) can also be regarded as a form of host-mediated assay in that macrophages are assessed for changes in their cloning properties *in vitro* after recovery from the peritoneum of treated rats. Three laboratories failed to discern any activity for 2AAF or BP, chemicals which were used to establish the test. Consequently, it was recommended in the miscellaneous assay workgroup report (Chapter 85) that use of this test system should cease.

The growing role of whole mammal genotoxicity assays for hazard assessment studies on in vitro genotoxins

The original demonstration that a high proportion of carcinogens are genotoxic *in vitro* (McCann and Ames, 1976) has subsequently been complemented by the knowledge that not all *in vitro* genotoxins are carcinogenic or mutagenic *in vivo* (Shelby and Stasiewicz 1984, Ashby, 1986). Against this background a role is emerging for *in vivo* genotoxicity assays to determine which new *in vitro* genotoxins are effective genotoxins in whole mammals. This role is referred to as hazard assessment herein, and it is already well-established that access to at least 2 assays in different tissues is required. The first part of this section is limited to a discussion of the performance of assays which are currently used routinely in several laboratories, for this purpose, i.e., the three classes of rodent bone-marrow chromosomal assay [sister-chromatid exchange (SCE), chromosomal aberrations (CA) and micronucleus formation (MN)] and the rodent liver assay for unscheduled DNA synthesis (UDS). Following these, assays are discussed which may, after suitable standardisation, provide useful additional assays for routine hazard assessment studies.

The mouse micronucleus assay is the simplest method of assessing a large number of cells for chromosome damage in the bone marrow (1.2.1–16 in Table 1). It is also sensitive to the effects of metaphase spindle damage, but that is unlikely to be relevant to the present chemicals. The manner in which this assay resolved BP from PYR, and 2AAF from 4AAF, is shown by the total database plots presented in Figures 4 and 5. The resolution of BP and PYR was dramatic and absolute while that of 2AAF and 4AAF was less marked, albeit qualitatively absolute. Only one investigator (1.2.6 in Table 1) recorded weak activity for 4AAF and PYR and none failed to detect 2AAF or BP as positive. This is a particularly encouraging finding given the current widespread use of this assay. Results were not qualitatively compromised by the choice of route of exposure, strain or sex of mouse, or dosing regimen adopted. The mouse bone marrow micronucleus assay therefore appears to be confirmed as a robust and sensitive *in vivo* genotoxicity test system.

If metaphase spindle interaction is neglected, it would be expected that assays for CA in the bone marrow would produce results which correlate with those for MN, and, perhaps, also with results from SCE. Reference to Table 1 indicates that the CA assays (1.1.1–9) were generally less 'sensitive' than the MN tests, while those for SCE (1.3.1–7 and 9) not only found the two carcinogens positive, but also produced a number of weak positive results for 4AAF and PYR. These relative assay sensitivities are shown diagrammatically in Figure 6. Absence of knowledge regarding the true carcinogenic status of 4AAF and PYR precludes use of the term 'false positive' response for these chemicals in some of the SCE assays, however, two general points can be made. First, it would be inappropriate to form conclusions regarding the absolute relative sensitivity of these three classes of assay due to their differing statistical design and positive response criteria. Thus, the apparent 'insensitivity' of the CA assays to 2AAF, as compared to the MN assay, may be associated with the fact that only ~ 100 metaphases are scored per animal in the former assay while 1000 cells are scored in the latter. Likewise, the relative ease with which the unit of observation in the SCE assays can be determined may lead to more extensive sampling of slides, and thus to the apparent higher sensitivity of these assays. Also, reference to Fig. 7 indicates that both 4AAF and PYR elicited some very weak effects in some of the MN assays, and while these were not classed as positive, they may be associated with the weak SCE effects for these chemicals. Second, evidence exists in the literature suggesting that there may be mechanisms of SCE induction that are unrelated to clastogenic changes. Therefore, although the *in vivo* SCE results closely parallel the relative potency of the two chemical pairs as genotoxins *in vitro*, the possibility exists that the very weak SCE responses observed for 4AAF and PYR may be due to bone marrow toxicity, as commented on in the SCE workgroup report (Chapter 37). This toxicity may be subtle and specific – thus, Montaldi *et al.* (1985) recently suggested that chemically-induced modulation of topoisomerase enzyme activity may lead to small increases in SCE which are not definitive of a genotoxic effect. Likewise, ultrasound produces unique SCE effects both *in vitro* and *in vivo* (Stella *et al.*, 1984). These combined considerations indicate that the mouse MN assay probably provides the simplest and most effective measure of genotoxicity in the bone marrow, and as such it is recommended here as a primary *in vivo* genotoxicity test.

A particular danger of a database as large as the present one is the loss of information that accompanies the simplification of test results to crude measures such as pluses and minuses, as used in Table 1. To partially overcome this problem, the mouse bone marrow micronucleus assay data for the 2 pairs of chemicals are shown in 3-dimensional projection in Figure 7. In these projections it is evident that BP and 2AAF are effective *in vivo* genotoxins while neither PYR or 4AAF are, irrespective of individual anguishing over the definitions of weak positive responses. Further, it will be recalled that Ivanov *et al.* (assay 1.2.6 in Table 1) recorded both 4AAF and

Figure 4. Mouse bone marrow micronucleus assay data for BP and PYR plotted on the same scales. These data are those shown qualitatively in Table 1 (1.2.1–1.2.16) and include the dose-response lines for all experiments, i.e., 24 h, 48 h; i.p. or oral, etc. Individual data sets are discussed elsewhere in this volume; these figures are intended to represent the gross activity of these two chemicals in this assay system. An additional way of viewing these data is shown in Figure 7.

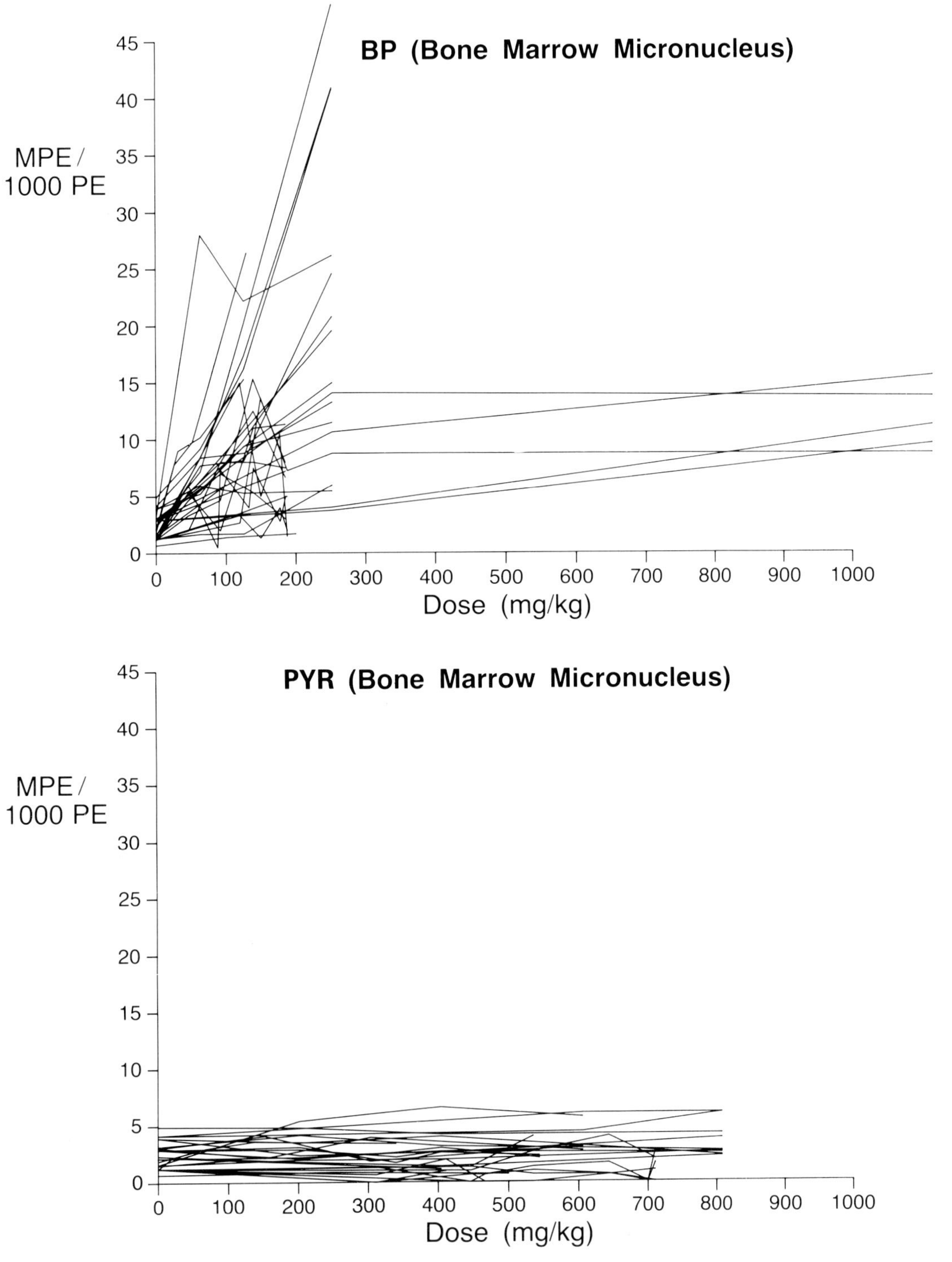

PYR as weak genotoxins and were thus apparently in conflict with the conclusions of the 15 other participating laboratories in that group. However, when all data are compared it is clear that although the Ivanov data for 4AAF and PYR can be individually identified in Figure 7, these weak results are in no way unique. Irrespective of whether or not 4AAF and PYR are weak bone marrow clastogens, some method which transcends + and − conclusions must be derived for transferring the critical difference in the genotoxicity *in vivo* of each member of these 2 pairs. Thus, the lower overall clastogenic potency in the bone-marrow of 2AAF, as compared to BP, is evident in Figure 7, but not in Table 1.

Despite the utility of the mouse bone-marrow micronucleus assay in detecting the 2 carcinogens in this study, sole reliance upon it or any other bone-marrow genotoxicity test for hazard assessment studies would lead to a range of genotoxic mammalian carcinogens being prematurely regarded as non-hazardous *in vivo*. A group of particular importance are liver carcinogens such as the nitrosamines and azo compounds. It is for this reason that a range of liver genotoxicity assays have been developed (2.1.5 in Table 1). Reference to Figure 8 illustrates that the liver carcinogen 2AAF was a potent inducer of unscheduled DNA synthesis (UDS) in the rat liver, as well as being a clastogen in the bone marrow (Figure 7). Most of

Figure 5. Mouse bone marrow micronucleus assay data for 2AAF and 4AAF plotted on the same scale. These data are those shown qualitatively in Table 1 (1.2.1–1.2.16) and include the dose-response lines for all experiments, i.e., 24 h, 48 h; i.p. or oral, etc. Individual data sets are discussed elsewhere in this volume; these figures are intended to represent the gross activity of these two chemicals in this assay system. An additional way of viewing these data is shown in Figure 7.

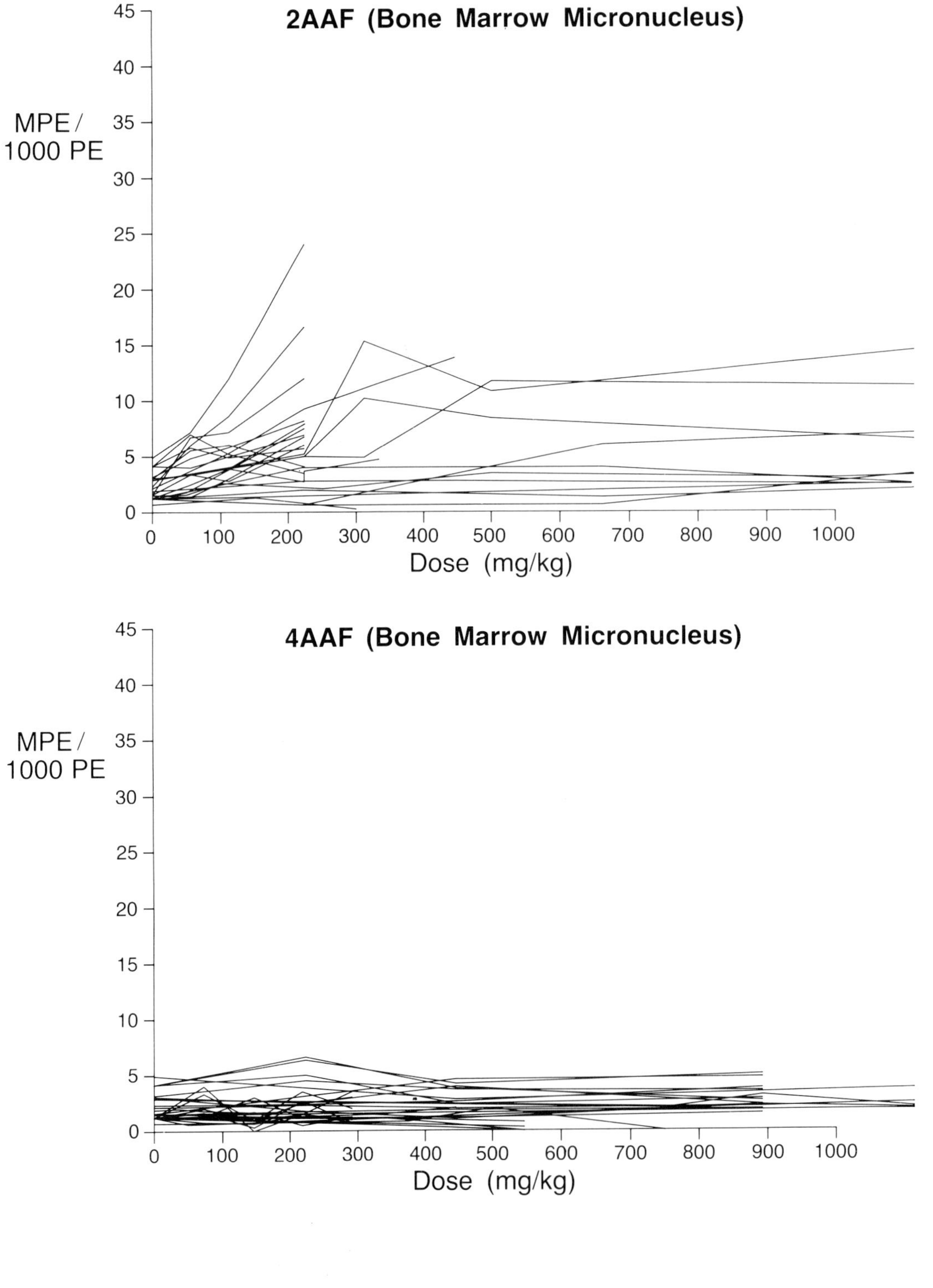

the rat liver UDS tests defined 4AAF as a weak *in vivo* genotoxin. In stark contrast to the bone marrow assays, BP was devoid of activity in both the UDS assays (Figure 8) and the majority of the other liver specific assays. This happens to be consistent with the lack of carcinogenicity of BP to the rat liver, but of greater importance, it underlines the danger of relying on any single *in vivo* test system for hazard assessment studies. Fewer laboratories studied BP/PYR in the liver assays because its carcinogenic inactivity in this organ was known at the start of the study, in fact, this was one of the reasons for the selection of this pair.

With only 2 carcinogens it is difficult to illustrate the importance of not relying on a single negative *in vivo* response when assessing the hazard of a new *in vitro* genotoxin. Thus, both 2AAF and BP were active in the mouse MN assay here, but only 2AAF was positive as a liver genotoxin in the UDS test. Had we elected to study the carcinogens diethylnitrosamine and 2, 4-dinitrotoluene, then negative results would have been observed in the bone marrow assays with positive results in the liver genotoxicity tests. For reasons of ease of conduct the MN assay is favoured as an initial *in vivo* assay, but the literature indicates that negative results in this test should be supplemented with evidence of non-genotoxicity to the liver (Ashby, 1986). The failure to endorse that principal in this study is merely a penalty of having to restrict the number of chemicals studied.

The data shown in Figure 9 illustrate the problems posed by the weak UDS activity of 4AAF in the liver, a topic discussed in greater detail by Margolin and Risko (Chapter 3) and which is probably of equal conceptual relevance to the bone marrow SCE activity of both 4AAF and PYR. From Figure 9 it is evident that, in the liver UDS assays, 2AAF showed strong activity at dose-levels between 5–100 mg/kg. In contrast, 4AAF showed only very weak activity at dose-levels between 50–1000 mg/kg, and then only in some experiments. Of great importance, this activity of 4AAF was essentially contained within the region of test response occupied by control variability, and this led the UDS workgroup (Chapter 55) to question its biological significance. In contrast, Margolin and Risko present evidence to confirm the high statistical significance of these data (Chapter 3). In fact, a conflict

Figure 6. Qualitative results of the three bone marrow cytogenetics assays presented as proportions of the total results. The heights of the bar have been normalized to enable comparisons to be made. The qualitative results employed are those discussed in the bone marrow assay group report (Allen *et al.*, this volume, Chapter 6). The trend of importance is that for the chromosomal aberration assays (CA), the lower sensitivity to 2AAF is accompanied by a large proportion of inconclusive results for both 2AAF and 4AAF. In contrast, the sister chromatid exchange assays (SCE) showed high sensitivity accompanied by a large proportion of positive responses for 4AAF and PYR and no inconclusive results. The micronucleus assays (MN) gave the clearest qualitative separation between the two carcinogens and their structurally-related non-carcinogens. These trends are discussed further in the text, and it must be emphasized that they suffer from a lack of quantitative definition beyond positive, weak positive, inconclusive, and negative. The distinction between a weak positive response and an inconclusive response is, on occasion, subjective, given that the reproducibility of data was not always ascertained.

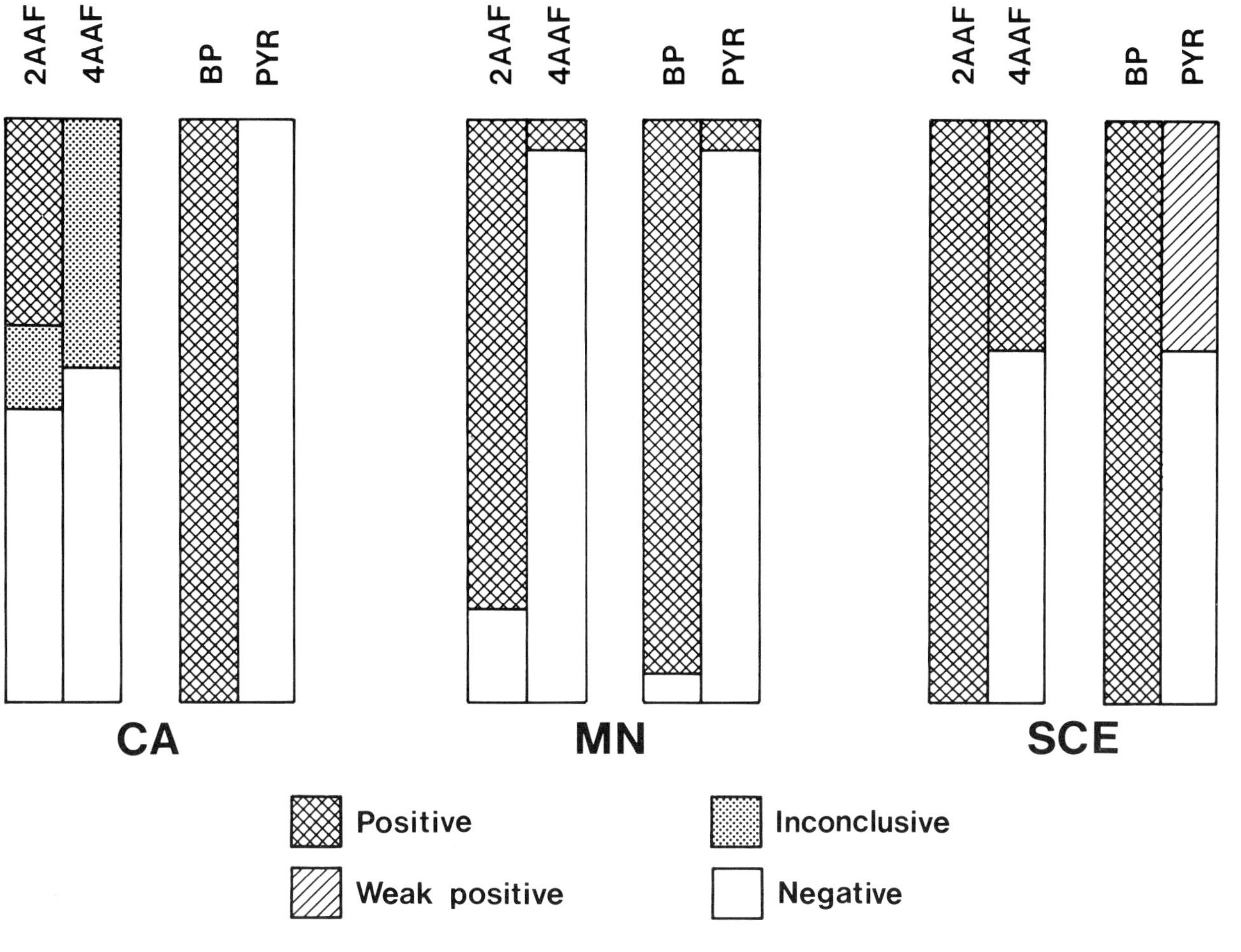

Figure 7. Mouse bone marrow micronucleus data for the four test chemicals plotted in 3-dimensional projection. Horizontal black lines on individual red or blue columns represent individual observations at a particular dose-level and time. These data are derived from the individual investigator chapters. The colouring adopted is red for 2AAF and BP and blue for 4AAF and PYR, i.e., colour does not indicate individual conclusions regarding activity or inactivity in the assay. Control incidence of MPE have been subtracted from data columns in these figures: zero or negative values appear as a dot on the figure.

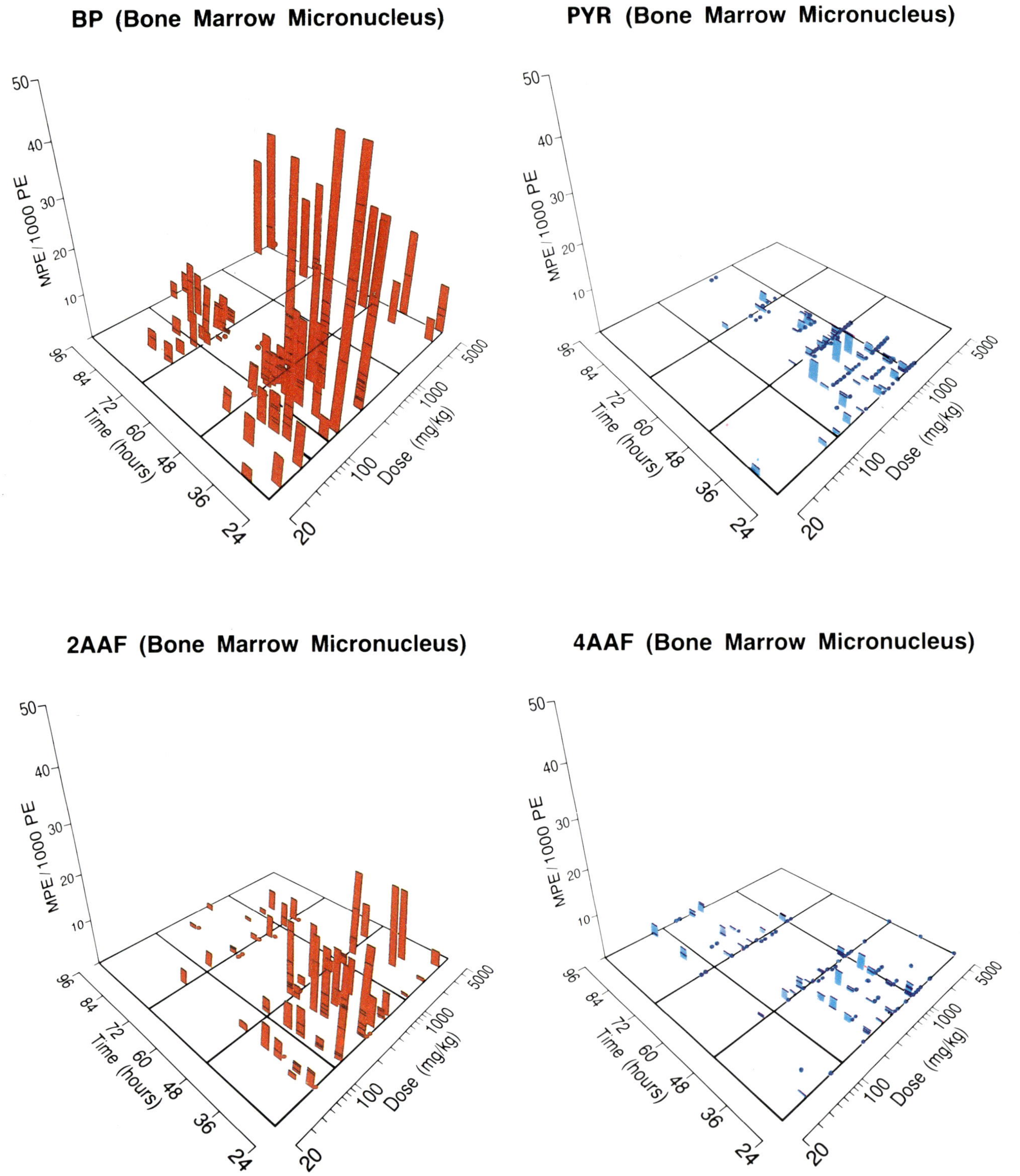

Figure 8. Rat liver UDS data for the four test chemicals plotted in 3-dimensional projection. Horizontal black lines on individual red or blue columns represent individual observations at a particular dose-level and time. These data are derived from the individual investigator chapters. The colouring adopted is red for 2AAF and BP and blue for 4AAF and PYR, i.e., colour does not indicate individual conclusion regarding activity or inactivity in the assay. Control values have been subtracted from the data columns in these figures:zero or negative values appear as a dot on the figure.

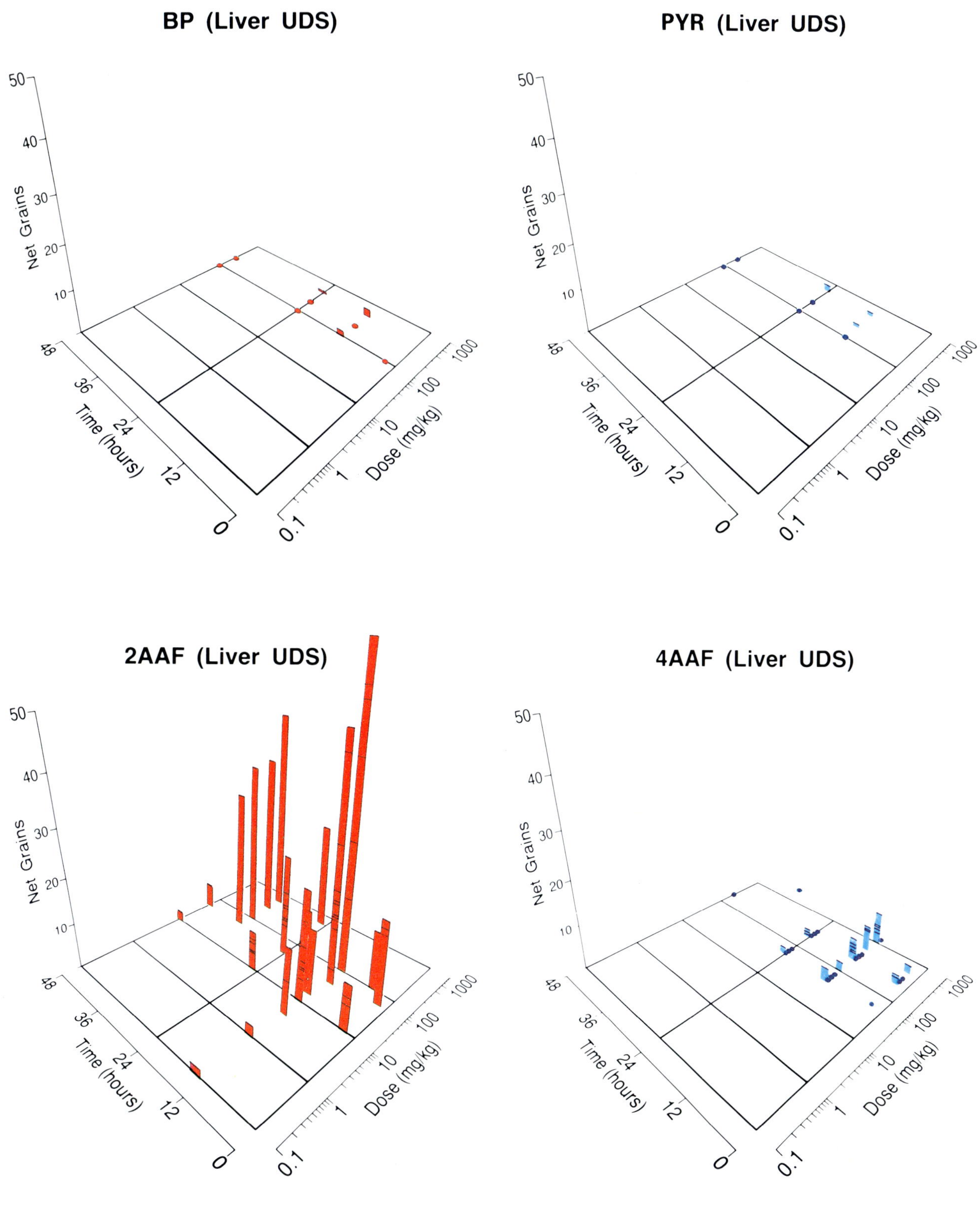

exists between Chapters 3 and 55 regarding the most appropriate positive response criteria for the rat liver UDS assay. However, when seeking evidence of the likely hazard presented by a chemical *in vivo*, the liver UDS test has adequately discriminated between 2AAF and 4AAF (Figure 8); what has to be resolved is whether 4AAF is formally active in the assay. That question must be settled, but this case lends support to the concept that + or − test response criteria are too simplistic for hazard assessment studies. Perhaps one of the most useful things to derive from the present study is the attention it has focussed on the need for general agreement on positive test criteria for widely used *in vivo* assays (see Chapter 3).

A final consideration regards which, among the many other test systems in this study could, or should be further developed as routine test systems for hazard assessment purposes. First, it seems sensible to base the development of additional *in vivo* assays on specific and well-defined needs or purposes. Thus, working down Table 1, the following assays were shown by this study to be less sensitive than the more widely used bone marrow or liver counterparts – their possible further development must therefore be based upon considerations beyond this study: mouse ascites cytogenetics (1.1.8), mouse blood micronuclei (1.2.17–20), UDS in weanling or neonate rats (1.2.10, 1.2.11), cytogenetic or SCE analysis in liver epitheloid cells (2.4.1 and 2.4.2) and the DNA strand break and cross-link assays (2.5.1–4). Of particular importance in this list is that the alkaline elution liver DNA strand break assay (2.5.1–3) produced such discordant results that the assay work group recommended its suspension from use pending an understanding of these problems (Chapter 67). In contrast, the mouse peripheral lymphocyte SCE assay (1.3.8) gave data consistent with the

Figure 9. Data for 2AAF and 4AAF in the rat liver UDS assays. Dose-response effects at all time points are included, as are repeat experiments from individual laboratories. To the left of each graph is a histogram of the recent and concurrent control data accrued in the seven participating laboratories. Concurrent control data are available from the individual chapters for assays 2.2.2–2.2.4 and 2.2.6–2.2.9 (Table 1). The additional control data were provided by courtesy of individual laboratories and are discussed more fully in the chapter by Margolin and Ricko (this volume, Chapter 5). N.B. Different dose scales for each figure.

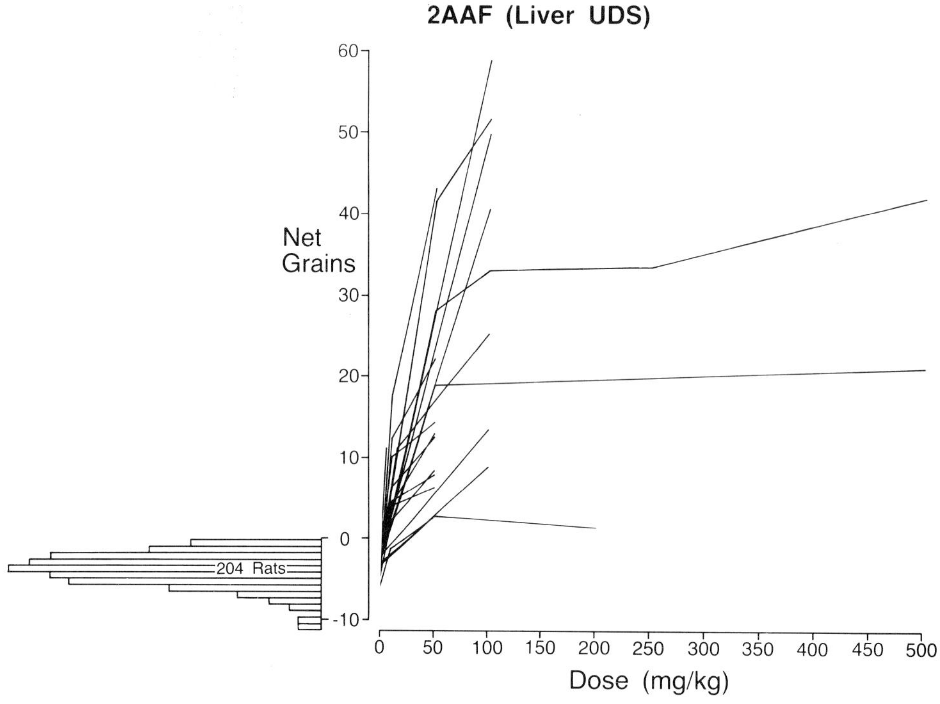

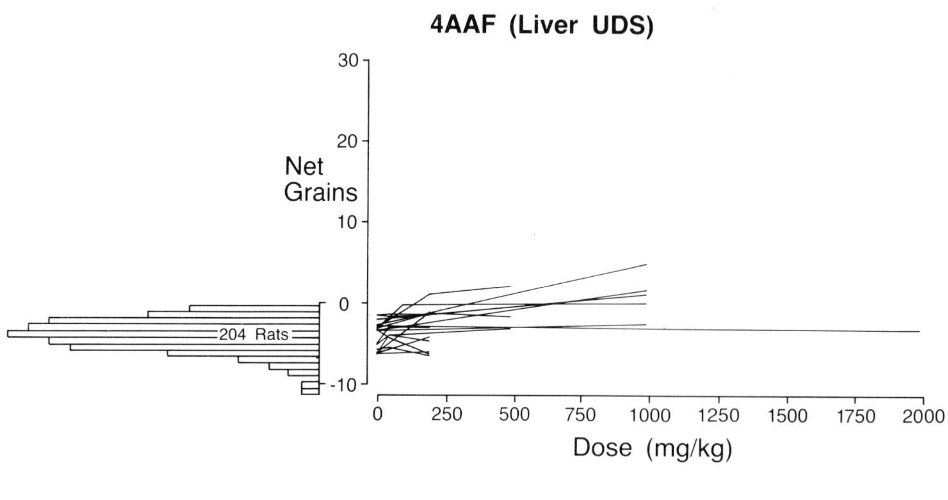

assumed carcinogenic status of the 4 test chemicals and therefore presents the prospect of a highly useful assay for hazard assessment purposes – the more so because blood is the vehicle of systemic exposure. Nonetheless, it is interesting that 2 of the 4 mouse blood micronucleus assays detected BP but not 2AAF as positive. Work to develop further a peripheral blood chromosomal genotoxicity assay seems to be justified, but the problem of S-phase specific clastogens must not be neglected (see Newton and Lilly, 1986). The liver micronucleus assay (2.4.3) showed no advantage over the UDS test in this study and is technically more complex to conduct. Nonetheless, a hidden advantage of this assay is that the expression of micronuclei does not occur until the mitotic surge following the partial hepatectomy. This means that several months may elapse between exposure to the genotoxin and the assessment for micronuclei, and this may prove useful in combined *in vivo* hazard assessment studies. The protocol of the liver DNA-unwinding tests (2.5.5 and 6) is still evolving and will require reappraisal at a later stage.

This section of the overview will be of particular relevance to those wishing to discern a minimum number of *in vivo* assays for routine hazard assessment studies. However, the use of only 4 test chemicals precludes definitive advice. Nevertheless, this study has inadvertently focussed attention on an underlying issue which must be resolved before further progress can be made on this topic. This relates to the possible existence of a class of carcinogens which is not characterised by an initial reaction between the test chemical and genetic material. If 'non-genotoxic' carcinogens exist, then their inactivity in *in vivo* genotoxicity assays is automatically justified. If, however, agreement cannot be gained concerning their separate existence, then their inactivity in genotoxicity assays becomes evidence of the 'insensitivity' of the latter. Previous discussion of this issue has centered mainly on liver carcinogens which are non-genotoxic both *in vitro* and *in vivo*, agents such as diethylhexylphthalate [DEHP; (Butterworth *et al.*, 1984)]. However, this topic can be approached here *via* the inactivity of 2AAF in the mouse liver genotoxicity assays as discussed later in this chapter. The imminent danger is that, in the absence of agreement on the reality of epigenetic mechanisms of carcinogenesis, the reaction may be to increase the number and types of *in vivo* genotoxicity assay in order to try to detect *all* carcinogens as *in vivo* genotoxins. This same problem has delayed progress in the effective utilization of *in vitro* assays and led to the IPCS study of *in vitro* assays (Ashby *et al.*, 1985). Thus, in order to devise a minimum battery of *in vivo* genotoxicity assays for hazard assessment studies it must first be decided what the assays should be sensitive to, *all* carcinogens, or just those which are DNA-reactive (see later).

Assays for the determination of germ cell mutagens

This discussion commences with consideration of all assays conducted using germ cell tissues from mice and rats and concludes with attention focussed on the results from the dominant lethal assay which is generally considered to be the most practical germ cell mutagenicity test for routine use.

First, considering the complete rodent germ cell database (Table 1; assays 5.1–3) which comprises assays for altered sperm morphology, DNA damage and dominant lethal mutations. Ten positive and 1 weak positive response were reported out of 25 determinations on BP and 2AAF. Only 1 weak positive response was reported for PYR out of a total of 23 experiments conducted on it and 4AAF. The simplest statement concerning the germ cell assay data, is therefore, that 40% of the responses were positive for the two carcinogens and 4% for the two non-carcinogens. These figures speak, superficially, of a moderate discrimination by the germ cell assays between the carcinogens and the non-carcinogens – a conclusion which is only worth stating so that it can be immediately devalued, and thereby, provide an example of the dangers inherent in the simplistic averaging of responses from several distinct assays. The information provided by these data are probably more as follows. The germ cell assays were conducted using both the oral and ip injection routes of administration, 20 and 28 determinations in each category, respectively. Ten of the 11 positive responses with BP/2AAF were associated with use of the ip injection route. Without pre-judging which route is the more appropriate, it is important to appreciate that this protocol variable can affect the outcome of germ cell assays in such a profound manner. Similar effects are reflected elsewhere in this database and are discussed later in this chapter. The second most striking observation is provided by comparison of the performance of the sperm morphology assays with that of the DNA-damage tests in their role as surrogates for definitive mutagenicity tests. BP was uniformly positive in the sperm morphology tests, and 2 of 5 determinations of 2AAF were also positive. In contrast, both carcinogens were non-reactive to germ cell DNA as measured by 2 independent techniques in the mouse and rat. With the exception of a weak positive sperm morphology response for PYR (assay 5.2.3), the 2 non-carcinogens were negative in both classes of assay, and by both routes of exposure. It is relevant to mention here that Russell (1977) has recommended use of the mouse spot test as a pre-screen (surrogate ?) for germ-cell mutagens, and in that case, both of the carcinogens showed activity while the non-carcinogens were inactive (assays 4.1.1–5, Table 1). A major conflict of evidence therefore surrounds the predicted capacity of BP and 2AAF to induce germ-cell mutations in rodents – the spot test and sperm morphology assays indicate a mutagenic potential which appears not to involve damage to germ cell DNA.

Turning now to that which was predicted by the above assays concerning the ability of BP and 2AAF to elicit germ-cell mutations in rodents, the data are clear: BP is a germ-cell mutagen when administered *via* ip injection while 2AAF is non-mutagenic by either route of administration. It is relevant to note that BP is known from the literature to be at most only marginally active as a mutagen (following ip injection) in the mouse heritable translocation (HT) and specific locus (SL) assays (reviewed in Chapter 3). The combined data from both this study and the literature for BP, therefore, confirm that although the HT assay has a higher statistical sensitivity (Russell and Shelby, 1984), the DL assay may, on occasion, be more sensitive to the induction of germ cell mutations than either the HT and SL assays. It is perhaps also worth emphasizing that although care is required in the interpretation of DL data, in particular the separate consideration of reduced fertility and late deaths, this assay is not prone to false (toxicity mediated) positive responses. If conducted according to an appropriate protocol it can give unequivocal evidence of germ cell mutagenicity [see reviews by Russell and Shelby (1985) and Anderson (1984)].

Further experiments will be required to confirm the following rationalization of the germ cell data from this study, itself derived mainly from the chapters by Generoso (101), Salamone (107), Leonard (100) and Forster (4).

1. PYR and 4AAF can be regarded as devoid of male rodent germ cell mutagenicity.
2. BP, but not 2AAF, is a germ-cell mutagen.
3. The mouse spot test represents an inadequate pre-screen for germ cell mutagens because it does not involve the study of germ cell tissues. However, it found both BP and 2AAF active, thus defining their somatic cell mutagenicity, a property which may be more readily determined using the bone marrow assays (Holden, 1982, and this Volume).
4. By finding both BP and 2AAF positive, the sperm morphology assays demonstrated their inability to discriminate the germ cell mutagen BP. The prospect that the activity of BP and 2AAF in these assays may have been mediated by non-genetic toxic effects is not a novel concept. This possibility, together with the fact that damage to Sertoli cells may lead to secondary morphological changes in sperm (Russell and Shelby 1985) suggest that this class of assay should not be used to predict a phenomenon as specific and important as the germ cell mutagenicity of a chemical.
5. Shelby (1986) has recently suggested that use of the ip injection route should be reserved for primary *in vivo* studies and that positive results observed in such cases are subject to subsequent qualification by results from tests employing more natural routes of administration of the test chemical. BP would not, therefore, be regarded as a significant germ-cell mutagenic hazard based on the negative results following its oral administration.
6. It is essential that the true heritable mutagenic status of BP to rodents be defined, including an explanation for the activity in the DL assay not being accompanied by activity in the HT and SL assays. This may require the conduct of specific chromosomal aberration and protamine binding assays in spermatocytes, oocytes and the corresponding zygotes. In the absence of such data it is difficult to reach a firm conclusion regarding the significance of the negative results for BP in the germ cell DNA damage assays. The complexity of the present data for BP indicates that attention should remain focussed on the necessary separation of studies designed to evaluate the mechanisms of germ-cell mutagenicity from those aimed at screening for possible germ-cell mutagens. To achieve this, a restatement in more specific terms of the exact role of rodent germ-cell mutation assays in the hazard definition or assessment process will be required. Thus, the recent GENE-TOX review (Green *et al.*, 1985) specifically suggests that the DL assay be used to confirm positive results from lower tier (*in vitro*?) chromosomal aberration detecting systems, and that it should not be used at a risk assessment method. Presumably, the HT or SL assays would be reserved for this purpose. In contrast, Brockman and de Serres (1983) place this test in the final of their 3 categories, that for risk estimation. Caught between these two extremes are the UKEMS mutagenicity testing guidelines (1983) which place this assay within its 'basic package' of tests to be conducted on new chemicals (p. 2). In fact, it was placed in an either/or position against the rodent bone marrow cytogenetic assays (micronucleus or, 'preferable' metaphase analysis) which were also regarded as 'basic' test systems. Later in that document (p. 170) the conduct of supplementary tests (i.e. beyond the basic test battery) is suggested to have the single purpose of ascertaining if positive *in vitro* results accurately reflect the potential of the compound for genotoxicity in the mammal. These two statements are obviously contradictory. Later still (p. 176) it is suggested that if a compound induces mutagenic changes in somatic cells *in vivo*, then there seems little practical value in studying its effects in germ cells. This latter statement implies that germ cell mutagens will also be somatic cell mutagens. Brusick (1980) and Holden (1982) have considered the relationship between somatic and germ cell mutagens. They took a similar but not identical position and proposed that there is little value in conducting germ cell tests on chemicals that are non-mutagenic in somatic cells. The results of the present study add further weight to the concept that germ cell genotoxicity and mutagenicity assays should be reserved for specific hazard assessment studies on established whole mammal somatic cell genotoxins and/or mutagens. As such, germ cell assays would appear to have no legitimate position in routing screening programmes for *in vivo* genotoxins.

In summary, the germ cell activities of BP have served to focus several important, but unresolved issues associated with conduct of, and the interpretation of data derived from, germ cell genotoxicity and mutagenicity assays.

Use of in vivo assays for research into the stages of chemically induced carcinogenesis

A fundamental pre-condition for tissue or organ-specific genotoxic chemical carcinogenesis will be the selective accumulation, metabolic activation and subsequent DNA-reactivity of the test chemical. However, these preliminary requirements (initiatory events) may apply to several tissues, but subsequent, and again tissue specific responses (referred to as promotional events) may determine in which organs tumours will develop. It follows, therefore, that short-term genotoxicity assays will only be capable of determining those tissues which may be *subject* to carcinogenesis, i.e., that they can only define initiated tissues or those at risk of developing tumours. Which, if any, of these tissues will eventually present tumours is beyond the scope of assays conducted within a few days of exposure to the test chemical. With the exception of the liver initiation/promotion assays (Table 1, 2.1.1–6) this restraint precludes any of the assays in this study from providing accurate predictions regarding the organotropy of a chemical carcinogen. In fact, the literature already contains sufficient data to establish this point. Nonetheless, the fact that the observed organotropic activity of a carcinogen will be determined by a process of selection from among the several initiated tissues, suggests that short-terms tests should give positive responses in assays sited in the target tissue for carcinogenesis. For example, it must be beyond reasonable doubt that the potent UDS activity of 2AAF in the rat liver presents evidence of it initiating carcinogenesis in this, its target organ in the rat. Equally, the activity of BP in assays associated with its sites of carcinogenesis, the skin and lung in the mouse and the forestomach in the rat (Figure 10), is probably significant, as is the corresponding genotoxic inactivity of PYR, 2AAF and 4AAF in these assays.

Forster (Chapter 4), when discussing the genotoxic profile of BP and PYR, emphasizes the dangers inherent in interpreting short-term *in vivo* genotoxicity data in terms of carcinogen organotropy, and his cautions are essentially endorsed here. Thus, it is recommended that only when sustained research effort is to be devoted to the study of a particular carcinogen should acute genotoxicity data be interpreted beyond providing evidence that the chemical is genotoxic *in vivo*. In particular, the earlier suggestion regarding the rat liver UDS test providing a complement to the mouse bone marrow MN assay was aimed at providing an efficient screen for *in vivo* genotoxins, rather than at providing a means to predict mouse leukaemogens or rat hepatocarcinogens, respectively. This statement has the potential to be subtlely reversed to the detriment of *in vivo* genotoxicity assays, as emphasized by Forster in Chapter 4. A clearer example is provided in a recent paper by Hellemann *et al.* (1984). These authors evaluated the colon-specific carcinogen 3,2′-dimethyl-4-aminobiphenyl in the *in vivo* rat liver UDS test, and found it to be active. Viewed from the present position this was a successful outcome. However, Hellemann *et al.*, counted this as evidence for the non-specifity of the liver assay. Fortunately, the inactivity of the mouse liver carcinogen 2AAF in the mouse liver UDS assay has provided a perfect vehicle for the future study of this topic, and this will be returned to later in this chapter.

Due to the protracted time required for their conduct, assays which monitor the increased incidence of enzyme altered foci, or the production of pre-neoplastic nodules (the liver initiation/promotion assays; Table 1, 2.1.1–6) cannot be considered 'short-term' and are therefore unsuitable for screening purposes. However, they provide a means of studying the progression of liver carcinogenesis. The only determination made following a single dose of the test chemical to animals which had not been compromised by surgery (assay 2.1.1) gave a negative response for 2AAF, while the other 5 tests found it

Figure 10. Qualitative responses observed for the four test chemicals in the five assays selected specifically for the evaluation of BP and PYR. These assays are entered under the miscellaneous heading in Table 1 because they are not employed routinely for screening purposes. Each of these assays was either known to be sensitive to polycyclic aromatic hydrocarbon genotoxins, or such sensitivity was expected, at the start of the study. However, their sensitivity to other classes of genotoxins was considered worthy of study. The reduced sensitivity observed for 2AAF indicates that assay sensitivity may be influenced strongly by the class of chemical under study. The exact reverse of the relative sensitivities shown in this Figure is observed when considering assays in the rodent liver; there, sensitivity is high to 2AAF and absent for BP.

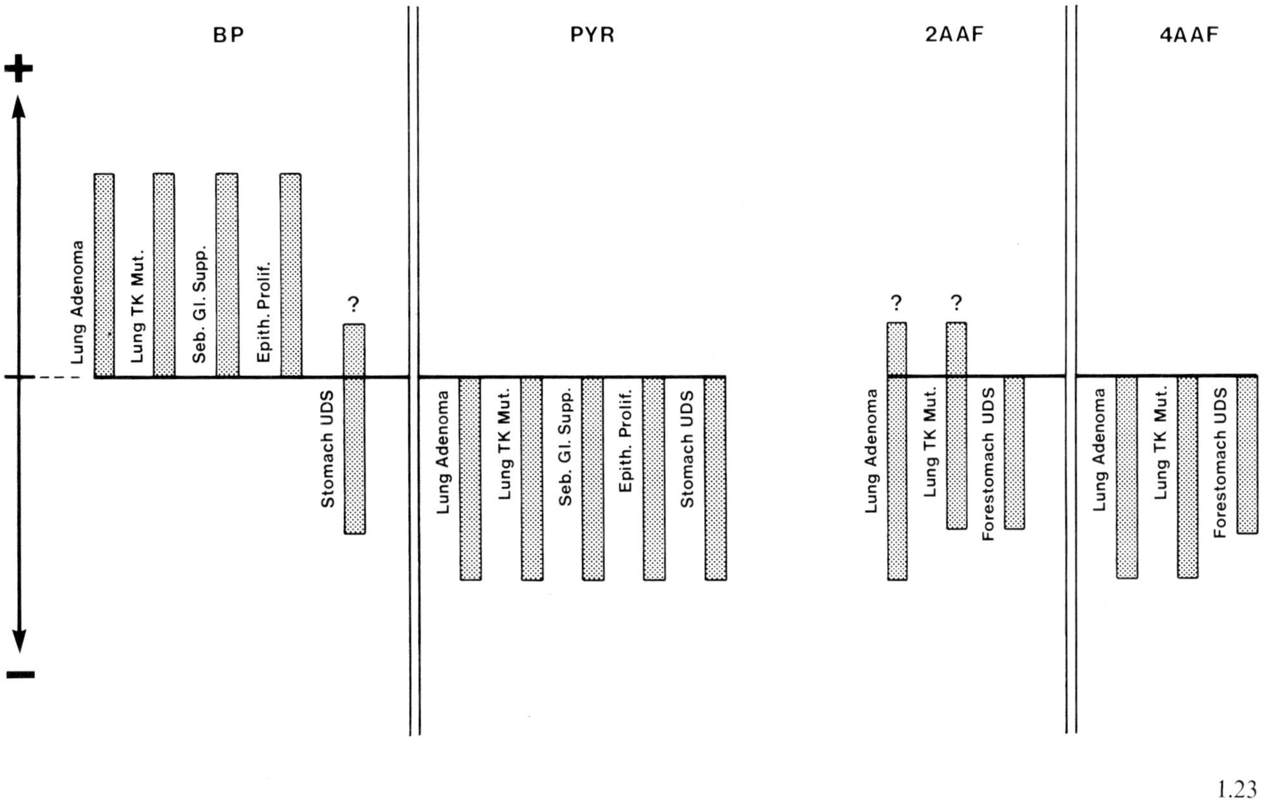

positive. This underlines a second problem of this class of assay, namely, that the dramatic effects of surgical intervention make it difficult to interpret such data for hazard assessment purposes. Thus, a single dose of 500 mg/kg of 2AAF followed by 60d of observation yielded a negative result (assay 2.1.1; Table 1) while a single dose of only 10 mg/kg, followed by a partial hepatectomy and promotion with phenobarbitone gave a clear positive response for this chemical after a similar period of observation (assay 2.1.2; Table 1). These considerations apart, this class of test is generally regarded as having a role to play in research into the development of liver cancer. The data derived from assays 2.1.2 – 2.1.5 involved submitting the test animals to a partial hepatectomy, and each found 2AAF positive. It is of interest that in 2 of these cases, 4AAF also showed activity. This may be consistent with its weak activity in the liver UDS tests (see Chapter 3). Of greater interest, two assays detected BP as positive which is inconsistent with all other genotoxicity data for this chemical in this tissue of the rat, as seen in this study. Perhaps the presence of an actively dividing liver encourages the expression of an activity not apparent in the quiescent tissue. It was for this reason that the UDS activity of BP was studied in weanling rats, but no activity was seen (assay 2.2.10). Enzyme altered foci do not provide definitive evidence of a genotoxic event, but the fact that a low level of DNA-BP adducts was detected in the rat liver following its ip injection (assay 3.2.7a) suggests that BP may have a very low level of genotoxicity in that tissue by that route of exposure. However, this level seems to be too low to elicit other genotoxic changes or tumours following its oral administration. As such, these data may provide an interesting example of low level genotoxic events which do not express themselves as cancer within the animal's lifetime. Hicks was able to discriminate clearly between 2AAF and 4AAF in an enzyme altered foci study conducted as part of her cancer bioassay of these two chemicals (assay 2.1.6; itself a part of assay 3.1.1). Clearly, the eventual outcome of that cancer bioassay of 4AAF will influence an appraisal of the performance of the assays in this group.

The definition of non-genotoxic mechanisms of carcinogenesis

This overview concludes with a discussion of the single most important issue requiring solution at present – the reality or otherwise of non-genotoxic mechanisms of cancer induction. The situation can be stated, using the extremes of 2AAF and diethylhexylphthalate (DEHP), as follows. 2AAF is clearly genotoxic both *in vitro* and *in vivo*, culminating in its potent carcinogenicity to rodents. Links among all of these separate activities are assumed to exist. In summary, 2AAF is a genotoxic carcinogen to rodents. At the other extreme are compounds such as the hepatocarcinogen DEHP which is essentially inactive in primary *in vitro* genotoxicity tests as well as in those conducted *in vivo*. In particular, it does not interact covalently with rodent hepatic DNA, neither does it elicit UDS in this, its sole target for carcinogenesis. DEHP has therefore been described as a non-genotoxic (or epigenetic) liver carcinogen. All of the current genotoxicity screening strategies, whether they involve assays conducted *in vitro* or *in vivo*, would, therefore, fail to alert to this carcinogen. Thus, so long as 2AAF and DEHP (for example) are regarded as unqualified 'carcinogens', the predictivity for carcinogenicity of genotoxicity tests will continue to be debated (see, for example, Legator and Harper, 1986). Two explicit formulations of this problem are stated here to justify the concern being expressed in this section of the Overview. Namely, that non-genotoxic mechanisms of carcinogenesis are only invoked by apologists for genotoxicity assays as a convenient hypnotic when faced with an apparently non-genotoxic carcinogen; or, that the perceived need to define 2 classes of carcinogen merely endorses what some have long held to be self-evident, namely, that chemically induced cancer is incapable of being reduced to a common denominator, and that genotoxicity is at best but parallel phenomena observed for a sub-group of electrophilic carcinogens. Alternatively, the viewpoint exists which accepts that some carcinogens are non-genotoxic, and that as such they present a reduced hazard as compared to genotoxic carcinogens. These extremes must be reconciled before genotoxicity assays can be employed with confidence. This will necessitate specific studies to evaluate the underlying assumptions of these arguments. What should probably be avoided is the current tendency to convenient rationalizations which change with each new problem and with each new observation. A vivid example of such a situation is provided in the recent literature by Ziegler-Skylakakis *et al.* (1985) who have presented the first limited evidence for the genotoxicity of thiourea, and in so doing, implicitly challenged Rubin's (1985) concomitant and considered acceptance of a hormonal mechanism of action for this carcinogen. One such situation occurred in this study and is raised here as it provides an optimum example for the further study of non-genotoxic mechanisms of carcinogenesis.

2AAF is a potent and general genotoxin to the rat liver. In particular, it elicits UDS in this tissue at dose-levels as low as 15 mg/kg. This is consistent with its potent carcinogenicity to the rat liver as confirmed by Hicks in the present study (assay 3.1.1). 2AAF is also a genotoxin in certain tissues of the mouse, as evidenced by its clastogenicity to both the bone marrow and peripheral lymphocytes of mice. It is carcinogenic to the mouse bladder and liver. 2AAF would, therefore, appear to be a genotoxic carcinogen to both the rat and the mouse. However, it failed to elicit a UDS response in the mouse liver. Mirsalis (Chapter 58) has suggested, in order to explain this occurrence, that the mouse liver carcinogenicity of 2AAF may be mediated by an epigenetic mechanism which is associated with the appearance of an increase in S-phase cells in the liver. Cells undergoing 'scheduled' DNA-synthesis (S-phase cells) are distinct from those undergoing UDS, and provide implicit (but not definitive) evidence of a wave of cell division occurring in the liver (Mirsalis *et al.*, 1985). The hepatocarcinogenicity of DEHP has recently been rationalized in these terms (Butterworth *et al.*, 1984), but in that case the suggestion of a non-genotoxic mechanism was made within the context of the well established non-genotoxicity of this agent. However, the proposal of a non-genotoxic mechanism for the mouse hepatocarcinogenicity of 2AAF, while not being novel (Williams and Numoto, 1984), is all together more provocative. The implications of this proposal are formalized in Figure 11. The main arguments for and against it are suggested to be as follows.

The hepatocarcinogenic activity of 2AAF to the mouse is dramatically attenuated as compared to its activity in the rat liver, as evidenced by the results of the 'mega-mouse' carcinogenicity bioassay (discussed in Appendix 3 and shown graphically in Figure 12). Liver tumours appeared only late in the study, and then only when the spontaneous incidence in the control animals was itself beginning to rise sharply (Figure 12). This is in marked contrast to the rat where animals succumbed from liver tumours within 6 months of cessation of a 3 month weekly-dosing regimen (Hicks, assay 3.1.1). The mouse liver carcinogenicity of 2AAF is therefore consistent with the temporal advancement of 'spontaneous' liver tumours in the treated animals due to chemically-mediated cell turnover in the liver. It is obviously important that the supporting hepatocyte S-phase data reported by Mirsalis (assay 2.3.4) should be enlarged and independently reproduced. The negative UDS responses reported for 2AAF in the mouse liver assays (2.2.1 and 2.3.5) were paralleled in this study by its inactivity as a UDS-inducing agent in isolated mouse hepatocytes *in vitro* (Appendix 2), thus suggesting a basic inability of these cells to metabolize 2AAF to a DNA-reactive species. Were 2AAF to be an agreed non-genotoxin, this construction to the data could be accepted as a reasonable explanation of the available facts. But 2AAF is a genotoxin in extrahepatic tissues of the mouse, and this demands a high level of discernment in accepting this explanation. Points mitigating against the S-phase explanation are as follows. First, the tetraploid population of hepatocytes in both mice and rats was reduced by 2AAF-treatment. This was associated by Styles (Chapter 78) with the genotoxicity of this chemical to the liver, and although the relevance of ploidy changes to carcinogenicity was questioned by the workgroup (Chapter 72), the effects induced by 2AAF in the rat have recently been independently confirmed (Schwarze *et al.*, 1986). Second, the ^{32}P post-labelling technique identified a low but positive level of 2AAF-DNA adducts in the mouse liver (assay 3.2.7c). The value observed was 17 lesions per 10^7 bases, as compared to $117/10^7$ 2AAF-DNA adducts in the rat liver (assay 3.2.7a). The value of $17/10^7$ bases, while low in relation to the rat liver value, is higher than observed for BP in the lung, one of BP's target organs for (presumably) genotoxic carcinogenesis. In the liver assay group report by Hunt (Chapter 48) it is suggested that $17/10^7$ lesions may be insufficient to trigger a positive UDS response. That is certainly worthy of further evaluation, but new data will be required. This is because the ^{32}P post-labelling experiments were conducted following ip injection of the test chemical, a generally inappropriate route of exposure for liver assays, especially if, as here, the derived data are to be contrasted with effects observed following oral administration. A third implication of the proposal of a non-genotoxic mechanism is the extent to which liver S-phase activity is indicative of hepatocarcinogenicity for all species. Thus, 4AAF has a truly dramatic effect on the incidence of S-phase cells in the rat liver. In some assays (e.g. 3.1.1) levels equalling that observed following a partial hepatectomy were seen (i.e. ~ 40% cells in S-phase). Mirsalis *et al.* (1985) and Butterworth *et al.* (1984) have explained the rat hepatocarcinogenicity of DEHP using S-phase effects much lower than those seen here for 4AAF. Therefore, the proposal of a non-genotoxic mechanism of carcinogenic action for 2AAF in the mouse liver implies that 4AAF will be a clear hepatocarcinogen to the rat. The ongoing study by Hicks has so far given no indication that this will occur; the final report is thus awaited with great interest.

4 Conclusions

The *in vivo* genotoxicity data from this study are plotted in Figure 13 and it is apparent that the trend established in the earlier study (Figure 2) has been maintained, i.e. most of the *in vivo* somatic cell genotoxicity assays were able to discriminate the two non-carcinogens from the related carcinogens. In some instances, weak activity was also evident for the 2 non-carcinogens, particularly for 4AAF. The study has focussed the general need for the development of adequate positive test response criteria. Some assays were insensitive to one or the other of the pairs of chemicals which underlines a trend already evident in the literature, namely, that negative *in vivo* data must be produced in at least two assays sited in different tissues if they are to be used for hazard assessment. The mouse bone marrow micronucleus assay is confirmed as a robust and sensitive assay and should probably be conducted first on new *in vitro* genotoxins. The reproducibility of the data from the rat liver UDS assay suggests that it could act as a second *in vivo* genotoxicity assay, but further attention should be

Figure 11. Philosophical assumptions which underly the rationalization presented for the inactivity of the mouse liver carcinogen 2AAF in the mouse liver UDS assay. On the left of the figure is presented what is assumed to be a classical genotoxicity profile (i.e. genotoxicity for a carcinogen *in vitro* and in the target organ of carcinogenesis *in vivo*). On the right is the presumed profile of a non-genotoxic (epigenetic) carcinogen such as diethylhexylphthlate (cf. Butterworth *et al.*, 1984; Mirsalis *et al.*, 1985). The question mark applies for the genotoxin 2AAF in the mouse liver *in vivo*. Its inactivity there is either evidence of assay insensitivity or of an epigenetic aetiology of cancer in the mouse liver (see text and Figure 12).

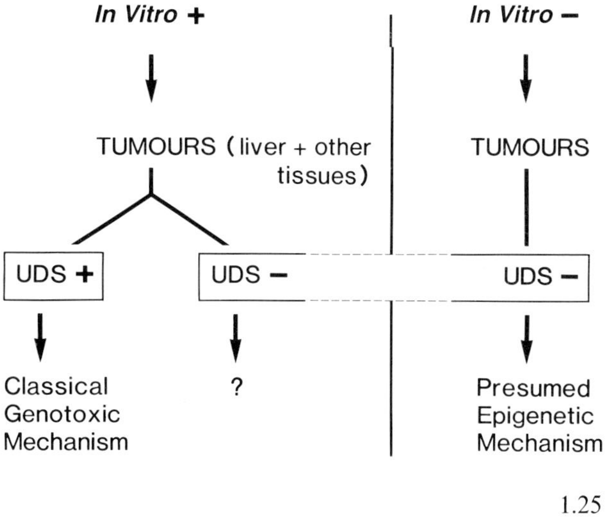

focussed on which liver genotoxicity test is most appropriate for general use; at present, the UDS assay is the only generally available option. The role of the rodent host-mediated and urine mutagenicity assays, and of assays in Drosophila, for routine hazard determination studies has been called into question by the results of this study.

Only BP gave evidence of germ cell mutagenicity, and then only when tested using the ip injection route. This selective activity demonstrates that the route of administration selected can affect the outcome of experiments. Further, this activity of BP was adequately predicted by the somatic cell mutagenicity assays, and this confirms earlier suggestions that germ cell tests should only be conducted to gain further information on the mutagenic spectrum of established *in vivo* somatic cell mutagens. The present two genotoxic carcinogens showed genotoxic activity in their target organs of carcinogenesis, but their additional activity in a range of other issues precludes the use of short-term *in vivo* assay data for making predictions of organ-specific carcinogenic activity. The majority of the positive responses observed in this study were obtained using the oral route and dose-levels substantially below the lethal levels. These facts, together with the observation of weak genotoxicity for 4AAF, indicate that short-term *in vivo* assays are not intrinsically insensitive, as is sometimes inferred. The suggestion made in this study that 2AAF may be a non-genotoxic carcinogen to the mouse liver should be pursued urgently in order to confirm or refute the validity of the concept of non-genetic mechanisms of cancer induction.

It is therefore concluded that short-term *in vivo* genotoxicity assays have a critical role to play in the hazard assessment process, though not as primary screens. This role is to define which new *in vitro* genotoxins are active *in vivo*, and thereby to identify those most likely to present a carcinogenic/mutagenic hazard to mammals. The genotoxic events monitored by the *in vivo* assays in current use may not precisely represent causative steps in tumour development, therefore, a clear distinction should be made between hazard assessment

Figure 12. Dose-response data for the carcinogenicity bioassay of 2AAF in the mouse (the mega-mouse study, J. Environ. Path. Tox. *3*, 17, 1979). Liver tumors were observed only late in the study and then in both the treated and control groups. In contrast, 2AAF is a potent and rapid carcinogen to the rat (see assay 3.1.1, this volume). These murine carcinogenicity data are discussed in Appendix 3 and were invoked in the liver genotoxicity assay workgroup report (Hunt *et al.*, this volume, Chapter 48). The blue hatched area represents the increasing incidence of tumor bearing animals (TBA) with time, and the red area the increase in TBA with increasing dose-levels of 2AAF at 33 months into the study.

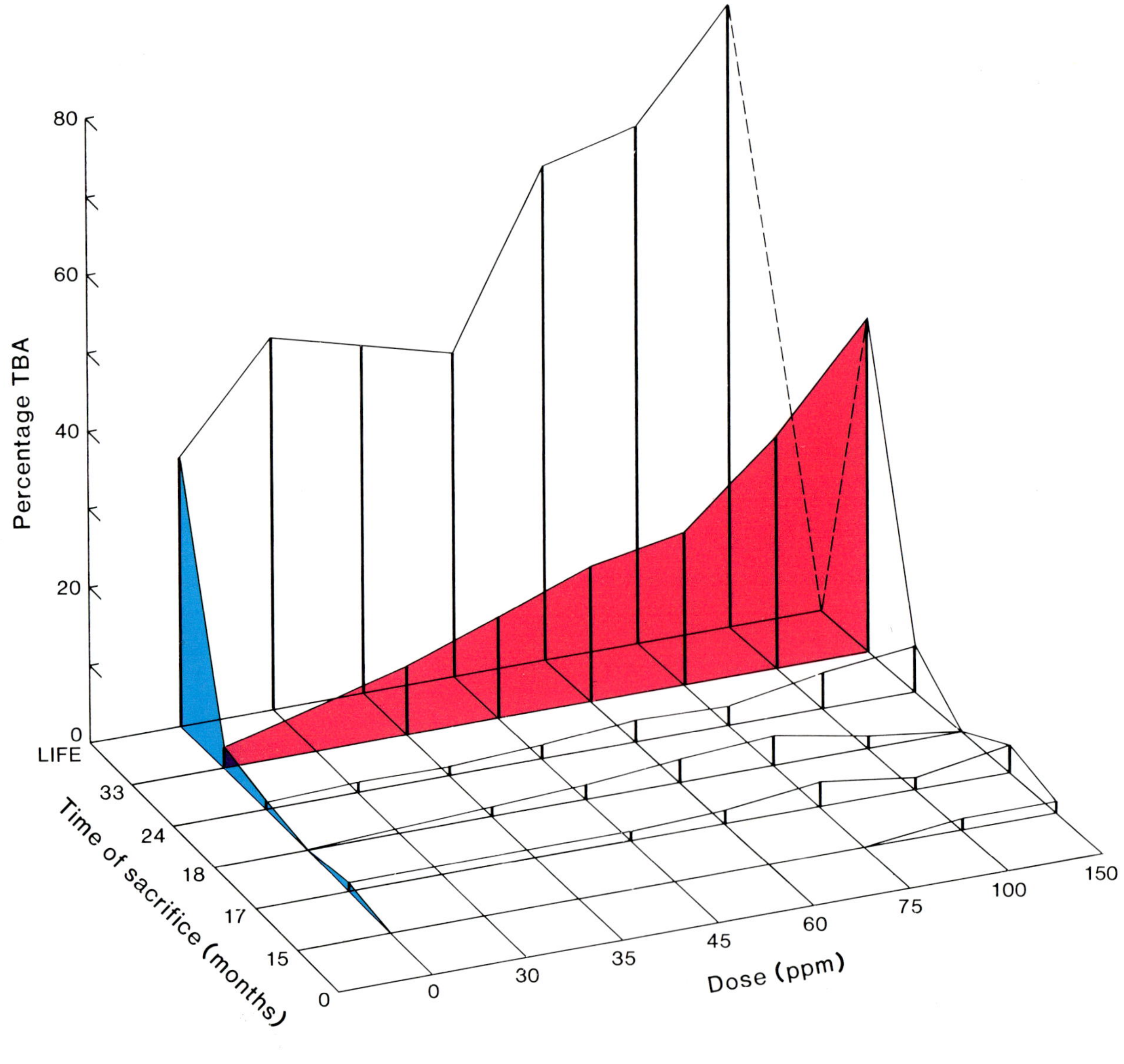

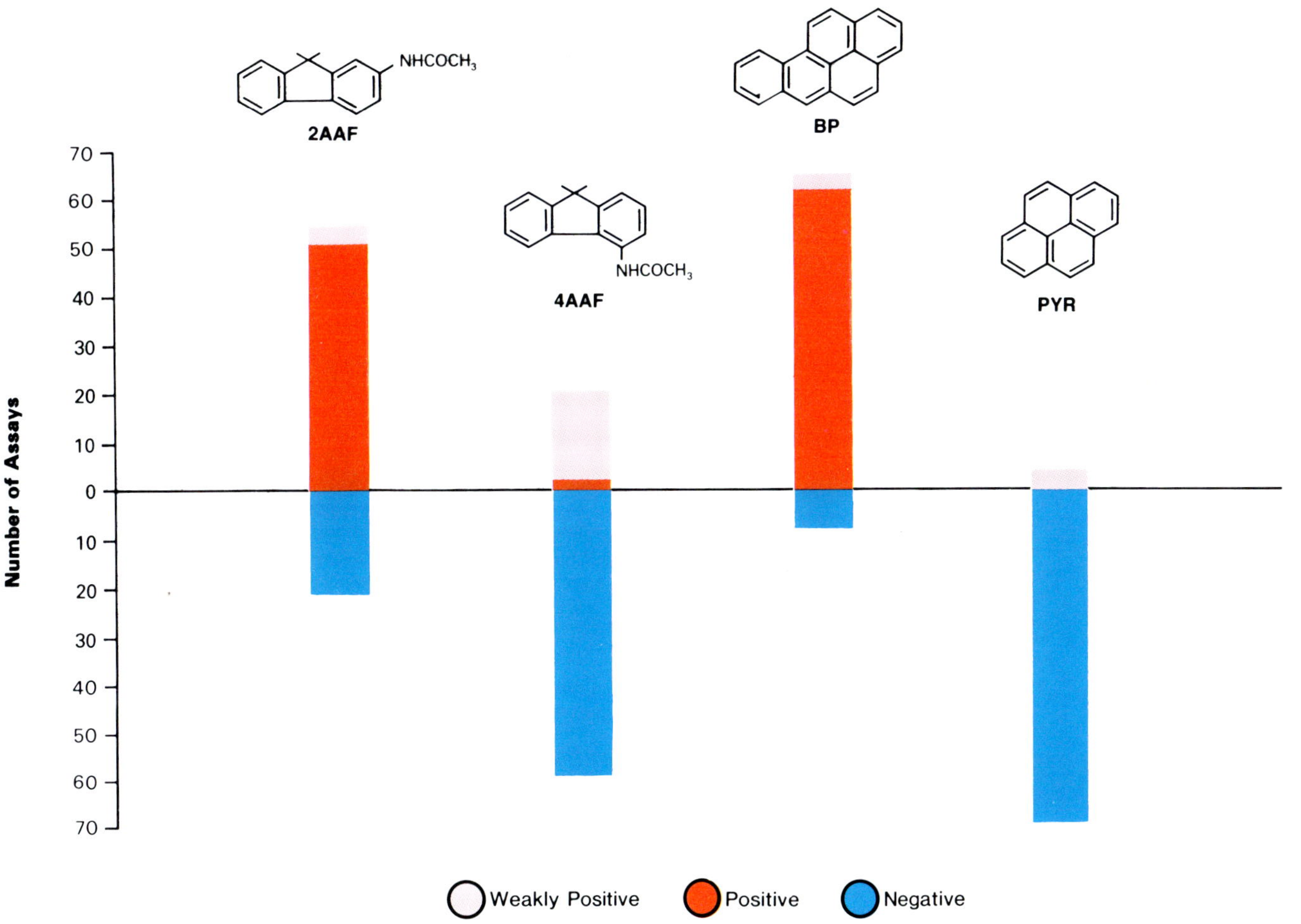

Figure 13. Activity of the 4 test chemicals in the *in vivo* genotoxicity assays shown in Table 1, with the following assays omitted: 2.3 and 3.3 (not evidence of genotoxicity), 3.1.1 and 3.1.3 (data awaited) and 6 (Drosophila). In addition, data from the following assays have been omitted based on comments made in the workgroup reports: 2.5.1–3; 3.3–3.6 and 5.2. The remaining 295 assays provide data on the interaction of the test chemicals with DNA *in vivo*, following either ip or oral administration. The 8 negative results for BP were tissue-specific (6 liver, 2 gonads) and the 4 weak positive responses for PYR were observed in the bone marrow (3 being for SCE). In contrast, 21 negative results for 2AAF were spread over 4 tissues and indicate a generally reduced genotoxicity *in vivo* when compared to BP. The weak positive responses for 4AAF were almost all in the bone marrow or liver assays. Both members of each pair must have been tested for inclusion in this figure, and a questionable response led to the neglection of a pair of responses. These histograms, when compared with those evident at the initiation of the study (Figure 1), indicate the general ability of the *in vivo* assays employed to resolve the two pairs of test chemicals.

studies and research into the mechanisms of chemical carcinogenesis.

Two of the most crucial issues in environmental mutagenesis and carcinogenesis are suggested to be the question of non-genotoxic mechanisms of chemical carcinogenesis and the distinct roles *in vitro* and *in vivo* short-term genotoxicity assays. The resolution of these issues will facilitate the central task of identifying agents which might reasonably be expected to cause cancer or mutation in exposed populations.

5 References

Anderson, D. (1984) The dominant lethal test in rodents. In *Mutagenicity Testing*, Ed. S. Venitt and J.M. Parry. IRL Press, Oxford p. 307.

Ashby, J., F.J. de Serres, M. Draper, M. Ishidate Jr., B.H. Margolin, B.E. Matter and M.D. Shelby Eds (1985) Evaluation of Short-term Test for Carcinogens: Report on the International Programme of Chemical Safety's Collaborative Study on *in vitro* Assays. *Progress in Mutation Research*, Vol. 5, Elsevier/North Holland, Amsterdam.

Ashby, J. (1986) The prospects for a simplified and internationally harmonized approach to the detection of possible human carcinogens and mutagens, *Mutagenesis*, **1**, 3–16.

Bridges, B. (1974) The three-tier approach to mutagenicity screening and the concept of radiation equivalent dose. *Mutat Res.*, **26**, 335–340.

Bridges, B. (1976) Evaluation of mutagenicity and carcinogenicity using a three-tier system. *Mutat Res.*, **41**, 71–72.

Brockman, H.E. and F.J. de Serres (1983) Short-term tests for genetic toxicity. In *Short-term Tests for Environmentally Induced Chronic Health Effects*. EPA-600/8-83-002, Office of Research and Development, US. Environmental Protection Agency, Washington, p. 29.

Brusick, D. (1980) *Principles of Genetic Toxicology*. Plenum Press, New York, p. 113.

Butterworth, B.E., E. Bermudez, T. Smith-Oliver, L. Earl, R. Cattley, J. Martin, S. Strom, R. Jirtle and G. Michalopoulos (1984) Lack of genotoxic activity of di(2-ethylhexyl)phthalate (DEHP) in rodent and human hepatocytes, *Carcinogenesis*, **5**, 1329–1335.

de Serres, F.J. and J. Ashby (Eds) (1981) Evaluation of Short-term Tests for Carcinogens: Report of the International Collaborative Study. *Progress in Mutation Research*, Vol. I, Elsevier/North Holland, Amsterdam.

Gocke, E., M.J. King, K. Eckhardt and D. Wild (1981) Mutagenicity of cosmetics ingredients licenced by the European Communities, *Mutation Res.*, **90**, 91–109.

Hellemann, A., C.J. Malansky, M. Bosland and G.M. Williams (1984) Rat liver damage by the nonhepatocarcinogen 3, 2′-dimethyl-4-aminobiphenyl, *Cancer Lett.*, **22**, 211–218.

Holden, H.E. (1982) Comparison of somatic and germ cell models for cytogenetic screening, *J. Appl. Toxicol.*, **2**, 196–200.

Jenssen, D. and C. Ramel (1980) The micronucleus test as part of a short-term mutagenicity test program for the prediction of carcinogenicity evaluated by 143 agents tested. *Mutat Res.*, **75**, 191–202.

Legator, M.S. and D. Williams (1975) The inadequacies of sub-mammalian tests for mutagenicity. In *Molecular and Environmental Aspects of Mutagenesis*. Ed, L. Prakash, Thomas Springfield.

Legator, M.S. and B. Harper (1988) Mutagenicity screening and *in vitro* testing: The end of an era; Animal and Human Studies: Directions for the future, *Ann NY Acad. Sci* (*in press*).

McCann, J. and B.N. Ames (1976) Detection of carcinogens as mutagens in the Salmonella/Microsome Test: Assay of 300 Chemicals: Discussion. *Proc. Natl. Acad. Sci. USA* **73**, 950–954.

Mirsalis, J.C. and B.E. Butterworth (1980) Detection of unscheduled DNA synthesis in hepatocytes isolated from rats treated with genotoxic agents. *Carcinogenesis*, **1**, 621–625.

Mirsalis, J.C., C.K. Tyson, E.N. Loh, K.L. Steinmetz, J.P. Bakke, C.M. Hamilton, D.K. Spak and J.W. Spalding (1985) Induction of hepatic cell proliferation and unscheduled DNA synthesis in mouse hepatocytes following *in vivo* treatment, *Carcinogenesis*, **6**, 1521–1524.

Newton, M.F. and L.J. Lilly (1986) Tissue-specific clastogenic effects of Chromium and Selenium salts *in vivo*, *Mutation Res.*, **169**, 61–69.

Rubin, H. (1985) Cancer as a Developmental Disorder, *Cancer Research*, **45**, 2935–2942.

Russell, L.B. and M.D. Shelby (1985) Tests for heritable genetic damage and for evidence of gonadal exposure in mammals. *Mutat. Res.*, **154**, 69–84.

Schwarze, P.E., E.O. Pettersen and P.O. Seglen (1986) Characterization of hepatocytes from carcinogen-treated rats by two-parametric flow cytometry, *Carcinogenesis*, **7**, 171–173.

Shelby, M.D. and S. Stasiewicz (1984) Chemicals showing no evidence of carcinogenicity in long-term two species rodent studies. The need for short-term test data. *Environ. Mutagen*, **6**, 871–878.

Shelby, M.D. (1986) A case for the continued use of the intraperitoneal injection route of exposure, *Mutation Res.* **170**, 169–171.

Stella, M., L. Trevison, A. Montaldi, G. Zaccaria, G. Rossi V. Bianchi and A.G. Levis (1984) Induction of SCE in human lymphocytes exposed *in vitro* and *in vivo* to therapeutic ultrasound, *Mutation Res.*, **138**, 75–85.

Styles, J.A. and M.P. Penman (1985) The mouse spot test: an appraisal of its performance in detecting chemical carcinogens and mutagens. *Mutat. Res.*, **154**, 183–204.

Vogel, E.W., A.A. van Zeeland, C.A. Raaymakers-Jansen Verplanke and J.A. Zijlstra (1985) Analysis of HMPA-induced genetic alterations in relation to DNA damage and DNA repair in Drosophila melanogaster, *Mutation Res.*, **150**, 241–260.

Williams, G.M. and S. Numoto (1984) Promotion of mouse liver neoplasms by the organochlorine pesticides chlordane and heptachlor in comparison to dichlorodiphenyltriochloroethane, *Carcinogenesis*, **5**, 1689–1696.

Zeigler-Skylakakis, K., S. Rossberger and V. Andrae (1985) Thiourea induces DNA repair synthesis in primary rat hepatocyte cultures and gene mutations in V79 Chinese hamster cells, *Arch Toxicol.*, **58**, 5–9.

Barry H. Margolin and Kenneth J. Risko
National Institute of Environmental Health Sciences, Biometry Risk Assessment Program, Research Triangle Park, North Carolina, USA

3

The statistical analysis of in vivo genotoxicity data: case studies of the rat hepatocyte UDS and mouse bone marrow micronucleus assays

1 Introduction

The vast amount of experimentation constituting the backbone of the present study provides a unique and invaluable database for examining the behavior of a wide variety of *in vivo* assays. The decision of the IPCS Steering Committee to computerize these data at the National Institute of Environmental Health Sciences greatly facilitated the analyses to be described in this chapter; pressures of timely publication of this volume, however, dictated that discussion of analyses be limited to two of the four assays presently judged appropriate for routine screening, the mouse bone marrow micronucleus assay and the measurement of unscheduled DNA synthesis (UDS) in rat hepatocytes. The UDS assay is considered first and in greatest detail because its need for criteria of evaluation appears paramount; however, the need to address such issues as evaluative criteria, measurement of potency, utility of historical controls and assessment of interlaboratory variability transcends any one *in vivo* assay, and certainly applies to all three mouse bone marrow cytogenetics assays highlighted in the Overview (Chapter 2).

2 UDS in rat hepatocytes

One of the major conclusions of the current IPCS study is that *in vivo* measurement of UDS in rat hepatocytes is an appropriate assay for further routine screening of *in vivo* genotoxins, and as such, UDS serves as a suitable complementary assay to the well-established primary *in vivo* screen for this purpose – the set of mouse bone marrow cytogenetics assays. In light of this conclusion, it is important to note that the criteria in common use for an unambiguous positive response in this assay, at least one dose that yields $\geqslant 5$ NG [net grains] (rat mean) coupled with evidence of a dose response, are conservative. These criteria are colored to a large extent by the criteria postulated by Williams (1977) for individual cells in his *in vitro* assay, namely, if a hepatocyte yielded $+5$ or more NG, it was deemed in repair.

Ashby *et al.* (1985) recently traced the history of the adoption and subsequent attempts at modification of the $+5$ NG criterion for an *individual* animal to be in repair for the *in vivo* assay. These authors proposed their own criterion of a mean of $+3$ or more NG, which by their description was arbitrary, but less conservative. Even this weaker criterion is deserving of a goodly measure of skepticism when one learns that none of the investigators participating in the current study has ever observed a single vehicle control animal with an average net grain response greater than zero. A + 5 NG criterion extended to the mean of a treated group of animals is an even more conservative criterion. Based on biological considerations, the final UDS workgroup report (Chapter 55) indicates that NG values greater than zero is one of the criteria necessary to define activity in the assay. In as much as this criterion does not reflect individual laboratories' experiences with both concurrent and historical control values, it still is somewhat arbitrary; nonetheless, it will be accepted here as reflective of biological significance. This colors the following discussion because recourse to control values in some form must be made in a statistical analysis, and this has led to the classification of some data, e.g. Butterworth's results for 4AAF (Chapter 56), as statistically positive even though only negative NG values were observed. Thus, while there may be good biological reasons for questioning the statistical decisions presented here, they are included to focus the issue: these biological considerations need further substantiation so that it is fully documented that only NG values >0 evidence nuclear UDS. If the UDS assay is to be employed widely in routine screening, there is a clear need for objective criteria for a positive response, criteria that would be appropriate for the weakly acting *in vivo* genotoxins that one envisions encountering. Based on present knowledge of carcinogen predictivity, there is no justification for devising assay evaluative criteria to detect only strongly acting *in vivo* genotoxins. The final UDS workgroup report (Chapter 55) reflects the progress that has been made during this study in constructing appropriate evaluative criteria.

Sources of variability

In devising objective evaluative criteria for *in vivo* UDS data, one must be cognizant of three potential sources of variability in control responses for a laboratory performing this assay:

(i) cell to cell within an animal,
(ii) animal to animal within a day, and
(iii) day to day, or experiment to experiment.

Despite the fact that there are UDS control data now available on over 350 rats, no published data appear to exist that permit one to distinguish between that portion of variability among control animals tested on different days that is due to uncontrolled variations in day-to-day experimental conditions, and that portion due to inherent animal-to-animal differences. The usual practice in laboratories performing this assay is to employ a single control animal per day (Butterworth, personal communication), a practice that clearly precludes resolution of these two sources of response variability. The inferential dilemma this poses for statistical analysis is substantial, for it leaves open the question of how to utilize most effectively a laboratory's historical control data (Margolin and Risko, 1984). Consideration of the variance of a control group mean in the next paragraph elucidates the dilemma.

If there is negligible day-to-day variability in control responses compared with animal-to-animal variability within a common day, then the variance of the mean response among n control animals is approximately $1/n$ times the variance of the response of a single control animal. In this case, the historical control distribution would give a clear picture of animal-to-animal variability, and would readily permit a statistical assessment of the evidence for genotoxicity in the response of a group of treated animals. This assumption is consistent with the position adopted by Ashby *et al.* (1985), and partially explains why there is typically only one concurrent control animal in an experiment. While this practice is surely expedient, it is based primarily on intuition, there being little published empirical evidence for its support. If, on the other hand, there is significant day-to-day variability reflected in the historical control responses, then the variance of the mean of n control animal responses obtained on a common day may be substantially greater than $1/n$ times the variance of a single animal's response when viewed from the perspective of the historical control database. For example, if the day-to-day variability is equal in magnitude to the animal-to-animal variability within a common day, then the variance, day-to-day, of the mean response of 5 control animals that have been studied on a common day is $3/5$, not $1/5$, times the variance of a single control animal's response because the day effect has not been eliminated. Were this magnitude of day-to-day variability to prevail, it would have direct implications for the design and analysis of UDS experiments. In the absence of evidence to the contrary, it will be assumed here that the intuitive position of most UDS investigators is correct and that day-to-day variability in control animal responses is relatively insignificant when compared to animal-to-animal variability for laboratories in this study. The need to resolve this matter formally and expeditiously is pressing; it will necessitate experiments with multiple replicate control animals on each of a number of days, preferably in more than one laboratory.

The role for historical control data

Precisely how is one to employ historical control data in the analysis of a UDS experiment? Historical rat control data for a number of UDS investigators were made available by Dr. Butterworth for purposes of this discussion. The data on an individual animal basis are presented graphically in Figure 1. Summary statistics and descriptions of these data are presented in Table 1, with the order of investigator determined by decreasing rat mean NG per animal value. Two immediate observations are that: (i) across laboratories, the inter-animal standard deviation generally increases with decreasing animal mean, and more important, (ii) there are substantial differences in the mean NG per animal values among the various historical control datasets in Table 1, the range of means varying from -5.7 to -2.0 NG per animal. These mean NG per animal differences may be attributable to either inter-laboratory or inter-strain differences. The reader will note, for example, that the two highest means in Table 1 were obtained by laboratories working with strains of rat other than the Fischer 344, which constituted the bulk of the historical control database examined. The recommendation of Ashby *et al.* (1985) that each laboratory should develop a sizeable control database before commencing routine use of the UDS assay thus becomes an imperative. A historical control database of 15 seems an absolute minimum before a laboratory commences routine UDS experimentation; employment of another laboratory's control data is presently unjustified and could be seriously misleading. Moreover, based on the substantial interlaboratory differences in control means and variances, it can be argued that no simple numerical criterion, e.g., + 5 NG, can apply universally to determine the significance of the mean response of a group of treated animals. Suitable criteria need tailoring to each laboratory's control results and the number of treated animals involved in an experiment. To this end, it would be most desirable if, in general, UDS data were published on an individual animal basis.

The statistical analysis to be developed below rests heavily on the historical control data available for a given laboratory and the assumption that day-to-day variability in control animals is negligible. Should future evidence prove the contrary, then the statistical analysis here will need modification. In any event, the uncertainties associated with the sources of variability suggest that: (i)

Table 1. *Historical control male rat data for a number of the UDS investigators*[a]

Investigator	Strain	Sample Size	Mean NG per animal	Standard Deviation among animals
Kornbrust	Sprague-Dawley	18	− 1.99	0.77
Ashby/Beije	Alderly Park	34	− 2.28	1.08
Doolittle	Fischer 344	9	− 3.71	2.12
Mirsalis and Butterworth	Fischer 344	52	− 4.26	1.37
Butterworth	Fischer 344	22	− 4.93	1.19
Mirsalis (corn oil)	Fischer 344	35	− 5.13	1.99
Mirsalis (water)	Fischer 344	26	− 5.66	2.68

[a] The vehicle was not specified, save for Mirsalis' two datasets; the names of the individual investigators generating data were provided by Dr. Butterworth.

3 *Figure 1.* Seven distinct sets of historical control data. Each box represents a single animal. The aggregate numbers of animals are indicated below the investigators' names. Mirsalis A and Mirsalis B represent corn oil and water controls, respectively. Summary statistics and strain information are given in Table 1.

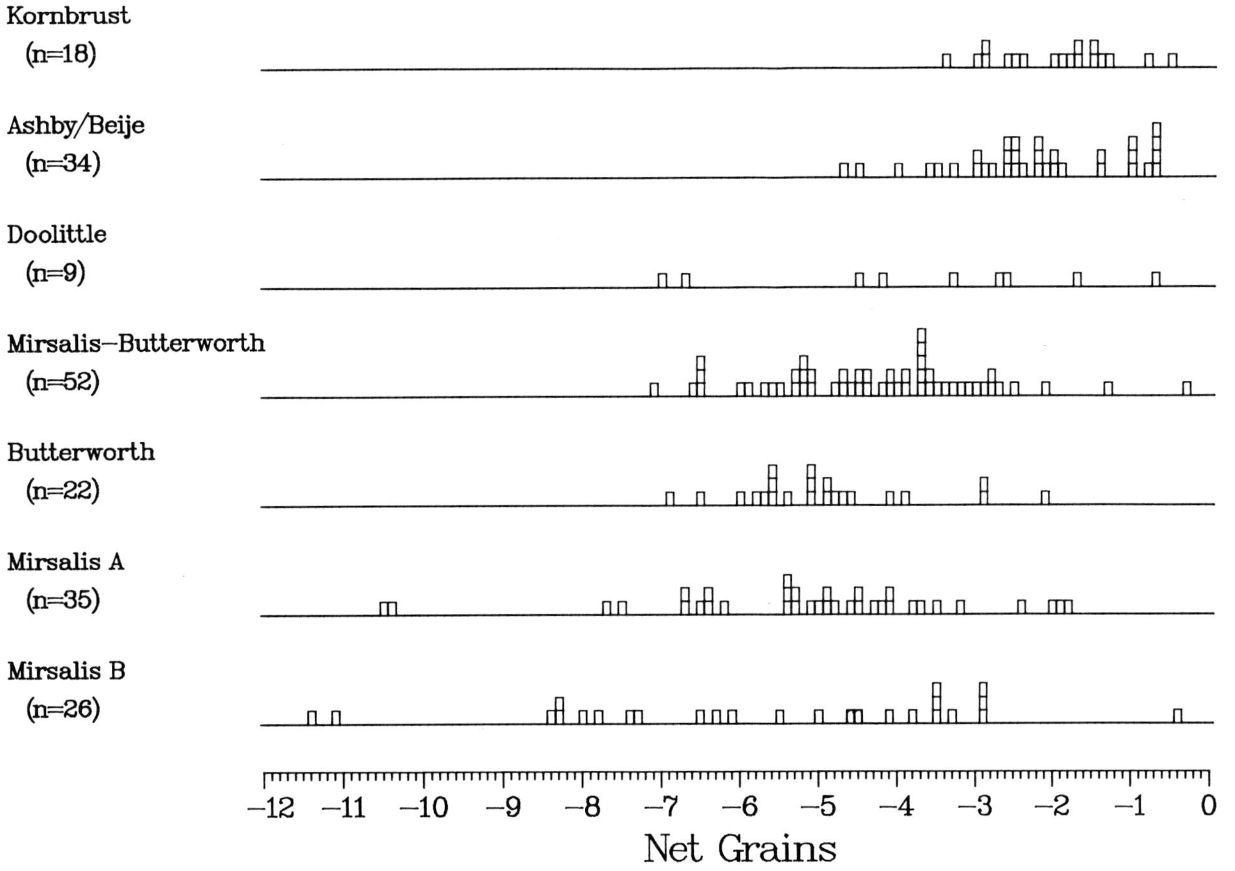

no statistical analysis should be performed before the relevant historical control database has grown to 15 animals or more, and (ii) weak, statistically positive findings are in need of replication. To focus the discussion of the role of historical control data further and to elucidate the main points of the statistical analysis, consider the 4AAF results with Fischer 344 rats obtained by Ashby *et al.* (Chapter 57) in the current study; these data, from three separate experiments, are reproduced in Table 2. The first difficulty encountered is that Ashby has an extensive historical control experience with Alderly Park (AP) rats (Table 1 and Figure 1), but apparently only three other F344 control animals (−2.16, −2.21, −3.4 NG) have been studied for UDS beyond the three in Table 2. All six F344 control rats means appear consistent with the AP rat distribution of historical controls reported for Ashby in Figure 1, but the evidence for no difference between strains within that laboratory is limited. If one did not appeal to the AP history, one would be hard-pressed to feel comfortable with any conclusion regarding 4AAF based solely on the data in Table 2. One way to proceed is by adoption of an additional assumption that inter-strain differences in mean and standard deviation are modest within this laboratory, an assumption in need of further supporting data.

As the use of 2–5 animals per dose is routine in UDS experiments, otherwise commendable nonparametric statistical methods, such as the Mann-Whitney test (Snedecor and Cochran, 1967), are of limited value because the permissible observed significance levels are too restricted; this can then result in diminished statistical power. The relatively rich historical control database, however, permits development of an appropriate parametric analysis based on the normal distribution. Consider the histogram of the frequencies of AP control animals presented by Ashby *et al.* (1985). Figure 2A presents a somewhat modified version of this plot, putting the frequencies on a proportional basis and dividing the X-axis somewhat more finely. In addition, the best fitting normal distribution, i.e., common means and variances, is superimposed. The fit of the normal distribution is not impressive, there being a mild suggestion of bimodality in these controls. From the standpoint of developing a statistical analysis, however, the key point is that one is more interested in the sampling behavior of the mean of 2–5 treated animals and whether one can apply normal sampling theory to that distribution. If the chemical does not induce UDS and there is insignificant day-to-day variability, then the behavior of the mean of the treated animals would be similar to that for the mean of 2–5 control animals drawn from the historical database. Figure 2B illustrates, in histogram form, the randomization sampling behavior of the mean of 3 animals drawn at random from the distribution of individual control animals represented in Figure 2A. Means for all possible triples, 5984 in number, are represented in Figure 2B. The best fitting normal distribution is again superimposed. The mean of the histogram in Figure 2B is of necessity the same as the mean of the histrogram in Figure 2A, whereas the variance has been reduced by the averaging process to approximately 1/3 the original variance. To illustrate the quality of the normal approximation in the tails of the randomization distribution of means of three responses, consider that in the right tail the empirically estimated probabilities associated with the nominal 0.1, 0.05 and 0.01 values are 0.106, 0.041 and 0.007, respectively, whereas in the left tail the corresponding values are 0.109, 0.055 and 0.011. The normal distribution affords a superb approximation for the randomization distribution of the mean of 3 control animals. The approximation is better still for the mean of more than 3 animals, and is acceptable even for the mean of 2 animals (Figure 2C). Similar results hold for all laboratories whose data are presented in Figure 1, confirming that mean UDS scores for two or more control animals from these laboratories may be considered

Table 2. *UDS results with Fischer 344 rats for 4AAF (Ashby et al., current study)*

Experiment	Dose (mg/kg)	Individual animal NG means	NG Group Mean	SD	*p*-value[a]
1	0	−2.67	−2.67		
	1000	6.06			
		3.45			
		7.74	5.08	2.22	$<10^{-4}$
		3.07			
2	0	−3.36	−3.36		
	100	0.31			
		−0.79	−0.12	0.59	.0003
		0.13			
	1000	0.12			
		0.09	0.18	0.13	.0002
		0.33			
3	0	−1.98	−1.98		
	1000	−0.17			
		1.9			
		1.9			
		1.4	1.26	1.00	$<10^{-4}$
		2.3			
		0.25			

[a] *p*-values obtained by a one-tailed normal test against the aggregated historical plus concurrent control (see Appendix 1).

Figure 2. Histogram and best-fitting normal distribution for: (A) original historical control data of Ashby/Beije, (B) means of all possible triples drawn from the original data, (C) means of all possible pairs drawn from the original data.

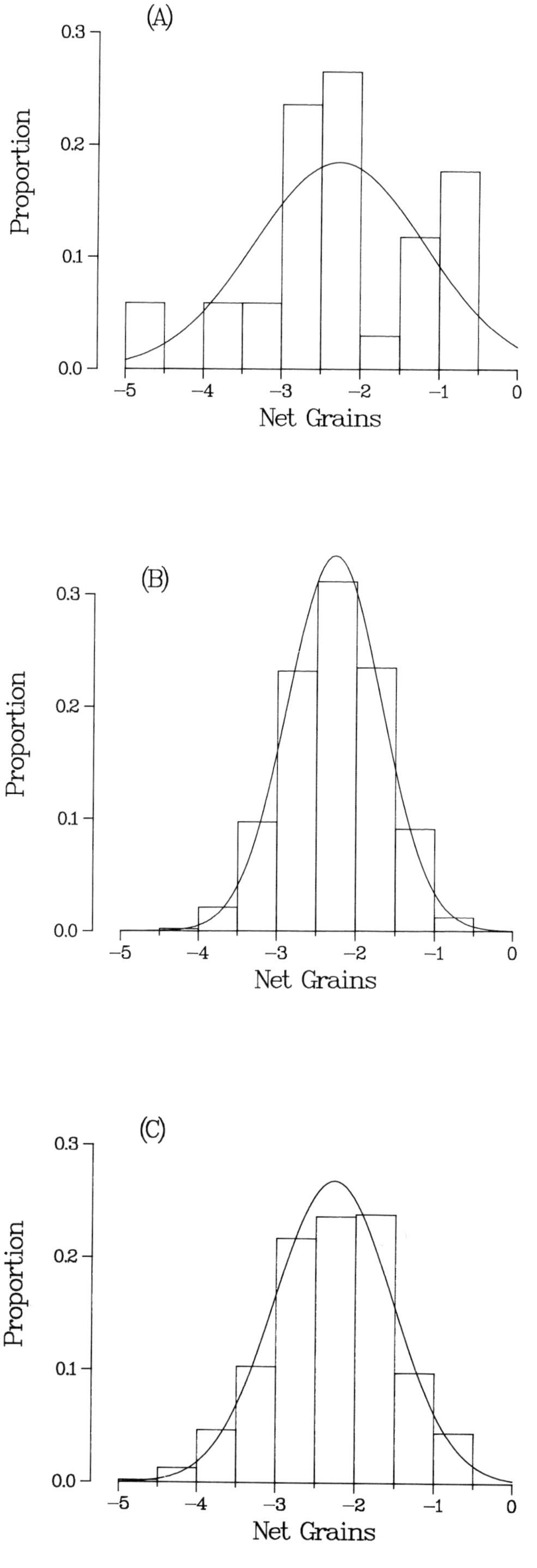

as being normally distributed, even though individual animal measurements may not be well-approximated by a normal distribution.

Methods for assessing statistical significance
Using the normality model for the mean of a set of animal UDS responses, one can readily assess each of the 4AAF group means in Table 2 for statistical significance. Each treatment group mean is first adjusted for a combined historical and concurrent control mean by differencing; this difference is then judged against its estimated standard deviation. The technical details of this statistical analysis are presented in Appendix 1, together with an illustration of the computation for the case of 4AAF at 100 mg/kg in Table 2. Applying this analysis, which ignores possible day-to-day variability, one obtains the statistical significance listed for each of the dose groups in Table 2. Although the inter-experiment variability in the magnitude of the UDS induction in Table 2 is impressive, the collective evidence supports the conclusion that for Ashby's laboratory, 4AAF is a weak but reproducible inducer of UDS.

The results of applying the proposed statistical analysis to all 2AAF and 4AAF data from those laboratories in the study that used the standard *in vivo* rat hepatocyte UDS assay are presented in Table 3. This table is a partial parallel to Table 1 in the UDS workgroup report (Chapter 55) and indicates the gains achieved by applying a common statistical procedure to all data generated with a comparable protocol. In certain cases, e.g., Ashby's 4AAF, the "lowest dose which gave a positive response" has been lowered by an order of magnitude; in other cases, e.g., Butterworth's 4AAF, a response reported to be negative by the investigator and the UDS workgroup is shown to be statistically significant. It is interesting to observe from Table 3 that if a laboratory's results with either chemical were judged statistically significant, the effect was invariably demonstrable at the lowest dose tested.

Multiple times, multiple doses
A number of UDS experiments reported in this volume involve multiple dose groups and multiple times of measurement post-treatment. Regarding the latter, each time presents an opportunity for any statistical analysis to produce a false positive result. If the judging of a response to be positive at any given time is sufficient to declare the chemical positive in the UDS assay, then the multiplicity of times causes the overall chance of falsely declaring a chemical positive to be elevated beyond that for any one time. This problem of inflation of the false-positive rate is well known in the statistics literature, and is labeled "the problem of multiple comparisons". It is not unique to UDS experiments, or even to genotoxicity assays; for example, it has been addressed in the context of the chronic rodent carcinogenicity assay, where the mutliplicity of comparisons is due to the variety of tumor types studied. Although there are various solutions to the problem posed by "multiple comparisons", the most common and most easily effected one is to use a Bonferroni correction to the significance level of each com-

parison. If, for a given chemical, there are n times of UDS experimentation, the results of each of which would be sufficient to label the chemical positive, then the Bonferroni adjustment of any one p-value, say of magnitude p, is to view that p-value loosely as having value equal to n times p. Under such an adjustment, for example, a p-value of 0.04 that might be suggestive of a UDS effect if it were the result of a sole planned comparison, would not justify a positive finding if it were the smallest p-value obtained from three to five different times of measurement post treatment.

The issues presented by multiple dose groups for a given time of measurement extend beyond those of multiple times of experimentation. If one were merely to contrast the response at each individual dose with the control, then again one would face the issue of multiple comparisons. Moreover, one would still be unable to make pronouncements about the evidence for dose response. A better approach that simultaneously eliminates the multiple comparisons among doses and addresses the issue of dose response is to perform a test of linear regression (Snedecor and Cochran, 1967) of the UDS response on dose. This procedure entails computation of a UDS regression coefficient, b, as well as an estimated standard deviation of b. Again, if the day-to-day variability is insignificant, the historical control data can be treated as if they were concurrent controls. The ratio of the regression coefficient b to its standard deviation can be referred to a table of the normal distribution to obtain the level of significance or p-value. As long as the true underlying dose response does not exhibit a substantial downturn at high doses, this statistical approach to assessing significance is presently as good as one can do in assessing significance. For UDS, although the dose response was commonly concave, there was no evidence of such a downturn for either 2AAF or 4AAF.

A measure of potency

The regression coefficient b estimates the *average* slope or amount of increase in NG per unit increase in dose of the test chemical within the range of doses tested. Most measures of potency proposed in toxicology in general are functions of a theoretical slope that describes the impact on response of a unit increase in dose of the test agent. In certain circumstances, e.g., Margolin *et al.* (1981), a mechanistic model relates the slope and the response; other times the relationship is purely empirical. The regression coefficient b is an appropriate empirical measure of potency as long as the observed response is essentially linear. If, however, there is substantial curvature, the use of b may be misleading. Figure 3 contains two hypothetical UDS datasets, both of which yield the same value of the coefficient of linear regression, $b = 0.43$. While b would still be appropriate for assessing statistical significance in these two cases, its use as a measure of potency misrepresents reality. In the case of the concave dose response, represented by triangles, the slope of the response initially is substantially greater than $b = 0.43$, and remains greater until a dose of 50 mg/kg is exceeded. Conversely, for the convex dose response, represented by o's, the slope of the response initially is substantially smaller than $b = 0.43$, and remains smaller until the same 50 mg/kg dose is exceeded.

As far as is known, no mechanistic models have been constructed for *in vivo* genotoxicity assays; as such, measures of potency need to be based entirely on empiricism. The proposed measure of potency to be presented derives from discussions with *R.* Tice, and is similar in

Table 3. *Statistical significance of the trend test for the in vivo hepatocyte UDS data*[a]

Investigator	2AAF Statistical Decision	2AAF p-value	4AAF Statistical Decision	4AAF p-value
Ashby	+[b] (0.10)[c]	<.001	w^+ (100)	<.001
Butterworth	+ (10)	<.002	?[d] (50)	.007
Mirsalis	+ (10)	<.001	w^+ (200)	<.001
Salock	+ (10)	<.001	w^+ (200)	<.001
Beije	+ (50)	<.001	− (2000)	0.92
Kornbrust	+ (10)	<.001	− (200)	0.09
Doolittle	+ (10)	<.001	− (500)	0.99

[a] All seven investigators evaluated 2AAF as positive in UDS; Ashby, Mirsalis and Salocks considered 4AAF as weakly positive, while the other four concluded it negative (see text).
[b] + is positive; − is negative; w^+ is a positive of limited magnitude (NG).
[c] Either the lowest dose that led to a positive decision, or the highest dose that led to a negative decision. For the positive decisions, the lowest dose that led to a positive decision was invariably the lowest dose tested.
[d] A statistically significant NG response, but one that failed to exhibit elevated % cells in repair (Chapter 56). The investigator concluded that this was negative.

spirit to, but technically advanced beyond, his SCE-5 approach (Chapter 37). Consider a simple quadratic dependence of UDS response on dose; the UDS response to dose D, Y, can be approximated as follows:

$$Y \doteq \alpha + \beta D + \gamma D^2, \qquad (1)$$

where α represents the response at zero dose, i.e., the spontaneous or control level, and β and γ are the linear and quadratic coefficients for the dose response.

The proposal here is to measure potency by the dose needed to induce a unit increase in response, i.e., 1 NG, over the control level α. Estimation of the *d*ose for a *u*nit *i*ncrease (DUI) can be achieved with the aid of the fitted version of equation (1), which is obtained by standard regression methods (Snedecor and Cochran, 1967). Appendix 2 describes the possible mathematical solutions of such a potency estimation equation. Note that in an experiment with a single dose group and a control group, one cannot detect curvature of response; only the simple regression coefficient, b, is available. For experiments with multiple times post-treatment for an individual chemical, the maximum potency (or minimum DUI) observed appears as good a summary measure as one can derive. Although it is potentially biased upwards, the maximum potency observed over different experimental conditions is not compromised by those experiments conducted at times post treatment when essentially no induction of UDS is present; this would be the case if one attempted an averaging of potencies across experimental conditions.

Table 4 presents estimates of potency for the 2AAF and 4AAF results from the *in vitro–in vivo* UDS assay. Note that for the seven laboratories, 2AAF exhibits an average relative DUI of approximately 10^{-2} compared with 4AAF, or equivalently, a relative potency approximately 100 times greater than that of 4AAF. The agreement among the first four laboratories is most reassuring. Although this agreement does not extend to all seven laboratories, there is still more to be gleaned from Table 4 regarding the UDS results. Recall from Table 3 that all seven investigators found evidence of UDS attributable to 2AAF, but only the first four in Table 4 produced statistically significant evidence of a weak 4AAF effect. The lack of such a response for Beije is understandable; the choice of 36 h, dictated by her desire to study

Table 4. *DUI[a] for the in vivo hepatocyte UDS assay*

	2AAF	4AAF	Relative DUI[b]
Ashby	0.35	43.2	123
Butterworth	0.38	20.9	55
Mirsalis	0.48	47.2	98
Salock	0.63	36.7	58
Beije	0.82	∞[c,d]	∞
Kornbrust	1.01	119.5[c]	118
Doolittle	1.34	401.7[c]	299

[a] DUI = dose (mg/kg) for a unit increase over controls.
[b] 4AAF/2AAF
[c] Results judged not statistically significant.
[d] No absolute increase seen above control.

Figure 3. Two hypothetical datasets with common linear regression coefficient $b = 0.43$.

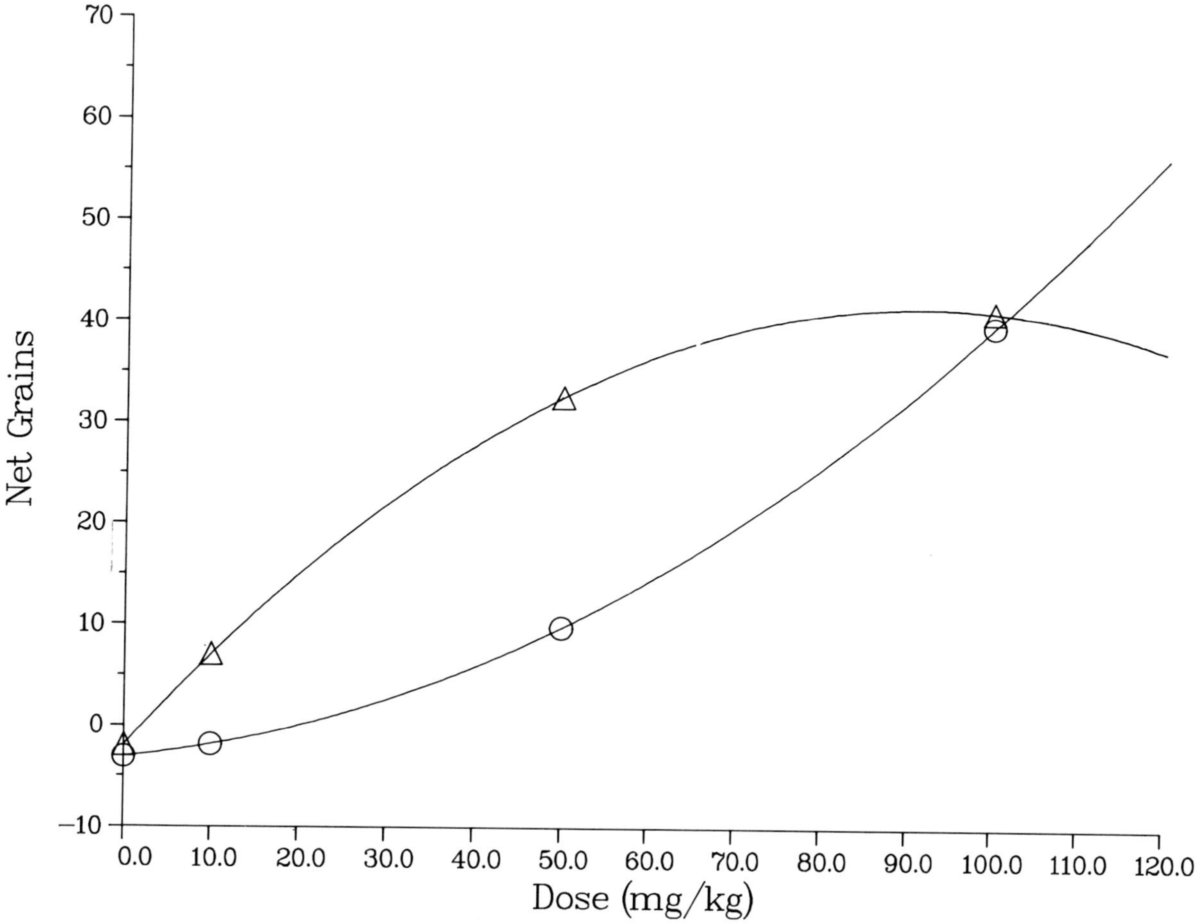

rat bone marrow micronucleus concomitantly, was an inappropriate time at which to try to elicit the weak UDS response to 4AAF. Table 4 offers a possible explanation as to why the other two investigators, Kornbrust and Doolittle, were unable to see evidence of a 4AAF effect. These two obtained substantially less potent responses to 2AAF as well. This is suggestive of a laboratory-specific phenomenon that has lowered the general sensitivity of the assays in these laboratories.

Concluding remarks regarding the UDS assay

A. Although the need to resolve the issue of whether there is substantial day-to-day variability in control animal responses is important, it is not the only outstanding question regarding UDS response variability. Equally important is the need to understand why some of the seven laboratories listed in Table 4 did not observe a small but reproducible 4AAF response within the time-dose parameters for which other laboratories obtained unequivocal statistical significance. The conjecture of lower assay sensitivity, if true, is still a symptom; further study of intra- and inter-laboratory results for a number of coded chemicals is needed to confirm the symptom and to begin to learn its etiology.

B. Another interpretation can be put on the 4AAF results, i.e., that the weak response to 4AAF seen by four laboratories is reproducible and statistically significant, but artifactual, e.g., due to a disturbance of homeostasis. Mirsalis *et al.* (Chapter 58) dismiss their 4AAF response as being of "limited biological significance," but what precisely does this mean and how can the assertion be documented? Do weak UDS assay responses lack specificity, and if so, how can this be altered so that genuine weak effects may be detected by this assay with confidence of their biological credibility?

C. All investigators who employ this assay report the percentage of cells in repair (%IR), i.e., $\geqslant +5$ NG, a response that is usually monotonically related to the mean NG response. The % IR is often used by investigators as an additional criterion of evaluation. This has value as a quality control check, ensuring that a positive UDS response is not attributable to a large NG response from a small percentage of cells. While there are alternative statistical methods to protect against the impact of a small number of observations, the informal use of %IR seems preferred because of its ease of use and familiarity to experimenters. One may also be concerned that a single treated animal can have a disproportionate impact on the results from an experiment. An "influence analysis" in which one atypical animal is omitted from the formal evaluation is an effective way to determine if an inference turns dramatically on any single animal's UDS response.

D. Although the UDS investigators believe that day-to-day variability is negligible, this may not prove true. In that event, the role of the concurrent control will need reexamination. Addition of at least one more concurrent control may then be justified, as would a change in the statistical analysis.

E. Finally, while the analysis that has been proposed here and illustrated in the Appendix is based on a normal distribution for the difference between the treated and control animal means, any similarity to the analysis employed by Joachim and Decad (1984) ends there. The latter procedure involves a *t*-test based on individual cell NG scores, and is statistically inappropriate for *in vivo* data, for it ignores the substantial animal-to-animal variability in UDS responses and bases its error estimates solely on the far smaller intra-animal cellular variability in response. The resulting statistical analysis, particularly when applied to experiments with one treated and one control animal, is then characterized by seriously inflated false-positive and false-negative error rates.

3 Mouse bone marrow micronucleus

Research publications employing the mouse bone marrow micronucleus assay have experienced an explosive growth over the last decade (Ashby, 1985), possibly because of the importance attached to chromosome aberrations and the belief that assessment of chromosome damage is more easily achieved via indirect monitoring of micronuclei rather than direct observation of chromosome aberrations (Salamone and Heddle, 1983). This assay's popularity is further reflected in the current collaborative study, in which 15 laboratories contributed data from this assay. Table 5 contains a summary of the micronucleus response for control animals from the current study. Figure 4 is a graphical representation of these data and clearly indicates that there is substantial variability in these spontaneous rates of micronucleated polychromatic erythrocytes (MPEs).

Table 5 also includes the strain of mouse employed by each of the 15 laboratories. In this study, selection of strain was left to the laboratories. The resultant diversity of strains is duly impressive, there being 10 different ones chosen by the 15 laboratories. Equally impressive is the nearly four-fold difference in the observed rates of spontaneous micronucleus formation in polychromatic erythrocytes (PEs) exhibited in Table 5, the lowest being 1.2/1000 PE, the highest 4.9/1000 PE, and the median 2.1/1000 PE. While the replication of strain results across laboratories is limited, there are two instances of significant disparity in control rates that suggest that strain differences alone cannot account for the variation in results observed:

Strain	*Control rate* (MPE/1000 PE)
(a) NMRI	1.21 (Suter), 1.58 (Wild), 3.90 (Jensen)
(b) C57BL/6J	1.60 (Ivanov), 2.90 (Styles)

In retrospect, it would have been more informative to have encouraged a number of the laboratories either to use multiple strains in their tests, or to agree to use one of a somewhat smaller set of pre-selected strains. As matters stand, the extent to which the variability in control rates in Table 5 is attributable to inter-laboratory differences, as opposed to inter-strain differences, is decidedly unclear.

Sampling distribution of control data

Despite its widespread use, there is still no commonly accepted method of statistical analysis for data from the micronucleus assay. Indeed, on a superficial first inspection, there appears to be as much confusion and disarray

Figure 4. Concurrent spontaneous rates of micronucleated polychromatic erythrocytes; data generated by 15 laboratories during the collaborative study.

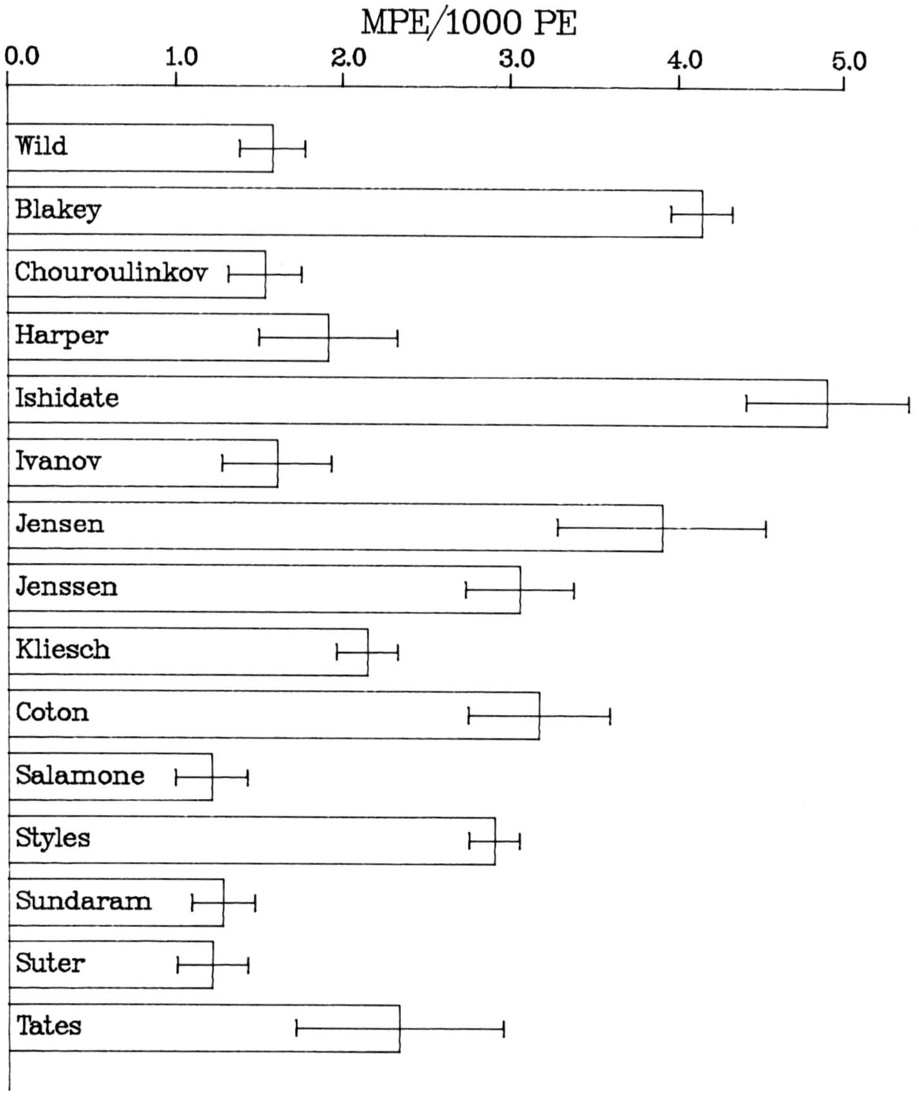

Table 5. *Concurrent control data for the mouse bone marrow micronucleus assay*

Investigator	Strain	Number of Animals	MPE[a] Count	PE[a] Count	MPE/ 1000 PE
Wild	NMRI	40	63	40,000	1.58
Blakey	HPB/b1	118	501	121,021	4.14
Chouroulinkov	Swiss	16	49	32,000	1.53
Harper	Ha/ICR	11	21	11,000	1.91
Ishidate	MS/Ae[b]	20	98	20,000	4.90
Ivanov	C57BL/6J	15	24	15,000	1.60
Jensen	NMRI	10	39	10,000	3.90
Jenssen	CBA	40	91	29,815	3.05
Kliesch	(101 × C3H)F1	64	137	64,000	2.14
Coton	Balb-c	18	57	18,000	3.17
Salamone	B6C3F1	53	32	26,500	1.21
Styles	C57BL/6J	131	380	131,000	2.90
Sundaram	Swiss	31	47	36,773	1.28
Suter	NMRI	28	34	28,000	1.21
Tates	(101 × C3H)F1	6	14	6,000	2.33

[a] PE = polychromatic erythrocyte; MPE = micronucleated PE
[b] Mutagen-sensitive

regarding evaluation of micronucleus data as there was previously with the *in vitro–in vivo* UDS assay. For example, various authors have proposed at least three different sampling distributions for the number of micronucleated cells/animal. Specifically, Mackey and MacGregor (1979) argued for the negative binomial, Hart and Engberg-Pedersen (1983) opted for the binomial, and Salamone and Heddle (1983) and Amphlett and Delow (1984) concluded that the Poisson distribution provided a satisfactory fit for micronucleus data. Hart and Engberg-Pedersen (1983), however, correctly noted that the binomial and Poisson models are indistinguishable when a large number of cells are scored per animal, e.g., 1000 or more, and the proportion of cells exhibiting micronuclei is small, say 1 to 3 per 1000. The negative binomial model was adopted by Mackey and MacGregor (1979) because their data exhibited variability in excess of that appropriate for a Poisson distribution. They reached this conclusion by including treatment data from strongly clastogenic agents, e.g., colchicine. The excess variability is not surprising, for animals seem to be moderately variable in their responses to strongly clastogenic agents. Had Mackey and MacGregor (1979) focused on control data only, it is conjectured that they, too, would have arrived at a binomial or essentially equivalent Poisson model. As Hart and Engberg-Pedersen (1983) pointed out, it is the behavior of control data that is relevant to devising tests of statistical significance. Variable response behavior when the treatment is effective can only impact on assay sensitivity; it cannot affect false-positive rates.

Hart and Engberg-Pedersen (1983) and Salamone and Heddle (1983) used the Pearson chi-square goodness of fit test to determine whether their data were adequately described by a Poisson distribution. The dispersion test employed by Amphlett and Delow (1984) for this same purpose, however, is more sensitive to excess variation above that of the binomial or Poisson models (Collings and Margolin, 1985), and can be employed with small samples. This procedure can also be employed to test for excess animal-to-animal variability among controls within a day or to test for excess day-to-day variability in experimental control groups. Appendix 3 contains the formal computations necessary for both of these tests. These two dispersion tests were applied to all micronucleus control data submitted for this study. Given the relatively small counts of micronuclei in 1000 (PE) control cells typical of this assay, it is not surprising that no excess variability among animals within a day was detected for any day and any of the 15 laboratories. What is more interesting is that only one laboratory (Jenssen) exhibited excess interday variability among its control groups. In that laboratory on one particular day apparently eight of 10 control animals were free of micronuclei among the roughly 500–700 cells scored per animal, a result substantially inconsistent with the observations on other days in this laboratory. The important conclusions then are that a binomial model, or, in the current context, its Poisson equivalent, is a good description of micronucleus control data from any one laboratory; that the rate within a laboratory stays relatively constant over moderate periods of time; and that there are significant rate differences among laboratories.

The second conclusion is fortuitous, for some laboratories did not run their control animals concurrently. In these cases, as much as a few months separated the testing of control and treated animals. Were there significant inter-day variability, this practice would be indefensible; omission of any truly concurrent controls still remains open to serious challenge on general principles and deserves reconsideration by those who adhere to it.

Statistical implications of the binomial model

The conclusion that a binomial (or Poisson) model provides an excellent description for all micronucleus control data from this sizeable study, as well as for most if not all of the published control data, yields a multitude of dividends. This point has been made previously by Margolin *et al.* (1986) in a tangentially related context, where the binomial model was empirically justified as an approximation to the control frequency of *in vitro* Chinese hamster ovary cells with chromosome aberrations. Margolin *et al.* (1986) point out that the applicability of this model has immediate implications for "(i) experimental quality control standards, (ii) methods of statistical analysis for a treatment effect, and (iii) sensitivity or 'power' assessments for alternative methods of analysis, or for alternative" protocol designs for the assay under discussion.

The implications for experimental quality control standards are simple and straightforward. For mouse bone marrow micronucleus control data, achievement of binomial sampling variability that is homogeneous both across and within days is readily attainable. Laboratories that exhibit persistent excess variability beyond that attributable to binomial sampling should examine their experimental procedures for the source of this presumably controllable variability.

The implications of the binomial model for a recommended method of statistical analysis are also immediate. The Cochran-Armitage test for trend in binomial proportions is the preferred way to assess the evidence for a dose response relationship (Collings *et al.*, 1981; Margolin *et al.*, 1986), the sole caveat being that there is no substantial downturn in the response at high doses. Mechanisms do exist that will produce a rising micronucleus dose response that exhibits a sharp downturn at high doses. For example, a classical spindle poison, such as vincristine, causes partial impairment of the spindle at low doses; thus, increasing numbers of micronuclei result from an increasing dose. At higher doses, however, the spindle impairment is complete, whence tetraploid nuclei are produced and totally expelled. This leaves behind erythrocytes devoid of any micronuclei (Schmid, 1976). All micronucleus experiments in the current study that consisted of a control and at least two treatment groups were checked for evidence of a substantial downturn, and none was found. For experiments in which no downturn in dose response was evident, the Cochran-Armitage test would be expected to outperform most statistical competitors, such as the chi-square test. This will be documented shortly.

Appendix 4 presents the formula for and an illustration of the computation of the Cochran-Armitage test. This test is similar in spirit to the test advocated here for the UDS data, involving again a linear regression of the observed endpoint measurement on the applied dose. The distinction is that the sampling model is now binomial. The results of applying the Cochran-Armitage trend test to the micronucleus data are entirely supportive of the micronucleus workgroup (Chapter 18). As this group reported, all 14 laboratories that tested BaP found it positive, and with the exception of Harper ($p > 0.25$), the other 12 laboratories that tested 2AAF had positive results with it as well. In each case of a positive finding, the inference was strongly supported by the observed significance level, it being less than 0.001. Results for 4AAF and PYR, on the other hand, were essentially negative throughout the micronucleus database, the exceptions being two clear positive results ($p < 0.001$ for both chemicals) obtained by Ivanov (Chapter 24). It is worth noting that Styles (Chapter 30), using the same C57BL/6J strain that Ivanov used, saw no effect for either 4AAF or PYR. An explanation for this anomalous situation is wanting.

The concept behind the earlier proposal for a measure of potency in UDS experiments, the DUI, is generic in its formulation; it can be applied straightforwardly to micronucleus data. For this application, equation (1) is adopted and fitted to the micronucleus data by ordinary least squares (Snedecor and Cochran, 1967). Then the values of β and γ that one obtains are inserted into equation (A.1) and the resultant equation is solved for the DUI.

Table 6 contains the mouse bone marrow micronucleus DUIs for 2AAF and BaP for each laboratory. Meaningful comparison of the DUIs for the two chemicals requires adjustment for the respective molecular weights; the relative DUIs in Table 6 reflect this adjustment. The variation in the absolute and relative DUIs appears to be due to a complex mixture of laboratory, mouse strain and chance effects.

For the thirteen laboratories in which the relative DUI is computable for both 2AAF and BaP, the median relative DUI is 1.9. Ten of these laboratories, including those that replicated strains NMRI and (101 × C3H) F1, produced data that indicated that BaP was more potent than 2AAF as measured by the DUI. Interestingly, the three laboratories whose data indicated greater potency for 2AAF relative to BaP consisted of the two laboratories (Ivanov, Styles) that employed C57BL/6J and the only laboratory (Sundaram) that used Swiss mice and tested both chemicals. Thus, the three strains that were replicated in pairs of laboratories exhibited complete intra-strain concordance regarding the relative potency of 2AAF and BaP. Moreover, although the number of replicated strains was small, there is a suggestion in the relative DUIs of Table 6 that BaP is more potent than 2AAF for most mouse strains, but the reverse may occur in a few.

The final dividend from validation of a binomial model for data from the mouse bone marrow micronucleus assay is an ability to study the dependence of assay sensitivity (or power) on the statistical analysis employed, or on protocol features, such as the number of cells scored/animal. As an illustration, Table 7 contains results of a small Monte Carlo study of the sensitivity of the micronucleus assay when there is a control group and three treated groups. It has been assumed that there are five animals/group and the doses are in the ratios of 1:2:3. No historical control data are incorporated in the analysis. In each case, either 500 or 1000 cells are scored per animal, and the statistical analysis is performed first with the Cochran-Armitage trend test and then again with the common Pearson chi-square test. For this exercise, the chemical was assumed first to have no effect, from which one can estimate the false-positive rate corresponding to nominal 0.05 and 0.01 levels for the two statistical procedures. Next, the chemical was assumed to induce micronuclei in a weak linear dose response that at the highest dose added either 2 or 4 micronuclei/1000 PE to the spontaneous rate for each animal so exposed. When the chemical had no effect, the estimated false-positive rates for both the trend test and the chi-square test were in

Table 6. *DUI[a] for the mouse bone marrow micronucleus assay*

Investigator	Strain	2AAF	BaP	Relative[b] DUI
Wild	NMRI	63.8	10.3	7.0
Blakey	HPB/b1	30.0	19.7	1.7
Chouroulinkov	Swiss	100.1	NT[c]	—
Harper	Ha/ICR	450.1[d]	6.5	78.3
Ishidate	MS/Ae	24.1	21.0	1.3
Ivanov	C57BL/6J	13.8	23.3	0.7
Jensen	NMRI	NT	24.8	—
Jenssen	CBA	21.1	17.9	1.3
Kliesch	(101 × C3H)F1	29.7	5.4	6.2
Coton	BALB/c	37.4	3.4	12.4
Salamone	B6C3F1	132.0	3.3	45.2
Styles	C57BL/6J	51.5	69.9	0.8
Sundaram	Swiss	70.6	135.5	0.6
Suter	NMRI	72.3	9.0	9.1
Tates	(101 × C3H)F1	18.7	10.9	1.9

[a] DUI = dose (mg/kg) at unit increase over control
[b] Corrected for molecular weights: (2AAF DUI ÷ 223.26)/(BaP DUI ÷ 252.30)
[c] NT = not tested
[d] Not statistically significant result

excellent agreement with the nominal false-positive rates under all conditions studied. (These results are not shown here.)

Turning to the sensitivity issue, if the chemical is assumed to have an additive, linear effect on the micronucleus rate, the results in Table 7 demonstrate that the Cochran-Armitage trend test consistently outperforms the chi-square test, often providing double the sensitivity. This table also provides quantitative evidence of the increased difficulty in detecting an additive chemical effect as the hypothetical underlying spontaneous rate increases. For example, if the spontaneous micronucleus rate is 1/1000, the induction of 4 micronuclei/1000 PEs is detected 87% of the time with 500 cells/animal for each of five animals and a 0.05 level trend test but this drops to 62% if the spontaneous micronucleus rate is 4/1000. Finally, based on this Monte Carlo, with an experimental design consisting of five animals per dose group, it seems advisable to use 100 cells/animal if one has a mouse strain with a moderate to high spontaneous rate of micronucleus formation or if one is desirous of detecting weakly clastogenic agents capable of inducing on average 2 micronuclei/1000 PE at the highest dose to be tested.

Concluding remarks regarding the micronucleus assay

1. The micronucleus analyses discussed here have all been directed to mouse bone marrow. While they form a basis upon which one can approach analysis of micronucleus data in other organs or other species, there still exists the need to verify applicability of the binomial model to these other contexts.

2. As experience accumulates with the proposed use of the Cochran-Armitage test for trend in assessing micronucleus data, an appreciation will develop as to how serious a concern is the issue of dose-response turndowns at high dose. If the vincristine phenomenon is a rarity, the analysis will stand as proposed here. If downturns resulting in errors of inference are frequent enough to be of concern, a modification of the analysis will be needed to preserve sensitivity. Two directions for such a modification are represented by the point-rejection approach adopted by Bernstein *et al.* (1982) for analyzing Ames test data and the recursive analysis developed by Simpson and Margolin (1986) for nonparametric testing for dose response.

3. Animals with aberrant measurements are always a possibility in *in vivo* assays. For micronucleus control data that are binomially distributed, statistical methods of Collings *et al.* (1981) can be employed as protection against aberrance. These methods would also apply to dose groups in which the test chemical had little or no effect, so that the binomial model was still applicable. If an aberrant value occurred in a dose group for which the test chemical was having a sizeable effect, statistical methods of analysis are relatively unimportant and the aberrant measurement would have limited impact.

4. A most intriguing and extremely important question is to what extent can one demonstrate empirically that the mouse bone marrow micronucleus assay is capable of substantially greater sensitivity for comparable cost when compared to the chromosome aberration assay? This question is clearly in need of further study, for *in vivo* cytogenetics assays have become regular components of regulatory screens for genetic toxicity. The much needed definitive answer to this question will provide regulators with important guidance in refining their requirements for testing.

4 Concluding remark

This chapter has focused exclusively on two of the many *in vivo* assays included in the current collaborative study. The discussion in this chapter has covered a number of statistical issues, including:

A. assessment of the sources of variability that impinge upon on assay response;

B. construction, where feasible, of adequate statistical

Table 7. *Estimated effects of the number of cells/animal and the method of statistical analysis on the sensitivity (power) of the micronucleus assay*[a]

			Nominal false-positive rate (p-value)			
			0.05		0.01	
Case	Mean MPE/1000 PE Ordered by Dose	Cells/ Animal	Trend	Chi-Square	Trend	Chi-Square
1	1.0, 2.0, 3.0, 5.0	500	0.87	0.64	0.69	0.38
		1000	0.99	0.93	0.95	0.81
2	2.0, 3.0, 4.0, 6.0	500	0.78	0.49	0.52	0.26
		1000	0.96	0.83	0.86	0.63
3	4.0, 5.0, 6.0, 8.0	500	0.62	0.34	0.34	0.15
		1000	0.86	0.63	0.66	0.38
4	1.0, 1.5, 2.0, 3.0	500	0.51	0.24	0.26	0.10
		1000	0.77	0.49	0.52	0.26
5	2.0, 2.5, 3.0, 4.0	500	0.38	0.18	0.16	0.06
		1000	0.61	0.33	0.33	0.15
6	4.0, 4.5, 5.0, 6.0	500	0.27	0.12	0.10	0.03
		1000	0.43	0.21	0.20	0.07

[a] The assumption is that a binomial model is applicable. The Monte Carlo study consists of 5000 replications of a control and three doses in the ratio 1:2:3, with 5 animals/group.

distributions as models that describe the behavior of control data;

C. appreciation of the utility of historical control data;

D. recognition of multiple comparisons and their impact on false-positive rates;

E. documentation of the sizeable strain/species/laboratory differences possible in control rates;

F. consideration of statistical tests for trend as sensitive indicators of dose response; and

G. exploration of a simple measure of potency, the DUI, as a useful research tool.

These issues are, in fact, widely applicable, and form a suitable basis for an in-depth study of methods to analyze most other *in vivo* assays for genetic toxicity. The need to proceed with these studies is strong and clear in view of the heightened interest in *in vivo* assay systems.

5 References

Amphlett, G.E. and G.F. Delow (1984) Statistical analysis of the micronucleus test, *Mutation Res.*, **128**, 161–166.

Ashby, J. (1985) Is there a continuing role for the intraperitoneal injection route of exposure in short-term rodent genotoxicity assays?, *Mutation Res.*, **156**, 239–243.

Ashby, J., P.A. Lefevre, B. Burlinson and M.G. Penman (1985) An assessment of the *in vivo* rat hepatocyte DNA-repair assay, *Mutation Res.*, **156**, 1–18.

Ashby, J., B. Burlinson and P.A. Lefevre (1988) Activity of 2AAF and 4AAF in an *in vitro* and an *in vivo/in vitro* assay for UDS in the rat liver, *in Chapter 57.*

Bernstein, L., J. Kaldor, J. McCann and M.C. Pike (1982) An empirical approach to the statistical analysis of mutagenesis data from the Salmonella test, *Mutation Res.*, **97**, 267–281.

Collings, B.J., B.H. Margolin and G.W. Oehlert (1981) Analyses for binomial data, with application to the fluctuation test for mutagenicity, *Biometrics*, **37**, 775–794.

Collings, B.J. and B.H. Margolin (1985) Testing goodness of fit for the Poisson assumption when observations are not identically distributed, *J. Amer. Statist. Assoc.*, **80**, 411–418.

Hart, J.W. and H. Engberg-Pedersen (1983) Statistics of the mouse bone-marrow micronucleus test: Counting, distribution and evaluation of results, *Mutation Res.*, **111**, 195–207.

Joachim, F. and G.M. Decad (1984) Induction of unscheduled DNA synthesis in primary rat hepatocytes by benzidine congener-derived azo dyes in the *in vitro* and *in vivo* assays, *Mutation Res.*, **136**, 147–152.

Mackey, B.E. and J.T. MacGregor (1979) The micronucleus test: Statistical design and analysis, *Mutation Res.*, **64**, 195–204.

Margolin, B.H., N. Kaplan and E. Zeiger (1981) Statistical analysis of the Ames Salmonella/microsome test, *Proc. Natl. Acad. Sci., (U.S.A.)*, **78**, 3779–3783.

Margolin, B.H. and K.J. Risko (1984) The use of historical data in the analysis of laboratory studies, *Proc. Intl. Biometrics Conf.*, **12**, 21–30.

Margolin, B.H., M.A. Resnick, J.Y. Rimpo, P. Archer, S.M. Galloway, A.D. Bloom and E. Zeiger (1986) Statistical analyses for *in vitro* cytogenetic assays using Chinese hamster ovary cells, *Environ. Mutag.*, **8**, 183–204.

Mirsalis, J.C., C.K. Tyson, E.K.N. Loh, J.P. Bakke, C.M. Hamilton and K.L. Steinmetz (1988) An evaluation of the ability of benzo(a)pyrene, pyrene, 2- and 4-acetylaminofluorene to induce unscheduled DNA synthesis and cell proliferation in the livers of male rats and mice treated *in vivo*, *in Chapter 58.*

Salamone, M.F. and J.A. Heddle (1983) The bone marrow micronucleus assay: Rationale for a revised protocol. In *Chemical Mutagens*, Vol. 8, ed. F.J. de Serres, Plenum Press, N.Y., pp. 111–149.

Schmid, W. (1976) The micronucleus test for cytogenetic analysis. In *Chemical Mutagens*, Vol. 4, ed. A. Hollaender, Plenum Press, N.Y., pp. 31–53.

Simpson, D.G. and B.H. Margolin (1986) Recursive nonparametric testing for dose-response relationships subject to downturns at high doses, *Biometrika*, **73**, 589–596.

Snedecor, G.W. and W.G. Cochran (1967) *Statistical Methods*, 6th Edition, Iowa State Press, Ames, Iowa.

Williams, G.M. (1977) Detection of chemical carcinogens by unscheduled DNA synthesis in rat liver primary cell cultures, *Cancer Res.*, **37**, 1845–1851.

Appendix 1. Assessment of the statistical significance of an individual UDS treatment group mean

The following analysis is predicated on an assumption that day-to-day variability is negligible. Let $\bar{Y}_c$ denote the mean NG from n_c combined historical and concurrent control animals. Similarly, let $\bar{Y}_t$ denote the mean NG from n_t animals treated with a given dose of the test chemical. Interest centers on the estimate of induced UDS given by

$$\Delta = \bar{Y}_t - \bar{Y}_c.$$

The variance of Δ, if the chemical is assumed to be without effect, is expressible as:

$$\text{var}(\Delta) = \left\{ \begin{matrix} \text{variance of the NG response} \\ \text{for a single control animal} \end{matrix} \right\} \times \left(\frac{1}{n_t} + \frac{1}{n_c} \right)$$

The variance V of the NG response for a single control animal is commonly estimated from the standard deviation of the combined historical and concurrent control data, say S_c, and the standard deviation of the treated group, say S_t. The estimate V is given by the following expression:

$$V = \frac{(n_c - 1)S_c^2 + (n_t - 1)S_t^2}{n_c + n_t - 2}$$

In view of the applicability of the normality model for means of UDS responses and the large historical databases anticipated, it follows that the ratio of the estimated induced UDS, Δ, divided by its estimated standard deviation $[\text{var}(\Delta)]^{1/2}$, can be referred to a one-tailed table of the standard normal distribution to assess its statistical significance. If the individual UDS measurements were normally distributed, it would be appropriate to compare this ratio with a *t*-distribution. This would permit analysis in smaller samples and yield somewhat smaller false-positive rates. The evidence in support of normality for the distribution of individual UDS observations is limited. Hence, it is recommended that statistical analyses be performed only when there is a historical control set of at least 15 animals. To illustrate the computation, consider the response for 100 mg/kg in the second experiment of Table 2, together with the historical control information for Ashby in Table 1. Here $n_c = 34 + 1 = 35$, which includes the single concurrent control for the experiment in question. In addition, $\bar{Y}_c = -2.31$ and $S_c = 1.08$, where these reflect, again, the addition of the concurrent control to the set of historical controls. For the dose of 4AAF in question, $n_t = 3$, $\bar{Y}_t = -0.12$ and $S_t = 0.59$.

The induced UDS is estimated to be

$$\Delta = -0.12 - (-2.31) = 2.19$$

For Ashby, the variance of a single animal's UDS response is estimated to be

$$V = \{34(1.08)^2 + 2(0.59)^2\}/(35 + 3 - 2) = 1.12$$

The variance of Δ is then

$$\begin{aligned} \text{var}(\Delta) &= V \cdot \left(\frac{1}{n_t} + \frac{1}{n_c}\right) \\ &= 1.12\ (1/3 + 1/35) \\ &= 0.41 \end{aligned}$$

and its standard deviation is $[\text{var}(\Delta)]^{1/2} = 0.64$. The ratio of Δ to its standard deviation is $2.19/0.64 = 3.42$. From a table of the normal distribution one obtains $p < 0.001$, implying that the observed mean UDS response of -0.12 based on three animals dosed at 100 mg/kg of 4AAF, even though negative in magnitude, has a probability of less than one in a thousand that it arose by chance.

Appendix 2. Solving for the empirical DUI

Let $\hat{\alpha}, \hat{\beta}, \hat{\gamma}$ denote the regression estimates of α, β and γ in equation (1), respectively. The DUI is defined as the dose such that

$$\hat{\alpha} + 1 = \hat{\alpha} + \hat{\beta}(\text{DUI}) + \hat{\gamma}(\text{DUI})^2,$$

or equivalently,

$$\hat{\beta}(\text{DUI}) + \hat{\gamma}(\text{DUI})^2 = 1 \qquad \text{(A.1)}$$

The following outcomes are possible:

(i) $\hat{\gamma} = 0$, i.e., no estimated curvature: If $\hat{\beta}$ is negative, then the DUI does not exist, there being no evidence that any dose will induce a unit increase in response. This will be indicated by setting DUI to ∞. On the other hand, if $\hat{\beta}$ is positive, the DUI is simply $1/\hat{\beta}$, the reciprocal of the estimated slope.

(ii) $\hat{\beta}^2 + 4\hat{\gamma}$ is negative: There is no finite DUI; again, the DUI is set to ∞.

(iii) If neither (i) nor (ii) obtain, then the solutions of (A.1) are given by

$$[-\hat{\beta} \pm (\hat{\beta}^2 + 4\hat{\gamma})^{1/2}]/2\hat{\gamma}$$

The DUI is taken to be the positive solution of smallest magnitude, provided one exists; if none does, then the DUI is set to ∞.

Appendix 3. Dispersion tests for excess variability in micronucleus control data

The two analyses to be presented attempt to resolve excess animal-to-animal variability under control conditions into components that are attributable to intra- and inter-day sources. Consider first the intra-day variability among animals. Let $Y_1, \ldots, Y_r$ denote the number of MPE scored per n PE cells for each of r control animals tested within a common day. The dispersion statistic D for that day is given by

$$D = \sum (Y_j - \hat{Y})^2/[\hat{Y}(n - \hat{Y})/n]$$

where $\hat{Y} = (\sum_{j=1}^{r} Y_j)/r$. For data from multiple days d in number, as would likely be the case, one must use an index to denote those values of D and r common to a day, i.e., D_i, r_i for day i. Assuming the r_i are relatively constant, then the overall test for excess variability is based on computing $\sum_{i=1}^{d} D_i$, which is referred to a table of critical values for a χ^2 distribution with $\sum_{i=1}^{d} (r_i - 1)$ degrees of freedom.

If there is no excess intra-day animal-to-animal variability among controls, one can then proceed to study inter-day variability. If S_i denotes the number of MPE cells scored among a total of k_i PE cells evaluated in a given control group, and if there are g such groups, then a test of excess inter-day variability can be based on $\sum_{i=1}^{g} [(S_i - k_i\bar{S})^2/k_i\bar{S}(1 - \bar{S})]$, where $\bar{S} = \sum S_i/(\sum k_i)$. This statistic is referred to a table of critical values for a χ^2 distribution with $g - 1$ degrees of freedom.

In the case of both dispersion tests, excess variability is indicated by large values of the test statistic.

Appendix 4. Computation of the Cochran-Armitage test for trend in binomial proportions

Assume the experiment consists of r dose groups plus a control group, where the doses are denoted by $d_0 = 0 < d_1 < d_2 < \cdots < d_r$. The Cochran-Armitage test only considers the total numbers of PEs and MPEs scored per dose group irrespective of the number of animals involved. According to the statistical model described here the number of animals and their patterns of response only influence assay sensitivity. Therefore, assume that n_i PEs have been scored at the ith dose, and that Y_i MPEs have been observed. The overall rate of MPE formation in the experiment is given by $\bar{p} = \sum_{i=0}^{r} Y_i/\sum_{j=0}^{r} n_j$ and the trend test statistic is specified by

$$Z = \sum_{i=0}^{r} (Y_i - n_i\bar{p})d_i/[\bar{p}(1 - \bar{p}) \sum_{j=0}^{r} n_j(d_j - \bar{d})^2]^{1/2},$$

where $\bar{d} = \sum n_j d_j/\sum n_k$. The value of Z is referred to a one-tailed table of the standard normal distribution to obtain a p-value. To illustrate, consider the data of Ivanov for 4AAF (Chapter 24):

Dose (d_i)	Number of PE (n_i)	Number of MPE (Y_i)
0	15,000	24
223	7,000	22
446	5,000	23
892	7,000	33

The value of $\bar{p}$ is 0.003. Z is evaluated to be 4.22. Referring to a standard normal table, one obtains a significance level $p < 0.001$ for these 4AAF data.

PART II

Biological profiles of the four test chemicals

Roy Forster (Work Group Chairman)[1]
Life Science Research Roma Toxicology Centre Pomezia, Roma, Italy

4

Overview: benzo(a)pyrene and pyrene

1 Introduction

The purpose of this chapter is to draw together the various strands of the IPCS II study from the perspective of the two test chemicals benzo(a)pyrene (BP) and pyrene (PYR). What is the spectrum of genetic activity of these two chemicals? Do the assay system responses fit together to form a coherent story? And how well do the different assay types succeed in predicting the relative carcinogenic activity of the pair?

As a by-product, IPCS II will certainly have produced the largest ever data-base of *in vivo* short-term tests on BP and PYR, thus contributing to the fifty-year old history of these chemicals (Phillips 1983). BP was first synthesised by J.W. Cook and C.L. Hewett, (working in the group of Sir Ernest Kennaway) who published their work in 1933, identifying BP with a highly carcinogenic fraction derived from pitch. The isolation of PYR was described by Kruber in 1931. In the subsequent years BP has become an archetypal carcinogen; PYR is believed to be a compound almost devoid of carcinogenic activity.

The massive literature on BP offers a background of understanding against which to discuss the results of the present study. Accordingly, this chapter includes a brief review of the fate of BP in the body after administration, its acute toxicity, metabolism and carcinogenicity. Regrettably there is little to say about PYR. The spectrum of genetic activity of the two chemicals *in vitro* is then described. The results of the present study have been divided into two broad headings: first, those assays having non-mammalian endpoints are discussed, and then those assays in which the genetic endpoint is mammalian (i.e. involving cellular material derived from treated mammals).

Finally, an attempt is made to draw together the conclusions with regard to the two chemicals, and to the predictive power of the different assays.

Throughout the chapter, the term "screening mode" refers to the use of short-term tests to indicate whether a chemical has the potential to cause genetic damage or cancer, in the way that the Ames test is routinely used to screen novel chemicals. "Hazard assessment", on the other hand, denotes the use of an assay to indicate more

[1] Workgroup participants: Jeffrey A. Allen, Janos Arany, Ivan Chouroulinkov, Chie Furihjata, Ulrich Graf, Barbara L. Harper, Jorn C. Jensen, Andrew D. Kligerman, Claude Lasne, Carol A. Luke, Michael A. Pereira, Claes Ramel, Charles B. Salocks, and Paul A. Watkins

concretely the hazard posed by *in vivo* exposure. The question that we hope to answer when employing an assay in the "hazard assessment" mode is "Will the Ames-positive chemical turn out to be a potent carcinogen?" The IPCS II study was undertaken in part to assess the proposal that *in vivo* studies should be deployed in a hazard-assessment mode (Ashby *et al.*, 1983).

2 BP: Some background

Absorption, excretion and disposition

BP is a pale yellow material, forming needle-like crystals. It is almost insoluble in water, but readily soluble in organic solvents and lipids (including cellular membrances). After oral administration to experimental animals, absorption is believed to be poor, and this may contribute to its relatively low acute oral toxicity. Once absorbed, BP binds to the lipoprotein fraction of plasma (Shu and Bymun 1983). First pass removal by the liver is not high, and both BP and reactive metabolites enter the systemic circulation. After ip administration to rats, peak levels of BP in venous blood were found after 22 hours, and decreased slowly over the next 28 hours (Lutz *et al.*, 1978).

The high rate of metabolism of BP results in fairly rapid elimination from the body – for example, almost all of a dose of 100 mg/kg administered ip to mice was lost over a seven day period, following first order kinetics (Camus *et al.*, 1984).

The bile is a major excretory pathway; estimates of the amount of biliary excretion within the first few hours after administration vary, but a large proportion of administered BP is excreted in the bile in the first 6 to 10 hours (Boroujerdi *et al.*, 1981, Shu and Bymun 1983, Schlede *et al.*, 1970).

BP entering the bile is almost entirely present in the form of conjugated metabolites. These compounds may then be subject to metabolism by the gut bacterial flora (Renwick and Drasar 1976) and by cells in the gut wall (Autrup *et al.*, 1978), both of which have relevant metabolic capabilities. There is evidence for enterohepatic circulation in both the rat and the rabbit (Chipman 1981, 1982). A range of conjugated metabolites can be detected in the urine, but the major pathway for removal from the body is via the faeces, even after exposure via inhalation (Mitchell 1982).

Once it has entered systemic circulation, BP is distributed throughout the body, including the brain, testes (Mitchell 1983) and saliva (Guttenplan *et al.*, 1980). As a result of the widespread ability of tissues to metabolise BP to reactive forms (Juchau *et al.*, 1979), DNA binding can also be detected in a range of tissue types, for example in the liver, lung, spleen, kidneys and skin after ip administration to mice (Prodi *et al.*, 1970, Stowers and Anderson 1984). Thus BP has the ability, not only to form adducts locally after administration (e.g. Brookes and Lawley 1964) but at also distant sites, with relevance to its ability to induce tumours in organs remote from the site of administration.

Metabolism

As noted in Figure 1, the metabolism of BP presents dismaying complexity. A number of mechanisms are involved in the production of the so-called "primary" metabolites (epoxides, quinones, dihydrodiols, phenols). Epoxides can be produced at five different positions by cytochrome P450-mediated oxidative metabolism. These epoxides may then be subject to spontaneous rearrangement to phenols, or may be reduced to dihydrodiol forms by epoxide hydrase enzymes. Some phenols however, may be formed directly, without intermediate epoxides. BP-quinones are produced by spontaneous oxidation of 6-OH-BP.

Once formed, these primary metabolites may be further metabolised, either by conjugation (primarily to sulphates, glucuronides and glutathione conjugates) or by further oxidative metabolism. The latter is of some importance; the majority of the metabolites binding to DNA after incubation with mouse liver microsomes were found to require more than one mono-oxygenation (Boobis *et al.*, 1978).

A series of the phenols thus formed have been shown to be mutagenic to *Salmonella typhimurium* (Glatt and Oesch 1976) and some dihydrodiols are also mutagenic (Malaveille *et al.*, 1975).

BP-quinones, on the other hand, have been implicated in cellular injury (Lorentzen *et al.*, 1979) and the inhibition of microsomal metabolism (Shen *et al.*, 1979). Epoxide metabolites are sufficiently reactive to bind to DNA directly, and BP-4, 5-oxide has been the subject of particular interest. This epoxide, formed at the electronically-favourable K-region of the molecule was at one time believed to be the ultimate carcinogen. It is stable under physiological conditions, can be detected after metabolism *in vitro* and *in vivo*, is mutagenic (Wood *et al.*, 1975) and weakly carcinogenic (Levin *et al.*, 1976).

Over the last decade, however, attention has been turned to the further metabolites of BP-7, 8-oxide. Reduction of this compound to the dihydrodiol can produce two

Figure 1. Metabolic pathways of BP to primary metabolites and to the major metabolite, BP-7, 8-oxide

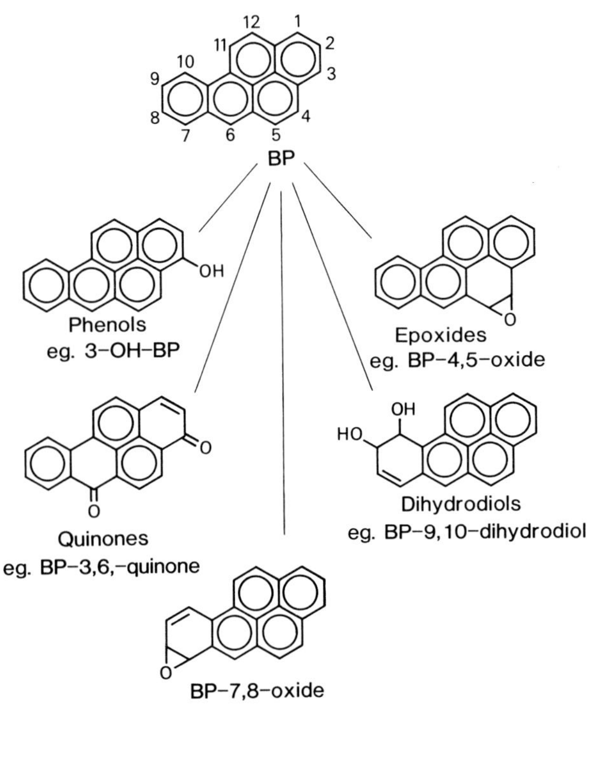

enantiomeric forms, but enzyme stereospecificity ensures that the (−)-enantiomer is produced. Further oxidative metabolism can then produce a pair of 7, 8-dihydrodiol-9, 10-epoxide enantiomers, in which the 7-hydroxy substituent is in the *syn* (BPDE I) or *anti* (BPDE II) position, relative to the epoxide ring (see Figure 2).

Predominantly BPDE II (correctly termed the (+)-anti-diol-epoxide) is produced, and, as luck would have it, this form appears to be the more harmful structure, and the ultimate carcinogenic metabolite of BP. Both are extremely reactive chemicals, but while BPDE I is the more potent mutagen to *S. typhimurium*, BPDE II is more mutagenic to mammalian cells (Newbold and Brookes 1976) and is the more potent carcinogen. Evidence from DNA binding in mammalian cells *in vitro* (Koreeda *et al.*, 1978) and *in vivo* (Grover *et al.*, 1976), mutagenicity and carcinogenicity (Buening *et al.*, 1978, Slaga *et al.*, 1979) (summarised in Pelkonen and Nebert 1982) supports this argument. These findings have further led to the development of the 'bay region' hypothesis, according to which the unique structural feature of BPDE II is the presence of the epoxide on a saturated angular benzo-ring, next to the sterically hindered region between carbons 10 and 11. This makes the epoxide highly susceptible to nucleophilic attack (Jerina *et al.*, 1977). The idea has been extended to the prediction of the carcinogenic activity of other polycyclic aromatic hydrocarbons.

While one strand of evidence that supported the BPDE hypothesis at the outset was the finding that BP bound to cellular DNA was present as adducts identical to those obtained from BP-7, 8-dihydrodiol (Sims *et al.*, 1974), the extent to which BPDE contributes to DNA binding *in vivo* is not yet clear. While some *in vivo* studies have indicated that the majority of adducts formed are derived from BPDE II and BPDE I (e.g. Adriaenssen *et al.*, 1983, Kulkarni *et al.*, 1984), in other studies these adducts have been of minor significance (Dunn 1983, Boroujerdi *et al.*, 1981).

BPDE adducts have, however, been shown to make a large contribution to DNA binding in a wide range of *in vitro* and *in vivo* studies. Included among these are studies with *S. typhimurium* (Santella *et al.*, 1979, Fahl *et al.*, 1981), and these have shown that the suggestion that the *in vitro* mutagenicity of BP to *S. typhimurium* results solely from BP-4, 5-oxide is due for revision. In some cases, such metabolites play a predominant role after *in vitro* metabolism; for example, BP-4, 5-oxide is a major metabolite of metabolism by mouse microsomes (Santella *et al.*, 1979, Oesch and Glatt 1976). Similarly, 9-OH-BP is a major metabolite in 3-methylcholanthrene-induced hepatocytes (Jernstrom *et al.*, 1978). The spectrum of *in vitro* metabolites is also be influenced by the ratio of BP to microsomal material.

Figure 2. Major pathway of metabolic activation of BP to BPDE II

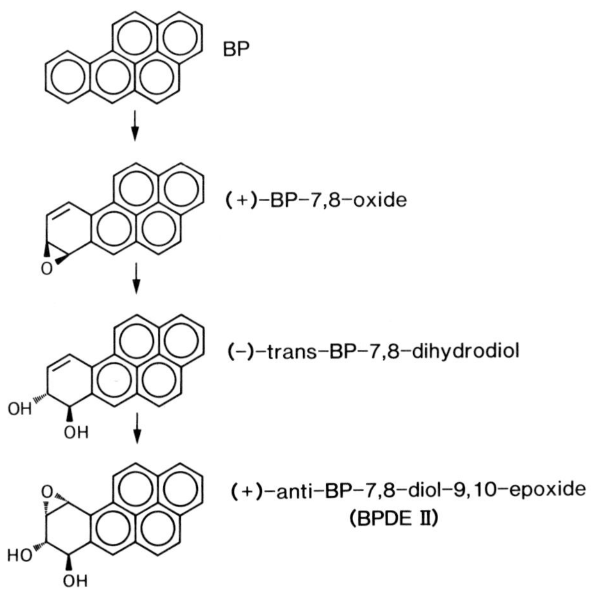

Carcinogenic activity

The carcinogenicity of BP and PYR is discussed elsewhere in this volume (Appendix 3). The principal features, relevant to this discussion are as follows:

(i) BP is a potent local carcinogen: where it is applied, it causes tumours.

(ii) Few tumours have been detected which are not local; these are leukaemias and lung adenomas after dietary administration to the mouse, and mammary tumours after oral or ip administration to the rat. Regrettably, little attention was given to these organotropisms in the present study.

(iii) In the special case of the newborn mouse, subcutaneous administration of BP induces hepatomas and lung adenomas with a very high incidence, in mice of different strains. These effects have been attributed variously to localisation of metabolites in these organs, to the different metabolic capability of newborn mice, or to their immunologically compromised status (Shimkin and Stoner 1975).

(iv) Few species-differences in BP carcinogenicity have been noted between rats and mice, the two principal species used in this study. Rats may be more resistant to skin carcinogenesis than mice (which vary tremendously in sensitivity according to strain), but this conclusion does not emerge clearly from the published data (IARC 1973) The hamster appears to be quite resistant to the carcinogenicity of BP to the skin.

(v) PYR appears to be a noncarcinogen but there are few studies of an acceptable standard (Appendix 3, this volume). There is some evidence for the induction of papillomas in rat skin. Regrettably, no short-term tests in the present study concerned effects in rat skin.

3 *In vitro* predictive tests

When we turn to the responses of BP and PYR in short-term *in vitro* assays for carcinogens, PYR becomes the focus of interest. For the results shown in Table 1, taken from the International Collaborative Program (de Serres and Ashby 1981), interpretations of the findings obtained with BP is straightforward; the results are overwhelmingly positive, and reference to the original data will confirm that strong effects were observed in the majority of cases. Many similar examples will also be found in the literature.

PYR presents a more complex situation. The Study

Overview concluded that PYR gave a "similar spread of positive results (to BP), albeit of a lower incidence and relative potency". Table 1 shows clearly, in fact, that the positive responses with PYR were mostly in the premutational and point mutation categories, and chromosomal, effects were not induced.

It is tempting at this point to speculate on the role of PYR-4, 5-oxides in producing these results. For example, MacDonald (1981) found that the addition of an epoxide hydrolase inhibitor enhanced the effects of PYR seen in *S. typhimurium* TA 1537. The case for the involvement of such epoxides is strong, but PYR also induced genetic effects in the absence of S9 metabolism in UDS (Martin and McDermid 1981) and bacterial repair assays (Rosenkranz *et al.*, 1981).

In vitro UDS assays gave particularly confusing results with PYR. It was negative in one investigation (Agrelo and Amos, 1981), positive in the absence of metabolism in another (Martin and McDermid 1981) and positive in the presence of S9 metabolism in a third laboratory (Robinson and Mitchell 1981).

PYR gave quite convincing evidence of mitotic gene conversion in *S. cerevisiae* in one investigation (Zimmermann and Scheel, 1981), but failed to induce a response in another laboratory (Sharp and Parry 1981). A third laboratory obtained negative results for mitotic gene conversion in *S. cerevisiae* but employed a rather insensitive protocol (Jagganath *et al.*, 1981).

Equivocal results with PYR were also obtained in assays for the induction of SCE's. For two investigators (Evans and Mitchell 1981, Perry and Thompson 1981), reproducible increases in SCE's were obtained, which never reached double the control values; a slight but not statistically significant increase was observed in a third laboratory (Natarajan and van Kesteren-van Leeuwen).

The evidence from the bacterial repair assays, UDS, SCE induction and mitotic gene conversion all indicates that PYR has the ability to induce DNA damage and repair *in vitro*; further support for this idea is supplied by the finding that PYR induces *rec-lex* dependent repair functions in *E. Coli* (Quillardet *et al.*, 1985).

Evidence for the activity of PYR as a gene mutagen is drawn from *S. tyhimurium* and mouse lymphoma cell studies. PYR was described in the bacterial assay summary, as "a mutagen which is not detected in the majority of laboratories" (Bridges *et al.*, 1981). Recent studies have confirmed PYR's activity toward *S. typhimurium* TA97 (Matijasevic and Zeiger 1985, Sakai *et al.*, 1985). 1, 6- and 1, 8-pyrene quinones are also mutagenic both in the absence and presence of S9 (Sakai *et al.*, 1985). Small increases have also been observed after treatment of mouse lymphoma L5178Y cells with PYR in the presence of S9 metabolic activation (Jotz and Mitchell, 1981).

In summary PYR demonstrated evidence for the induction of mitotic gene conversion and UDS but these results were poorly reproducible between laboratories. Reproducible increases in SCE and mammalian cell mutation (L5178Y), and reproducible mutagenic activity in *S. typhimurium* have been observed. No evidence for the induction of chromosomal aberrations was observed in the limited number of assays presented in the International Collaborative Program.

Table 1. *Summary of the in vitro genetic activity of BP & PYR*

	BP	PYR
Permutational		
Microbial repair assays	+	–
Mitotic gene conversion/yeasts	+	– (a)
Unscheduled DNA synthesis	+	+
Sister chromatid exchange	+	– (a)
Gene Mutation		
S. typhimurium	+	+
Mouse lymphoma L5178Y	+	+
Chromosomal		
Chromosomal aberrations	+	–
Mitotic aneuploidy in yeast	+	–

[a] Overall negative, but some positive results were reported as described in text.
Data derived from investigator or consensus views in de Serres and Ashby (1981)

4 *In vivo* tests: non-mammalian endpoints

Without too much apology, those assays with *in vivo* non-mammalian endpoints are here treated separately from those with mammalian endpoints. Three rather diverse nonmammalian assay types have been presented in the present study; *Drosophila melanogaster*, hostmediated and urinary microbial mutation assays. These are considered together in this section.

As described by Vogel elsewhere in this volume, there is ample published data to demonstrate that *D. melanogaster* is insensitive to the mutagenic effects of BP. It has been suggested that this results from the preferential metabolism of BP to quinones, rather than to mutagenic forms, by *D. melanogaster* microsomes (Hallstrom and Grafstrom 1981). For other reasons, but with the same result, BP is poorly detected in host-mediated microbial mutation assays (see Forster, this volume). In contrast, the urinary bacterial mutation assays give positive results with BP. While these findings result from the presence of potentially reactive metabolites in the urine (which may be present as conjugates), they are probably not concerned with the detection of proximate or ultimate carcinogens.

These considerations are reflected in the observed performance of the assays in the present study (shown in

Table 2. *The in vivo genetic activity of BP & PYR: Nonmammalian endpoints*

	BP	PYR
D. melanogaster		
– SLRL	–	–
– Chromosome loss	–	–
– SMART	+	–
Host-mediated assay		
– Mouse/*S. typhimurium*	–	–
– Mouse/*S. cerevisiae*	+	+
Urinary assay/S. typhimurium		
– Rat	+	+
– Induced rat	+	+
– Mouse	+	+

Summary of the findings reported in this volume.

Table 2). Sex-linked recessive lethal (SLRL) and chromosome loss assays in *D. melanogaster* continued to give negative responses with BP despite administration to larval forms. The Somatic Mutation And Recombination Tests (SMART) discriminated between BP and PYR, but the positive effects observed for BP were weak. In view of previous explanations of the inactivity of BP in the *D. melanogaster* germ-cell assays, it is not clear what meaning should be given to the positive results observed here.

In the host-mediated assay, negative responses with both compounds were obtained with *S. typhimurium* TA100 using mice as host organisms at dose-levels up to 250 mg/kg (a dose-level which was clearly effective in the induction of some other in vivo endpoints such as micronuclei). *S. cerevisiae* D4 recovered from mouse liver, on the other hand, failed to discriminate between carcinogen and noncarcinogen and produced positive results for mitotic gene conversion with both BP and PYR. PYR is active in *S. cerevisiae* D4 *in vitro* in the presence of S9 metabolism, and the *in vivo* assay thus confirms the sensitivity of the mitotic gene conversion endpoint to this chemical. The result with BP is more difficult to interpret, in view of the negative *S. typhimurium* responses reported in this study (Pagano, Chapter 124, Volume 2) and elsewhere (Glatt *et al.*, 1985). It is worth noting, however, that mitotic gene conversion seems to be a non-specific response to DNA damage; the endpoint is induced by a wide range of mutagens without marked mutagen specificity (Zimmermann *et al.*, 1984, Murthy 1979). It may be anticipated therefore that mitotic gene conversion is induced by a wider range of BP and PYR metabolites than those that are able to induce point mutations.

In the urinary assays several different patterns of results were observed, but the greatest responses were obtained in the presence of deconjugating enzymes and S9 metabolism. These assays were not able to distinguish between BP and PYR; published data indicates that, for BP at least, the reactive metabolites present in the urine are not related to proximate carcinogenic forms (Camus *et al.*, 1984).

Overall, despite the ability of the *D. melanogaster* SMART assays and some rat urine assays to distinguish between BP and PYR, Table 2 shows that the findings supplied by these non-mammalian tests would not help to discriminate between the carcinogen and noncarcinogen.

The reasons in each case appear to be related to metabolism, or to the sensitivity of the target organisms to diverse metabolites.

The inability of such assays to function in a screening mode precludes their use in the hazard-assessment of these two chemicals. There is no particular reason to imagine that the two assays which here distinguish between BP and PYR are, in a more general context, any better able to discriminate between carcinogens and noncarcinogens than their companion assays in this group.

5 *In vivo* tests: mammalian endpoints

Introduction

The results from the *in vivo* assays with mammalian endpoints constitute the major part of the present study. To clarify the presentation of the results, the various assay types have been categorised by endpoint. Those assays concerned with evidence of DNA damage and repair have been placed together under the heading of "*Premutational*" endpoints. The induction of SCEs which may be associated with the repair of DNA damage (Perry *et al.*, 1984) has been included in this category together with UDS and strand breakage tests. None of these assays offers formal evidence of mutational change, although all are associated with DNA damage and repair. Those assays which do measure *gene mutation* are conveniently classed together, as are those measuring structural or numerical *chromosomal* damage. Finally, a number of assays measure non-genetic endpoints which might be related to later steps in the carcinogenic sequence; these include assays for immunotoxicity, liver initiation or promotion, cell proliferation and hyperplasia. These assays have been placed together under the title "*Carcinogenesis-related*". Since the sperm morphology assay is believed to measure not only gene mutation but also misdifferentiation (Komatsu *et al.*, 1982, Wyrobek *et al.*, 1983) it has been tentatively placed in this category. It is not our purpose here to argue strongly for the classification of any particular assay, but to bring an element of order into the diverse array of results.

The complete data-base is presented in Table 3. Where only one laboratory undertook an assay, their findings are presented; if a group of laboratories performed the assay, the group consensus is given. This table summarises the responses obtained in the assays. No attempt has been made to quantify the level of the response seen in individual assays.

In the discussion below this data set is considered, first under a number of general headings (e.g. germ-cell assays, bone-marrow) which draw together particular groupings of assays. The findings are then briefly reviewed in terms of the endpoint classification in Table 3. Finally, some consideration is given to general features of assay conduct.

Germ-cell assays

This family of assays is represented here by sperm-cell UDS, the dominant lethal test and the sperm morphology assay. The mouse spot test is not a germ-cell assay and measures somatic cell mutation; nonetheless, since it has been proposed as a pre-screen for germ cell mutagens (Russell 1977) the results from this test are considered here. The assays are drawn from all four categories of endpoint defined above, and we may therefore imagine *a priori* that there will be some differences in the responses observed. Data are summarized in Table 4.

Pyrene presents a straightforward situation; it is uniformly negative. Only in the sperm morphology assay was there a weak positive response in one of six assays (Arany, Chapter 110, Volume 2); this finding resulted from a unique combination of strain, route and vehicle and thus, although it contrasted with the consensus finding of the investigators in the group, it was felt that the result could not justifiably be over-ruled.

When we ask if BP is a germ-cell mutagen, the available results demand more careful interpretation. The only evidence available is oblique, being derived from

Table 3. *The in vivo genetic activity of BP and PYR: Mammalian endpoints Summary table*

	Mouse		Other		
	BP	PYR	BP	PYR	Species
1. *Permutational*					
DNA strand breaks	NT	NT	—	—	Rat
– with repair inhibitor	+	–	NT	NT	—
UDS – liver	–	–	–	–	Rat
UDS – forestomach	NT	NT	±	–	Rat
UDS – sperm-cell stages	–	–	–	–	Rat
SCE – bone marrow	+	w^+	+	w^+	Rat CH
SCE – liver epith. cells	NT	NT	w^+	–	Rat
SCE – circ. lymphocytes	+	–	NT	NT	–
SCE – intestinal epithelium	NT	NT	+	–	CH
2. *Gene mutation*					
In vivo/in vitro cell mutation	NT	NT	+	–	SH
Mouse spot test	+	–	NT	NT	–
3. *Chromosomal*					
Micronucleus – bone marrow	+	–	+	–	Rat
Micronucleus – peripheral erythrocytes					
– maternal	+	–	NT	NT	–
– foetal	+	–	NT	NT	–
Micronucleus – hepatocytes	NT	NT	–	–	Rat
Chromosome aberrations					
– bone marrow	+	–	+	–	Rat, CH
– liver epith. cells	NT	NT	w^+	–	Rat
Dominant lethal test	+	–	NT	NT	–
4. *Carcinogenesis-related*					
T-cell cytotoxicity	NT	NT	+	–	Rat
Liver initiation/promotion	NT	NT	+	–	Rat
Hepatocyte ploidy	+	–	+	–	Rat
Hepatocyte S-phase	–	NT	–	–	Rat
Mouse skin hyperplasia/ sebaceous gland suppression	+	–	NT	NT	–
Sperm morphology	+	–	NT	NT	–

Summary of the findings reported in this volume.

Key to Table 3

NT :not tested
± :equivocal result
w^+ :weak positive result
CH :Chinese hamster
SH :Syrian hamster

Table 4. *Germ-cell assays: reported results with BP*

Assay	Species	Route	Dose range (mg/kg)	Result
Specific locus test	Mouse	ip	500	– (i)
Sperm morphology	Mouse	ip/po	> 200	+
Dominant lethal	Mouse	ip	500	+
	Mouse	po	500	–
Heritable translocation test	Mouse	ip	500	– (ii)
Chromosomal aberrations	Mouse Oocytes	–	450 × 2	+ (iii)
Spermatocyte UDS	Rat	po	100–500	—
	Mouse	ip/po	100–500	—
Testes DNA binding	Mouse	ip		– (iv)

Data from this study except (i) Russell *et al.* (1981)
(ii) Generoso *et al.* (1982)
(iii) Basler and Rohrborn (1976)
(iv) Sega (1982)

ip :intraperitoneal administration
po :oral administration

assays (the mouse spot test and the sperm morphology assay) which do not formally measure germ-cell mutation.

Positive data was obtained in both assays, suggesting, albeit tentatively, that on further investigation it may be formally demonstrated that BP induces gene mutations in germ-cells. The mouse specific locus test (SLT) was not undertaken as part of the present study, but this test has previously given inconclusive results with BP (Russell *et al.*, 1981).

The positive results obtained with BP in the dominant lethal assay both in the present study and elsewhere (Generoso *et al.*, 1982), indicate the ability of BP to cause germ-cell chromosomal damage. The effect is very specific to mating intervals between 4.5 and 9 days after treatment, indicating that only late spermatids and mature spermatozoa are susceptible to these effects. BP was effective after ip administration but negative results were obtained in experiments using oral administration. Positive findings were also obtained after treatment of females. Again, no heritable translocation test was undertaken in the present study, but published data indicate a negative response using a dose and route of BP administration that is effective in inducing dominant lethals (Generoso *et al.*, 1982). This difference may be related to the clastogenic spectrum of BP; bone marrow studies indicate that BP induces predominantly chromatid deletions, and that very few exchanges are induced (Adler, this study; Kliesch *et al.*, 1982). Clearly, results from a spermatogonial cell or spermatocyte cytogenetic study would be valuable, but to our knowledge there is no published information. It has been reported, however, that BP induces chromosome aberrations in the oocytes of treated female mice (Basler and Rohrborn, 1976).

Evidence for the ability of BP to induce DNA damage and repair in germ-cells (in contrast to gene or chromosomal mutations) may be supplied by sperm-cell UDS. The dominant lethal response was obtained by the treatment of cells in the late spermatid and mature spermatozoal stages; these stages appear to be completely incapable of UDS, perhaps because the DNA is highly condensed and the lesions become inaccessible to repair enzymes (Sega and Sotomayor, 1982).

Examination of cells treated with BP as early spermatids (after the final scheduled DNA synthesis in primary spermatocytes, but before the UDS block in late spermatids) gave no evidence of a UDS response (Sega, this study). However, the presence, absence or amount of spermatid UDS cannot readily be related to germ-cell mutagenic effects for a number of agents (Sotomayor *et al.*, 1978). Furthermore, UDS assays appear to be relatively insensitive to BP, as discussed below.

Given the positive dominant lethal and sperm morphology assay responses, it is more difficult to reconcile the reported lack of germ-cell DNA binding by BP (Sega and Sotomayor 1982, Sega 1982). It has not been possible to demonstrate binding to germ-cell DNA, even when BP was present in the testes; the ability of testicular tissues to metabolise BP has also been shown (Lee and Nagayama 1980). It would seem desirable to obtain confirmation of the lack of binding.

In conclusion, it appears that BP is at best a poor germ-cell mutagen as measured by the authoritative specific locus and heritable translocation tests. Dominant lethals are, however, induced after treatment of either sex; the route of administration is critical. Spermatocyte UDS and DNA binding have produced some unexpected negative results, which may require reconsideration of the mechanisms of action of BP in these tissues.

Bone marrow

Acute toxicity studies with BP have demonstrated damage in the bone marrow, a well vascularised tissue, indicating the accessibility and sensitivity of the tissue of systemic BP. In contrast to the testes, the systemic availability of BP results in readily measureable genetic damage in this tissue. In the present study both bone marrow clastogenicity assays (chromosome aberration analysis and the micronucleus test) discriminated well between BP and PYR, the investigators unanimously finding BP positive and PYR negative.

BP was found to induce deletions but not exchanges, and it was concluded that "BP is a weak clastogen" compared with classical agents such as cyclophosphamide. This is consistent with the analysis of Aust *et al.* (1984), who showed that BPDE has the mutational specificity of a point mutagen, with only modest ability to induce mutations by mechanisms other than base-pair substitutions.

In part, this clear discrimination may have been assisted by the stringent requirements for a positive result. The CA assay required approximately seven-fold increases to achieve statistical significance (Adler, Chapter 7, Volume 1) and thus PYR remained negative despite a tendency to induce slightly elevated CA yields at high doses. The micronucleus test was only considered positive if 5% significance was achieved at two dose-levels.

The other assay which used bone marrow as the target tissue was the induction of SCEs; this is an endpoint indicative of DNA damage (Perry *et al.*, 1984) but not known to have deleterious genetic effects. This assay was responsive to BP, producing uniformly positive results in mice, rats and Chinese hamsters. Although a great deal less potent than BP, PYR also induced SCE's at high dose-levels in mice after oral or ip administration, and in male rats by oral administration. Positive findings were obtained in five out of fifteen trials (considering sexes separately).

These findings have been explained in terms of high-dose effects on cellular homeostasis, but where similar dose-levels were used in CA or micronucleus studies there is little evidence of corresponding artefactual induction of responses, which might be expected. On the contrary, in view of the positive *in vitro* SCE responses obtained with PYR (see Section 3), it seems reasonable to accept these results at face value. The positive result in male but not female rats after the oral administration of PYR, may reflect the greater capacity of males for xenobiotic metabolism (Chung *et al.*, 1973).

Unscheduled DNA synthesis

One of the most striking contrasts highlighted by the present study, is that BP gives a clear negative response in the *in vivo* UDS and single strand breakage assays, but positive responses in the same target cells *in vitro*

(Table 5). BP has been shown to be an inducer of UDS in rat, mouse and hamster hepatocytes *in vitro*. *In vivo*, however, using dose-levels which are effective in other *in vivo* assays, hepatocyte UDS assays are clearly negative.

Furthermore, while BP is able to induce UDS in a variety of different cell-types *in vitro* (e.g. HeLa, WI-38), it is equally true that it fails to induce UDS in a range of tissues *in vivo*, as exemplified in this study by hepatocytes, rat forestomach mucosal cells and sperm-cell stages.

It may be that these results reflect the organotropism of BP; this is consistent with the finding that chemically-induced *in vivo* strand breakage (as measured by alkaline elution) is greatest in the target organs for tumour induction (Petzold and Swenberg 1978). But in the liver, for example, the same major DNA-BP adduct can be found as in target organs such as the skin. The untested hypothesis is that target organs for tumours (e.g. mouse skin or Strain A mouse lung) would demonstrate UDS *in vivo*. Discussion has not focussed on *in vitro*/*in vivo* differences that would cause the same cells to react differently inside or outside of the body. Assuming the *in vivo* dose-levels to have been adequate, it may be that repair *in vivo* is inhibited by pathophysiological responses to the acute *general* rather than *genetic* toxicology. Replicative DNA synthesis is inhibited in the testes (Seiler 1977) and several other organs (Hellman *et al.*, 1984) after treatment with BP, and the same may be true for unscheduled synthesis.

Nucleotide pool imbalances, or hyperthermia, for example, are both responses which might reasonably be expected to result from the acute toxicity of BP, both of which may inhibit UDS (Bianchi *et al.*, 1982, Ben-Hur *et al.*, 1974).

Regardless of the reasons, it is true that all UDS endpoints (and similarly strand breakage, except in the presence of a repair inhibitor) have demonstrated a lack of sensitivity to BP *in vivo*. Accordingly, we should consider the PYR assays to be 'no test' since the endpoint appears to be insensitive to the structurally-related carcinogen. *In vitro* studies gave some indications of PYR activity, but *in vivo* responses were uniformly negative.

Liver assays

This group of assays is discussed in detail elsewhere (Hunt, Chapter 48, Volume 1). A large number of the assays are still in an early stage of development and only modest data-bases of results are available. The rationale and interpretation of many of these assays rest on features of hepatic biology rather than genetic toxicology (qv. the identity of the proliferating epithelioid cells used for cytogenetic studies? The role of GGT-foci? Or the meaning of hepatocyte ploidy values?) and thus are best discussed in that context.

The overall performance of this group of assays was not very good although there were noteable exceptions. An immediate explanation for the dissappointing performance of some of the hepatic assays is that BP is a poor hepatocarcinogen, and the assays are indicative of organotropic effects. The bone marrow, however, is not inhibited by considerations of organotropism, and bone-marrow metaphase analysis and micronucleus tests are responsive to a wide range of agents. At least at the level of metabolism, the general effectiveness of S9 preparations argues against highly specific mechanisms in the liver.

One should be careful in the use of arguments based on organotropism to justify negative results. One immediate implication of designating an assay 'organotropic' is to remove it from a role as a screening-mode assay. Screening assays, after all, should be widely responsive assays, and not highly specific. If screening assays were organotropic in their responses we would be obliged to examine an impossibly wide range of tissues in order to detect, for example, potential pancreatic or zymbal gland carcinogens.

An alternative explanation for the negative results may be that the liver, as the primary site of metabolism, has a more developed set of protective mechanisms than for example, bone marrow, or it may simply be that as a non-proliferating tissue the sensitivity of the genetic material is reduced. Whatever the explanation, it would appear from the overall performance of this very diverse group that some of these assays are at present poorly adapted for use in a screening mode.

Protocol features

Despite the wealth of data generated in the present study, few points emerge about the influence of experimental parameters on the outcome of the studies.

Table 5. *Unscheduled DNA synthesis: reported results with BP and PYR*

	BP	PYR	Ref
In vitro			
HeLa cells	+	+	Martin (1981)
Rat hepatocytes	+	−	Probst *et al.* (1981)
Rat hepatocytes		−	Tong *et al.* (1981) (a)
Rat hepatocyte nuclei	+	+	Althaus *et al.* (1982)
Mouse hepatocytes	+	−	McQueen *et al.* (1983)
Hamster hepatocytes	+	+	McQueen *et al.* (1983)
In vivo			
Rat hepatocytes	−	−	This study
Rat forestomach	−	−	This study
Rat spermatocytes	−	−	This study
Rat hepatocytes	−	NT	Mirsalis *et al.* (1982)

[a] Pyrene induced an increase in net grains/nucleus although considered negative.
NT :not tested.

No marked species differences were noted; where comparisons can be made, the results obtained in the rat and Chinese hamster are generally in agreement with those obtained in the mouse. Insufficient data are available to comment on the relative sensitivity of different animal strains, beyond the findings reported by the individual assay study groups.

The study design does not permit careful evaluation of the effect of route of administration; at best there are occasional examples of differences in findings where different routes were used. Conclusions about the relative value of different routes of administration would therefore be out of place.

Only one pronounced difference was noted; in the dominant lethal test BP gave a positive result after intraperitoneal administration but negative results after oral administration. Published data indicates that the levels of DNA binding of BP are approximately five-fold greater after intraperitoneal administration than oral, and this difference was presumed to result from the efficiency of absorption of BP through the gut (Lutz 1979). Similar findings have been obtained for the induction of lung adenomas in A-strain mice (Shimkin and Stoner 1984).

Overview by endpoint

Returning now to the classification of results by endpoint, we can draw some general conclusions. These conclusions are valid only for the two test chemicals BP and PYR. It may be of interest to generalise them to closely related compounds, or to the family of polycyclic aromatic hydrocarbons; it would be imprudent to generalise them further, as is shown by the different conclusions to emerge from the companion chapter on 2AAF and 4AAF (Combes, this volume).

The most mixed set of results, with both negative calls for BP and positives for PYR, comes from the premutational endpoints.

Whether this is the result of an unfortunate coincidence of special circumstances which affected the UDS and SCE assays or whether it is intrinsic for such endpoints may be debated. What has been established here is that for BP and PYR this group of assays gives poor discrimination, on the basis of the conventional interpretation of their carcinogenicity.

Only two assays fall clearly into the gene mutation category, the mouse spot test and an *in vivo/in vitro* mutation assay; each undertaken by one laboratory only. The mouse coat-colour spot test, assays for somatic mutation in foetal melanocytes. In the *in vivo/in vitro* mutation assay the lung cells of treated male Syrian hamsters were isolated and plated *in vitro* to determine the mutation frequency to thioguanine resistance. Both assays clearly distinguished between the carcinogen and noncarcinogen. BP produced positive findings, confirming published data for each assay (Fahrig, McGregor, this volume). PYR, despite evidence for its activity as a point mutagen *in vitro*, was negative in both assays. The sperm morphology assay which responds, *inter alia*, to germ-cell gene mutations produced similar results. Despite the limited number of assays, the evidence thus indicates that BP induces gene mutations *in vivo* and that PYR does not. Discrimination between carcinogen and noncarcinogen, on the basis of the conventional interpretation, is good.

The chromosomal endpoint assays similarly discriminated between the carcinogen and noncarcinogen, despite the conclusion that BP is a weak clastogen, and the slightly discordant negative result in the published heritable translocation test. The assays that failed to discriminate convincingly were those utilising proliferating liver-derived epithelioid cells. The other assays were not only able to discriminate between BP and PYR, but were also highly reproducible (see for example, the micronucleus test data).

The carcinogenesis-related endpoint data, despite the lack of a common rationale, also discriminated well between BP and PYR. These assays have not been developed, standardised and exploited as routine short-term assays to the same degree as many of the other kinds of assays under discussion. These findings therefore augur well for the future of such assays.

The epidermal hyperplasia/sebaceous-gland suppression assay was particularly sensitive, and only three applications of 50 ug BP per mouse were required to obtain positive results, while PYR was negative despite three dosings of 1200 ug/mouse (Lasne, Chapter 77, Volume 2). The majority of the other assays reported in Table 3 required dose-levels between 25 and 250 mg/kg BP to obtain positive results, and some used dose-levels considerably higher. Lower dose-levels can be effective in the production of tumours; for example, a single intragastric administration of 200 ug BP per mouse produced 14 tumours in five mice out of eleven (Pierce 1961, reported in IARC 1973). This dose level is in the order of 8 to 10 mg/kg; other similar examples of the production of tumours by relatively low doses of BP exist. It seems that the order of sensitivity of the assays for BP is *in vitro* > cancer bioassays > *in vivo* short-term tests.

6 Conclusions

We have seen that BP is widely distributed in the body after administration, and binds to a wide range of tissues. Its principal carcinogenic action is the causation of local tumours, although some remote tumours (e.g. leukaemias and lung adenomas) can be detected after dietary administration. *In vitro* short-term assays give uniformly positive results with BP. In non-mammalian *in vivo* assays, *D. melanogaster* is generally unresponsive to BP, and mixed results are obtained with urinary and host-mediated assays. *In vivo*, BP is a gene mutagen and a clastogen, although the clastogenic activity is limited to breaks and few complex exchanges are induced. *In vivo* UDS and strand breakage assays, for reasons yet to be defined, do not respond to BP. It is at best a weak germ-cell mutagen.

For PYR on the other hand little or no information is available concerning tissue distribution, metabolism and DNA binding. Available data suggests that it is a noncarcinogen, but studies of an adequate standard are lacking. *In vitro* studies give negative results for chromosomal damage but positive results for a variety of permutational endpoints (UDS, SCE, mitotic gene conversion and SOS induction). *In vitro* activity as a point mutagen is observed in *S. typhimurium* and in the L5178Y mutation system. Positive results were obtained *in vivo* with the yeast mitotic gene conversion endpoint in a host mediated assay. *In vivo* mammalian assays are all negative with the exception of the induction of SCE's (at very high dose levels).

It would appear that the present results support the proposal that *in vivo* short-term tests have the resolving power to distinguish carcinogens from those noncarcinogens which have some *in vitro* mutagenic activity. It is a pity that no definitive cancer bioassay with PYR is planned to vindicate (or contradict) this conclusion. For these two compounds only the SCE and UDS/strand breakage assays seem to be out of accord with this interpretation, and attention will no doubt be turned to examine the reasons for this.

In the *S. typhimurium* data, there may be a clue which prefigures the conclusions which were finally drawn for PYR after much further testing. The activity of PYR is principally towards tester strains TA1537 and TA97. These strains respond primarily towards intercalating agents, which are usually poor carcinogens. At least two of the other *in vitro* assays responding to PYR have endpoints which do not necessarily indicate covalent binding to DNA (mitotic gene conversion and sister chromatid exchange). The UDS and L5178Y mutation assays gave suggestive but not definitive results *in vitro* (see section 3).

Thus it is possible that further definition of the nature and mechanisms of the *in vitro* responses would have obviated the need for *in vivo* testing, or at the least have given strong indications of the likely outcome. The two assays which gave positive results *in vivo* with PYR have endpoints which are non-specific responses to DNA damage; these were mitotic gene conversion in the host mediated assay, and sister chromatid exchange, Randerath (this volume) failed to detect any DNA binding by PYR, using the sensitive post-labeling technique.

6 Acknowledgements

The author is grateful to Miss Sheridon Lummis for her patient assistance in the preparation of this chapter, and to any readers who have reached this point.

7 References

Adriaenssens, P.I., C.M. White and M.W. Anderson (1983) Dose-response relationships for the binding of benzo(a)pyrene metabolites to DNA and protein in lung, liver and forestomach of control and butylated hydroxyanisole-treated mice, *Cancer Res.*, **43**, 3712–3719.

Agrelo, C. and H. Amos (1981) In de Serres and Ashby (1981).

Althaus, F.R., S.D. Lawrence, G.L. Sattler, D.G. Longfellow and H.C. Pitot (1982) Chemical quantification of unscheduled DNA synthesis in cultured hepatocytes as an assay for the rapid screening of potential chemical carcinogens, *Cancer Res.*, **42**, 3010–3015.

Ashby, J., F.J. de Serres, M. Draper, M. Ishidate Jr., B.E. Matter and M. Shelby (1983) The two IPCS Collaborative Studies on Short-Term Tests for Genotoxicity and Carcinogenicity, *Mutation Res.*, **109**, 123–126.

Ashby, J., F.J. de Serres, M. Draper, M. Ishidate Jr., B.H. Margolin, B.E. Matter and M.D. Shelby (1985) Evaluation of short-term tests for carcinogens, *Progress in Mutation Research*, Vol. 5, Elsevier/North Holland.

Ashby, J. and B.J. Kilbey (1981) In de Serres and Ashby (1981).

Aust, A.E., N.R. Drinkwater, K. Debien, V.M. Maher, and J.J. McCormick (1984) Comparison of the frequency of diptheria toxin and thioguanine resistance induced by a series of carcinogens to analyse their mutational specificities in diploid human fibroblasts. *Mutat. Res.*, **125**, 95–104.

Autrup, H., C.C. Harris, B.F. Trump and A.M. Jeffery (1978) Metabolism of benzo(a)pyrene and identification of the major benzo(a)pyrene-DNA adducts in cultured human colon, *Cancer Res.*, **38**, 3689–3696.

Basler, A. and G. Rohrborn (1976) Chromosome aberrations in oocytes of NMRI mice and bone marrow cells of Chinese hamster induced with 3, 4, Benzopyrene, *Mutation Res.*, **38**, 327–332.

Ben-Hur, E., M.M. Elkind and B.V. Bronk (1974) Thermal enhanced radioresponse of cultured Chinese hamster cells: Inhibition of repair of sublethal damage and enhancement of lethal damage. *Radiat. Res.*, **58**, 38.

Bermudez, J., J.C. Mirsalis and H.C. Eales (1982) Detection of DNA damage in primary cultures of rat hepatocytes following *in vivo* and *in vitro* exposure to genotoxic agents, *Environmental Mutagenesis*, **4**, 667–679.

Bianchi, V., F. Nuzzo, A. Abbondandolo, S. Bonatti, E. Capelli, R. Fiorio, E. Giulotto, A. Mazzaccaro, M. Stefanini, L. Zaccaro, A. Zantedeschi and A.G. Levis (1982) Scintillometric determination of DNA repair in human cell lines, *Mutation Res.*, **93**, 447–463.

Boobis, A.R., D.W. Nebert and O. Pelkonen (1978) Effects of Microsomal Enzyme Inducers *in vivo* and inhibitors *in vitro* on the covalent binding of Benzo(a)pyrene metabolites to DNA catalyzed by liver microsomes from genetically responsive and nonresponsive mice, *Biochemical Pharmacology*, Vol. 28, 111–121.

Boroujerdi, M., H.-O. Kung, A.G.E. Wilson and M.W. Anderson (1981) Metabolism and DNA Binding of Benzo(a)pyrene *in vivo* in the rat, *Cancer Res.*, **41**, 951–957.

Bridges, B.A., E. Zeiger and D.B. McGregor (1981) In de Serres and Ashby (1981).

Brookes, P. and P.D. Lawley (1964) Evidence for the binding of polynuclear aromatic hydrocarbons to the nuclei acids of mouse skin. Relation between carcinogenic power of hydrocarbons and their binding to DNA, *Nature*, **202**, 781–784.

Buening, M.K., P.G. Wislock, W. Levin, H. Yagi, D.R. Thakker, H. Akagi, M. Koreeda, D.M. Jerina and A.H. Conney (1978) Tumorigenicity of the optical enantiomers of the diastereomeric benzo(a)pyrene 7, 8-diol, 9, 10-epoxides, in newborn mice: Exceptional activity of (+)-7, 8 -dihydroxy-9, 10 -epoxy-7, 8, 9, 10-tetrahydrobenzo(a)pyrene, Proc. *Natl. Acad. Sci. U. S.A.*, **75**, 5358–5361.

Camus, A.-M., A. Aitio, N. Sabadie, J. Wahrendorf and H. Bartsch (1984) Metabolism and urinary excretion of mutagenic metabolites of benzo(a)pyrene in C57 and DBA mice strains, *Carcinogenesis* Vol. 5, 35–39.

Chipman, J.K., P.C. Hirom, G.S. Frost and P. Milburn (1981) The biliary excretion and enterohepatic circulation of benzo(a)pyrene and its metabolites in the rat. *Biochem. Pharmacol*, **30**, 937–944.

Chipman, J.K., N.A. Bhave, P.C. Hirom and P. Milburn (1982) Metabolism and excretion of benzo(a)pyrene in the rabbit. *Xenobiotica*, **12**, 397–404.

Chung, L.W.K., G. Raymond and S. Fox (1973) Role of neonatal androgen in the development of hepatic microsomal drug-metabolising enzymes. *J. Pharm. Exp. Ther.*, **193**, 621–630.

Cook, J.W., C.L. Hewett and I. Hieger (1933) The

isolation of a cancer-producing hydrocarbon from coal-tar, *J. Chem Soc.*, 395–405.
Dunn, B.P. (1983) Wide-range linear dose-response curve for DNA binding of orally administered benzo(a)pyrene in mice, *Cancer Res.*, **43**, 2654–2658.
Evans, E.L. and A.D. Mitchell (1981) In de Serres and Ashby (1981).
Fahl, W.E., D.G. Scarpelli and K. Gill (1981) Relationship between benzo(a)pyrene-induced DNA base modification and frequency of reverse mutations in mutant strains of *Salmonella*, *Cancer Res.*, **41**, 3400–3406.
Generoso, W.M., K.T. Cain, C.S. Hellwig and N.L.A. Cacheiro (1982) Lack of association between induction of dominant-lethal mutations and induction of heritable translocations with benzo(a)pyrene in postmeiotic germ cells of male mice, *Mutation Res.*, **94**, 155–163.
Glatt, H.R. and F. Oesch (1976) Phenolic benzo(a)pyrene metabolites are mutagens, *Mutagen Res.*, **36**, 379–384.
Glatt, H. *et al.* (1985) Host mediated mutagenicity experiments with benzo(a)pyrene and two of its metabolites. *Mutation Res.*, **156**, 163–169.
Grover, P.L., A. Hewer, K. Pak and P. Sims (1976) The involvement of a diol-epoxide in the metabolic activation of benzo(a)pyrene in human bronchial mucosa and in mouse skin. *Int. J. Cancer*, **18**, 1–6.
Guttenplan, J.B., R. Bonomo, J.H. Fliessbach (1980) Recovery of benzo(a)pyrene metabolites and dimethylnitrosamine (DMN) in rat saliva (meeting abstract), *Cancer Res.*, **21**, 124.
Hallstrom and R. Grafstrom (1981) The metabolism of drugs and carcinogens in isolated subcellular fractions of *Drosophila melanogaster*, II. Enzyme induction and metabolism of benzo(a)pyrene, *Chem. Biol. Interact.*, **34**, 145–159.
Hellmann, R., R. D'argy and S. Ullberg (1984) The *in vivo* uptake of tritiated thymidine as a potential short-term test of toxic effects of polycyclic aromatic hydrocarbons in different organs. *Toxicology*, **29**, 183–194.
Hubbard, S.A., M.H.L. Green, B.A. Bridges, A.J. Wain and J.W. Bridges (1981) In de Serres and Ashby (1981).
IARC Monographs on the evaluation of carcinogenic risk of chemicals to man (1973). *Certain polycyclic aromatic hydrocarbons and heterocyclic compounds*, **3**, 91–136.
Jagannath, D.R., D.M. Vultaggio and D.J. Brusick (1981) In de Serres and Ashby (1981).
Jerina, B.M., R. Lehr, M. Schaefer-Ridder, H. Yagi J.M. Karle, D.R. Thakker, A.W. Wood, A.Y.H. Lu, D. Ryan, S. West W. Levin and A.H. Conney (1977) Bay region epoxides of dihydrodiols: A concept which explains the mutagenic and carcinogenic activity of benzo(a)pyrene and benzo(a)anthracene. In *Origins of Human Cancer*, ed. by J.D. Watson, and H. Hiatt, pp. 639–658.
Jernstrom, B., S. Orrenius, O. Undeman, A. Graslund and A. Ehrenberg (1978) Fluorescence study of DNA-binding metabolites of benzo(a)pyrene formed in hepatocytes isolated from 3-methylcholanthrene-treated rats. *Cancer Res.*, **38**, 2600–2607.
Jotz, M.M. and A.D. Mitchell (1981) In de Serres and Ashby (1981).
Juchau, M.R., J. DiGiovanni, M.J. Namking and A.M. Jones (1979) A comparison of the capacity of fetal and adult liver, lung and brain to convert polycyclic aromatic hydrocarbons to mutagenic and cytotoxic metabolites in mice and rats, *Toxicology and Applied Pharmacology*, **49**, 171–178.
Kliesch, U., I. Roupova and I.-D. Adler (1982) Induction of chromosome damage in mouse bone marrow by benzo(a)pyrene, *Mutation Res.*, **102**, 265–273.
Komatsu, H., T. Kakizoe, T. Nijima, T. Kawachi and T. Sugimura (1982) Increased sperm abnormalities due to dietary restriction, *Mutation Res.*, **93**, 439–446.
Koreeda, M.N., P.D. Moore, P.G. Wislocki, W. Levin, A.H. Conney and D.M. Jerina (1978) Binding of benzo(a)pyrene 7,8-diol-9,10-epoxides to DNA, RNA and protein of mouse skin occurs with high stereoselectivity. *Science*, **199**, 778–781.
Kruber (1931) *Chemische Berichte*, **64**, 84.
Kulkarni, M.S. and M.W. Anderson (1984) Persistence of benzo(a)pyrene metabolite:DNA adducts in lung and liver of mice, *Cancer Res.*, **44**, 97–101.
Lee, I.P. and J. Nagayama (1980) Metabolism of benzo(a)pyrene by the isolated perfused rat testis, *Cancer Res.*, **40(9)**, 3297–3303.
Levin, W., A.W. Wood, H. Yagi, P.M. Dansette, D.M. Jerina and A.H. Conney (1976) Carcinogenicity of benzo(a)pyrene 4,5-, 7,8- and 9,10- oxides on mouse skin. *Proc. Natl. Acad. Sci. U.S.A.*, **73**, 243–247.
Lorentzen, R.J., S.A. Lesko, K. McDonald and P.O.P. Ts-o, (1979) Toxicity of metabolic benzo(a)pyrenediones to cultured cells and the dependence upon molecular oxygen, *Cancer Res.*, **39**, 3194–3198.
Lutz, W.K., A. Viviani and C. Schlatter (1978) Nonlinear dose-response relationship for the binding of the carcinogen benzo(a)pyrene to rat liver DNA *in vivo*, *Cancer Res.*, **38**, 575–578.
Lutz, W.K. (1979) *In vivo* covalent binding of organic chemicals to DNA as a quantitative indicator of chemical carcinogenesis. *Mutat. Res.*, **65**, 289–356.
Malaveille, C., H. Bartsch, P.L. Grover and P. Sims (1975) Mutagenicity of non-K-region diols and diol-epoxides of benzo(a)anthracene and benzo(a)pyrene in *S. typhimurium* TA100, *Biochem. Biophys. Res. Commun.*, **66**, 693–700.
Matijasevic, Z. and E. Zeiger (1985) Mutagenicity of pyrene in *Salmonella, Mutation Res.*, **142**, 149–152.
MacDonald, D.J. (1981) In de Serres and Ashby (1981).
Martin, C.N. and A.C. McDermid (1981) In de Serres and Ashby (1981).
McQueen, C.A., B.M. Kreiser and G.M. Williams (1983) The hepatocyte primary culture/DNA repair assay using mouse or hamster hepatocytes, *Environmental Mutagenesis*, **5**, 1–8.
Mitchell, C.E. (1983) The metabolic fate of benzo(a)pyrene in rats after inhalation, *Toxicology*, **28**, 65–73.
Mohn, G.R., P.R.M. Kerklaan, W.P.C. ten. Bokkum-Coenradi and T.E.M. ten Hulscher (1983) A differential DNA-repair test using mixtures of E. coli K12 strains in liquid suspension and animal-mediated assays, *Mutation Res.*, Vol. 113, No. 5, 403–415.
Mirsalis, J.C., C.K. Tyson and B.E. Butterworth (1982) Detection of genotoxic carcinogens in the *in vivo–in vitro* hepatocyte DNA repair assay, *Environmental Mutagenesis*, **4**, 553–562.
Murthy, M.S.S. (1979) Induction of gene conversion in diploid yeast by chemicals: correlation with mutagenic action and its relevance in genotoxicity screening. *Mutation Research*, **64**, 1–17.
Natarajan, A.T. and A.C. van Kesteran-van Leeuwen (1981) In de Serres and Ashby (1981).
Newbold, R.F. and P. Brookes (1976) Exceptional mutagenicity of a benzo(a)pyrene diol epoxide in cultured mammalian cells. *Nature*, **261**, 52–54.
Parry, J.M. and D. Sharp (1981) In de Serres and Ashby (1981).
Pelkonen O. and D.W. Nebert (1982) Metabolism of Polycyclic Aromatic Hydrocarbons: Etiologic role in carcinogenesis. *Pharm Rev.*, **34**, 190–222.
Perry P., L. Henderson and D. Kirkland (1984) Sister chromatid exchange in cultured cells in: Report of the UKEMS sub-committee on guidelines for mutagenicity testing: Part II. United Kingdom Environmental Mutagen Society, Swansea.
Perry, P.E. and E.J. Thompson (1981) In de Serres and Ashby (1981).
Petzold G.L. and J.A. Swenberg (1978) Detection of DNA

damage induced *in vivo* following exposure of rats to carcinogens, *Cancer Res.*, **38**, 1589–1594.
Phillips, D.H. (1983) Fifty years of benzo(a)pyrene, *Nature*, Vol. 303, 468–472.
Probst, G.S., R.E. McMahon, L.E. Hill, C.Z. Thompson, J.K. Epp and S.B. Neal (1981) Chemically-induced unscheduled DNA synthesis in primary rat hepatocyte cultures: A comparison with bacterial mutagenicity using 218 compounds, *Environmental Mutagenesis*, **3**, 11–32.
Prodi, G., P. Rocchi and S. Grilli (1970) Binding of 7,12-Dimethylbenz(a)anthracene and benzo(a)pyrene to nucleic acids and proteins of organs in rats, *Cancer Res.*, **30**, 1020–1023.
Quillardet, P., C. de Bellecombe and M. Hofnung (1985) The SOS chromotest, a colorimetric bacterial assay for genotoxins: validation study with 83 compounds, *Mutation Res.*, **147**, 79–85.
Renwick, A.G. and B.S. Drasar (1976) Environmental carcinogens and large bowel cancer, *Nature*, Vol. 263, 234–235.
Robinson, D.E. and A.D. Mitchell (1981) In de Serres and Ashby (1981).
Rosenkranz, H.S., J. Hyman and Zev. Leifer (1981) In de Serres and Ashby (1981).
Russell, L.B., (1977) Validation of the *in vivo* somatic mutation method in the mouse as a prescreen for germinal point mutations, *Arch. Toxicol.*, **38**, 75–85.
Russell, L.B., P.B. Selby, E. von Halle, W. Sheridan and L. Valcovic (1981) The mouse specific-locus test with agents other than radiations, *Mutation Res.*, **86**, 329–354.
Salamone, M.F. (1981) In de Serres and Ashby (1981).
Santella, R.M., D. Grunberger and B. Weinstein (1979) DNA-benzo(a)pyrene adducts formed in a *Salmonella typhimurium* mutagenesis assay system, *Mutation Res.*, **61**, 181–189.
Sakai, M., D. Yoshida and S. Mizusaki (1985) Mutagenicity of polycyclic aromatic hydrocarbons and quinones on *Salmonella typhimurium* TA97.
Schlede, E., R. Kuntzman and A.H. Conney (1970) Stimulatory effect of benzo(a)pyrene and phenobarbital pretreatment on the biliary excretion of benzo(a)pyrene metabolites in the Rat. *Cancer. Res.*, **30**, 2898–2904.
Sega, G.A. (1982) DNA repair in spermatocytes and spermatids of the mouse. In *Banbury Report 13. Indication of Genotoxic exposure.* Bridges, Butterworth and Weinstein (Eds), Cold spring harbour NY, 503–513.
Sega G.A. and Sotomayor (1982) Unscheduled DNA synthesis in mammalian germ cells – its potential use in mutagenicity testing. In *Chemical Mutagens: Principles and methods for their detection*, Vol. 7, de Serres F.J. and A. Hollaender (Eds), Plenum Press, New York pp. 421–445.
Seiler, J.P. (1977) Inhibition of testicular DNA synthesis by chemical mutagens and carcinogens. Preliminary results in the validation of a novel short term test. *Mutation Res.*, **46**, 305.
de Serres, F.J. and J. Ashby (1981) Evaluation of Short-Term Tests for Carcinogens: Report on the International Collaborative Program. Elsevier/North Holland, New York, 1981.
Sharp D.C. and J.M. Parry (1981) In de Serres and Ashby (1981).
Shen, A.L., W.E. Fahl, S.A. Wrighton and C.R. Jefcoate (1979) Inhibition of benzo(a)pyrene and benzo(a)pyrene 7,8-Dihydrodiol metabolism by benzo(a)pyrene quinones, *Cancer Res.*, **39**, 4123–4129.
Shimkin M.B. and G.D. Stoner (1975) Lung tumours in mice: application to carcinogenesis bioassay. *Adv. Cancer Res.*, **21**, 1–58.
Shu, H.P. and E.N. Bymun (1983) Systemic excretion of benzo(a)pyrene in the control and microsomally induced rat: the influence of plasma lipoproteins and albumin as carrier molecules, *Cancer Res.*, **43**, 485–490.
Sims, P., P.L. Grover, A. Swaisland, K. Pal and A. Hewer (1974) Metabolic activation of benzo(a)pyrene proceeds by a diol-epoxide. *Nature*, **252**, 326–328.
Slaga, T.J., W.M. Bracken, G. Gleason, W. Levin, H. Yagi, D.M. Jerina and A.H. Conney (1979) Marked differences in the skin tumour-initiating activities of the optic enantiomers of the diastereometic benzo(a)pyrene 7,8-diol-9,10-epoxides, *Cancer Res.*, **39**, 67–71.
Sotomayor, R.E., G.A. Sega and R.B. Cumming (1978) Unscheduled DNA synthesis in spermatogenic cells of mice treated *in vivo* with the indirect alkylating agents cyclophosphamide and mitomen, *Mutation Res.*, **50**, 229–240.
Stoner, G.D., E.A. Greisiger, H.A.J. Schut, M.A. Pereira, T.R. Loeb, J.E. Klaunig and D.G. Branstetter (1984) A comparison of the lung adenoma response in strain A/J mice after intraperitoneal and oral administration of carcinogens. *Toxicol. Appl. Pharm.*, **72**, 313–323.
Scott, W.T., R.H. Reitz, A.M. Schumann and P.G. Watanabe (1981) Genetic and nongenetic events in neoplasia, *Fd. Cosmet. Toxicol.*, **19**, 567–576.
Stowers, S.J. and M.W. Anderson (1984) Ubiquitous binding of benzo(a)pyrene metabolites to DNA and protein in tissues of the mouse and rabbit. *Chem. Biol. Interact.*, **51**, 151–166.
Tong, C., M.F. Laspia, S. Telang and G.M. Williams (1981) The use of adult rat liver cultures in the detection of the genotoxicity of various polycyclic aromatic hydrocarbons. *Environ. Mutagenesis*, **3**, 477–487.
Wood, A.W., R.L. Goode, R.L. Chang, W. Levin, A.H. Conney, H. Yagi, P.M. Dansetto and D.M. Jerina (1975) Mutagenic and cytotoxic activity of Benzo(a)pyrene 4,5-, 7,8-, and 9,10-oxides and the six corresponding phenols, *Proc. Natl. Acad. Sci. U.S.A.*, **72**, 3176–3180.
Wyrobek, A.J., L.A. Gordon, J.G. Burkhart, M.W. Francis, R.W. Kapp Jr., G. Letz, H.V. Malling, J.C. Topham and M.D. Whorton (1983) An evaluation of the mouse sperm morphology test and other sperm tests in nonhuman mammals, *Mutation Res.*, **115**, 1–72.
Zimmermann, F.K., R.C. von Borstel, E.S. von Halle, J.M. Parry, D. Siebert, G. Zetterberg, R. Barale and N. Loprieno (1984) Testing of chemicals for genetic activity with Saccharomyces cerevisiae: a report of the U.S. EPA Gene-Tox program. *Mutation Research*, **133**, 199–244.
Zimmermann, F.K. and I. Scheel (1981) In de Serres and Ashby (1981).

Robert D. Combes (Workgroup Chairman)[1]
School of Biological Sciences, Portsmouth Polytechnic Portsmouth, UK

5

Overview: 2-acetylaminofluorene and 4-acetylaminofluorene

1 Introduction

Scope of chapter

This chapter summarises comparative genotoxicities of 2AAF and 4AAF detected by *in vivo* short term assays conducted in the collaborative trial. Using consensus results from the respective working groups, the different end points induced by the 2 compounds in each organism and tissue are classified and discussed. The performance of assays with respect to their abilities to distinguish between 2AAF and 4AAF and to predict correctly animal carcinogenicity is also considered. The efficacy and utility of the *in vivo* tests are assessed by comparing the data with results of analogous assays carried out *in vitro*. The data base compiled from individual investigators' reports for responses of the chemicals in the same systems also is evaluated as far as possible for the extent of interlaboratory differences. The degree of variation in responses and the qualitative nature of the different end points induced are discussed with respect to species of host animal, target organ, route of compound administration and several other experimental variables. In addition some of the factors such as the nature of DNA adducts, metabolic activation and *in vivo* distribution which may account for divergent data obtained for 2AAF and 4AAF are debated.

Previous literature

2AAF and related compounds have been reviewed on numerous occasions and Weisburger and Weisburger (1958) should be consulted for a detailed discussion of early literature especially on metabolism and carcinogenesis. A more recent account of the latter topic in relation to the activities of other aromatic amines has been written by Clayson and Garner (1976). A comprehensive and contemporary treatise, including genotoxicity, can be found in Weisburger (1981) and Thorgiersson *et al.* (1983) whilst detailed information on metabolism

[1] Workgroup participants: Ilse-Dore Adler, Robert A. Baan, Brita Beije, Eldiberto Bermudez, Hyman C. Birnboim, Jack B. Bishop, Rolf Braun, Georgio Bronzetti, Pawan S. Chauhan, Rudolf Fahrig, R. Marion Hicks, Dag Jenssen, Ian Kimber, Douglas Kornbrust, Naguib Nashed, Fabrizio Palitti, Michael F. Salamone, Jörg Schöeneich, Antony G. Searle, Tohru Shibuya, Reggie H. Stevens, Hiroyuki Tsuda, Ruby Valencia, Ekkehart W. Vogel, and Dieter Wild

and DNA adduct formation is presented in the NCI Monograph No. 58 (1981).

Summary of metabolic activation pathways
It is agreed generally that conversion of 2AAF to an electrophilic reactive species requires cytochrome P-450 – mediated hydroxylation at the N-acetyl N atom, whilst ring hydroxylations at the 1-, 3-, 5-, 7-C and the 9-N positions are mainly detoxifying (Miller and Miller, 1969). Subsequent conjugations of ring hydroxylated metabolites (e.g., *O*-glucuronidation) are also deactivating (Wang *et al.*, 1985). Also it is thought that N-hydroxylation is the rate-limiting step and that the efficiency of this reaction can be modulated by the presence of various ring hydroxylated metabolites (Razzouk and Roberfroid, 1983). Species and organ variation in sensitivity to 2AAF may in part be due to differences in the levels of N-hydroxylation (Lotlikar *et al.*, 1984; and see p. 1.68). Glucuronidation of N-OH-2AAF in the liver may be important for generating metabolites which exert carcinogenic effects at extrahepatic sites (Malejk-Giganti and Ryzewski, 1985).

There have been few metabolism studies on analogues of 2AAF and none on 4AAF. However, it is noteworthy that the weak carcinogen 3AAF was N-hydroxylated at a 50-fold slower rate than 2AAF by a rat liver microsomal preparation (Weisburger, 1981). Presumably one reason for the weaker activities of analogues of 2AAF is their intrinsic resistance to N-hydroxylation (p. 1.70). Mechanistically this will be a function of the ease with which the N-atom proximal to the ring can interact with the fluorene system in order to reduce its charge density as a pre-requisite for O addition. It is also feasible that the resulting metabolite is considerably less stable or amenable to activating transformations. It has been shown (Leow *et al.*, 1985) that relative mutagenic potencies for 2AAF and 4AAF agree with the respective values calculated for various molecular electronic parameters considered by these authors to reflect ease of N-hydroxylation rather than ring hydroxylation of these compounds.

Numerous studies have suggested that the NOH metabolite is a substrate for several different pathways leading to a number of candidate reactive species. Moreover, both microsomal and cytosolic enzymes are implicated variously in these reactions which comprise: 1) deacetylation to form NOH-2AF; 2) peroxidation giving nitrosofluorene (a potent direct-acting mutagen) via a nitroxyl radical (Reigh and Floyd, 1981); 3) transfer/conjugation yielding esterified N-O conjugates mediated by sulphotransferases especially, and 4) N, O-acyl transferase to give N-acetoxy-2AAF (see Faustmann-Watts *et al.*, 1984). Depending upon the particular pathway followed, the ultimate electrophile is considered to be either an N-acetyl-N-aryl-nitrenium ion or the N-aryl-nitrenium ion. The predominant mode of activation may depend upon the capacity of a species or organ to N-hydroxylate AAF or AF, thus deacetylation may occur before or after N-hydroxylation (Aune *et al.*, 1985a, b; Weber and Hein, 1985).

4-nitrosofluorene is a weak direct-acting mutagen, 4AF and NOH-4AAF are activated to mutagens by S9 mix and N-acetoxy-4AAF was a direct-acting mutagen but considerably less potent than N-acetoxy-2AAF (Ashby, Paton and Callander, this volume). These data suggest that, although 4AAF may be metabolized in a similar fashion to 2AAF, the resulting intermediates are less electrophilic and reactive to bacterial DNA. Furthermore, different metabolites are responsible for the induction of particular end points observed *in vitro* and *in vivo*. For example, deacetylation is not required for formation of certain DNA adducts since 80% of bound forms of 2AAF retain the N-acetyl group (p. 1.70; Poirier, 1984). However, deacetylation is believed to be necessary for the formation of mutagenic metabolites, at least *in vitro* (Stout *et al.*, 1976; Aune *et al.*, 1985a) and the mutagenicity of 4AAF may be a consequence of its susceptibility to deacetylation by rat liver homogenate (Weisburger and Weisburger, 1958). However, the need for deacetylation in the production of mutagenic metabolites can vary according to the tissue from which S9 is prepared (Aune *et al.*, 1985a). In contrast, sulphotransferase catalysed N-esterification is not thought to be important for the generation of mutagenic metabolites from 2AAF (Thorgiersson *et al.*, 1982). Nevertheless, this enzyme plays a significant role in the production of active species *in vivo* since inhibition of sulphation reduces any such activity. Indeed the sulphate ester of NOH-2AAF is believed to be the ultimate carcinogen in the rat. The dependence of different genotoxic effects *in vitro* and *in vivo* on the generation of varying reactive metabolites could be a significant cause for the existence of discrepancies between species susceptibility to cancer and the genotoxicity of 2AAF (p. 1.70).

The relative contributions of cytosolic and microsomal enzymes to the activation of 2AAF is unclear. Thus 2AAF can be activated to a mutagen by rat S105 fraction (Forster *et al.*, 1982). The activity of 4AAF under the same conditions is worthy of investigation. Although deacetylation is a microsomal reaction, the formation of the putative highly reactive metabolites, N-acetoxy-2AAF and its sulphate ester respectively is mediated by cytosolic enzymes (Weeks *et al.*, 1978). Further work is required due to problems of contamination of the various fractions with free or membrane-bound enzymes and due to the complex relationship between deacylases and acylases with regard to production of metabolites with varying activities (Vaught *et al.*, 1985).

Pharmacokinetics, distribution and elimination
Whilst 2AAF has been subjected to numerous studies on its metabolic fate, there is a paucity of information for 4AAF due to the lack of radiolabelled material. No integral overview of the pharmacokinetics of 2AAF is available.

Early studies with radiolabelled samples suggest that, after oral administration to rats, 2AAF is absorbed efficiently from the small intestine and is rapidly disseminated to the liver. Hepatic metabolites are distributed in the blood to most major organs including the kidneys from which a significant proportion of the applied dose is voided in urine. Also many liver metabolites are excreted in bile although faecal labelling is lower than expected due to some enterohepatic circulation. After i.p. dosing, label persists for longer in the blood and there is the possibility that a more general distribution of the compound occurs such that the proportion of unmetabolised 2AAF reaching extra-hepatic sites may be increased. Moreover, the relative amount of label eliminated in the urine, as opposed to bile, is greater than after p.o. dosing. Excretion of label occurs rapidly so that only a small fraction of the administered dose remains in the body 4 days after a single application. Exhalation of label has not been detected (Weisburger and Weisburger, 1958; Combes *et al.*, Volume 1).

There have been few distribution studies with other species although it is known that, when compared with rats, guinea pigs excrete approximately 3 times the proportion of label in their urine at the expense of the faecal route. In mice there is evidence for transplacental transfer of 2AAF and its NOH metabolite (Takahashi and Yasuhira, 1975; p. 1.70). Considerable species variation in the urinary excretion of 2AAF and its metabolites (NCI, 1981) has been attributed to differential complements of activating/detoxification enzymes in organs (pp. 1.58 and 1.70).

After single i.p. injections of rats with suspensions of ^{3}H-2AAF or ^{3}H-4AAF, the compounds appeared to reach the liver to the same extent whereas more 4AAF label was associated with blood and the remaining carcass (Combes *et al.*, this volume). A correspondingly higher level of faecal elimination of 2AAF occurred. Further experiments with 4AAF especially will assist an understanding of the reasons for divergent *in vivo* responses of the 2 compounds.

It is clear from both chemical analysis (Weisburger and Weisburger, 1958) and from the testing of body fluids (Combes *et al.*, 1984) that large amounts of 2AAF in bile and urine are present as promutagenic conjugates. Furthermore, administration of 4AAF to rats and mice results in urine with mutagenicity similar to that exhibited by urines from animals treated with 2AAF (McGregor *et al.*, this volume). However, chemical analysis of such urines is necessary before attributing these observations to production and excretion *in vivo* of metabolites analogous to those derived from 2AAF.

2 *In vitro* activities of 2AAF and 4AAF

Assays used

Information on the genotoxicities of 2AAF and 4AAF was derived from de Serres and Ashby (1981), Weisburger (1981), Kada and Ishidate (1980), and individual investigators' reports in this volume.

The test systems that have been used are listed in Table 1.

Both compounds have been assayed for induction of the major end points except that 4AAF has not been assessed for its mutagenicity or clastogenicity towards cultured mammalian cells. Although aneuploidy has been scored this was in fungi and the *in vitro* activity of 4AAF in the micronucleus assay is unknown. In fact SCE represents the only test system sensitive to chromosome effects in cultured mammalian cells to which 4AAF has been subjected. Therefore, for certain test systems more data are available for *in vivo* responses of 4AAF, a fact which restricts comparisons between *in vitro* and *in vivo* short-term assays (p. 1.73).

Although rat liver S9 has been the most frequently used source of metabolic activation, homogenates from several other organs and organisms have been utilized especially with the Salmonella mutation assay.

Summary of activities

The qualitative data presented in Table 1 reveal that in only 3 of 8 assay systems in which both compounds have been tested has there been a distinction in the activities of 2AAF and 4AAF observed. All 3 of these tests are specific for premutational damage. Both UDS in cultured mam-

Table 1. *Summary of qualitative responses of 2AAF and 4AAF in in vitro genotoxicity assays*[a]

End point/assay	Qualitative Result[b] 2AAF	4AAF
Premutational damage		
DNA binding/adducts/strand breaks	●	nt
DNA repair (bacterial)	●/◐/○	●/○
(fungal)	○	●
(mammalian, UDS)	●	○
Recombination (mitotic, fungal)	●	○
(conversion, fungal)	●	●/○
Gene mutation		
Point mutation (yeast)	●	nt
(mammalian)	●	nt
(bacterial)	●	◐
Chromosomal effects		
Chromosome loss (fungal)	●	●
Structural aberrations	●/○	nt
Micronucleus	nt	nt
SCE	◐/○	○
Cell transformation	●	●/○

[a]Data from de Serres and Ashby, (1981) and Weisburger (1981).
[b]● = positive response; ◐ = weak positive response; ○ = negative response; where more than 1 response is recorded indicates conflicting results; nt = not tested.

malian cells and mitotic recombination in fungi correctly predict positive and negative responses respectively for 2AAF and 4AAF. However, when stimulation of repair was scored in fungi the reverse order of response was observed. 2AAF induces premutational damage, gene mutation, chromosomal effects and cell transformation in Syrian Hamster kidney cells. 4AAF consistently was negative for 3/8 end points studied (Table 1). Conflicting results were obtained for 3 assays and 4AAF was unequivocally positive for fungal DNA repair, bacterial mutation and aneuploidy in fungi. Although no weak positive responses were recorded for 4AAF, a comparison with Minimal Effective Dose (MED) values (de Serres and Ashby, 1981) together with dose responses suggests that, in the majority of *in vitro* tests, 4AAF is the less genotoxic compound by a substantial margin. For example in the author's laboratory using the Salmonella mutagenicity assay, 4AAF was more than 100 fold less potent (Dashwood and Combes, these volumes).

Role of metabolic activation

As expected with aromatic amines, 2AAF and 4AAF genotoxicity *in vitro* is dependent upon the presence of activating enzymes. In the case of bacterial repair assays, apparent direct activity has been detected for reasons which have been discussed elsewhere (Combes, 1983). 2AAF is converted to genotoxic metabolites by enzymes from several different organs and species including those which are refractory to extensive carcinogenic effects by this chemical (McGregor, 1975; Walters and Combes, 1985; Weisburger, 1981, Aune *et al.*, 1985a, b, p. 1.73).

Putative, ultimate electrophilic forms of 2AAF such as N-acetoxy 2AAF exhibit potent genotoxicity in the absence of S9. Negative responses for 2AAF have been obtained for several end points scored in cultured mammalian cells without addition of enzymes. However, under these conditions, and especially with some primary cell cultures, 2AAF exhibits some activity which appears to be independent of S9. In other experiments, although the parent compound is inactive without metabolism, 2AF or N-OH-2AAF variously are genotoxic implying that different cell lines possess varying capacities for metabolic activation (p. 1.58) (see Weisburger, 1981 for review). In certain systems, for example transformation in hamster embryo and kidney cells, 2AAF may be inactive even in the presence of S9 despite the fact that Nacetoxy 2AAF was positive. Detoxification reactions may predominate in such circumstances, perhaps due to a combination of S9 and cell metabolism. The capacity of cultured mammalian cells themselves to activate and detoxify chemicals requires further consideration.

3 Carcinogenicity of 2AAF and 4AAF

4AAF in relation to 2AAF

Previously (de Serres and Ashby, 1981) it was concluded that, whilst adequate evidence existed to suggest that 2AAF is carcinogenic in several species, 4AAF was left unclassified. This was due to the fact that in 2 studies (Weisburger *et al.*, 1952; Schintz *et al.*, 1955) involving oral administration to rats, data evaluation was precluded owing to the use of low dosages, lack of controls and inadequate numbers of animals. Moreover, the maximum mean duration of dosing was approximately 13 months only. Thus, despite the fact that 4AAF is quoted as a non-carcinogen (Weisburger, 1981) on the basis of these data there is no way of distinguishing between weak and negative carcinogenicity (see Hicks and Ashby, these volumes). The rodent carcinogenicity has been re-assessed therefore using orally-dosed rats (Hicks *et al.*, these volumes). The experiments are thus far incomplete but already it is apparent that after 26 weeks no tumours have been detected following chronic administration of doses 4 times those capable of producing many initiating/promoting events (p. 1.67) and liver tumours under the same conditions when 2AAF was the test compound. These results do not preclude the possibility that tumours may arise after longer dosing times. The working group concluded that if 4AAF proves to be a carcinogen under these circumstances it will be significantly less potent than 2AAF (McGregor *et al.*, these volumes). Of the 2 other cancer studies in the trial, results from the Japanese quail embryo assay are incomplete although a single preneoplastic change was induced by 2AAF. In contrast, the completed data from the mouse lung adenoma system are equivocal. Stoner *et al.* (these volumes) reported a significant increase in tumour incidence occurring with 2AAF but not 4AAF. However, the consensus result (McGregor *et al.*, these volumes) was that, although statistically valid, the rise was sufficiently small to require further verification.

2AAF

2AAF generally is considered to be a potent carcinogen in a wide range of species (Weisburger and Weisburger, 1955; Weisburger, 1981; Miller, 1978). 2AAF is described as a complete hepatocarcinogen, functioning as both an initiator and a promotor in this organ (Stevens and Peraino, 1983). There is less evidence for this dual activity in other tissues, although this depends on species.

Effect of species with 2AAF. For 2AAF-induced hepatocarcinogenesis the following order of rodent species-sensitivity has been observed: rat hamster > mouse > guinea pig the last species being resistant. In the rat, 2AAF induces liver tumours earlier than in the bladder, whereas in the mouse a reversal in sensitivity was observed. This is likely to be due to the fact that, in the mouse, 2AAF is an initiating carcinogen for the liver. Whereas, in the rat 2AAF is thought to be a strong promoter as well as an initiator for the liver (Kuchlbauer *et al.*, 1985). Moreover, unlike the situation in the rat, exposed mice show no signs of tumours of zymbal's (ear duct) gland. Also 2AAF induces tumours in mammary glands of rats and in stomach and intestine of hamsters, although cancers at all sites in this species are less frequent.

Effect of route with 2AAF. Except for the hamster, p.o. administration either in the diet or by gavage results in a wider spectrum of organ sensitivity. Although there is 1 report of induction of lung tumours after i.p. dosing (Fischer, 1959), no induced liver tumours have been detected in any species as a consequence of injection of 2AAF. However, administration of 2AAF by this route results in the production of mammary tumours but males

Table 2. *Classification of end points detected in non-mammalian cells*

Assay	Number of tests 2AAF	4AAF
Urine analysis (bacteria)[a]	7	6
Host-mediated (bacteria)[a]	1	1
Host-mediated (yeast)[a.b.e.]	5	5
Drosophila– sex-linked recessive lethal (SLRL)[a.b.c.]	5	5
–Somatic recombination and Mutation (SMART)[a.b.c.d.f.]	3	3
–Chromosome Loss[e]	2	1
Total tests considered	23	21

[a]gene mutation; [b]detections; [c]chromosomal structural aberrations; [d]mitotic recombination; [e]gene conversion; [f]aneuploidy

Table 3. *Classification of end points detected in mammalian cells*

End point	Assay	Number of tests 2AAF	4AAF
Premutational	Covalent DNA binding	1	1
	DNA adduct formation	8	8
	single strand breakage (alkaline elution)	4	3
	single strand breakage (unwinding)	1	1
	UDS (stomach)	1	1
	(liver, rat weaning/adult)	10	9
	(spermatogonial)	2	2
Gene Mutation	Mammalian somatic cell (lung)	1	1
	(spot test)[a]	5	5
Chromosomal	*SCE (circulating erythrocytes)*	1	1
	(bone marrow)	10	10
	(liver cell)	1	1
	(intestinal epithelium)	1	1
	Micronucleus (liver)	1	1
	(bone marrow)	14	14
	(foetal blood)	3	3
	Structural aberrations (ascites tumour cells)	1	1
	(bone marrow)	10	10
	(liver)	1	1
	Dominant lethal (mouse)	5	5
Carcinogenesis related	Liver cell initiation/promotion	6	6
	Immunotoxicity	2	2
	Changes in liver cell ploidy	1	1
	S-phase liver cell proliferation	6	6
	Skin hyperplasia	0	0
	Sebaceous gland suppression	0	0
	Abnormal differentiation (sperm morphology)[a]	5	5
Cancer Bioassays	Lung adenoma (mouse)	1	1
	Embryo lethality/carcinogenesis (Japanese Quail)	1	1
	F344 lifetime bioassay (rat)	1	1
	Total tests considered =	104	102

[a] assay may be sensitive to several other end points (see text)

are apparently resistant. This contrasts with the observation that male rats generally are more susceptible to induction of liver tumours after feeding with 2AAF. Seven different mouse tissues are susceptible to tumour induction when the compound is applied p.o., whereas tumours are restricted to the ovary after s.c. injection. In the hamster, i.p. dosing of 2AAF resulted in more tumour induction in the gastrointestinal tract. The bladder was equally sensitive to either p.o. or i.p. routes but the liver was susceptible only using the former method of dosing.

4 *In vivo* short term assays for detection of 2AAF and 4AAF

Classification of end points and assays

The *in vivo* assays used for the detection of genotoxicity of 2AAF and 4AAF have been classified according to the different types of end points involved (Tables 2 and 3). The relative abilities of the 2 compounds to induce these end points are summarised in Table 4, the data for which are derived from consensus results.

Assays giving unresolved activities for 2AAF and 4AAF

Several assays (Table 5) were unable to distinguish between the 2 compounds which were either both active or inactive according to host or route of administration.

5 End points induced *in vivo* by 2AAF and 4AAF in non-mammalian cells

In microbial cells (urine)

Both these types of system attempt to bridge the gap between *in vitro* and *in vivo* testing by involving the assay of activity following compound administration to animals

Table 4. *Qualitative results with in vivo assays*[a]

End point/assay	Consensus result 2AAF	4AAF
Premutational (mammalian)		
DNA covalent binding	●	◐
adducts	●	○
strand breakage	●	○
UDS (liver)	●	○
(sperm)	○[b]	○
(stomach)	○	○
Gene mutation (mammalian)		
Somatic (lung)	○	○
(foetal, spot test)	●	○
Chromosomal effects (mammalian)		
SCE (blood)	◐	○
(bone marrow)	●	◐
(intestinal epithelium)	●	○
(liver)	◐	○
Micronucleus (liver)	●	○
(bone marrow)	●	○
(foetal blood)	●	○
Structural aberrations (ascites tumour)	○	○
(bone marrow)	●	○
(liver)	◐	○
Dominant Lethal	○	○
Carcinogenesis related effects		
Initiation/Promotion	●	◐
Immunology	●	◐
Liver cell ploidy	●	○
Liver cell proliferation	●	●[c]
Skin hyperplasia	nt[d]	nt
Sebaceous gland suppression	nt	nt
Abnormal differentiation (sperm)	●	○
Cancer		
Lung adenoma	○	○
Quail egg	○[b]	○[b]
F344 lifetime rodent bioassay	●	○[b]
Non-mammalian (organism, end point)		
Urine (bacteria, gene mutation)	●	●
Host-mediated (bacteria, gene mutation)	○	○
(yeast, various)	◐	○
Somatic cell (SMART, Drosophila, various)	●	◐
Germ cell (SLRL, Drosophila, various)	○(♂)●(♀)	○(♂)◖(♀)
(Drosophila, chromosome loss)	○	○

[a] does not include macrophage transformation
[b] preliminary data
[c] 4AAF was significantly more active
[d] not tested; ● = positive; ◐ = weak positive; ○ = negative

and subsequent detection of genotoxicity in microbial cells *in vitro* (Legator *et al.*, 1982; Combes *et al.*, 1984). The tests differ in that the urine assays rely on *in vitro* exposure of indicator organisms to excreted metabolites or unchanged chemical whilst the other methods include exposure in situ at different sites after which detection of end points is facilitated by isolation of tester strain cells and growth *in vitro*. Both methodologies have their limitations discussed in the reviews above and in McGregor *et al.* (these volumes).

The relevance of the data from the urine assays is difficult to assess since none of the experiments undertaken was able to detect any consistent differences in the activities of 2AAF and 4AAF. Collectively the data show that both compounds resulted in excretion of urine samples with mutagenicity detectable without any further exogenous exposure to activating enzymes. With both analogues, since mutagenicity was potentiated by further metabolism with S9 and/or deconjugating enzymes it is clear that a considerable proportion of the dose was excreted in the form of promutagenic metabolites. In the author's laboratory detection of urinary mutagenicity after i.p. injection of rats with either chemical was in agreement with comparable proportions of injected dose in the urines from animals of the same species using radiolabelled samples (Dashwood and Combes, these volumes; Combes *et al.*, these volumes). Although some interlaboratory variation was apparent, the results from urine assays imply that neither 2AAF nor 4AFF was detoxified *in vivo* to forms that can no longer be converted by enzymes into metabolites capable of mutating bacteria. It is still unclear from the combined observations whether or not activation of 4AAF to ultimate reactive moieties can occur *in vivo*. More research is required concerning the possibilities for extrahepatic metabolic activation of urinary promutagens for example in the bladder mucosa. No useful conclusions can be gleaned from the urine results with respect to relative potencies of 2AAF and its isomer.

Host-mediated assays

Three host-mediated assays were undertaken each differing with respect to indicator organism, end point detected and the variety of organs from which the indicator cells were recovered (McGregor *et al.*, this volume). All the protocols involved dosing mice mainly by the oral route. Conflicting data were obtained from these assays. When gene mutations were scored using bacterial tester strains isolated from liver no activity of either chemical was detected. This implies the predominant effect of *in vivo* liver metabolism is detoxification despite the fact that *in vitro* 2AAF and 4AAF are activated to mutagens. However, 2AAF and 4AAF produced weak positive results when tested for their abilities to induce gene conversion in yeast cells isolated from liver but not lung. However, the consensus view was that these results require repetition since the responses were marginal, despite the fact that they met the stringent requirements for a positive effect used by the working group (McGregor *et al.*, these volumes).

Yeast was also the indicator organism in another host mediated assay involving the detection of gene mutation, mitotic recombination and gene conversion. With single dosings all experiments were negative for the 2 chemicals. However, the investigator claimed that 2AAF, but not 4AAF, specifically induced mutations in cells recovered from the liver and kidney after chronic administration (see Bronzetti *et al.*, these volumes). The working group suggested caution in the interpretation of host mediated assay results in view of several complicating factors (McGregor *et al.*, these volumes). Thus it is not possible to conclude that either chemical is capable of inducing gene mutation or premutational changes in microbial cells using these protocols.

Table 5. *In vivo assays not distinguishing between 2AAF and 4AAF*

Host-mediated, gene mutation in bacteria
Urine (all species and routes)
Tissue distribution (rat, i.p.)
UDS – liver (mouse i.p. or p.o.)
– stomach
– sperm (rat and mouse, p.o.)
Micronucleus – bone marrow (rat, i.p.)
Chromosomal aberrations – bone marrow (mouse, p.o.)
– ascites tumour cells
Dominant Lethal
Gene mutation – Spot test (mouse, p.o.)
– Lung
S-phase proliferation, liver (all species and routes)
Sperm morphology (rat and mouse, p.o.)
Drosophila – SLRL in males
– Chromosome loss

In Drosophila

Several different Drosophila assays were used which were capable of detecting a variety of endpoints (Table 2). The sex-linked recessive lethal test is the only method in the collaborative trial designed to detect specifically gene mutations in germ cells. It was shown (Table 4) that both 2AAF and 4AAF induced gene mutations in germ cells of Drosophila and both chemicals were active in somatic cell tests sensitive to mutational and other related phenomena such as deletions. However, 4AAF was less than half as active as 2AAF. Thus for germ cells 2AAF was clearly positive at 0.5 mM whereas 4AAF was only weakly active at twice this concentration. For somatic cells 4AAF gave a weak positive response at almost a 10-fold higher concentration than the lowest level of 2AAF which resulted in definite activity.

The importance of using the larval feeding technique to obtain positive data with these aromatic amines in the sex-linked recessive lethal assays and the apparent sex effect on response (Table 4) are discussed on page 1.69.

6 End points induced *in vivo* by 2AAF and 4AAF in mammalian cells

Premutational effects

Covalent binding and adduct formation. Only preliminary data are available for covalent binding to rat liver DNA. After i.p. administration of 2AAF a covalent binding index (CBI) of 78.5(+/− 1.7) was obtained. Thus far, only one determination has been undertaken for 4AAF resulting in a value of 12.2. Both compounds were administered i.p. to male Wistar rats and the DNA was isolated 24 hr after dosing (Combes *et al.*, these volumes). These data are tentative in view of the limited number of determinations. In these experiments both compounds reached the liver to the same extent. The CBI for 2AAF is lower than other published values of 83–245 obtained using the same species and dosing regime (Lutz, 1979). The fact that 4AAF in some form reaches the liver also is apparent from data showing that this compound causes S-phase in liver cells. (p. 1.67).

Both 2AAF and 4AAF form adducts with DNA *in vivo*. However, quantitatively the numbers of adducts formed by 4AAF in rat liver were 400 fold less than those induced by 2AAF. This large differences was obtained after i.p administration, a route which is known to result in equal distribution of either compound to rat liver, as revealed by experiments with radiolabelled samples. The difference in extent of DNA binding was even greater (almost 800 fold) in rat lung due to the extremely low level of 4AAF binding to lung DNA and the greater extent of interaction of 2AAF with liver DNA. The CBI result predicts that 4AAF will exhibit very weak carcinogenicity (Lutz, 1979). In view of the reported genotoxicity of 4AAF *in vitro* (de Serres and Ashby, 1981) it would be of interest to assess whether this difference in DNA binding between the 2 chemicals occurs *in vitro*. Nevertheless, the data show that 4AAF is capable of forming adducts in DNA *in vivo* in rat lung and liver albeit to a very low extent. This very reduced level of binding could be due to: (a)-lack of activation enzymes in the target tissues, and (b)-qualitative differences in adduct formation such that any DNA lesions induced by 4AAF are subject to error-free repair. Since 4AAF is genotoxic *in vitro* in the presence of an activation system it is evident that the molecule can be metabolically converted to an electrophile. However, reactivity of the ultimate form of the chemical may be substantially lower than for 2AAF. The contribution of (a) above is impossible to assess until the relative binding efficiencies of the 2 compounds to DNA *in vitro* have been compared in detail (p. 1.70).

Evidence was obtained from both the ^{32}P-post-labelling assay and from the use of monoclonal antibodies for chemical differences in the lesions induced by 2AAF and 4AAF. However, further chemical analysis of adducts is required to characterise these differences further.

Strand breaks. Three different methods were used for the detection of strand breaks. Two of these were conducted in the rat and 1 in the mouse. In all cases liver DNA was analysed. Results obtained with the alkaline elution assay from different sources were conflicting and the investigators agreed that they should not be considered due to problems with interpretation (Hunt *et al.*, these volumes).

2AAF and 4AAF were equally efficient at inducing strand breaks in the DNA-protein cross-linking assay, whilst the alkali unwinding fluorimetric system distinguished between the 2 compounds. Thus 2AAF, but not 4AAF, induced single strand breaks in mouse liver DNA after p.o. administration (Hunt *et al.*, these volumes). The activity of 2AAF was weak but could be potentiated by at least 5-fold to a clear positive response if a repair inhibitor was present *in vivo* during treatment. This suggests that such strand breaks are subject to repair. The breaks themselves are due to removal of adducts initially formed in the DNA. Inability to detect breaks following treatment with 4AAF could be due to the very low numbers of adducts induced (p. 1.63) and/or to the fact that any such adducts are not removed in a manner which leads to strand breakage.

Both of the above tests for strand breakage were undertaken in only 1 laboratory. Further studies are required to establish the relative activities of the 2 chemicals in these assays especially as the tentative data imply that rat liver is more susceptible, a result which agrees with the order of sensitivity to 2AAF-induced carcinogenesis.

Unscheduled DNA synthesis. UDS measures repair synthesis which has been stimulated by the presence of DNA damage. Presumably such synthesis is a manifestation of the repair of single strand breaks. UDS was scored in 3 different tissues: spermatogonia, stomach and liver. The majority of experiments involved the liver. 2AAF induced UDS in liver cells at the lowest dose tested in every experiment with rat. However, all investigators failed to detect any activity in the mouse (see p. 1.68). 4AAF was mostly inactive or was weakly active on a few occasions in the rat and was devoid of activity in the mouse (Table 4). 2AAF did not induce UDS in either spermatogonia or stomach. The latter result (Furihata *et al.*, these volumes) should be regarded as preliminary. Both chemicals were negative in dominant lethal and sperm morphology tests using rats and mice under similar conditions where they were inactive for sperm UDS. Thus it is impossible to ascertain from the available data whether or not they cross the blood-testis barrier. However, 2AAF or a metabolite reaches the testis after i.p. administration to mice (p. 1.64) but unfortunately sperm UDS was not analysed using this route.

The UDS data predict hepatocarcinogenicity for 2AAF and negative or at best weak liver carcinogenicity for 4AAF in the rat. In this species no carcinogenicity in stomach or developing tissue is predicted and in the mouse the results imply that both 2AAF and 4AAF are non-carcinogens for all 3 organs in the mouse. The possibility that for 2AAF the differences in adduct formation with mouse and rat liver DNA could account for the negative UDS data in the mouse has been discussed elsewhere (Hunt *et al.*, these volumes). It is difficult to compare the relative sensitivities of the UDS and strand breakage assays since conflicting data were obtained. However, it is noteworthy that 2AAF induced strand breaks in the mouse (p. 1.58) at the same dose at which it was negative for UDS.

Gene mutation

Two assays for gene mutation undertaken in mammalian cells were the mouse coat colour spot test and the *in vivo-in vitro* cells mutation assay (see McGregor *et al.*, these volumes). The former system detects somatic mutations induced in the foetus but may also be sensitive for several other end points (Fahring, 1985). It is thus specific for compounds or their metabolites capable of transplacental distribution. The lung assay was the only protocol used that was designed to detect mutations in adult somatic cells. Although some sperm abnormalities are likely also to be caused by point mutations the relative contribution of other phenomena to the formation of aberrant sperm is unknown but may be considerable (Wyrobek *et al.*, 1983). Likewise dominant lethal mutations are considered to be due primarily to chromosome damage (Anderson, 1984). Consequently no reliable indication of the ability of 2AAF or 4AAF to induce germ cell mutations *in vivo* in mammalian cells are forthcoming from the trial.

Only 2AAF was active in the mouse spot test implying that 2AAF induces mutations in foetal tissue. It is unkown whether this is due to the passage of 2AAF itself across the placenta followed by activation in situ within the foetus or to transport of metabolites which were activated at some site within maternal tissue. Indeed human foetal liver microsomes activate NOH-2AAF to mutagenicity (Aune *et al.*, 1985b). However transplacental transfer of labelled 2AAF in the mouse occurs (Takahashi and Yasuhira, 1975).

After i.p. injection into baby Syrian hamsters, 4AAF failed to induce mutation in lung cells, when 6-TG resistance was scored. 2AAF apparently gave conflicting results in the same assay. In 2 experiments it was inactive whilst in 1 test a statistically significant number of resistant cells was observed. The investigator recorded a weak positive (Inui *et al.*, these volumes) but since there was no dose-related effect the consensus opinion was that 2AAF was unable to induce mutations in this system (Table 4, McGregor *et al.*, these volumes), despite the fact that it was positive on a previous occassion when 8-AG resistant lung cells from Chinese hamster were selected (Dean and Senner, 1977).

Thus 4AAF cannot induce gene mutations in mammalian adult somatic cells *in vivo*. However, 2AAF is active in this respect when mutations are scored in foetal tissue. This divergent result may be due to species and/or organ specific effects. For example baby Chinese hamster may be incapable of activating 2AAF either at any site or specifically in the lung. Furthermore, it is noteworthy that when administered p.o., 2AAF induces lung tumours in hamsters. Thus route as well as age of host may be important in determining the results obtained (p. 1.70). Unfortunately no other information is available from, for example, DNA adduct analysis to indicate the level of genotoxic metabolites capable of reaching the lung in Syrian hamsters.

From the results of the Drosophila assays (p. 1.63) it maybe surmised that the inability of 4AAF to induce gene mutations in mammalian cells *in vivo* is likely to be due to a cause other than an intrinsic inability of this chemical to mutate eukaryotic DNA. Unfortunately the capacity of 4AAF to induce mutations in cultured mammalian cells has not been assessed. However, the information available from other assays for premutational damage implies that the extent of primary DNA damage induced *in vivo* by 4AAF may be below the threshold necessary to give rise to detectable levels of somatic mutation in organs of exposed animals.

Chromosomal effects (including SCE)

Chromosomal effects discussed in this section include SCE, micronuclei, structural aberrations and the dominant lethal test. Chromosomal effects were induced by 2AAF in mammalian somatic cells *in vivo* as revealed by positive responses in the 3 different cytogenetic systems used (Table 4). However, 2AAF was inactive in all dominant lethal assays with either sex and using both rat and mouse. This implies that 2AAF is unable to induce chromosomal effects in mammalian germ cells *in vivo* (Green *et al.*, 1985). The positive results obtained in the sperm abnormality assay after i.p. dosing of mice with 2AAF show that in males, at least, 2AAF affects the germ cells after administration using this route. However, it is possible that, after p.o administration, inactivity of 2AAF in the dominant lethal assay in males as well as females is due to a deficiency in the transport of the compound to the germ cells as both the sperm morphology and the sperm UDS assays were negative.

4AAF was likewise inactive in all the dominant lethal assays conducted and is thus incapable of causing chromosomal effects in germ cells. This compound was in fact negative in all the assays for chromosomal effects in somatic cells, except for the induction of SCE in bone marrow in which a weak positive response was recorded (Table 4). 2AAF elicited clear positive effects in the majority of these assays. However, it gave weak positive results for SCE in blood and liver cells and when aberrations were scored in liver cells. 2AAF was negative for aberrations in ascites tumour cells. The ability of 2AAF to induce aberrations in the bone marrow was dependent on the organism used and other factors (section 7). 4AAF, unlike 2AAF, is unable to induce definite positive effects on mammalian chromosomes *in vivo* (Table 6a). Furthermore, a distinction is apparent between the induction of weak positive effects by the 2 compounds. Thus whilst very few assays detected 2AAF as weakly active, 4AAF was weakly positive in several laboratories but only for induction of SCE's in bone marrow. 2AAF was a weak positive for SCE induction in blood on 1 occasion. However, even in this experiment the 2 compounds were distinguished, as 4AAF was inactive. There was also 1 weak positive response out of 10 tests for induction of micronuclei following p.o. administration of 4AAF to the mouse. Although this result was obtained using 223 mg/kg, applications of higher doses up to 5000 mg/kg were definitely inactive in 9 other replicate experiments. Thus the consensus was that 4AAF does not induce micronuclei in mammalian cells *in vivo* (Wild, *et al.*, these volumes) since it was inactive in all other micronucleus assays.

Where a weak positive result was obtained for induction of SCE's by 4AAF a direct comparison of potency with 2AAF is feasible using the respective minimum SCE-5 determinations (Tice *et al.*, these volumes; and see Table 6b). Thus 4AAF was 6–13 fold less active than 2AAF in the mouse depending on route of administration and in the rat 2AAF was just over 3 fold more potent than 4AAF. The effects of route and species are discussed on page 1.70.

Overall the SCE and micronucleus assays (Table 6c) were more sensitive than chromosome aberration tests for the detection of induction of chromosomal effects by the 2 chemicals. The micronucleus test was the most sensitive when scored in liver cells since 2AAF was a clear positive but only a weak positive for SCE's and aberrations in this tissue (Meyer and Tates, this volume). Likewise, detection of SCE's in bone marrow was more sensitive than with other tissues since 2AAF was positive and 4AAF a weak positive. A clear positive/negative distinction was obtained when SCE's were scored in intestinal epithelium and the only weak positive response for 2AAF was obtained using blood cells. Apart from this result no weak positive or negative results were obtained for 2AAF in SCE tests whereas 4AAF was either incapable of inducing SCE's or exhibited weak activity in almost half of the experiments undertaken. In the micronucleus assays 2AAF was definitely active in nearly all experiments whilst 4AAF was entirely negative except in 1 laboratory, and a weak positive in another.

Chromosome aberrations were the most insensitive of the tests for chromosomal effects induced by either chemical. Thus only approximately 1/4 of these assays indicated definite clastogenicity of 2AAF whilst 1/2 of the experiments was negative (Adler *et al.*, these volumes). 4AAF was devoid of any activity in all of the aberration assays. When the MED values for 2AAF in the 3 types of assays for chromosomal effects are compared (Table 5) it can be seen that the sensitivities of the assays vary according to species, route of administration and organ. This is discussed more fully on pages 1.68–1.70. Apart from when effects are scored in the liver after dosing rats, the MED values for clear positive results for SCE are either lower or equivalent to the values obtained using the micronucleus assays. In the mouse after oral dosing, induction of aberrations is the least sensitive assay. However, although 300 mg/kg administered i.p. was sufficient to give a weak positive effect using the rat, this dose did not result in the production of micronuclei in the bone marrow of this species.

The limited data available for the activities of these 2 compounds in the *in vivo* assays conducted in this study suggest that the SCE tests undertaken show more similarity in response to other assays for premutational damage, especially the UDS tests, rather than to the micronucleus and aberration assays. This implies that, for these chemicals at least, SCE assays are sensitive to the induction of premutational damage. This conclusion is supported by the fact that qualitative and quantitative differences occur in the frequency and distribution of SCE's observed.

It was suggested by the working group that the greater sensitivity of SCE assays to the 2 compounds may be due to the fact that higher numbers of cells are scored routinely for the presence of effects on chromosomes than the numbers screened when using the 2 other cytogenetic test systems. If this is a valid reason for increased sensitivity, then SCE assays should be more susceptible to other chemicals tested under similar conditions. The data from the trial show that the aberration, micronucleus and SCE tests conducted *in vivo* were almost equally sensitive to BP although the SCE assays were the most responsive to PYR which was recorded as a weak positive in 33% of experiments (Forster, these volumes). The underlying reason for the greater sensitivity of SCE tests in this study requires further investigation.

If 4AAF is found to be devoid of carcinogenicity in the compelete rodent bioassay then all 3 micronucleus tests will exhibit the best predictive response for carcinogenicity of both compounds. However, if the chemical proves to be a weak carcinogen then detection of SCE's in bone marrow is necessary for an indication of weak *in vivo* genotoxicity. Even though approximately 1/4 of the experiments involving detection of chromosomal effects indicated non-genotoxicity for 2AAF, it is clear from the

Table 6. *Analysis of assays detecting chromosomal effects*

(a) Summary of overall performance of assays[a]

Compound	total tests[a]	%−	%+	%w+
2AAF	49	26.5	59.2	12.2
4AAF	49	83.7	0	14.2

[a] 2 assays gave inconclusive results

(b) Potencies of 2AAF and 4AAF in SCE

Compound	Min SCE − 5(mg kg^{-1}) in: mouse	rat
2AAF	183	133
4AAF	2012	448

(c) Potencies of 2AAF and 4AAF in micronucleus[a]

Compound	species	tissue[b]	min dose positive (mg kg^{-1})	max fold increase over background	max dose negative (mg kg^{-1})
2AAF	mouse	bm	10	312	—
4AAF	mouse	bm	446	2.5	5000
		blood	446	?	892

[a]from individual investigators' reports; [b]bm – bone marrow.

proportion of results showing a weak positive response for either compound that collectively the *in vivo* cytogenetic assays are predicting a much greater potency *in vivo* for 2AAF. However it is surprising that this analogue was only a weak positive for aberrations and SCE's when scored in liver cells, the target organ.

Carcinogenesis-related effects

Although 7 assays for carcinogenesis-related effects were conducted in the trial, 2AAF and 4AAF were not tested for induction of skin hyperplasia or sebaceous gland suppression.

Depression of nuclear ploidy in rat liver. 2AAF and 4AAF unequivocally were distinguished as active and inactive respectively by the liver polyploidy assay using both rats and mice. Moreover, 2AAF was active at a dose which was half the maximum concentration of 4AAF which proved negative.

The nuclear ploidy assay is based upon previous observations, (Styles *et al.*, 1985) that a reduction in the ratio of 4N to 2N nuclei in liver cells is concomitant with early stages of chemically induced liver cancer using several hepatocarcinogens. However, the mechanism causing such a change is unknown although it is suggested that positive chemicals are selectively toxic to the binucleate 2N cell population. The relationship between this end point and carcinogenesis is obscure and interpretation of the assay is further confounded by the fact that complex changes in the liver ploidy ratio can be induced by a variety of chemicals some of which are not known carcinogens or hepatocarcinogens, including BP (McGregor *et al.*, this volume). However, if the link between activity in this assay and putative hepatocarcinogenicity is assumed then the results at least for these 2 compounds correctly predict the liver carcinogenicity of 2AAF. Clearly more research into the significance of liver cell ploidy status and carcinogenesis is required before the results of this assay can be fully assessed with respect to *in vivo* activity.

Pre-neoplastic changes (initiation/promotion) assays. Several assays designed to detect early pre-neoplastic stages of hepatocarcinogenesis were undertaken. These were based upon histochemical identification of altered liver foci detected either on morphological and staining criteria or by the presence of elevated levels of gamma glutamyl trans peptidase (GGT). Although it is known that such altered foci appear before tumour formation it was concluded by the working group (Hunt *et al.*, this volume) that specific induction of initiation or promotion by any particular chemical could not be distinguished. Consequently the results for these assays have been grouped together as "initiation/promotion". All these assays were undertaken in the rat and, apart from 1 negative result, all tests were positive for 2AAF, out of a total of 5 experiments, irrespective of whether observations were made after 70 days or 6 months. The reasons for the negative result are obscure and cannot be due to administration of insufficient chemical since a single dose of 500 mg/kg was inactive compared with doses of 10 or 12 mg/kg giving positive results in the other experiments. Also, although no pre-inducer was used and observations were made after 8 weeks, 2AAF was active without a necessity for pre-induction and after shorter time periods. Moreover, inactivity in this assay is unlikely to be due to the fact that foci were scored by histological appearance only since, in another study involving chronic administration in the diet, foci were detected by this method and also using the GGT method. In the same study liver tumours were observed eventually in the same batch of animals (Hicks *et al.*, these volumes).

With 4AAF, 3/5 initiation/promotion assays were negative whilst 2 indicated activity. This was considered by the respective working group (Hunt *et al.*, this volume) to be a weak positive response. The result does not necessarily depend on the duration of the experiment since although no activity was detected at 70 days at a maximum dose of 500 mg/kg, 4AAF was weakly active at 6 weeks or 6 months at a dose as low as 12 mg/kg. The significance of these 2 observations is difficult to assess especially as 1 result was obtained using a single dose whilst the other was observed using chronic dosing in the diet. Furthermore, another experiment involving similar administration of the compound was negative. It is noteworthy that in 1 of the tests where 4AAF was active, the MED was 2.5 fold higher than the value for 2AAF. However, in the other experiment comparable levels of the 2 chemicals were active.

Thus the data from this group of assays are conflicting. Overall the assays predict correctly that 2AAF is a liver carcinogen in the rat. Despite the fact some activity for 4AAF was seen, it should be remembered that these results were considered to be weak effects. Thus these assays collectively infer that either 4AAF will be a non-carcinogen or at best will exhibit weak hepatocarcinogenicity.

Liver cell S-phase proliferation. The liver cell proliferation assay detects induction of S-phase DNA synthesis in liver cells (Mirsalis *et al.*, this volume) 4AAF was an extremely potent mitogen for liver cells in all tests carried out. When cells were scored 24 h after dosing this large effect was obtained with a dose of 50 mg/kg. At 48 h 2AAF was positive, although still less potent than 4AAF.

The relevance to carcinogenesis of liver cell S-phase DNA synthesis needs to be established. Several carcinogens, devoid of other genotoxic activities are known to induce proliferation.

Immunotoxicity tests. Two different assays, were undertaken to determine the abilities of the chemicals to suppress the immunological activity of respectively natural killer lymphocytes (NK cell assay) and T lymphocytes (T cell assay) (see McGregor *et al.*, this volume). In both cases the end point measured was cytotoxicity although different immune functions are affected. Both of the assays were undertaken in rats, albeit different strains, but they varied in respect of route of compound administration, and also in the manner in which immune function was scored (the so-called direct and indirect methods). In both assays the compounds were dosed chronically over a 2 week period. The rationale for using the NK cell assay is the observation that several chemical carcinogens transiently depress the activities of natural killer cells thus increasing the incidence of tumour formation due to a prolongation of the survival of malignant tissue. In the T cell assay an increase in the activity of immunocompetent T cells is detected as a result of induction of damage altering the antigenic characteristics of host tissues.

After oral administration, 2AAF was a clear positive on 2 separate occasions in the NK cell assay using the direct method of scoring. 4AAF was inactive in this system at a maximum dose which was approximately 2-fold higher than the MED for 2AAF. In the T cell assay, after i.p. injection and the indirect scoring procedure, 2AAF again was positive whilst 4AAF was weak positive (20-fold increase in activity). Further comparison of relative potencies in this assay is complicated by the fact that the same MED values are quoted by the authors for both compounds. It is noteworthy that no definite dose responses were established in either system, although assay sensitivity to low concentrations of chemical was high.

The T cell protocol and method of scoring would seem to be the most sensitive test for the 2 chemicals in question, although the reason for the differential susceptibilities is obscure. However, the efficacy of the weak positive response for 4AAF will become apparent when the carcinogenicity data for this chemical are available.

Sperm morphology assays. The effects on sperm morphology were assessed in the rat and mouse using the oral route only in the former species but after administration p.o. or i.p. using the mouse. When applied orally to rats both 2AAF and 4AAF were inactive. This may be due to failure of both compounds to reach the developing germ cells as the results for both dominant lethal mutations and sperm UDS in rats were negative. The same problem of interpretation arises with results showing inactivity of 2AAF and 4AAF after their administration to the mouse p.o. However, the 2 chemicals could be distinguished when injected i.p. since, in 2 separate experiments, 2AAF induced abnormal sperm whilst 4AAF was unable to do so. However, it should be noted that the MED for 2AAF was approximately 3.5 fold greater than the highest dose of 4AAF tested. Moreover, in the negative assays with rats the highest dose of 2AAF used was 10 fold lower than the maximum level of 4AAF administered. It is possible that 2AAF may have been active at this higher concentration. The differential responses obtained using different routes may be due to effects on the ability of the compounds to traverse the blood-testis barrier.

7 Analysis of *in vivo* data

Comparison of end points induced by 2AAF and 4AAF

Based upon consensus results for the test systems (Table 5) it is seen that 2AAF induced the 3 main forms of genotoxic damage: premutational effects, gene mutation (in mammalian and non-mammalian cells) and chromosomal effects. In assays for premutational changes, 2AAF was negative for UDS in the stomach only but this result must be considered premature (p. 1.64). Activity of 2AAF in the spot test has been assumed to indicate induction of gene mutation and/or recombination in foetal tissue (p. 1.64). This result contrasts with the negative data obtained in the lung cell mutation assay. Thus 2AAF failed to induce mutation in adult somatic cells *in vivo* (p. 1.65) a process which is presumed to be required for tumour initiation. Either the mutation system used was too insensitive or the premutational damage caused *in vivo* does not lead to mutation in such tissue.

2AAF gave positive or weak positive results in most tests for chromosome effects. It should be noted that weak positive results for SCE and aberrations were obtained in liver. Ascites tumour cells were the only somatic cells refractory to induction of structural aberrations by 2AAF. However, absence of dominant lethals following treatment with 2AAF implies that this carcinogen is unable to induce chromosomal effects in germ cells.

In contrast, 4AAF did not produce a clear positive response in any 1 of these 3 types of assays. This compound was unable to induce gene mutations *in vivo* but gave weak positive results for DNA binding and adduct formation. Some activity of 4AAF was seen in sperm but not in liver or stomach. Negative data are most likely a cause of the very low level of interaction of this chemical with DNA such that strand breaks and concomitant DNA repair went undetected. 4AAF was devoid of any activity in all tests for chromosome effects, except weak positive responses for SCE's in bone marrow.

2AAF was positive in all assays for carcinogenesis-related effects whereas 4AAF gave a clear positive response in the liver cell proliferation system only. In fact this result was unique since 4AAF was considerably the more active chemical. 4AAF gave negative responses in 2 other tests in this group of assays and weak positive scores for initiation/promotion and immunotoxicity.

Interpretation of end points induced in non-mammalian cells is not straightforward. Both chemicals induced gene mutation in bacteria as indicated by urine assays. However, urines required further metabolism *in vitro* before clear positive responses were obtained in all participating laboratories. Furthermore, this activity for either compound was not confirmed in host-mediated assays where metabolism is *in vivo* only. The weak positive activity of 2AAF with yeast needs to be corroborated before it is assumed that this system is more sensitive for distinguishing between the carcinogen and putative non-carcinogen.

It is concluded from the Drosophila data that neither 2AAF nor 4AAF causes aneuploidy although 2AAF was positive and 4AAF weak positive in assays sensitive to other forms of genetic damage in this organism.

Overall the consensus results for 31 different assays, excluding those specific for cancer (Table 7) show that the same proportion of end points was positive and negative respectively for 2AAF and 4AAF. Also, 4AAF gave weak positive responses for twice the number of end points. 2AAF clearly was incapable of inducing 1/5 of the end points scored.

Species – and strain – specific effects

Insufficient data from the *in vivo* assays were generated to enable identification of any particular animal species as any more sensitive overall to genotoxic end points.

The organotropic pattern of response detected by

the ^{32}P-post-labelling technique for DNA adduct formation after exposure of rats 2AAF (Randerath *et al.*, these volumes and p. 1.64) was not observed when the mouse was used. Also when the rat was utilized for the detection of micronuclei in the bone marrow it was found necessary to use the oral route in order to obtain a clear positive/negative distinction between 2AAF and 4AAF respectively. However with the mouse the same distinction was achieved whether or not dosing was by injection or p.o.

When 2AAF in DMSO was injected into mice, the following strain susceptibilities with respect to induction of chromosomal aberrations in bone marrow were apparent:
C57 BL/6 (negative) < $B6C3F_1$ (weak positive) < 101/Elx (positive) (the respective activities with each strain are indicated in parentheses afterwards). When 2AAF was administered p.o. in corn oil, strain CBA marginally was the most sensitive.

Organ-specific effects

In lifetime cancer studies using 2AAF, organ-specificity varies with species such that in the rat the liver is more sensitive than the bladder but in the mouse this order of susceptibility is reversed. Also in the rat, Zymbal's glands are particularly prone to undergo tumorigenesis. Little direct verification of these findings can be gleaned from the data provided by the short-term *in vivo* assays. Generally livers from both rat and mouse were sensitive to a wide range of different genotoxic end points induced by 2AAF (Hunt *et al.*, this volume). For example, in the liver cell ploidy assay, 2AAF was equally potent in both species (Styles, this volume and Volume 2) but this chemical was less efficient by a factor of almost 7 fold at forming adducts with DNA in livers of mouse compared with rat. Also, in the UDS assay mouse liver was less sensitive than rat liver to 2AAF (Hunt *et al.*, this volume). Thus 10 times the minimal effective dose values obtained using rat liver were negative in the mouse. It is also of interest that, in the rat, the liver was the most susceptible organ to induction of UDS as no effect was discernible in spermatogonia or stomach.

In the ^{32}P-post-labelling DNA adduct assay, injection of 2AAF into mice resulted in the following order of organ sensitivity: liver > lung = kidney. The factor increase in binding between liver and lung was 2 whereas in the rat it was 3.

With regard to the induction of chromosomal effects, when the same species, strain, route and vehicle were used, both 2AAF and 4AAF were unable to induce aberrations irrespective of whether these were scored in the bone marrow or in ascites tumour cells (Adler *et al.*, these volumes). However, 2AAF was weakly positive when aberrations were second in rat liver cells (Meyer and Tates, these volumes).

Table 7. *Consensus results for 2AAF and 4AAF for 31 end points*[a]

Compound	Number (%) with responses +	w$^+$	−
2AAF	21(68)	4(13)	6(19)
4AAF	2(7)	8(26)	21(68)

[a] See Table 6 for assays and separate responses, does not include results of cancer tests.

Other factors

Effects due to sexual status of host. Little meaningful data on the effects of sexual status of host were forthcoming from the study since in most laboratories male animals were used for dosing. However, several different responses were recorded when the sex of the test organism was the varying factor. For example, 1 investigator (Vogel *et al.*, these volumes) reported that female Drosophila larvae were more sensitive than males to the induction of sex-linked recessive lethal mutations by both 2AAF and 4AAF. The consensus opinion (Vogel *et al.*, these volumes) was that this apparent sex difference was due to variation between males and females in the efficiency of repair of DNA damage. The possibility of differential metabolic activation between the sexes was rejected (Vogel *et al.*, 1983). Bone marrow in rats (but not mice) seemed to be more sensitive to SCE induction by 4AAF when males were dosed p.o. such that in order to obtain a clear positive/negative distinction between the 2 compounds it was necessary to treat females (Tice *et al.*, these volumes). No evidence for a sex-related effect was obtained using 2AAF.

Effects of solvent/vehicle. Within individual experiments little variation in the use of solvents for dissolving test compounds and for obtaining negative control values was apparent. A solvent-related effect was recorded in the chromosome aberration assays since in bone marrow 2AAF was active when prepared in oil but was negative when made up in DMSO (Adler *et al.*, these volumes). However, although these experiments were conducted with the same mouse strains, routes of application varied and may have affected the result (p. 1.97). A solvent effect was very noticeable when using the mouse coat spot test (Fahrig *et al.*, these volumes), since activity of 2AAF was seen only if the chemical was prepared in DMSO before injection.

Effects of dosing schedule. In the majority of experiments, chemicals were administerd as single doses. Apart from the rodent cancer bio-assay little attention has been paid in this trial or indeed in other studies to the use of chronic treatments, which are normally applied in the diet. This is surprising in view of the fact that this is a major mode of human exposure.

Although a mixture of chronic and acute exposures was utilized in the assays for pre-neoplastic (initiation/promotion) changes, comparison of the results is precluded since other variables were not controlled. Only in one test system as undertaken in a single laboratory (Bronzetti *et al.*, these volumes) was any effect of dosing schedule apparent. Thus in the host mediated assay when end points were scored in yeast it was claimed that chronic administration of 2AAF was necessary to obtain a positive result with this chemical. Moreover, the number of organs from which altered yeast cells were obtained was increased when chronic dosing was utilized. This phenomenon deserves further study in this and in a larger range of test systems.

Effects of route of exposure. As in previous sections, insufficient data are available from the collaborative trial to facilitate any direct comparisons between route effects on carcinogenicity and genotoxic response *in vivo*. It should be added that the study was not designed initially to investigate this potential modulating factor although clearly more information is required.

It is possible to recommend the avoidance of administration of either 2AAF or 4AAF by particular routes if a distinction between their respective genotoxicities in individual *in vivo* short term tests is to be obtained (see Table 8).

A route effect for induction of chromosome aberrations in mouse bone marrow is apparent since 2AAF was active when injected i.p. but was ineffective if applied p.o. However, interpretation of these results is confounded by the fact that the solvents used varied although the mouse strain was maintained constant (Adler *et al.*, these volumes). No consistent modifying effect of route of exposure were observed when micronuclei were scored in bone marrow. Styles *et al.*, (these volumes) observed no difference in qualitative responses of 2AAF or 4AAF using either route. 4AAF was negative with either route. When immunotoxicity was assayed in the rat a positive/negative result for 2AAF and 4AAF respectively was observed when dosing was oral whereas when the chemicals were injected the response for 4AAF was reduced to a weak positive. This effect may nevertheless have been caused by the use of different mouse strains.

In the mouse coat colour spot test a definite requirement for i.p. administration was detected in the same laboratory under the same experimental conditions (Fahrig *et al.*, these volumes) in order for a positive and negative distinction between 2AAF and 4AAF respectively to be obtained. If dosing was p.o. then both chemicals were inactive. This mandatory requirement was coupled with a solvent effect and is further evidence that in this assay transplacental transport of 2AAF and 4AAF may be dependent on these dosing parameters. The same requirement for the i.p. route was observed in the sperm abnormality assays which may also be due to effects on absorption of chemicals across the blood-testis barrier.

Thus overall in 2 different short-term *in vivo* genotoxicity assays it was necessary to dose i.p. in order to distinguish between the activities of 2AAF and 4AAF. In 1 assay the p.o. method was required. It has been argued (Ashby, 1985a) that injection into the peritoneum is an artificial route of entry with respect to modes of human exposure and thus complicates hazard evaluation. It is well established that diffusion of molecules of test agent *in vivo* to target organs following injection can vary substantially depending upon the physico-chemical nature of the formulation. The other main choice of route used in this study is p.o. However, although this takes account of a major mode of human exposure its use in genotoxicity assays also is not without problems. Firstly, dietary feeding techniques are unsatisfactory if precise levels of uptake of chemicals from animal feeds are to be calculated. This problem is caused by dietary avoidance. Secondly, although known amounts of sample can be delivered to the stomach by intubation techniques, subsequent uptake and distribution of the large levels of chemical often administered may differ significantly from the pharmacokinetic profiles which are likely to occur as a consequence of less artificial exposures.

Table 8. *Routes to avoid when testing 2AAF and 4AAF*[a]

Assay	Species	Route
Micronucleus (bone marrow)	rat	i.p.
Sperm morphology	mouse	p.o
Sperm morphology	rat	p.o.
Spot test	mouse	p.o.

[a] Routes indicated are those required to get +, – or +, w + responses.

It is the author's opinion that no single route of dosing should be used at the expense of any other for use in short-term *in vivo* genotoxicity tests. The utilization of inhalation for volatile or gaseous chemicals is used routinely where appropriate, but the adoption of dermal exposure, for example resulting in mutagenic urine after application of certain hair dyes to rats (Ammenheuser and Warren 1979), is rare in the literature and was not a part of any protocol in the present collaborative trial. It is likely that results obtained using several routes of exposure for these assays together with information concerning uptake, organ distribution and excretion of samples applied likewise would facilitate hazard evaluation.

Potential mechanisms for differential activities of 2AAF and 4AAF

Divergent activities of chemicals *in vivo* and *in vitro* and in various organs of the same and different species may be due to a plethora of factors, especially when metabolic activation is necessary. Inactivity *in vitro* can be a consequence of inefficient activation perhaps caused by a structural feature of the molecule (p. 1.58). Alternatively there may be a lack of the requisite activation enzyme in the organ homogenate. 4AAF as well as the non-carcinogen 3AAF apparently are not extensively N-hydroxylated, at least *in vivo* (Roberfroid *et al.*, 1982). Negative responses *in vivo* can be caused by similar reasons, perhaps giving rise to an organ specific effect (Reigh and Floyd, 1981; Malejka-Giganti and Ryzewski, 1985). Moreover, inactivity in a particular tissue may be attributable to failure of transport of sufficient chemical from the portal of entry to the target site.

The data presented in the collaborative trial show that, in contrast with 2AAF, 4AAF at best binds weakly to DNA as revealed by 3 different methods. The low extent of adduct formation by 4AAF also may be a result of reduced interaction between the reactive metabolite and DNA or because of adduct instability for stereochemical reasons. Adducts may go undetected owing to repair which presumably must be more rapid and error-free than the rapair of 2AAF induced adducts. In this connection it should be noted that the minimum time period after which DNA was analysed for adducts following dosing was 2 hrs and 24 hrs in 2 other tests (McGregor *et al.*, these volumes). Moreover, no strand breakage was induced by 4AAF and only weak responses in UDS were seen, exclusively in spermatogonia.

Adducts formed by 4AAF were sufficiently different from those induced by 2AAF to be distinguishable by both the ^{32}P post-labelling and radioimmunoassay techniques. It is thought that at least 3 different types of adducts are formed by 2AAF (Singer and Grunberger, 1983; Poirier, 1984). These arise by covalent interactions with Guanine (G) and include: the C8 adduct N-(deoxyguanosin8-yl)-AAF (the major product); the N^2-G adduct 3-(deoxyguanosin-N2-yl) AAF (a minor form). These 2 adducts arise after reaction of N-acetoxy 2AAF with DNA. A third deacetylated adduct due to interaction with C-8 of G has been detected in rat liver DNA after *in vivo* exposure and is N(deoxyguanosin-8-yl)-AF. This last adduct is probably the form detected by the ^{32}P post-labelling system whilst the G C8-AAF adduct was the predominant product recognised by the immune assay (McGregor *et al.*, these volumes).

After covalent binding of 2AAF to Guanine it is believed that the fluorene ring becomes accommodated within the helix by rotation of the phosphodiester bond (Grunberger, 1981). The concomitant stacking between adjacent base pairs distorts the helix and is a potential cause of the subsequent premutational damage detected. 2AAF produces 2 further G adducts with labile glycosidic bonds resulting in depurination (Tarpley, *et al.*, 1982). The apurinic site gives rise to the strand breaks detected. After interaction of 4AAF with G the proximity of the N-acetyl moeity to the ring system of the purine may preclude the occurrence of any prolonged association or rotation of the adduct.

Further experiments with antibodies specific for other potential 4AAF adducts are required to study the basis for the reduced interaction of this compound with DNA. Quantitative variation in adduct formation in different tissues may be a reflection of qualitatively different adducts arising at different sites (Poirier, 1984).

Differential metabolism (page 1.58) between tissues and species is a possible factor contributing to organotropic and species-specific effects of 2AAF (p. 1.58). Sulphotransferase and N-hydroxylase are more active or possess higher substrate affinity in species with an elevated susceptibility to 2AAF-induced carcinogenesis (Table 9). Although guinea-pig microsomes efficiently N-hydroxylate and deacylate 2AAF activating the compound to a mutagen *in vitro* (Takeishi *et al.*, 1979; Vaught *et al.*, 1985), resistance to carcinogenesis of this species could be due to the rearrangement of the NOH metabolite to a ring hydroxy form which arises *in vivo* only (Roberfroid *et al.*, 1982).

Recently species differences in cancer sensitivity could not be accounted for by using hepatocytes from rats, mice, hamsters or guinea-pigs in place of S9 in the Salmonella mutagenicity assay nor by comparing the sensitivity of hepatocytes to cytotoxicity or UDS induction when using 2AAF (Holme and Soderland, 1985).

Metabolites generated in the liver may be too labile to be active after transport to extrahepatic sites and organ-specificity is then likely to be due primarily to differential metabolism. However, glucuronidation of N-OH-2AAF in the liver, with subsequent transport of the resulting metabolites to the mammary gland may contribute significantly to any tumorigenesis at the latter site in the rat (Malejka-Giganti and Ryzewski; 1985). Differences in organ-specificity have been related to variations in tissue complement of mutiple forms of cytochrome P-450 which are differently induced by 2AAF itself and which vary in their capacities to activate the chemical (Astrom and DePierre, 1985 a and b; Lotlikar *et al.*, 1984). Data obtained *in vitro* using different tissue homogenates in the Salmonella assay often do not agree with organ susceptibility to the chemical being tested. Nevertheless, rat liver S9 generally is more effective for the activation of 2AAF (Walters and Combes, 1985). With respect to the *in vivo* genotoxicity assays, it is noteworthy that a high proportion of positive responses for both 2AAF and 4AAF was obtained when the end points were scored in rat liver (Table 10).

Comparison of organ specificity of genotoxic response with *in vivo* distribution of compound is restricted by lack of metabolism data. Positive responses in liver, blood and urine of rats after i.p. or p.o. dosing with 2AAF all agree with the distribution of the compound to these tissues. With 4AAF, positive or weak positive responses in the liver, blood and urine are compatible with the presence of 4AAF at these sites following i.p. dosing.

8 Appraisal of *in vivo* short term assays

Analysis and predictability of in vivo assays

A total of 134 experiments was analysed for responses of *in vivo* genotoxicity assays to 2AAF and 4AAF. For 2 assays, epidermal hyperplasia and sebaceous gland suppression, neither compound was tested. Also either 2AAF or 4AAF

Table 9. *Species differences in metabolism and susceptibility to 2AAF*[a]

Species	Relative activity of 2AAF for: carcinogenicity	mutagenicity	N-hydroxylase activity	affinity	Sulpho-transferase activity
Rat	+++	+	++	+++	+++
Hamster	++	+++	+++	+	
Mouse	+	++	++++	+	+
G-pig	–	+++	+++		
Monkey	–	+++	+++[b]		
Man	?	+	+++		

[a] data from McGregor (1975); Roberfroid *et al.* (1982); Boobis *et al.* (1982); Razzouk and Roberfroid (1983); Weisburger (1981); Holme and Soderland (1985); Vaught *et al.* (1985).
[b] determined by urinary excretion of NOH-2AAF.

was recorded as not tested variously in several experiments.

The results of the analyses are presented in Tables (10,11) and Figure (1). Due to the differing members of assays in each group analysed, it is difficult to compare the data between groups. For example differences are apparent between results due to i.p. and p.o. administration but less than 1/2 the number of experiments were conducted using the former route.

It can be seen from analysis of the total experiments (Table 10) that almost 2/3 of responses were positive for 2AAF contrasted with less than 1/5 of the responses being positive for 4AAF. The frequency of weak positive results was only marginally higher for 4AAF whereas the amount of equivocal data was low in both cases. It is noteworthy that the combined *in vivo* results show that 2AAF was negative in 1/3 of all experiments. This figure remains unaltered when results obtained using the p.o. route of administration are analysed separately. However, there is a decrease in the frequency of negative responses for 2AAF when data obtained using the i.p. route (irrespective of species, organ) or obtained with rat are considered individually. This decline is due to a rise in the number of positive responses and this effect is sustained or is even more pronounced when the results for rat liver are used alone in the analysis. In both these cases more than 3/4 of all experiments recorded 2AAF as positive. When data for rat liver are analysed there is little effect on the frequency of weak positive responses but there is some decline in the proportion of weak positive results when the use of the i.p. route is considered alone. Moreover, compared with the total data base, almost 1/3 of all experiments conducted in rat liver detected 4AAF as positive with a small increase in the number of weak positive responses. In contrast, when data using the mouse are compared with the complete data base, whilst there is little effect on the activity of 2AAF, the proportion of negative results for 4AAF increases by 14%. This is due to a decrease in both the positive and weak positive responses. Comparison of these data with results obtained with mouse liver is hindered by the relatively low numbers of experiments. Nevertheless, when these data are considered it is noteworthy that there is little effect on the spectrum of activity of 2AAF, except for a decline in the frequency of weak positive scores. However, 4AAF now registers positive more frequently. When the results of scoring end points in mouse bone marrow are considered separately, 3/4 of all experiments registered 2AAF as active whilst none gave any positive response with 4AAF. Since few experiments were conducted with rat bone marrow, data comparison is precluded.

In conclusion it would appear that assays conducted in rat liver or in both species using i.p., are more likely to yield positive results for both 2AAF and 4AAF with an additional increase in the proportion of weak positive results for 4AAF in the liver. This result would be

Table 10. *Analysis of assay responses under different conditions*[a]

		Number (%) of experiments[b] which were:				
Compound	Conditions	+	W⁺	−	+/−	Total
2AAF	All	76(58.5)	11(8.4)	42(32.1)	1(0.8)	130
4AAF	All	17(13.5)	19(15.1)	89(70.6)	1(0.8)	126
2AAF	I.P	27(73.0)	1(2.7)	8(21.6)	1(2.7)	37
4AAF	I.P	7(20.0)	4(11.4)	23(65.7)	1(2.9)	35
2AAF	P.O.	47(56.6)	8(9.6)	28(33.7)	0(0.0)	83
4AAF	P.O.	10(12.2)	12(14.6)	60(73.2)	0(0.0)	82
2AAF	RAT	34(65.4)	3(5.8)	14(26.9)	1(1.9)	52
4AAF	RAT	14(28.0)	10(20.0)	26(52.0)	0(0.0)	50
2AAF	MOUSE	35(58.3)	5(8.3)	20(33.3)	0(0.0)	60
4AAF	MOUSE	3(5.0)	5(8.3)	51(85.0)	1(1.6)	60
2AAF	RAT, LIVER	22(71.0)	3(9.7)	6(19.4)	0(0.0)	31
4AAF	RAT, LIVER	8(27.6)	6(20.7)	15(51.7)	0(0.0)	29
2AAF	MOUSE, LIVER	5(55.6)	1(11.1)	3(33.3)	0(0.0)	9
4AAF	MOUSE, LIVER	2(22.2)	0(0.0)	7(77.8)	0(0.0)	9
2AAF	MOUSE, BM.	19(73.1)	3(11.5)	4(15.4)	0(0.0)	26
4AAF	MOUSE, BM.	0(0.0)	5(19.2)	20(76.9)	1(3.8)	26

[a] Derived from data submitted by investigators at Cape d'Agde and in respective Group A reports. Wherever possible any alterations made subsequently have been incorporated where these have been corroborated. However, some differences may exist between the data base used and the overall results presented elsewhere in this volume (which reflects the responses given in the individual investigators' reports). The detailed information is available from the work group coordinator; [b]excludes experiments where compound was not tested.

expected in view of the target organ status of the liver for carcinogenesis in this species. When using the mouse, where this organ is not the primary target, the bone marrow would seem to be the most suitable tissue for distinguishing between 2AAF and 4AAF. The majority of these bone marrow experiments involved the micronucleus assay which exhibited a high degree of resolution between the 2 compounds.

The proportions of different responses for both 2AAF and 4AAF recorded by the same investigator are presented in a 2 × 2 contingency table (Table 11). The total number of reports of responses for the chemicals together is 126 when the numbers of reports where either compound is missing have been subtracted. It can be seen that, whilst 2AAF was recorded as positive and 4AAF as negative on 42 occassions (approx. 1/3 of the time), 4AAF was positive also in 14 experiments (11% of the time). Furthermore, there were just 3 reports (all S-phase DNA synthesis) where 2AAF was negative and 4AAF positive, and no instances where 4AAF was weak positive or equivocal when 2AAF was negative. On 38 occasions (30% of the time) both compounds were found to be inactive.

Table 11. *Summary of dual activities of 2AAF and 4AAF in 126 experiments*

	Activity of 2AAF: +	W^+	−	+/−	(Total)
ACTIVITY OF 4AAF: +	14	0	3	0	17
W^+	17	2	0	0	19
−	42	8	38	1	89
+/−	1	0	0	0	1
(Total)	74	10	41	1	126

Repeatability of individual assays *conducted under the same conditions* is summarised for the rat and mouse and the 2 different routes of administration (Figure 1). The data base for other species was too small for comparative purposes. Assays for which only 1 result for either or both compounds is available have been ommited. This has resulted in the inclusion of 17 different assays systems. The main conclusion from this analysis are:

1. 2AAF was uniformly active in 6 assays and completely inactive in 2 assays.
2. In the rat, 2AAF showed inactivity not observed in every experiment in 3/5 assays: proliferation, single strand breaks and initiation/promotion tests. 2AAF showed some inactivity in the mouse in 6 assays: chromosomal aberrations, host mediated (yeast), micronucleus (p.o.) and spot tests.
3. 4AAF was uniformly active in 2 assays both in the rat and completely inactive in 7 tests or 40% of the assays used. In 2 further assays, most of the experiments with 4AAF were negative.
4. Consistent negative results for 4AAF were found using the mouse irrespective of dosing route.
5. 2AAF and 4AAF were both uniformly non-genotoxic in 2 assays (dominant lethal and sperm morphology, both i.p.) neither of which was in the rat.
6. Only 2 assays (micronuclei in bone marrow and sperm morphology in the mouse after i.p. dosing) exhibited 100% ability to distinguish 2AAF as positive and 4AAF as negative. No assay in Fig. 1 resolved the compounds as having the reverse activities.

Performance of in vivo assays compared with in vitro tests
The spectra of end points and assays for which 2AAF and 4AAF have exhibited *in vitro* and *in vivo* activities are presented in Figure 2. It is evident that whenever 2AAF has proved positive for a particular assay *in vivo* it has also

Figure 1. Repeatability of short-term *in vivo* assays with 2AAF and 4AAF

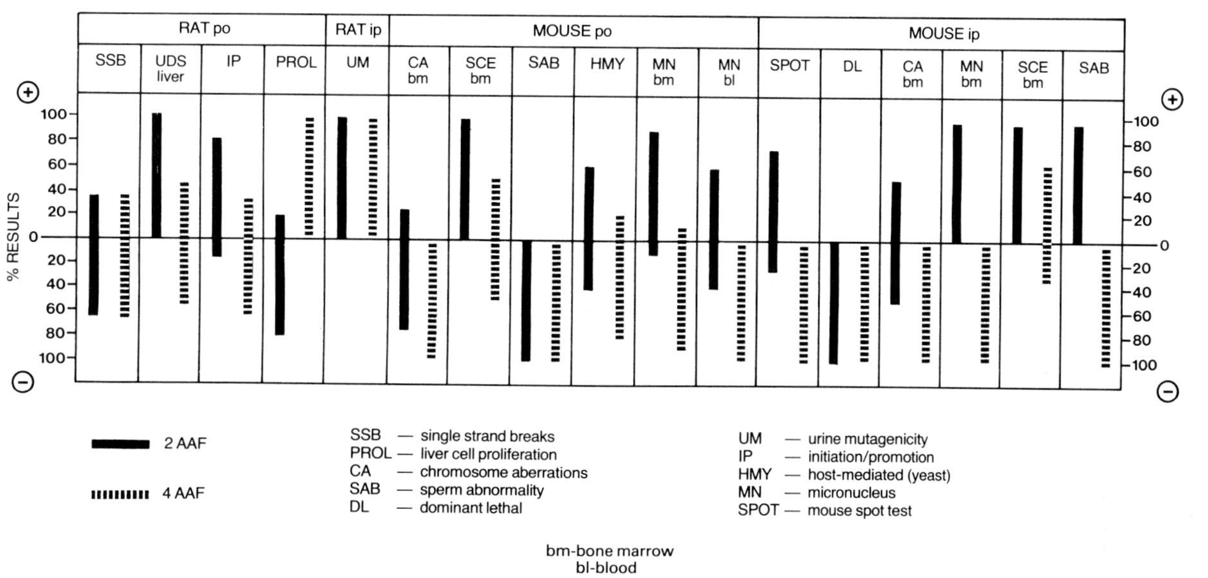

been registered as positive for a similar end point *in vitro*. For SCE-induction both weak positive and negative responses have been detected *in vitro*, whereas *in vivo* this end point was especially sensitive to 2AAF. Conflicting *in vitro* data have been obtained also for the induction of chromosomal aberrations in mammalian cells and premutational damage and gene mutations in microorganisms. In the case of chromosome aberrations, discrepancies in response are most likely due to use of varying levels of additional metabolic activation (p. 1.60). With primary DNA damage it is noteworthy that *in vivo* at best only weak positive responses were recorded.

Whilst 2AAF is a bacterial mutagen *in vitro*, its ability to mutate microbial cells *in vivo* as detected by the host-mediated assays using yeast or bacteria, remains to be confirmed. No activity was detected using Salmonella tester strains in this assay. However, it is clear that 2AAF unequivocally induces gene mutation in mammalian cells both *in vitro* and *in vivo*.

With 4AAF the data base is less extensive since the ability of this chemical to interact with DNA *in vitro* apparently has not been tested. Moreover its activities in gene mutation and clastogenesis assays with cultured mammalian cells have not been published. However, in 2 assays, UDS and SCE, 4AAF gives weak positive responses *in vivo* but has proved negative *in vitro*. Thus, unlike the situation with 2AAF, these results alone suggest that conducting an assay *in vivo* with 4AAF can result in increased sensitivity. 4AAF was negative *in vivo* with microorganisms and the combined assays for micronuclei in mammalian cells and aneuploidy in fungi. Unfortunately this compound has not been tested for induction of micronuclei *in vitro* so a more accurate comparison of the data is precluded. Conflicting results have been obtained for the induction by 4AAF of premutational damage *in vitro*. It is noteworthy also that less activity of 4AAF is generally detected by the assays irrespective of whether they have been undertaken *in vivo* or *in vitro*.

In conclusion, it is clear that, for 2AAF, the *in vivo* assays conducted in the trial, apart from the host-mediated tests for microbial mutation and some tests for premutational damage, have exhibited a similar spectrum of response as the corresponding tests conducted *in vitro*. With regard to distinguishing between 2AAF and 4AAF the relative predictive values of assays conducted *in vivo* and *in vitro* will become more apparent when the animal cancer data are complete. Should 4AAF be found to exhibit weak carcinogenicity then several assays conducted *in vivo* would be providing false negative information whilst other tests (UDS and SCE) would be more predictive than if they had been used *in vitro*. Some might consider also that the results of the Salmonella mutagenicity assay alone could have been used to predict weak positive responses including carcinogenicity for 4AAF and therefore other compounds *in vivo*. Although this might be the case for 4AAF, if should not be assumed that Salmonella mutagenicity assay would be as predictive for carcinogenicity of other chemicals as the short-term *in vivo* assays. A larger data base obtained using short term *in vivo* assays with structurally related carcinogen/suspect carcinogen pairs of chemicals is required before their predictivity in general can be assessed.

9 Summary and conclusions

The majority (71%) of *in vivo* assays/end points in mammalian cells (Table 4) distinguished between 2AAF and 4AAF. However, rather less (35%) were able to register a clear positive/negative distinction whilst the rest recorded either similar responses or 1 of the chemicals was a weak positive. When these data are compared with those obtained from *in vitro* testing only tentative conclusions can be drawn due to the limited *in vitro* data base for 4AAF.

Discussion of the overall significance of the *in vivo* short-term test data for 2AAF and 4AAF is complicated by several factors. First, the lifetime cancer bioassay is incomplete for 4AAF. Secondly, 4AAF was active in

Figure 2. Comparison of *in vivo* and *in vitro* activities of 2AAF and 4AAF in genotoxicity assays.

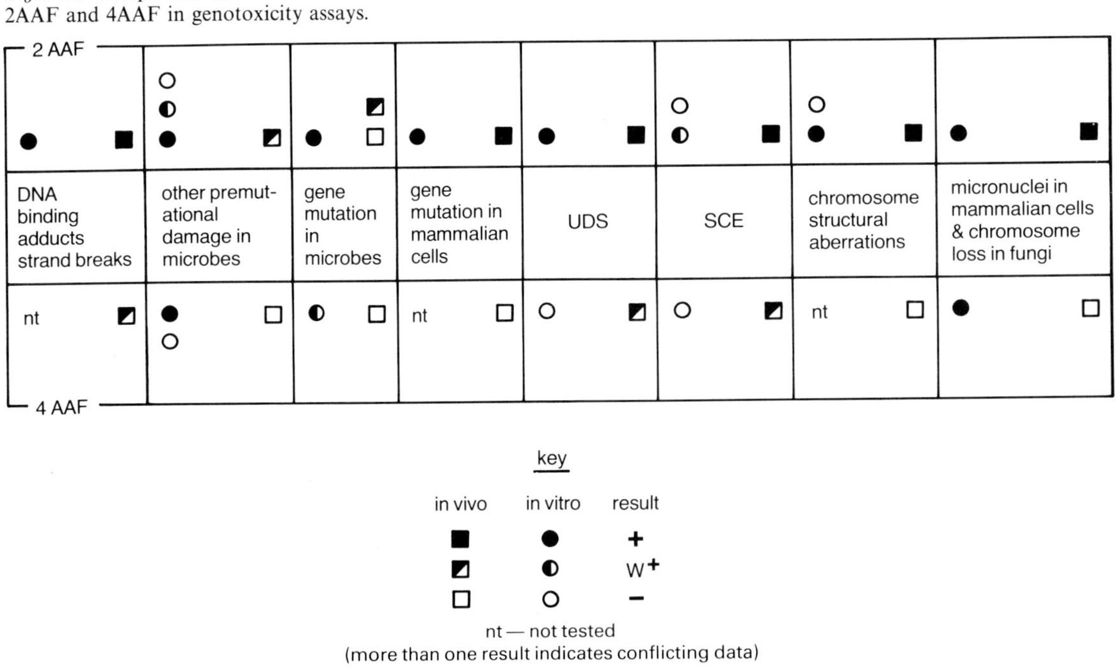

several of the *in vivo* tests used. Thirdly, the Salmonella mutagenicity assay is able to distinguish between the relative potencies of 2AAF and 4AAF as positive and weak positive, respectively. Lastly, detailed distribution studies on the analogues were not conducted. Knowledge of the amounts of administered compound and its metabolites reaching any organ can facilitate interpretation of data, especially negative results obtained with an end point measured in that organ.

Clearly the potential of 4AAF to induce DNA damage as shown by *in vitro* tests is realized *in vivo*. However, when compared with 2AAF, 4AAF was active in a smaller number of tests conducted *in vivo*. Where activity was observed, 4AAF was, with one exception, always the less genotoxic of the two analogues. Since 2AAF and 4AAF can be distinguished as positive and weak positive in the Salmonella assay, the contribution of the results obtained from the *in vivo* assays to their genotoxicity profiles with respect to potential carcinogenicity could be questioned. However, it is not possible to make quantitative predictions about carcinogenicity from the results of *in vitro* tests. A major reason for this is the likelihood that the balance between activating and detoxification enzymes present *in vivo* is different in exogenous metabolising systems.

The main classes of assays in the trail include those used currently to predict human/animal carcinogenicity (Garrett *et al.*, 1984). Overall it is apparent that these tests provide no evidence to suggest that 4AAF is equipotent to 2AAF as a carcinogen, rather they imply that at best 4AAF is a weak carcinogen at any site.

Recently (Ashby, 1985b) it has been suggested that *in vivo* short term genotoxicity assays will facilitate the definition of potential carcinogenicity of chemicals with *in vitro* genotoxic effects. Moreover, it was recommended that a bone marrow micronucleus assay followed by a liver specific test such as UDS (if the micronucleus test is negative) will prove the most useful in this respect. This recommendation is supported by the data base (Table 9). It is noteworthy that 4AAF was inactive in the micronucleus assays whereas it was a weak positive in several UDS assays in the liver. If 4AAF turns out to be a weak carcinogen then the above recommendation for the additional use of UDS will have been endorsed.

Further experiments are required to verify that 2AAF is unable to induce somatic mutations in lung cells, especially as a clear positive/negative distinction; respectively, for 2AAF and 4AAF was obtained in the spot test. In this context it is noteworthy that, using the micronucleus test with the mouse, foetal blood cells were less sensitive to induction of micronuclei by 2AAF.

Of the tests for chromosomal effects, those for structural aberrations were the least sensitive, when the responses of both chemicals are considered. This is not unexpected in view of the good but incomplete concordance of aberration tests with animal cancer data (Preston *et al.*, 1981). It is not possible yet to recommend any one assay for chromosomal effects in bone marrow or in any other tissue for testing of other compounds. Until such time as data become available for a wider range of chemicals it is advisable to use several systems as suggested elsewhere (Tice and Ivett, 1985).

The data base of the *in vivo* study provides insufficient information to assess whether or not short term *in vivo* systems are useful for predicting organotropic effects or potentially hazardous routes of exposure in relation to carcinogenesis.

Several of these assays may be useful for determining heritable effects without the need for using other specific tests for induction of germ line mutations (Ashby, 1985c). 2AAF was active in somatic cells including the bone marrow but was negative in dominant lethal assays. However, sperm abnormalities were induced implying that gonadal exposure to this chemical occurs. On the contrary there is no evidence that 4AAF traverses the blood testis barrier, since it was inactive for all end points scored in germ cells. Therefore the available information suggests that 4AAF is unlikely to induce inheritable effects whereas 2AAF may do so.

10 References

Ammenheuser, M.M. and W.E. Warren (1979) Detection of mutagens in the urine of rats following topical application of hair dyes, *Mutation Research*, **66**, 241–245.

Anderson, D. (1984) The dominant lethal test in rodents. In *Mutagenicity Testing, A Practical Approach*, S., Venitt and J.M. Parry, eds, IRL press, Oxford, pp. 307–335.

Armuth, V. and I. Berenblaum (1977) Possible two-stage transplacental liver carcinogenesis in C57 BL/6 mice, *Int. J. Cancer*, **20**, 292–295.

Ashby, J. (1985a) Is there a continuing role for the intraperitoneal route of exposure in short-term rodent genotoxicity assays? *Mutation Research*, **156**, 239–243.

Ashby, J. (1985b) Genetic toxicology in industry: perspectives and initiatives, *Food Chem. Toxicology*, **23**, 43–45.

Ashby, J. (1985c) Gonadal genotoxicity assays as practical surrogates for germ-cell mutagenicity assays, *Environ. Mutagenesis*, **7**, 263.

Astrom, A. and J.W. DePierre (1985a) Metabolism of 2-acetylaminofluorene by eight different forms of cytochrome P-450 isolated from rat liver, *Carcinogenesis*, **6**, 113–120.

Astrom, A. and J.W. DePierre (1985b) Identification of the forms of cytochrome P-450 induced in rat liver by 2-acetylaminofluorene using immunoblotting and partial purification, *Biochem. Biophys. Res. Comm.*, **128**, 578–585.

Aune, T., R.R. Vanderslice, J.E. Croft, E. Dybing, J.R. Bend and R.M. Philpot (1985a) Deacetylation to 2-aminofluorene as a major initial reaction in the microsomal metabolism of 2-acetylaminofluorene to mutagenic products in preparations from rabbit lung and liver, *Cancer Res.*, (in press).

Aune, T., A. Haugen and E. Dybing (1985b) Metabolism and mutagenicity of aromatic amines by human foetal liver, *Arch. Toxicol.*, **57**, 136–138.

Boobis, A.R., M.J. Brodie, M.E. McManus, N. Staiano, S.S. Thorgiersson and D.S. Davies (1982) Metabolism and mutagenic activation of 2-acetylaminofluorene by human liver and lung. In *Biological Reactive Intermediates-II, Chemical Mechanisms and Biological Effects*, R., Snyder *et al.*, eds, Plenum Press, pp. 193–201.

Clayson, D.B. and R.C. Garner (1976) Carcinogenic aromatic amines and related compounds. In *Chemical Carcinogens*, C.E. Searle, ed., pp. 366–461 ACS Monograph 173 Washington.

Combes, R.D. (1982) An analysis of the efficacy of

bacterial DNA repair assays for predicting genotoxicity, *Mutation Research*, **108**, 81–92.
Combes, R.D., D. Anderson, T.M. Brooks, S. Neale and S. Venitt (1984) The detection of mutagens in urine, faeces and body fluids. In *Guidelines for Mutagenicity Testing*. Pt II, Supplementary Tests, BJ. Dean *et al.*, eds, UKEMS, Swansea pp. 203–244.
Dean, B. and K.R. Senner (1977) Detection of chemically induced somatic mutation in Chinese hamsters, *Mutation Research*, **46**, 403–407.
De Serres, F.J. and J. Ashby (1981) *Evaluation of Short Term Tests for Carcinogens* – Report of the International Collaborative Program, Elsevier/North Holland, Amsterdam pp. 827.
Fahrig, R. (1984) Genetic mode of action of cocarcinogens and tumour promoters in yeast and mice, *Molecular Gen. Genetics*, **194**, 7–14.
Faustman-Watts, E.M., H.Y.L. Yang, M.J. Namkung, J.C. Greenaway, A.G. Fantel and M.R. Juchau (1984) Mutagenic, cytotoxic and teratogenic effects of 2-acetylaminofluorene and reactive metabolites *in vitro*, *Teratogenesis, Carcinogenesis and Mutagenesis*, **4**, 273–283.
Forster, R., M.H.L. Green, A. Priestly and J.W. Gorrod (1981) Apparent activation of 2-acetylaminofluorene and some aromatic amines by cytosolic preparations, *Mutation Research*, **85**, 187–194.
Garrett, N.E., H.F. Stack, M.R. Gross and M.D. Waters (1984) An analysis of the spectra of genetic activity produced by known or suspected human carcinogens, *Mutation Research*, **134**, 89–111.
Green, S., A. Auletta, J. Fibricant, R. Kapp, M. Manandhar, C.J. Sheu, J. Springer and B. Whitfield (1985) Current status of bioassays in genetic toxicology – the dominant lethal assay, *Mutation Res.*, **154**, 49–67.
Grunberger, D. (1981) N-2-fluorenylacetamide-induced conformational and functional damage to DNA, In *NCI Monograph No 58 Carcinogenic and Mutagenic N-Substituted Aryl Compounds*, US. Dept. Health and Human Services, Bethesda, pp. 193–199.
Holme, J.A. and E.J. Soderland (1985) Species differences in the cytotoxic and genotoxic effects of 2-acetylaminofluorene and its primary metabolites 2-aminofluorene and N-OH-2-acetylaminofluorene, *Carcinogenesis*, **6**, 421–425.
Kada, T. and M. Ishidate, Jr. (1980) *Environmental Mutagens Data Book*, Scientist Inc., Tokyo.
Kuchlbauer, J., W. Romen and H.G. Neumann (1985) Syncarcinogenic effects on the initiation of rat liver tumours by trans-4-acetylaminostilbene and 2-acetylaminofluorene, *Carcinogenesis*, **6**, 1337–1342.
Legator, M.S., E. Bueding, R. Batzinger, T.H. Connor, E. Eisenstadt, M.G. Farrow, G. Ficsor, A. Hsie, J. Seed and R.S. Stafford (1982) An evaluation of the host-mediated assay and body fluid analysis, *Mutation Research*, **98**, 319–374.
Lotlikar, P.D., P.N. Pandy, M.S. Clearfield and S.M. Paik (1984) Oxidative activation and inactivation of the carcinogen 2-acetylaminofluorene by various purified cytochromes P-450 from 3-methylcholanthrene pretreated rats, *Toxicology Letters*, **21**, 111–118.
Lowe, G.H., M. Poulson, E. Kirkjian, J. Ferrell, B.S. Sudhindra and M. Rebagliati (1985) Computer-assisted mechanistic structure-activity studies: Application to diverse classes of chemical carcinogens, *Environ. Hlth. Persp.* **61**, 69–96.
Lutz, W.K. (1979) *In vivo* covalent binding of organic chemicals to DNA as a quantitative indicator in the process of chemical carcinogenesis, *Mutation Research*, **65**, 289–356.
Malejka-Giganti, D. and C.N. Ryzewski (1985) Microsomal metabolism of the carcinogen, N-2-fluorenyl-acetamide, by the mammary gland and liver of female rats. II. Glucuronidation of ring- and N-hydroxylated metabolites of N-2 fluorenylacetamide, *Carcinogenesis*, **6**, 687–692.
McGregor, D.B. (1975) The relationship between 2-acetamidofluorene in plate tests with its *in vivo* liver cell component distribution and its carcinogenic potential, *Mutation Research*, **30**, 305–316.
Miller, E.C. (1978) Some current perspectives on chemical carcinogenesis in humans and experimental animals, *Cancer Research*, **38**, 1479–1496.
Miller, J.A. and E.C. Miller (1969) The metabolic activation of carcinogenic aromatic amines and amides, *Progr. exp. Tumor Res.*, **11**, 273–301.
NCI Monograph No 58, (1981) *Carcinogenic and Mutagenic N-Substituted Aryl Compounds*, US. Dept Health and Human Services, Bethesda.
Pitts, C.M. and R. Lindahl (1979) Transplacental and perinatal carcinogenicity of 2-acetylaminofluorene in rats. *Proc. Amer. Assoc. Cancer Research*, **20**, 1.
Poirier, M.C. (1984) The use of carcinogen-DNA adduct antisera for quantitation and localization of genomic damage in animal models and the human population, *Environmental Mutagenesis*, **6**, 879–887.
Preston, R.J., W. Au, M.A. Bender, J.G. Brewen, A.V. Carrano, J.A. Heddle, A.F. McFee, S. Wolff and J.S. Wassom (1981) Mammalian *in vivo* and *in vitro* cytogenetic assays: A report of the US EPA's gene-tox program, *Mutation Res.*, **87**, 143–188.
Razzouk, C. and M. Roberfroid (1983) Comparative study of rat, dog and human liver microsomal arylamide N-hydroxylases, *Toxicology Letters*, **16**, 339–345.
Reigh, D.L. and R.A. Floyd (1981) Evidence for a cytochrome P-420 catalyzed mechanism of activation of N-hydroxy-2-acetylaminofluorene, *Cancer Biochem. Biophys.*, **5**, 213–217.
Roberfroid, M., C. Razzouk, E. Agazzi-Leonard, M. Batardy-Gregoire, F. Poncelet and M. Mercier (1982) Microsomal metabolism of 2-acetylaminofluorene: role in mutagenicity and carcinogenicity. In *Biological Reactive Intermediates-II, Chemical Mechanisms and Biological Effects*, R. Snyder *et al.*, (eds), Plenum Press, pp. 921–929.
Schintz, H.R., H. Fritz-Niggli, T.W. Campbell and H. Schmid (1955) Krebsbildung durch aminofluorene und verwandte korpe, *Oncologia*, **8**, 233–245.
Singer, B. and D. Grunberger (1983) Reactions of metabolically activated carcinogens with nucleic acids. In *Molecular Biology of Mutagens and Carcinogens*, B. Singer and D. Grunberger, Plenum Press, New York pp. 143–197.
Stevens, F. and C. Peraino (1983) Liver as a model system for analyzing mechanisms of tumor initiation and promotion. In *Organ Specificity in Chemical Carcinogenesis*, R. Langenbach *et al.* (eds) Plenum Press, New York, pp. 231–252.
Stout, D.L., J.N. Baptist, T.S. Matney and C.R. Shaw (1976) N-hydroxy-2-aminofluorene: the principal mutagen produced from N-hydroxy-2-acetylaminofluorene by a mammalian enzymic preparation supernatant, *Cancer Letters*, **1**, 269–274.
Styles, J., B.M. Elliott, P.A. Lefevre, M. Robinson, N. Pritchard, D. Hart and J. Ashby (1985) Irreversible depression in the ratio of tetraploid: diploid liver nuclei in rats treated with 3′-methyl-4-dimethyl-aminoazobenzene (3′M), *Carcinogenesis*, **6**, 21–28.
Takahashi, G. and K. Yasuhira (1975) Macroautoradiographic assays in pregnant mice and their foetuses given N-acetyl-9-[^{14}C]-2-aminofluorene and p-[^{14}C] dimethylaminoazobenzene, *Cancer Research*, **35**, 2126–2131.
Takeishi, K., S. Okuno-Kaneda and T. Seno (1979) Mutagenic activation of 2-acetylaminofluorene by guinea pig liver homogenates: essential involvement of cytochrome P-450 mixed-function oxidases, *Mutation Res.*, **62**, 425–437.

Tarpley, W.G., J.A. Miller and E.C. Miller (1982) Rapid release of carcinogen guanine adducts from DNA after reaction with N-acetoxy-2-acetylaminofluorene or N-benzoyloxy-N-methyl-4-aminoazobenzene, *Carcinogenesis*, **3**, 81–88.

Thorgeirsson, S.S., P.J. Wirth, N. Staino and C.L. Smith (1982) Metabolic activation of 2-acetylaminofluorene, in: *Biological Reactive Intermediates-II, Chemical Mechanisms and Biological Effects.* Plenum Press, R. Snyder *et al.*, (eds), pp. 897–919.

Thorgeirsson, S.S, I.B. Glowinski and M.E. McManus (1983) Metabolism, mutagenicity, carcinogenicity of aromatic amines. In *Review in Biochemical Toxicology*, Vol 5, E. Hodgson, J. Bend and R. Philpot (eds), Elsevier/N. Holland, New York, pp. 349–386.

Tice, R.R. and J.L. Ivett Cytogenetic analysis of bone marrow damage. In *Toxicology of the Blood and Bone Marrow*, R.D. Irons, (ed.), Raven Press, New York, pp. 119–140.

Vaught, J.B., I.B. Glowinski and C.M. King (1985) Metabolism of N-hydroxy-2-acetylaminofluorene and N-hydroxyphenacetin by guinea pig liver microsomal enzymes, *Carcinogenesis*, **6**, 495–499.

Walters, J.M. and R.D. Combes (1985) Characterisation of a microsomal fraction from rat small intestine for metabolic activation of some promutagens, *Carcinogenesis*, **6**, 1415–1420.

Wang, C.Y., K. Zukowski and M.S. Lee (1985) Glucuronidation of carcinogenic arylamine metabolites by rat liver microsomes, *Biochem. Pharmacol*, **34**, 837–841.

Weber, W.W. and D.W. Hein (1985) N-acetylation pharmacogenetics, *Pharmacol Revs.*, **37**, 25–79.

Weeks, C.E., W.T. Allaben, S.C. Louie, E.J. Lazear and C.M. King (1978) Role of arylhydroxamic acid acyltransferase in the mutagenicity of N-hydroxy-N-2-fluorenylacetamide. In *Salmonella typhimurium, Cancer Research*, **38**, 613–618.

Weisburger, E.K. (1981) N-2-fluorenylacetamide and derivatives. In *Carcinogens in Industry and the Environment*, J.M. Sontag, (ed.), Ch. 14, pp. 583–666, M. Dekker Inc., New York.

Weisburger, E.K. and J.H. Weisburger (1958) Chemistry, carcinogenicity and metabolism of 2-fluorenamine and related compounds, *Advs. Cancer Res*, **5**, 331–431.

Weisburger, J.H., E.K. Weisburger and H.P. Morris (1952) Analogues of the carcinogen 2-acetylaminofluorine: the isometric 4-acetylaminofluorine, *J. Am. Chem. Soc.*, **74**, 4540–4543.

Wyrobek, A.J., L.A. Jordon, J. (G) Burkhart, M.C. Francis, R.W. Kapp, T. Letz, H.V. Malling, J.C. Doplan and M.A. Whorton (1983) An evaluation of human sperm as indicators of chemically induced alterations of spermatogenic function, *Mutation Research*, **115**, 73–148.

PART III

In vivo assays: bone marrow

James W. Allen (Workgroup Chairman)[1]
Environmental Protection Agency, Research Traingle Park, North Carolina, USA

6

Summary report on the performance of sister chromatid exchange, micronucleus, and chromosome aberration tests in rodent bone marrow cells

1 Introduction

The task of this work group was to compare the results obtained in rodent bone marrow using different cytogenetic assays of 2-acetylaminofluorene (2AAF)/4-acetylaminofluorene (4AAF) and benzo(a)pyrene (BP)/pyrene (PYR) chemical pairs. Any relationships or patterns among the various endpoints – sister chromatid exchange (SCE), micronucleus (MN), and chromosomal aberration (CA) – and among the major experimental variables (e.g., species, dose, route of exposure) were to be evaluated.

In many important ways, the separate endpoint data sets are not directly comparable. Interpretations of relative sensitivities are limited by the general lack of coordination for individual doses and dose ranges and by a failure to test for the lowest doses that could elicit a positive response. Further, the criteria used for concluding positive and negative responses were not standardised. These discrepancies limit comparison of endpoint capabilities to discriminate (by positive and negative responses) between presumed carcinogenic and noncarcinogenic chemicals.

Review of the summary papers prepared separately for each cytogenetic endpoint (Adler *et al.*, 1986; Tice *et al.*, 1986; Wild *et al.*, 1986) also reveals a general lack of uniformity in testing procedures. Although all investigators within an endpoint group followed a common set of test guidelines, these were loosely structured and allowed for considerable latitude in protocol design. Experimental variables included the following: (1) animals – species, strain, age, sex, source, maintenance conditions; (2) treatments – doses, solvents, route of exposure, number of exposures, duration of exposure; (3) experimental designs – controls, sample sizes, statistical treatments, criteria used to interpret results. That these protocol differences usually occurred in various combinations places constraints on judging the importance of any one variable by itself, either within or between endpoints.

Regardless of the limitations involved in comparing results obtained with different cytogenetic assays, a

[1] Workgroup participants: Ole Andersen, Hyman C. Birnboim, David H. Blakey, Pawan S. Chauhan, Riccardo Crebelli, Howard P. Glauert, Makoto Hayashi, Vladimir I. Ivanov, Ulf Kliesch, Carol Luke, James T. MacGregor, Jeremy A. Styles, Guy Thomas, Raymond R. Tice, Barbara Westbrook-Collins, Viacheslav S. Zhurkov

general assessment of overall test performances, as well as a determination of whether or not any notable relationships exist among endpoints, seems warranted. This report compares the conclusions (summarised investigators' calls) for each chemical and cytogenetic endpoint evaluated in rodent bone marrow and examines the similarities and differences in doses and responses. Also, two studies carried out in rodent peripheral blood lymphocytes and foetal liver red blood cells are considered. A perspective on experimental variables for each endpoint is briefly discussed. This report was prepared from data provided by the chairpersons of the SCE, MN, and CA work groups at the time of the IPCS-CSSTT (phase II) meeting (May 1985).

2 Assessments

Results defined by investigator calls

Table 1 summarises the multi-cytogenetic endpoint test results for 2AAF, 4AAF, BP, and PYR. Entries represent the number and percentage of responses classified by individual investigators as either positive, weak positive, negative, or inconclusive. These relative calls should be considered rough estimates in view of the lack of standard criteria used in performing the tests and interpreting the results. Comparative endpoint performances to provide positive results for carcinogens and negative results for noncarcinogens are as follows. The frequencies of positive results for 2AAF were obtained in descending order by SCE > MN > CA endpoints; negative results for 4AAF by MN > SCE ⩾ CA; positive results for BP by CA and SCE ⩾ MN; and negative results for PYR by CA ⩾ MN > SCE.

For both 4AAF and PYR, the CA endpoint (when conclusive) was always negative, while positive or weak positive responses were observed occasionally with the MN endpoint and more often with the SCE endpoint. It cannot be concluded here as to whether these elevated MNs and SCEs are accurately reflecting weak carcinogenic potential. Similar uncertainties surround the question of whether CAs are too insensitive to detect weak carcinogens or are most accurate in revealing an actual lack of activity for PYR and 4AAF. For BP and 2AAF, SCE and MN responses were always, or nearly always, positive. Although CA responses to BP exposure were always positive, they were clearly positive with 2AAF in less than half of the exposure instances. In view of the mechanistic similarities between MN and CA endpoints, the latter's weak responsiveness to 2AAF may reflect some decreased sensitivity due to a smaller sample size.

As a visual aid to perceiving relative endpoint capabilities to discriminate (by + and − responses) between the presumed carcinogen and noncarcinogen of each chemical pair, the responses from Table 1 are graphed in Figure 1. Although only the SCE endpoint provided 100% positive results for the two carcinogens, mixed (+, weak +, −) SCE results were obtained for the presumed noncarcinogens. CAs completely discriminated between BP and PYR; results were mixed or inconclusive for the 2AAF/4AAF chemical pair. The MN endpoint was most consistent in discriminating fairly well for both chemical pairs.

Results defined by dose-response spectra

Figures 2–7 are spectral presentations of the data for each investigator, grouped by chemical pairs for the different endpoints. These graphs display the dose range, number of doses, lowest effective dose tested (LEDT), and highest ineffective dose tested (HIDT). In comparing the graphs, differences in individual study designs, as well as patterns and deviations in test responses, become apparent. Clearly, the largest data base (number of tests, number of doses tested) derives from the MN assay; the smallest from the CA assay. LEDTs and especially HIDTs were often variable both within the between endpoints. A dose of approximately 1 mmole/kg, which was common to all the endpoints, usually gave a positive response for the carcinogens and a negative response for the supposed noncarcinogens.

It may be concluded from the spectral graphs the MN and SCE assays were quite sensitive to BP and 2AAF at low doses and, in general, were rather similar in their

Table 1. *Cytogenetic test results for 2AAF/4AAF and BP/PYR*

Chemical	Endpoint	Total number of tests	Positive no.	Positive (%)	Weak positive no.	Weak positive (%)	Negative no.	Negative (%)	Inconclusive no.	Inconclusive (%)
2AAF	CA[a]	14	5	(36)	0	(0)	7	(50)	2	(14)
	MN[b]	19	16	(84)	0	(0)	3	(16)	0	(0)
	SCE[c]	10	10	(100)	0	(0)	0	(0)	0	(0)
4AAF	CA	14	0	(0)	0	(0)	8	(57)	6	(43)
	MN	19	1	(5)	0	(0)	18	(95)	0	(0)
	SCE	10	4	(40)	0	(0)	6	(60)	0	(0)
BP	CA	14	14	(100)	0	(0)	0	(0)	0	(0)
	MN	20	19	(95)	0	(0)	1	(5)	0	(0)
	SCE	10	10	(100)	0	(0)	0	(0)	0	(0)
PYR	CA	13	0	(0)	0	(0)	13	(100)	0	(0)
	MN	20	1	(5)	0	(0)	19	(95)	0	(0)
	SCE	10	0	(0)	4	(40)	6	(60)	0	(0)

[a]Chromosomal aberrations.
[b]Micronuclei.
[c]Sister chromatid exchanges.

response patterns for all the chemicals. Although no quantitative information pertaining to dose-response relationships is provided, MN and SCE endpoints tend to exhibit comparable numbers of positive and negative results over close dose ranges. As noted earlier, the actual lowest effective doses may be very different from those reported here as the lowest dose tested and found to be positive. Thus, relative endpoint sensitivities cannot be reliably judged by comparing the low dose responses. For similar reasons, the derivation of discrimination indices from LEDT/HIDT ratios is considered unreliable. While these ratios have proven useful for comparing chemical activities across different dose ranges for the same genetic endpoint (Garrett *et al.*, 1984), we consider them to be potentially misleading for such comparisons across different endpoints.

Overall, BP consistently induced significant increases in all 3 endpoints at doses much lower than the highest PYR doses that were tested and found to yield negative results. A similar although less consistent relationship is evident for 2AAF and 4AAF. Further, when BP and 2AAF are compared at equimolar doses, BP is clearly the more potent chemical for causing cytogenetic damage under these experimental conditions.

Results influenced by experimental variables

In no case where dose-response curves were generated did the choice of species, strain, sex, route of exposure, or solvent appear to affect the results such that they changed from overall positive to negative or negative to positive. In general, there was an insufficient amount of data on any one experimental variable to draw firm conclusions in regard to its singular influence upon the results. The species-specific responses presented in Figures 2–7 suggest that mice may be more sensitive than rats and hamsters to the test carcinogens, especially BP. There are exceptions and too few studies in rats and hamsters were carried out to allow meaningful analyses of between-endpoint similarities and differences. Within endpoints, the individual importance of certain experimental variables may be evaluated, to some extent, as indicated below.

MN assay. Because of differences in experimental protocols, there were only 2 sets of data in which all the variables of species, strain, vehicle, sex, and route were held constant. Whenever one of these variables was different in another study, the response was not considered specific for that variable unless it exceeded the range of responses among laboratories with identical conditions. With this standard of measure, no significant differences in responses were associated with the use of different solvents. The data pertaining to species and sex variables were considered insufficient to draw conclusions. One laboratory used the MS strain of mouse specifically because of its reported high sensitivity to

Figure 1. Comparison of results obtained in chromosomal aberration (CA), micronucleus (MN), and sister chromatid exchange (SCE) assays of 2AAF/4AAF and BP/PYR. Differentially patterned bars reflect the relative percentages of specific responses (investigator calls from Table 1). Although total bar dimensions are similar, the sample sizes they represent vary among endpoints.

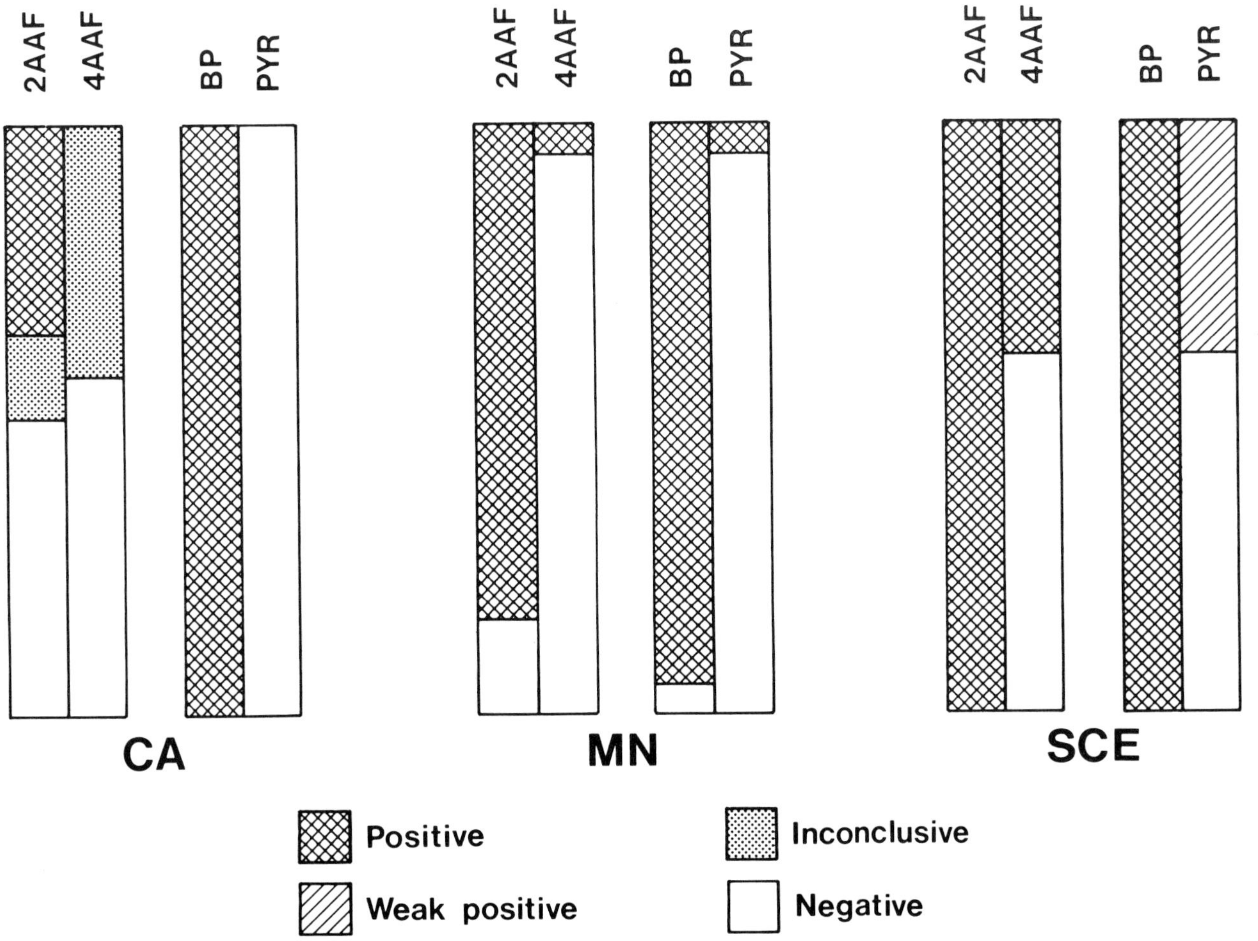

micronucleus induction. The highest BP- and 2AAF-induced micronucleus frequencies were observed in this strain. Finally, based on data from 2 laboratories which conducted both foetal liver and adult/weanling marrow assays, BP was equipotent for inducing the MN response in these hematopoietic cell types. 2AAF was less active in the foetal cells, possibly as a result of transplacental pharmacokinetic factors.

CA assay. Experimental conditions among the different investigators carrying out the CA assay were particularly variable. Only 2 studies used the same strain, and the route of administration, solvent, dose range, and sampling times were all different. Three studies using DMSO as the solvent consistently revealed lower levels of aberrations in controls as compared with the corn oil controls within each laboratory. DMSO and corn oil were never used for the same chemical; thus, induced CA levels cannot be compared. In 1 study, 5 mouse strains were tested at 1 dose for each of the 4 chemicals. BP, PYR, and 4AAF caused similar effects across strains. However, 2AAF was positive for 2 strains and negative for 3 strains (Figure 7). No further comparisons appeared justified.

SCE assay. SCE studies were also typically discordant for several experimental variables. However, the following observations were made for each chemical considered separately.

BP was positive by the 2 routes tested (ip, gavage) in all 3 species (Chinese hamster, mouse, rat), in every strain (mouse = 4, rat = 1), in both sexes, and using 3 solvents (corn oil, corn oil + DMSO, olive oil). It was not possible to determine whether differences in the potency of BP was solvent dependent; however, by comparing SCE-5 values (the dose of BP that would induce an increase of 5 SCEs above the corresponding control values; Tice *et al.*, 1986), several experimental factors seemed important. It appeared that the ip route was more potent than the gavage route of exposure. Furthermore, greater activities were evident in mice > rats ≥ Chinese hamsters. Marked sex differences were not observed, although male rats generally showed ≥ the responsiveness of female rats. The dose-response curve saturated at about one-fourth molar in mice.

PYR was negative in most experiments. However, where SCEs were induced by this chemical, the shape of the dose-response increase was nonlinear, fluctuating significantly across dose. In addition, there were significant differences between male and female rats (+ in males, − in females). The data were insufficient to determine whether route, species, or strain was preferentially invol-

Figure 2. Spectral display of SCE results for BP and PYR. Each investigator's study is described by a vertical bar, the boundaries of which define dose ranges tested. Horizontal cross bars mark individual doses. Heavy shading is used to denote a positive response ($p < 0.05$), whether at one dose or connecting multiple doses. Light shading indicates a negative response. Abbreviations of route, cell type, or mouse strain are sometimes noted. These graphs show a "breakout" of all the results obtained by each investigator for a chemical test. This information may or not appear consistent with the investigators' discretionary calls overall, noted in Table 1 and Figure 1 for the chemicals.

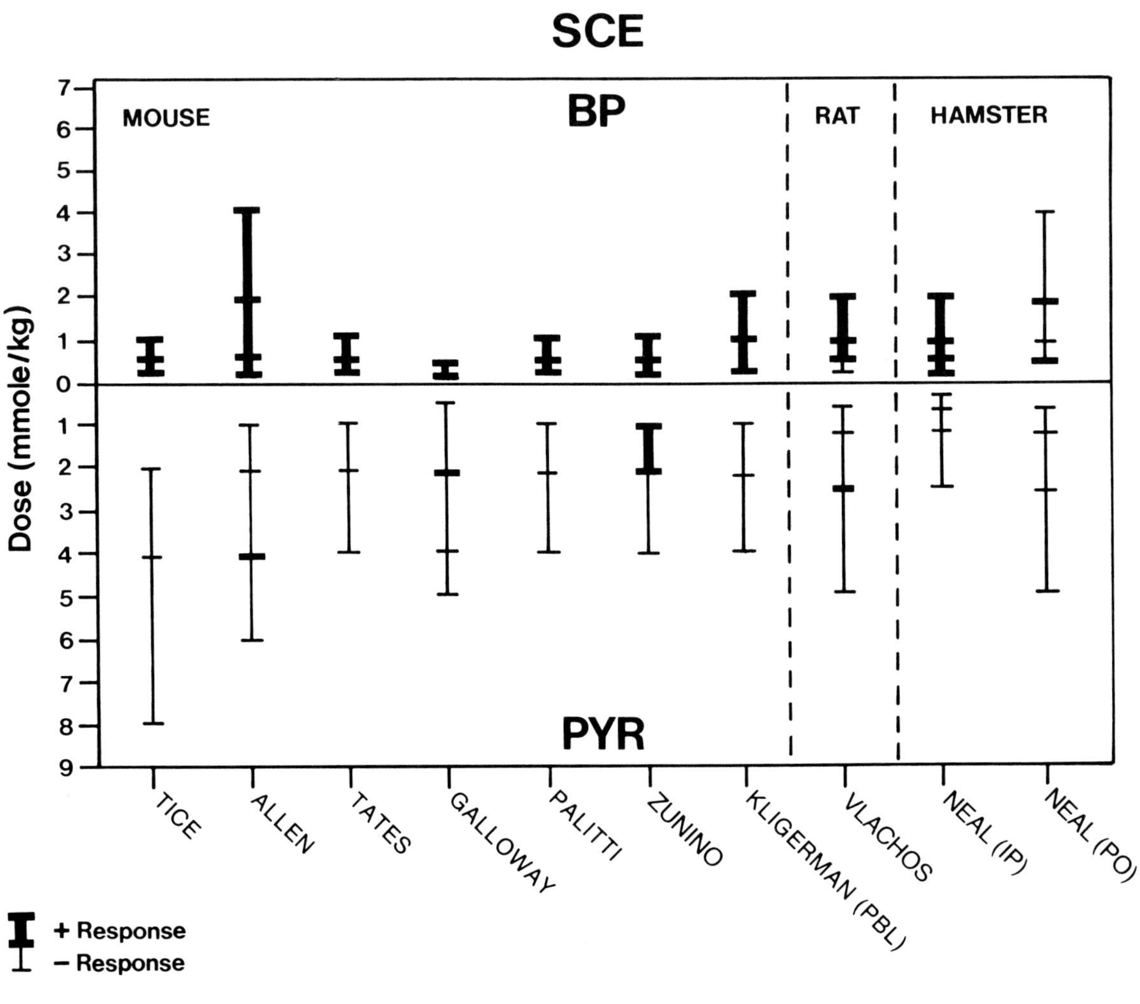

Figure 3. Spectral display of SCE results for 2AAF and 4AAF. (See Figure 2 legend for further explanation.)

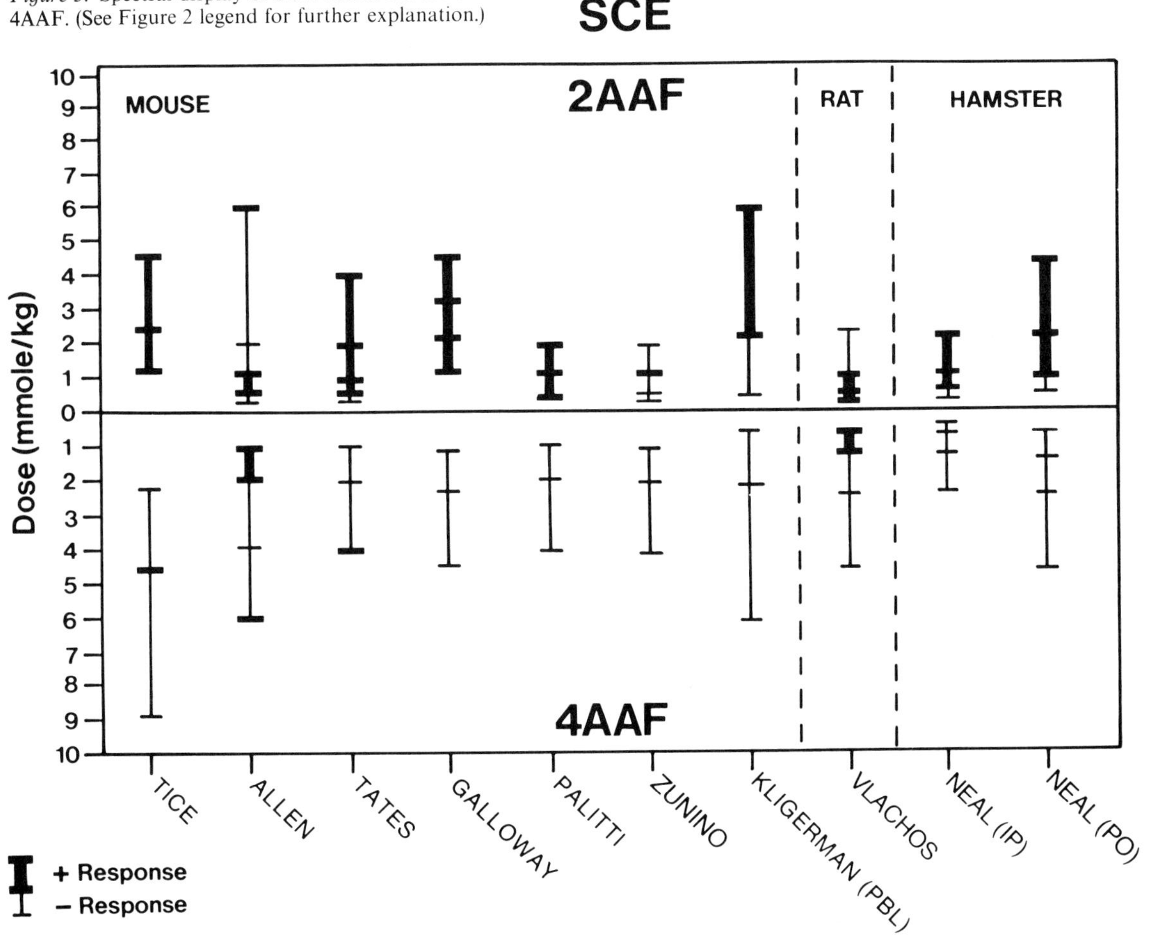

Figure 4. Spectral display of MN results for BP and PYR. (See Figure 2 legend for further explanation.)

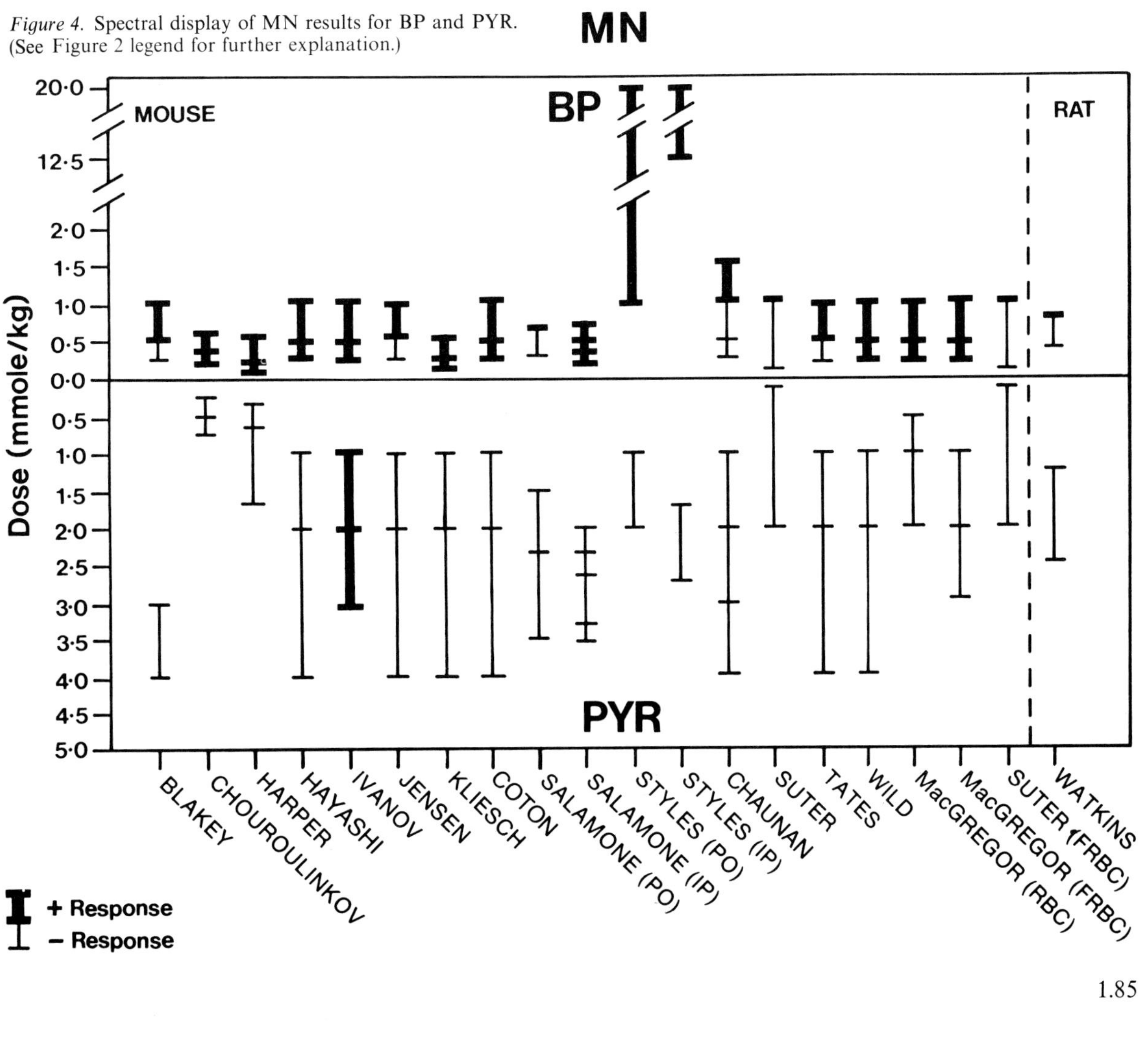

Figure 5. Spectral display of MN results for 2AAF and 4AAF. (See Figure 2 legend for further explanation.)

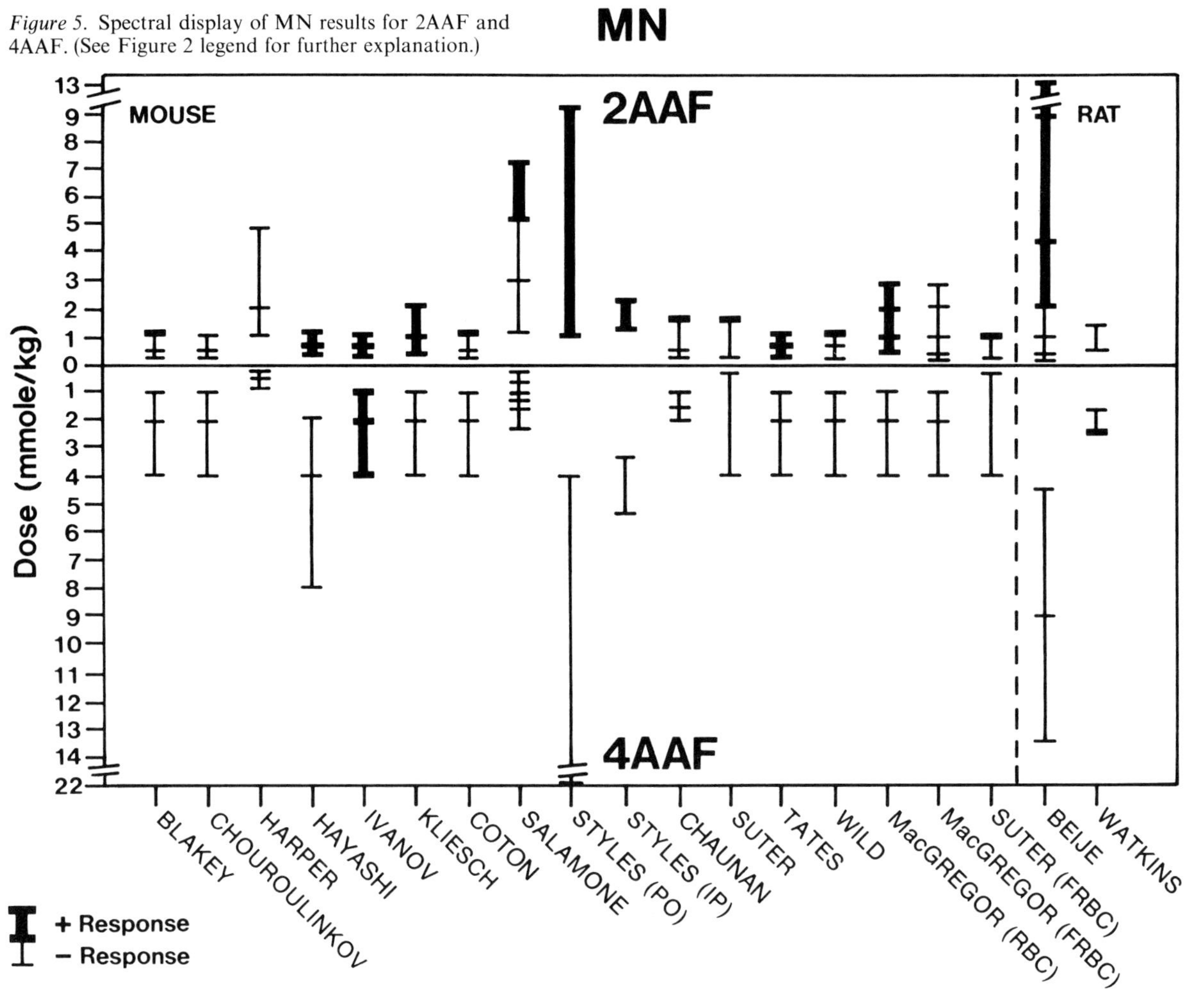

Figure 6. Spectral display of CA results for BP and PYR. (See Figure 2 legend for further explanation.)

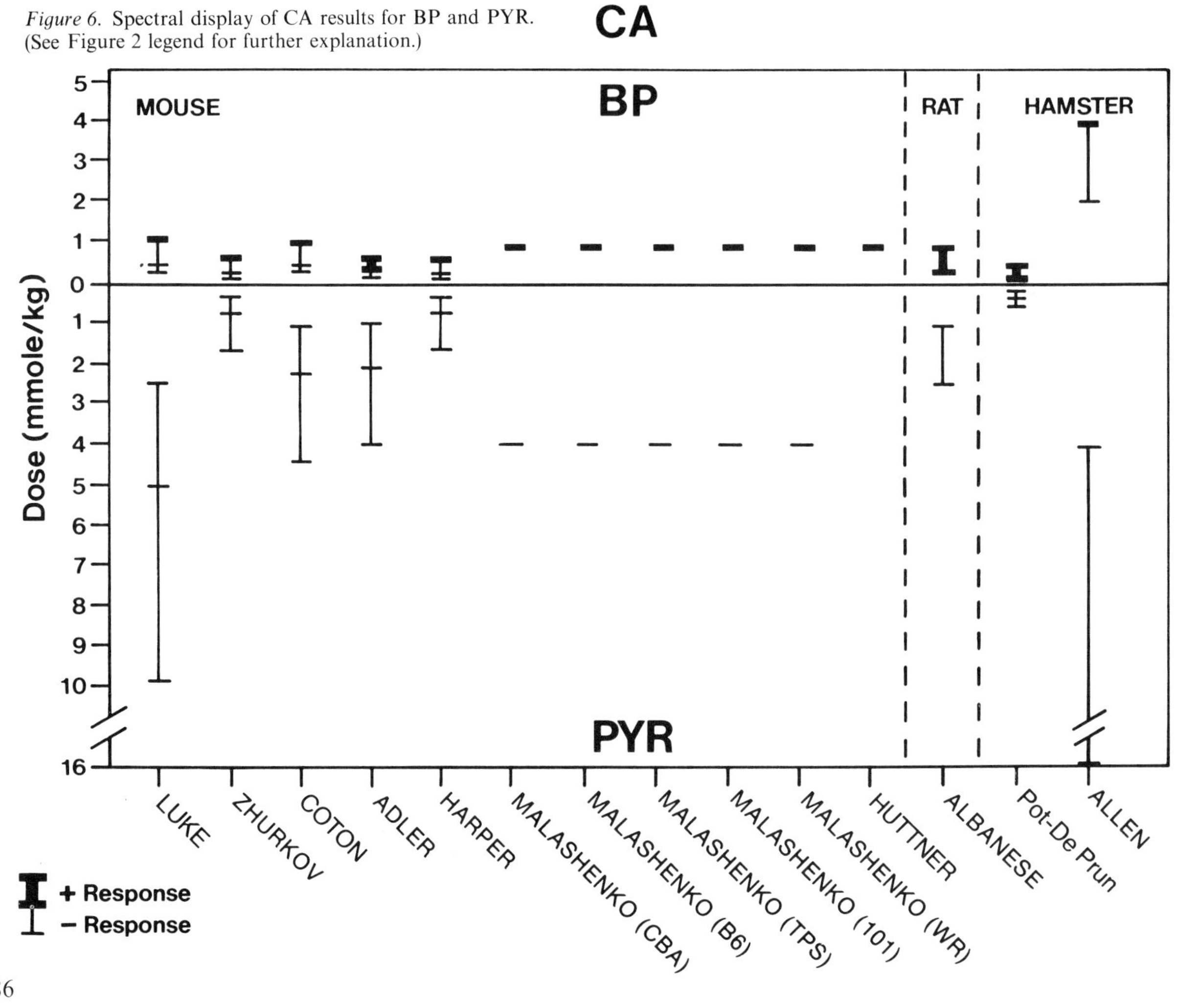

involved in the responses, although the results in CD-1 mice appeared positive in 2 or 3 trials.

2AAF induced a positive SCE response in 3 species (Chinese hamster, rat, mouse) by gavage and ip routes in all strains and in both sexes. Significant species, strain, or sex differences in activity were not apparent. The ip route in Chinese hamsters was slightly more active than the gavage route, while in the mouse, gavage exposures were often associated with greater activity. Apparently, different dose-response curves resulted from ip as compared with those from gavage routes of exposure. With the former, approximately linear responses were obtained. With the latter, more complex responses with an initial rise followed by a decline, and then a second rise tended to occur.

4AAF was positive for SCE induction in a few trials with mice and rats; it was negative in Chinese hamsters. After ip exposures, activity was noted in male but not female rats and in female but not male CD-1 mice. When administered by gavage, 4AAF was active in male 101 × $C3HF_1$ mice and in male and female C57B1/6 mice. The shape of the dose-response curve, similar to that observed for 2AAF, depended on the route of exposure. Linear responses were observed with ip exposures and more complex responses with gavage exposures.

3 Summary and conclusions

All assays discriminated between the paired carcinogens and noncarcinogens on the basis of relative cytogenetic activity. In general, BP and PYR were more efficiently separated than 2AAF and 4AAF by all the endpoints. The induction of MN provided the most consistent criterion of clear positive results after exposure to the carcinogens and negative results after exposure to the presumptive noncarcinogens.

Some summarised observations pertaining to each chemical are as follows: For BP, clustering of positive responses at relatively low doses was a typical finding. All of the CA and SCE and most of the MN trials were positive for this chemical. PYR was usually negative, although weak positive results were frequently reported with the SCE assay. 2AAF was always or nearly always detected as positive with SCE and MN tests; however, it was more often negative or inconclusive with the CA test. 4AAF was usually negative for MN induction; yet, it was sometimes inconclusive for CAs and sometimes positive for SCEs.

In some instances, the data were suggestive of experimental conditions having a significant impact on the results. Isolated studies showed species, strain, sex, and maternal/foetal differences affecting the strength or

Figure 7. Spectral display of CA results for 2AAF and 4AAF. (See Figure 2 legend for further explanation.)

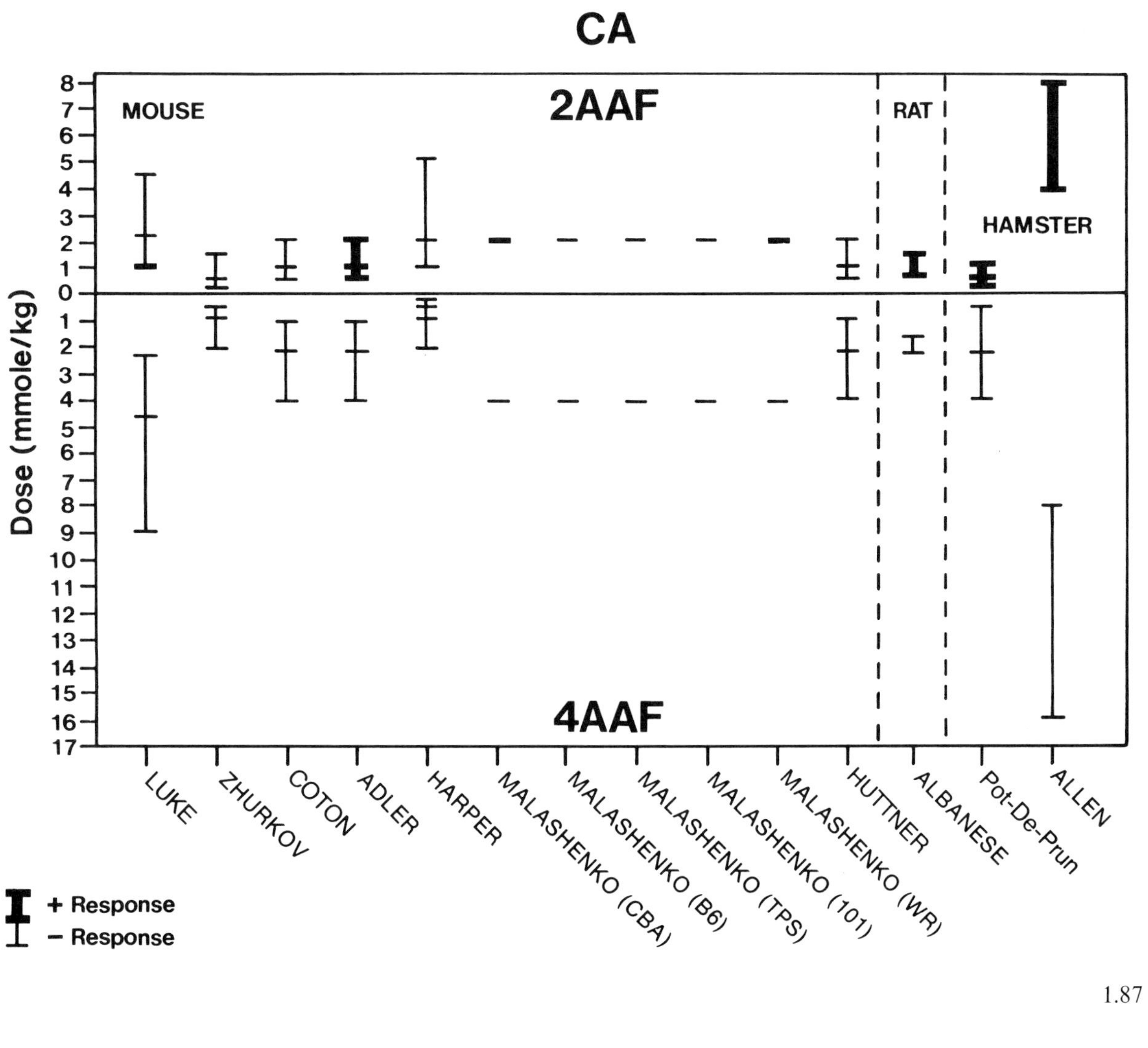

pattern of responsiveness to the carcinogens. More impressive were the several instances where slopes and shapes of dose–response curves appeared dependent on exposure modes. In this regard, perhaps the linear vs biphasic increases in SCE after respective ip vs gavage exposures to 2AAF were most pronounced.

The purpose of the *in vivo* studies with 2AAF, 4AAF, BP, and PYR was to generate comprehensive activity profiles under varied experimental conditions (Ashby *et al.*, 1983). These multi-cytogenetic endpoint data have contributed significantly to that aim. In particular, the carcinogenic chemicals proved positive under very diverse conditions.

The hope that these *in vivo* assays might act as "arbiters" of potential carcinogenicity for agents proven *in vitro* to be genotoxic has not been fully realised. The cytogenetic data discussed here are clearly consistent with earlier observations of reduced activity *in vivo* for 4AAF and PYR, as compared with *in vitro* test results for those chemicals (Ashby, 1981). Unequivocal negative results for the presumed noncarcinogens would have supported the contention that the *in vitro* responses were false positives. Yet, this was not found. In only 1 of 6 trials with the noncarcinogens (CA assay of PYR) were the results considered to be uniformly negative. As mentioned earlier, it cannot be concluded here whether the apparent cytogenetic toxicities of the presumptive noncarcinogens are also false positive, or indeed, show evidence of low-level carcinogenic potential. The latter is reasonably suspect from the *in vivo* cytogenetic data. Metabolism of the paired chemicals may differentiate them with regard to their actual capabilities to induce tumours, despite close structure and activity relationships between the parent compounds. By the same account, the relative accuracies of the individual assays for predicting carcinogenic potential cannot be determined without further information on the true carcinogenicity of these chemicals.

Throughout this report, conclusions regarding the relationships between MN, CA, and SCE assays have been qualified on the basis of different experimental designs employed. While these differences have in some ways strengthened interpretations of assay discriminating powers, they have weakened abilities to compare results and assess inter-endpoint relationships. Any future studies aimed at achieving greater definition of these relationships should incorporate tight standardisation of experimental protocol and evaluation procedures. This would make it possible to compare meaningfully the low-dose sensitivities; slopes of dose–response curves, and LEDT/HIDT ratios (as discrimination indices).

4 Acknowledgements

We are particularly grateful to Frank Stack for his counsel and help concerning the generation of spectral graphs. Thanks are also due to Frank Stack and Andrew Kligerman for reviewing the manuscript and to Shirley Milton for typing it.

5 Disclaimer

The research described in this article has been reviewed by the Health Effects Research Laboratory, U.S. Environmental Protection Agency, and approved for publication. Approval does not signify that the contents necessarily reflect the views and policies of the Agency nor does mention of trade names or commercial products constitute endorsement or recommendation for use.

6 References

Adler, I.D., J. Allen, E. Kristiansen, A. Leonard, C. Luke, A. McFee, M. Richold, J. Schoneich and V. Zhurkov (1986) Summary report on mammalian cell cytogenetics *in vivo: Chromosome aberrations – metaphase analysis* (this volume).

Ashby, J. (1981) Overview of study and test chemical activities, in: F.J. de Serres and J. Ashby (Eds) Evaluation of Short-Term Tests for Carcinogens, *Progress In Mutation Research*, Vol. 1, Elsevier, Amsterdam, pp. 112–171.

Ashby, J., F.J. de Serres, M. Draper, M. Ishidate Jr., B. Matter and M. Shelby (1983) The two IPCS collaborative studies on short-term tests for genotoxicity and carcinogenicity, *Mutation Res.*, **109**, 123–126.

Garrett, N.E., H.F. Stack, M.R. Gross and M.D. Waters (1984) An analysis of the spectra of genetic activity produced by known or suspected human carcinogens, *Mutation Res.*, **134**, 89–111.

Tice, R., J.W. Allen, S.M. Galloway, A.D. Kligerman, S.B. Neal, B.G. Ormiston, F. Palitti, A. Zunino, G.S. Probst, A.D. Tates and D. Vlachos (1986) *Interlaboratory Collaborative Study on Sister Chromatid Exchange Induction* (this volume).

Wild, D., B. Beije, D.H. Blakey, P.S. Chauhan, I. Chouroulinkov, C. Coton, B.L. Harper, M. Hayashi, V.I. Ivanov, J.C. Jensen, D. Jenssen, U. Kliesch, J.T. MacGregor, M.F. Salamone, J.A. Styles, K.E. Suter, A.D. Tates and P. Watkins (1986) *Summary Report on the IPCS-CSSTT Micronucleus Assays Performed on BP, PYR, 2AAF and 4AAF* (this volume).

Part III
In vivo assays: bone marrow

A
Chromosome aberrations

Ilse-Dore Adler (Workgroup Chairman)[1]
Institut für Säugetiergenetile, Gesellschaft für Strahlen- und Umweltforschung, Neuherberg, W, Germany

7

Summary report on the performance of the in vivo mammalian cell cytogenetics chromosome aberrations – metaphase analysis assays

1 Summary

Ten laboratories analysed bone marrow cells for structural chromosomal aberrations. One laboratory additionally used ascites tumour cells as target. Seven laboratories performed the experiments with mice, one used rats and two used Chinese hamsters. BP was found positive by all investigators and PYR gave negative results throughout. The results with 2AAF were contradictory; 2 investigators found a positive response, 3 called their data weakly positive, one regarded his data as equivocal, and 4 studies were negative. The analogue 4AAF gave negative results throughout. Sex, strain, and species differences, routes of application, vehicles, scoring criteria, sample size, and expression times were compared by the workgroup to elucidate the discrepancies in response to 2AAF.

2 Introduction

Cytogenetics data submitted by ten different laboratories are summarised. Rodent bone marrow preparations were used to analyse chromosomal aberrations in cells at mitosis. Cytogenetic analysis is regarded as an established procedure since it is described for the base set of testing dangerous substances both in the European Community Directives, Annex V, and the Organization for Economic Cooperation and Development Guidelines.

The principle of the test is to analyse cells at the first posttreatment mitosis. With conventional staining procedures, chromatid-type aberrations are more readily recognised than chromosome-type aberrations (particularly in the uniform karyotype of the mouse). Because of the possibility of preparational artifacts, numerical aberrations are not included in the evaluation. The test procedure and application has been described by Dean (1969) and recently in more detail by Adler (1984).

The current understanding of the mechanisms of aberration formation was reviewed by Brewen and Stetka (1982) and dealt with specifically in the proceedings of a meeting on DNA Repair, Chromosome Alterations and

[1] Workgroup participants: Rosario Albanese (absent, but data considered), Jeffrey A. Allen, Ivan Chouroulinkov (absent, but data considered), Colette Coton (absent, but data considered), Barbara L. Harper (absent, but data considered), Edith Hüttner (absent, but data considered), Eva Kristiansen, Alain Léonard, Carol Luke, Alexei M. Malashenko (absent, but data considered), Alfred McFee, Margaret Richold, Jaques Pot-Deprun (absent, but data considered), Jörg Schöneich, and Viacheslav S. Zhurkov

Chromatin Structure held in The Netherlands in 1981 (Natarajan *et al.*, 1982).

A survey of chemicals tested in the mammalian bone marrow chromosome assay, which included publications up to 1979, is given in a report of the U.S. EPA's Gene-Tox Program (Preston *et al.*, 1981). That survey contains none of the chemicals studied in the present trial. Chromosomal aberration induction in mouse bone marrow by BP has been studied extensively by Kliesch *et al.* (1982). Therefore, BP was regarded as the positive control in the present experiments.

Before the start of the study it was suggested that only acute treatments should be used, followed by multiple sampling of bone marrow cells. A range of doses was proposed based on the molarity of the test chemicals and/or LD 50 data (Salamone, 1981). Reference was made to a specific classification of aberrations (Savage, 1976), and reporting sheets were supplied to ensure a common

Table 1. *Experimental animals*

Investigator	Species	Strain	Age (w)	Weight (g)	Sex	Tissue
Zhurkov	mouse	C57BL/6	8	17–20	males + females	bone marrow
Luke	mouse	B6C3F1	10–16	25–30	males	bone marrow
		1C3F1	15	23–26	males	bone marrow
Coton	mouse	BALBc	8	20	males + females	bone marrow
Adler	mouse	1C3F1	12–14	25–28	males	bone marrow
Harper	mouse	Ha/ICR	15	25	males	bone marrow
Malashenko	mouse	CBA/LacY C57BL/6JY TPS/Y 101/HY WR/Y	7–8	20	males	bone marrow
Hüttner	mouse	NMRI	6–8	24–32	males	bone marrow
Hüttner	mouse	NMRI	7	30	males	ascites tumours
Albanese	rat	Wistar (ALPK)	5–6	100–120	females	bone marrow
Pot-Deprun	Chin h[a]		8–12	30–32.5	males	bone marrow
Allen	Chin h[a]		8–9	20	males + females	bone marrow

[a]Chin h = Chinese hamster

Table 2. *Treatment details*

Investigator	BP Dose (mg/kg)	Vehicle	Route[a]	PYR Dose (mg/kg)	Vehicle	Route[a]	2AAF Dose (mg/kg)	Vehicle	Route[a]	4AAF Dose (mg/kg)	Vehicle	Route[a]
Zhurkov	130 52 26	sun flower seed oil	po	340 136 68	sun flower seed oil	po	331 132 66	DMSO	ip	446 178 90	DMSO	ip
Luke	250 125 62.5	corn oil	ip	2000 1000 500	corn oil	ip	1000 500 250	corn oil	ip	2000 1000 500	corn oil	ip
							446	DMSO	ip			
Coton	252 126 63	olive oil	po	900 450 225	olive oil	po	446 223 111	olive oil	po	892 446 223	olive oil	po
Adler	126 63 31.5	corn oil	po	808 404 202	corn oil	po	446 223 111.5	DMSO	ip	892 446 223	DMSO	ip
							446	corn oil	po			
Harper	130 52 26	olive oil	po	340 136 68	olive oil	po	1110 446 223	olive oil	po	446 182 73 36	olive oil	po
Malashenko	252	olive oil	po	808	olive oil	po	446	olive oil	po	892	olive oil	po
Hüttner	252	corn oil	po	nt[b]			446 223 111	corn oil	po	892 446 223	corn oil	po
Albanese	200 100	corn oil	po	500 250	corn oil	po	300 150	corn oil	ip	500 375	corn oil	ip
Pot-Deprun	100 50 25	olive oil	po	100 50 25	olive oil	po	223 111.5 55.8	olive oil	po	892 446 223	olive oil	po
Allen	1008 504	corn oil	po	3232 808	corn oil	po	1784 892	corn oil	po	3568 1784	corn oil	po

[a]ip = intraperitoneal injection; po = per os (gavage).
[b]nt = not tested.

format of data presentation for computerisation of the data.

Many variables were left to the investigators discretion since all participants had previous experience with the assay system and had established standardised laboratory protocols. Statistical treatment of the data varied from laboratory to laboratory. At present it is difficult to identify a statistical procedure that is optimal for use with chromosome aberration data. The raw data of all investigators were computerised at the National Institute of Environmental Health Sciences; at the workgroup meeting, access to these data and statistical treatment was not available. Therefore, the workgroup decided to accept the individual investigator's final conclusions. Four categories of judgement of the investigators were distinguished: negative (−), positive (+), weak positive (+w) and equivocal (±), (i.e., further testing is required).

3 Protocols

Details of the experimental animals used in the study are given in Table 1. Seven laboratories used mice, one used rats and two used Chinese hamsters. Only 2 investigators used the same mouse strain (C57BL/6). Of the 10 different types of mice, 6 were inbred, 2 were outbred and 2 were F_1 hybrids. Ages varied for mice from 6–16 weeks and weights varied from 17–30 g. In 7 studies the experimental animals were males only; the rat study used females only, and 3 investigators used both sexes in their experiments. While bone marrow was the target tissue in all laboratories, one group provided additional data on ascites tumour cells with 2AAF and 4AAF.

Details of doses, vehicles and route of applications are presented in Table 2. Six laboratories used 3 doses for all 4 chemicals. One laboratory used only one dose of each chemical but compared 5 different strains of mice. Two

Table 3. *Sampling details*

Investigator	Highest dose/Lower dose sampling intervals	No. of animals	Cells per animal	Total cells/ dose-time	Colchicine dose/duration	Statistics
Zhurkov	18, 24 30 h/18 h	5♂ + 5♀	50	500	2.5 mg/kg, 1.5 h	Student's test
Luke	17 h/17 h, with BrdUrd	8♂	50	400	2 mg/kg, 2 h	Kruskall-Wallis
	24 h, without BrdUrd	5♂	100	500		chi-square
Coton	18, 24, 30 h/24 h	5♂ + 5♀	50	500	0.1 mg/animal, 1 h	chi-square
Adler	BP 30 h/30 h PYR 6, 24,48 h/24 h AAF 18, 24, 30, 48 h/24 h	5♂	100	500	2 mg/kg, 1 h	chi-square
Harper	24, 48 h/24 h	5♂	50	250	1 mg/kg, 1.5–2.0 h	chi-square
Malashenko	BP, PYR 30 h 2AAF, 4AAF 24 h	5♂	50	250	2 mg/kg, 1 h	Fisher's exact test
Hüttner	12, 24, 30, 42 h all	5♂	50	250	0.13 mg/animal, 2.5 h	chi-square
Albanese	24 h/24 h	10♀	50	283–500	5 mg/kg, 1 h (vinblastin)	Fisher's exact test
Pot-Deprun	30 h/30 h	4♂	125	500	0.14 mg/animal, 1 h	Mann-Whitney-Wilcoxon
Allen	12, 18, 24 h all	3♂ + 3♀	50	300	4 mg/kg, 2 h	Kruskall-Wallis

Table 4. *Author's statements*[a]

Investigators	Species	BP	PYR	2AAF	4AAF
Zhurkov	mouse	+w	−	−	−
Luke	mouse	+	−	+w	−
Coton	mouse	+w	−	−	−
Adler	mouse	+	−	+	±
Harper	mouse	+	−	−	−
Malashenko	mouse	+	−	+w	−
Hüttner	mouse	+	−	−	−
Albanese	rat	+	−	±	−
Pot-Deprun	Chin h[b]	+	−	+	−
Allen	Chin h[b]	+w	−	+w	−

[a] + = positive; +w = weakly positive; ± = equivocal; − = negative.
[b] Chin h = Chinese hamster.

laboratories used 2 doses of all chemicals and one laboratory tested 2AAF and 4AAF at 3 dose levels and BP at one dose. Vehicles for application were sunflower seed oil, corn oil, olive oil, or dimethylsulphoxide (DMSO). Routes of application were mostly gavage (po) and in some instances, intraperitoneal (ip) injection.

Details of sampling intervals, sample sizes, mitotic arrest, and statistical treatment of the data are given in Table 3. Six laboratories used multiple sampling intervals at least for the highest dose. Four laboratories sampled bone marrow only at one interval. One of these investigators determined first posttreatment mitoses by BrdUrd incorporation and scored SCEs in the same experiment (Luke).

Two investigators used the same animals for the micronucleus test and the chromosome analysis by sampling bone marrow from one femur for each assay (Albanese, Harper).

The minimum number of animals per dose-time group was 4, the maximum 10. A range of 50 and 125 cells was scored per animal. For mitotic arrest, colchicine was injected 1–2.5 h prior to sacrifice of the animals except in one instance where vinblastine was used. All investigators coded their slides. Statistical treatment of the data varied from chi-square and Fisher's exact test to tests that used individual animals as the unit of treatment (i.e., Kruskall-Wallis or Mann-Whitney-Wilcoxon). Analysis of variance was applied to test correlations, such as sex differences or strain differences.

4 Clastogenic activity of each chemical

Table 4 gives the author's conclusions on the clastogenicity of the 4 chemicals. BP was uniformly called positive, though weakly so, in 3 instances. PYR was uniformly called negative. Disagreement existed among the authors on 2AAF, 2 called it positive, 3 called it weakly positive, 4 called it negative, and one called the data inconclusive. On 4AAF all authors came to the conclusion that it was not clastogenic.

BP

The general tendency in the overall positive response to BP treatment was that a dose-related increase in aberration frequencies was observed wherever more than one dose had been tested. Table 5 compares the minimum positive dose with the maximum negative dose, the increases in aberrant cells (excluding gaps) over those of the concurrent control levels and the conclusions of the authors. The multiples of control levels, whenever a weak positive response was stated, ranged from 3.2 to 8.5. Positive calls ranged from 2.3–18.0 times control level. Compared with classical clastogens such as cyclophosphamide, which was used as positive control in several laboratories, it is certain that BP has to be called a weak clastogen.

PYR

The uniformity of results is striking in the case of PYR. Table 6 compares the increase in aberrant cells at the highest dose tested over those of the concurrent control levels and the conclusions of the authors. The range of multiples over control is 1–5.0. The range of the highest doses tested is from 2.5–16 times the molecular weight in mg/kg body weight. Looking at the individual sets of data, there was a tendency toward an elevation of the frequencies of aberrant cells at the highest dose in 2 studies, however, it was not significant at the 0.05 level (Luke, Adler). The work group came to the conclusion that PYR did not show clastogenic properties in the present study.

2AAF

The response to 2AAF was quite variable. Table 7 compares doses of positive and negative responses and increases of aberrant cell frequencies above control level. Only 2 authors called 2AAF clearly positive with 3.4 and 22 times the control level at the lowest dose positive. One of the 2 studies (Pot-Deprun) found a linear dose response. In the other experiment (Adler), the aberration rates were elevated significantly at the lowest dose tested

Table 5. *BP: Range of doses tested, increases in multiples of control levels and results (molecular weight 252)*

Investigator	Species	Optimum sampling time	Minimum dose pos. (mg/kg)	Maximum dose neg. (mg/kg)	Increase above control[a]	Result[b] (+, +w ±, −)
Zhurkov	mouse	18	130	52	M[c]3.8	+w
			na[g]	130	F[c]1.3	−
Luke	mouse	17	250	125	11.0	+
Coton	mouse	24	250	125	8.5	+w
Adler	mouse	na	63	31.5	7.7	+
Harper	mouse	24	130	52	3.2	+w
Malashenko	mouse	na	252[d]	na	2.4–16.0[e]	+
Hüttner	mouse, bm[f]	24	252[d]	na	18.0	+
Albanese	rat	na	100	na	4.0	+
Pot-Deprun	Chin h[h]	na	25	na	2.3	+
Allen	Chin h	18	1008	504	4.0	+w

[a]Multiple over control level at the minimum dose positive.
[b]Author's conclusion, not based on the factor of increase calculated by the workgroup.
[c]F = females; M = males.
[d]One dose tested.
[e]Range given for five strains.
[f]bm = bone marrow cells.
[g]na = not applicable.
[h]Chin h = Chinese hamsters.

and did not increase further to the highest dose. In this study, the oral route of an oil suspension was compared with the ip route of a DMSO solution for the highest dose of 2AAF. While the ip results were positive, the oral results were negative. Three authors called the response to 2AAF weakly positive with multiples over control values of 2.6 to 10.0. One of these studies showed a positive response with 2 out of 5 strains of mice (Malashenko). Analysis of variance allowed the authors to conclude that the 2AAF result overall was weakly positive but did not allow the conclusion that there was a strain difference in response. In the other study only the lower of 2 doses tested produced a significant increase in aberrant cells at 24 h (Allen). The third study found a significant increase only with the lowest dose (Luke). The authors repeated the 2AAF experiment with different mice, a different solvent, and a different dose. They again obtained a weak positive response.

One author called the results inconclusive due to lack of a dose response even though in one case the multiple over control level was 9.0 and the increase in aberrant cells was significant at both doses tested (Al-

Table 6. *PYR: Range of doses tested, increases in multiples of control levels and results (molecular weight 202)*

Investigator	Species	Optimum sampling time	Minimum dose pos. (mg/kg)	Maximum dose neg. (mg/kg)	Increase above control[a]	Result[b] (+, +w ±, −)
Zhurkov	mouse	na[c]	na	340	M[d]2.3	−
					F[d]1.3	−
Luke	mouse	na	na	2000	5.0	−
Coton	mouse	na	na	900	2.1	−
Adler	mouse	na	na	808	1.8	−
Harper	mouse	na	na	340	2.2	−
Malashenko	mouse	na	na	808[e]	1.0–3.0[f]	−
Albanese	rat	na	na	500	2.0	−
Pot-Deprun	Chin h[g]	na	na	100	1.0	−
Allen	Chin h[g]	na	na	3232	1.7	−

[a]Multiple of control level at the maximum dose negative.
[b]Author's conclusion, not based on the factor of increase calculated by the workgroup.
[c]na = not applicable.
[d]F = females; M = males.
[e]One dose tested.
[f]Range given for five strains.
[g]Chin h = Chinese hamsters.

Table 7. *2AAF: Range of doses tested, increases in multiples of control levels and results (molecular weight 223)*

Investigator	Species	Optimum sampling time	Minimum dose pos. (mg/kg)	Maximum dose neg. (mg/kg)	Increase above control[a]	Result[b] (+, +w ±, −)
Zhurkov	mouse	30	na[c]	331	M[d]4.0	−
					F[d]1.5	−
Luke	mouse, I	17	250	1000	7.0	+w
	mouse, II	24	446	na	2.6	+w
Coton	mouse	24	na	446	2.5	−
Adler	mouse	24	111.5	na	22.0	+
Harper	mouse	na	na	1110	2.4	−
Malashenko	mouse					
	CBA	na	446	na	7.0	+ } +w
	WR	na	446	na	10.0	+ } +w
	B6	na	na	446	1.6	− } +w
	TPS	na	na	446	2.5	− } +w
	101	na	na	446	2.7	− } +w
Hüttner	mouse, bm[e]	na	na	446	2.0	−
	mouse, at[f]	na	na	446	0.0	−
Albanese	rat	na	150	na	9.0	±
Pot-Deprun	Chin h[g]	na	55.8	na	3.4	+
Allen	Chin h	24	892	na	5.0	+w

[a]Multiple of control level at the minimum dose positive or the maximum dose negative.
[b]Author's conclusion, not based on the factor of increase calculated by the workgroup.
[c]na = not applicable.
[d]F = females, M = males.
[e]bm = bone marrow cells.
[f]at = ascites tumour cells.
[g]Chin h = Chinese hamster.

banese). The author suggests that the result may be "due to systemic toxicity as a result of exposure to peri-lethal doses".

The remaining 4 investigators found no increase in aberration frequencies after 2AAF treatment, with multiples over control levels ranging from 0–4.

The workgroup concluded that 2AAF has weak clastogenic properties, and the response may depend on species, strain, solvent, and route of exposure.

4AAF

The responses in the 4AAF experiments were negative. The multiples over control values (Table 8) ranged from 1.3 to 8.0. In one study the highest 4AAF dose was significantly above the control (Adler). The authors regarded the result as equivocal because of the exceptionally low control value. The highest doses tested represented 2–16 times the molecular weight in mg/kg body weight. The workgroup came to the conclusion that 4AAF did not show clastogenic properties in the present study.

5 Discussion

The workgroup discussed a variety of comparisons and correlations using the data available. These included dose levels and response, strain differences, species differences, control levels, routes of administration, sample sizes, expression times, and dose responses.

BP was positive at the one molar dose level in 4 of 5 experiments and negative at 1/8th of the molar dose in 3 of 4 experiments. In contrast, 2AAF was only positive in 2 of 6 experiments at the one molar dose. Three experiments were negative at the same dose level, and one experiment gave inconclusive results. While BP showed a dose response in all but one experiment, 2AAF only gave a linear dose response in one experiment (Pot-Deprun) and significantly elevated aberration levels that were very similar at all three doses tested in another study (Adler). Both investigators called 2AAF positive. A lack of a dose response was obvious in 2 studies which called 2AAF weakly positive (Luke, Allen).

Strain differences could only be assessed in one study (Malashenko). No strain difference was found for BP. With 2 strains giving a positive response and 3 strains giving a negative response, the analysis of variance for 2AAF did not detect strain differences, and the authors suggest that the sample sizes may have been too small.

Species differences were difficult to assess since most laboratories used the mouse. Rats and Chinese hamsters were used in only one and 2 laboratories, respectively. However, a limited comparison was attempted. Consistency in response was obtained for BP, PYR, and 4AAF while inconsistencies were obvious for 2AAF:

Mouse	5 pos.	4 neg.	1 inconclusive
Rat			1 inconclusive
Chinese hamster	2 pos.		

Both experiments with Chinese hamsters were positive and this may suggest that the karyotype of the Chinese hamster of $2n = 22$ with metacentric as well as acrocentric chromosomes is better suited for analysis of chromosomal aberrations than the karyotype of the mouse of $2n = 40$ with only acrocentric chromosomes. However, it is also possible that metabolic differences exist between the mouse and Chinese hamster.

Control levels were compared and found to range from 0 to 2.4% cells with aberrations (excluding cells with gaps only). No correlation of the control levels was found with strain, age, or route of administration. In one study (Zhurkov) the control values for males were uniformly lower than that for females, but analysis of variance did not reveal a significant difference. A correlation, however, seemed to exist with solvents used, in that the oil

Table 8. *4AAF: Range of doses tested, increases in multiples of control levels and results (molecular weight 223)*

Investigator	Species	Optimum sampling time	Minimum dose pos. (mg/kg)	Maximum dose neg. (mg/kg)	Increase above control[a]	Result[b] (+, +w ±, −)
Zhurkov	mouse	na[c]	na	446	M[d]8.0 F[d]2.5	−
Luke	mouse	na	na	2000	0.35	−
Coton	mouse	na	na	892	1.7	−
Adler	mouse	na	892	446	10.0	±
Harper	mouse	na	na	446	1.6	−
Malashenko	mouse	na	na	892[e]	1.0–4.0[f]	−
Hüttner	mouse, bm[g]	na	na	892	1.6	−
	mouse, at[h]	na	na	892	0.6	−
Albanese	rat	na	na	500	2.5	−
Pot-Deprun	Chin h[i]	na	na	446	1.9	−
Allen	Chin h	na	na	3568	1.3	−

[a] Multiple of control level at the maximum dose positive.
[b] Author's conclusion, not based on the factor of increase calculated by the workgroup.
[c] na = not applicable.
[d] F = females; M = males.
[e] One dose tested.
[f] Range given for five strains.
[g] bm = bone marrow cells.
[h] at = ascites tumour cells.
[i] Chin h = Chinese hamster.

controls tended to be higher than the DMSO controls in 2 laboratories (Adler, Zhurkov).

	Zhurvov	Adler
Oil	♂ 1.2%; ♀ 2.4%;	♂ 0.8% cells with aberrations
DMSO	♂ 0.4%; ♀ 0.8%;	♂ 0.07% cells with aberrations

Another correlation was found through a calibration excercise. Slide samples had been sent to the coordinator by all investigators using mice. The slides were rescored by Brian Dean (BD), UK, to see whether low reading or high reading consistencies could be determined.

No. of cells	Cells with gaps		Cells with breaks		
174	2	1.1%	1	0.06%	BD
50	2	4%	3	6%	Author
200	2	1%	0	0	BD
50	1	2%	2	4%	Author
200	3	1.5%	1	0.5%	BD
50	0	0	3	6%	Author

This limited comparison suggests major differences in cell selection and/or scoring criteria. The work group discussed the possibility that the difference may be due specifically to different criteria applied to the discrimination of gaps and breaks. Some authors used very stringent discrimination criteria in that they required a dislocation or misalignment of the fragment for breaks. These criteria were adopted by Brian Dean in his calibration exercise. Other authors required only an unstained region longer than the width of a chromatid to characterise a break. The tendency emerged that whenever the less stringent criteria were applied, the control levels were higher. The workgroup came to the conclusion that the discrimination criteria were not critical for the overall outcome of the experiment as long as they had been applied consistently to controls and treated groups. This was ensured by the fact that all investigators had coded their slides before scoring and only decoded the results at the end of the experiment.

Only 2 routes of application were used in the present experiments, and it was not possible to determine any correlation between routes and responses. One laboratory gave BP by ip injection of a corn oil solution; all other laboratories used gavage for BP treatments with no obvious difference in response. Similarly, PYR was negative, regardless of the route of administration. In the case of 2AAF, where the responses differed, it did not seem to be correlated to the route. One ip experiment was positive (Adler), one was weakly positive (Luke), one was negative (Zhurkov), and one was inconclusive (Albanese). Of the oral experiments one was positive (Pot-Deprun), 2 were weakly positive (Malashenko, Allen), and three were negative (Coton, Hüttner, Harper). 4AAF was negative regardless of the route of application. Likewise, there did not seem to be any correlation between the vehicle and the results obtained.

Sample sizes seem to influence the results. Without computer aided analysis of the present data, the correlation between sample size and sensitivity of the test could not be determined. The work group discussed the need to optimise the test protocol both in regard to the minimum number of animals required and the minimum number of cells to be scored per animal, factors that are necessarily interrelated. To determine these criteria, a detection level has to be fixed. In the present set of data with sample sizes between 4 and 10 animals and 250 and 500 cells per group, a 7-fold increase above control level seemed to emerge as the minimum detection level (Tables 5–8). Weak positive calls were based on increases between 2.5 and 4.0 times the control level.

Expression times and dose-response curves were also discussed in the workgroup. The general consensus was that multiple samples are necessary to determine the optimum expression time. A single sampling interval may be justified if cell cycle kinetics have been determined under the influence of the test chemical by BrdUrd incorporation prior to the onset of the actual study. However, the water insolubility of the chemicals used in the present experiment, particularly of 2AAF, may have interfered with the distinction of a single optimum expression time and a clear dose response. Resorption of the test chemicals by the animal was a slow process, not necessarily related directly to the applied dose but to the physical status of the experimental animals. In some of the ip experiments traces of 2AAF or 4AAF were found in the peritoneum at the time of sacrifice, as late as 48 h after treatment (Adler). Thus, the lack of time response and/or dose response could be partly explained by the degree of water insolubility for the chemicals being tested.

6 Conclusions and recommendations

The workgroup was of the unanimous opinion that the analysis of metaphase cells for chromosomal aberrations is a valuable test to determine the clastogenic potential of a chemical. It is not the most sensitive test with the presently used sample sizes. The negative results with PYR and 4AAF were attributed to the inactivity of the chemicals, while the differing results with 2AAF were thought to be due to the limited power of detection (7 times control level) with the given sample sizes. The negative results with 2AAF also may be due to strain and/or species differences in metabolism, vehicle, and route of administration.

7. References

Adler, I.-D. (1984) Cytogenetic tests in mammals. In S. Venitt and J.M. Parry (Eds) *Mutagenicity Testing: A Practical Approach*, IRL Press, Oxford, Washington, DC, pp. 275–306.

Brewen, J.G. and D.G. Stetka (1982) Cytogenetic events *in vivo*. In J.A. Heddle (Eds) *Mutagenicity: New Horizons in Genetic Toxicology*, Academic Press, N.Y., London, pp. 351–384.

Dean, B.J. (1969) Chemical-induced chromosome damage, *Lab. Animal*, **3**, 157–174.

Natarajan, A.T., G. Obe and H. Altmann (Eds) (1982)

DNA Repair, Chromosome Alterations and Chromatin Structure, *Progress in Mutation Research*, Vol. 4, Elsevier, NY.
Kliesch, U., I. Roupova and I.-D. Adler (1982) Induction of chromosome damage in mouse bone marrow by benzo-a-pyrene, *Mutation Res.*, **102**, 265–273.
Preston, R.J., W. Au, M.A. Bender, J.G. Brewen, A.V. Carrano, J.A. Heddle, A.F. McFee, S. Wolff and J.S. Wassom (1981) Mammalian *in vivo* and *in vitro* cytogenetic assays. A report of the US EPA's Gene-Tox program, *Mutation Res.*, **87**, 143–188.
Salamone, M.F. (1981) Toxicity of 41 carcinogens and non-carcinogenic analogs. In F.J. de Serres and J. Ashby (Eds), *Evaluation of Short-Term Tests for Carcinogens, Progress in Mutation Research*, Vol. 1, Elsevier, NY, pp. 682–685.
Savage, J.R.K. (1976) Classification and relationships of induced chromosomal structural changes, *J.Med. Genet.*, **13**, 103–122.

Ilse-Dore Adler and Ingrid Ingwersen
Institut für Säugetiergenetik, Gesellschaft für Strahlen- und Umweltforschung, Neuherberg, W. Germany

8

Evaluation of chromosomal aberrations in bone marrow of male IC3F1 mice

1 Summary

The 4 chemicals BP, PYR, 2AAF, and 4AAF supplied by ICI (Macclesfield, UK) were tested for induction of chromosomal aberrations in bone marrow of male IC3F1 mice. None of the chemicals was water-soluble. BP and PYR were dissolved in corn oil and applied by stomach intubation. The AAF compounds were dissolved in pure DMSO and injected intraperitoneally. The time dependent increase in aberration yields was studied with PYR, 2AAF, and 4AAF by sampling bone marrow at 3 or 4 intervals after treatment (6–48 h). BP was tested at the 30 h interval only. Each of the chemicals was tested at 3 dose levels.

The results show that of the 4 compounds BP and 2AAF have clastogenic potential while PYR gave a negative result and the data with 4AAF were inconclusive. The fact that the chemicals were water-insoluble influenced the results. No clear-cut time maximum in the clastogenic response could be recognized. Traces of 2AAF and 4AAF were found in the peritoneum at the time of bone marrow sampling even as late as 48 h after treatment, indicating a slow rate of absorption. Similarly, the dose-response curves were hampered by the insolubility of the compounds and the resulting difficulties of dosing and resorption.

2 Introduction

The principle of the test is to analyse cells at first posttreatment mitosis. Due to its proliferation activity bone marrow is generally the tissue of choice to evaluate mitotic chromosomes for the presence of aberrations such as gaps, breaks, and reunion configurations. The test procedure and application has been described by Dean (1969) and recently in more detail by Adler (1984). Chromosomal aberration induction in mouse bone marrow by BP has extensively been studied by Kliesch *et al.* (1982). Therefore, BP was regarded as the positive control in the present experiments.

3 Materials and methods

Animals

Male IC3F1 (101/ElxC3H/El) mice were used and randomly distributed to treated and control groups. The animals were 10 to 14-weeks old. Their weight ranged from 25 to 28 g. The animals were housed under conventional conditions with a 12 h light-dark cycle in air-conditioned quaters.

Chemicals, solvents and doses

All 4 test chemicals were supplied by ICI, Macclesfield (UK). BP and PYR were dissolved in corn oil and orally applied by stomach intubation in volumes of 0.1 or 0.2 ml/10 g body weight. 2AAF and 4AAF were dissolved in pure DMSO and intraperitoneally (ip) injected in a volume of 0.12 ml/mouse.

Doses of BP corresponded to 1/2, 1/4, and 1/8 of the molecular weight (i.e., 126, 63 and 32 mg/kg). Doses of PYR corresponded to 4, 2, and 1 times the molecular weight (i.e., 808, 404, and 202 mg/kg). Doses of 2AAF corresponded to 2, 1 and 1/2 of the molecular weight (i.e., 446, 223, and 111.5 mg/kg). Doses of 4AAF corresponded to 4, 2 and 1 times the molecular weight (i.e., 892, 446, and 223 mg/kg).

Sampling intervals and sample size

Bone marrow of BP treated animals was sampled after 30 h only, and the whole experiment was repeated. Sampling intervals for bone marrow of animals treated with the highest dose of PYR were 6, 24, and 48 h. For the two lower doses bone marrow was sampled at the 24 h interval. Bone marrow of animals treated with the highest dose of 2AAF and 4AAF, respectively, was sampled after 18, 24, 30, and 48 h. Animals treated with the 2 lower doses were sampled at the 24 h interval. The experiment with the highest dose of 2AAF was repeated as an ip injection dissolved in DMSO and by oral intubation suspended in corn oil.

Each dose-time group consisted of 5 treated and one or 2 solvent control males. One hundred cells at mitotic metaphase were scored microscopically per male (i.e., 50 cells per slide) independently by 2 investigators. Mitotic indices were determined by counting the number of mitoses among 500 cells in a given field on each slide (i.e., a total of 1000 cells per male).

Bone marrow preparation

The animals were dosed with 2 mg/kg of colchicine exactly 1 h prior to sacrifice. Twelve animals were sampled at one time. The bone marrow from both femurs of each male was flushed into a test tube containing 5 ml HBSS. After centrifugation the cells were hypotonically treated with 0.56% KCl solution for 15 min at room temperature. Centrifugation was repeated before and after each of 3 steps of fixation with 3 parts methanol + 1 part glacial acetic acid. Air dried slides with 3 drops of the cell suspension were stained with 2% acetic orcein solution, cleared in ethanol and xylene and mounted with Eukitt. All slides were coded before scoring.

Microscopic analysis

The slides were scored under 1250 X magnification and phase contrast. All aberrant cells were photographed. The aberrations found were of the chromatid or isochromatid type (i.e., gaps, breaks, fragments, intrachanges and interchanges of various kinds). Numerical aberrations were omitted from scoring for their susceptibility to preparational artifacts, which means that only cells with 40 centromeres were included in the evaluation. Furthermore, scoring was restricted to cells with proper chromatid alignment, and cells with apparent colchicine effect (split centromeres, split and contracted chromatids) were excluded, as well as cells with a surplus of chromosomal overlap.

The slides were decoded when the reading of an entire experiment was completed.

Statistics

The main parameter for statistical evaluation was the frequency of cells with aberrations excluding cells with gaps only. To test for inhomogeneity among animals within or between groups, the binomial dispersion test (Snedecor and Cochran, 1967) was applied. For calculation of statistically significant differences between treated and control groups, the chi-square test (with Yates' correction, if appropriate) (Sachs, 1974) was used.

4 Results

BP

The results of both experiments with BP are shown in Table 1 and the upper part of Figure 1. The frequencies of gaps were not elevated but breaks and exchanges occurred more often in the treated than in the control sample. The combined data (aberrant cells) from the control animals of both experiments plus the combined data of the animals treated with BP showed inhomogeneity ($P < 0.05$), while each combined group (controls or BP-treated) within itself showed no inhomogeneity ($P > 0.05$).

Figure 1. Frequencies of cells with chromatid aberrations in bone marrow of male lC3Fl mice after oral treatment with BP (top) and PYR (bottom). Points represent mean values of 5 animals and bars represent the 95% confidence limits.

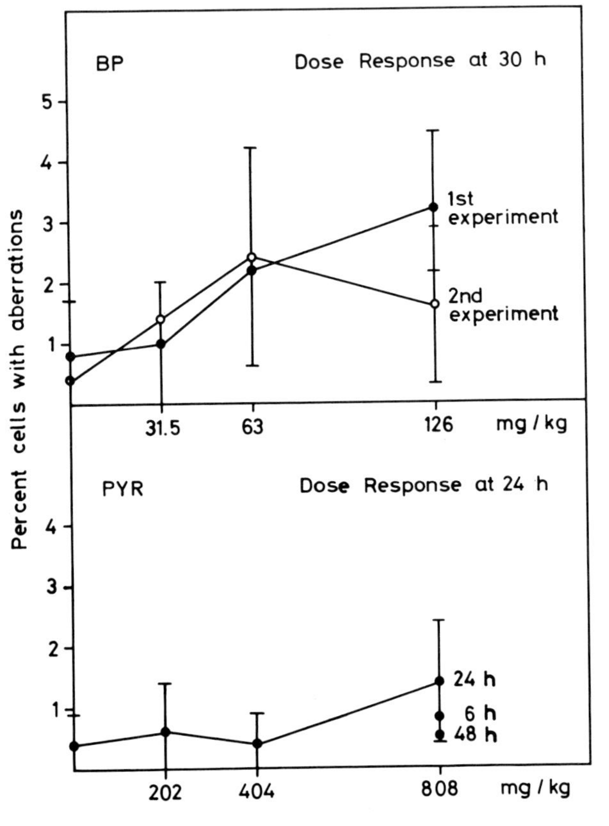

Comparing single treatment groups and concurrent controls, an increase in cells with aberrations was found to be significant at the 2 higher doses of the first experiment and the middle dose of the second experiment ($P < 0.05$). When both sets of data were combined the aberration frequencies at the 2 higher doses were significantly above the combined control ($P < 0.001$). BP was classified as clastogenic.

PYR

The results for PYR are shown in Table 1 and the lower part of Figure 1. The frequency of gaps was within control range throughout the experiment. Exchange configurations occurred sporadically; one in the control, 2 in the treated groups. The frequency of breaks was increased at the highest dose sampled 24 h after treatment. The binomial dispersion test revealed no inhomogeneity

Table 1. *Chromatid aberrations in bone marrow after oral treatment of male 1C3F1 mice with BP and PYR*

Treatment	Dose (mg/kg)	Interval (h)	No. of males/cells	No. of aberrations: Gaps	Breaks	Exchanges	Aberrant cells[a]: n	% ± SE	Mitotic index[b]
First experiment									
BP	126	30	5/500	26	14	2	16	3.2 ± 0.6[c]	2.8
	63	30	5/500	37	8	3	11	2.2 ± 0.8[c]	2.0
	31.5	30	5/500	20	5	1	5	1.0 ± 0.6	1.3
Control (corn oil)		30	5/500	25	4	0	4	0.8 ± 0.5	2.6
Second experiment									
BP	126	30	5/500	14	7	2	8	1.6 ± 0.7	nd[d]
	63	30	5/500	20	10	4	12	2.4 ± 0.9[c]	nd
	31.5	30	5/500	25	6	1	7	1.4 ± 0.5	nd
Control (corn oil)		30	5/500	18	1	1	2	0.4 ± 0.3	nd
PYR	808	6	5/500	8	3	1	4	0.8 ± 0.6	3.7
	808	24	5/500	14	6	1	7	1.4 ± 0.5	2.0
	808	48	4/400	15	2	0	2	0.5 ± 0.5	2.5
	404	24	5/500	11	2	0	2	0.4 ± 0.3	nd
	202	24	5/500	5	3	0	2	0.6 ± 0.4	nd
Control (corn oil)			8/800	23	2	1	3	0.4 ± 0.3	3.5

[a]Excluding cells with gaps only.
[b] Per cent cells at mitosis (based on 100 cells counted per animal).
[c] $P < 0.05$ (chi-square test with Yates' correction).
[d]nd = not determined.

Table 2. *Chromatid aberrations in bone marrow after ip treatment of male 1C3F1 mice with 2AAF and 4AAF*

Treatment	Dose (mg/kg)	Interval (h)	No. of males/cells	No. of aberrations: Gaps	Breaks	Exchanges	Aberrant cells[a]: n	% ± SE	Mitotic index[b]
2AAF	446	18	5/500	56	22	0	22	4.4 ± 0.4[a]***	1.4
	446 po[c]	24	5/500	10	1	1	2	0.4 ± 0.2	nd
	446	24	5/500	37	10	1	11	2.2 ± 1.4[a]*	2.4
	446 R[d]	24	5/500	30	17	1	16	3.2 ± 0.8[a]**	2.4
	446	30	5/500	33	23	2	20	4.0 ± 1.3[a]***	2.7
	446	48	5/500	21	8	2	9	1.8 ± 0.7[a]*	3.7
	223	24	5/500	23	12	1	13	2.6 ± 0.7[a]**	2.3
	111.5	24	5/500	22	10	1	11	2.2 ± 1.0[a]*	2.6
4AAF	892	18	5/500	17	2	0	2	0.4 ± 0.2	1.8
	892	24	5/500	11	4	1	5	1.0 ± 0.5[a]**	2.3
	892	30	5/500	10	3	0	3	0.6 ± 0.3	2.0
	892	48	5/500	13	3	1	4	0.8 ± 0.2[a]*	3.2
	446	24	5/500	12	8[e]	1	3	0.6 ± 0.3	2.3
	223	24	5/500	8	3	0	3	0.6 ± 0.4	2.6
Control (DMSO)			15/1500	28	1	0	1	0.07 ± 0.07	2.9

*$P < 0.05$; **$P < 0.01$; ***$P < 0.001$ (chi-square test with Yates' correction).
[a] Excluding cells with gaps only.
[b] Per cent cells at mitosis (based on 1000 cells counted per animal).
[c] po = 2AAF suspended in corn oil and orally intubated.
[d]R = repeat.
[e]One cell with 7 fragments.

among the data (aberrant cells) of the combined corn oil control animals plus the data of all animals treated with the highest dose of PYR ($P > 0.05$). Likewise, neither the control nor the PYR-treated group was inhomogeneous in itself ($P > 0.05$). Comparing the highest experimental value (808, 24 h; 1.4%) with the concurrent control, the increase was not significant ($P = 0.083$). PYR was classified as nonclastogenic.

2AAF

The results with 2AAF are shown in Table 2 and Figure 2. Four of 30 animals treated with the highest dose died within 48 h after treatment. The aberration rates were significantly ($P < 0.05$) elevated at all 3 sampling intervals after treatment with the highest dose. The number of aberrant cells was also significantly elevated at all 3 doses ($P < 0.05$) without showing a clear dose response. The orally treated group sampled after 24 h gave a negative response. 2AAF was classified as clastogenic if applied ip in DMSO.

4AAF

The results with 4AAF are shown in Table 2 and Figure 3. Three of 20 animals treated with the highest dose of 4AAF died within 48 h after treatment. Gaps were not increased in any of the treated groups when compared with the DMSO control ($P > 0.1$). Cells with aberrations were significantly increased at the highest dose 24 h after treatment ($P = 0.004$) and 48 h after treatment ($P = 0.02$) when compared with the low DMSO control. The samples at the different times of those animals treated with the highest dose of 4AAF showed no inhomogeneity and were pooled. Compared with the DMSO control, the mean of aberrant cells at the highest dose was significantly increased ($P = 0.01$).

The DMSO control value is extremely low (1/1500; 0.07%) and determined the significant increase in aberration rates rather than the actually observed number of events in the treated groups. The historical control value is ~0.5%. The corn oil control was also 0.5% (9/1800). The two control groups do not differ significantly ($P = 0.052$). It could be argued that the DMSO control is low due to the specific property of DMSO. It was reported that DMSO scavenges hydroxyl radicals and could thus have antimutagenic properties (Cohen and Cederbaum, 1979). However, in a recently conducted series of experiments with 80% DMSO as solvent the control value was 0.56% (14/2500). The present low control value is attributed to chance. Compared with the combined corn oil and DMSO control of 0.3% (10/3300), neither the 24 h nor the 48 h group was significantly different ($P = 0.06$ and $P = 0.19$, respectively). The 4AAF results are regarded as inconclusive because of the low control value.

5 Discussion

The 4 compounds of the present collaborative trial were tested in the cytogenetic bone marrow assay by analysing chromatid-type aberrations in mitotic metaphases. Sampling intervals were varied for the highest doses of 3 of the 4 chemicals. For BP, published information (Kliesch *et al.*, 1982) from our own laboratory suggested that the optimal time for aberration scoring was 30 h after treatment, which was used for all 3 doses of the present experiments. Since the aberration yields did not reach the level previously reported, the experiment was repeated. In the repeated test the rates of aberrant cells showed a slight though not significant decline against the first experiment ($P = 0.1$). The dose-response curve of the

Figure 2. Frequencies of cells with chromatid aberrations in bone marrow of male IC3F1 mice after intraperitoneal treatment with 2AAF. Points represent mean values of 5 animals and bars represent the 95% confidence limits.

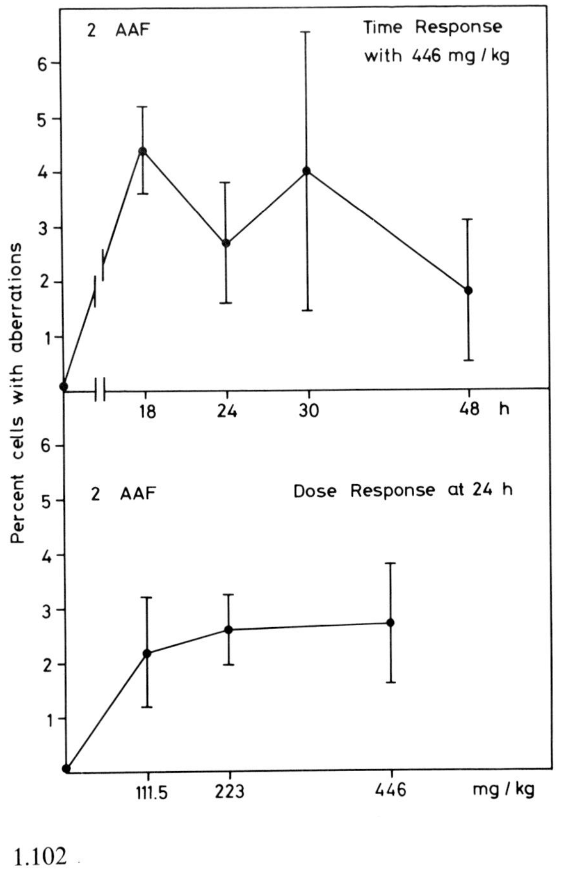

Figure 3. Frequencies of cells with chromatid aberrations in bone marrow of male IC3F1 mice after intraperitoneal treatment with 4AAF. Points represent mean values of 5 animals and bars represent the 95% confidence limits.

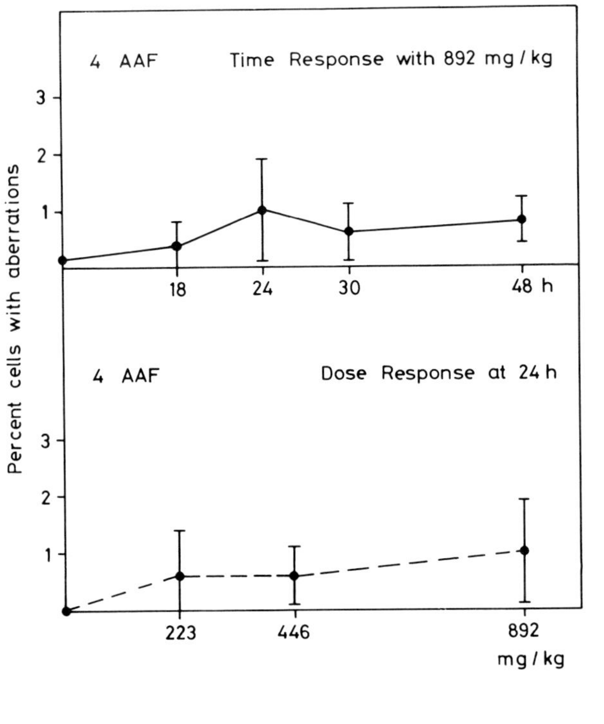

combined experiments showed a plateau between 63 and 126 mg/kg.

The results of these experiments demonstrate the difficulties encountered with water-insoluble test chemicals. Precipitated BP was found in the peritoneum when the compound was dissolved in corn oil and injected ip. It is conceivable that BP also precipitates in the stomach. Resorption may not necessarily correspond directly to the injected dose and may take place over an extended period of time, thus disturbing both the time-response and the dose-response curves. The same arguments apply to 2AAF, traces of which were found in the peritoneum 48 h after ip treatment and in the stomach 24 h after oral treatment. The frequencies of aberrant cells were increased at all 3 2AAF doses 24 h after treatment without showing a dose response. The oral experiment was negative, indicating an influence of the route of application and/or the solvent.

In an 8-day LD50 trial conducted according to Weil (1952), the value obtained for PYR was 900 mg/kg; for 2AAF 600 mg/kg; and for 4AAF 900 mg/kg with the lC3Fl mice used in present experiments. Thus, the highest doses applied for the chromosome study corresponded closely to the maximum tolerated doses for PYR and 4AAF. A slight reduction in mitotic indices after treatment corroborates the fact that the doses should not have been increased. PYR was classified as nonclastogenic. The experiment with 4AAF was regarded as inconclusive because of the low control value.

Summarising, it should be said that under the present experimental conditions the carcinogen BP gave a positive clastogenic response, while the noncarcinogen PYR was negative. The carcinogen 2AAF responded in inducing chromosomal aberrations, while the homologous chemical 4AAF gave inconclusive results.

6 Acknowledgement

The skilled technical assistance of Ruth Schmöller and Claudia Zeller is gratefully acknowledged.

7 References

Adler, I.-D. (1984) Cytogenetic tests in mammals. In S. Venitt and J.M. Parry (Eds) *Mutagenicity Testing: A Practical Approach*, IRL Press, Oxford, Washington, D.C., pp. 275–306.

Cohen, G. and A.I. Cederbaum (1979) Chemical evidence for production of hydroxyl radicals during microsomal electron transfer, *Science*, **204**, 66–68.

Dean, B.J. (1969) Chemical-induced chromosome damage. *Lab. Animal*, **3**, 157–174.

Kliesch, U., I. Roupova and I.-D. Adler (1982) Induction of chromosome damage in mouse bone marrow by benzo(a)pyrene, *Mutation Res.*, **102**, 265–273.

Sachs, L. (1974) *Angewandte Statistik*, Springer, Berlin, pp. 269–273.

Snedecor, G.W. and W.G. Cochran (1967) *Statistical Methods*, 6th ed., Iowa State Univ. Press, Ames, Iowa, pp. 240–242.

Weil, C. (1952) Tables for convenient calculation of median-effective dose (LD_{50} or ED_{50}) and instructions in their use, *Biometrics*, **8**, 249–263.

Viacheslav S. Zhurkov and
Irena F. Vyskubenko
A.N. Sysin Research Institute of General and Community Hygiene of the USSR AMS, Moscow, USSR

9

Analysis of chromosomal aberrations in bone marrow cells of C57BL/6 mice treated with benzo(a)pyrene, pyrene, 2-acetylaminofluorene and 4-acetylaminofluorene

1 Summary

The frequency of chromosomal aberrations in bone marrow cells of C57BL/6 male and female mice was studied 18, 24, and 30 h after a single exposure to BP and 2AAF and to their structurally related analogues PYR and 4AAF. BP and PYR were administered per os, 2AAF and 4AAF intraperitoneally at doses corresponding to 1/2, 1/5, and 1/10 of the LD50. A significant increase in the frequency of metaphases with chromosomal aberrations compared with the control was produced only with BP at the maximum dose of 130 mg/kg.

2 Introduction

Analysis of chromosomal aberrations in mammalian bone marrow cells is one of the common methods for assessing clastogenic effects of chemicals. Increased chromosomal aberrations due to chemical exposure are observed only in cases when a substance or its metabolite have clastogenic activity, and when the concentration of the clastogen or its metabolite can reach the target cells under study. Hence, *in vivo* tests with bone marrow cells are not always reliable for mutagenicity assessment.

In this study, the induction of chromosomal aberrations was evaluated in bone marrow cells of mice after a single exposure to the chemicals.

3 Materials and methods

The experiments were conducted using C57BL/6 Sto mice (males 18–20 g, females 17–19 g). In each experimental group there were 5 males and 5 females. BP, PYR, 2AAF, and 4AAF were supplied by ICI (UK). BP and PYR were dissolved in sterile sunflower seed oil. Their concentrations in stock solution were calculated to give the required dose of 0.2 mg/10 g weight. 2AAF and 4AAF were dissolved in pure DMSO. The concentrations in stock solution were calculated to give the required doses of 0.12 ml per mouse.

Each chemical was tested in 3 single doses approximating 1/2, 1/5, and 1/10 of the LD50 (BP at 130, 52, and 26 mg/kg; PYR at 340, 136 and 60 mg/kg; 2AAF at 331, 132; and 66 mg/kg; 4AAF at 446, 178 and 90 mg/kg). BP and PYR were administered per os (po), 2AAF and 4AAF intraperitoneally (ip). Control animals were treated with sunflower seed oil, 0.2 ml per 10 g body weight po, or with 0.12 ml DMSO per mouse ip.

Colchicine was injected ip 1.5 h prior to sacrifice.

Table 1. *Chromosomal aberrations in bone marrow of C57BL/6 mice after a single oral treatment with BP*[a]

Dose (mg/kg)	Interval (h)	Sex[b]	Per cent cell showing: Breaks	Chromatid exchanges	Chromosome exchanges	All aberrations	Frequence per cell of: Breaks	Chromatid exchanges	Chromosome exchanges	All aberrations
130	18	M	3.2	0.0	0.0	3.2	0.032	0.000	0.000	0.032
130	24	M	3.2	0.0	0.0	3.2	0.032	0.000	0.000	0.032
130	30	M	4.4	0.0	0.0	4.4	0.044	0.000	0.000	0.044
52	18	M	1.2	0.0	0.0	1.2	0.012	0.000	0.000	0.012
26	18	M	2.8	0.0	0.0	2.8	0.028	0.000	0.000	0.028
0[c]	24	M	1.2	0.0	0.0	1.2	0.012	0.000	0.000	0.012
130	18	F	8.0	0.0	0.0	8.0	0.084	0.000	0.000	0.084
130	24	F	3.2	0.0	0.0	3.2	0.040	0.000	0.000	0.040
130	30	F	2.4	0.0	0.0	2.4	0.028	0.000	0.000	0.028
52	18	F	1.2	0.0	0.0	1.2	0.012	0.000	0.000	0.012
26	18	F	2.0	0.4	0.0	2.4	0.020	0.004	0.000	0.024
0[c]	24	F	2.4	0.0	0.0	2.4	0.024	0.000	0.000	0.024

[a]250 cells from each of 5 mice were examined at every dose.
[b]M = males; F = females.
[c]Sunfower seed oil, 0.2 ml/10 g body weight of mouse.

Table 2. *Chromosomal aberrations in bone marrow of C57BL/6 mice after a single oral treatment with PYR*[a]

Dose (mg/kg)	Interval (h)	Sex[b]	Per cent cells showing: Breaks	Chromatid exchanges	Chromosome exchanges	All aberrations	Frequency per cell of: Breaks	Chromatid exchanges	Chromosome exchanges	All aberrations
340	18	M	2.8	0.0	0.0	0.8	0.028	0.000	0.000	0.028
340	24	M	2.4	0.0	0.0	2.4	0.024	0.000	0.000	0.024
340	30	M	1.2	0.0	0.0	1.2	0.012	0.000	0.000	0.012
136	18	M	2.0	0.0	0.0	2.0	0.020	0.000	0.000	0.020
68	18	M	0.4	0.0	0.0	0.4	0.004	0.000	0.000	0.004
0[c]	24	M	1.2	0.0	0.0	1.2	0.012	0.000	0.000	0.012
340	18	F	1.6	0.0	0.0	1.6	0.016	0.000	0.000	0.016
340	24	F	3.2	0.0	0.0	3.2	0.036	0.000	0.000	0.036
340	30	F	2.4	0.0	0.0	2.4	0.024	0.000	0.000	0.024
136	18	F	0.4	0.0	0.0	0.4	0.004	0.000	0.000	0.004
68	18	F	0.8	0.0	0.0	0.8	0.008	0.000	0.000	0.008
0[c]	24	F	2.4	0.0	0.0	2.4	0.024	0.000	0.000	0.024

[a]250 cells from each of 5 mice were examined at every dose.
[b]M = males; F = females.
[c]Sunflower seed oil, 0.2 ml/10 g body weight of mouse.

Table 3. *Chromosomal aberrations in bone marrow of C57BL/6 mice after a single intraperitoneal injection with 2AAF*[a]

Dose (mg/kg)	Interval (h)	Sex[b]	Per cent cells showing: Breaks	Chromatid exchanges	Chromosome exchanges	All aberrations	Frequency per cell of: Breaks	Chromatid exchanges	Chromosome exchanges	All aberrations
331	18	M	1.2	0.0	0.0	1.2	0.012	0.000	0.000	0.012
331	24	M	0.8	0.0	0.0	0.8	0.008	0.000	0.000	0.008
331	30	M	1.6	0.0	0.0	1.6	0.016	0.000	0.000	0.016
132	18	M	0.0	0.0	0.0	0.0	0.000	0.000	0.000	0.000
66	18	M	0.4	0.0	0.0	0.4	0.004	0.000	0.000	0.004
0[c]	24	M	0.4	0.0	0.0	0.4	0.004	0.000	0.000	0.004
331	18	F	1.2	0.0	0.0	1.2	0.012	0.000	0.000	0.012
331	24	F	0.4	0.0	0.0	0.4	0.004	0.000	0.000	0.004
331	30	F	0.0	0.0	0.0	0.0	0.000	0.000	0.000	0.000
132	18	F	1.2	0.0	0.0	1.2	0.012	0.000	0.000	0.012
66	18	F	0.4	0.0	0.0	0.4	0.004	0.000	0.000	0.004
0[c]	24	F	0.8	0.0	0.0	0.8	0.008	0.000	0.000	0.008

[a]250 cells from each of 5 mice were examined at every dose.
[b]M = males; F = females.
[c]DMSO – 0.12 ml per mouse intraperitoneally.

Preparations of bone marrow cells for metaphase analysis were made by the standard method. Slides were coded before microscopic analysis. Chromosomal aberrations were analysed for each high dosage group at 18, 24, and 30 h after treatment. For all other doses one sample (18 h) was used. 50 well-spread metaphases were analysed. Breaks, chromatid and chromosome exchanges, and cells with multiple aberrations and with pulverisation were scored. Gaps, Robertsonian translocations, and polyploidies were not evaluated.

Statistical analysis was made only for the index "per cent of cells with total aberrations". For the stabilisation of variances we used the following transformation:

$$z = 1/2\left(\arcsin\sqrt{\frac{m}{n+1}} + \arcsin\sqrt{\frac{m+1}{n+1}}\right),$$

where m = number of metaphases with aberreations in a mouse and n = total number of metaphases for the same mouse. Statistical significance of differences between treated and control groups was evaluated by Student's test. Analysis of variance was used for evaluating doses of chemical, time interval after treatment, and sex in relation to the frequency of cells with chromosome aberrations.

4 Results

The frequency of cells with chromosomal aberrations in the control group treated with sunflower seed oil averaged 1.2% in males and 2.4% in females. In the control animals given DMSO, this index was 0.4% in males and 0.8% in females. Analysis of variance did not reveal any significant effects of sex and solvent on the frequency of cells with chromosomal aberrations in the controls. Only breaks were observed.

Cytogenetic data on BP exposure are presented in Table 1. The frequency of chromosomal aberrations in female mice was 8.0% at 18 h after administration of the highest BP dose, significantly exceeding the control level. At other intervals after exposure, this index did not differ from the control. The frequencies of chromosomal aberrations in male mice was approximately 3 times higher than in the controls at any time following treatment with the maximum BP dose. This increase was not statistically significant. Analysis of variance did not indicate a significant effect of sex and time of sacrifice after the maximum BP dose on the frequency of metaphases with chromosomal aberrations in the bone marrow.

The dose-effect relationship for BP was analysed only at 18 after treatment. BP doses of 52 and 26 mg/kg did not increase the number of aberrant metaphases either in female or male mice. Analysis of variance showed a marked effect of BP dose ($P < 0.05$) and no effect of sex on the frequency of aberrant metaphases.

Breaks were the basic type of aberrations in all experiments with BP. The dose of 26 mg/kg was shown to produce 1 asymmetric chromatid exchange in one female. On the whole, 1 aberration per aberrant metaphase cell was observed.

The results of cytogenetic analysis with PYR, 2AAF, and 4AAF are presented in Tables 2,3, and 4, respectively. The frequencies of metaphases with chromosomal aberrations did not differ from the control after treatment with any of these 3 chemicals. Aberrations were manifested only by breaks.

5 Discussion

Of the 2 tested carcinogens whose mutagenicity was proved experimentally in *in vitro* studies, only BP at a dose close to the LD50 caused a significant increase in the frequency of metaphases with chromosomal aberrations in bone marrow cells of mice. 2AAF in the dose range tested did not cause changes in the frequency of chromosomal aberrations. It is quite likely that the clastogenic metabolites of 2AAF do not reach the bone marrow cells at effective concentrations. Their structurally related analogues, PYR and 4AAF, did not demonstrate cytogenetic activity. The experimental design described is based on a single exposure to the chemical and on a preparation of slides at fixed time intervals after treat-

Table 4. *Chromosomal aberrations in bone marrow of C57BL/6 mice after a single intraperitoneal injection with 4AAF*[a]

Dose (mg/kg)	Interval (h)	Sex[b]	Per cent cells showing: Breaks	Chromatid exchanges	Chromosome exchanges	All aberrations	Frequency per cell of: Breaks	Chromatid exchanges	Chromosome exchanges	All aberrations
446	18	M	2.0	0.0	0.0	2.0	0.020	0.000	0.000	0.020
446	24	M	3.2	0.0	0.0	3.2	0.032	0.000	0.000	0.032
446	30	M	0.4	0.0	0.0	0.4	0.004	0.000	0.000	0.004
178	18	M	1.2	0.0	0.0	1.2	0.012	0.000	0.000	0.012
90	18	M	1.2	0.0	0.0	1.2	0.012	0.000	0.000	0.012
0[c]	24	M	0.4	0.0	0.0	0.4	0.004	0.000	0.000	0.004
446	18	F	2.0	0.0	0.0	2.0	0.020	0.000	0.000	0.020
446	24	F	1.2	0.0	0.0	1.2	0.012	0.000	0.000	0.012
446	30	F	1.6	0.0	0.0	1.6	0.016	0.000	0.000	0.016
178	18	F	0.8	0.0	0.0	0.8	0.008	0.000	0.000	0.008
90	18	F	1.2	0.0	0.0	1.2	0.012	0.000	0.000	0.012
0[c]	24	F	0.8	0.0	0.0	0.8	0.008	0.000	0.000	0.008

[a] 250 bells from each of 5 mice were examined at every dose.
[b] M = males; F = females.
[c] DMSO – 1.12 ml per mouse intraperitoneally.

ment. For future studies, it may be expedient to use 5 daily exposures and to prepare slides 6 h after the final administration; this might enhance the sensitivity of the method, demonstrate minimum dose effect, and reduce the number of cells under study (Šrám *et al.*, 1981).

6 Reference

Šrám, R.J., V.S. Zhurkov, J. Navakova and I. Kodytkova (1981) Changes in the frequency of chromosomal aberrations in the bone marrow of cells of mice examined at various intervals single dose and continual exposures to cyclophosphamide, *Folia, Biol.* (Praha), **27**, 58–65.

Colette Coton, Alain Léonard, and
Gilbert Decat
Mammalian Genetics Laboratory, Department of Biology, C.E.N. S.C.K., Mol, Belgium

10

Induction of chromosomal aberrations in BALB/c mouse bone marrow cells

1 Summary

The analysis of metaphase bone marrow cells has been utilized to study the ability of BP, PYR, 2AAF, and 4AAF to induce structural chromosomal aberrations *in vivo* in mammalian somatic cells.

Treatment with BP increased weakly, but significantly the yield of structural aberrations, whereas the three other compounds yielded negative results.

2 Introduction

Two techniques, metaphase analysis and micronucleus induction, are currently used to assess the capability of chemicals to induce chromosomal aberrations *in vivo* in mammalian somatic cells.

In the present study, the clastogenic properties of BP and 2AAF, and their analogues, PYR and 4AAF, were evaluated by metaphase analysis of bone marrow cells.

3 Materials and methods

The preparations were made according to the method used routinely by our laboratory.

Test agents
The compounds were obtained from ICI (UK). They were suspended in olive oil (pharmacopoiea grade).

Animals
The experiments were carried out on 8-weak-old inbred BALB/c mice weighing 19–21 g. Each group consisted of 5 male and 5 female animals that received the substance per os through a stomach tube.

Treatment
The compounds were administered in a single treatment at the following doses:

BP – 63, 126 and 252 mg/kg;
PYR – 150, 225, 300, 450, 600 and 900 mg/kg;
2AAF – 111, 223 and 446 mg/kg;
4AAF – 223, 446 and 892 mg/kg.
Negative controls were given olive oil only.

Chromosomal analysis
One h before killing the mice were injected with 0.4 ml colchicine (0.025 mg/100 ml). The bone marrow preparations were stained with toluidine blue, and 50 metaphase cells were analysed for each animal. Three intervals after treatment (18, 24 and 30 h) were used for the highest dose

Table 1. *Results of the chromosome analysis on negative controls (olive oil), using BALB/c mice*

						Chromatid type exchanges				Chromosome type exchanges			
Sex	Preparation interval (h)	Animals analysed	Cells scored	Gaps	Breaks	I	S	A	C	I	S	A	R
M	18	4	200	—	—	—	—	—	—	—	—	—	—
	24	4	200	—	—	—	—	—	—	—	—	—	—
	30	5	250	2	—	—	—	—	—	—	—	—	—
F	18	5	250	1	—	—	—	—	—	—	—	—	—
	24	4	200	—	—	—	—	—	—	—	—	—	—
	30	5	243	—	—	—	—	—	—	—	—	—	—

Abbreviations: M = Male; F = Female; I = Intrachanges; S = Symmetrical interchanges; A = Asymmetrical interchanges; C = complex configurations; R = Robertsonian translocations.

Table 2. *Results of the chromosome analysis of BP in BALB/c mice*

						Chromatid type exchanges				Chromosome type exchanges			
Treatment (mg/kg)	Preparation interval (h)	Animals analysed	Cells scored	Gaps	Breaks	I	S	A	C	I	S	A	R
63	24	5	250	—	—	—	—	—	—	—	—	—	—
126	24	5	227	—	2	—	—	—	—	—	—	1	—
252	18	5	250	5	5	1	—	—	—	1	—	—	—
	24	5	250	2	1	1	—	—	—	—	—	—	—
	30	5	250	2	—	—	—	—	—	—	—	—	—
63	24	4	200	1	2	—	—	—	—	—	—	—	—
126	24	5	250	—	—	—	—	—	—	—	—	—	—
252	18	5	250	—	1	—	—	—	—	—	—	—	1
	24	5	250	—	5	—	—	—	—	—	—	—	—
	30	5	250	—	1	1	—	—	—	—	—	—	—

Abbreviations, see footnotes, Table 1.

Table 3. *Results of the chromosome analysis on BALB/c mice treated with PYR*

						Chromatid type exchanges				Chromosome type exchanges			
Treatment (mg/kg)	Preparation interval (h)	Animals analysed	Cells scored	Gaps	Breaks	I	S	A	C	I	S	A	R
225	24	5	250	—	5	—	—	—	—	—	—	—	—
450	24	5	250	—	—	—	—	—	—	—	—	—	—
900	18	5	250	—	—	—	—	—	—	—	—	—	—
	24	5	250	1	—	1	—	—	—	—	—	—	—
	30	5	250	2	—	—	—	—	—	—	—	—	—
150	24	5	243	—	—	—	—	—	—	—	—	—	—
300	24	5	250	—	—	—	—	—	—	—	—	—	—
900	18	5	250	1	—	—	—	—	—	—	—	—	—
	24	5	250	2	—	—	—	—	—	—	—	—	—
	30	5	250	—	—	—	—	—	—	—	—	—	—

Abbreviations, see footnotes, Table 1.

of each compound and a single one (24 h) for the lower doses.

Statistical evaluation

The influence of the sex on aberration yields was estimated by Friedman's two-way analysis of variance. Since no difference was observed between the sexes, the data were pooled, and the Kruskal-Wallis one-way analysis of variance was employed to investigate the time- and dose-response relationships.

For each compound the pooled data from the treated animals were compared with the negative controls by the Chi-square test with Yates' correction.

4 Results

Tables 1 to 5 show the results obtained for different doses of the 4 compounds and at various intervals of time after treatment.

2AAF, PYR and 4AAF failed to increase significantly the yield of structural chromosomal aberrations above control values. The animals given BP showed a statistically higher number of structural aberrations, but no clear dependence on dose- or time-response could be detected.

5 Discussion

Confirming the results reported in the literature (Kliesch *et al.*, 1981, 1982), exposure to BP produced a significant increase in the yield of aberrations, the numbers of anomalies was rather small and did not allow us any conclusion as to the dependence on dose or time. Furthermore, most structural changes were either breaks or gaps.

It should be mentioned that the quality of the preparations after treatment of the mice with 2AAF, and chiefly with 4AAF, was generally very poor and some experiments had to be repeated.

6 Conclusions

The observations performed on bone marrow cells of animals treated with BP, PYR, 2AAF and 4AAF do not

Table 4. *Results of the chromosome analysis on BALB/c mice treated with 2AAF*

						Chromatid type exchanges				Chromosome type exchanges			
Treatment (mg/kg)	Preparation interval (h)	Animals analysed	Cells scored	Gaps	Breaks	I	S	A	C	I	S	A	R
111	24	5	230	—	—	—	—	—	—	1	—	—	—
223	24	5	250	—	—	—	—	—	—	—	—	—	—
446	18	5	250	—	2	—	—	—	—	—	—	—	—
	24	5	250	2	—	—	—	—	—	—	—	—	—
	30	5	250	—	—	—	—	—	—	—	—	—	—
111	24	5	250	—	—	—	—	—	—	—	—	—	—
223	24	5	250	—	1	—	—	—	—	—	—	—	—
446	18	1	> 50	—	1	—	—	—	—	—	—	—	—
	24	5	250	2	—	—	—	—	—	—	—	—	—
	30	5	250	—	—	—	—	—	—	—	—	—	—

Abbreviations, see footnotes, Table 1.

Table 5. *Results of the chromosome analysis on BALB/c mice treated with 4AAF*

						Chromatid type exchanges				Chromosome type exchanges			
Treatment (mg/kg)	Preparation interval (h)	Animals analysed	Cells scored	Gaps	Breaks	I	S	A	C	I	S	A	R
223	24	5	230	1	2	—	—	—	—	—	—	—	—
446	24	5	200	—	2	—	—	—	—	—	—	—	—
892	18	5	250	1	2	—	—	—	—	—	—	—	—
	24	5	250	—	—	—	—	—	—	—	—	—	—
	30	4	200	1	—	—	—	—	—	—	—	—	—
223	24	5	250	—	—	—	—	—	—	—	—	—	3
446	24	5	250	—	1	—	—	—	—	—	—	—	—
892	18	5	250	—	—	—	—	—	—	—	—	—	1
	24	4	200	—	—	—	—	—	—	—	—	—	—
	30	3	150	—	—	—	—	—	—	—	—	—	—

Abbreviations, see footnotes, Table 1.

allow us to establish a relation between the carcinogenic properties of these chemicals and their clastogenic properties.

7 Acknowledgement

This research was supported by a grant from the "Ministère Belge de la Santé."

8 References

Kliesch, U., N. Danford, and I.-D. Adler (1981) Micronucleus test and bone-marrow chromosome analysis, A comparison of 2 methods *in vivo* for evaluating chemically induced chromosomal alterations, *Mutation Res.*, **80**, 321–332.

Kliesch, U., I. Roupova, and I.-D. Adler (1982) Induction of chromosome damage in mouse bone-marrow by benzo(a)pyrene, *Mutation Res.*, **102**, 265–273.

Barbara L. Harper and Marvin S. Legator
Department of Preventive Medicine and Community Health, Division of Environmental Toxicology, The University of Texas Medical Branch, Galveston, Texas, USA

11

Metaphase analysis test results in male ICR Swiss mice

1 Summary

Each of the 4 test chemicals (BP, PYR, 2AAF, and 4AAF) was administered orally in olive oil to male ICR Swiss mice. Femurs were harvested 24 h after treatment with an additional high-dose treatment group harvested 48 h after treatment. BP caused a dose-related increase in chromosome aberrations that was statistically significant at both harvest times of the highest dose. PYR and 2AAF caused increased but not significant effects.

2 Introduction

We participated in this collaborative study by using a combined testing protocol in which several endpoints of genotoxicity can be measured in the same animal (Legator and Harper, 1982). The metaphase analysis is presented here.

3 Materials and Methods

Male Ha/ICR Swiss mice were obtained from Timco/Harlan Sprague-Dawley, Houston, TX. They were housed on corn-cob bedding with food and water ad libitum and were acclimatized 7 days before treatment. Treatment groups consisted of 5 mice weighing 22–26 g (15 weeks of age). Femurs were harvested 24 h after treatment and 1 to $1\frac{1}{2}$ h after colchicine (Fisher Chemical Co.) was injected intraperitoneally (ip) at 1 mg/kg. Metaphases were harvested by standard methods except that the hypotonic 0.075 M KCl solution contained 0.002% sodium N-lauroyl sarcosine (Sigma Chemical Co.) to facilitate spreading. Fifty metaphases were scored from each animal, and aberrations were classified according to the method of Savage (1976). Data are presented as both percentage of cells with aberrations (Kliesch *et al.*, 1982) and frequency of aberrations per cell (Preston *et al.*, 1981). Gaps were scored and are presented separately. The chi-square test was used for statistical analysis.

Test chemicals were supplied for this study from ICI (UK). Cyclophosphamide (CP) was purchased from Mead Johnson, Evansville, Indiana. BP and PYR were dissolved in olive oil (Fisher Chemical Co.) with heating, and 2AAF and 4AAF were suspended in olive oil without DMSO. All chemicals were administered orally at 0.1 ml per 10 g body weight except CP, which was dissolved in saline and injected ip in the same volume per weight. Doses were established by the individual test coordinators.

Table 1. *Metaphase analysis*

Compound and dose (mg/kg)		Interval (h)	No. animals	No. cells scored	Total No. cells with:[a] Gaps	Breaks	Other	Total + gaps	Total − gaps	% cells showing: Gaps	Breaks	Other	All aberr. − gaps	Frequency per cell of: Gaps	Breaks	Other	All aberr. + gaps	All aberr. − gaps
BP	(130)	48	5	250	3	12	2	14	14	1.2	4.8	0.8	5.6	0.012	0.048	0.008	0.068	0.056
	(130)	24	5	250	4	16	0	20	16	1.6	6.4	0	6.4	0.024	0.080	0	0.104	0.080
	(52)	24	5	250	5	9	0	14	9	2.6	3.6	0	3.6	0.028	0.044	0	0.072	0.044
	(26)	24	5	250	0	7	0	7	7	0	2.8	0	2.8	0	0.036	0	0.056	0.056
PYR	(340)	48	4	250	9	9	1	17	10	3.6	3.6	0.04	2.5	0.044	0.040	0.004	0.088	0.044
	(340)	24	5	250	3	8	3	13	11	1.2	3.2	1.2	4.4	0.012	0.036	0.012	0.060	0.048
	(136)	24	5	250	2	2	0	4	2	0.8	0.8	0	0.8	0.008	0.008	0	0.016	0.008
	(68)	24	5	235	0	4	0	4	4	0	1.6	0	1.6	0	0.016	0	0.016	0.016
2AAF	(1110)	48	5	241	0	8	2	10	10	0	3.3	0.8	4.1	0	0.033	0.008	0.041	0.041
	(1110)	24	5	250	1	9	2	12	11	0.4	3.6	0.8	4.8	0.008	0.044	0.008	0.060	0.052
	(446)	24	5	250	1	5	7	13	12	0.4	0.2	2.8	4.8	0.004	0.020	0.032	0.056	0.052
	(223)	24	5	250	4	3	0	7	3	1.6	1.2	0	1.2	0.016	0.012	0	0.028	0.012
4AAF	(446)	48	5	250	2	3	5	9	8	0.8	1.2	2.0	3.2	0.012	0.012	0.020	0.044	0.032
	(446)	24	5	250	0	1	3	4	4	0	0.4	1.2	1.6	0	0.004	0.012	0.016	0.016
	(182)	24	5	250	3	3	2	7	5	1.2	1.2	0.8	2.0	0.012	0.012	0.008	0.032	0.020
	(73)	24	5	233	5	6	1	12	7	2.1	2.6	0.4	3.0	0.021	0.034	0.004	0.060	0.038
	(36)	24	5	250	6	12	2	19	13	2.4	4.8	0.8	5.2	0.028	0.060	0.008	0.096	0.068
Untreated		—	5	250	2	3	2	7	5	0.8	1.2	0.8	2.0	0.008	0.012	0.008	0.028	0.020
CP	(50)[a]	24	7	266	0	28	15(M) + 36(P)	43(− P)	43(− P)	0	10.5	0 + 19.2(M + P)	16.2	0	0.548	0	0.548	0.548

[a] M = multiple (> 10) aberrations per cell;
P = pulverisations, shattered chromosomes;
− P = pulverised cells not included.

4 Results and discussion

As shown in Table 1, BP was the only chemical to cause a significant ($p < 0.05$) increase in aberration frequency, although the higher doses of 2AAF at both time points and the 24 h harvest of the highest dose of PYR show increases but are slightly below significance. 2AAF has been found to be positive in other cytogenetic endpoints when administered ip in DMSO (Salamone *et al.*, 1981; Tsuchimoto and Matter, 1981). It is possible that administration as a suspension in olive oil rather than DMSO decreased absorption below the levels needed to cause a more significant effect. The dosage of PYR that caused a slight increase in aberrations was considerably higher than the highest dose of BP. It is possible that semiquinone radicals and reactive oxygen species generated during the metabolism of PYR could be contributing to this effect, (Chesis *et al.*, 1984) if this effect is confirmed.

5 References

Chesis, P.L., D.E. Levin, M.T. Smith, L. Ernster and B.N. Ames (1984) Mutagenicity of quinones: pathways of metabolic activation and detoxification, *Proc. Natl. Acad. Sci.*, **81**, 1696–1700.

Kliesch, U., I. Rupova and I.D. Adler (1982) Induction of chromosome damage in mouse bone marrow by benzo(a)pyrene, *Mutation Res.*, **102**, 265–273.

Legator, M.S. and B.L. Harper (1982) Animal human studies: an integrated approach to identifying hazardous chemicals in the workplace. In M. Sorsa and H. Vainio, (Eds), *Mutagens in our environment*, Alan R. Liss, NY, pp. 181–202.

Preston, R.J., W. Au, M.A. Bender, J.G. Brewen, A.V. Carrano, J.A. Heddle, A.F. McFee, S. Wolff and J.S. Wassom (1981) Mammalian *in vivo* and *in vitro* cytogenetic assays: a report of the U.S. EPA's Gene-Tox Program, *Mutation Res.*, **87**, 143–188.

Salamone, M.F., J.A. Heddle and M. Katz (1981) Mutagenic activity of 41 compounds in the *in vivo* micronucleus assay. In F.J. de Serres and J. Ashby, (Eds), *Evaluation of short-term tests for carcinogens,* Elsevier/North Holland, NY, pp. 686–697.

Savage, J.R.K. (1976) Classification and relationships of induced chromosomal structural changes, *J. Med. Genet,* **13**, 103–122.

Tsuchimoto, T. and B.E. Matter (1981) *Activity of coded compounds in the micronucleus test.* In F.D. de Serres and J. Ashby, (Eds), Elsevier/North Holland, NY, pp. 705–711.

Alexei M. Malashenko and Tatijana B. Uvarova
Scientific Research Laboratory of Experimental Biological Models of the Academy of Medical Sciences of the USSR, Moscow Region, USSR

12

A study of the ability of benzo(a)pyrene, pyrene, 2-acetylaminofluorene and 4-acetylaminofluorene to damage bone marrow cell chromosomes in five inbred strains of mice

1 Summary

The cytogenetic effects of BP, PYR, 2AAF and 4AAF were studied in bone marrow cells of male mice using 5 inbred strains. The chemicals were dissolved or suspended in olive oil and administered by gavage.

In experiments with BP and PYR, PYR was not clastogenic. In the 2AAF/4AAF pair, 2AAF was found to be a weak clastogen. The method applied did not give a distinct answer to the question: is 4AAF a clastogen *in vivo*?

2 Introduction

The aim of this work was an attempt to reveal possible cytogenetically determined strain differences among the 4 chemical compounds. Mammalian populations are genetically heterogenous in their response to mutagenic and carcinogenic agents, and it is desirable to elucidate what influences the sensitivity and variability in populations of mammals, including humans.

Five inbred strains of mice were used. The origin and genetic specificity of these strains have been summarised previously (Malashenko, 1983a, b; Malashenko and Kosarev, 1984). The TPS strain was selectively developed for sensitivity to thio-TEPA. The WR strain is also sensitive to thio-TEPA but relatively resistant to the mutagens cyclophosphamide and mitomycin C. WR mice may have a defective mutant enzyme. Strain 101 and, possibly, B6 have a defect in genetic repair. Strain CBA, included as "normal", is characterised by low spontaneous and induced levels of chromosomal damage.

3 Materials and methods

Chemicals: BP, PYR, 2AAF and 4AAF were supplied by ICI (UK).

The choice of doses, solvent, route of administration, intervals between treatment, and preparation of bone marrow were performed in accordance with the programme coordinator's recommendations. All chemicals were dissolved or suspended in warm olive oil using a magnetic stirrer immediately before the animals were treated. They were administered with a stomach tube at a volume of 10 ml/kg body weight.

Animals: The male mice used in these experiments were 7–8 weeks old and weighed about 20 g. The 5 inbred strains used were: CBA/LacY (CBA), C57BL/6JY (B6), TPS/Y (TPS), 101/HY (101), and WR/Y (WR). For each of the 5 strains, 5 animals were used. Five control mice of each strain were treated in the same way with olive oil only.

Treatment: For BP and PYR, the sampling interval was 30 h and for 2AAF and 4AAF it was 24 h. The air-dried bone marrow cell preparations were made 1 h after colchincine injection. For each animal 50 metaphases were analysed.

Fisher's exact test was used to compare experimental and control data. These and other calculations were made by using "Biometria" (Plochinski, 1970).

4 Results

The results of scoring metaphase cell preparations are summarised in Tables 1 to 5. Relatively weak clastogenic effects were found. Only the frequencies of gaps and chromatid breaks were elevated in the experimental groups over those in the controls. The frequencies of cells with chromosomal damage (FCD, gaps + breaks + exchanges) were compared statistically (Table 6). The FCD increased in all strains after BP treatment. It varied from 6.4% in WR and TPS mice to 2.8% in CBA mice.

The differences between control and BP-treated mice are statistically significant for the TPS, WR, and 101 strains. It is clear that BP induced chromosomal damage in mice of all strains, but the FCD is so low that the samples used are insufficient for statistical corroboration (Table 6). If the BP treatment data are compared with those of the untreated controls (when 1000 cells per strain are analysed), a significant difference is shown with all strains ($P < 0.01$, Table 6). In the BP experiment, differences between strains are not significant, while previous experiments with other chemicals showed distinct strain differences. PYR did not increase the FCD in any strain (Table 3).

2AAF increased the FCD in all groups by an average of 3 times background (Table 4). Increases in the FCD were observed with all strains but are only statistically significant with WR ($P < 0.01$) and CBA ($P < 0.05$) when compared with the solvent controls. A two-factor analysis of variance, in which the effect of combined 2AAF data vs control data for all genotypes (TPS, WR, 101, B6, CBA) was compared, showed a significant difference

Table 1. *Chromosome damage in controls (interval – 30 h; treated with olive oil; po)*

			Per cent cells showing			Frequency per cell of			
Strain	No. of animals	No. of cells scored	Gaps	Breaks	All aberrations excl. gaps	Gaps	Breaks	All aberrations incl. gaps	All aberrations excl. gaps
TPS	5	250	0.8 ± 0.56	0.0	0.0	0.008	0.0	0.008	0.0
WR	5	250	0.4 ± 0.40	0.0	0.0	0.004	0.0	0.004	0.0
101	5	250	0.4 ± 0.40	0.8 ± 0.56	0.8 ± 0.56	0.004	0.008	0.012	0.008
B6	5	250	1.2 ± 0.69	0.4 ± 0.40	0.8 ± 0.56	0.004	0.004	0.02	0.008
CBA	5	250	0.4 ± 0.40	0.0	0.0	0.004	0.0	0.004	0.0

Table 2. *BP Effect in mouse bone marrow cells (interval – 30 h; dose – 252 mg/kg; po)*

			Per cent cells showing			Frequency per cell of			
Strain	No. of animals	No. of cells scored	Gaps	Breaks	All aberrations excl. gaps	Gaps	Breaks	All aberrations incl. gaps	All aberrations excl. gaps
TPS	5	250	3.6 ± 1.80	2.8 ± 1.04	2.8 ± 1.04	0.044	0.028	0.072	0.028
WR	5	250	3.6 ± 1.80	2.4 ± 0.97	2.8 ± 1.04	0.068	0.024	0.092	0.024
101	5	250	2.1 ± 0.91	3.1 ± 1.10	3.5 ± 1.16	0.04	0.032	0.076	0.036
B6	5	250	1.2 ± 0.69	3.6 ± 1.80	3.6 ± 1.80	0.012	0.044	0.056	0.044
CBA	5	250	1.6 ± 0.80	1.2 ± 0.69	1.2 ± 0.69	0.016	0.012	0.028	0.012

Table 3. *PYR effect in mouse bone marrow cells (interval – 30 h; dose – 808 mg/kg; po)*

			Per cent cells showing			Frequency per cell of			
Strain	No. of animals	No. of cells scored	Gaps	Breaks	All aberrations excl. gaps	Gaps	Breaks	All aberrations incl. gaps	All aberrations excl. gaps
TPS	5	250	0.4 ± 0.40	0.4 ± 0.40	1.6 ± 0.80	0.004	0.004	0.008	0.004
WR	5	250	0.0	1.2 ± 0.69	1.2 ± 0.69	0.0	0.012	0.012	0.012
101	5	250	1.2 ± 0.69	0.0	0.0	0.012	0.0	0.012	0.0
B6	5	250	0.4 ± 0.40	0.4 ± 0.40	0.4 ± 0.40	0.004	0.004	0.008	0.004
CBA	5	250	0.0	0.0	0.0	0.0	0.0	0.0	0.0

Table 4. *2AAF Effect in mouse bone marrow cells (interval – 24 h; dose 446 mg/kg; po)*

Strain	No. of animals	No. of cells scored	Per cent cells showing: Gaps	Breaks	All aberrations excl. gaps	Frequency per cell of: Gaps	Breaks	All aberrations incl. gaps	All aberrations excl. gaps
TPS	5	250	0.4 ± 0.40	1.6 ± 0.80	1.6 ± 0.80	0.004	0.016	0.02	0.016
WR	5	250	1.6 ± 0.80	2.4 ± 0.97	2.4 ± 0.97	0.016	0.024	0.04	0.024
101	5	250	1.6 ± 0.80	1.6 ± 0.80	1.6 ± 0.80	0.02	0.02	0.04	0.02
B6	5	250	2.5 ± 0.97	0.8 ± 0.56	0.8 ± 0.56	0.024	0.008	0.032	0.008
CBA	5	250	2.0 ± 0.89	0.8 ± 0.56	0.8 ± 0.56	0.02	0.008	0.028	0.008

Table 5. *4AAF effect in mouse bone marrow cells (interval – 24 h; dose – 892 mg/kg; po)*

Strain	No. of animals	No. of cells scored	Per cent cells showing: Gaps	Breaks	All aberrations excl. gaps	Frequency per cell of: Gaps	Breaks	All aberrations incl. gaps	All aberrations excl. gaps
TPS	5	250	0.8 ± 0.56	0.8 ± 0.56	0.8 ± 0.56	0.008	0.008	0.016	0.008
WR	5	250	0.8 ± 0.56	0.8 ± 0.56	0.8 ± 0.56	0.008	0.008	0.016	0.008
101	5	250	0.0	1.6 ± 0.80	2.0 ± 0.86	0.0	0.02	0.024	0.024
B6	5	250	0.8 ± 0.56	2.0 ± 0.89	2.0 ± 0.86	0.008	0.02	0.028	0.02
CBA	5	250	0.4 ± 0.40	0.0	0.0	0.004	0.0	0.004	0.0

Table 6. *Chromosome damage induced in bone marrow cells*

Chemical (Dose; mg/kg)	Strain	Cells scored	Frequency of cells (% + SE) with: Gaps	Breaks and exchanges	FCD[a]	P values
Control (Olive oil)	TPS	250	0.80 ± 0.56	0.0	0.80 ± 0.56	
	WR	250	0.40 ± 0.40	0.0	0.40 ± 0.40	
	101	250	0.40 ± 0.40	0.80 ± 0.56	1.20 ± 0.69	
	B6	250	1.20 ± 0.69	0.80 ± 0.56	2.0 ± 0.88	
	CBA	250	0.40 ± 0.40	0.0	0.40 ± 0.40	
Control (untreated)	TPS	1000	0.50 ± 0.22	0.40 ± 0.20	0.90 ± 0.30	
	WR	1000	0.40 ± 0.22	0.30 ± 0.17	0.70 ± 0.26	
	101	1000	0.50 ± 0.22	0.60 ± 0.24	1.10 ± 0.33	
	B6	1000	0.50 ± 0.22	0.80 ± 0.28	1.30 ± 0.36	
	CBA	1000	0.20 ± 0.14	0.30 ± 0.17	0.50 ± 0.22	
BP (252)	TPS	250	3.60 ± 1.80	2.80 ± 1.04	6.40 ± 1.55	< 0.001
	WR	250	3.60 ± 1.80	2.80 ± 1.04	6.40 ± 1.55	< 0.001
	101	250	2.10 ± 0.91	3.50 ± 1.16	5.60 ± 1.46	< 0.01
	B6	250	1.20 ± 0.69	3.60 ± 1.80	4.80 ± 1.35	
	CBA	250	1.60 ± 0.80	1.20 ± 0.69	2.80 ± 1.04	
PYR (808)	TPS	250	0.40 ± 0.40	1.60 ± 0.80	2.00 ± 0.88	
	WR	250	0.0	1.20 ± 0.69	1.20 ± 0.69	
	101	250	1.20 ± 0.69	0.0	1.20 ± 0.69	
	B6	250	0.40 ± 0.40	0.40 ± 0.40	0.80 ± 0.56	
	CBA	250	0.0	0.0	0.0	
2AAF (446)	TPS	250	0.40 ± 0.40	1.60 ± 0.80	2.00 ± 0.88	
	WR	250	1.60 ± 0.80	2.40 ± 0.97	4.00 ± 1.24	< 0.001
	101	250	1.60 ± 0.80	1.60 ± 0.80	3.20 ± 1.11	
	B6	250	2.40 ± 0.97	0.80 ± 0.56	3.20 ± 1.11	
	CBA	250	2.00 ± 0.89	0.80 ± 0.56	2.80 ± 1.04	< 0.05
4AAF (892)	TPS	250	0.80 ± 0.56	0.80 ± 0.56	1.60 ± 0.79	
	WR	250	0.80 ± 0.56	0.80 ± 0.56	1.60 ± 0.79	
	101	250	0.0	2.00 ± 0.89	2.00 ± 0.89	
	B6	250	0.80 ± 0.56	2.00 ± 0.89	2.80 ± 1.04	
	CBA	250	0.40 ± 0.40	0.0	0.40 ± 0.40	

[a] FCD = Frequencies of cells with chromosomal damage (gaps + breaks + exchanges).

between treatment groups and controls (0.25, $P < 0.001$) but not between strains.

The analysis of chromosomal aberrations after treatment with 4AAF gave inconclusive results (Table 5). Mean FCD for experimental groups increased 1.7 times, but that for the CBA males remained at the control level. All differences between groups of treated and control males are statistically insignificant. With the B6 strain, the increase in FCD was 2.8%, but this was still not significantly different from the control. It is concluded that 4AAF tends to increase FCD. A more definitive conclusion may be possible if only B6 mice were used in an extensive investigation (more than 50 cells per mouse and more than one dose used). We cannot conclude with confidence the clastogenic inactivity of 4AAF.

5 Discussion and conclusions

The clastogenic activity of BP found in this experiment is low compared with that found by Kliesch *et al.* (1982). It is unlikely that this difference is due to genetic characteristics of the mice. Differences in diet, pH of drinking water (in our laboratory, HCl is added to the water, pH = 4.5), and metabolism by intestinal microflora are probably the reasons for the difference. It is known that BP, a typical aromatic hydrocarbon, exerts its clastogenic and carcinogenic effects after metabolic activation controlled by the *Ah* gene complex. Because BP was clastogenic in all 5 strains, it is possible that they carry the Ah^b gene, as is known to be the case with the B6 strain.

High sensitivity of some strains compared with the normal CBA strain in this experiment may also be due to other reasons. WR mice have metabolic peculiarities. Excision repair defects are present in strain 101. An increased number of bone marrow cells with greater sensitivity to some chemicals can be observed in TPS mice. The strain ranking for sensitivity to BP is similar to that for thio-TEPA (Malashenko, 1983a).

Clastogenic metabolites are probably not formed by the metabolism of PYR, and this compound is not mutagenic *in vivo*.

2AAF demonstrated weak clastogenic activity. WR mice were more sensitive to 2AAF than were the other strains. WR mice are also more sensitive to the pesticide captan (Malashenko and Kosarev, 1984). The WR strain may be useful for studying the biological effects of xenobiotics because of its metabolic peculiarities.

Comparison of 2AAF and 4AAF effects does not allow as clear a conclusion as for BP and PYR. We cannot reject the possibility of a very weak clastogenic activity of 4AAF based on these data. The sensitivity of the assay is not sufficiently powerful for the complete evaluation of 4AAF.

In our opinion, this experiment demonstrated the expediency of testing chemical compounds, using an array of strains. The next stage of experimentation would be a more complete investigation using a more sensitive strain. For example, WR mice should be used in a study of BP and 2AAF and B6 mice for the further evaluation of 4AAF.

In conclusion, the assay allows us to discriminate between BP and PYR. PYR shows no clastogenic effect *in vivo*. With 2AAF/4AAF, it is not possible to claim that 4AAF is not clastogenic. More extensive investigations with 4AAF are desirable.

6 References

Kliesch, U., I. Roupova and I.-D. Adler (1982) Induction of chromosome damage in mouse bone marrow by benzo(a)pyrene, *Mutation Res.*, **102**, 256–273.

Malashenko, A.M. (1983a) Use of inbred mice in studies of chemical mutagenesis, *Vestn. Akad. Med. Nauk SSR*, **9**, 39–42.

Malashenko, A.M. (1983b) Specificity of inbred TPS, WR and CBA/LacY mice strains to the action of chemical mutagens, *Genetika (USSR)*, **19**, 1463–1468.

Malashenko, A.M. and A.V. Kosarev (1984) Choice of mice lines for the testing of chemical mutagens, *Genetika (USSR)*, **20**, 1289–1294.

Plochinski, N.A. (1970) *Biometria*, Moscow Univ. Publ., Moscow.

Edith Hüttner and Jörg Schöneich
Zentralinstitut für Genetik und Kulturpflanzenforschung der Akademie der Wissenschaften der DDR, 4325 Gatersleben, German Democratic Republic

13

Induction of chromosomal aberrations in bone marrow cells and ascites tumour cells of mice by 2-acetylaminofluorene, 4-acetylaminofluorene and benzo(a)pyrene

1 Summary

Male mice were treated per os with 2AAF, 4AAF, and BP. Bone marrow cells or ascites tumour cells excluding BP treatment were examined for chromosomal aberrations 12, 24, 30, and 42 h after treatment. This analysis was carried out on 50 metaphase cells per animal (5 animals per group). A significant increase in the percentage of cells with chromosomal aberrations in bone marrow cells was obtained with animals treated with BP. In the 2AAF- and 4AAF-treated group there were no increases in the frequencies of aberrations compared with those of the corn oil treated control, neither in bone marrow cells nor in ascites tumour cells.

2 Materials and methods

Animals and tumours

Male Gat: NMRI mice (outbred), 6 to 8-weeks old, were used in all experiments. The animals were maintained under barrier conditions with water and diet ad libitum. The light-dark cycle was 12 h, room temperature 21 ± 1 °C, relative humidity 65 ± 5%. Ehrlich ascites tumour cells with a modal number of 44 chromosomes were routinely transplanted (5-day intervals) and used for the experiments (Schöneich, 1967; Wobus *et al.*, 1978).

Test compounds

2AAF, 4AAF, BP, and PYR were supplied by ICI (UK). 2AAF and 4AAF were suspended and BP was dissolved in corn oil (Mazola Keimöl).

Experimental design

Each mouse received 0.2 ml of a freshly prepared test solution by gavage. 2AAF was administered at 111, 223, and 446 mg/kg; 4AAF at 223, 446, and 892 mg/kg; and BP at a single dose of 252 mg/kg. The concentrations of the mutagenic solutions were adjusted to the body weight of the animals. Each dose and interval was analysed with a group of 5 animals. A control group of 5 animals received 0.2 ml corn oil.

Ascites tumour cells

24 h before dose administration, the animals were injected with 0.5 ml ascites fluid. At 12, 24, 30, and 42 h after treatment, the animals were killed by cervical dislocation. At 2 h before sacrifice, 0.5 ml = 12/μg/mouse colchicine was injected intraperitoneally (ip). 0.5 ml ascites fluid was

Table 1. *Aberration frequency in bone marrow cells of mice after administration of 0.2 ml corn oil (control)*

Dose (mg/kg)	Interval (h)	No. of animals	No. of cells scored	% Cell showing	
				Gaps	Breaks
Corn oil	12	5	250	0.40	0
	24	5	250	0	0
	30	5	250	0	0
	42	5	250	1.20	0

Table 2. *Induction of chromosomal aberrations in bone marrow cells of mice by BP (252 mg/kg; 250 cells from 5 animals per interval)*

	% Cells showing							Frequency per cell of					
Interval (h)	Gaps	Breaks	Chromatid exchanges	Chromosome exchanges	Multiple abberations	Poly-ploidy	All aberrat. ex. gaps	Gaps	Breaks	Chromatid exchanges	Chromosome exchanges	All aberrat. inc. gaps	All aberrat. ex. gaps
12	2.4	0.8	0	0	0	2.8	3.6	0.024	0.008	0	0	0.032	0.008
24	4.8	4.4	0.4	0	0	4.8	9.6	0.076	0.048	0.004	0	0.128	0.052
30	1.6	3.2	1.2	1.2	0.4	4.0	10.0	0.024	0.044	0.012	0.012	0.092	0.068
42	1.2	0.8	0	0	0	0.8	1.6	0.012	0.008	0	0	0.020	0.008

recovered from each animal and was incubated with 8 ml prewarmed, hypotonic 1.12% sodium citrate solution for 30 min at 37 °C. After centrifugation, the cells were fixed twice with ethanol: acetic acid (3:1). Air-dried slides were prepared and stained with 2% orcein in 60% acetic acid.

Bone marrow cells

Bone marrow cells were sampled 12, 24, 30, and 42 h after treatment; 2 h before sacrifice, the animals were treated with colchicine as described above. Bone marrow cells were flushed from both femurs with prewarmed Hanks balanced salt solution. After centrifugation, the cells were resuspended in a 0.075 mol KCl solution, incubated 10 min at 37 °C, and fixed twice in methanol: acetic acid (3:1). Slides were air dried and stained with 2% orcein in 60% acetic acid.

All slides were coded, and 50 well-spread metaphases from each mouse were analysed for chromosomal aberrations according to the schedule of Savage (1976).

Table 3. *Induction of chromosomal aberrations in bone marrow cells of mice by 2AAF*

Dose (mg/kg)	Interval (h)	No. of animals	No. of cells scored	% Cells showing	
				Gaps	Breaks
111	12	5	250	0	0
	24	5	250	0.4	0
	30	5	250	0.4	0
	42	5	250	0	0
223	12	4	200	0	0
	24	5	200	0	0
	30	5	200	0.5	0
	42	5	250	0	0.4
446	12	5	250	0.8	0.8
	24	5	250	0	0.4
	30	5	250	0.4	0
	42	5	250	0.8	0

Table 4. *Induction of chromosomal aberrations in bone marrow cells of mice by 4AAF*

Dose (mg/kg)	Interval (h)	No. of animals	No. of cells scored	% Cells showing	
				Gaps	Breaks
223	12	5	250	0	0
	24	5	250	0	0
	30	5	185	0	0
	42	5	200	0	0
446	12	5	220	0.45	0
	24	5	250	0.40	0
	30	5	250	0.40	0
	42	5	250	0.40	0
892	12	5	250	1.20	0
	24	4	200	0.50	0
	30	5	250	0.40	0
	42	5	200	0.50	0

Table 5. *Induction chromosomal aberrations in ascites tumour cells of mice by 2AAF*

Dose (mg/kg)	Interval (h)	No. of animals	No. of cells scored	% Cells showing	
				Gaps	Breaks
111	12	5	250	0	0
	24	5	205	0	0
	30	5	220	0	0
	42	4	140	0	0
223	12	5	250	0	0
	24	5	250	0	0
	30	4	200	0	0
	42	5	250	0	0
446	12	5	170	0	0
	24	4	200	0	0
	30	5	200	0	0
	42	5	170	0	0

Table 6. *Induction of chromosomal aberrations in ascites tumour cells of mice by 4AAF*

Dose (mg/kg)	Interval (h)	No. of animals	No. of cells scored	% Cells showing		
				Gaps	Breaks	Chromatid exchanges
223	12	5	250	0.40	0	0
	24	5	250	0.40	0	0
	30	5	245	0	0	0
	42	5	120	0	0	0
446	12	5	250	0.40	0	0
	24	5	250	0	0	0
	30	5	200	0	0	0
	42	5	250	0.40	0	0
892	12	5	220	0	0	0.45
	24	3	150	0	0	0
	30	5	250	0	0	0
	42	5	200	0	0	0

3 Results and discussion

Control results of bone marrow cells are shown in Table 1. As shown in the Table 2 oral administration of 252 mg/kg BP resulted in a significant increase of aberrant bone marrow metaphase cells. The peak of aberration frequencies was found 24 and 30 h after treatment. The main aberration types induced by BP were breaks and polyploid cells. The results observed are in general agreement with those published by Kliesch *et al.* (1982). The increase of polyploid cells indicates an effect that BP has on the mitotic spindle first observed by von Moellendorff (1939) in tissue cultures of rabbit fibroblasts.

In contrast to BP, 2AAF (Table 3) and 4AAF (Table 4) did not induce significant increases of aberrations compared with control group.

Control results of ascites tumour cells are shown in Table 7. Neither 2AAF (Table 5) nor 4AAF (Table 6) showed significant increases of aberrations in ascites tumour metaphase cells compared as those with the corn oil treated control.

4 References

Kliesch, U., I. Roupova and I.D. Adler (1982) Induction of chromosomal damage in mouse bone marrow by benzo(a)pyrene. *Mutation Res.*, **102**, 265–273.

von Moellendorff, W. (1938) *Z. Zellforsch. u. mikroskop. Anal.*, **28**, 512.

Savage, I.R.K. (1976) Classification and relationships of induced chromosomal structural changes, *J. Med. Genetics*, **13**, 103–122.

Schöneich, J. (1967) The induction of chromosomal aberrations by hydrogen peroxide in strains of ascites tumors in mice. *Mutation Res.*, **4**, 384–388.

Wobus, A.M., J. Schöneich and R. Thieme (1978) The effect of the mode of administration of nitrogen mustard and cytosine arabinoside on the production of chromosomal aberrations in mouse bone marrow and ascites tumour cells. *Mutation Res.*, **58**, 67–77.

Table 7. *Aberration frequency in ascites tumour cells of mice after administration of 0.2 ml corn oil (control)*

Dose (mg/kg)	Interval (h)	No. of animals	No. of cells scored	% Cells showing Gaps	% Cells showing Breaks
Corn oil	12	6	275	0	0.36
	24	6	240	0	0
	30	6	250	0.40	0
	42	5	200	0	0

Carol. A. Luke, Raymond R. Tice, Valerie M. Miller, Brian G. Ormiston, Anita C. Wilton and Michael D. Shelby[1]

Medical Department, Brookhaven National Laboratory, Upton, New York (USA) and[1] *National Institute of Environmental Health Sciences, Research Triangle Park, North Carolina, USA*

14

A comparative analysis of the induction of chromosomal aberrations in the bone marrow cells of B6C3F1 mice

1 Summary

Using the bromodeoxyuridine-dependent differential staining technique to analyse first generation post-treatment metaphase cells, the ability of two established carcinogens, benzo(a)pyrene and 2-acetylaminofluorene, to induce chromosomal aberrations and to inhibit cellular proliferation in bone marrow cells of male B6C3F1 mice was compared with that of two purported noncarcinogenic analogues, pyrene and 4-acetylaminofluorene, respectively. Based on an analysis of individual animal aberration data by the Kruskal-Wallis nonparametric test using an α of 0.05: (1) treatment with BP induced a significant increase in the percentage of damaged cells ($P = 0.0115$) and in the number of aberrations per cell ($P = 0.0216$), (2) treatment with 2AAF induced a weak but significant increase in both measures of clastogenicity ($P = 0.0365$), and (3) exposure to PYR and 4AAF did not induce a significant increase in the percentage of damaged cells or in the number of aberrations per cell. Also, bone marrow cellular proliferation kinetics was inhibited significantly by BP, PYR, and 4AAF ($P < 0.05$) but not by 2AAF ($P = 0.1711$), as determined by a one-way analysis of variance of average generation time data.

2 Introduction

Benzo(a)pyrene (BP) and 2-acetylaminofluorene (2AAF) are widely recognised as animal carcinogens (Kriek, 1974; NCI, 1951–1978; IARC, 1982, 1983) while pyrene (PYR) and 4-acetylaminofluorene (4AAF), structural analogues of BP and 2AAF, respectively, have not demonstrated carcinogenic activity (Ashby, this volume). Both carcinogens are active in a variety of short-term *in vivo* and, with metabolic activation, *in vitro* assays, including cytogenetic, DNA damage and repair, cell transformation, and prokaryotic/eukaryotic mutational tests (IARC, 1983; Ashby, 1981). In several of the *in vitro* assays, PYR and 4AAF were also found to induce a positive genotoxic response (Ashby, 1981). Because of these *in vitro* findings, collaborative studies focusing on different biological endpoints were undertaken to compare, as extensively as possible, the *in vivo* genotoxicity of these chemicals. In this report, we compare the abilities of 2AAF, 4AAF, BP, and PYR to induce chromosomal aberrations (CAs), a well-established indicator of genotoxic damage (Preston *et al.*, 1981) and, as a secondary measure of toxicity, to inhibit cellular proliferation (Tice and Ivett, 1985) in mouse bone marrow cells *in vivo*.

3 Materials and methods

The experimental design and methods used in this study follow the U.S. National Toxicology Program standard protocol for *in vivo* mouse bone marrow cytogentics evaluation (McFee *et al.*, 1986). This methodology is presented in part in the following sections.

Animals

Male B6C3F1 mice, received from Frederick Cancer Research Center as weanlings, were used throughout these studies. Until selected for experimentation, mice were housed at a population density no greater than one animal per 90 sq cm under controlled environmental conditions [i.e., 25 °C; 50% relative humidity; 20 air exchanges/h and a 12 h photoperiod (light 0700–1900 hr)]. All mice were bedded on corn cob litter (Bed O'Cobs, J.R. Nielsen and Sons, Inc.) with food (Purina Laboratory Rodent Chow 5001) and water available ad libitum. Within each experiment, animals used were of the same date of birth (between 10 to 16 weeks of age) and weighed within a 2 g range of a mean weight of 25 to 30 g. For maximum tolerated dose determinations, generally 3 animals per exposure group were used; for CA analysis, 10 to 12 animals per dose were used, depending on the expected rate of lethality, to ensure that 8 animals per exposure group could be analysed.

Chemicals

BP (WHO 12), PYR (WHO 11), 2AAF (WHO 14), and 4AAF (WHO 13) were obtained from Lancaster Synthesis Ltd. (UK). Mitomycin C (MMC; Lot No. 12F-0600), 5-bromodeoxyuridine (BrdUrd; Lot No. 73F-0548), and reagent grade corn oil (Lot No. 131-076) were obtained from Sigma; colchicine from Eli Lilly; Hoechst 33258 from American Hoechst; and Giemsa from Harleco. The 4 test chemicals and MMC were stored at 4 °C. Within one-half hour of administration, MMC was dissolved in phosphate buffered saline (PBS; pH 7.4) and the 4 test chemicals were dissolved or suspended in corn oil. When necessary (e.g. 2AAF, 4AAF, PYR), sonication (5 min, 37 °C) and/or syringing through an 18 g needle were used to prepare a homogenous suspension prior to injection. All injections were intraperitoneal (ip) at a volume of 0.4 ml.

4 Experimental protocol

Each animal was anaesthetised lightly with Metofane® (Pitman-Moore, Inc.), a small incision (~6 mm) made on the right lateral aspect of the abdomen, a 50 mg paraffin-coated BrdUrd tablet implanted subcutaneously, and the incision closed with wound clips (Clay-Adams). The BrdUrd tablets were made on a Parr pellet press using a die (0.178 in diam) and 55 ft lbs of pressure (Allen *et al.*, 1979) and coated (~70%) with paraffin (McFee *et al.*, 1983). Thirty minutes after BrdUrd tablet implantation (1500–1600 h), each animal was injected ip with the test chemical or the appropriate solvent. Fifteen and one-half hours later (16 h after BrdUrd tablet implantation), each surviving animal was injected ip with colchicine (2 mg/kg). Two hours later (18 h after BrdUrd tablet implantation), the animals were killed by cervical dislocation, and the bone marrow from both femurs removed by flushing with PBS. This sample time was selected based on previous studies demonstrating a maximal yield of chemically-induced aberrant first division cells at this time (McFee *et al.*, 1984). The aspirated bone marrow was pelleted by centrifugation, incubated in 0.075 M KCl (25 min; 37 °C), repelleted, fixed with ice cold 3:1 methanol:glacial acetic acid, and stored at 4° to −5 °C until used for slide preparation.

The MMC positive control experiments, which used a dose of 1.0 mg/kg, were conducted using the same protocol.

Slide preparation, staining, and analysis

Each fixed bone marrow sample was washed twice in fixative and coded flame-dried slides were prepared. All slides were differentially stained for CA and cell kinetic analysis using a modified (Tice *et al.*, 1978) fluorescence-plus-Giemsa technique (Goto *et al.*, 1975). Slides were scored without knowledge of the treatment. Slides for PYR and 2AAF were scored concurrently. In each set of slides, similarly coded MMC positive control and MMC solvent control slides were placed randomly. To examine individual animal bone marrow proliferation kinetics, 100 randomly selected metaphase cells were scored for replicative history (i.e. the number of S phases completed since the time of BrdUrd tablet implantation) Tice *et al.*, 1976; Tice and Ivett, 1985). For CA analysis, 50 first generation metaphase cells per animal were selected for scoring based on good chromatid differential staining properties, lack of overlapping chromosomes, and chromosome number (40 ± 3 chromosomes). In each CA study, 2 individuals scored one-half of each animal's data (i.e. 25 cells). Individual types of aberrations (i.e. chromatid vs chromosome; achromatic lesion (or gap) vs break vs chromosome rearrangement, including rings, triradials, quadriradials, Robertsonian translocations, etc.) were recorded separately. Chromosomal lesions were scored as gaps if the length of the pale staining region was equal to or less than the width of the chromatid and if no displacement occurred. Chromosomal lesions were scored as breaks if separation was greater than the width of a chromatid and/or if displacement was apparent.

Maximum tolerated dose (MTD) determinations

For BP and 2AAF, the test doses used in these studies were the same as those used in previous CA studies with the same chemicals (Ivett *et al.*, 1984). For PYR and 4AAF, preliminary studies were conducted to determine the optimal dose range to be tested for each chemical. The purpose of the MTD determination was to identify a range of doses that would evaluate, as completely as possible, the ability of a test chemical to induce CAs in bone marrow cells *in vivo*. Thus, selection of the MTD was made on the basis of one or more of the following criteria: over a 24 h period and when necessary, including the maximum dose possible consistent with obtaining a homogenous injection suspension, (1) animal lethality was apparent and/or, (2) first generation, post-treatment, metaphase cells only were present (Ivett *et al.*, 1984; McFee *et al.*, 1985). In the absence of available lethality data, groups of animals (generally 3 per group) were injected ip following the standard protocol for CA analysis at doses of 10, 100, and 1000 mg/kg (Ivett *et al.*,

1984; Tice *et al.*, this volume). Where lethality information was available, doses around the LD_{50} were tested. Cell proliferation kinetics were determined in bone marrow material from all survivors. Based on the lethality and/or bone marrow kinetic response, higher, intermediate or lower doses were tested in the same manner until an MTD was determined. For CA and proliferative inhibition analysis, the MTD, half of the MTD and a quarter of the MTD were tested.

Statistical analysis

For cell proliferation kinetic analysis, the proportion of cells among the first, second, and third generations in each animal was transformed into an average generation time (AGT):

AGT = BrdUrd exposure duration/Replicative index,

Where the Replicative Index (RI) = 1 times the frequency of first generation metaphase cells + 2 times the frequency of second generation metaphase cells + 3 times the frequency of third generation metaphase cells...(Ivett and Tice, 1982). AGT data were analysed by one-way ANOVA, using an α of 0.05 to determine whether the treatment had any effect on bone marrow cell kinetics. Individual animal data were omitted from the analysis if differential staining quality was considered inadequate.

For statistical analysis of CA data, information of individual aberrations within an animal were combined and analysed by the Kruskal-Wallis nonparametric test as: (i) percentage of aberrant cells (including and excluding gaps) and (ii) the number of aberrations per cell (including or excluding gaps). For the latter analysis, all aberrations were considered equal, regardless of the number of probable breakage events involved. Significance was based on an α of 0.05. If a significant treatment effect was observed, a pairwise comparison of each treatment group against the control group were conducted using Mann-Whitney simultaneous test procedure (including a Bonferroni correction) for multiple comparisons. For comparative purposes, alternative analytic approaches were conducted on each chemical. The cell-by-cell data were pooled among animals within treatment groups for aberrant vs normal cells to form two-way [exposure ($a = 4$) versus normal cell category ($b = 2$)] frequency tables. The null hypothesis of independence of exposure dose vs cell type factors (i.e. the equality of proportions of damaged cells among doses) was tested using the chi-square and log-likelihood ratio (G-statistic) methods (Sokal and Rohlf, 1969). In addition, a test of linear trend of proportions (Cochran, 1954) of aberrant cells was conducted. Statistical comparisons of the total number of aberrations observed per exposure group were conducted using the one-way goodness of fit test and calculating chi-square and G. The null hypothesis for these analyses was that the frequency of aberrations was equal among treatment groups.

Additional experiments with 2AAF

An additional experiment was conducted to assess laboratory and protocol variability in regard to 2AAF by following the protocol of Adler (Adler, this volume). Male C31F1 mice, kindly supplied by Dr. W. Generoso (Oak Ridge National Laboratory, Oak Ridge, TN), were received as weanlings. Housing and environmental conditions were as described above. At 15 weeks of age, mice (5 per exposure group; 23–26 g in body weight) were injected ip (volume 0.1 ml) with 2AAF (446 mg/kg) dissolved in DMSO (Aldrich) or with DMSO only. BrdUrd tablets were not implanted in these animals. Twenty-three hours after treatment, each animal was injected ip with colchicine (2 mg/kg). One h later (24 h after treatment) the animals were killed, bone marrow obtained, processed and slides made. Slides were stained for 5 min in 4% Giemsa, rinsed with distilled water and air dried. One hundred metaphase cells per animal were analysed for chromosomal aberrations.

5 Results

Maximum tolerated dose (MTD) determinations

To allow a direct comparison between this study and an earlier clastogenic study on BP and 2AAF (Ivett *et al.*,1984; Tice *et al.*, this volume), an MTD of 250 mg/kg for BP (based on lethality) and of 1000 mg/kg for 2AAF (based on cell cycle delay) were used. Treatment with PYR at doses of 500, 1000, and 1500 mg/kg did not kill any animals nor did it cause any delay in bone marrow cellular proliferation. A dose of 2000 mg/kg was selected as the MTD because of difficulty in obtaining a homogenous suspension at higher doses (Table 1). As with PYR, treatment with 4AAF did not cause any animal lethality nor inhibit bone marrow proliferation when tested at doses of 250, 500, and 1000 mg/kg (Table 1). In a second study, conducted at 1500 and 2000 mg/kg, 4AAF again induced no animal lethality nor inhibited cell proliferation kinetics. Because of difficulty in achieving a homo-

Table 1 *Maximum tolerated dose determinations*[a]

Chemical	Dose (mg/kg)	AGT (h)	Chemical	Experiment	Dose (mg/kg)	AGT (h)
PYR[b]	0	11.4	4AAF[b]	I	0	12.4
	500	11.1			250	12.2
	1000	10.6			500	12.6
	1500	10.9			1000	12.4
				II	0	11.4
					1500	13.6
					2000	12.3

[a] Three animals were treated and survived for 24 h at each dose.
[b] MTD of 2000 mg/kg selected based on suspension characteristics.

genous suspension at higher doses, an MTD of 2000 mg/kg was selected.

Chromosomal aberration (CA) and average generation time (AGT)

BP Data on the CA and AGT response to treatment with BP are presented in Table 2. A dose-related increase in gaps and breaks resulted in a statistically significant increase in the number of aberrations per cell and in the percentage of damaged cells, excluding and including gaps ($P < 0.5$). The increase in aberrations was entirely due to chromatid breaks and gaps; an increase in complex rearrangements was not observed. The lowest effective dose was 250 mg/kg. Cellular proliferation kinetics, as measured by AGT, was significantly affected with the greatest effect (+2.2 h) occurring at the highest dose tested (250 mg/kg). No lethality occurred at any of the doses tested. The positive control, MMC, induced a significant increase ($P < 0.05$) in both measures of clastogenicity.

PYR Following treatment with PYR, regardless of

Table 2. *CA and AGT for BP*

BP[a] (mg/kg)	Gaps[b]	Breaks[b]	Complex aberrations	Aberrations per cell[c] (+Gaps)	Aberrations per cell[c] (−Gaps)	% Damaged cells (+Gaps)	% Damaged cells (−Gaps)	AGT (h)[d]	n
0	0	1	0	0.002 (0.000–0.020)[e]	0.002 (0.000–0.020)	0.25 (0.00–2.00)	0.25 (0.00–2.00)	10.6 ± 0.4	9
62.5	3	2	1	0.015 (0.000–0.040)	0.008 (0.000–0.020)	1.50 (0.00–4.00)	0.75 (0.00–2.00)	10.9 ± 0.2	10
125	5	5	0	0.025 (0.000–0.100)	0.012 (0.000–0.080)	1.75 (0.00–6.00)	1.00 (0.00–6.00)	11.0 ± 0.2	10
250	7	11	0	0.045 (0.020–0.080)	0.027 (0.000–0.060)	4.25 (0.00–8.00)	2.75 (0.00–6.00)	12.8 ± 0.4	10
P value[f]				0.0057[g]	0.0216[g]	0.0039[g]	0.0115[g]	0.0000[g]	
MMC positive control data									
0	1	0	0	0.002 (0.000–0.020)	0.000	0.25 (0.00–2.00)	0.00		
1.0	3	15	3	0.050 (0.000–0.160)	0.042 (0.000–0.120)	4.25 (0.00–12.00)	3.75 (0.00–10.00)		
P value[f]				0.0105[g]	0.0036[g]	0.0103[g]	0.0037[g]		

[a] Fifty cells were scored for each of eight animals (total of 400 cells) per dose.
[b] Induces chromatid and isochromatid events.
[c] Number of chromosomal aberrations per cell (± gaps). Simple aberrations (chromatid and isochromatid events) and complex aberrations (chromatid interchanges) are combined.
[d] Group mean AGT value ± standard error of the mean among n animals.
[e] Numbers in parentheses indicate the range of individual animal values.
[f] CA data analysed by the Kruskal-Wallis test using individual animal values; AGT data analysed by one-way ANOVA.
[g] Significantly different at $\alpha = 0.05$.

Table 3. *CA and AGT data for PYR*

PYR[a] (mg/kg)	Gaps[b]	Breaks[b]	Complex aberrations	Aberrations per cell[c] (+ Gaps)	Aberrations per cell[c] (− Gaps)	%Damaged cells (+ Gaps)	%Damaged cells (− Gaps)	AGT (h)[d]	n
0	1	1	0	0.005 (0.000–0.020)[e]	0.002 (0.000–0.020)	0.50 (0.00–2.00)	0.25 (0.00–2.00)	12.4 ± 0.5	9
500	0	1	0	0.002 (0.000–0.020)	0.002 (0.000–0.020)	1.25 (0.00–2.00)	0.25 (0.00–2.00)	11.2 ± 0.3	10
1000	0	3	0	0.007 (0.000–0.040)	0.007 (0.000–0.040)	0.75 (0.00–4.00)	0.75 (0.00–4.00)	14.1 ± 0.6	10
2000	4	4	1	0.022 (0.000–0.100)	0.012 (0.000–0.040)	2.00 (0.00–8.00)	1.25 (0.00–4.00)	14.6 ± 0.6	12
P value[f]				0.1408	0.2626	0.1608	0.2626	0.0002[g]	
MMC positive control data									
0	3	4	0	0.017 (0.000–0.060)	0.010 (0.000–0.040)	1.75 (0.00–6.00)	1.00 (0.00–4.00)		
1.0	15	14	0	0.072 (0.040–0.200)	0.040 (0.000–0.120)	6.00 (4.00–10.00)	3.25 (0.00–6.00)		
P value[f]				0.0077[g]	0.0678	0.0077[g]	0.0671		

See Table 2 for explanation of footnotes.

whether gaps were included or excluded in the analysis, a significant increase in either aberrations per cell or in the percentage of damaged cells was not observed. However, there was slight increase in chromatid breaks and gaps at the highest dose tested (i.e. 2000 mg/kg) (Table 3). The AGT was significantly increased, with the largest increase (+2.2 h) occurring at the highest dose tested (2000 mg/kg). No lethality occurred among any of the treated animals. MMC induced a significant increase in aberrations per cell and in the percentage of damaged cells when gaps were included in the analysis ($P = 0.0077$), but MMC induced a marginally negative response when gaps were excluded from the analysis ($P = 0.068$).

2AAF Data on the CA and AGT response to treatment with 2AAF appear in Table 4. No complex aberrations were observed in any of the analysed material. There was a significant increase ($P = 0.0365$) in the number of breaks per cell and in the percentage of damaged cells (excluding gaps), but the response was not dose-dependent. Treatment with 2AAF had no significant effect on AGT, and no lethality occurred among any of the treatment groups. MMC was found to be a positive clastogen, when gaps were including in the analysis (see PYR).

To exclude the possibility that the apparent lack of a positive response for 2AAF-induced CAs at the higher doses occurred because of cell sample size (i.e. 50 cells per animal), all slides were recoded and reanalysed and 100 cells were scored per animal (Table 5). In this analysis, although the level of CAs at an intermediate dose (500 mg/kg) was elevated, a significant increase in aberrations per cell or in the percentage of damaged cells was not observed.

Data from male C31F1 mice treated with 2AAF

Table 4. *CA and AGT data for 2AAF*

2AAF[a] (mg/kg)	Gaps[b]	Breaks[b]	Complex aberrations	Aberrations per cell[c] (+Gaps)	Aberrations per cell[c] (−Gaps)	% Damaged cells (+Gaps)	% Damaged cells (−Gaps)	AGT (h)[d]	*n*
0	3	1	0	0.010	0.002	0.75	0.25	117 ± 0.4	9
				(0.000–0.060)[e]	(0.000–0.020)	(0.00–4.00)	(0.00–2.00)		
250	5	7	0	0.030	0.017	3.00	1.75	12.8 ± 0.5	10
				(0.000–0.060)	(0.000–0.040)	(2.00–6.00)	(0.00–4.00)		
500	4	2	0	0.015	0.005	1.50	0.50	13.3 ± 0.6	10
				(0.000–0.040)	(0.000–0.020)	(0.00–4.00)	(0.00–2.00)		
1000	4	2	0	0.015	0.005	1.50	0.50	12.6 ± 0.5	9
				(0.000–0.040)	(0.000–0.020)	(0.00–4.00)	(0.00–2.00)		
P value[f]				0.0662	0.0365[g]	0.0497[g]	0.0365[g]	0.1711	
MMC positive control data									
0	3	4	0	0.017	0.010	1.75	1.00		
				(0.000–0.060)	(0.000–0.040)	(0.00–6.00)	(0.00–4.00)		
1.0	15	14	0	0.072	0.040	6.00	3.25		
				(0.040–0.200)	(0.000–0.120)	(4.00–10.00)	(0.00–6.00)		
P value[f]				0.0077[g]	0.0678	0.0077[g]	0.0671		

See Table 2 for explanation of footnotes.

Table 5. *CA data for 2AAF – Second analysis*

2AAF[a] (mg/kg)	Gaps[b]	Breaks[b]	Complex aberrations	Aberrations per cell[c] (+Gaps)	Aberrations per cell[c] (−Gaps)	% Damaged cells (+Gaps)	% Damaged cells (−Gaps)
0	5	4	0	0.011	0.005	1.00	0.50
				(0.000–0.040)[d]	(0.000–0.020)	(0.00–3.00)	(0.00–2.00)
250	8	7	0	0.019	0.009	1.50	0.88
				(0.000–0.050)	(0.000–0.020)	(0.00–4.00)	(0.00–2.00)
500	16	13	0	0.036	0.016	3.38	1.50
				(0.000–0.047)	(0.000–0.030)	(0.00–7.00)	(0.00–3.00)
1000	12	7	0	0.024	0.007	2.38	0.75
				(0.000–0.080)	(0.000–0.020)	(0.00–8.00)	(0.00–2.00)
P value[e]				0.0752	0.1330	0.0508[f]	0.1603

[a] One hundred cells were scored for each of eight animals (total of 800 cells) per dose.
[b] Includes chromatid and isochromatid events.
[c] Number of chromosomal aberrations per cell (± gaps). Simple aberrations (chromatid and isochromatid events) and complex aberrations (chromatid interchanges) are combined.
[d] Numbers in parentheses indicate the range of individual animal values.
[e] CA data analyzed by the Kruskal-Wallis test using individual animal values.
[f] Significantly different at $\alpha = 0.05$.

(446 mg/kg) following the protocol of Adler (this volume) appear in Table 6. Treatment with 2AAF resulted in an increased number of gaps and breaks and of complex rearrangements. These data, when analysed by Kruskal-Wallis, are significant (at $\alpha = 0.05$) only when aberrations per cell (including gaps) are considered ($P = 0.0276$). Probability values for the other measures of clastogenic activity ranged from 0.0562 for the percentage of damaged cells (including gaps) to 0.1105 for aberrations per cell (excluding gaps). When the same data are analysed by the same statistical method (chi-square analysis) used by Adler, the percentage of damaged cell data (± gaps) becomes highly significant (+ gaps, $P = 0.0014$; – gaps, $P = 0.0259$).

4AAF Chromosomal aberration and AGT summaries for 4AAF can be found in Table 7. A significant treatment-related difference was not detected in any of the categories analysed. There was a significant effect on cellular proliferation kinetics with the maximum delay in generation time (+ 3.5 h) occurring at the highest tested dose (2000 mg/kg). No lethality occurred among any of the treated animals. Exposure to MMC resulted in an increase in all 3 classes of aberrations and, although the percentage of damaged cell data (excluding gaps) was not statistically significant ($P = 0.0631$), it was significant when gaps were included ($P = 0.0217$).

6 Discussion

The inclusion of the BrdUrd methodology for identifying the replicative history of individual metaphase cells examined in *in vivo* clastogenic studies permits (1) evaluation of the ability of the test chemical to affect the rate of cellular proliferation in the target tissue, and (2) restriction of the assessment of CAs to the cell population most likely to exhibit clastogenic damage. An analysis of cell cycle kinetics provides an independent measure of treatment-induced cellular toxicity. In the absence of clastogenic damage, the presence of cell cycle delay can be an indication that the chemical has reached the target

Table 6. *CA data for 2AAF in C31F1 mice*

2AAF[a] (mg/kg)	Number animals	Number cells	Gaps[b]	Breaks[b]	Complex aberrations	Aberrations per cell[c] (+ Gaps)	Aberrations per cell[c] (– Gaps)	% Damaged cells (+ Gaps)	% Damaged cells (– Gaps)
0	5	500	8	7	0	0.030 (0.000–0.050)[c]	0.016 (0.000–0.030)	2.80 (0.00–5.00)	1.40 (0.00–3.00)
446	5	483[d]	19	17	3	0.080 (0.040–0.120)	0.040 (0.010–0.060)	7.20 (4.00–12.00)	3.60 (1.00–6.00)
Kruskal-Wallis P value[e]						0.0278[f]	0.1105	0.0562	0.0897
Chi-square P value								0.0014[f]	0.0258[f]

[a] Includes chromatid and isochromatid events.
[b] Number of chromosomal aberrations per cell (± gaps). Simple aberrations (chromatid and isochromatid events) and complex aberrations (chromatid interchanges) are combined.
[c] Numbers in parentheses indicate the range of individual animal values.
[d] 83 cells were scored in one animal.
[e] Data analyzed by the Kruskal-Wallis test using individual animal values.
[f] Significantly different at $\alpha = 0.05$.

Table 7. *CA and AGT data for 4AAF*

4AAF[a] (mg/kg)	Gaps[b]	Breaks[b]	Complex aberrations	Aberrations per cell[c] (+ Gaps)	Aberrations per cell[c] (– Gaps)	% Damaged cells (+ Gaps)	% Damaged cells (– Gaps)	AGT (h)[d]	n
0	3	3	0	0.015 (0.000–0.060)[e]	0.007 (0.000–0.040)	1.50 (0.00–6.00)	0.75 (0.00–4.00)	12.7 ± 0.9	10
500	3	5	0	0.020 (0.000–0.040)	0.012 (0.000–0.020)	2.00 (0.00–4.00)	1.25 (0.00–2.00)	13.6 ± 0.8	10
1000	2	1	1	0.010 (0.000–0.020)	0.005 (0.000–0.020)	1.00 (0.00–2.00)	0.50 (0.00–2.00)	16.0 ± 0.8	10
2000	4	4	1	0.012 (0.000–0.020)	0.002 (0.000–0.020)	1.25 (0.00–2.00)	0.25 (0.00–2.00)	16.2 ± 0.7	8
P value[f]				0.6916	0.2924	0.5549	0.2059	0.0094[g]	
MMC positive control data									
0	5	3	0	0.020 (0.000–0.040)	0.007 (0.000–0.040)	2.00 (0.00–4.00)	0.75 (0.00–4.00)		
1.0	9	12	5	0.095 (0.000–0.200)	0.065 (0.000–0.160)	6.25 (0.00–12.00)	4.00 (0.00–10.00)		
P value[f]				0.0220[g]	0.0558	0.0217[g]	0.0631		

See Table 2 for explanation of footnotes.

tissue (Tice and Ivett, 1985). The selection of the optimal metaphase cell population to examine for CAs is based on 2 competing forces. For the induction of CAs by most chemicals, the cell being examined must have progressed through one (or most of one) round of DNA replication since the induction of the majority of the DNA damage. However, if sampling occurs at late times after the induction of CAs (i.e. after enough time that multiple cell generations would have occurred), the frequency of aberrant cells will decrease due to selective toxicity and/or to the dilution of damage among daughter cells. Unless pharmacokinetic considerations limit the formation of the clastogenic species, the optimal cell population to analyse for CAs are metaphase cells that have undergone only one round of DNA replication since treatment. When kinetic analysis based on BrdUrd-dependent differential staining patterns of the tissue being scored for CAs is included in the clastogenicity assay system, the selective scoring of first generation, post-treatment metaphase cells is achieved easily, regardless of the extent of agent-induced proliferative delay. Further, it has been observed that first generation, post-treatment metaphase cells appearing at late times after treatment exhibit more damage than first generation, post-treatment metaphase cells appearing at early times after treatment (McFee *et al.*, 1983). This pattern would be expected if it takes longer for a more heavily damaged cell to divide than a less damaged cell. Thus, by selectively sampling late-appearing first generation metaphase cells, increased sensitivity in the assay is achieved. The occurrence of a confounding interaction between BrdUrd-substituted DNA and the test chemical remains a possibility (Schvartzman and Tice, 1982). However, CA yields for the limited number of clastogenic agents that have been tested *in vivo*, using the present protocol or in the absence of BrdUrd, are not significantly different when a sampling time which ensures 100% first generation, post-treatment metaphase cells in the non BrdUrd tablet containing animals is used (McFee *et al.*, 1986). The methodology used in the present studies appear to offer increased resolving power at less effort then the traditional cytogenetic approach using multiple sampling times.

Exposure to BP and 2AAF significantly increased (at an $\alpha = 0.05$) the frequency of aberrant cells and the number of aberrations per cell. The marginal, dose-independent response for 2AAF was reappraised by re-examining the slides and scoring twice the original number of metaphase cells for aberrations. Although an increase in the number of gaps and breaks was observed, the treatment-related response was not statistically significant ($P > 0.05$). The ability of 2AAF to induce CAs was examined further by attempting to duplicate the positive response observed by Adler in male C31F1 mice, sampled at 24 h after an ip administration of 2AAF at a dose of 446 mg/kg. When the resulting data were analyzed using the same statistical approach of Adler (i.e., combining individual cell data and analyzing by chi-square), the response was clearly positive. However, when individual animal values were analysed by Kruskal-Wallis, the results were not statistically significant. It is concluded from these various studies that the carcinogen 2AAF is weakly clastogenic in bone marrow cells of mice following ip administration and that the magnitude of the clastogenic activity may be strain specific. The lack of a dose-related increase in CAs following treatment with 2AAF could suggest an experimental artifact related to administering doses exceeding the solubility of the chemical in the solvent. However, since the SCE response to 2AAF increased linearly over the same dose range (Tice *et al.*, this volume), this explanation must be considered unlikely. Exposure to the two purported noncarcinogens, PYR and 4AAF, up to doses that induced significant cell cycle delay, did not induce a significant increase in CAs. The clastogenic response to PYR, although statistically negative, was suggestive as there was an increasing absolute number of aberrant cells with increasing dose.

It is interesting to compare the dose-response curves for BP-induced CAs with that of BP-induced SCEs (Tice *et al.*, this volume). The dose-response curve for SCE induction saturates at the lowest dose tested (62.5 mg/kg). However, at identical doses, the induction of CAs by BP increases with increasing dose. This difference in dose-response curves can be explained if, as it has been suggested for benzene-induced SCEs and CAs in bone marrow cells *in vivo* (Tice *et al.*, 1980), different reactive metabolites of BP (having different pharmacokinetics) are responsible for the induction of one or the other, but not both genotoxic endpoints. This possibility requires further consideration and analysis.

A variety of statistical methods have been used to analyse *in vivo* CA data. For *in vivo* studies, the unit of exposure is the animal, and unless statistically proven

Table 8. *Statistical analysis of in vivo mouse bone marrow chromosomal aberration data using aberrant cell information*

	+ Gaps				− Gaps			
Chemical	KW[a]	Chi-square[b]	G Test[c]	Trend test[d]	KW[a]	Chi-square[b]	G Test[c]	Trend test[d]
2AAF	0.0497[e]	0.0920	0.1011	0.7946	0.0365[e]	0.0605	0.0906	0.7955
4AAF	0.5549	0.6722	0.6786	0.5109	0.2059	0.3613	0.3581	0.2233
BP	0.0039[e]	0.0005[e]	0.0004[e]	0.0001[e]	0.0115[e]	0.0071[e]	0.0099[e]	0.0014[e]
PYR	0.1680	0.0391[e]	0.0543	0.0164[e]	0.2626	0.2188	0.2225	0.0470[e]

[a] P values for the Kruskal-Wallis test using individual animal values.
[b] P values for the Chi-square analysis (2 × 2 tests of independence) using total cell data.
[c] P values for the G test of independence using total cell data.
[d] P values for the Trend test using a linear model and total cell data.
[e] Significantly different at $\alpha = 0.05$

otherwise, *in vivo* CA data should be analysed statistically using individual animal responses. Due to the low frequency of CAs within each classification group, data on individual aberrations within an animal (including and excluding gaps) are combined into 2 measures of clastogenic activity: (1) the percentage of damaged cells and (2) the mean number of aberrations per cell. Use of the number measure avoids the issue of how to analyse seriously damaged cells, while the latter measure attempts to take into account the occurrence of such cells. However, when calculating the mean number of aberrations per cell, all aberrations are considered to have equal weight, and no attempt is made to estimate the number of breakage events involved. Achromatic lesions (gaps) are included and excluded in these analyses because, although gaps may not represent true chromosomal mutations, their classification vs that of a chromosome break is somewhat arbitrary and differs considerably among different laboratories. Data on the individual types of CAs observed can be used to give weight to the decision as to the chemical's clastogenicity and/or to indicate the need for additional studies.

To compare the statistical methods used in analysing these studies vs other more traditional approaches (e.g. chi-square analysis), these data were analysed also after combining all cell data within each exposure group. Tables 8 and 9 present the *P* values determined for the different statistical approaches used to analyse the percentage of damaged cell data and the total number of CAs scored for each of the 4 chemicals. When considering the total number of damaged cells (Table 8), the data on BP become more highly significant, the data on PYR become significant (especially if the analysis by the trend test is considered) the data on 2AAF become less significant, and the data on 4AAF remain statistically nonsignificant. For total aberrations scored (Table 9), the data on BP remain significant, the data on PYR become significant when data on gaps are included, the data on 2AAF becomes insignificant, and the data on 4AAF remain nonsignificant. As discussed earlier, the 2AAF data on the C31F1 mice become highly significant when individual animal data are combined. Thus, depending on the statistical approach used, the same *in vivo* CA data can be concluded to be significantly or nonsignificantly different. The use of individual animal responses is probably a more biologically valid approach for the analysis of *in vivo* CA data, but a more rigorous statistical analysis of such data is required before the most appropriate statistical methods can be determined.

7 Conclusion

When administered ip, BP induced a significant increase in CAs in bone marrow cells of male B6C3F1 mice. Treatment with 2AAF was marginally clastogenic, but the response was not dose dependent. When 2AAF was administered ip to male C31F1 mice, the clastogenic response was more dramatic but not statistically significant when analysed by the Kruskal-Wallis test. When analysed by the chi-square test for independence, these 2AAF data were highly significant. Exposure to 4AAF and PYR was ineffective in inducing CAs up to doses which induced significant cell cycle delay. The difference in *P* values for each CA assay, depending on whether the total cell data are combined among animals within an exposure group or animal-to-animal variability is considered a component in the analysis, indicates the importance of the statistical approach in assessing the clastogenicity of a chemical.

8 Acknowledgements

The authors would like to thank Ms. L. Wasson and Ms. G. Callister for typing this contribution. The work in this contribution was supported by the National Toxicology Program under National Institute of Health Sciences Interagency Agreement No. YO1-ES-20098-05 and by the U.S. Department of Energy under Contract DE-AC02-76CH00016. Accordingly, the U.S. Government retains a nonexclusive, royalty-free license to publish or reproduce the published form of this contribution, or allow others to do so, for U.S. Government purposes.

9 References

Adler, I-A. *et al.*, this volume.

Allen, J.W., C.F. Schuler, R.W. Mendes, and S.A. Latt (1979) A simplified technique for *in vivo* analysis of sister chromatid exchanges using 5-bromodeoxyuridine tablets, *Cytogent. Cell Genet.*, **18**, 231–237.

Ashby, J. (1981) Overview of study and test chemical activities. In F.J. de Serres and J. Ashby (Eds)

Table 9. *Statistical analysis of in vivo mouse bone marrow chromosomal aberration data using aberrations per cell (KW) or total aberration data*[a] *(chi-square and G test)*

	+ Gaps			− Gaps		
Chemical	KW[b]	Chi-square[c]	G Test[d]	KW[b]	Chi-square[c]	G Test[d]
2AAF	0.0662	> 0.05	> 0.05	0.0365[g]	> 0.05	> 0.05
4AAF	0.6916	> 0.05	> 0.05	0.2924	> 0.05	> 0.05
BP	0.0057[g]	< 0.01[g]	< 0.01[g]	0.0216[g]	< 0.05[g]	< 0.05[g]
PYR	0.1408	< 0.05[g]	< 0.05[g]	0.2626	> 0.05	> 0.05

[a] Simple and complex aberrations added together without taking into consideration the number of breakage events involved.
[b] P values for the Kruskal-Wallis test using individual animal values.
[c] P values for the Chi-square analysis (one-way goodness of fit) using total cell data.
[d] P values for the G test using total cell data.
[g] Significantly different at $\alpha = 0.05$.

Evaluation of Short-Term Tests for Carcinogens, Report of the International Collaborative Program, Elsevier/North-Holland, Amsterdam, pp. 112–171.

Ashby, J. (1986) this volume.

Cochran, W.G. (1954) Some methods for strengthening the common Chi-square test, *Biometrics*, **10**, 417–441.

Goto, K., T. Akematasa, H. Shimazu, and T. Sugiyama (1975) Simple differential Giemsa staining of sister chromatids after treatment with photosensitive dyes and exposure to light and the mechanism of staining, *Chromosoma*, **53**, 223–230.

IARC (1982) *Evaluation of the Carcinogenic Risk of Chemicals to Humans*, Suppl. 4, Chemicals, Industrial Processes and Industries Associated with Cancer in Humans, World Health Organization International Agency for Research on Cancer, Lyon, France.

IARC (1983) *Evaluation of the Carcinogenic Risk of Chemicals to Humans*, Vol. 32, Polynuclear Aromatic Compounds, Part I, Chemical, Environmental, and Experimental Data, World Health Organization International Agency for Research in Cancer, Lyon, France.

Ivett, J.L. and R.R. Tice (1982) Average Generation Time: A new method of analysis and quantitation of cellular proliferation kinetics, *Environ, Mutagen.*, **4**, 358.

Ivett, J.L., C.A. Luke and R.R. Tice (1984) Range finding studies for determining the optimal doses and termination times for evaluating chromosomal aberrations and sister chromatid exchanges *in vivo*, *Environ. Mutagen.*, **6**, 468.

Kriek, E. (1974) Carcinogenesis by aromatic amines. *Biochim. Biophys. Acta*, **355**, 177–203.

McFee, A.F., K.W. Lowe, and J.R. San Sebastion (1983) Improved sister chromatid differentiation using paraffin-coated bromodeoxyuridine tablets in mice, *Mutation Res.*, **119**, 83–88.

McFee, A.F., K.W. Lowe, and M.N. Sherrill (1984) Influence of time on the *in vivo* induction of sister chromatid exchanges and chromosomal aberrations, *Environ. Mutagen.*, **6**, 469.

McFee, A.F., R.R. Tice and M.D. Shelby (1986) Interlaboratory standardization of procedures for the *in vivo* mouse bone marrow cytogenetics assay. In preparation.

NCI (1951–1978) *Survey of Compound Which Have Been Tested for Carcinogenic Activity*. U.S. Dept. of Health and Human Services, Wasington, DC.

Preston, R.J., W. Au, M.A. Bender, J.G. Brewen, A.V. Carrano, J.A. Heddle, A.F. McFee, S. Wolff, and J.S. Wassom (1981) Mammalian *in vivo* and *in vitro* cytogenetic assays: A report of the Gene-Tox Program. *Mutation Res.*, **87**, 143–188.

Schvartzman, J.B. and R.R. Tice (1982) 5-Bromodeoxyuridine and its role in the production of sister chromatid exchanges. In A.A. Sandberg (Ed.), *Sister Chromatid Exchanges*, Alan R. Liss Press, NY, pp. 123–134.

Sokal, R.R. and F.J. Rohlf (1969) *Biometry*, Second Edition, W.H. Freemand and Co., San Francisco, CA.

Tice, R.R. and J.L. Ivett (1985) Cytogenetic analysis of bone marrow damage. In R.D. Irons (Ed.), *Toxicology of the Blood and Bone Marrow*, Raven Press, NY, pp. 119–140.

Tice, R.R., E.L. Schneider, and J.M. Rary (1976) The utilization of bromodeoxyuridine incorporation into DNA for the analysis of cellular kinetics, *Exp. Cell Res.*, **102**, 232–236.

Tice, R.R., M.A. Bender, J.L. Ivett and R.T. Drew (1978) Cytogenetic effects of inhaled ozone, *Mutation Res.*, **58**, 293–304.

Tice, R.R., D.L. Costa and R.T. Drew (1980) Cytogenetic effects of inhaled benzene in murine bone marrow. Induction of sister chromatid exchanges, chromosomal aberrations and cellular proliferation inhibition in DBA/2 mice, *Proc. Natl. Acad. Sci. USA*, **77**, 2148–2152.

Tice, R.R., C.A. Luke, V.M. Miller, B.G. Ormiston, A.C. Wilton and M.D. Shelby (1986) *A comparative analysis of the induction of sister chromatid exchanges in the bone marrow cells of B6C3F1 mice*, this volume.

Rosario Albanese
Imperial Chemical Industries PLC, Pharmaceuticals Division, Alderley Park, Cheshire, UK

15

An in vivo cytogenetic test in bone-marrow cells of the Alderley Park Wistar-derived rat

1 Summary

BP, PYR, 4AAF and 2AAF were administered to Alderley Park Wistar-derived female rats. The compounds were formulated as ball-milled suspensions in corn oil. Corn oil and cyclophosphamide (CY) were used as the vehicle and reference positive controls, respectively. Corn oil, BP, and PYR were administered as single oral doses (by gavage) and 4AAF, 2AAF, and CY were administered as single intraperitoneal (ip) doses. Bone marrow cells from each animal were analysed for structural chromosome damage, 24 hrs after dosing.

The incidence of chromosomally aberrant cells in the bone marrow of rats dosed with PYR was similar to the vehicle control. The incidence of chromosomally aberrant cells in the BP, 2AAF, 4AAF (aberrations excluding gaps), and CY dosed groups was statistically significantly increased compared to the vehicle control. However, the increased number of chromosomally aberrant cells was dose-related in the BP dosed groups only. Thus, only the results obtained with BP were considered to be indicative of a true clastogenic response.

2 Introduction

The *in vivo* cytogenetic test is a short-term mutagenicity test and detects the potential of a compound to induce structural chromosome aberrations. Bone marrow cells are examined as the tissue contains a large population of dividing cells with a relatively short cell-cycle time. A single expression time (24 hrs after dosing) was used. For screening purposes, this time interval provides a good indication of clastogenic potential (Topham *et al.*, 1983). Each compound was dosed at the approximate maximum tolerated dose (MTD) and $\frac{1}{2}$ MTD. The doses and route of administration for each compound were selected after consideration of the toxicity data provided by the IPCS group organiser. The method described in this paper allows the scoring of chromosomal aberrations and micronuclei in bone marrow cells from the same animal (the "cytonucleus test", Albanese and Watkins, 1983; Albanese, 1985). This paper discusses the results of the chromosomal aberration assay only. The results of the parallel micronucleus study are reported by Watkins (this volume).

3 Materials and methods

BP, PYR, 4AAF, and 2AAF were supplied by ICI UK and were formulated as ball-milled (overnight) suspensions in corn oil. Corn oil (Kraft Wesson) and cyclophosphamide (in saline; Koch Light Laboratories) were used as the vehicle and reference positive controls, respectively. Corn oil (5 ml/kg), BP (100 and 200 mg/kg), and PYR (250 and 500 mg/kg) were administered as single oral doses (by gavage). 4AAF (375 and 500 mg/kg), 2AAF (150 and 300 mg/kg), and cyclophosphamide (CY; 20 mg/kg) were administered as single ip doses. (Doses of 1000 and 750 mg/kg 4AAF were not tolerated and the doses were reduced to 500 and 375 mg/kg, see Table 1).

Healthy female Alderley Park Wistar-derived rats weighing 100–120 g were used. Animals were acclimatized in the animal room for 7 days prior to the start of the study. The animals were individually identified by an ear mark and were fed pelleted PCD diet with water available ad libitum. For convenience of handling, dosing was subdivided over 4 consecutive days with animals from each dose group represented in each of these parts. The rats were housed (up to 4 per cage) in galvanised steel mesh cages (dimensions 46 × 28 × 21 cm). The animals were randomly allocated to cages and groups using a randomisation plan provided by the statistician.

Chromosome preparations were made 24 hrs after dosing. One and a half hrs prior to necropsy, 5 mg/kg vinblastine sulphate was administered ip to each animal. Cells were flushed out of a single dissected femur from each animal (using 3 ml foetal calf serum) into 10 ml centrifuge tubes. The resulting cell suspensions were centrifuged at 1000 rpm for 5 mins. The supernatant was discarded and the pellets resuspended in the remaining fluid. One drop of this cell suspension was spotted onto a clean, grease-free microscope slide and then drawn across the slide, leaving a "bone-marrow cell film". These slides were used to assess the incidence of micronuclei (Watkins, this volume). 5.0 ml 0.075 M KCl was added to each of the remaining cell suspensions and left at room temperature for 15 mins. Chromosome preparations from each animal were made following conventional methods (i.e. the cells were centrifuged at 1000 rpm for 5 mins; the supernatant was discarded and the cells fixed using 3:1 methanol: glacial acetic acid fixative). This centrifugation/fixative process was repeated 2 more times and air-dried chromosome preparations were made from each animal (Cohen and Hirschhorn, 1971). The cells were stained using 2% Giemsa solution (Gurrs improved R66 in buffer, pH 6.8) for 2 mins. When dry, the slides were dipped in xylene and mounted using DPX (BDH Ltd.). Before analysis, the slides were coded using a series of random numbers. Where possible 50 metaphase cells from each animal were analysed for structural chromosome damage. The number of cells with aberrations (including and excluding gaps) and the number of polyploid cells in each dose group was compared to the vehicle control group, using Fisher's exact test for 2 ≃ 2 tables and single-sided significance level.

4 Results

The results are summarised in Table 1. The incidence of chromosomally aberrant cells in rats dosed with PYR was similar to that of the vehicle control. In the 4AAF-dosed groups there was a statistically significant increase in the incidence of aberrant cells excluding gaps (Table 1), but this was within the historical control range for this laboratory (range 0–1.0%). The incidence of chromosomally aberrant cells (including and excluding gaps) was statistically significantly increased in the BP- and 2AAF-dosed groups compared to the vehicle control. For aberrations (gaps included) there was a 4- and 9-fold increase at 100 and 200 mg/kg BP, respectively and an approximate 9-fold increase at both 150 and 300 mg/kg 2AAF. CY, the positive control, caused the expected significant increase (34-fold) compared to the vehicle control.

5 Discussion

The chromosomal aberrations recorded in the BP-dosed groups included not only gaps and breaks but also chromosomal fragments and exchange figures. The in-

Table 1. *Incidence of chromosome aberrations in Alderley Park female rats*

Compound and dose (mg/kg)		Number of animals	Number of cells: Analysed	With gaps only	With breaks	With exchanges	With multiple damage	With Polyploidy	% of cells with: Gaps only	Aberrations excluding gaps[b]	Aberrations including gaps[b]
Corn oil		20	963	4	0	0	0	13	0.4	0	0.4
BP	100	10	500	2	6	0	0	6	0.4	1.2**	1.6*
	200	10	443	4	11	1	0	10	0.9	2.7***	3.6***
PYR	250	10	440	2	0	0	0	2	0.4	0	0.4
	500	10	426	2	2	0	0	0	0.4	0.4	0.8
4AAF	375	7[c]	283	0	3	0	0	1	0	1.0*	1.0
	500	8[d]	365	0	4	0	0	0	0	1.0**	1.0
2AAF	150	10	419	7	8	0	0	0	1.7*	1.9***	3.6***
	300	8	301	6	6	0	0	0	1.9*	1.9***	3.8***
CY	20	10	436	15	24	6	15	8	3.4**	10.1***	13.5***

a = > 10 aberrations in one cell.
b = Statistical significant increase compared to vehicle control, corn oil, * $P < 0.05$, ** $P < 0.01$, *** $P < 0.001$.
c = 3 animals died after dosing (2 after 1000 mg/kg and 1 after 750 mg/kg).
d = 2 animals died after dosing.

cidence of chromosomally aberrant cells was also dose related, criteria indicative of a true clastogenic response. In contrast, the aberrations recorded in the 2AAF-dosed groups were gaps and breaks only, and the incidence was not dose related (Table 1). Thus, the results obtained with 2AAF may not be due to the chromosome damaging potential of the compound but due to systemic toxicity as a result of exposure to near-lethal doses.

6 Conclusions

BP induced chromosomal damage in the bone marrow cells of the rat but PYR did not.

Equivocal results were obtained with the 2AAF. 4AAF did not induce chromosomal abnormalities in bone marrow cells of the rat.

7 Acknowledgements

I would like to thank Dr. A. Ferguson for the statistical analysis, and I would also like to acknowledge the technical assistance provided by Dr. R. Dhadial and Mr I.J. Thompson.

8 References

Albanese, R. (1985) United Kingdom Environmental Mutagen Society Second Collaborative Study. 4-dimethylaminoazobenzene (DAB) and 4-cyano-dimethylaniline (CDA) – an *in vivo* cytogenetic study in the Alderley Park (Wistar) rat. In J.M. Parry and C.F. Arlett (eds) *Comparative Genetic Toxicology.* MacMillan, London pp. 475–481.

Albanese, R. and P.A. Watkins (1983). Is the "cytonucleus test" a possibility? *Environ. Mutagen.*, **5**, 945.

Cohen, M.M. and K. Hirschhorn (1971). Cytogenetic Studies in animals. In A. Hollaender (ed.) Chemical Mutagen. *Principles and Methods for their detection.* Vol. 2. Plenum Press, NY, pp. 515–534.

Topham, J.C., R. Albanese, J. Bootman, D. Scott and D. Tweats (1983). *In vivo* cytogenetic assays. In B.J. Dean (Ed.) *UKEMS Sub-Committee on Guidelines for Mutagenicity Testing Report.* Part I. Basic Test Battery. UKEMS Society Swansea, pp. 119–141.

Watkins, P.A. (1987) *Micronucleus Test in the Alderley Park Wistar-Derived Albino Rat* (this volume).

Jacques Pot-Deprun and Ivan Chouroulinkov
Unité de Cancérogénèse Expérimentale et Toxicologie Génétique, I.R.S.C.-CNRS, Villejuif, France

16

Mutagenic activity of four compounds in chromosomal aberrations assay using Chinese hamster bone marrow cells

1 Summary

The frequency of chromosomal aberrations induced in Chinese hamster bone marrow cells was evaluated for BP, PYR, 2AAF, and 4AAF. The treatments were done *in vivo* on 8 to 12-week-old male Chinese hamsters. BP and 2AAF gave positive responses in the chromosomal aberration test. 4AAF was weakly positive at the highest dose, while PYR was negative.

2 Materials and methods

Animals

Male Chinese hamsters were originally obtained from the specific pathogen-free breeding colony at the Institut de Recherches Scientifiques sur le Cancer, Villejuif, (France). The animals weighed 30–33 g, and were 8–12 weeks old. For each dose there were 4 hamsters in the series treated and 2 hamsters in the controls.

Test chemicals, handling and dilution

PYR was provided by Sigma Laboratories (USA), BP (Figure 1, 2) was synthesized and supplied by Laboratoire P. Jacquignon, Gif sur Yvette (France). 2AAF and 4AAF were supplied by ICI (UK). All the compounds were stored at −20 °C until used. Olive oil (extra pure, Codex) was chosen for the administration of the test chemicals. The insoluble compounds (2AAF and 4AAF) were administered as a suspension, while BP and PYR were dissolved in the same vehicle.

Treatment administration was by gavage. There were 3 dose groups for each chemical and doses were selected (as previously recommended) on the basis of the molecular weight except for BP and PYR which had been previously tested.

A single administration was chosen for these experiments, and the volume administered was 0.3 ml per animal.

The optimal expression time of 30 h after dosing was used for the 4 compounds.

Suspension medium and bone marrow cells harvest

One h before preparation the animals were injected with 0.2 ml of Colchicine (Colcinéos – Laboratories Houdé, France, 6 mg/10 ml of distilled water).

Bone marrow cells were harvested using the method of Kliesch *et al.* (1981). Cells were sampled from the femur

30 h after treatment in Hank's balanced salt solution (Gibco-Europe).

Hypotonic treatment was performed using a 1% sodium citrate solution (Merck) for 15 min at 37 °C.

Fixation, staining and cell count

The cells were suspended (3 times) in the fixation medium (methanol, acetic acid, 3:1), dropped on microscope slides, and gently flame dried. The cells were stained with 5% Giemsa in buffer pH 7.2.

For each treated animal 125 metaphase cells were analyzed, and for each control animal 200 cells were analyzed. Cells were examined under code and checked by double-blind examination.

Cells were analysed for chromatid type aberrations such as gaps, breaks, and fragments.

Table 1. *Chromosomal aberrations in Chinese hamster bone marrow cells induced by BP and PYR*

Compound (mol. wt.)	Dose (mg/kg)	Cells Scored (n)	Deletions *n*	Deletions % ± SE	Mitotic index % ± SE
	0	1200	22	1.8 ± 0.2	4.5 ± 0.1
BP	25	500	21	4.2 ± 0.5[a]	4.1 ± 0.2
(250)	50	500	30	6.0 ± 0.2	3.9 ± 0.3
	100	500	63	12.6 ± 0.6	2.7 ± 0.1
	0	1200	19	1.6 ± 0.1	4.6 ± 0.2
PYR	25	500	7	1.4 ± 0.3	4.0 ± 0.1
(202)	50	500	10	2.0 ± 0.5	3.8 ± 0.1
	100	500	8	1.6 ± 0.3	3.9 ± 0.1

[a] Lowest dose at which a significant elevation was found ($p < 0.02$).

Figure 1. Spectrophotometric UV analysis of BP: BP was dissolved in ethanol and the spectra were recorded in spectrophotometer Varian series 634. Wavelengths changed from 400 to 200 nm.

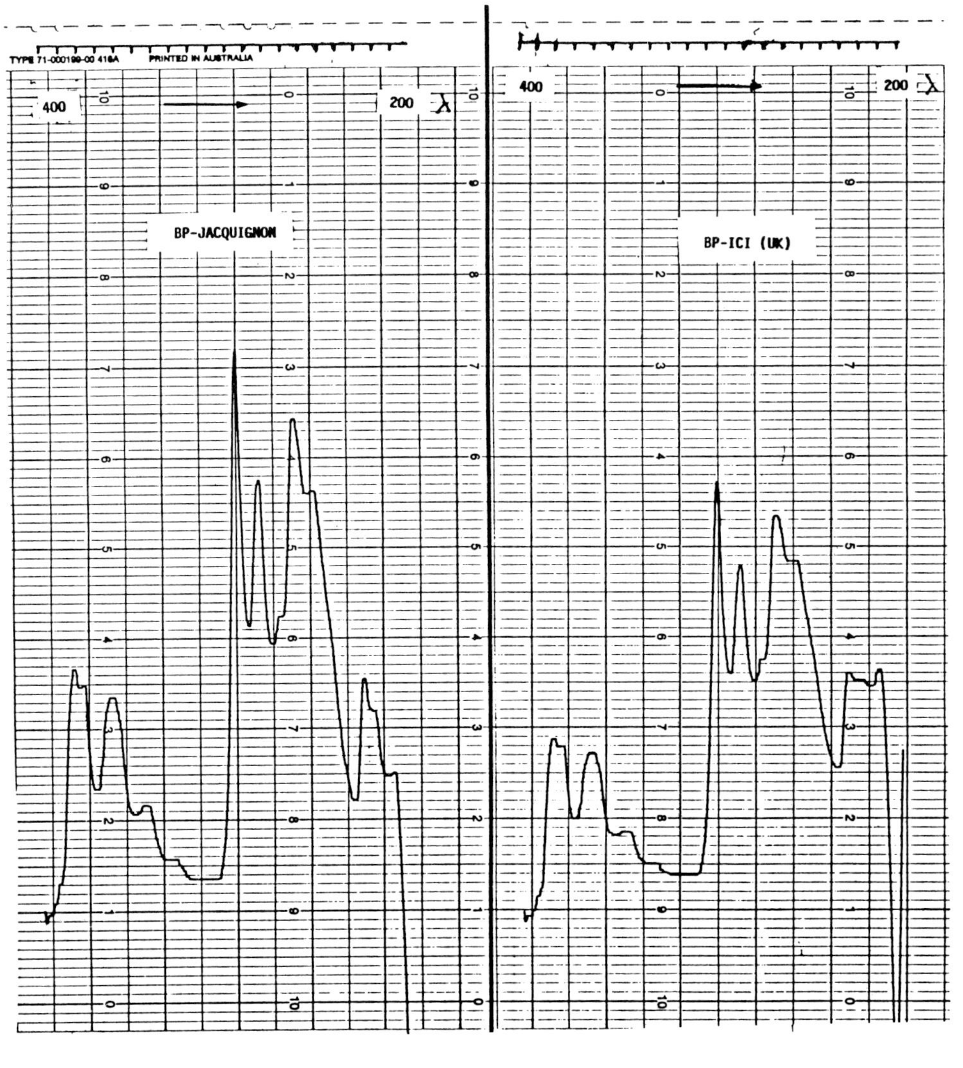

Statistics
The Mann-Whitney-Wilcoxon test (U test) was used to compare the data obtained. The possible dose–response relationship was evaluated by linear regression analysis.

3 Results and discussion

BP and PYR
For BP clear increases in aberration frequencies were obtained (Table 1), which shows a dose-dependent relationship (Figure 3). After a dose of 25 mg/kg the aberrations were 4.2% vs 1.8% for the control ($p < 0.02$). The equation of the dose-response curve was $y = 0.016X + 1.546$. These results confirm those obtained by Kliesch *et al.* (1982). PYR was found to be negative in this test. No increase in the number of aberrations, was observed at any dose (Table 1). PYR was also negative in the micronucleus test (Kirkhart, 1981) but positive in the SCE test (Kirkhart, 1981).

2AAF and 4AAF
For 2AAF the frequencies of chromosomal aberrations were increased (Table 2), and a dose–response effect was observed (Figure 4). The equation of the dose–response

Table 2. *Chomosomal aberrations in Chinese hamster bone marrow cells induced by 2AAF and 4AAF*

Compound (mol. wt.)	Dose (mg/kg)	Cells scored (n)	Deletions		Mitotic index % ± SE
			n	% ± SE	
	0	1200	23	1.9 ± 0.1	4.5 ± 0.1
2AAF	55.8	500	32	6.4 ± 0.6[a]	4.1 ± 0.1
(223)	115.5	500	52	10.4 ± 0.6	3.1 ± 0.2
	223	500	77	15.4 ± 0.6	2.4 ± 0.2
	0	1200	28	2.3 ± 0.1	4.1 ± 0.1
4AAF	223	500	9	1.8 ± 0.2	4.1 ± 0.2
(223)	446	500	14	2.8 ± 0.2	4.0 ± 0.1
	892	500	22	4.4 ± 0.4	3.1 ± 0.3

[a] Lowest dose at which a significant elevation was found ($p < 0.01$).

Figure 2. BP HPLC analysis: Beckman 344 gradient liquid chromatograph. Absorbance Detector Model 160 (λ 254 Lamp Mercury). Column: Ultrasphere-ODS′ (C_{18}) 4,6 ≧ 150 mm. Mobile phase: H_{20}/Methanol. Flow rate: 1 ml/min. Prog.: 60: 70% → 10′, 7′; 70: 90 → 25′; 90: 100% → 3′; 100 → 3′. Range UV: 60 → 100% = 0.05, 100% = 2.

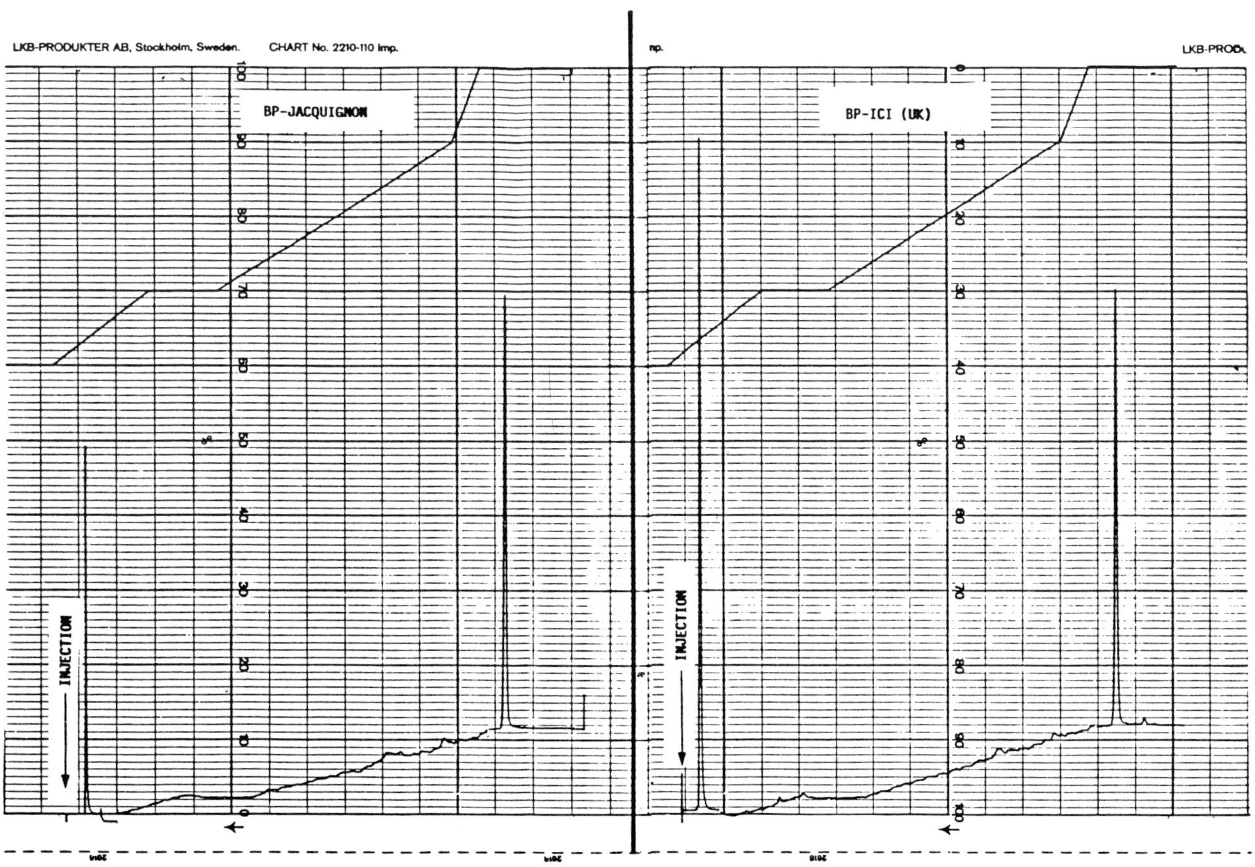

curve was $y = 0.0607X + 2.526$ ($p < 0.001$). The lowest dose at which a significant effect was found is 55.8 mg/kg.

For 4AAF a statistically significant increase was found only at the highest dose (892 mg/kg) (Table 2). However, a significant linear trend with increasing dosage was only observed if the control value (0 dose) was not taken into account. Thus, the evidence for an actual dose–response relationship is not convincing, and further experiments are necessary to confirm this point.

Figure 3. Dose–response curve ($y = 0.106X + 1.546$) of aberrations (%) in Chinese hamster bone-marrow cells after treatment with BP.

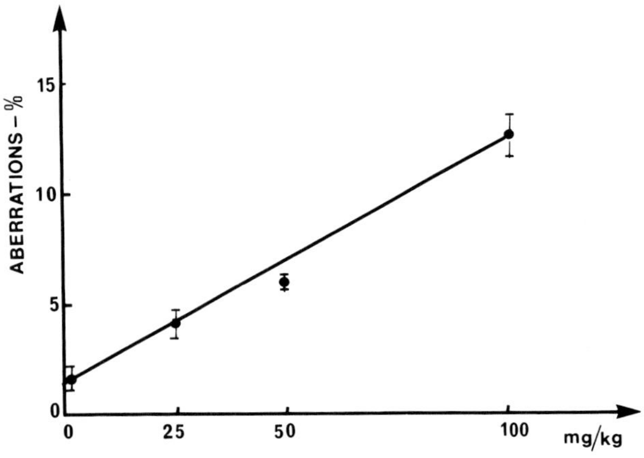

Figure 4. Dose–response curve ($y = 0.0607X + 2.526$) of aberrations (%) in Chinese hamster bone-marrow cells after treatment with 2AAF.

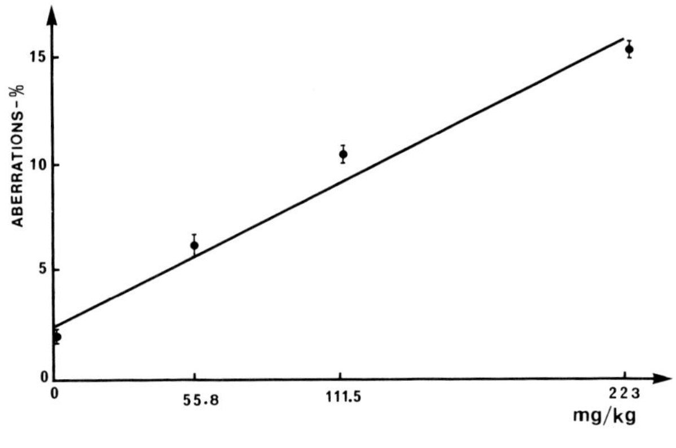

Our results obtained seem to be similar those obtained by Tsuchimoto and Matter (1981) in the micronucleus test.

4 Conclusion

Of 4 chemicals tested in this study, BP and 2AAF were positive in the chromosomal aberration assay on Chinese hamster bone marrow cells.

PYR was negative while 4AAF showed a slight response only at the highest dose tested.

It would be untimely to assess the data on 4AAF without further study. Although the chromosomal aberration assay is more time-consuming than the micronucleus test, it remains a sensitive and valuable cytogenetic test *in vivo* for genotoxic studies.

6 References

Kirkhart, B. (1981) Micronucleus test on 21 compounds. In F.J. de Serres and J. Ashby (Eds), *Progress in Mutation Research*, Vol. 1, Evaluation of Short-Term Tests for Carcinogens, pp. 698–704, Elsevier/North Holland.

Kliesch, U., N. Danford and I.-D. Adler (1981) Micronucleus test and bone-marrow chromosomes analysis. A comparison of 2 methods *in vivo* for evaluating chemically induced chromosomal alterations, *Mutation Res.*, **80**, 321–332.

Kliesch, U., I. Roupova, and I.-D. Adler (1982) Induction of chromosome damage in mouse bone-marrow by benzo(a)pyrene, *Mutation Res.*, **102**, 265–273.

Tsuchimoto, T., and B.E. Matter (1981) Activity of coded compounds in the micronucleus test. In F.J. de Serres and J. Ashby (Eds), *Progress in Mutation Research*, Vol. 1., Evaluation of Short-Term Tests for Carcinogens, pp. 705–711, Elsevier/North Holland.

Jeffrey A. Allen and Raymond J. Proudlock
Department of Mutagenesis and Cell Biology, Huntingdon Research Centre, Ltd, Huntingdon, Cambridgeshire, UK

17

Examination of metaphase chromosomes obtained from the bone marrow of Chinese hamsters (Cricetulus griseus)

1 Summary

Bone marrow cells in metaphase obtained from Chinese hamsters have been examined for chromosomal damage after a single administration by gavage of PYR, BP, 2AAF, or 4AAF in corn oil. None of the 4 compounds acted as a potent clastogen, but BP and 2AAF both caused low levels of chromosome damage, while PYR and 4AAF caused little or no damage.

2 Introduction

Chemicals that induce mutation in an eukaryote genome also often cause structural damage to the chromosomes (Evans, 1976). Consequently, assessment of chromosomal damage forms an integral part of the evaluation of the genotoxic profile of a chemical (Brusick, 1982; OECD, 1984a). For *in vivo* cytogenetic test systems, the cells examined are usually obtained from the bone marrow of the treated animal, normally a rodent. These test systems have been validated by their ability to detect chemicals that induce cancer in animals and, thus, are useful in predicting risk to the individual.

In this study, an examination has been made of the clastogenic effect of PYR, its carcinogenic derivative BP, the carcinogen 2AAF, and its 4AAF isomer on the bone marrow of Chinese hamsters, which are particularly suitable for cytogenetic analysis because of their large chromosomes and favourable karyotype.

3 Materials and methods

Animals

Chinese hamsters were obtained from Bantin and Kingman, Hull, U.K. They were supplied with tap water and standard laboratory rodent diet (Spratt's Laboratory Rodent Diet No. 1) *ad libitum* except that the diet was removed for the overnight period immediately prior to dosing.

Chemicals

PYR, BP, 2AAF, and 4AAF were supplied by ICI Central Toxicology Laboratory, U.K. Each chemical was prepared as a suspension immediately before dosing in corn oil warmed to 37 °C.

Procedure

Groups of 5 male and 5 female (control) and 3 male and 3 female (test material) Chinese hamsters weighing 20 g (8 to 9 weeks old) were dosed by gavage with 0.4 ml of corn oil/suspension. Two dose levels of each chemical were administered. The upper dose levels were chosen on the basis of a preliminary toxicity test and were, as recommended by the OECD (1984b), approximately the maximum tolerated dose, which for 4AAF was the maximum administrable dose. The upper and lower dose levels were chosen so that there was a common dose, on a moles/kg bodyweight basis, within each pair of compounds.

Twelve, eighteen, and twenty-four hours after dose administration, one group of 10 control and 6 test animals per dose level were sacrificed by cervical dislocation. Two hours prior to sacrifice, colchicine in 0.9% saline was administered at a concentration of 4 mg/kg body weight by intraperitoneal injection.

After sacrifice a femur was excised, the proximal epiphysis removed, and the marrow flushed into a plastic universal container with 10 ml Hanks' balanced salts solution, using a syringe fitted with a 25 g needle. Each bone marrow cell suspension was centrifuged at 70 times *g* for 7 min. The resultant cell sediment was resuspended in 10 ml hypotonic Hanks' solution (water: Hanks' solution 3:1 v/v) for 10 min at room temperature, and then centrifuged at 70 times *g* for 7 min. The cell pellet was fixed overnight at 4 °C in 5 ml methanol: glacial acetic acid (3:1 v/v, freshly prepared). The fixative was then decanted and the pellet resuspended in 5 ml fresh fixative, sedimented by centrifugation at 70 times *g* for 7 min and finally resuspended in 0.5 ml fresh fixative. Two or three drops of this suspension were dropped onto wet, ice-cold microscope slides, which were air-dried, washed in pH 6.8 buffered water, and stained with 10% Giemsa solution in pH 6.8 buffered water for 20 min. The slides were dried, mounted with coverslips, and 50 metaphase figures per animal were examined, under code, for the presence of structural chromosomal abberrations as classified by Savage (1976).

4 Results and discussion

The percentages of cells exhibiting chromosomal damage are presented in Table 1. Because the significance of gap damage is controversial, the results are divided into values including and excluding gap damage.

Although a wide range of aberrant chromosomal figures can be produced by clastogens, the abnormalities seen in this study were restricted to gaps and breaks/fragments except for one exchange figure that was observed in an animal sacrificed 24 h after treatment with 2AAF at 892 mg/kg.

None of the chemicals caused extreme amounts of chromosomal damage and, as such, none can be classified as potent clastogens.

2AAF and 4AAF No statistically significant increase in the proportion of cells containing chromosomal aberrations was observed at any time after treatment with 4AAF, and the maximum response observed was 2.3 times the background level. Similarly, no significant increase in chromosomal damage was detected 12 or 18 h after treatment with 2AAF, but after 24 h the proportion of cells exhibiting chromosomal damage was approximately 3 times the control value at a dose of 1784 mg/kg and 5 times the control value at a dose of 892 mg/kg ($p < 0.05$ excluding gap damage, $p < 0.01$ including gap damage). The lack of an obvious dose relationship may be due to a low effective level of clastogen in the bone marrow, since the group mean mitotic indices at both dose levels at 24 h (2.6% at 892 mg/kg and 3.1% at 1784 mg/kg) were similar to the control volue of 2.7%.

PYR and BP At the lower dose of PYR, slight but statistically significant ($p < 0.05$) increases in the proportion of cells with chromosomal damage were observed 12 h after dosing when gap damage is excluded but not when gap damage is included in the results and at 18 h when gap damage is included but not when excluded. However, as with 4AAF, the maximum level of response was only 2.3 times the control value. Eighteen hours after dosing, BP provoked an approximately 4-fold increase above control values ($p < 0.01$ including and excluding gap damage) in the proportion of cells with chromosome damage at the top dose level of 1008 mg/kg, which, on a molar basis, is equivalent to the lower dosage of PYR. However, by 24 h the proportion of aberrant cells had reverted to control levels. No substantial difference in the group mean mitotic index was observed at this time after treatment with BP (2.8% at 504 mg/kg and 2.3% at

Table 1. *Percentage of metaphase figures with structural chromosomal aberrations*

	Dose			Time after dose (h)		
Chemical	mg/kg	moles/kg	N[a]	12	18	24
Corn oil	—	—	500	1.0 (2.0)[b]	1.0 (1.6)	1.0 (1.8)
2AAF	892	4	300	0.7 (1.7)	0.3 (2.3)	5.0*(8.7**)
2AAF	1784	8	300	0.7 (2.3)	1.3 (3.0)	3.3 (5.0)
4AAF	1784	8	300	0.3 (0.3)	2.3 (3.3)	1.0 (1.7)
4AAF	3568	16	300	1.0 (1.0)	1.3 (2,0)	1.0 (1.7)
BP	504	2	300	1.3 (2.7)	1.3 (3.0)	0.7 (2.0)
BP	1008	4	300	1.7 (2.3)	4.0[c]**(7.0**)	0.3 (1.3)
PYR	808	4	300	2.3*(3.0)	2.0 (3.7*)	2.3 (2.3)
PYR	3232	16	300	1.0 (3.0)	1.0 (1.3)	0.7 (2.0)

[a]N, number of cells examined per group.
[b]Figures in parentheses refer to values which include gap damage and those outside parentheses exclude gap damage.
[c]Kruskal-Wallis test, *$p < 0.05$, **$p < 0.01$.

1008 mg/kg) compared with the control value of 2.7%. The low level and apparent specificity in time of response is probably a reflection of the kinetics of absorption, metabolism, and detoxification of BP in the Chinese hamster, which may have been dependent on the dose formulation of a suspension in corn oil.

To conclude, none of the 4 chemicals were potent clastogens in the Chinese hamster after administration of the maximum tolerated dose. The maximum response obtained for PYR and 4AAF was 2.3 times the negative control value while BP and 2AAF both caused responses equal to or in excess of 4 times the background levels, indicating weak clastogenic activity.

5 References

Beusick, D. (1982) Genetic Toxicology. In A.W. Hayes (Ed.), *Principles and Methods of Toxicology*, 1st ed., Raven Press, NY, pp. 223–272.

Evans, H.J. (1976) Cytological methods for detecting chemical mutagens. In A. Hollaender (Ed.), *Chemical Mutagens, Principles and Methods for Their Detection*, Vol. 4, Plenum Press, NY, pp. 1–29.

OECD (1984a) DIG 18, Mutagenicity. In *Data interpretation guidelines for initial hazard assessment of chemicals*, Organisation for Economic Co-operation and Development, Paris, pp. 58–60.

OECD (1984b) "Genetic toxicology: *In vivo* mammalian bone marrow cytogenetic test-chromosomal analysis." In *OECD Guideline for testing of chemicals*, Organisation for Economic Co-operation and Development, Paris, No. 475.

Savage, J.R.K. (1976) Annotation: Classification and relationships of induced chromosomal structural changes, *J. Med. Genetics*, **13**, 103–122.

PART III
In vivo assay: bone marrow

B
Micronuclei

Dieter Wild (Workgroup Chairman)[1]
Institut für Pharmakologie und Toxikologie der Universität Würzburg Würzburg, Federal Republic of Germany

18

Summary report on the performance of the in vivo micronucleus assays

1 Introduction

A total of 18 laboratories conducted micronucleus assays for the collaborative study; 16 laboratories performed assays using mice and 2 used rats. Two laboratories performed the recently developed assays on peripheral blood and/or foetal blood: one conducted peripheral and foetal blood assays, the other foetal blood and maternal bone marrow assays (Table 1).

2 Co-ordination of experimental conditions

Prior to the beginning of the study, a common experimental protocol was suggested to all investigators in the group; this is given in Appendix 1.

3 Mouse bone marrow micronucleus assay

Principle of the test
Micronuclei are fragments of a cell nucleus; they can arise by the formation of acentric chromosomal fragments which are not incorporated into one of the daughter nuclei during the next mitosis. Alternatively, they can arise by a misdistribution of intact chromosomes at mitosis, for example due to an inhibitor of the normal spindle function.

Micronuclei have been observed *in vitro* and *in vivo*, in numerous types of eukaryotic cells. For the purpose of *in vivo* mutagenicity testing, young erythrocytes of the bone marrow have been most frequently studied. The main advantage of the micronucleus assay on mammalian erythrocytes is the prominent appearance of micronuclei in these normally anucleate cells.

Test protocol
The individual test protocols are described in the investigators' reports. The recommendations for the protocols are represented in Appendices 1 and 2. They are based on literature reviews by Schmid (1976) and Salamone and Heddle (1983).

Data analysis
Statistical methods to determine the significance of induced effects were chosen and performed by each

[1] Workgroup participants: Brita Beije, David H. Blakey, Pawan S. Chauhan, Ivan Chouroulinkov, Colette Coton, Barbara L. Harper, Makoto Hayashi, Vladimir I. Ivanov, Jorn C. Jensen, Dag Jenssen, Ulf Kliesch, James T. MacGregor, Michael F. Salamone, Jeremy A. Styles, Kurt E. Suter, Ad D. Tates, and Paul A. Watkins

investigator independently. The working group agreed that the classification of a result as positive in the summary report required at least 2 data points showing a significant effect at $p \leqslant 0.05$ and that the effect be significant compared with both the concurrent control value and composite control value calculated from all vehicle control animals used by that investigator throughout this study. This requirement for at least 2 significant data points has been used previously as a criterion for a significant positive response (Heddle *et al.*, 1983). The comparison with both concurrent and composite control values was adopted to help eliminate "false" interpretations caused by incidentally high experimental or low vehicle control values.

Results

The mouse bone marrow assay was carried out in 15 laboratories using 10 strains of mice. Spontaneous frequencies of micronucleated polychromatic erythrocytes (MPE) were in the range of 0.12 – 0.41% (mean: 0.24 ± 0.10%) (Table 1).

BP registered positive in all 15 laboratories. Table 1 presents the maximal fold increase in MPE over the corresponding vehicle control frequency. These values ranged from 3.3 to 30 with an average of 11.7 ± 8.2. The lowest doses detected as positive in these laboratories were typically between 0.1 and 1 m mol/kg (25–252 mg/kg).

PYR was reported inactive in 16 sets of experiments performed in 14 laboratories. The experimental conditions for these assays were essentially the same in each laboratory as those used in the positive BP tests, except that the doses of PYR were generally higher, with the maximum dose tested being 4 m mol/kg (808 mg/kg). At this dose, lethality was observed in several laboratories. The highest doses of PYR tested were 10–40 times higher than the lowest doses of BP found to induce a significant increase of MPE. This difference in activity between the 2 compounds can provide a rough estimate of the maximal activity of PYR in relationship to BP. By this criterion, the hypothetical clastogenic activity of PYR is less than 1/10 to 1/40 the clastogenic activity of BP. PYR was found positive in one laboratory, with a 3.5-fold maximal increase. This effect was not dose dependent.

Table 1. *Overview on the micronucleus assay working group and results*

Investigator	Mouse or rat strain	Sex	VEH	Route	Spont. freq. %	Maximum fold-Increase * (m mol/kg) BP	PYR	2AAF	4AAF
Mouse bone marrow									
Bla	HPB/B1	M	OO[a]	GAV	.41	3.32 (.5–1)	0 (4)	2.2 (1)	0 (4)
Cho	SW	M	OO	GAV	.18	5 (.2–.6)	0 (4)	3.7 (1)	0 (4)
Har	ICR	M	OO	GAV	.19	13.7 (.1–.5)	0 (1.7)	0 (5)	0 (.8)
Ish	MS	M	OO	GAV	.40	30 (.25–1)	0 (4)	6 (25–1)	0 (8)
Iva	C57/B1/6J	M	OO	GAV	.16	12.9 (.25–1)	3.5 (1–2)	6.2 (.25–1)	2.5 (1–4)
JJe	NMRI	M, F	OO	GAV	.39	5.9 (.5–1)	0 (4)	NT	NT
DJe	CBA	M	CO	GAV	.30	4.0 (.25–1)	0 (3)	3.1 (1)	0 (4)
Kli	(101 × C3H)F1	M	CO	GAV	.22	7 (.125–.5)	0 (4)	NT	NT
Kli	(101 × C3H)F1	M	DMSO	IP	.22	NT	NT	6.3 (.5–2)	0 (4)
Leo	Balb/c	M	OO	GAV	.32	13.3 (.25–1)	0 (4)	5.8(1)	0 (4)
Sal	B6C3F1	M, F	DMSO	IP	.14	7.9 (.18–.73)[d]	0 (3.5)[d]	8.6 (3–7)[d]	0 (2.4)[d]
Sal	B6C3F1	M, F	CO	GAV	.11	5.5 (.6–.75)[d]	0 (3.5)[d]	NT	NT
Sty	C57BL/6J	M, F	OO	GAV	.29	7.7 (1–20)	0 (2)	6.8 (1–9)	0 (22)
Sty	C57BL/6J	M, F	OO	IP	.29	9.6 (12.5–19.8)	0 (2.7)	5.3 (1.4–2.2)	0 (5.4)
Sun	SW	M	CO[b]	IP	.12	6.3 (1–1.5)	0 (4)	3.9 (1.5)	0 (2)
Sut	NMRI	PF	OO	GAV	.12	16.3 (1)	0 (2)	6.3 (1)	0 (4)
Tat	(101 × C3H)F1	M	OO[c]	GAV	.23	21 (.25–1)	0 (4)	7.8 (.25–1)	0 (4)
Wil	NMRI	M	OO	GAV	.16	29 (.25–1)	0 (4)	3.9 (1)	0 (4)
Mouse peripheral blood									
Mac	SW	M	OO	GAV	.15	16.5 (.25–1)[e]	0 (2)[e]	4.8 (.5–2)[e]	0 (4)[e]
Mouse foetal blood									
Mac	SW	—	OO	GAV	.36	8.7 (.25–1)	0 (2)	0 (2.8)	0 (4)
Sut	NMRI	—	OO	GAV	.50	21 (1)	0 (2)	2.1 (1)	0 (4)
Rat bone marrow									
Bei	AP	M	CO	GAV	.15	NT	NT	3.2 (2.2–9)	0 (13)
Wat	AP	F	CO	GAV	.07	2.4 (.4)	0 (2.5)	NT	NT
Wat	AP	F	CO	IP	.07	NT	NT	0 (1.3)	0 (2.2)

[a] olive oil or olive oil + DMSO
[b] corn oil or corn oil + 40% DMSO
[c] olive oil + TWEEN-80
[d] dose administered on each of 2 consecutive days
[e] dose administered on each of 4 consecutive days

Abbreviations:
CO – corn oil
OO – olive oil
F – female animals
PF – pregnant females
M – male animals
NT – not tested

* Positive response (range of concentrations giving a positive response), negative response (maximum concentration giving a negative response)

2AAF was reported positive by 13 of 14 laboratories. The maximal fold increase ranged from 2.2 to 8.6 (mean: 5.4 ± 1.8). These values indicate a considerably lower activity for 2AAF when compared with the respective values for BP. In addition, the minimal effective molar doses of 2AAF (0.25–3 m mol/kg; 56–669 mg/kg) appear to be higher than the corresponding minimal effective doses for BP. The differences between 2AAF and BP in the maximal detected effect as well as in the lowest positive doses indicate lower clastogenic activity for 2AAF as compared with that of BP.

4AAF was found inactive in 13 laboratories and positive in one laboratory in which PYR also was found to be positive. The positive result was found at 3 doses, but there was no dose effect relationship, and the result was considered by the investigators to be of uncertain significance. The highest inactive doses of 4AAF were 4–22 times higher (highest doses: 8 and 22 m mol/kg; 1784 and 5000 mg/kg) than those used of the carcinogenic isomer 2AAF. An approximate maximal value for the potential genotoxic activity of 4AAF can be calculated from these data in a way analogous to that described in the section on PYR. This maximal value ranges from 1/4 to 1/88 of the activity of 2AAF.

4 Micronucleus assay on mouse peripheral blood

One laboratory used mouse peripheral blood, rather than bone marrow, to assess the frequency of MPE induced by the four test agents. Test agents were administered on 4 consecutive days and peripheral blood was sampled on the fifth day.

BP and 2AAF were positive at the lowest dose tested (0.25 and 0.5 m mol/kg/day, respectively) and induced 16.5- and 4.8-fold increases over control values at 1 and 2 m mol/kg/day, respectively. PYR and 4AAF were negative by the criteria adopted for a positive response. These results were comparable with those obtained by other investigators in the bone marrow assay.

5 Micronucleus assay on mouse foetal blood

In the foetus, haematopoesis occurs in the liver, and the transplacental foetal micronucleus assay has been shown to be more sensitive than the bone marrow assay to certain agents known to be activated by the liver.

Two laboratories conducted both the foetal assay and the adult or the weanling assay in which the target cells are bone marrow haematopoetic cells. Both laboratories used the same vehicle (olive oil), route (po), and species (mouse). In the foetal assays, both laboratories examined foetal blood from mice exposed on day 15 of pregnancy and used the critical sampling times of 30 and 48 h. In the adult assay, one laboratory administered the test material in a single dose and sampled bone marrow erythrocytes at various times after dosing, while the second laboratory sampled peripheral blood at a single time after 4 consecutive doses.

Results from both laboratories were consistent for all 4 compounds. BP was found to be highly active, and to have approximately equal potency in both the bone marrow and foetal liver assays. PYR was inactive. 2AAF was found to induce a 5 to 6-fold increase over the spontaneous micronucleus frequency in the bone marrow-based assay. The foetal assay was less sensitive to 2AAF, with one laboratory finding a doubling of the spontaneous frequency and the other finding no increase. 4AAF was inactive.

6 Micronucleus assays on rat bone marrow

Two laboratories conducted assays on rat bone marrow. Both performed the micronucleus assays in conjunction with another assay (UDS in liver and chromosome analysis in bone marrow, respectively), and the protocols used for the micronucleus assay were partly dictated by the other assays.

One laboratory used males of the AP rat strain and studied the AAFs only, using expression times of 24 and 36 h. 2AAF was inactive at 24 h but was significantly active at 36 h posttreatment, with a maximal fold increase of 3.2 and a lowest effective dose of 2.2 m mol/kg (500 mg/kg). Although these values lie within the range of the respective values found in mouse bone marrow, they suggest a weaker response of this rat assay compared with that of the mouse assays. 4AAF was found inactive in this laboratory. Using female rats of the same strain, the other laboratory found BP weakly active and PYR, 2AAF, and 4AAF inactive. The very weak increase reported for 4AAF was observed at one data point only. Therefore, 4AAF was classified as negative (see earlier section on data analysis). The negative result obtained with 2AAF may be spurious because: (1) the dose administered was only 60% of the lowest dose found to be active in the first laboratory; (2) the route of administration was ip as opposed to gavage in the first laboratory; (3) the single expression time of 24 h used in this laboratory is probably inadequate in the light of the expression data of the first laboratory; (4) a lower response of female compared with male rats cannot be excluded.

7 Influence of specific experimental factors

A special chapter (Group Report on Bone Marrow Assays, these volumes) compares the result of SCE, micronucleus and chromosome aberration assays in rodent bone marrow and deals with the influence of particular experimental variables.

8 Conclusions

Thirteen of 15 laboratories (87%) which performed micronucleus assays on mouse bone marrow reported a significant increase in micronucleated cells with the carcinogens 2AAF and BP and a lack of such effects with the presumed noncarcinogens 4AAF and PYR. This clear "+/− response" was found with 9 different strains of mice and with male and female, including pregnant, animals. The clearcut discrimination was obtained not only in assays using administration of the compounds in olive or corn oil by gavage but also in assays performed using ip administration of DMSO solutions and ip administration of olive or corn oil solutions or suspensions. These latter conditions required relatively high doses for a detectable effect. The apparent insensitivity of the assays using oil and ip injection probably is due to slow resorption of oil and BP from the peritoneum.

One of the two laboratories which failed to demonstrate "+/− responses" for both pairs of chemicals could at least discriminate between BP and PYR, while the other laboratory found quantitative but not qualitative differences between the carcinogens and the noncarcinogens.

Based on equimolar doses, BP induced clearly higher effects than AAF.

The assay on peripheral blood of mice performed in one laboratory also discriminated between the 2 carcinogens and the presumed noncarcinogens.

The transplacental assay on foetal blood revealed a difference between the 2 pairs of compounds: the BP/PYR pair yielded the "+/− response" as observed in adult bone marrow and peripheral blood. With 2AAF one laboratory found no effect in the foetal assay, and the other found an effect clearly lower than in the adult bone marrow. Therefore, this assay was less sensitive to 2AAF than was the adult bone marrow assay and could not unambiguously distinguish 2AAF and 4AAF.

The assays in bone marrow of rats both were performed as research assays under conditions which were partly dictated by other assays and were conducted simultaneously in the same animals. One of the 2 laboratories reported results with the AAFs analogous to those obtained in mouse bone marrow, whereas the other laboratory found a weak effect of BP and no effect with PYR, 2AAF, and 4AAF. The lack of concordance between the AAF data of the two laboratories is probably attributable to specific differences in the experimental protocols.

In summary, the micronucleus assays on mouse bone marrow and peripheral blood show a consistent concordance of the results and a clear potential to discriminate qualitatively between the BP/PYR and 2AAF/4AAF pairs. These results confirm and extend the data obtained in the previous international collaborative program (Kirkart, 1981; Salamone *et al.*, 1981; Tsuchimoto and Matter, 1981).

9 References

Heddle, J.A., M. Hite, B. Kirkart, K. Mavournin, J.T. MacGregor, G.W. Newell and M.F. Salamone (1983) The induction of micronuclei as a measure of genotoxicity. A report of the U.S. Environmental Protection Agency Gene-Tox Program, *Mutation Res.*, **123**, 61–118.

Kirkart, B. (1981) The micronucleus test on 21 compounds. In F.J. de Serres and J. Ashby (Eds), Evaluation of Short-Term Tests for Carcinogens, *Progress in Mutation Research*, Vol. 1, Elsevier, Amsterdam, pp. 698–704.

Salamone, M.F. and J.A. Heddle (1983) The bone marrow micronucleus assay: rationale for a revised protocol. In F.J. de Serres (Ed.), *Chemical Mutagens*, Vol. 8, Plenum, New York, pp. 111–149.

Salamone, M.F., J.A. Heddle and M. Katz (1981) The mutagenic activity of 41 compounds on the *in vivo* micronucleus assay. In F.J. de Serres and J. Ashby (Eds), Evaluation of Short-Term Tests for Carcinogens, *Progress in Mutation Research*, Vol. 1, Elsevier, Amsterdam, pp. 686–697.

Schmid, W. (1976) The micronucleus test for cytogenetic analysis. In A. Hollaender (Ed.), *Chemical Mutagens*, Vol. 4, Plenum, New York, pp. 31–53.

Tsuchimoto, T. and B. Matter (1981) Activity of coded compounds in the micronucleus test. In F.J. de Serres and J. Ashby (Eds), Evaluation of Short-Term Tests for Carcinogens, *Progress in Mutation Research*, Vol. 1, Elsevier, Amsterdam, pp. 705–711.

Appendix 1

Vehicle for administration of test chemicals: olive oil (pharmacopoeia grade); the AAFs are insoluble and must be administered as suspensions.

Mouse strain: any defined strain.

Number of replicate animals: 5 or more replicates.

Sex of animals: groups of 5 animals should be composed of males; if males and females are used, more animals should be included in each group.

Age of mice: 8–12 weeks.

Number of doses: 3 (or more) doses of each chemical, and a positive and a negative (vehicle) control.

Appointment of doses: doses should be selected on the basis of the molecular weight of each chemical (Appendix 2).

Number of treatments: a single treatment.

Route of application: based on pilot experiments, peroral treatment by gavage is recommended for all 4 compounds.

Volume administered: 1% of the body weight.

Expression time: it is recommended to use 3 expression times in the range of 24–48 h after dosing for the highest dose of each compound. This time-effect experiment will provide the *optimal* expression time (in the case of a mutagenic compound). For the lower doses the use of the optimal expression time is considered adequate. Inactive compounds are studied at the optimal expression time found for their mutagenic/carcinogenic paired compound.

Suspension medium for bone marrow: fetal bovine serum.

Fixation and staining of bone marrow smears: May-Grünwald and Giemsa, according to Schmid (1976). It is advisable to filter the Giemsa prior to use.

Scoring of slides: 1000 polychromatic erythrocytes (PE) per animal on coded slides.

Appendix 2. *Doses suggested for the CSSTT micronucleus assays*

Compound	Molecular weight	Doses in mg/kg (m mol/kg) high	intermediate (m mol/kg)	low
BP	252	252 (1)	126 (0.5)	63 (0.25)
PYR	202	606 (3)	404 (2)	202 (1)
2AAF	223	223 (1)	111.5 (0.5)	55.8 (0.25)
4AAF	223	892 (4)	446 (2)	223 (1)

Dieter Wild
Institut für Pharmakoligie und Toxikologie Universität Würzburg, Würzburg, Federal Republic of Germany

19

Bone marrow micronucleus tests on NMRI mice

1 Summary

Male NMRI mice were treated by gavage with single doses of 2-acetylaminofluorene (2AAF; 56, 112, 223 mg/kg), 4-acetylaminofluorene (4AAF; 223, 446, 892 mg/kg), benzo(a)pyrene (BP; 63, 126, 252 mg/kg) and pyrene (PYR; 202, 404, 808 mg/kg). Animals were sacrificed 24, 30, and 48 hours after treatment. Bone marrow smears were prepared and analysed for micronucleated erythrocytes.

Chemical dose- and time-dependent induction of micronuclei was observed following 2AAF and BP treatments, whereas 4AAF and PYR were inactive in this respect.

This study is a part of the IPCS Collaborative Study on *in vivo* short-term tests and the data, in conjunction with those of the other studies, allow multiple inter-laboratory and interassay comparisons, the results provide a deeper understanding of the genotoxicity of the test compounds and create the basis for an improved assessment of the assay systems.

2 Introduction

Radiation-induced micronuclei in fertilized mouse eggs were recognized and reported as early as 1937 (Brenneke, cited in Hertwig, 1957). More than 30 years later, the micronucleus test on bone marrow was developed as an economic screening test for cytogenetic damage (Boller and Schmid, 1970; Matter and Schmid, 1971). In the meantime, this test has been put to use in many laboratories and included in guidelines for short-term tests for mutagens and carcinogens. The cytogenetic basis of the assay, the technique, and the data obtained have been described and reviewed repeatedly (Schmid, 1976; Jenssen and Ramel, 1980; de Serres and Ashby, 1981; Salamone and Heddle, 1983; Heddle *et al.*, 1983).

This study utilizes the bone marrow micronucleus test for the assay of two pairs of compounds consisting of a carcinogen and a presumed noncarcinogen. The rationale for the selection of the pairs of compounds and the structure of the entire study are described elsewhere in this volume (chapter 1).

3 Materials and methods

Male NMRI mice, 8–12 weeks old, were purchased from Ivanovas (Kisslegg, FRG). The animals were treated in groups of five. The test compounds 2AAF, 4AAF, BP, and PYR were samples from Lancaster Synthesis Ltd.

Table 1. *Results of micronucleus assays with BP and PYR, administered by gavage*

Compound	Dose (mg/kg)	Expression time (h)	Animals treated/evaluated	$\frac{PE}{PE + NE}$	MPE/PE (%) $\pm s$	P
BP	0	30	5/5	0.56 ± 0.08	0.18 ± 0.04	
	252	24	5/5	0.50 ± 0.07	2.22 ± 1.01	*
	252	30	5/5	0.55 ± 0.05	3.48 ± 1.08	*
	252	48	5/5	0.44 ± 0.07	1.24 ± 0.58	*
PROC[a]	6.4	30	5/5	0.59 ± 0.07	0.46 ± 0.05	*
BP	0	30	5/5	0.58 ± 0.05	0.16 ± 0.11	
	63	30	5/5	0.56 ± 0.07	0.84 ± 0.38	*
	126	30	5/5	0.54 ± 0.05	1.74 ± 1.38	*
	252	30	5/5	0.50 ± 0.02	4.70 ± 1.77	*
PROC	6.4	30	5/5	0.53 ± 0.06	0.66 ± 0.05	*
PYR	0	30	5/5	0.50 ± 0.05	0.16 ± 0.05	
	808	24	5/4	0.43 ± 0.09	0.13 ± 0.05	
	808	30	5/4	0.46 ± 0.08	0.30 ± 0.14	
	808	48	5/3	0.45 ± 0.02	0.17 ± 0.15	
PROC	6.4	30	5/5	0.50 ± 0.04	0.72 ± 0.13	*
PYR	0	30	5/5	0.56 ± 0.04	0.14 ± 0.05	
	202	30	5/5	0.54 ± 0.02	0.20 ± 0.07	
	404	30	5/5	0.53 ± 0.04	0.20 ± 0.07	
	808	30	5/4	0.39 ± 0.13	0.23 ± 0.19	
PROC	6.4	30	5/5	0.50 ± 0.03	0.86 ± 0.26	*

[a] PROC: Procarbazine
*: $P \leq 0.01$.

Table 2. *Results of micronucleus assays with 2AAF and 4AAF, administered by gavage*

Compound	Dose (mg/kg)	Expression time (h)	Animals treated/evaluated	$\frac{PE}{PE + NE}$	MPE/PE (%) $\pm s$	P
2AAF	0	30	5/5	0.44 ± 0.05	0.18 ± 0.08	
	223	24	5/5	0.44 ± 0.11	0.20 ± 0.10	
	223	30	5/5	0.53 ± 0.06	0.62 ± 0.13	*
	223	48	5/5	0.49 ± 0.14	0.28 ± 0.13	
PROC[a]	6.4	30	5/5	0.48 ± 0.06	1.10 ± 0.19	*
2AAF	0	30	5/5	0.58 ± 0.07	0.12 ± 0.08	
	55.8	30	5/5	0.54 ± 0.07	0.34 ± 0.17	+
	111.5	30	5/5	0.47 ± 0.09	0.28 ± 0.15	
	223	30	5/5	0.56 ± 0.08	0.58 ± 0.15	*
PROC	6.4	30	5/5	0.54 ± 0.04	1.20 ± 0.28	*
4AAF	0	30	5/5	0.51 ± 0.08	0.18 ± 0.04	
	892	24	5/5	0.54 ± 0.05	0.14 ± 0.05	
	892	30	5/5	0.51 ± 0.09	0.16 ± 0.05	
	892	48	5/5	0.48 ± 0.11	0.14 ± 0.05	
PROC	6.4	30	5/5	0.49 ± 0.04	0.90 ± 0.26	*
4AAF	0	30	5/5	0.50 ± 0.07	0.14 ± 0.09	
	223	30	5/5	0.55 ± 0.04	0.12 ± 0.08	
	446	30	5/5	0.48 ± 0.06	0.12 ± 0.13	
	892	30	5/5	0.53 ± 0.04	0.14 ± 0.05	
PROC	6.4	30	5/5	0.53 ± 0.05	1.14 ± 0.21	*

[a] PROC: Procarbazine.
*: $P \leq 0.01$.
+: $P \leq 0.05$.

distributed for the IPCS collaborative study. The compounds were dissolved or suspended in olive oil DAB7 (C. Roth, Karlsruhe, FRG) with sonication. In addition to the test chemical groups, a negative, olive oil-treated (0.1 ml per 10 g mouse weight), and a positive, procarbazine-treated control group were included in each experiment. Procarbazine (Hoffmann-LaRoche, Basle, Switzerland) was dissolved in 0.9% sterile saline solution. All compounds were administered by gavage, in volumes adjusted according to the body weight of the mice (0.1 ml per 10 g mouse body weight). Mice had continuous access to food and water.

For each compound, a time-response experiment was performed first, using the highest dose and expression times of 24, 30, and 48 h; from this experiment the optimal expression time – in the case of a mutagen – was deduced. This was followed by the dose-response experiment using three doses and the optimal expression time. For inactive compounds, an "optimal expression time" in the true sense of the word cannot be derived. Therefore, the quasi-optimal expression time in tests with inactive compounds was adopted from the paired mutagen/carcinogen.

Each mouse received a single treatment and was sacrificed after an expression time of 24, 30, or 48 hours. All olive oil control animals were sacrificed 30 h after treatment, because previous experiments had shown the control values to be independent of the time interval between olive oil treatment and bone marrow preparation. Bone marrow from both femora was homogeneously suspended in fetal calf serum (1.2 ml per mouse in Eppendorff conical plastic tubes) and concentrated by centrifugation and resuspension; smears were prepared, air-dried, fixed, and stained with May-Grünwald and Giemsa stains according to Schmid (1976).

Slides were coded before scoring. On each slide 1000 polychromatic (PE) and the concurrent normochromatic erythrocytes (NE) were analysed for micronuclei. In the Tables 1–2, the frequencies of micronucleated polychromatic erythrocytes (MPE/PE) are presented. The proportion of PE among total erythrocytes (PE/PE + NE) also given in the tables is a measure of toxic damage to the erythropoesis: clastogenic compounds at high doses frequently decrease PE/PE + NE strongly.

MPE/PE values in treated groups were compared with the values in the concurrent negative control groups by means of the tables of Kastenbaum and Bowman (1970). Significance is marked by *, $P \leq 0.01$ and $^{+}$, $P \leq 0.05$.

4 Results and discussion

In the course of this study, 40 olive oil-treated control mice were evaluated; the mean frequency of MPE in the bone marrow of these animals was 0.158%. This value is close to the spontaneous frequencies of 0.205% and 0.185% published previously which were means of 313 and 240 control animals, respectively (Wild *et al.*, 1980; Wild *et al.*, 1983).

BP was soluble in olive oil at all concentrations tested; PYR was soluble at the low and the medium concentration; at the highest concentration it was not dissolved completely. 2AAF and 4AAF were suspended in olive oil, sonication yielded homogenous, fine suspensions. They were stirred continuously during animal treatment to avoid sedimentation and distortion of the doses administered.

BP induced micronuclei in all experimental groups (Table 1). The optimal expression time was 30 h, with much lower MPE/PE-values at 24 and 48 h. At 30 h after treatments, the micronucleus induction was proportional to the BP dose. The MPE/PE values in the three dose groups were increased 5-, 11-, and 29-fold over the control.

PYR, a presumed noncarcinogen, was dosed to 808 mg/kg (4 mmol/kg). At the highest dose it was lethal for 5 out of 20 mice; no losses were induced at the lower doses. The proportion of PE among total erythrocytes was slightly reduced at the highest dose. With regard to micronucleus induction, PYR was clearly inactive at all expression times and at all doses (Table 1). (The quasi-optimal expression time for PYR was adopted from BP.)

With 2AAF at the highest dose (223 mg/kg, 1 mmol/kg), a significant increase of MPE was seen at an expression time of 30 h; no significant effects were seen at 24 and 48 h (Table 2). At the lower doses and 30 h expression, 2AAF was not significantly active; the slight increase with borderline significance, $P = 0.05$, in the lowest dose group appears to be of questionable relevance. Thus, 2AAF induced a reproducible increase of MPE at the highest dose and 30 h expression time. The MPE frequencies in these groups were increased over the corresponding controls by factors of 3.4 and 4.8, respectively.

4AAF was dosed 4 times higher than 2AAF (up to 4 mmol/kg), because 4AAF is a presumed noncarcinogen and had been found nonmutagenic in previous micronucleus tests (Salamone *et al.*, 1981; Tsuchimoto and Matter, 1981). In the high dose experiments with variable expression times, 4AAF was inactive (Table 2). The dose-response experiment was performed using the optimal expression time found for 2AAF (30 h). This procedure is based on the assumption that the mechanism of a possible 4AAF effect, if there is any, is at least similar to that of 2AAF. The dose–response experiment indicated no mutagenic effect of 4AAF (Table 2).

The frequency of micronucleated normochromatic erythrocytes (MNE) was 0.08% in the control mice (mean of 40); it was increased in one group of mice only, 48 h after the highest dose of BP and reached 0.22% ($P = 0.05$, data not shown). This result illustrates that the high sensitivity of the micronucleus test can be achieved only if PE and NE are differentiated and if scoring is restricted to PE.

5 Conclusions

The spontaneous frequency of MPE in this study was 0.158% (mean of 40 mice). The carcinogenic compounds 2AAF and BP increased the frequency of MPE significantly. For these test compounds, optimal expression times were 30 h. The comparison of the induced frequencies of MPE following 2AAF and BP, at doses of 1 mmol/kg, shows a 9-fold higher activity of BP. High doses, up to 4 mmol/kg, of the presumed noncarcinogens 4AAF and PYR did not induce micronuclei.

6 References

Boller, K. and W. Schmid (1970) Chemische Mutagenese beim Säuger. Das Knochenmark des Chinesischen Hamsters als *in vivo*-Testsystem. Hämatologische Befunde nach Behandlung mit Trenimon, *Humangenetik*, **11**, 35–54.

de Serres, F.J. and J. Ashby (1981) *Evaluation of Short-Term Tests for Carcinogens: Report of the International Collaborative Program*, Elsevier, NY.

Heddle, J.A., M. Hite, B. Kirkhart, K. Mavournin, J.T. MacGregor, G.W. Newell and M.F. Salamone (1983) The induction of micronuclei as a measure of genotoxicity; a report of the USEPA Gene-Tox Program, *Mutation Res.*, **123**, 61–118.

Hertwig, P. (1957) Strahlenschäden und Strahlenschutz im zellulären Bereich, *Berichte über die Verhandlungen der sächsischen Akademie der Wissenschaften zu Leipzig*, **102(6)**, 1–43.

Jenssen, D. and C. Ramel (1980) The micronucleus test as part of a short-term mutagenicity test program for the prediction of carcinogenicity evaluated by 143 agents tested, *Mutation Res.*, **75**, 191–202.

Kastenbaum, M.A. and K.O. Bowman (1970) Tables for determining the statistical significance of mutation frequencies, *Mutation Res.*, **9**, 527–549.

Matter, B. and W. Schmid (1971) Trenimon-induced chromosomal damage in bone-marrow cells of six mammalian species, evaluated by the micronucleus test, *Mutation Res.*, **12**, 417–425.

Salamone, M.F. and J.A. Heddle (1983) The bone marrow micronucleus assay: rationale for a revised protocol. In F.J. de Serres (Ed.), *Chemical Mutagens* Vol. 8, Plenum Press, NY. pp. 111–149.

Salamone, M.F., J.A. Heddle and M. Katz (1981) Mutagenic activity of 41 compounds in the *in vivo* micronucleus assay. In F.J. de Serres and J. Ashby (Eds), *Evaluation of Short-Term Tests for Carcinogens, Report of the International Collaborative Program*, Elsevier, NY, pp. 686–697.

Schmid, W. (1976) The micronucleus test for cytogenetic analysis. In A. Hollaender, (Ed.), *Chemical Mutagens*, Vol. 4, Plenum Press, NY, pp. 31–53.

Tsuchimoto, T. and B.E. Matter (1981) Activity of coded compounds in the micronucleus test. In F.J. de Serres and J. Ashby (Eds), *Evaluation of Short-Term Tests for Carcinogens, Report of the International Collaborative Program*, Elsevier, NY, pp. 705–711.

Wild, D., M.-T. King and K. Eckhardt (1980) Cytogenetic effect of ortho-phenylenediamine in the mouse, Chinese hamster, and Guinea pig and of derivatives, evaluated by the micronucleus test, *Arch. Toxicol.*, **43**, 249–255.

Wild, D., M.-T. King, E. Gocke and K. Eckhardt (1983) Study of artificial flavouring substances for mutagenicity in the Salmonella/microsome, Basc and micronucleus tests, *Food Chem. Toxicol.*, **21**, 707–719.

David H. Blakey, George R. Douglas, Ron D.L. Bell and Michael Goddard
Mutagenesis Section, Environmental Health Centre, Tunney's Pasture, Ottawa, Ontario, Canada

20

Bone marrow micronucleus induction in HPBBL mice

1 Summary

The clastogenic activity of two carcinogens, benzo (a)pyrene (BP) and 2-acetylaminofluorene (2AAF), and structurally related noncarcinogens, pyrene (PYR) and 4-acetylaminofluorene (4AAF), was determined using the mouse bone marrow micronucleus assay. Significant increases in micronucleated polychromatic erythrocytes were induced by BP and 2AAF but not by PYR or 4AAF. The bone marrow micronucleus assay was able to discriminate between the carcinogens and their noncarcinogenic analogues that were examined in this study.

2 Introduction

The bone marrow micronucleus assay has been used to measure the *in vivo* clastogenic activity of a broad spectrum of compounds (Heddle *et al.*, 1983). Micronuclei are small nuclear fragments that are not included in either daughter nucleus following cell division. In the case of polychromatic erythrocytes (PE) in the bone marrow, micronuclei may remain in the cell's cytoplasm following the expulsion of the primary nucleus. Although not all chromosome aberrations are manifested as micronuclei (Carrano and Heddle, 1973), the ease with which thousands of cells can be analysed and the persistence of micronuclei following cell division offset this disadvantage.

In this study, the induction of micronuclei by BP, PYR, 2AAF and 4AAF was determined in the bone marrow of the mouse. BP and 2AAF are carcinogenic in several species, while PYR and 4AAF are structurally related compounds that are noncarcinogenic (Purchase *et al.*, 1981). In a previous international collaborative study, several *in vitro* short-term mutagenicity tests gave positive results for both of these carcinogens as well as their noncarcinogenic analogues (Ashby, 1981), while the *in vivo* tests included in the study were able to discriminate between the carcinogens and noncarcinogens (Purchase and Ray, 1981). The present collaborative study was undertaken to characterize further genotoxic effects of these 4 compounds using several *in vivo* short-term tests.

3 Materials and methods

Animals

8 to 12 week old male HPBBL mice were used throughout this study. HPBBL is a strain derived from C57BL mice, and all mice were supplied by the Animal Resources

Division, HPB, Health and Welfare Canada. The animals were given Purina laboratory rodent chow and water ad libitum. Five animals were included in each treatment group.

Chemicals

The 4 test compounds were supplied by the International Program on Chemical Safety (IPCS). Cyclophosphamide was obtained from Sigma, USP grade olive oil from Eastman, and dimethylsulphoxide (DMSO) from Aldrich. Whenever possible, the test compounds were dissolved or suspended directly in olive oil. This procedure often produced a suspension that contained undissolved test compound. Since small volumes of test compound were to be administered to the animals, these non-uniform suspensions could cause non-uniform dosing of animals. In some experiments, the compounds were first dissolved in a small volume of DMSO and then added to the olive oil. This method produced suspensions that appeared to be homogeneous. The volumes of DMSO used in each experiment and the doses of test compounds are given in Table 1.

Treatment

The treatment doses, route of administration and expression times used in this study were recommended by the group coordinators. Animals received the test compounds by gavage, and the volume administered was equal to 1% of the animal's body weight. The solvent control consisted of olive oil containing the same proportion of DMSO as the olive oil/test compound solution. The positive control animals received 75 mg/kg cyclophosphamide in saline by intraperitoneal (ip) injection.

Animals were sacrificed by cervical dislocation at 24, 36 and 48 h following treatment. Femurs were removed and the marrow flushed out with approximately 0.05 ml fetal calf serum (GIBCO) containing EDTA (1.67 mg/ml). Cell suspensions were smeared onto a clean glass slide, air-dried, and fixed with absolute methanol for 5 min. The slides were stained with 8% Giemsa (Gurr R66 or Fisher) pH 6.8 for 10 – 20 min. One thousand PEs were scored for each animal.

Statistical analysis

The data were transformed using the weighted Freeman Tukey transformation, then tested for significance using the t test (1-tailed). Preliminary analysis was also performed (by D. Blakey) using the chi-square analysis recommended by Hart and Engberg-Pedersen (1983). These two methods of analysis led to essentially the same data interpretation. Throughout this report, unless otherwise stated, significance was determined at $p < 0.05$.

4 Results

Assay performance

To establish the characteristics of the bone marrow micronucleus assay as it was performed in these experiments, the control data collected throughout this study were pooled and composite control values were cal-

Table 1. *Treatment doses, expression times, and solvent conditions*[a]

Agent	Trial	Expression times (h)	Doses Tested (mg/kg) High	Medium	Low	%DMSO
BP	1	24, 36, 48	252			0
	2	36	252	126	63	0
PYR	1	24, 36, 48	606			25
	2	24, 36	808			6
2AAF	1	24, 36, 48	223			12
	2	24, 36, 48		111.5	55.8	13
	3	24, 36, 48	223			0
4AAF	1	24, 36, 48	892			13
	2	24, 36, 48		446	223	13

[a] In some experiments, the test compound was dissolved in DMSO prior to being added to the olive oil. %DMSO indicates the percentage of DMSO in the olive oil.

Table 2. *Micronucleus test composite vehicle and positive control data in HPBBL mice*[a]

Treatment	Exp. time (h)	%MPE	%MNE	PE/PE + NE
Vehicle Control	24	0.390 ± 0.030	0.237 ± 0.038	0.618 ± 0.014
	36	0.435 ± 0.030	0.204 ± 0.030	0.589 ± 0.019
	48	0.402 ± 0.036	0.215 ± 0.027	0.564 ± 0.024
	overall	0.410 ± 0.018	0.214 ± 0.018	0.592 ± 0.011
Cyclophosphamide	24	1.91 ± 0.068	0.225 ± 0.025	0.403 ± 0.016
	36	3.57 ± 0.149	0.264 ± 0.022	0.328 ± 0.016
	48	3.57 ± 0.170	0.377 ± 0.020	0.212 ± 0.141

[a] ± Standard error of the mean.

culated for both vehicle control and positive control animals (Table 2). The overall vehicle control micronucleated polychromatic erythrocyte (MPE) frequency was 0.410% which is within the limits proposed for acceptable control data (Heddle *et al.*, 1983). The overall micronucleated normochromatic erythrocyte (MNE) frequency was 0.214% and the PE/PE + NE ratio was 0.592. Although the PE/PE + NE ratio in vehicle control animals decreased slightly with time, no increase in micronuclei was associated with this marginal effect. It is possible that treatment with the vehicle alone was slightly toxic.

In positive control animals, a significant increase in MPE frequency was induced by cyclophosphamide in

Table 3. *Summary of micronucleus test data in HPBBL mice*[a]

Agent	Trial	Dose (mg/kg)	Exp. time (h)	PE/PE + NE	%MNE	%MPE
BP	1	0	24	0.549 ± 0.022	0.23 ± 0.10	0.28 ± 0.06
			36	0.541 ± 0.033	0.07 ± 0.04	0.38 ± 0.12
			48	0.597 ± 0.018	0.17 ± 0.12	0.50 ± 0.12
		252	24	0.557 ± 0.040	0.18 ± 0.05	0.84 ± 0.21
			36	0.549 ± 0.037	0.15 ± 0.08	1.32 ± 0.24
			48	0.466 ± 0.033	0.39 ± 0.09	0.98 ± 0.23
BP	2	0	36	0.422 ± 0.020	0.30 ± 0.07	0.42 ± 0.08
		63	36	0.517 ± 0.017	0.21 ± 0.08	0.84 ± 0.26
		126	36	0.563 ± 0.043	0.31 ± 0.10	0.88 ± 0.20
		252	36	0.546 ± 0.037	0.29 ± 0.11	1.34 ± 0.29
PYR	1	0	24	0.612 ± 0.058	0.25 ± 0.05	0.56 ± 0.13
			36	0.796 ± 0.057	0.22 ± 0.12	0.30 ± 0.06
			48	0.814 ± 0.037	0.05 ± 0.06	0.54 ± 0.11
		606	24	0.337 ± 0.036	0.22 ± 0.03	0.60 ± 0.09
			36	0.424 ± 0.116	0.33 ± 0.17	0.47 ± 0.15
			48	0.410 ± 0.081	0.16 ± 0.08	0.30 ± 0.11
	2	0	24	0.652 ± 0.053	0.36 ± 0.18	0.42 ± 0.10
			36	0.582 ± 0.019	0.10 ± 0.05	0.50 ± 0.08
		808	24	0.313 ± 0.099	0.24 ± 0.02	0.60 ± 0.23
			36	0.239 ± 0.051	0.19 ± 0.02	0.60 ± 0.16
2AAF	1	0	24	0.582 ± 0.050	0.08 ± 0.06	0.40 ± 0.08
			36	0.619 ± 0.028	0.44 ± 0.09	0.62 ± 0.09
			48	0.526 ± 0.038	0.26 ± 0.07	0.34 ± 0.06
		223	24	0.587 ± 0.052	0.19 ± 0.11	0.46 ± 0.09
			36	0.593 ± 0.057	0.34 ± 0.16	0.90 ± 0.08
			48	0.567 ± 0.042	0.39 ± 0.11	0.84 ± 0.14
	2	0	24	0.563 ± 0.062	0.29 ± 0.12	0.36 ± 0.06
			36	0.499 ± 0.048	0.10 ± 0.02	0.42 ± 0.15
			48	0.476 ± 0.056	0.25 ± 0.02	0.46 ± 0.08
		55.8	24	0.587 ± 0.040	0.23 ± 0.12	0.70 ± 0.13
			36	0.535 ± 0.016	0.16 ± 0.15	0.40 ± 0.11
			48	0.402 ± 0.019	0.14 ± 0.03	0.58 ± 0.17
		111.5	24	0.471 ± 0.026	0.10 ± 0.04	0.52 ± 0.07
			36	0.593 ± 0.036	0.37 ± 0.16	0.50 ± 0.17
			48	0.432 ± 0.039	0.13 ± 0.02	0.50 ± 0.13
	3	0	24	0.649 ± 0.032	0.41 ± 0.17	0.28 ± 0.10
			36	0.630 ± 0.051	0.18 ± 0.14	0.30 ± 0.12
			48	0.549 ± 0.022	0.15 ± 0.07	0.22 ± 0.11
		223	24	0.463 ± 0.102	0.22 ± 0.03	0.24 ± 0.09
			36	0.563 ± 0.021	0.31 ± 0.06	0.48 ± 0.11
			48	0.557 ± 0.031	0.11 ± 0.06	0.30 ± 0.15
4AAF	1	0	24	0.678 ± 0.024	0.07 ± 0.05	0.50 ± 0.07
			36	0.566 ± 0.032	0.22 ± 0.13	0.50 ± 0.09
			48	0.400 ± 0.027	0.25 ± 0.06	0.34 ± 0.05
		892	24	0.651 ± 0.022	0.17 ± 0.05	0.28 ± 0.09
			36	0.583 ± 0.054	0.11 ± 0.06	0.26 ± 0.08
			48	0.429 ± 0.049	0.22 ± 0.05	0.50 ± 0.09
	2	0	24	0.654 ± 0.011	0.21 ± 0.11	0.32 ± 0.10
			36	0.674 ± 0.020	0.19 ± 0.10	0.50 ± 0.06
			48	0.603 ± 0.016	0.36 ± 0.07	0.46 ± 0.12
		223	24	0.624 ± 0.022	0.36 ± 0.14	0.50 ± 0.07
			36	0.632 ± 0.038	0.17 ± 0.09	0.66 ± 0.06
			48	0.628 ± 0.030	0.23 ± 0.08	0.63 ± 0.15
		446	24	0.711 ± 0.028	0.10 ± 0.07	0.24 ± 0.07
			36	0.632 ± 0.020	0.44 ± 0.06	0.38 ± 0.16
			48	0.563 ± 0.043	0.26 ± 0.09	0.42 ± 0.10

[a] Values are given as group means ± standard error.

every experiment. At the optimal expression time of 48 h, the composite MPE frequency for cyclophosphamide-treated animals was 3.57%. The dose of cyclophosphamide used in this experiment induced toxicity in the bone marrow as indicated by a reduction in the PE/PE + NE ratio.

Test results

The data for the 4 test compounds are summarized in Table 3.

BP and PYR

In the first trial, BP (252 mg/kg) induced a significant increase in MPE frequency at 24, 36 and 48 h with the highest MPE frequency detected 36 h following treatment. The PE/PE + NE ratio was significantly reduced by BP in animals sacrificed 48 h following treatment. In the second trial, animals were treated with BP at 63, 126 and 252 mg/kg and samples were collected 36 hr following treatment. In this experiment, animals treated with 126 and 252 mg/kg had an increased MPE frequency and a dose response was evident (Table 3).

PYR did not increase the MPE frequency at either 606 or 808 mg/kg. In both trials, PYR reduced the PE/PE + NE ratio. Following treatment with PYR (808 mg/kg) in the second trial, none of the animals survived to the 48 h sacrifice time. Thus, in these experiments, PYR was toxic but not clastogenic in the bone marrow.

2AAF and 4AAF

In the first trial, 2AAF (223 mg/kg) induced a small but significant increase in MPE frequency at 36 and 48 h. In the second 2AAF trial, 111.5 mg/kg did not induce an increase in MPE frequency, but 55.8 mg/kg did induce a significant increase at 24 h. However, since the effect was not dose related, it is listed as a questionable response in Table 4.

Since only a small significant increase in MPE frequency was induced by 2AAF in the first trial, the experiment was repeated, with one important change. In the first trial, to achieve a more homogeneous suspension, 2AAF was dissolved in DMSO before its addition to olive oil. When this experiment was repeated, (Trial 3) 2AAF (233 mg/kg) was added directly to the olive oil to avoid any additional factors that might affect the clastogenic activity of 2AAF. In the third trial (Tables 3 and 4), 223 mg/kg of 2AAF did not induce a significant increase in MPE frequency ($p > 0.05$) (Tables 2 and 3). We were therefore unable to confirm the positive response for MPE frequency detected in the first trial.

4AAF, the noncarcinogenic analogue of 2AAF, did not induce an increase in MPE frequency at 223, 446 or 892 mg/kg.

5 Discussion

BP and PYR

Previous studies demonstrated that BP is carcinogenic in several animal species including sub-human primates (IARC, 1973). Furthermore, BP is mutagenic in a wide variety of *in vitro* and *in vivo* short-term tests (Brookes and Preston, 1981; Purchase and Ray, 1981). It is not surprising that in this study BP induced a significant, dose-related increase in MPE frequency (Tables 3 and 4). The maximum BP-induced increase in MPE frequency was detected 36 h following treatment, indicating that there was little mitotic delay and that the compound was taken up and metabolized rapidly.

PYR was included in this study as the noncarcinogenic analogue of BP. However, PYR has a poorly defined carcinogenic potential (Purchase *et al.*, 1981). Furthermore, in a previous IPCS collaborative study, PYR was genotoxic for several *in vitro* endpoints, although it was determined that PYR induced neither sister chromatid exchanges nor chromosome aberrations (Brookes and Preston, 1981). In this study, PYR did not induce an increase in the MPE frequency in mouse bone marrow even after oral administration of doses which were lethal within 48 h or toxic in the bone marrow as indicated by a reduction of the PE/PE + NE ratio. These negative *in vivo* results are in agreement with a previous study in which

Table 4. *Interpretation of micronucleus test results in HPBBL mice*[a]

Agent	Trial	Dose (mg/kg)	MPE[b] 24	36	48	MNE 24	36	48	PE/PE + NE 24	36	48
BP	1	252	+	+++	+	−	−	+	−	−	++
	2	252		+++			−			−	
		126		+			−			−	
		63		−			−			−	
PYR	1	606	−	−	−	−	−	−	+++	++	+++
	2	808	−	−		−	−		+++	+++	
2AAF	1	223	−	+	+++	−	−	−	+++	−	−
	3	223	−	−	−	−	−	−	+	−	−
	2	111.5	−	−	−	−	+	−	−	−	−
		55.8	++?	−	−	−	−	−	−	−	−
4AAF	1	892	−	−	−	−	−	−	−	−	−
	2	446	−	−	−	−	+	−	−	−	−
		223	−	−	−	−	−	−	−	−	−

[a] Data analysis was performed using the t test on wieghted, Freeman Tukey transformed data.
[b] + indicates significance at $p < 0.05$,
++ indicates significance at $p < 0.01$,
+++ indicates significance at $p < 0.005$,
− indicates no significant effect ($p > 0.05$),
? indicates a questionable result which was significant but not dose related.

PYR administered by ip injection in mice did not induce bone marrow micronuclei (Salamone *et al.*, 1981; Kirkhart, 1981; Tsuchimoto and Matter, 1981) or sperm abnormalities (Wyrobek *et al.*, 1981).

2AAF and 4AAF

2AAF has been shown to exhibit carcinogenic activity in several animal species (Clayson and Garner, 1976). In short-term mutagenicity tests, 2AAF induced chromosome aberrations *in vitro* (Brookes and Preston, 1981) and micronuclei in the mouse bone marrow (Salamone *et al.*, 1981; Tsuchimoto and Matter, 1981).

In this study, 2AAF (223 mg/kg) induced a small but significant increase in MPE frequency at 36 and 48 h in the first trial. In Trial 2, the low dose of 2AAF (55.8 mg/kg) increased the MPE frequency at 24 h ($p < 0.05$). However, when the first trial was repeated (Trial 3), 2AAF (223 mg/kg) did not induce an increase in MPE frequency. This discrepancy may be attributed to the fact that the highest 2AAF dose (223 mg/kg) recommended for use in this study was too low to induce consistent increases in MPE frequencies. In previous studies, the lowest effective dose for bone marrow micronucleus induction was 280 mg/kg administered by ip injection (Tsuchimoto and Matter, 1981). It is also possible that the discrepancy between the trials was caused by the use of different vehicles. In the first trial, 2AAF was dissolved in DMSO before it was added to the olive oil, but in the repeat experiment, the 2AAF was added directly to the olive oil. The use of DMSO in the first trial may have enhanced the clastogenic activity of 2AAF because of increased solubility or uptake. For example, it has been shown in Salmonella that BP was nonmutagenic when dissolved in mineral oil alone but mutagenic when dissolved in mineral oil plus a variety of solvents including DMSO, ethanol, acetone, or benzene (V.F. Simmon, personal communication).

4AAF was included in this study as the noncarcinogenic analogue of 2AAF (Purchase *et al.*, 1981). 4AAF has also been shown to exhibit genotoxic activity for at least one endpoint *in vitro* (Ashby, 1981). In a previous IPCS collaborative study, 4AAF was positive in the baby hamster kidney cell *in vitro* transformation assay (Daniel and Dehnel, 1981; Styles, 1981). In this study, 4AAF (892 mg/kg) given by gavage did not induce an increase in MPE frequency. These data are supported by a previous study indicating that 4AAF (600 mg/kg) given by ip injection was negative in the mouse bone marrow micronucleus assay (Tsuchimoto and Matter, 1981).

6 Conclusions

In these experiments, the mouse bone marrow micronucleus assay was able to discriminate between the carcinogens (BP and 2AAF) and their noncarcinogenic analogues (PYR and 4AAF).

However, since the positive response for 2AAF evident in the first trail could not be confirmed in the repeat experiment, the data for the carcinogen/noncarcinogen pair 2AAF and 4AAF are somewhat equivocal and require further study using higher doses.

7 References

Ashby, J. (1981) Overview of Study and Test Chemical Activities. In F.J. de Serres and J. Ashby (Eds), *Evaluation of Short-Term Tests for Carcinogens*, Elsevier/North Holland, NY, pp. 112–172.

Brookes, P. and R.J. Preston (1981) Summary Report on the Performance of *In Vitro* Mammalian Assays. In F.J. de Serres and J. Ashby (Eds), *Evaluation of Short-Term Tests for Carcinogens*, Elsevier/North Holland, NY, pp. 77–85.

Carrano, A.V. and J.A. Heddle (1973) The fate of chromosome aberrations, *J. Theor. Biol.*, **38**, 289–304.

Clayson, D.B. and R.C. Garner (1976) Carcinogenic Aromatic Amines and Related Compounds. In C.E. Searle (Ed.), *Chemical Carcinogens*, American Chemical Society, Washington, pp. 366–461.

Daniel, M.R. and J.M. Dehnel (1981) Cell Transformation Test with Baby Hamster Kidney Cells. In F.J. de Serres and J. Ashby (Eds), *Evaluation of Short-Term Tests for Carcinogens*, Elsevier/North Holland, NY, pp. 626–637.

Hart, J.W. and H. Engberg-Pedersen (1983) Statistics of the mouse bone-marrow micronucleus test: counting, distribution and evaluation of results, *Mutation Res.*, **111**, 195–207.

Heddle, J.A., M. Hite, B. Kirkhart, K. Mavournin, J.T. MacGregor, G.W. Newell and M.F. Salamone (1983) The induction of micronuclei as a measure of genotoxicity, A report of the U.S. Environmental Protection Agency Gene-Tox Program, *Mutation Res.*, **123**, 61–118.

International Agency for Research on Cancer (1973) *IARC Monographs on the Evaluation of Carcinogenic Risk of Chemicals to Man*, Vol. 3, Certain polycyclic aromatic hydrocarbons, International Agency for Research on Cancer, Lyon, pp. 91–136.

Kirkhart, B. (1981) Micronucleus Test on 21 Compounds. In F.J. de Serres and J. Ashby (Eds), *Evaluation of Short-Term Tests for Carcinogens*, Elsevier/North Holland, NY, pp. 698–704.

Purchase, I.F.H. and V. Ray (1981) Summary Report on the performance of *In Vivo* Assays. In F.J. de Serres and J. Ashby (Eds), *Evaluation of Short-Term Tests for Carcinogens*, Elsevier/North Holland, NY, pp. 86–95.

Purchase, I.F.M., D.B. Clayson, R. Preussmann and L. Tomatis (1981) Activity of 42 Compounds in Animal Carcinogenicity Studies. In F.J. de Serres and J. Ashby (Eds), *Evaluation of Short-Term Tests for Carcinogens*, Elsevier/North Holland, NY, pp. 21–32.

Salamone, M.F., J.A. Heddle and M. Katz (1981) Mutagenic Activity of 42 Compounds in the *In Vivo* Micronucleus Assay. In F.J. de Serres and J. Ashby (Eds), *Evaluation of Short-Term Tests for Carcinogens*, Elsevier/North Holland, NY, pp. 686–697.

Styles, J.A. (1981) Activity of 42 Coded Compounds in the BHK-21 Cell Transformation Test. In F.J. de Serres and J. Ashby (Eds), *Evaluation of Short-Term Tests for Carcinogens*, Elsevier/North Holland, NY, pp. 638–646.

Tsuchimoto, T. and B.E. Matter (1981) Activity of Coded Compounds in the Micronucleus Test. In F.J. de Serres and J. Ashby (Eds), *Evaluation of Short-Term Tests for Carcinogens*, Elsevier/North Holland, NY, pp. 705–711.

Wyrobek, A., L. Gordon and G. Watchmaker (1981) Effect of 17 Chemical Agents Including 6 Carcinogen/Noncarcinogen Pairs on Sperm Shape Abnormalities in Mice. In F.J. de Serres and J. Ashby (Eds), *Evaluation of Short-Term Tests for Carcinogens*, Elsevier/North Holland, NY, pp. 712–717.

Jacques Pot-Deprun and Ivan Chouroulinkov
Unité de Cancérogénèse Expérimentale et Toxicologie Génétique, I.R.S.C.-CNRS, Villejuif France

21

Micronuclei induction in mouse bone marrow cells with benzo(a)pyrene, pyrene, 2-acetylaminofluorene and 4-acetylaminofluorene

1 Summary

Carcinogen/non carcinogen pairs BP/PYR and 2AAF/4AAF were studied in the mouse bone marrow micronucleus assay. Each compound dissolved in olive oil was administered orally to male mice. Polychromatic erythrocytes with micronuclei were scored after 24 hours. Only BP and 2AAF were found active with dose–response relationships.

2 Introduction

The *in vivo* micronucleus assay detecting chromosome breaks or chromosome loss is recommended by OECD and CEE as a cytogenetic test for chemical genetic hazard control. The validation of such a test for carcinogenic potential prediction is of great importance. This report is a small contribution to such a validation. It concerns the two carcinogen/non carcinogen pairs BP/PYR and 2AAF/4AAF and their capacity to induce micronuclei in mouse bone marrow cells.

3 Materials and methods

Animals

For the *in vivo* micronucleus assay, male Swiss mice were obtained from specific pathogen-free breeding colony at the Institut de Recherches Scientifiques sur le Cancer, Villejuif, France.

The animals were 8–12 weeks old at the beginning of the experiments. For each dose the number of replicates were 4 animals (treated) and 2 for the controls.

Test chemicals, handling and dilution

BP was synthetized by Laboratoire P. Jacquignon, CNRS, Gif sur Yvette, France (quality "purissim", Figures 1, 2) for the main assay, and Lancaster synthesis Ltd. Eastgate, Morecombe, (ICI-UK) for repetition of dose 100 mg/kg.

PYR was obtained from Sigma Lab. U.S.A. for the main assay, and from Lancaster Synthesis Ltd. Eastgate, Morecombe, (ICI-UK) for repetition of dose 100 mg/kg.

2AAF and 4AAF were furnished by ICI-UK.

All compounds were stored at $-20\,^{\circ}C$ until use. Olive oil (extra pure, Codex) solutions were prepared just before treatment.

Treatment administration was by gavage.

Micronucleus test procedure

Bone marrow cells were harvested according to Kliesch *et al.* (1981) procedure. Cells were sampled from the femur 24 hr after treatment in fetal calf serum (Flow Laboratories, Paris). Smears were prepared according to Schmid (1977). For staining slides, Sörensen buffer (pH 6.8) was used instead of distilled water.

Figure 1. Spectrophotometre UV. Analysis of BP:BP dissolved in ethanol and the spectra were recorded in spectrophotometer Varian series 634. Wavelengths changed from 400 to 200 nm.

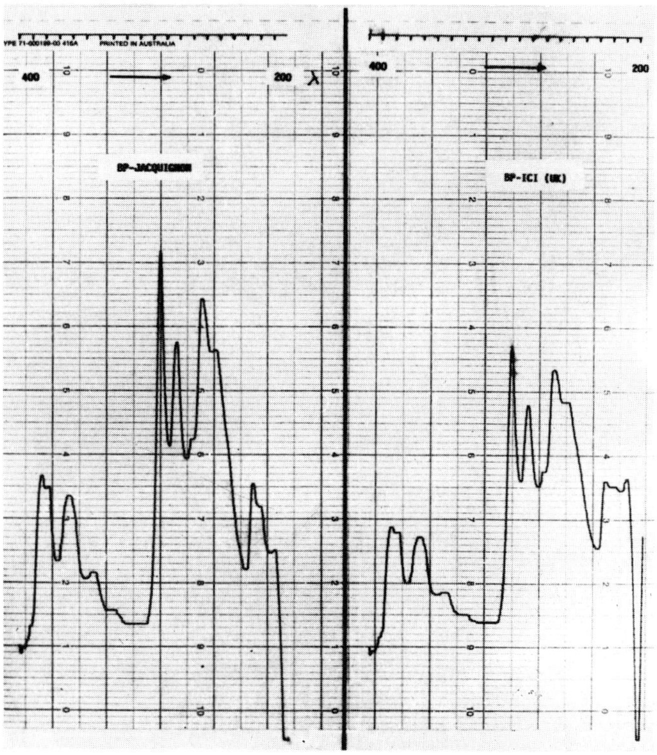

Analysis of bone marrow smears

Slides were screened using a Leitz-Orthoplan photomicroscope under × 1000 magnification. 2000 polychromatic erythrocytes per animal were counted and checked for the presence of micronuclei.

Micronuclei were counted in polychromatic erythrocytes. The ratio poly- to normochromatic erythrocytes was calculated based on 4000 cells for the controls and 8000 cells for the treated animals.

Statistical analysis

The Mann-Whitney-Wilcoxon test (U-test) was used to compare the data obtained. The possible dose–response relationship was calculated by means of linear regression after pooling the particular outcomes obtained for each animal.

4 Results and discussion

BP

BP was found positive in the micronucleous test and the summary results are presented in Table 1 and Figure 3.

For each tested dose we observed a significant enhancement in the frequency of micronucleated polychromatic erythrocytes. The dose–response relationship in our experimental conditions is linear (Figure 3) with equation $y = 0.0501X + 0.146$, $r = 0, 967$, $p < 0.001$.

The results are generally in agreement with those reported by Kirkart (1981), Kliesch (1981), and Salamone (1981). We should note that for the choosen interval (24 h) after treatment our results are higher (Salamone, 1981). The strain susceptibility may be at the basis for this observation. However, in our opinion, for unknown chemicals 2 or 3 intervals (24, 30 and 36 h) should be included in the experimental protocol. The determination of the doses and the intervals for such new compounds

Figure 2. BP HPLC analysis: Beckman 344 gradient liquid chromatograph. Absorbance Detector Model 160. (254 Lamp Mercury). Column: Ultrasphere-ODS′ (C_{18}) 4,6 × 150 mm. Mobile phase: H_{20}/Methanol. Flow rate: 1 ml/min. Prog.: 60:70% → 10′, 70% → 7′; 70:90 → 25′; 90:100% → 3′; 100 → 3′. Range UV: 60 → 100% = 0.05, 100% = 2.

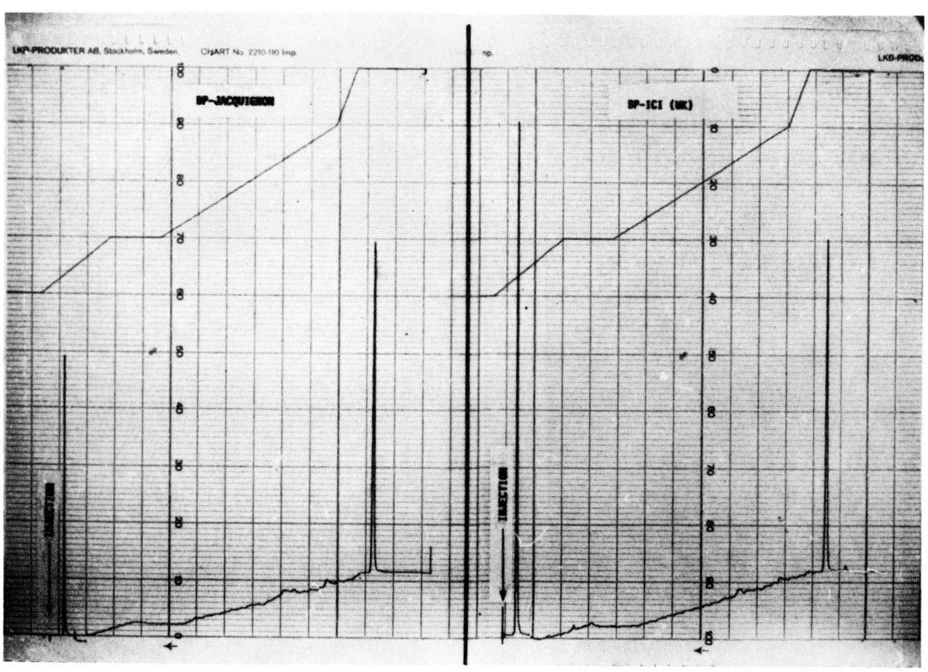

Table 1. *Micronuclei induced in mouse bone marrow cells of mice 24 h after single oral administration of BP and PYR*

Compound	Dose (mg/kg)	Polychromatic erythrocytes scored (n)	with MN n	with MN % ± SE	Ratio of poly/normo-chromatic erythrocytes
	0.0	12 000	16	0.13 ± 0.01	0.45
BP	50.0	8 000	36	0.45 ± 0.03[a]	0.45
(Jacq.)	100.0	8 000	46	0.60 ± 0.04	0.35
	150.0	8 000	73	0.91 ± 0.06	0.31
BP	0.0	4 000	8	0.20 ± 0.01	0.51
(ICI-UK)	100.0	8 000	48	0.60 ± 0.03	0.46
	0.0	12 000	17	0.14 ± 0.07	6.45
PYR	50.0	8 000	15	0.18 ± 0.02	0.38
(Sigma)	100.0	8 000	14	0.17 ± 0.02	0.41
	150.0	8 000	14	0.17 ± 0.02	0.40
PYR	0.0	4 000	6	0.15 ± 0.02	0.46
(ICI-UK)	100.0	8 000	16	0.20 ± 0.02	0.50

[a] Lowest dose at which a significant effect was found: $p < 0.01$ (Mann-Whitney-Wilcoxon test).

Table 2. *Micronuclei induced in mouse bone marrow cells 24 h after single oral administration of 2AAF and 4AAF*

Compound	Dose (mg/kg)	Polychromatic erythrocytes scored (n)	with MN n	with MN % ± SE	Ratio of poly/normo-chromatic erythrocytes
	0.0	12 000	17	0.14 ± 0.01	0.32
2AAF	55.8	8 000	11	0.14 ± 0.02	0.36
	111.5	8 000	24	0.30 ± 0.03[a]	0.37
	223.0	8 000	54	0.67 ± 0.02	0.33
	0.0	12 000	18	0.15 ± 0.02	0.30
4AAF	223.0	8 000	14	0.17 ± 0.01	0.36
	446.0	8 000	15	0.18 ± 0.01	0.37
	892.0	8 000	15	0.18 ± 0.02	0.36

[a] Lowest dose at which a significant effect was found; $p < 0.001$ (Mann-Whitney-Wilcoxon test).

Figure 3. Dose–response relationship *in vivo* (mouse) micronucleus assay with BP (single oral administration and 24 hours expression time). ($y = 0.0501\,X + 0.146$, $r = 0.967$, $p < 0.001$).

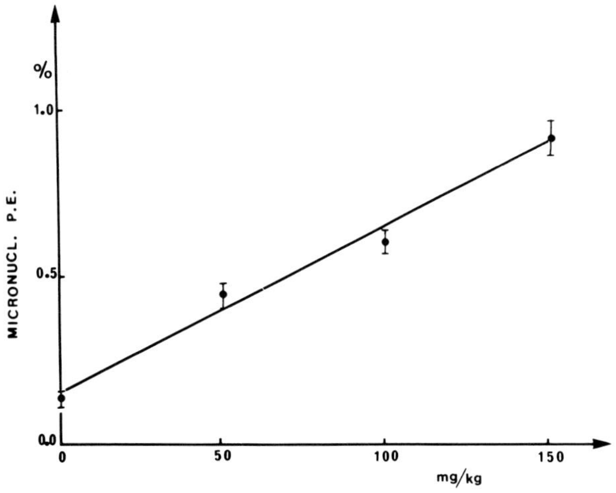

Figure 4. Dose–response relationship *in vivo* (mouse) micronucleus assay with 2AAF (single oral administration and 24 hours expression time).

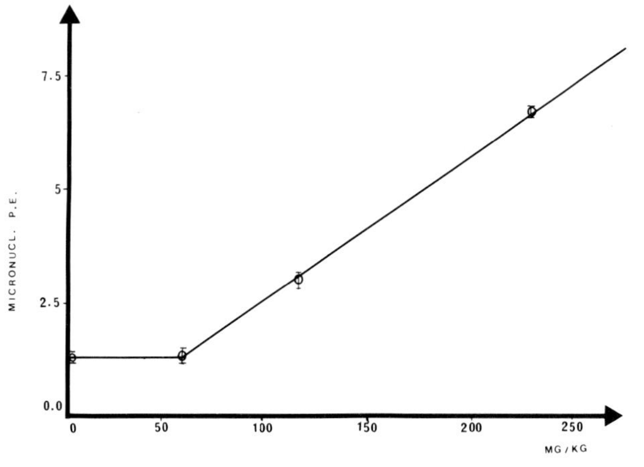

should be the decision of the expert who designs the protocol.

PYR

This compound gave negative results in the micronucleus test in our experimental conditions. Despite positive results reported in other test systems (SCE, UDS) our results confirm those obtained by others (Kirkart, 1981, and Salamone, 1981).

2AAF

2AAF was found positive in the micronucleus test (Table 2, Figure 4). The lowest dose with significant effect tested is 111.5 mg/kg. The dose–response relationship exhibits clear evidence of non-linearity.

4AAF

4AAF did not show a positive effect in the micronucleus test. The calculated dose–response regression curve ($y = 0.000419X + 1.576$, $r = 0.408$) is not significant.

5 Conclusion

The micronucleus test was positive for BP and 2AAF and negative for PYR and 4AAF. These results correlate well with the known carcinogenic potential of the studied compounds. Moreover the micronucleus test remains one of the most rapid and inexpensive *in vivo* mutagenicity test which can give quantitative data.

6 References

Kirkart, B. (1981) Micronucleus test on 21 compounds. In F.J. de Serres and J. Ashby (Eds), *Progress in Mutation Research*, Vol. 1, Evaluation of Short-Term Tests for Carcinogens, pp. 698–704, Elsevier/North Holland.

Kliesch, U., N. Danford and I.-D. Adler (1981) Micronucleus test and bone-marrow chromosomes analysis. A comparison of 2 methods *in vivo* for evaluating chemically induced chromosomal alterations, *Mutation Res.*, **80**, 321–332.

Salamone, M.F., J.A. Heddle and I. Katz (1981) Mutagenic activity of 41 compounds in the *in vivo* micronucleus assay. In F.J. de Serres and J. Ashby (Eds), *Progress in Mutation Research*, Vol. 1, Evaluation of Short-Term Tests for Carcinogens, pp. 686–697, Elsevier/North Holland.

Schmid, W. (1977) The micronucleus test. In *Handbook of Mutagenicity test procedure*, pp. 235–242, Elsevier/North Holland.

Barbara L. Harper and Marvin S. Legator
Department of Preventive Medicine and Community Health, Division of Environmental Toxicology, The University of Texas Medical Branch, Galveston, Texas, USA

22

Micronucleus test results for two carcinogen/noncarcinogen chemical pairs in male ICR Swiss mice

1 Summary

Bone marrow micronuclei were evaluated in male ICR Swiss mice after single oral treatments with BP, PYR, 2AAF and 4AAF. Chemicals were dissolved or suspended in olive oil and administered at doses approaching toxicity. Animals were sacrificed 24 h after treatment, with an additional 48 h harvest for the highest dose. A confirmatory experiment with 2AAF was harvested at 24, 36 and 48 h. 4AAF and PYR gave negative results, while BP was positive at both 24 and 48 hours after treatment in a dose–responsive manner. 2AAF was negative at 24 and 48 hours at all doses, while a repeat experiment at 1 mM/kg showed a moderate response in 2 out of 5 animals at 36 hours, the most sensitive time point as described in the summary report. Outbred Swiss mice may be less sensitive to 2AAF than are other strains and may be more variable in their response.

2 Introduction

As part of the collaborative study, we participated by using a combined testing protocol (Legator and Harper, 1982) in which one femur of each treated mouse was harvested for micronucleus analysis and one for metaphase analysis, and the pooled urine from the entire treated group was examined for the presence of bacterial mutagens. The micronucleus results are presented here.

3 Materials and methods

Male Ha/ICR Swiss mice, an outbred strain, were obtained from Timco (Harlan Sprague-Dawley, Houston, TX). They were housed on corn-cob bedding with food and water ad libitum, and were acclimatized 7 days before treatment. Treatment groups consisted of 5 mice or 10–11 mice in the case of control animals, all weighing 22–26 g. Femurs were harvested 24 h after treatment, with an additional group at the highest dose of each chemical also harvested at 48 h and one 2AAF-treated group at 36 h. Femurs were flushed into fetal calf serum, and slides were stained with Wright's and Giemsa stains (Heddle *et al.*, 1984; Salamone and Heddle, 1983). One thousand polychromatic erythrocytes (PE) were scored for micronuclei (MPE) from each animal, 500 cells from each of 2 duplicate slides. The Mann Whitney/Wilcoxon U test was used for statistical analysis.

Test chemicals were supplied for this study from ICI (UK). Benzene (thiophene-free) was purchased from Fisher Chemical Co., Houston, TX, and cyclophosphamide (CP) from Mead Johnson, Evansville, IN. BP, PYR and benzene were dissolved in U.S. pharmaceutical grade olive oil (Fisher Chemical Co.) with heating, and 2AAF and 4AAF were suspended in olive oil without DMSO. All chemicals were administered orally at 0.1 ml per 10 g body weight except CP, which was dissolved in saline and injected intraperitoneally (ip) in the same volume per weight.

4 Results and discussion

As shown in Table 1, BP was the only test chemical that caused significantly elevated levels of micronuclei and gave a dose–responsive increase for the 24-hour harvest times. The lowest dose of BP correlates well with the lowest oral dose detected by Kliesch *et al.* (1982) and in the papers reviewed by Heddle *et al.* (1983) for the Gene-Tox Program. In addition, the International Collaborative Program found BP to be positive in the micronucleus test when given ip in DMSO (Salamone *et al.*, 1981; Kirkhart, 1981; Tsuchimoto and Matter, 1981).

In other studies, 2AAF was shown to be positive in mice when administered ip in DMSO (Heddle *et al.*, 1983), with the lowest positive dose being 275–280 mg/kg. It was also found to be positive in the International Collaborative Study when given ip in DMSO (Salamone *et al.*, 1981; Tsuchimoto and Matter, 1981). Results presented by other authors in this volume indicate no advantage in using DMSO and ip administration. Inbred strains were used by other investigators, while outbred mice were used here; this outbred strain is of comparable sensitivity to the other strains with respect to BP (see summary report, this volume) but appears to be less sensitive to 2AAF. Since 36 hours was found to be the most sensitive time point by the overall study, a confirmatory experiment was done using 2AAF at 1 mM/kg (232 mg/kg) with 5 mice harvested at 24, 36, and 48 h. There was no indication of response at 24 and 48 hours, as before, but at 36 h 2 out of 5 animals showed a moderately elevated response (0.8% MPE; MPE/1000 PE = 0, 3, 4, 8, 9). This observation may indicate a variable or polymorphic metabolism of 2AAF in outbred mice, and the amount of variability within a group reduces the ability to obtain a statistically significant increase. Species differences in 2AAF metabolism are well known, and there are some indications that there may be quantitative and qualitative differences in cytochrome P450-mediated N-hydroxylation of 2AAF. Primary hepatocytes from C57B1/6 mice are more sensitive to alkali-labile DNA damage caused by 2AAF than are DBA/2 hepatocytes, and the difference is more pronounced after TCDD induction of the host (Miller *et al.*, 1984). Thus, differences in metabolic activation of 2AAF between mouse strains plus a low potency, which makes the optimal harvest time more critical, appeared to have a combined effect on the ability of outbred mice to detect the effect of 2AAF in the micronucleus test.

5 References

Heddle, J.A., E. Stuart and M.F. Salamone (1984) The Bone Marrow Micronucleus Test. In B.J. Kilbey, M.S. Legator, W. Nichols and C. Ramel (Eds), *Handbook of Mutagenicity Test Procedures*, second edition, Elsevier, NY, pp. 441–457.

Heddle, J.A., M. Hite, B. Kirkhart, K. Mavournin, J.T. MacGregor, G.W. Newell and M.F. Salamone (1983) The induction of micronuclei as a measure of genotoxicity. A report of the U.S. Environmental Protection Agency Gene-Tox program. *Mutation Res.*, **123**, 61–118.

Kirkhart, B. (1981) Micronucleus Test of 21 compounds. In F.J. de Serres and J. Ashby (Eds), *Evaluation of Short-Term Tests for Carcinogens*, Elsevier/North Holland, NY, pp. 698–711.

Kliesch, U., I. Rupova and I.D. Adler (1982) Induction of chromosome damage in mouse bone marrow by benzo(a)pyrene. *Mutation Res.*, **102**, 265–273.

Table 1. *Micronucleus test results*

Chemical	Dose (mg/kg)	Time of harvest (h)	(n)	MPE/PE (%) ($x \pm$ S.D.)	
BP	130	48	5	12.0 ± 3.2[a]	$p < 0.001$[b]
BP	130	24	5	26.4 ± 10.1	$p < 0.001$
BP	52	24	5	9.8 ± 3.1	$p < 0.001$
BP	26	24	5	7.8 ± 2.4	$p < 0.001$
PYR	340	48	5	2.4 ± 0.5	NS[c]
PYR	340	24	5	3.4 ± 2.5	NS
PYR	136	24	5	3.6 ± 2.4	NS
PYR	68	24	5	3.0 ± 1.6	NS
2AAF	1110	48	4	1.5 ± 1.3	NS
2AAF	1110	24	5	2.4 ± 1.3	NS
2AAF	446	24	5	2.8 ± 0.4	NS
2AAF	223	24	5	2.8 ± 1.3	NS
4AAF	182	48	5	2.2 ± 1.1	NS
4AAF	182	24	5	1.6 ± 1.3	NS
4AAF	73	24	5	1.8 ± 0.8	NS
4AAF	36	24	5	1.4 ± 2.1	NS
CP	50	24	10	18.1 ± 6.3	$p < 0.001$
Benzene	440	24	10	28.5 ± 9.0	$p < 0.001$
Untreated	—	—	11	2.9 ± 1.6	—

[a] Per cent micronucleated polychromatic erythrocytes.
[b] With respect to untreated animals,
[c] NS = not significant.

Legator, M.S. and B.L. Harper (1982) Animal-Human Studies: An Integrated Approach to Identifying Hazardous Chemicals in the Workplace. In M. Sorsa and H. Vainio (Eds), *Mutagens in Our Environment*, Alan R. Liss, NY, pp. 181–602.

Miller, M.E., I.B. Glowinski and S.S. Thorgeirsson (1984) The genotoxicity of aromatic amines in primary hepatocytes isolated from C57B1/6 and DBA/2 mice. *Carcinogenesis*, **5**, 797–804.

Salamone, M.F. and J.A. Heddle (1983) The Bone Marrow Micronucleus Assay: Rationale for a Revised Protocol. In F.J. de Serres (Ed.), *Chemical Mutagens: Principles and Methods for Their Detection*, Plenum Press, NY, Vol. 8, pp. 111–149.

Salamone, M.F., J.A. Heddle and M. Katz (1981) Mutagenic Activity of 41 Compounds in the *In Vivo* Micronucleus Assay. In F.J. de Serres and J. Ashby (Eds), *Evaluation of Short-Term Tests for Carcinogens*, Elsevier/New Holland, NY, pp. 686–697.

Tsuchimoto, T. and B.E. Matter (1981) Activity of Coded Compounds in the Micronucleus Test. In F.J. de Serres and J. Ashby (Eds), *Evaluation of Short-Term Tests for Carcinogens*, Elsevier/North Holland, NY, pp. 705–711.

Makoto Hayashi, Toshio Sofuni and Motoi Ishidate, Jr
Division of Mutagenesis, Biological Safety Research Center, National Institute of Hygienic Sciences, Tokyo, Japan

23

The micronucleus test in mutagen sensitive MS/Ae mice

1 Summary

The micronucleus test was carried out on 2 carcinogens, benzo(a)pyrene (BP) and 2-acetylaminofluorene (2AAF), and 2 analogues, pyrene (PYR) and 4-acetylaminofluorene (4AAF), using a mutagen-sensitive inbred strain (MS/Ae) of mice. Preparations were made from bone marrow cells after a single oral administration of each chemical. The treatment dose and the sampling time were chosen after a range-finding test by the multiple sampling at multiple dose levels method.

Both carcinogens were positive; the frequency of micronucleated polychromatic erythrocytes increased significantly showing a dose-effect relationship. Both PYR and 4AAF were negative in the present experiment.

2 Introduction

The micronucleus test has been widely used as an *in vivo* short-term test for the detection of chemicals with clastogenic effects and of spindle poisons. Our recent study suggests that the formation of micronuclei in polychromatic erythrocytes (PE) is associated with chromosomal aberrations induced by the chemical in erythroblasts in the bone marrow (Hayashi *et al.*, 1984a). In this study, we used the mutagen sensitive mouse strain MS/Ae (Hayashi *et al.*, 1982) established from the CD-1 strain of Aeschbacher *et al.* (1979). For the micronucleus test, the estimation of treatment doses and the sampling time after treatment are critical. We applied a pilot experiment procedure called a multiple sampling at multiple dose levels method to find appropriate dose levels and the optimal sampling time after administration of each chemical (Hayashi *et al.*, 1984b). For the evaluation of experimental results, both negative and positive control groups were included; comparisons were made between test data and our own historical background data for the MS/Ae mouse micronucleus test.

3 Materials and methods

Animals

MS/Ae mice were obtained from Dr H.U. Aeschbacher, Nestle Products Technical Assistance Company Ltd. (Switzerland) and maintained at our laboratory. They were fed with pellets (OA-2, Japan Clea), and water was given *ad libitum* throughout the experiments. For the pilot experiment, one male and one female 8 to 11-week old mouse were assigned to each group. For the full-scaled

Figure 1. Summary of results from the pilot experiments on four chemicals (D, dead).

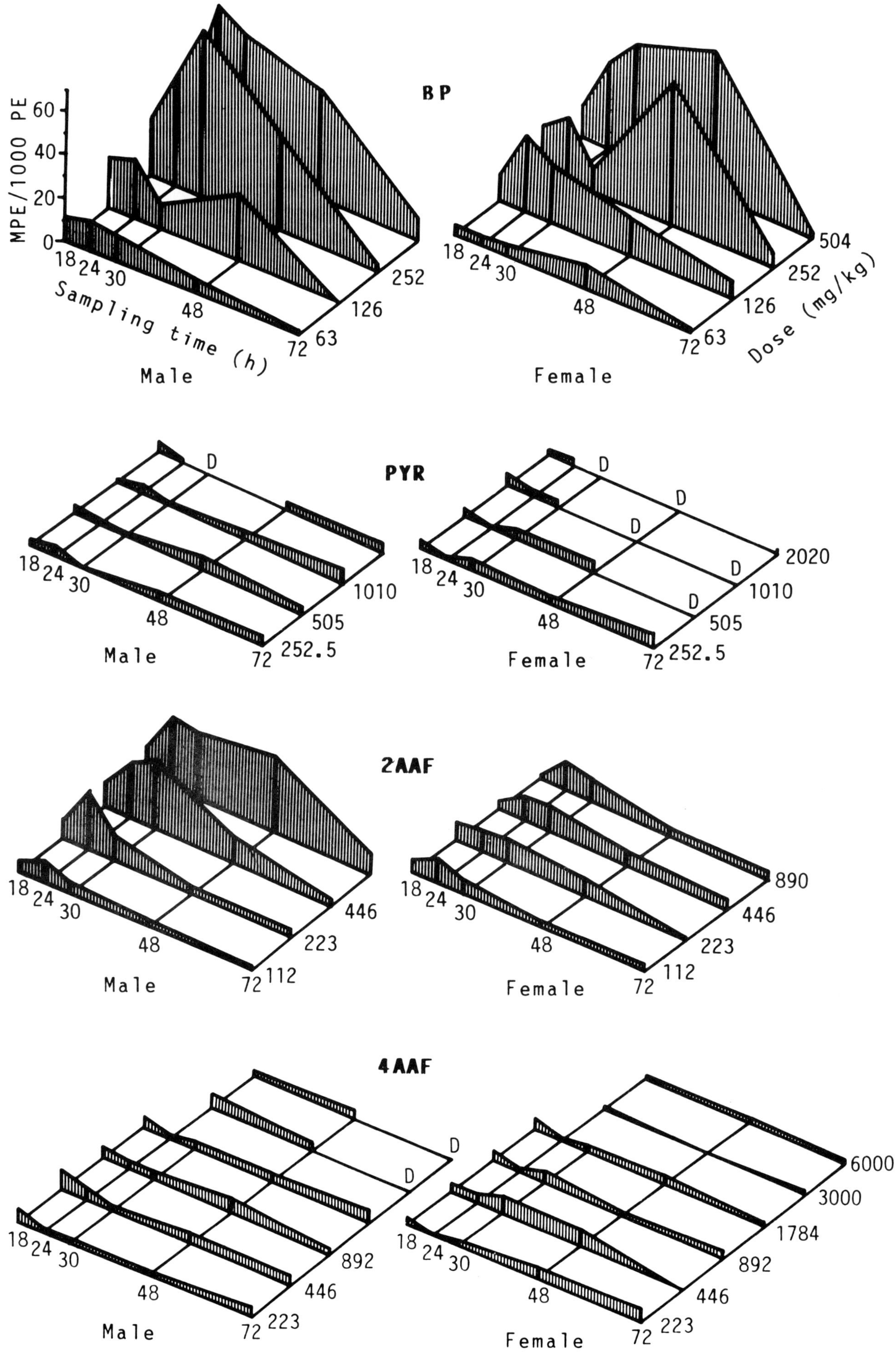

experiment, five male 8 to 11-week old mice were randomly assigned to each group.

Chemicals

The four chemicals tested (BP, PYR, 2AAF and 4AAF) were supplied from the IPCS committee. All these chemicals were dissolved or suspended in olive oil (Yoshida Chemicals Co., Japan) just prior to use. Olive oil was used for the negative control, and mitomycin C (Kyowa Hakko, Japan) dissolved in saline was used for the positive control.

Pilot experiment

The multiple sampling at multiple dose levels method (Hayashi *et al.*, 1984b) was applied to find the most suitable sampling times and treatment dose levels for each chemical before starting the full-scaled experiments. To test for any sex differences in the response, both sexes were used for this pilot experiment. Animals were treated with test chemicals at 4 different dose levels by gavage and killed at 18, 24, 30, 48 and 72 h after administration. Femoral marrow cells were smeared, fixed with methanol, and stained with Acridine Orange according to the technique described by Hayashi *et al.* (1983). To determine sex differences, data were analysed by means of the analysis of variance.

Full-scaled experiment

Five animals were randomly assigned to each group. Sex of the animals, dose levels and sampling time for each chemical were selected according to the results of the pilot experiments. Chemicals were administered as in the pilot experiment, and mice were killed at the most suitable sampling time or the same time as for the analogue compound. According to the common framework proposed by the standing committee, animals were also killed at an additional 2 sampling times between 24 and 48 h after single administration of the highest dose.

Femoral marrow cells were smeared on clean glass slides, fixed with methanol for 5 min, and stained with 3% Giemsa solution (Merck, diluted with 1/150 M phosphate buffer at pH 6.8) for 30 min. After a rinse in the same buffer, the slides were bleached with 0.004% citric acid aqueous solution for a few seconds to differentiate PE and normochromatic erythrocytes (NE) and rinsed in deionized water. Slides were coded and observations were made without knowledge of the treatment. The frequencies of micronucleated polychromatic erythrocytes (MPE) per 1000 PE and of micronucleated normochromatic erythrocytes (MNE) per 1000 NE were scored at a magnification of 1000 X. The ratio of PE to the total 1000 erythrocytes was also scored.

4 Data evaluation

The data obtained were compared not only with concurrent negative control groups (Kastenbaum and Bowman, 1970) but also with our historical background data for MPE in the negative control group of the MS mice. When the negative (vehicle) and positive control group showed the expected frequency of MPE, the experiment was considered to be technically acceptable. Data from treatment groups were then compared with the historical background data. The background frequency of MPE followed a binomial distribution with a mean of 0.44%. The distribution of values with a mean of 0.88% (2 × 0.44% an expected mean of the lowest positive control)

Table 1. *BP micronucleus test results in MS mice*

Dose mg/kg (m mole/kg)	Expression time (h)	PE/(PE + NE) (%)	MNE/NE (%)	MPE/PE (%)
0(0)	30	45.34 ± 13.15	0.20 ± 0.16	0.42 ± 0.27
63(0.25)	30	57.26 ± 5.67	0.20 ± 0.10	0.92 ± 0.27[a]
126(0.5)	30	58.52 ± 2.90	0.22 ± 0.22	1.62 ± 0.33[a]
252(1.0)	30	61.56 ± 3.97	0.26 ± 0.17	4.08 ± 0.99[a]
252(1.0)	24	57.60 ± 8.37	0.30 ± 0.23	4.06 ± 2.09[a]
252(1.0)	48	45.54 ± 10.39	0.60 ± 0.38	2.82 ± 1.42[a]
MMC[b]	30	37.98 ± 10.33	0.52 ± 0.15	11.98 ± 2.13[a]

[a] $\alpha < 0.01$, $\beta < 0.05$ (historical control); $\alpha < 0.01$ (concurrent control).
[b] As positive control, mitomycin C was injected intraperitoneally at a dose of 2.0 mg/kg.

Table 2. *PYR micronucleus test results in MS mice*

Dose mg/kg (m mole/kg)	Expression time (h)	PE/(PE + NE) (%)	MNE/NE (%)	MPE/PE (%)
0(0)	30	61.96 ± 5.44	0.40 ± 0.35	0.60 ± 0.14
202(1.0)	30	56.98 ± 13.55	0.36 ± 0.21	0.48 ± 0.13
404(2.0)	30	59.96 ± 9.72	0.46 ± 0.31	0.42 ± 0.28
808(4.0)	30	51.88 ± 6.60	0.46 ± 0.21	0.42 ± 0.38
808(4.0)	24	52.56 ± 4.54	0.26 ± 0.11	0.54 ± 0.19
808(4.0)	48	51.34 ± 14.49	0.30 ± 0.16	0.42 ± 0.24
MMC[b]	30	54.12 ± 6.34	0.28 ± 0.22	13.24 ± 1.96[b]

[a] As positive control, mitomycin C was injected intraperitoneally at a dose of 2.0 mg/kg.
[b] $\alpha < 0.01$, $\beta < 0.05$ (historical control); $\alpha < 0.01$ (concurrent control).

was assumed to follow the same type of binomial curve. The probability of a false positive (first class error) (α) was set at 0.01 and the probability of a false negative (second class error) (β) was set at 0.05. Under these conditions, when the total number of MPE per group (5000 PE) was less than 34, equal to 34, or more than or equal 35, the result was judged negative, inconclusive, or positive, respectively.

5 Results

Results from the pilot experiments are summarised in Figure 1. BP and 2AAF induced elevated frequencies of MPE, but PYR and 4AAF did not increase the frequency of MPE at any dose or sampling times, even though some animals died in the high dose groups. Data from the pilot experiment suggested that 2AAF induced more MPE in male mice than in female mice, so that for the full-scaled experiment on 2AAF, male mice were used. The other 3 chemicals showed no significant differences between male and female, so only male mice were used in the full-scaled experiment.

BP

This chemical was clearly positive in this test system at doses of 63, 126, and 252 mg/kg when the mice were killed at 30 h after a single oral administration. The data in Table 1 show a dose-response relationship at 30 h. For the highest dose, samples taken at 24 and 30 h show more MPE than samples taken at 48 h. The frequency of MNE and the ratio of PE to (PE + NE) did not change at any points for all chemicals tested.

PYR

Because no sign of positive response was obtained in the pilot experiment, the sampling time was fixed at 30 h (the same as that of BP), and the maximum dose was set at 808 mg/kg. There was no statistical significant increment in the frequencies of MPE (Table 2). The test was also negative when the animals were killed at 24 and 48 h after the administration of the highest dose.

2AAF

This chemical induced MPE at doses of 56, 112, and 223 mg/kg when mice were killed at 24 h after a single oral administration (Table 3). The frequency of MPE increased with the treatment dose. Samples taken after the longer expression times (30 and 48 h) from the highest dose group showed relatively fewer MPE.

4AAF

The sampling time was chosen to be the same as that of 2AAF, 24 h after a single oral administration. 4AAF was negative in this test system at doses up to 1784 mg/kg. Animals given the highest dose were sampled at 30 and 48 h as well, and there was no induction of MPE (Table 4).

6 Discussion

Both carcinogens, BP and 2AAF, induced MPE at significant levels, while their presumed noncarcinogenic analogues, PYR and 4AAF, did not induce MPE under the present experimental conditions. In the pilot experiments, animals were treated with PYR and 4AAF at dose levels that were lethal to some animals. Therefore, the

Table 3. *2AAF micronucleus test results in MS mice*

Dose mg/kg (m mole/kg)	Expression time (h)	PE/(PE + NE) (%)	MNE/NE (%)	MPE/PE (%)
0(0)	24	49.76 ± 2.89	0.38 ± 0.16	0.52 ± 0.11
56(0.25)	24	52.78 ± 3.86	0.28 ± 0.23	0.72 ± 0.27[a]
112(0.5)	24	46.94 ± 6.59	0.24 ± 0.09	1.20 ± 0.55[b]
223(1.0)	24	52.64 ± 4.80	0.28 ± 0.13	2.40 ± 0.66[b]
223(1.0)	30	46.40 ± 3.54	0.24 ± 0.21	1.82 ± 1.17[b]
223(1.0)	48	47.26 ± 6.83	0.52 ± 0.16	0.68 ± 0.26
MMC[c]	30	26.52 ± 3.85	1.04 ± 0.58	13.34 ± 2.52[b]

[a] $\alpha < 0.01$; $\beta < 0.05$ (historical control); N.S. (concurrent control).
[b] $\alpha < 0.01$, $\beta < 0.05$ (historical control); $\alpha < 0.01$ (concurrent control).
[c] As positive control, mitomycin C was injected intraperitoneally at a dose of 2.0 mg/kg.

Table 4. *4AAF micronucleus test results in MS mice*

Dose mg/kg (m mole/kg)	Expression time (h)	PE/(PE + NE) (%)	MNE/NE (%)	MPE/PE (%)
0(0)	24	44.14 ± 2.82	0.24 ± 0.11	0.42 ± 0.18
446(2.0)	24	52.26 ± 7.69	0.22 ± 0.13	0.28 ± 0.31
892(4.0)	24	53.14 ± 3.90	0.28 ± 0.22	0.34 ± 0.23
1784(8.0)	24	46.58 ± 8.46	0.40 ± 0.12	0.44 ± 0.18
1784(8.0)	30	59.10 ± 3.79	0.42 ± 0.19	0.18 ± 0.15
1784(8.0)	48	40.64 ± 9.10	0.26 ± 0.11	0.32 ± 0.18
MMC[a]	30	40.22 ± 9.56	0.52 ± 0.19	14.02 ± 1.59[b]

[a] As positive control, mitomycin C was injected intraperitoneally at a dose of 2.0 mg/kg.
[b] $\alpha < 0.01$, $\beta < 0.05$ (historical control); $\alpha < 0.01$ (concurrent control).

highest doses of PYR (808 mg/kg, 4 mmole/kg) and 4AAF (1784 mg/kg, 8 mmole/kg) were considered as the maximum tolerated dose for the micronucleus test. There was little possibility of false negative results because animals were killed at 5 different sampling times from 18 to 72 h after a single administration for all dose groups in the pilot experiments. The data for PYR and 4AAF were judged to the negative when compared with our historical background data. Also identical results were obtained by comparing concurrent negative control and treatment groups.

The result of the pilot experiment on 2AAF indicated that more MPE were induced in males than in females. These results were confirmed using the ddY strain of mice (Hayashi *et al.*, unpublished data). In the micronucleus test, a sex difference has been reported with benzene (males were more sensitive; Meyne and Legator, 1980) and ethyl methanesulfonate (females were more sensitive; Henry, *et al.*, 1980), so the findings on 2AAF are of interest in relation to the metabolism of the substance.

7 Conclusions

Two carcinogens, BP and 2AAF, were positive in the micronucleus test using the mutagen sensitive mouse strain MS/Ae. The two presumed noncarcinogenic analogues, PYR and 4AAF, were negative under the experimental conditions used. Dose levels tested and sampling times for each chemical were fixed by the multiple sampling at multiple dose levels method. All data obtained were evaluated using concurrent control data and historical control data from our laboratory.

8 References

Aeschbacher, H.U., D. Gottwick, H. Meier and A.W. Poot (1979) Mutagen-sensitive strain of mice, *Mutation Res.*, **59**, 301–304.

Hayashi, M., T. Sofuni and M. Ishidate, Jr. (1982) High-sensitivity in micronucleus induction of a mouse strain (MS), *Mutation Res.*, **105**, 253–256.

Hayashi, M., T. Sofuni and M. Ishidate, Jr. (1983) An application of Acridine Orange fluorescent staining to the micronucleus test, *Mutation Res.*, **120**, 241–247.

Hayashi, M., T. Sofuni and M. Ishidate, Jr. (1984a) Kinetics of micronucleus formation in relation to chromosomal aberrations in mouse bone marrow, *Mutation Res.*, **127**, 129–137.

Hayashi, M., T. Sofuni and M. Ishidate, Jr. (1984b) A pilot experiment for the micronucleus test: The multi-sampling at multi-dose levels method, *Mutation Res.*, **141**, 165–169.

Henry, M., S. Lupo and K.T. Szabo (1980) Sex difference in sensitivity to the cytogenetic effects of ethyl methanesulfonate in mice demonstrated by the micronucleus test, *Mutation Res.*, **69**, 385–387.

Kastenbaum, M.A. and K.O. Bowman (1970) Tables for determining the statistical significance of mutation frequencies, *Mutation Res.*, **9**, 527–549.

Meyne, J. and M.S. Legator (1980) Sex-related differences in cytogenetic effects of benzene in the bone marrow of Swiss mice, *Environ. Mutagen.*, **2**, 43–50.

Vladimir I. Ivanov, Viacheslav S. Zhurkov,[1] Ludmila M. Konstantinova, Yulia V. Korogodina and Elena G. Feldt[1]

Institute of Medical Genetics, The USSR Academy of Medical Sciences, Moscow (USSR) and[1] *A.N. Sysin Institute of General and Communal Hygiene, The USSR Academy of Medical Sciences, Moscow, USSR*

24

The micronucleus test in strain C57BL/6J mice

1 Summary

The *in vivo* mutagenicity of 2 known carcinogens (BP and 2AAF) was compared with that of 2 closely related noncarcinogens (PYR and 2AAF). The test chemicals were administered orally to male mice of the C57BL/6J Sto strain. The incidence of micronucleated polychromatic erythrocytes was scored.

Distinct dose-dependent mutagenic effects were found for BP and 2AAF. Uncertain dose-independent increases in the incidence of the micronucleated polychromatic erythrocytes were observed after PYR and 4AAF treatment.

2 Introduction

The micronucleus test with mouse bone marrow polychromatic erythrocytes has become a popular short-term *in vivo* test on mammals for screening potential mutagens. The test reveals the consequences of chromosome aberrations, nondisjunction, etc. leading to the loss of chromosomal material from daughter nuclei during cell division.

In the present study the ability of two known carcinogens, BP and 2AAF, to induce micronuclei in mouse bone marrow polychromatic erythrocytes was compared with that of PYR and 4AAF, two reported noncarcinogens.

For the most part, we followed the experimental protocol recommended by the IPCS coordinator for the micronucleus group.

3 Materials and methods

Male mice of C57BL/6J Sto strain 2 to 3 months old (20 ± 1 g body weight) obtained from Stolbovaya centre of laboratory animal breeding were used in this assay.

Test chemicals were obtained from the IPCS organizers. The chemicals were thoroughly mixed with olive oil in proportions corresponding to the required doses of the compounds (in 0.2 ml of the ready-to-use suspension for BP and PYR and 0.4 ml for 2AAF and 4AAF). For the AAF compounds the volume of suspension was doubled to facilitate administration of the thick suspensions of the test substances especially at high doses.

Three doses of each of the chemicals were administered per os (63, 126 and 252 mg/kg BP; 202, 404 and 606 mg/kg for PYR; 55.8, 111.5 and 223 mg/kg for 2AAF; 223, 446 and 892 mg/kg for 4AAF). In parallel with the treatment, 2 concurrent controls were performed. Negative control mice received 0.2 or 0.4 ml olive oil per os. Positive control animals were either injected intraperitoneally with 46 mg/kg of cyclophosphamide or given the same amount per os in olive oil. Five to seven animals were used for an experimental group.

The animals were killed by cervical dislocation 40 h after treatment, and the femoral marrow was washed out with foetal calf serum. The cells were suspended, centrifuged, resuspended and smeared on slides. Dried smears were fixed with methanol for 5 min and stained for 10 min with Giemsa diluted in proportion 1:6 with distilled water (pH = 7.0). The methods employed were those suggested by W. Schmid (1973, 1975) with minor modifications.

The 40 h expression time was chosen because the maximum yield of micronucleated polychromatic erythrocytes (MPE) observed in preliminary experiments was between 30 and 40 h after treatment.

Separate slides were made for each test animal. 1000 polychromatic erythrocytes (PE) were scored on each slide, and the number of MPE was recorded.

Student's t test was used for comparisons between groups.

4 Results

All results are presented in Table 1.

Control: The results obtained for a group of five olive oil-treated mice examined concurrently with those treated with the test chemicals are given in the upper part of the table. Low and identical values (MPE/PE = 0.1%) were obtained for all animals. Therefore, an additional negative control using 10 mice was performed. This gave MPE/PE = 0.19%. As no heterogeneity was revealed in the controls, the data were pooled and gave a mean value of MPE/PE = 0.16%. The latter two figures are close enough to those reported for some other strains of mice (e.g., Raj *et al.*, 1983; Przybojewska *et al.* 1984; Rolandi *et al.* 1984).

We also had a historical negative control values obtained for an outbred stock of white mice. The overall mean value of MPE/PE for 6 groups including 38 males was 0.21%. Statistical calculations have shown that the comparison of data for treated mice with those obtained for any of the negative control groups of for pooled control results could not alter our conclusions on the significance of the differences obtained.

Positive control results are also shown in Table 1.40 h after ip injection or po administration of 46 mg/kg of cyclophosphamide, the observed MPE/PE ratio is similar for both routes of administration. These ratios are comparable with those reported for similar experiments and are much greater than the negative control values.

5 Treatment data

At all doses of BP, the incidence of MPEs was significantly higher than in the negative control. ($P < 0.001$). Especially numerous were the MPEs at the highest dose. The slope of the regression of the effect observed on the doses applied also significantly differed from zero ($P < 0.05$).

At any of the 3 doses of PYR, the incidence of MPEs was significantly higher ($P < 0.001$) than that in the negative control. However, no significant between-dose differences were revealed ($P < 0.05$). The overall mean value of MPE/PE = 0.59%, although high, is within the range of published data on spontaneous MPE/PE values in mice. At all doses of 2AAF, the incidence of MPEs was higher than in the negative control ($P < 0.001$) at the high and the low doses, and $P < 0.01$ at the intermediate dose. Slope of regression of MPE/PE on dose was found to be significant ($P < 0.05$). Especially great was the difference between the high dose and the two other doses.

Table 1. *Results of the micronucleus test in C57BL6J mice*

Compound		Dose (mg/kg)	Animals treated/evaluated	MPE/PE Number	MPE/PE % ± SD
Control:					
concurrent		—	5	5/5000	0.10 ± 0.00
additional		—	10	19/10,000	0.19 ± 0.12
Cyclophosphamide:					
	po	46	5	135/5000	2.70 ± 0.81
	ip	46	5	204/5000	4.08 ± 0.63
BP	po	63	7	54/7000	0.77 ± 0.15
		126	7	57/7000	0.81 ± 0.23
		252	7	172/7000	2.46 ± 0.53
PYR	po	202	5	27/5000	0.54 ± 0.09
		404	7	46/7000	0.66 ± 0.17
		606	6	34/6000	0.57 ± 0.25
2AAF	po	55.8	7	47/7000	0.67 ± 0.34
		111.5	5	36/5000	0.72 ± 0.36
		223	5	60/5000	1.20 ± 0.38
4AAF	po	223	7	22/7000	0.31 ± 0.12
		446	5	23/5000	0.46 ± 0.09
		892	7	33/7000	0.47 ± 0.16

At any of the three doses of 4AAF, the incidence of MPEs was significantly higher than in the negative control ($P < 0.01$ for the low dose and $P < 0.001$ for the 2 lower doses). However, no significant between-dose differences were revealed ($P < 0.05$) and the overall mean MPE/PE = 0.41% was well within the range of published data on spontaneous MPE/PE values in mice.

6 Discussion

BP and 2AAF showed distinct mutagenic action. A substantial increase in the incidence of MPE directly related to the dose applied was invariably observed in all groups of treated animals. PYR and 4AAF treatment also exceeded the negative control level, but in these cases neither was a direct dose-effect relation found nor did the level of MPE/PE reach the upper limit of a range of values reported for spontaneous MPE incidence.

We have no plausible explanation for the moderate but systematic increase of MPEs in PYR and 4AAF treated animals. We are unable to rule out, unequivocally, the mutagenicity of these 2 substances. The level of response to the noncarcinogens (if any) is a great deal lower than that to the carcinogens. At the highest dose of PYR (606 mg/kg) the MPE incidence was 4.3 times lower than that at the highest dose of BP (252 mg/kg) in spite of the fact that the dose of PYR was 3 times as large by mortality. This holds true for the 2AAF vs 4AAF comparison. Following 892 mg/kg 4AAF treatment, the MPE incidence was 2.6 times lower than after 223 mg/kg of 2AAF.

7 Conclusions

Per os treatment of male mice with BP or 2AAF resulted in distinct and substantial increases in the incidence of MPE directly related to the dose applied. These compounds are considered mutagenic.

After similar treatment of mice with PYR and 4AAF, a moderate dose-independent but regular increase in the incidence of MPE was observed. The yield of MPEs in these cases did not exceed the upper limit of the range of spontaneous MPE values in male mice. Although the mutagenicity (under conditions employed) of the noncarcinogens tested may not be ruled out completely, it must be a great deal lower than that of their carcinogenic counterparts.

8 References

Przybojewska, B., E. Dziubaltowska and Z. Kowalski (1984) Genotoxic effects of ethyl acrylate and methyl acrylate in the mouse evaluated by the micronucleus test, *Mutation Res.*, **135**, 189–191.

Rolandi, A., E. De Marinis and M.De Caterina (1984) Dithiocarbamate pesticides: Activity of Propineb in the micronucleus test in mice, *Mutation Res.*, **135**, 193–197.

Schmid, W. (1975) The micronucleus test, *Mutation Res.*, **31**, 9–15.

Raj, A.S., J.A. Heddle, H.L. Newmark and M. Katz (1983) Caffeic acid as an inhibitor of DNA-induced chromosomal breakage in mice assessed by bone-marrow micronucleus test, *Mutation Res.*, **124**, 247–253.

Jørn C. Jensen,
National Food Institute, Institute of Toxicology,
Mørkhøj Bygade 19, 2860 Soeborg, Copenhagen, Denmark

25

Micronucleus test on benzo(a)pyrene and pyrene performed in NMRI mice

1 Summary

The micronucleus test was carried out in 2 steps. The first step was aimed at finding the optimum expression time. For this purpose, 8 groups of 2 male and 2 female mice were dosed orally with the maximum tolerated dose (MTD) of pyrene (PYR) or benzo(a)pyrene (BP), and bone marrow samples taken after 24, 36, 48, and 60 h. As MTD, 252 mg/kg body weight (bw) was used for BP and 808 mg/kg bw for PYR.

The highest response was seen after 36 h with BP, while no increase in micronucleated erythrocytes was seen with PYR at any time.

Dose–response studies were then carried out using 36 h sampling time with the MTD, 1/2 × MTD and 1/4 × MTD for both compounds. Concurrent negative controls (olive oil) and positive controls (cyclophosphamide 46 mg/kg bw) were included. Each group consisted of 5 male and 5 female mice.

No effect was seen with PYR, while BP revealed a linear dose–response curve.

2 Introduction

The micronucleus test is used to detect chromosome damage in the erythropoetic bone marrow cells of rodents. The sampling time after dosing should be adjusted so that the polychromatic erythrocytes (PE) that are scored have been exposed to the test chemical in the G_2 phase of their second last cell cycle and the G_1 phase of their last cycle, (Cole *et al.*, 1981, 1983).

Although the sampling time is theoretically known, experiments have shown that because of differences in pharmacokinetic properties, the optimum expression time differs among chemicals, (Salamone *et al.*, 1980).

To find the optimum expression time for PYR and BP, an initial experiment was set up for both compounds using sampling times of 24, 36, 48, and 60 h. As the purpose of this experiment was to compare responses obtained with varying sampling times, no controls were included. Only 4 mice were used per group, because only major differences in time response were considered relevant, and because some knowledge of BP effects already known, (Kliesch *et al.*, 1982).

Results of this experiment showed clearly that 36 h was the optimum expression time for BP. Regarding PYR, all the groups showed a similar percentage of micronuclei that corresponded to our historical negative control level. On this basis 36 h was used as sampling time for PYR as well, because the route of administration and the vehicle were similar, as well as some of the chemical properties.

The dose–response study was performed with 3 dose groups plus a negative and a positive control group, using the doses proposed by the coordinator on these tests. All groups consisted of 5 Male and 5 female mice. The highest dose of each test compound seemed to be close to the MTD, because it resulted in a few deaths.

The schedule for dosing and sampling was designed to allow the time needed for handling the animals without interfering with the demand of running all groups, including negative and positive control groups, concurrently. All animals were dosed within 3 h and all bone marrow samples were taken within 3 h (36 h later) according to the time of dosing.

In a few cases, the dosing with stomach tube was not considered precise, because traces of the administered solution could be seen in the mouths or in the bedding material. Unfortunately, these observations were not ascribed to the individual mouse number, partly because the female mice were caged 5 in each cage. It is believed, that some of the variation in responses within groups could stem from this problem.

All other conditions were in agreement with the description in the materials and methods section.

3 Materials and methods

Test compounds: PYR and BP were received in 1983 from Lancaster Synthesis. Lancashire U.K. Olive oil (pure grade) and cyclophosphamide were purchased from Bie and Berntsen, Roedovre, Copenhagen, Denmark.

Animals: Outbred NMRI mice: Bom (SPF) were delivered from Bomholtgaard, Ry, Denmark. The animals were 9 and 11 weeks of age when received and 14 weeks of age at the time of each experiment. Males were caged singly, and females were kept 5 in each cage (Makrolon IV). The animals were distributed randomly to the test groups.

Housing conditions: Temperature 22 ± 1 °C; humidity 60% ± 5%; light period 9 p.m. to 9 a.m.; air change 8 times per hour; food IT-chow (nutritionally adequate); water acidified with citric acid to pH 3.5.

Dosing: PYR and BP were dissolved in olive oil shortly prior to use. BP solutions of 25.2 mg/ml, 12.6 mg/ml and 6.3 mg/ml were made, so that a mouse weighing 30 g received 0.3 ml of test solution corresponding to 252, 126 or 63 mg/kg bw, respectively. The same procedure was used with PYR, except that solutions were 80.8, 40.4 or 20.2 mg/ml. The mice were fasted overnight with access to water only, and weighed the following morning just before dosing to calculate the exact amount of test solution to be administered. The female mice weighed 27 g to 32 g and the males 34 g to 42 g. Dosing was carried out with a 1 ml syringe with a special stomach tube needle. Cyclophosphamide was dissolved in water to 4.6 mg/ml immediately before use, and given intraperitoneally (ip) in a dose of 46 mg/kg bw.

Bone marrow preparations: The mice were killed and the right femur removed. The proximal end was cut and the bone marrow flushed with a syringe into 5 ml of foetal calf serum at 0 °C. The cells were spun down and smears were made on glass slides.

Fixation and staining: Smears were fixed in methanol and stained in May-Grünwald/Giemsa according to Schmid (1975).

Scoring: 1000 polychromatic erythrocytes (PE) were counted from each animal, and the number of clearly distinguishable micronuclei recorded. The ratio of normochromatic erythrocytes (NE) to PE was found by counting approximately 200 PE and simultaneously recording the number of NE. Micronuclei in NE were recorded throughout until 1000 PE were reached.

4 Results

The first part of the assay, where the optimum expression time was to be found, showed a clear optimum response of BP at 36 h while no response at any time was seen with PYR (Table 1, Figure 1).

Dose–response studies using 36 h sampling time with BP at 252 mg/kg bw, 126 mg/kg bw and 63 mg/kg bw, corresponding to the maximum tolerated dose (MTD), 1/2 × MTD and 1/4 × MTD, showed an approximately linear dose–response curve (Table 2, Figure 2). The highest dose resulted in an approximately 4-fold increase over the control level. The control level in this study, 0.39% micronucleated PE, was about twice our historical control level.

Dose–response studies using 36 h sampling time with PYR at 808 mg/kg bw, 404 mg/kg bw, and 202 mg/kg bw, corresponding to the MTD, 1/2 × MTD and 1/4 × MTD, confirmed the negative findings in the initial test (Figure 2).

5 Discussion

At least two micronucleus studies in mice with BP in the same dose range as the present study have been

Figure 1. Percentage of micronuclei (MN) + S.D. in polychromatic erythrocytes at varying sampling times.

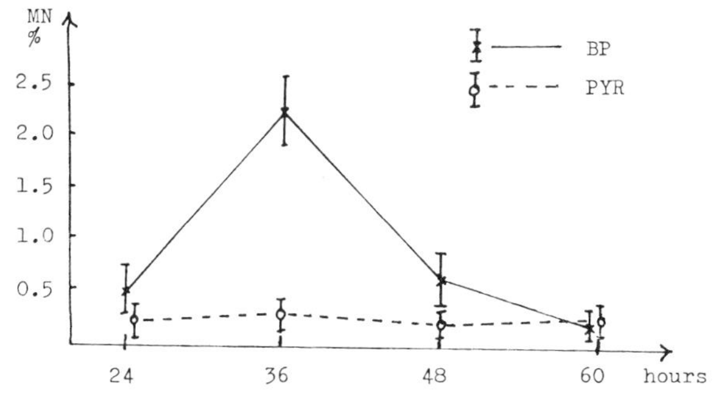

Figure 2. Dose–response of BP and PYR at 36 h sampling time. (MTD = maximum tolerated dose. For BP, MTD = 252 mg/kg bw, for PYR, MTD = 808 mg/kg bw)

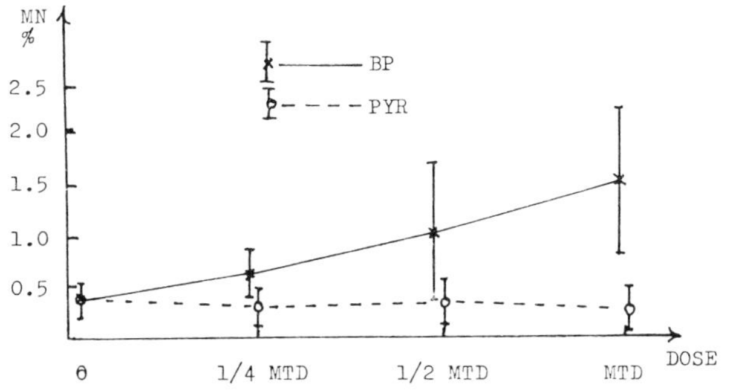

Table 1. *Induction of micronuclei at varying sampling times*

	Mouse No.	Hours[a]	Sex[b]	Survival	PE/total[c]	MN/NE[d]	MN/PE[e]	% MN/PE[f]
	9–10	24	M	2/2	50.5	2	12	
	11–12	24	F	2/2	40.5	6	8	0.5 ± 0.22
	13–14	36	M	2/2	41.0	11	41	
BP	15–16	36	F	2/2	46.5	17	40	2.28 ± 0.48
252 mg kg	17–18	48	M	2/2	42.5	4	13	
	19–20	48	F	1/2	41.0	0	4	0.57 ± 0.24
	21–22	60	M	1/2	37.0	0	2	
	23–24	60	F	1/2	42.0	2	2	0.2 + 0.0
	25–26	24	M	2/2	40.5	2	6	
	27–28	24	F	2/2	38.5	3	3	0.23 + 0.15
	29–30	36	M	2/2	47.5	1	6	
PYR	31–32	36	F	2/2	41.0	3	6	0.3 + 0.17
800 mg/ kg	33–34	48	M	1/2	56.0	0	1	
	35–36	48	F	2/2	38.5	5	4	0.17 ± 0.13
	37–38	60	M	2/2	50.0	1	4	
	39–40	60	F	1/2	47.0	0	4	0.27 ± 0.16

[a] Sampling time,
[b] M = male, F = female
[c] Polychromatic erythrocytes (PE) as a percentage of total erythrocytes.
[d] Total number of micronuclei (MN) registered in normochromatic erythrocytes (NE) during the scoring of 1000 PE per animal.
[e] Total number of MN in PE (1000 PE per animal).
[f] Percentage MN in PE ± S.D.

Table 2. *Induction of micronuclei: Dose–response at 36 h sampling time*[a]

	Mouse No.	Sex[b]	Survival	PE/total[c]	MN/NE[d]	MN/PE[e]	% MN/PE[f]
Olive oil	41–46	M	5/5	41.8	15	25	0.39 ± 0.16
control	46–51	F	5/5	42.0	3	14	
BP 63	71–76	M	4/5	49.8	4	24	0.56 ± 0.24
mg/kg	76–81	F	5/5	43.8	12	26	
BP 126	61–66	M	5/5	44.0	12	47	0.96 ± 0.72
mg/kg	66–71	F	3/5	44.0	10	30	
BP 252	51–56	M	3/5	40.7	4	34	1.47 ± 0.72
mg/kg	56–61	F	3/5	40.3	10	54	
PYR 202	101–106	M	4/5	39.8	6	14	0.29 ± 0.17
mg/kg	106–111	F	5/5	41.2	9	12	
PYR 404	91–96	M	5/5	41.2	9	20	0.33 ± 0.23
mg/kg	96–101	F	5/5	38.6	7	13	
PYR 808	81–86	M	5/5	37.4	11	10	0.29 ± 0.26
mg/kg	86–91	F	3/5	25.3	10	13	
CY 46	111–116	M	5/5	38.6	21	216	4.76 ± 0.99
mg/kg	116–121	F	5/5	33.8	25	260	

[a] See Table 1 for footnote.
[g] CY = cyclophosphamide.

reported earlier. Kliesch *et al.* (1982) finds the optimum expression time to be 36 h, with the responses at 24 and 48 h considerably weaker. Salamone *et al.* (1981) finds the optimum expression time to be 48 h, with a much broader time-response range. These differences may be due to the different strains of mice used and to the route of administration (oral vs ip).

The present study, which uses almost the same protocol as Kliesch *et al.* (1982), is in good agreement with this latter study with respect to the optimum expression time. The dose response seems to be somewhat weaker in the present study, but both studies find a linear dose response. The rather large variation within groups found in this study, is believed to some extent to result from difficulties in the dosing procedure (see Introduction), which indicates that a few mice actually received a smaller dose than anticipated. Furthermore, the response in the initial part of the study using MTD with 36-h sampling time was about 30% higher.

When tested in chi-square tests, the two highest doses were highly significant ($p < 0.001$), while the lowest dose was just above the 5% significance level ($p < 0.05$).

The number of micronucleated PE in all test groups with PYR was similar to the control level and is interpreted as a negative response.

6 Conclusion

The micronucleus test was carried out with BP and PYR in 2 steps. The optimum expression time was determined to be 36 h for BP while no response at any time was seen with PYR. Using a 36 h sampling time, dose–response studies revealed a linear dose response for BP while the lack of response with PYR was confirmed.

7 References

Cole, R.J., N. Taylor, J. Cole and C.F. Arlett (1981) Short term tests for transplacentally active carcinogens, I. Micronucleus formation in fetal and maternal mouse erythroblasts, *Mutation Res.*, **80**, 141–157.

Cole, R.J., J. Cole, L. Henderson, N.A. Taylor, C.F. Arlett and T. Regan (1983) Short Term tests for transplacentally Active carcinogens. A comparison of sister-chromatid exchange and the micronucleus test in mouse foetal liver erythroblasts, *Mutation Res.*, **113**, 61–75.

Kliesch, U., I. Roupova and I.D. Adler (1982) Induction of chromosome dammage in mouse bone Marrow by benzo(a)pyrene, *Mutation Res.*, **102**, 265–273.

Salamone, M., J. Heddle, E. Stuart and M. Katz (1980) Towards an Improved micronucleus test, studies on 3 model agents, Mitomycin C, cyclophosphamide and dimethylbenzanthracene, *Mutation Res.*, **74**, 347–356.

Schmid, W. (1975) The micronucleus test, *Mutation Res.*, **31**, 9–15.

Dag Jenssen
Department of Genetic Toxicology, Wallenberg Laboratory, University of Stockholm, Sweden

26

Micronucleus test in bone marrow cells of CBA male mice

1 Summary

Negative results were obtained by the 2 noncarcinogenic compounds pyrene and 4-acetylaminofluorene.

The 2 carcinogenic compounds benzo(a) pyrene and 2-acetylaminofluorene were found positive in the present investigation, which was in agreement with expectations.

2 Introduction

The four test compounds pyrene (PYR), benzo(a)pyrene (BP), 2-acetylaminofluorene (2AAF) and 4-acetylaminofluorene (4AAF) have been tested for their ability to induce micronuclei in bone marrow cells of mice. The validity of this assay as a test system for mutagenic activity has been reviewed elsewhere (Jenssen and Ramel, 1980; Jenssen, 1982; Heddle *et al.*, 1983).

3 Materials and methods

The test compounds were dissolved in corn oil just before they were given orally to CBA male mice 9 to 16 weeks old. Three doses of each compound were tested with 5 mice in each group. The mice were weighed before administration, and they received a single treatment of 0.4 ml solution per 25 g body weight. Control animals were treated with the solvent. Methyl methanesulphonate (MMS), also dissolved in corn oil, was used as a positive control. The mice were killed by cervical dislocation 30 hours after treatment. The bone marrow of the left femur was taken out, prepared on slides, and stained according to the procedure described earlier (Jenssen and Ramel, 1976). All slides were coded and examined under a microscope at a magnification of 1650X. One slide per mouse and 2000 erythrocytes, both normochromatic and polychromatic, were examined for the presence of micronuclei as defined by Schmid (1975). After that, an additional number of polychromatic erythrocytes, totalling 1000, were investigated for the presence of micronuclei. The observations were grouped according to polychromatic erythrocytes and normochromatic erythrocytes with or without micronuclei. The treatment groups were analysed versus control with Student's test.

4 Results

None of the 4 compounds dissolved completely in corn oil. However, each sample taken for treatment represented an unbiased estimate of the evenly dispersed crystals in the stock solution.

All results of the tests are summarised in table 1.

In the experiment with PYR, the dose of the positive control MMS was 52 mg/kg body weight (bw). This dose was later changed to 62 mg/kg in the experiments with the other 3 compounds.

In all the experiments, the induction of micronuclei by the positive control MMS was clearly demonstrated.

Concerning the results of the test with the 4 compounds, both BP and 2AAF showed a statistically significant and dose related effect on micronuclei induction after 30 hours of expression. PYR and 4AAF were negative in the test.

5 Discussion

The results of treatment with PYR and BP were as expected. The noncarcinogen PYR was found negative, but the carcinogen BP was positive in the micronucleus test system. Earlier reports on BP in this test system are both negative and positive (Heddle *et al.*, 1983).

In the case of the carcinogen 2AAF, we found a low but statistically significant increase of micronucleated polychromatic erythrocytes at 30 h. Both negative and positive results on 2AAF are also reported in the literature (Heddle *et al.*, 1983). In the report of Friedman and Staub (1977), an increase of micronuclei by 2AAF could be detected only at doses above 1000 mg/kg. In the present investigation, 224 mg/kg was the highest dose tested, and this was found to be the lowest detectable level. The noncarcinogenic analogue 4AAF was negative in the present investigation.

6 Conclusions

Negative results were obtained by the 2 noncarcinogenic compounds PYR and 4AAF in the micronucleus test.

BP and 2AAF were found positive in the micronucleus test.

7 Acknowledgement

We thank Mrs Birgitta Dillner for excellent technical performance and acknowledge Swedish Work Environment Fund for financial support.

8 References

Fruedman, M.A. and J. Staub (1977) Induction of micronuclei in mouse and hamster bone-marrow by chemical carcinogens, *Mutation Res.*, **43**, 255–262.

Heddle, J.A., M. Hite, B. Kirkhart, K. Mavournin, J.T. MacGregor, G.W. Newell and M.F. Salamone (1983) The induction of micronuclei as a measure of genotoxicity. A report of the U.S. Environmental Protection Agency Gene-Tox Program, *Mutation Res.*, **123**, 61–118.

Jenssen, D. (1982) The induction of micronuclei. In A.A. Sandberg (ed.), *Progress and topics in cytogenetics*, Vol. 2, Alan R. Liss, NY, pp. 47–63.

Jenssen, D. and C. Ramel (1976) Dose-response at low doses of X-irradiation and MMS on the induction of micronuclei in mouse erythroblasts, *Mutation Res.*, **41**, 311–320.

Jenssen, D. and C. Ramel (1980) The micronucleus test as a part of a short-term mutagenicity test program for the prediction of carcinogenicity evaluated by 143 agents tested, *Mutation Res.*, **75**, 191–202.

Schmid, W. (1975) The micronucleus test, *Mutation Res.*, **31**, 9–15.

Table 1. *Micronucleus test of PYR, BP 4AAF and 2AAF in CBA male mice, with expression time of 30 h.*

Chemical	Dose (mg/kg)	PE / PE + NE	MPE/PE (%)	S.E.	Student's test[a]
PYR	0	0.38	0.34	0.067	
	152	0.41	0.28	0.059	NS
	303	0.40	0.48	0.087	NS
	606	0.47	0.30	0.063	NS
MMS	52	0.42	1.00	0.251	*x*
BP	0	0.34	0.26	0.051	
	63	0.31	0.60	0.089	*x*
	126	0.27	1.06	0.157	*xx*
	252	0.30	1.20	0.158	*xxx*
MMS	62	0.25	2.40	0.399	*xxx*
4AAF	0	0.24	0.16	0.051	
	223	0.23	0.12	0.037	NS
	446	0.19	0.20	0	NS
	892	0.23	0.22	0.080	NS
MMS	62	0.30	2.32	0.291	*xxx*
2AAF	0	0.37	0.42	0.097	
	56	0.36	0.40	0.095	NS
	112	0.33	0.62	0.102	NS
	224	0.29	0.92	0.191	*xx*
MMS	62	0.27	2.98	0.429	*xxx*

[a] NS = not significant; $x = P < 0.05$; $xx = P < 0.01$: $xxx = P < 0.001$.

Colette Coton, Alain Leonard and Gilbert Decat
Mammalian Genetics Laboratory, Department of Biology, C.E.N.-S.CK., B-2400 Mol, Belgium

27

Induction of micronuclei in BALB/c mouse bone marrow cells

1 Summary

The clastogenicity of BP, PYR, 2AAF, and 4AAF has been evaluated using the micronucleus test on mouse bone marrow cells from BALB/c male mice. BP and 2AAF but not PYR and 4AAF, increased significantly the number of micronucleated polychromatic erythrocytes.

The present results indicate that carcinogenic potential and clastogenic properties correlate well for these compounds.

2 Introduction

The micronucleus test represents a fast and inexpensive *in vivo* screening system to evaluate the genetic hazards of environmental agents and is widely used as a short-term assay for the detection of carcinogens (Matter and Grauwhiller, 1974; Salamone *et al.*, 1980; Jenssen and Ramel, 1980; Heddle *et al.*, 1983). This test has been shown to have about the same specificity ($\pm$ 80%) and predictive value (74%) as the Ames test. Its sensitivity appears, however, to be somewhat lower.

The purpose of our study was to determine to which extent the outcome of the micronucleus test correlates with the carcinogenic potential of two carcinogens, BP and 2AAF, and their noncarcinogenic analogues, PYR and 4AAF. The experiments were performed according to the general protocol recommended by the coordinators of the IPCS Collaborative Study to see the influence of methodological factors on the qualitative and quantitative response obtained in the micronucleus test.

3 Material and methods

The experiments were carried out on 8-week-old male inbred BALB/c mice weighing 19–21 g. The compounds, provided by ICI (U.K.), were suspended in 0.25 ml olive oil (pharmacopoiea grade) and were administered per os in a single treatment at the following doses:

BP – 0.25, 0.50, and 1 m mole/kg (63, 126, 252, mg/kg, respectively);
PYR – 1.0, 2.0, and 4.0 m mole/kg (202, 404, 808 mg/kg, respectively);
2AAF – 0.25, 0.50, and 1 m mole/kg (55.8, 111.8, 223 mg/kg, respectively);
4AAF – 0.25, 0.5, and 4.0 m mole/kg (223, 446, 892 mg/kg, respectively).

Positive controls were given procarbazine per os at

a dose of 12.9 mg/kg (0.05 mmole/kg); negative controls received the vehicle (olive oil) only.

Bone marrow preparations were made according to the method described by Schmid (1975) at 3 intervals after treatment (24, 30, and 48 h) for the highest dose of each compound and for negative and positive controls. An interval of 30 h was used for lower doses. The preparations were stained by the May-Grünwald-Giemsa method. Six animals were treated per dose and 1,000 polychromatic erythrocytes (PE) per mouse were analysed for the presence of micronuclei.

Treated and control groups were compared with the Mann-Whitney U test. The dose- and time-effect relationship were analysed by the Kruskal-Wallis test.

4 Results

Controls (Table 1)
The frequency of micronucleated PEs in the controls receiving olive oil did not exceed 0.2% in agreement with the spontaneous rates (0.1 to 0.4%) generally reported (Heddle *et al.*, 1983).

As expected (Bruce and Heddle, 1979; Kliesch *et al.*, 1981; Cole *et al.*, 1981), procarbazine increased drastically the yield of PEs; the maximum yield was observed 36 h after treatment.

BP (Table 2) Treatment with BP resulted in a significant increase ($p \leqslant 0.001$; Mann-Whitney U test) of PEs

Table 1. *Results of the micronucleus test on control male BALB/c mice (olive oil)*

Treatment	Preparation interval (h)	PE[a] / PE + NE	MNE/NE[a] (mean% ± S.E.)	MPE/PE[a] (mean% ± S.E.)
	24	0.67 ± 0.009	0.21 ± 0.09	0.33 ± 0.03
Olive oil	30	0.67 ± 0.01	0.38 ± 0.06	0.35 ± 0.02
	48	0.66 ± 0.01	0.37 ± 0.06	0.26 ± 0.06
	24	0.61 ± 0.009	0.20 ± 0.06	1.32 ± 0.11
Procarbazine	30	0.63 ± 0.01	0.44 ± 0.06	2.4 ± 0.25
(12.9 mg/kg)	48	0.62 ± 0.01	0.53 ± 0.09	0.73 ± 0.06

[a] PE, polychromatin erythrocytes; NE, Normochromatin erythrocytes; MNE, micronucleated normochromatin erythrocytes; MPE, micronucleated polychromatin erythrocytes.

Table 2. *Results of the micronucleus test on male BALB/c mice treated with BP*

Treatment (mg/kg)	Preparation interval (h)	PE[a] / PE + NE (± S.E.)	MNE/NE[a] (mean % ± S.E.)	MPE/PE[a] (mean% ± S.E.)
63	30	0.65 ± 0.01	0.40 ± 0.06	2.8 ± 0.5
126	30	0.60 ± 0.03	0.45 ± 0.05	2.21 ± 0.27
	24	0.65 ± 0.01	0.34 ± 0.09	1.68 ± 0.07
252	30	0.62 ± 0.09	0.25 ± 0.03	2.61 ± 0.20
	48	0.57 ± 0.01	0.93 ± 0.17	4.28 ± 0.45

[a] For footnote, see Table 1.

Table 3. *Results of the micronucleus test on male BALB/c mice treated with PYR.*

Treatment (mg/kg)	Preparation interval (h)	PE[a] / PE + NE (± S.E.)	MME/NE[a] (mean % ± S.E.)	MPE/PE[a] (mean % ± S.E.)
202	30	0.59 ± 0.03	0.37 ± 0.07	0.41 ± 0.09
404	30	0.60 ± 0.01	0.21 ± 0.04	0.31 ± 0.06
	24	0.67 ± 0.01	0.31 ± 0.11	0.38 ± 0.06
808	30	0.65 ± 0.02	0.33 ± 0.10	0.26 ± 0.06
	48	0.66 ± 0.01	0.52 ± 0.08	0.43 ± 0.09

[a] For footnote, see Table 1.

with micronuclei for all dose administered and time interval tested. Following the highest dose (252 mg/kg), the yield of micronuclei increased in a statistically significant manner ($P < 0.005$; Kruskall-Wallis test) from 1.68% after 24 h to 4.28% after 48 h.

PYR (Table 3) The incidence of micronucleated PEs did not significantly increase in animals given PYR.

2AAF (Table 4) This compound increased significantly the yield of PEs with micronuclei. Differences between doses were not statistically significant and, for the highest dose the yield of micronucleated cells, was not related to the time interval between treatment and observation.

4AAF (Table 5) Treatment with 4AAF failed to increase the number of micronucleated PEs above the control values.

5 Discussion

Our findings indicate that the results obtained in the micronucleus test correlate well with the carcinogenic potential of the 4 chemicals: the 2 carcinogens 2AAF and BP showed a marked positive response, whereas, their noncarcinogenic analogues, 4AAF and PYR, were inactive. Our observations are in agreement with the large majority of the results reported in the literature of finding no excess of micronuclei after treatment with 4AAF or PYR (Tsuchimoto and Matter, 1981; Salamone *et al.*, 1981) and a marked increase with 2AAF (Friedman and Staub, 1977; Bruce and Heddle, 1979; Salamone *et al.*, 1981; Tsuchimoto and Matter, 1981) or BP (Cole *et al.*, 1981; Kirkhart, 1981; Kliesch *et al.*, 1982). Many mutagens or carcinogens have been detected in the bone marrow micronucleus assay only at doses of more than 50% of LD50. The negative findings obtained by Trzos *et al.* (1978) for 2AAF and by Bruce and Heddle (1979) for BP might be explained by the low dose used.

Confirming the results of Friedman *et al.* (1978), our observations indicate that, compared with BP, 2AAF is a relatively weak mutagen.

The yields of cells with micronuclei in animals given BP are remarkably similar to the values reported by Kliesch *et al.* (1982). Kliesch and his coworkers found that the cell-cycle length can be prolonged by increasing doses of BP in our experiments with BALB/c mice as well as in the study of Cole *et al.* (1981) on random bred albino mice (Porton strain), the maximum response to BP was observed 48 h after treatment. This fact suggests that those 2 strains could be more sensitive to BP than the 1C3F1 hybrid male mice used by Kliesch *et al.* (1982), who observed maximum response 30–36 h after treatment.

For the compounds tested, our results indicate that the carcinogenicity of chemicals and their *in vivo* clastogenic properties correlate well as estimated by the micronucleus test on bone marrow cells.

6 Acknowledgement

This research was supported by a grant from the "Ministère Belge de la santé."

7 References

Bruce, W.R. and J.A. Heddle (1979) The mutagenic activity of 61 agents as determined by the micronucleus, Salmonella and sperm abnormality assays, *Can. J. Genet. Cytol.*, **21**, 319–334.

Table 4. *Results of the micronucleus test on male BALB/c mice treated with 2AAF.*

Treatment (mg/kg)	Preparation interval (h)	PE[a] PE + NE (± S.E.)	MNE/NE[a] (mean % ± S.E.)	MPE/PE[a] (mean % ± S.E.)
55.8	30	0.63 ± 0.02	0.33 ± 0.09	0.48 ± 0.09
111.8	30	0.70 ± 0.009	0.38 ± 0.09	0.58 ± 0.10
223	24	0.74 ± 0.02	0.57 ± 0.07	0.90 ± 0.20
	30	0.77 ± 0.02	0.56 ± 0.08	0.81 ± 0.20
	48	0.69 ± 0.01	0.48 ± 0.09	1.86 ± 0.72

[a] For footnote, see Table 1.

Table 5. *Results of the micronucleus test on male BALB/c mice treated with 4AAF.*

Treatment (mg/kg)	Preparation interval (h)	PE[a] PE + NE (± S.E.)	MNE/NE[a] (mean % ± S.E.)	MPE/PE[a] (mean % ± S.E.)
223	30	0.65 ± 0.03	0.51 ± 0.09	0.45 ± 0.07
446	30	0.61 ± 0.03	0.31 ± 0.04	0.36 ± 0.03
892	24	0.70 ± 0.01	0.40 ± 0.07	0.50 ± 0.11
	30	0.73 ± 0.01	0.49 ± 0.07	0.45 ± 0.05
	48	0.67 ± 0.01	0.46 ± 0.09	0.36 ± 0.07

[a] For footnote, see Table 1.

Cole, R.J., N. Taylor, J. Cole and C.F. Arlett (1981) Short-term tests for transplacentally active carcinogens, *Mutation Res.*, **80**, 141–157.

Friedman, M.A., W.H. Carter, J. Staub and A. Segreti (1978) Relative mutagenic potencies determined in the micronucleus test, *J. Theor. Biol.*, **73**, 575–584.

Friedman, M.A. and J. Staub (1977) Induction of micronuclei in mouse and hamster bone-marrow by chemical carcinogens, *Mutation Res.*, **43**, 255–262.

Jenssen, D. and C. Ramel (1980) The micronucleus test as part of a short-term mutagenicity test program for the prediction of carcinogenicity evaluated by 143 agents tested, *Mutation Res.*, **75**, 191–202.

Heddle, J.A., M. Hite, B. Kirkhart, K. Mavournin, J.T. MacGregor, G.W. Newell and M.F. Salamone (1983) The induction of micronuclei as a measure of genotoxicity. A Report of the U.S. Environmental Protection Agency Gene-Tox Program, *Mutation Res.*, **123**, 61–118.

Kirkhart, B. (1981) Micronucleus test on 21 compounds. *Progr. Mutation Res.*, **1**, 698–704.

Kliesch, U., N. Danford and I.D. Adler (1981) Micronucleus test and bone-marrow chromosome analysis. A comparison of 2 methods *in vivo* for evaluating chemically induced chromosome alterations, *Mutation Res.*, **80**, 321–332.

Kliesch, U., I. Roupova and I.D. Adler (1982) Induction of chromosome damage in mouse bone marrow by benzo(a)pyrene, *Mutation Res.*, **102**, 265–273.

Matter, B.E. and J. Grauwhiller (1974) Micronuclei in mouse bone-marrow cells. A simple *in vivo* model for the evaluation of drug-induced chromosomal aberrations, *Mutation Res.*, **23**, 239–249.

Salamone, M.F., J.A. Heddle and M. Katz (1981) Mutagenic activity of 41 compounds in the *in vivo* micronucleus assay. In F.J. de Serres and J. Ashby (Eds), *Progress in Mutation Research*, Vol. 1, Elsevier, N.Y. pp. 686–697.

Salamone, N., J. Heddle, E. Stuart and M. Katz (1980) Towards an improved micronucleus test. Studies on 3 model agents, mitomycin C, cyclophosphamide and dimethylbenzanthracene, *Mutation Res.*, **74**, 347–356.

Schmid, W. (1975) The micronucleus test, *Mutation Res.*, **31**, 9–15.

Trzos, R.J., G.L. Petzold, M.N. Brunden and J.A. Swenberg (1978) The evaluation of sixteen carcinogens in the rat using the micronucleus test, *Mutation Res.*, **58**, 79–86.

Tsuchimoto, T. and B.E. Matter (1981) Activity of coded compounds in the micronucleus test, *Prog. Mutation Res.*, **1**, 705–711.

Ulf Kliesch and Ilse-Dore Adler
Institut für Säugatiekgenetik
Gesellschaft für Strahlen-und
Unwelt forschung Neuherberg,
Federal Republic of Germany

28

Induction of micronuclei in polychromatic erythrocytes in bone marrow of adult IC3F$_1$, hybrid male mice

1 Summary

The possible clastogenic effects of benzo(a)pyrene (BP), pyrene (PYR), 2-acetylaminofluorene (2AAF), and 4-acetylaminofluorene (4AAF) were studied in mice *in vivo*. Micronuclei were counted in polychromatic erythrocytes of mouse bone marrow after treatment with various doses of the test compounds. Except for the BP experiments, bone marrow from the femur was sampled at various intervals. For BP and PYR the solvent was corn oil, and the animals were treated by oral intubation. 2AAF and 4AAF were dissolved in DMSO and given as intraperitoneal (ip) injections. One experiment with 446 mg/kg 2AAF was carried out by treatment with oral intubation of an oil suspension.

BP gave a positive response, but no clear dose response in the micronucleus frequencies was observed 36 h after treatment. PYR gave a negative response; 2AAF gave a positive response. The maximum effect was observed 48 h after treatment with a dose dependent increase in the micronucleus frequency. 4AAF gave a negative result.

2 Introduction

The *in vivo* bone marrow micronucleus test (MNT) is used to identify the clastogenic potential of a compound as it measures clastogenic effects indirectly and can be carried out with a minimal amount of training. An advantage of the MNT is the fast comparison of test results in any number of different mammalian species, and the test is independent of the karyotype.

The test is based on the following principles: acentric chromatid and chromosome fragments lag in anaphase, while the centric elements move toward the spindle poles. The lagging materials is distributed to only one of the daughter cells. It may be included in the main nucleus or form one or more separate small nuclei (i.e., micronuclei). Similar events occur if a malfunction of the spindle apparatus causes lagging chromosomes. Micronuclei can be observed in any cell type of proliferating tissue. Most suitable for micronucleus analysis are the young erythrocytes.

Address correspondence to: Ulf Kliesch
Institute für Genetik
Gesellschaft für Strahlen- und
Umweltforschung
D-8042 Neuherberg
W. Germany

In mouse bone marrow, the erythroblasts go through 6 or 7 cell divisions with a cell-cycle length of about 10 h (Cole *et al.*, 1979). About 10 h after completion of the last mitosis, erythroblasts expel their main nucleus, micronuclei remain in the cytoplasm of the young erythrocytes and can be observed easily. With a combination of Giemsa and May-Grünwald staining, the young erythrocytes, which are named polychromatic (PE), stain bluish to purple due to the high content of ribonucleic acid in the cytoplasm. In contrast, the older erythrocytes, called normochromatic (NE), stain yellow-orange.

Ten h after the last mitotic division, the expulsion of the nucleus is completed and the resulting PE remains in the bone marrow for another 10 h. Data published by Hart and Hartley-Asp (1983) suggest that the interval between mitosis and nuclear expulsion may be as short as 4 to 5 h. Chemicals that are suspected to be spindle poisons or G_2 clastogens may require preparation of bone marrow as early as 6 h after treatment. Breakage-derived micronuclei produced by S-dependent clastogens may not appear in PE earlier than 10 h after treatment. But experiments with known chemical mutagens show that the maximum number of micronuclei appear 30–36 h after treatment.

Comparison of single and multiple treatment as recommended in the original test protocol (Boller and Schmid, 1970) suggested that all of the tested chemicals could be recognized in a single treatment schedule with multiple sampling times (Kliesch *et al.*, 1981). Depression of bone marrow proliferation affects the MN yields, and maximum frequencies vary with time after treatment from chemical to chemical. Therefore, the screening of unknown chemicals with a maximum chance of detection should start with single treatment and multiple sampling intervals.

Induction of micronuclei in mouse bone marrow by BP has extensively been studied by Kliesch *et al.* (1982). Therefore, BP was regarded as the positive control in the present experiments.

3 Materials and methods

Animals

Male (101/El × C3H/El)F_1 hybrid mice were used and randomly distributed to treated and control groups. The animals were 10- to 14-weeks old. Their weight ranged from 25–28 g. The animals were housed under conventional conditions in air-conditioned quarters with a 12 h light-dark cycle and fed pellet food and drinking water ad libitum.

Chemicals, solvents and doses

All 4 test chemicals were supplied by ICI, Macclesfield (England). BP and PYR were dissolved in corn oil and orally applied by stomach intubation in volumes of 0.1 ml/10 g body weight. 2AAF and 4AAF were dissolved in pure DMSO and injected ip in a volume of 0.12 ml/mouse. Doses of PYR corresponded to 4, 2, and 1 x the molecular weight (i.e., 808, 404, and 202 mg/kg). Doses of 2AAF corresponded to 2, 1, and 1/2 of the molecular weight (i.e., 446, 223, and 111.5 mg/kg). Doses of 4AAF corresponded to 4, 2, and 1 x the molecular weight (i.e., 892, 446, and 223 mg/kg).

Sampling intervals and sample size

Bone marrow of BP-treated animals was sampled after 36 h only. Sampling intervals for bone marrow of animals treated with the highest dose of PYR were 24, 36, and 48 h. For the 2 lower doses, bone marrow was sampled at the 24 h interval. Bone marrow of animals treated with the highest dose of 2AAF and 4AAF was sampled after 24, 36, 48, and 72 h. Animals treated with the two lower doses were sampled at the 48 h interval.

Each experimental group consisted of 5 treated males and 1 solvent control male. 2000 polychromatic erythrocytes were scored per male, 1000 each for 2 slides. Due to death of the animals, 4 groups consisted only of 4 treated males (223 mg/kg 2AAF/48 h; 446 mg/24 h and 36 h; and 892 mg/kg 4AAF/48 h) and one group consisted only of 3 treated males (892 mg/kg 4AAF/36 h).

One experiment with 446 mg/kg 2AAF was carried out by treatment with oral intubation (0.1 ml/10 g body weight), suspension in corn oil, and sampling 24 h after treatment. For 2AAF and 4AAF the 24 h sampling interval was repeated.

For calculation of statistical significances in micronucleus frequencies, groups of animals were compared by the nonparametric Mann-Whitney-Wilcoxon test (Mann-Whitney U test).

Bone marrow preparation

The method has been described in detail (Heddle, 1973; Matter and Schmid, 1971; Schmid, 1976), and only the basic steps are outlined here. The bone marrow from both femurs was flushed into a test tube containing 5 ml fetal calf serum. After centrifugation, the supernatant was discarded. The pellet was resuspended and one drop of the suspension was placed on one end of a dry glass slide and pulled behind a hemocytometer cover slip. The slides were air dried before staining. The staining procedure was a combination of May-Grünwald and Giemsa staining. All slides were coded before scoring.

Microscopic analysis

The slides were scored under 1 250 x magnification. The ratio of normochromatic (NE) to polychromatic erythrocytes (PE) was determined by counting a total of about 2 000 erythrocytes per slide so that all NEs were counted in the same field scored to count 1 000 PEs. The normal ratio is nearly 1:1. The number of micronuclei in both cell types was noted.

4 Results

BP

The results are summarised in Table 1. BP gave a positive response although no clear dose-dependent increase in the micronucleus rates was observed 36 h after treatment. The lowest dose tested (31.5 mg/kg) resulted in a micronucleus yield significantly above control level ($P < 0.01$).

PYR

PYR results are summarised in Table 2. PYR gave a negative response. None of the single dose-time groups or the pooled treated groups showed a micronucleus rate above control level ($P > 0.05$).

Table 1. *Micronucleus test with BP in mouse bone marrow*

Dose (mg/kg)	Sampling interval (h)	No. of animals	PE[a] / PE + NE	MN/NE	%	MN/PE	%MN ± SE
0.0	—	5	0.49	8/ 9970	0.08	15/10000	0.15 ± 0.02
31.5	36	5	0.48	6/ 9997	0.06	90/10000	0.90 ± 0.11[b]
63.0	36	5	0.46	7/10000	0.07	102/10000	1.02 ± 0.12[b]
126.0	36	5	0.48	4/ 9975	0.04	153/10000	1.53 ± 0.13[b]

[a] NE = normochromatic erythrocytes; PE = polychromatic erythrocytes; MN = micronuclei.
[b] $P < 0.01$ (Mann-Whitney-Wilcox on test).

Table 2. *Micronucleus test with PYR in mouse bone marrow*

Dose (mg/kg)	Sampling interval (h)	No. of animals	PE[a] / PE + NE	MN/NE	%	MN/PE	%MN ± SE
0.0	—	8	0.46	4/16000	0.02	26/16000	0.17 ± 0.02
202.0	24	5	0.49	5/10000	0.05	22/10000	0.22 ± 0.02
404.0	24	5	0.47	7/ 9871	0.07	27/10000	0.27 ± 0.04
808.0	24	5	0.48	3/10000	0.03	25/10000	0.25 ± 0.04
	36	5	0.46	7/ 9959	0.07	16/10000	0.16 ± 0.03
	48	5	0.48	2/10000	0.02	10/10000	0.10 ± 0.04

[a] NE = normochromatic erythrocytes; PE = polychromatic erythrocytes; MN = micronuclei.

Table 3. *Micronucleus test with 2AAF in mouse bone marrow*

Dose (mg/kg)	Sampling interval (h)	No. of animals	PE[a] / PE + NE	MN/NE	%	MN/PE	%MN ± SE[b]
0.0	—	10	0.49	18/19847	0.09	59/20000	0.27 ± 0.03
111.5	48	5	0.45	16/10000	0.16	53/10000	0.53 ± 0.10^{+}
223.0	48	4	0.46	10/ 8000	0.13	74/ 8000	0.93 ± 0.10^{++}
446.0	24[c]	5	0.48	4/10000	0.04	46/10000	0.46 ± 0.05^{+}
	24	4	0.49	6/ 8000	0.08	33/ 8000	0.41 ± 0.07^{+}
	24	5	0.51	9/ 9701	0.09	47/10000	0.47 ± 0.09^{+}
	36	4	0.47	5/ 8000	0.06	74/ 8000	0.93 ± 0.10^{++}
	48	5	0.46	10/10000	0.10	138/10000	1.38 ± 0.15^{+++}
	72	5	0.50	10/ 9823	0.10	33/10000	0.33 ± 0.05

[a] NE = normochromatic erythrocytes; PE = polychromatic erythrocytes; MN = micronuclei.
[b] $^{+} = P < 0.05$; $^{++} = P < 0.01$; $^{+++} = P < 0.001$ (Mann-Whitney-Wilcoxon Test).
[c] Experiment with stomach intubation of oil suspension.

Table 4. *Micronucleus test with 4AAF in mouse bone marrow*

Dose (mg/kg)	Sampling interval (h)	No. of animals	PE[a] / PE + NE	MN/NE	%	MN/PE	%MN ± SE
0.0	—	9	0.48	15/17817	0.08	42/18000	0.23 ± 0.03
223.0	48	5	0.50	8/ 9706	0.08	23/10000	0.23 ± 0.04
446.0	48	5	0.47	8/10000	0.08	20/10000	0.20 ± 0.02
892.0	24	5	0.50	13/10000	0.13	49/10000	0.49 ± 0.09
	24	5	0.47	12/10000	0.12	27/10000	0.27 ± 0.06
	36	3	0.46	2/ 6000	0.03	17/ 6000	0.28 ± 0.03
	36	5	0.49	11/10000	0.11	34/10000	0.34 ± 0.03
	48	4	0.48	7/ 8000	0.09	17/ 8000	0.21 ± 0.02
	72	5	0.48	15/10000	0.15	19/10000	0.19 ± 0.05

[a] NE = normochromatic erythrocytes; PE = polychromatic erythrocytes; MN = micronuclei.

2AAF

2AAF results are shown in Table 3. 2AAF gave a positive response. The study of the time dependence of micronucleus appearance in PE revealed the maximum at 48 h. The micronucleus rates increased with dose at the 48 h interval. The lowest tested dose (111.5 mg/kg) resulted in a micronucleus rate significantly above control level ($P < 0.05$).

4AAF

4AAF results are shown in Table 4. 4AAF gave a negative response. None of the single time groups or the pooled treated groups showed a micronucleus rate above control level ($P < 0.05$).

5 Discussion

The 4 compounds of the present collaborative trial were extensively tested in the bone-marrow assay micronucleus test. Sampling intervals were varied for the highest doses of 3 of the 4 chemicals.

For BP, preexisting information (Kliesch *et al.*, 1982) from our own laboratory suggested that the optimal time for micronuclei scoring was 36 h after treatment, the time used consequently for all 3 doses of the present experiments. In the previous experiment, the frequency of micronuclei in polychromatic erythrocytes increased linearly in the tested dose range of 12.5 to 250 mg/kg. With a dose of 25 mg/kg, the frequency of micronucleated erythrocytes was significantly different from the concurrent control ($P < 0.002$). Precipitated BP was found in the peritoneum (Adler, unpublished) when the compound was dissolved in corn oil and given ip. It is conceivable that BP also precipitates in the stomach. Resorption may not necessarily correspond directly to the injected dose. This may explain the small increase in micronucleus rates reported for the present experiments for the dose range of 31.5 to 63 mg/kg.

For 2AAF, the study of the time response revealed the maximum of micronuclei at 48 h after treatment. At this time interval, the micronucleus frequencies increased with dose. The lowest tested dose of 111.5 mg/kg proved an effect significantly above the control groups.

The other two chemicals, PYR and 4AAF, were classified as non-clastogens. None of the single time groups or the pooled treated groups showed a micronucleus frequency above control level. In a brief LD_{50} trial [according to Weil (1952)], the value obtained for PYR was 900 mg/kg, for 2AAF 600 mg/kg, and for 4AAF 900 mg/kg with the (101/El × C3H)F_1 hybrid mice used in present experiments. Thus, the highest doses applied for the micronucleus study corresponded closely with the maximum tolerated doses for PYR and 4AAF. After repetitive treatments, cell-killing effects may reduce the micronucleus frequencies. It seems justified to state that the 2 chemicals that gave a negative response were optimally tested.

It can be stated that, under the present experimental conditions, only the 2 known carcinogens, BP and 2AAF, induced micronuclei in mouse bone marrow. The two noncarcinogens PYR and 4AAF gave a negative result.

6 Acknowledgements

The authors are indebted to Miss C. Zeller for skilled technical assistance. The experiments were partly supported by the EG Contract No. ENV-637-D(B).

7 References

Boller, K. and W. Schmid (1970) Chemische Mutagenese beim Säuger. Das Knochenmark des Chinesischen Hamsters als *In-vivo*-Test-system, Hämatologische Befunde nach Behandlung mit Trenimon, *Humangenetik*, **11**, 35–54.

Cole, R.J., N.A. Taylor, J. Cole and C.F. Arlett (1979) Transplacental effects of chemical mutagens detected by the micronucleus test, *Nature* (London), **277**, 317–318.

Hart, J.W. and Beryl Hartley-Asp (1983) Induction of micronuclei in the mouse. Revised timing of the final stage of erythropoiesis, *Mutation Res.*, **120**, 127–132.

Heddle, J.A. (1973) A rapid *in vivo* test for chromosomal damage, *Mutation Res.*, **41**, 311–320.

Kliesch, U., N. Danford and I.-D. Adler (1981) Micronucleus test and bone-marrow chromosome analysis, A comparison of 2 methods *in vivo* for evaluating chemically induced chromosomal alterations, *Mutations Res.*, **80**, 321–332.

Kliesch, U., I. Roupova and I.-D. Adler (1982) Induction of chromosome damage in mouse bone marrow by benzo(a)pyrene, *Mutation Res.*, **102**, 265–373.

Matter, B. and W. Schmid (1971) Trenimon-induced chromosomal damage in bone marrow cells of six mammalian species, evaluated by the micronucleus test, *Mutation Res.*, **12**, 417–425.

Schmid, W. (1976) The micronucleus test for cytogenetic analysis. In A. Hollaender (Ed.), *Chemical Mutagens*, Vol. 4, Plenum, New York, pp. 31–53.

Weil, C.S. (1952) Tables for convenient calculation of median-effective dose (LD_{50} or ED_{50}) and instructions in their use, *Biometrics*, **8**, 249–263.

Michael F. Salamone, Susan Chiu, and David M. Logan[1]
MOE-Biohazard Laboratory, Rexdale, Ontario, (Canada) and [1]Department of Biology, York University, North York, Ontario, Canada

29

Tests on two pairs of structural analogues in the mouse in vivo micronucleus assay

1 Summary

As part of an international collaborative program, two pairs of structurally similar compounds, of which one member of each pair was a suspect carcinogen, were tested in the mouse *in vivo* micronucleus assay. Test were conducted using both an intraperitoneal and a gavage mode of treatment. In each trial the suspect carcinogens elicited positive micronucleus assay responses, while the supposed non-carcinogenic analogues did not.

2 Introduction

The *in vivo* micronucleus assay is a cytogenetic procedure that can be used to detect agents that either break chromosomes or interfere with the normal movement of chromosomes. Since either consequence poses a potentially serious genetic hazard, it is important that substances suspected of eliciting such a response be assessed by the most sensitive and reliable test systems available. In this regard, monitoring with *in vivo* systems offers a particular advantage over *in vitro* assessment. That is, *in vivo* assays better reflect the natural system. The potential involvement of natural barriers, activation enzyme systems, and other aspects of pharmacokinesis may all be expressed in an *in vivo* assay system. With *in vitro* assays, the interpretation of the results is often compromised by the use of exogeneous activation enzymes, and the application of the test agent directly to the cells does not reflect a real life situation.

In this report are presented *in vivo* micronucleus assay data on the 4 chemicals included in a large collaborative study. The study was established to assess the genotoxic behaviour of two pairs of structurally analogous and potentially dangerous chemicals in *in vivo* assay systems. Each of the 4 compounds had previously been shown to elicit positive responses in several *in vitro* genotoxic assays; however, only one member of each pair had been classified as carcinogenic. It was hoped that tests using *in vivo* genotoxic assays would provide a more critical estimate of the potential hazard of each compound to intact multicellular organisms. This expectation has been realised with the chemicals tested for this portion of the study.

3 Materials and methods

Chemicals

The 4 compounds tested, benzo(a)pyrene (BP); pyrene (PYR); 2-acetylaminofluorene (2AAF); and 4-acetylam-

Table 1. *Micronucleus assay results for PYR in B6C3F1 mice*

Treatment mode: ip   Solvent: DMSO   Age of mice at time of treatment: 18 wks

Treatment and dose mg/kg	Sex	46 hour Mean ± SE PE/NE + PE	46 hour MN/500 PE	70 hour Mean ± SE PE/NE + PE	70 hour MN/500 PE	94 hour Mean ± SE PE/NE + PE	94 hour MN/500 PE
PYR							
411 × 2	No. of mice treated = 23 (11M + 12F)				No. of survivors = 23		
	M	0.26 ± 0.03	1.50 ± 0.87	0.34 ± 0.05	1.00 ± 0.71	0.43 ± 0.05	0.33 ± 0.33
	F	0.42 ± 0.08	1.00 ± 0.00	0.46 ± 0.06	0.50 ± 0.29	0.45 ± 0.11	0.50 ± 0.29
	M + F	0.34 ± 0.05	1.33 ± 0.56[a]	0.40 ± 0.04	0.75 ± 0.37[a]	0.44 ± 0.06	0.50 ± 0.22[a]
Untreated controls							
0	No. of mice = 6 (3M + 3F)				No. of survivors = 6		
	M					0.33 ± 0.05	1.00 ± 0.58
	F					0.59 ± 0.01	1.05 ± 0.50
	M + F	No sample taken		No sample taken		0.43 ± 0.07	1.20 ± 0.37
DMSO (solvent control)							
0.1 ml × 2	No. of mice treated = 6 (3M + 3F)				No. of survivors = 6		
	M			0.26 ± 0.07	3.00[b]		
	F			0.34 ± 0.03	0.67 ± 0.67		
	M + F	No sample taken		0.31 ± 0.03	1.25 ± 0.75	No sample taken	
DMBA (positive control)							
25 × 2	No. of mice treated = 12 (6M + 6F)				No. of survivors = 12		
	M	0.18 ± 0.06	5.00 ± 3.00	0.40 ± 0.08	6.67 ± 2.40		
	F	0.52 ± 0.05	6.00 ± 0.58	0.43 ± 0.09	3.00 ± 1.00		
	M + F	0.35 ± 0.08	5.60 ± 1.03[c]	0.42 ± 0.05	4.83 ± 1.42[c]	No sample taken	

[a] Not significant.
[b] Due to a limited number of cells on the slide, the MN–PE level could be determined for only one mouse.
[c] Significant at $P \leqslant 0.01$.

Table 2. *Micronucleus assay results for PYR in B6C3F1 mice*

Treatment mode: ip   Solvent: DMSO   Age of mice at time of treatment: 14–15 wks

Treatment and dose mg/kg	Sex	47 hour Mean ± SE PE/NE + PE	47 hour MN/500 PE	56 hour Mean ± SE PE/NE + PE	56 hour MN/500 PE	73 hour Mean ± SE PE/NE + PE	73 hour MN/500 PE
PYR							
711 × 2	No. of mice treated = 18 (9M + 9F)				No. of survivors = 16 (8M + 8F)		
	M	0.37 ± 0.05	0.83 ± 0.17	0.43 ± 0.08	1.00 ± 1.00	No sample taken	
	F	0.30 ± 0.08	1.33 ± 0.67	0.50 ± 0.01	0.00 ± 0.00	0.40 ± 0.21	0.50 ± 0.50[a]
	M + F	0.34 ± 0.04	1.00 ± 0.24[a]	0.47 ± 0.03	0.40 ± 0.40[a]		
643 × 2	No. of mice treated = 18 (9M + 9F)				No. of survivors = 16 (9M + 7F)		
	M	0.27 ± 0.07	0.83 ± 0.40	0.45 ± 0.11	2.00 ± 2.00	No sample taken	
	F	0.40 ± 0.12	0.33 ± 0.33	0.42 ± 0.21	0.33 ± 0.33	0.59	1.00[a,b]
	M + F	0.31 ± 0.07	0.75 ± 0.31[a]	0.43 ± 0.12	1.00 ± 0.77[a]		
535 × 2	No. of mice treated = 18 (9M + 9F)				No. of survivors = 16 (8M + 8F)		
	M	0.43 ± 0.03	0.67 ± 0.33	0.55 ± 0.08	1.33 ± 0.67	0.39 ± 0.06	2.00 ± 1.00
	F	0.38 ± 0.05	0.50 ± 0.50	0.55 ± 0.04	0.33 ± 0.33	0.59	0.00[b]
	M + F	0.42 ± 0.03	0.50 ± 0.22[a]	0.55 ± 0.04	0.83 ± 0.40[a]	0.45 ± 0.08	1.33 ± 0.88[a]
446 × 2	No. of mice treated = 18 (9M + 9F)				No. of survivors = 17 (9M + 8F)		
	M	0.37 ± 0.12	0.00 ± 0.00	0.45 ± 0.10	0.50 ± 0.50	0.30 ± 0.09	0.67 ± 0.67
	F	0.52 ± 0.00	0.33 ± 0.33	0.47 ± 0.05	0.67 ± 0.33	0.43 ± 0.06	0.00 ± 0.00
	M + F	0.45 ± 0.07	0.17 ± 0.17[a]	0.46 ± 0.04	0.60 ± 0.24[a]	0.35 ± 0.06	0.40 ± 0.40[a]
DMSO (solvent control)							
0.1 ml × 2	No. of mice treated = 6 (3M + 3F)				No. of survivors = 6		
	M			0.62 ± 0.02	0.00 ± 0.00		
	F			0.63 ± 0.03	1.00 ± 0.58		
	M + F	No sample taken		0.63 ± 0.01	0.50 ± 0.34	No sample taken	
DMBA (positive control)							
25 × 2	No. of mice treated = 8 (4M + 4F)				No. of survivors = 8		
	M	0.27 ± 0.07	19.50 ± 1.50				
	F	0.36 ± 0.06	4.50 ± 1.44				
	M + F	0.32 ± 0.05	9.50 ± 3.31[c]	No sample taken		No sample taken	

[a] Not significant.
[b] Only one animal survived to sample time.
[c] Significant at $P \leqslant 0.01$.

Table 3. *Micronucleus assay results for PYR in B6C3F1 mice*

Treatment mode: gav		Solvent: corn oil		Age of mice at time of treatment: 14 wks	
Treatment and dose (mg/kg)	Sex	46 hour Mean ± SE		70 hour Mean ± SE	
		PE / NE + PE	MN / 500 PE	PE / NE + PE	MN / 500 PE
PYR					
	No. of mice treated = 14(6M + 8F)			No. of survivors = 5(4M + 1F)	
700 × 2	M	0.33 ± 0.03	0.00 ± 0.00		
700 × 1	F	0.44	0.00[a]		
	M + F	0.35 ± 0.03	0.00	No sample taken	
	No. of mice treated = 14(6M + 8F)			No. of survivors = 9(6M + 3F)	
467 × 2	M	0.35 ± 0.06	0.33 ± 0.33	0.42 ± 0.08	0.00 + 0.00
467 × 1	F	0.21 ± 0.05	0.33 ± 0.33	No sample taken	
	M + F	0.28 ± 0.05	0.33 ± 0.21		
311 × 2	No. of mice treated = 14(6M + 8F)			No. of survivors = 14	
	M	0.34 ± 0.10	0.67 ± 0.33[b]	0.47 ± 0.05	0.33 + 0.33
	F	0.41 ± 0.06	0.00 ± 0.00	0.45 ± 0.04	0.00 ± 0.00
	M + F	0.38 ± 0.05	0.29 ± 0.18	0.46 ± 0.03	0.14 ± 0.14
Corn oil (solvent control)					
0.3 ml × 2	No. of mice treated = 7(3M + 4F)			No. of survivors = 7	
	M	0.40 ± 0.03	0.67 ± 0.67		
	F	0.59 ± 0.03	0.50 ± 0.29		
	M + F	0.51 ± 0.04	0.57 ± 0.30	No sample taken	

[a] Only one animal survived to sample time.
[b] Not significant.

Table 4. *Micronucleus assay results for BP in B6C3F1 mice*

Treatment mode: ip		Solvent: DMSO		Age of mice at time of treatment: 18 wks			
Treatment and dose mg/kg	Sex	46 hour Mean ± SE		70 hour Mean ± SE		94 hour Mean ± SE	
		PE / NE + PE	MN / 500 PE	PE / NE + PE	MN / 500 PE	PE / NE + PE	MN / 500 PE
BP							
117 × 2	No. of mice treated = 24(12M + 12F)				No. of survivors = 24		
	M	0.31 ± 0.03	1.67 ± 1.20	0.28 ± 0.11	2.00 ± 2.00	0.49 ± 0.05	1.50 ± 0.29
	F	0.43 ± 0.01	2.00 ± 0.58	0.43 ± 0.10	5.25 ± 0.75	0.37 ± 0.05	2.75 ± 1.03
	M + F	0.37 ± 0.03	1.83 ± 0.60[a]	0.36 ± 0.08	4.17 ± 0.98[b]	0.43 ± 0.04	2.13 ± 0.55[b]
Untreated controls							
0	No. of mice = 6(3M + 3F)				No. of survivors = 6		
	M					0.33 ± 0.01	1.00 ± 0.58
	F					0.59 ± 0.01	1.50 ± 0.50
	M + F	No sample taken		No sample taken		0.43 ± 0.07	1.20 ± 0.37
DMSO (solvent control)							
0.1 ml × 2	No. of mice treated = 6(3M + 3F)				No. of survivors = 6		
	M			0.26 ± 0.07	3.00[c]		
	F			0.34 ± 0.03	0.67 ± 0.67		
	M + F	No sample taken		0.31 ± 0.03	1.25 ± 0.75	No sample taken	
DMBA (positive control)							
25 × 2	No. of mice treated = 12(6M + 6F)				No. of survivors = 12		
	M	0.18 ± 0.06	5.00 ± 3.00	0.40 ± 0.08	6.67 ± 2.40		
	F	0.52 ± 0.05	6.00 ± 0.58	0.43 ± 0.09	3.00 ± 1.00		
	M + F	0.35 ± 0.08	5.60 ± 1.03[b]	0.42 ± 0.05	4.83 ± 1.42[b]	No sample taken	

[a] Significant at $P \leqslant 0.05$.
[b] Significant at $P \leqslant 0.01$.
[c] Due to a limited number of cells on the slide, the MPE level could be determined for only one mouse.

Table 5. *Micronucleus assay results for BP in B6C3F1 mice*

Treatment mode: ip		Solvent: DMSO	Age of mice at time of treatment: 20 wks
Treatment and dose mg/kg	Sex	48 hour Mean ± SE PE/(NE + PE)	48 hour Mean ± SE MN/500 PE
BP			
	No. of mice treated = 5F		No. of survivors = 5F
177 × 1	F	0.48 ± 0.06	1.40 ± 0.51[a]
133 × 1	F	0.43 ± 0.04	2.00 ± 0.71[b]
88 × 1	F	0.48 ± 0.05	0.25 ± 0.25[a]
Untreated controls			
	No. of mice = 3F		No. of survivors = 3F
	F	0.51 ± 0.01	0.33 ± 0.33
DMSO (solvent control)			
	No. of mice treated = 5F		No. of survivors = 5F
0.1 ml × 2	F	0.50 ± 0.02	0.60 ± 0.40
DMBA (positive control)			
	No. of mice treated = 5F		No. of survivors = 5F
25 × 1	F	0.43 ± 0.03	2.40 ± 0.81[b]

[a] Not significant.
[b] Significant at $P \leqslant 0.01$.

Table 6. *Micronucleus assay results for BP in B6C3F1 mice*

Treatment mode: ip   Solvent: DMSO   Age of mice at time of treatment: 8 wks

Treatment and dose mg/kg	Sex	48 hour Mean ± SE PE/(NE + PE)	48 hour Mean ± SE MN/500 PE	72 hour Mean ± SE PE/(NE + PE)	72 hour Mean ± SE MN/500 PE	80 hour Mean ± SE PE/(NE + PE)	80 hour Mean ± SE MN/500 PE
BP							
185 × 2	No. of mice treated = 22(10M + 12F)				No. of survivors = 22		
	M	0.61 ± 0.09	1.50 ± 1.50	0.34 ± 0.08	3.33 ± 0.88	0.57 ± 0.09	5.67 ± 1.67
	F	0.35 ± 0.07	2.50 ± 0.50	0.26 ± 0.04	3.75 ± 1.31	0.52 ± 0.06	4.00 ± 0.82
	M + F	0.44 ± 0.07	2.17 ± 0.54[a]	0.29 ± 0.05	3.57 ± 0.78[a]	0.54 ± 0.05	4.71 ± 0.84[b]
139 × 2	No. of mice treated = 24(12M + 12F)				No. of survivors = 24		
	M	0.54 ± 0.04	4.33 ± 1.45	0.45 ± 0.08	6.25 ± 2.06	0.64 ± 0.04	5.50 ± 3.23
	F	0.53 ± 0.04	3.75 ± 0.85	0.29 ± 0.04	4.50 ± 1.71	0.48 ± 0.08	4.00 ± 0.41
	M + F	0.53 ± 0.02	4.00 ± 0.72[b]	0.37 ± 0.05	5.38 ± 1.28[b]	0.56 ± 0.05	4.75 ± 1.53[b]
92 × 2	No. of mice treated = 18 (9M + 9F)				No. of survivors = 18		
	M	0.53 ± 0.08	3.50 ± 2.50	0.65	4.00[c]	0.63 ± 0.05	1.00 ± 0.58
	F	0.39 ± 0.07	3.67 ± 1.45	0.49 ± 0.09	4.00 ± 0.58	0.50 ± 0.02	2.33 ± 0.88
	M + F	0.44 ± 0.06	3.60 ± 1.12[a]	0.53 ± 0.08	4.00 ± 0.41[a]	0.57 ± 0.04	1.67 ± 0.56[a]
46 × 2	No. of mice treated = 18(9M + 9F)				No. of survivors = 18		
	M	0.46 ± 0.05	2.33 ± 1.86	0.60 ± 0.02	2.67 ± 0.33	0.53 ± 0.04	2.67 ± 0.33
	F	0.52 ± 0.07	3.00 ± 0.58	0.44 ± 0.01	2.67 ± 1.20	0.62 ± 0.03	1.67 ± 1.20
	M + F	0.49 ± 0.04	2.67 ± 0.88[a]	0.52 ± 0.04	2.67 ± 0.56[a]	0.58 ± 0.03	2.17 ± 0.60[a]
Untreated controls							
	No. of mice = 6(3M + 3F)				No. of survivors = 6		
	M					0.48 ± 0.08	1.67 ± 1.20
	F					0.61 ± 0.05	1.00 ± 1.00
	M + F	No sample taken		No sample taken		0.54 ± 0.05	1.33 ± 0.71
DMSO (solvent control)							
0.1 ml × 2	No. of mice treated = 6(3M + 3F)				No. of survivors = 6		
	M	0.53	3.00[c]				
	F	0.62 ± 0.01	0.33 ± 0.33				
	M + F	0.60 ± 0.02	1.00 ± 0.71	No sample taken		No sample taken	
DMBA (positive control)							
25 × 2	No. of mice treated = 7(3M + 4F)				No. of survivors = 7		
	M			0.30 ± 0.07	4.67 ± 2.03		
	F			0.49 ± 0.02	3.50 ± 2.50		
	M + F	No sample taken		0.38 ± 0.06	4.20 ± 1.39	No sample taken	

[a] Significant at $P \leqslant 0.05$.
[b] Significant at $P \leqslant 0.01$.
[c] Slides for two animals could not be scored.

inofluorene (4AAF), were supplied in pure form by Imperial Chemical Industries (UK). Depending on the treatment mode, compounds were dissolved in either dimethylsulphoxide (DMSO) or a pharmaceutical grade (Fisher laboratory grade) of corn oil (CO). DMSO was used in all experiments in which animals were treated by intraperitoneal (ip) injections. The final volume of DMSO per treatment never exceeded 0.1 ml. Corn oil was used for all gavage treatments at a final volume of 0.3 ml per treatment.

Animals

For all experiments B6C3F1 hybrid mice from inbred parent strains C57BL/6 × C3H/HeB were used. The mice were obtained from Charles River of Canada (Quebec) at the age of 4–5 weeks and allowed to acclimate to the laboratory environment until they reached at least 7–8 weeks of age. The mice were maintained in a large clean animal facility where light, temperature and other environmental conditions were automatically controlled and where they were given Purina rodent chow and fresh water ad libitum.

In all but one experiment both male and female mice were used. Animals were age matched for each experiment (i.e., only animals of the same age from a single shipment were used in a given test). At the time of treatment the females generally weighed between 20 and 24 g, while the males weighed between 23 and 30g.

Assay procedure

The micronucleus assay procedure used here followed that described by Salamone and Heddle (1983) and Salamone *et al.* (1980). Mice were generally given 2 consecutive daily treatments of the test substance at a particular dose, initially and again at approximately 24 h. Two to 4 bone marrow samples were taken between 46 and 96 h after the initial treatment. Where possible, all results were confirmed. In some cases, the method varied from that of Salamone *et al.* (1980) in that 3 or 4 doses rather than 1 or 2 were tested at each sample time. Intraperitoneal treatments were given with a 26-gauge hypodermic needle attached to a one milliliter syringe, while gavage treatments were administered with 18-gauge steel 1.5-inch straight feeding needles (Popper and Sons, NY).

The harvesting of the bone marrow and the preparation, staining, and scoring of the slides were as a described by Heddle *et al.* (1984). The mountant used was DPX (BDH Chemical Co.). During the project one of the shipments of DPX was found to be faulty. After the DPX had hardened, the slides became unscorable. Remounting the slide did not help, and as a result, several confirmation studies, especially with 2AAF and 4AAF, had to be discarded.

Analysis of data

To facilitate uniformity in the analysis of the data among the different laboratories that participated in the *in vivo* micronucleus phase of this study, data were analysed using the tables of Kastenbaum and Bowman (1970). The control value used in this report was obtained by pooling the control frequencies for all the ip experiments reported here. The value obtained closely approximated the historical control level for this laboratory. As there was only one set of corn oil control data, it was used when analysing the gavage results. A response was considered positive if, when compared with the control value, it was significantly different at $p \leqslant 0.05$.

Table 7. *Micronucleus assay results for BP in B6C3F1 mice*

Treatment mode: gav — Solvent: corn oil — Age of mice at time of treatment: 14 wks

Treatment and dose mg/kg	Sex	47 hour Mean ± SE PE/(NE + PE)	47 hour Mean ± SE MN/500 PE	70 hour Mean ± SE PE/(NE + PE)	70 hour Mean ± SE MN/500 PE
BP					
188 × 2	No. of mice treated = 14(6M + 8F)			No. of survivors = 14	
	M	0.25 ± 0.06	3.67 ± 1.76	0.35 ± 0.07	1.00 ± 0.58
	F	0.43 ± 0.01	2.50 ± 0.96	0.32 ± 0.04	0.75 ± 0.48
	M + F	0.35 ± 0.04	3.00 ± 0.87[a]	0.33 ± 0.03	0.86 ± 0.34[b]
150 × 2	No. of mice treated = 14(6M + 8F)			No. of survivors = 14	
	M	0.45 ± 0.05	3.67 ± 0.33	0.36 ± 0.08	0.67 ± 0.33
	F	0.41 ± 0.06	1.75 ± 1.11	0.35 ± 0.05	0.25 ± 0.25
	M + F	0.43 ± 0.04	2.57 ± 0.72[a]	0.35 ± 0.04	0.43 ± 0.20[b]
120 × 2	No. of mice treated = 14(6M + 8F)			No. of survivors = 14	
	M	0.46 ± 0.04	1.33 ± 0.33	0.38 ± 0.07	1.67 ± 0.67
	F	0.56 ± 0.05	1.75 ± 1.18	0.45 ± 0.02	0.75 ± 0.25
	M + F	0.51 ± 0.04	1.57 ± 0.65[a]	0.42 ± 0.03	1.14 ± 0.34[a]
Corn oil (solvent control)					
0.3 ml × 2	No. of mice treated = 7(3M + 4F)			No. of survivors = 7	
	M	0.40 ± 0.03	0.67 ± 0.67		
	F	0.59 ± 0.03	0.50 ± 0.29		
	M + F	0.51 ± 0.04	0.57 ± 0.30	No sample taken	

[a] Significant at $P \leqslant 0.01$.
[b] Not significant.

Results are presented in tables, one table for each experiment. In the first column of each table is presented the drug treatment dose and the number of times it was given. Data are presented as the mean ± the standard error of the mean (SE) for both the frequency of micronuclei for 500 polychromatic erythrocytes (PE) and the ratio of PE to that of normochromatic erythrocytes (NE). With some sample times the words "No sample taken" are found in place of data. This designation indicates that no animals were sampled at this time because either the number of animals surviving the treatment was not sufficient to permit assessment at these times or, in the case of controls, samples were only taken at designated times.

PYR and BP

Initial test results with PYR administered by ip were negative at all 3 sample times from 46 to 96 h (Table 1). As the dose used did not induce a toxic effect in any of the animals, a confirmation test was conducted using much higher doses. After 2 treatments at these higher doses, the animals were all sick. To avoid losing too many animals due to the toxic effects of the chemical, the sample intervals were shortened and the final sample was taken at 73 h. The results of the test clearly indicate that PYR does not induce a significant increase in the number of micronucleated polychromatic erythrocytes (MPEs) compared with that found using solvent alone (Table 2). When administered by gavage at near toxic doses, PYR still did not induce a significant increase in the level of micronuclei (Table 3). These findings agree with previously published data on PYR from this and other laboratories (Salamone *et al.*, 1981; Tsuchimoto and Matter, 1981; Kirkhart, 1981).

The initial BP test used 2 ip treatments, each of 177 mg/kg. The results indicate that BP induced significant increases in the level of MPEs. This effect was found at all 3 sample times with the 70 h sample time showing

Table 8. *Micronucleus assay results for 4AAF*

Treatment mode: ip   Solvent: DMSO   Age of mice (wks) at time of treatment: 9–10 wks

Treatment and dose (mg/kg)	Sex	48 hour Mean ± SE PE/(NE + PE)	48 hour MN/500 PE	72 hour Mean ± SE PE/(NE + PE)	72 hour MN/500 PE	80 hour Mean ± SE PE/(NE + PE)	80 hour MN/500 PE	96 hour Mean ± SE PE/(NE + PE)	96 hour MN/500 PE
4AAF									
291 × 2		No. of mice treated = 48 (24M + 24F).			No. of survivors = 48				
	M	0.57 ± 0.04	0.63 ± 0.18	0.49 ± 0.04	0.63 ± 0.32	0.71 ± 0.04	1.00 ± 0.41	0.52 ± 0.06	1.75 ± 0.63
	F	0.57 ± 0.02	0.38 ± 0.18	0.50 ± 0.04	0.38 ± 0.26	0.60 ± 0.04	0.50 ± 0.29	0.49 ± 0.04	1.00 ± 0.41
	M + F	0.57 ± 0.02	0.50 ± 0.13[a]	0.49 ± 0.03	0.50 ± 0.20[a]	0.65 ± 0.02	0.75 ± 0.18[a]	0.50 ± 0.03	1.38 ± 0.38[a]
218 × 2		No. of mice treated = 48 (24M + 24F).			No. of survivors = 48				
	M	0.58 ± 0.01	0.50 ± 0.27	0.54 ± 0.04	0.25 ± 0.16	0.69 ± 0.05	1.75 ± 0.85	0.53 ± 0.04	0.75 ± 0.48
	F	0.59 ± 0.04	1.50 ± 0.38	0.60 ± 0.03	0.75 ± 0.16	0.61 ± 0.05	0.75 ± 0.48	0.46 ± 0.02	1.25 ± 0.25
	M + F	0.57 ± 0.02	1.00 ± 0.26[a]	0.57 ± 0.02	0.50 ± 0.13[a]	0.65 ± 0.03	1.25 ± 0.35[a]	0.50 ± 0.02	1.00 ± 0.27[a]
146 × 2		No. of mice treated = 36 (18M + 18F).			No. of survivors = 36				
	M	0.58 ± 0.04	1.50 ± 0.22	0.52 ± 0.06	0.86 ± 0.40	0.58 ± 0.03	0.50 ± 0.50	0.55	0.00
	F	0.53 ± 0.04	0.17 ± 0.17	0.54 ± 0.07	0.33 ± 0.21	0.62 ± 0.01	0.67 ± 0.33	0.47 ± 0.04	0.50 ± 0.50
	M + F	0.56 ± 0.03	0.83 ± 0.24[a]	0.53 ± 0.04	0.62 ± 0.24[a]	0.60 ± 0.01	0.60 ± 0.24[a]	0.49 ± 0.03	0.33 ± 0.33[a]
73 × 2		No. of mice treated = 36 (18M + 18F).			No. of survivors = 36				
	M	0.59 ± 0.03	0.67 ± 0.21	0.51 ± 0.08	0.67 ± 0.49	0.60 ± 0.06	1.67 ± 0.33	0.47 ± 0.07	2.00 ± 0.00[b]
	F	0.52 ± 0.02	0.83 ± 0.31	0.54 ± 0.07	0.33 ± 0.21	0.57 ± 0.05	1.00 ± 0.58	0.53 ± 0.05	0.67 ± 0.67
	M + F	0.55 ± 0.02	0.75 ± 0.18[a]	0.53 ± 0.05	0.50 ± 0.26[a]	0.59 ± 0.04	1.33 ± 0.33[a]	0.50 ± 0.04	1.33 ± 0.42[a]
Control (untreated)									
0		No. of mice treated = 12 (6M + 6F).			No. of survivors = 12				
	M					0.60 ± 0.04	0.00 ± 0.00	0.52 ± 0.05	0.67 ± 0.33
	F					0.60 ± 0.05	0.00 ± 0.00	0.50 ± 0.04	1.00 ± 0.58
	M + F	No sample taken		No sample taken		0.60 ± 0.03	0.00 ± 0.00	0.51 ± 0.03	0.83 ± 0.31
DMSO (solvent control)									
	D	No. of mice treated = 14 (7M + 7F).			No. of survivors = 14				
	M			0.54 ± 0.02	0.57 ± 0.20				
	F			0.55 ± 0.04	0.29 ± 0.18				
	M + F	No sample taken		0.54 ± 0.02	0.43 ± 0.14	No sample taken		No sample take	
DMBA (positive control)									
25 × 2		No. of mice treated = 16 (8M + 8F).			No. of survivors = 16				
	M	0.44 ± 0.03	5.00 ± 1.69						
	F	0.44 ± 0.07	3.75 ± 1.06						
	M + F	0.44 ± 0.03	4.38 ± 0.98[c]	No sample taken		No sample taken		No sample taken	

[a] Not significant.
[b] Significant at $P \leqslant 0.05$.
[c] Significant at $P \leqslant 0.01$.

the best response (Table 4). Two confirmation tests of these results were carried out using slightly different treatment regimens. In the first test, a single treatment of 3 different doses was administered and the bone marrow was sampled 48 h later. In this case the data were less convincing, and only the 133 mg/kg dose gave a clearly significant response (Table 5). In the second confirmation test, 2 treatments of 4 different doses were administered and mice were sampled at 3 different times. In this case the data were unequivocal. Significant clastogenic activity was noted for all doses at every sample time (Table 6).

BP also induced significant increases in the MPE level when administered by gavage (Table 7). These increases were observed only at the early sample time, and these data illustrate the effect of treatment mode on peak response time. A comparison of the ip and gavage BP data shows that there is a shift to an earlier peak response when BP is administered by gavage (Figure 1). This observation supports a previous report of Kliesch *et al.* (1982). In contrast, when BP is administered by ip, there is a shift to a later peak response, especially at higher doses, which is consistent with the general findings of Salamone *et al.* (1980). The different times of peak response reflect the complexity of uptake and metabolism of substances from the peritoneal cavity or digestive tract, making questionable any direct comparison of data obtained by use of different treatment modes.

Overall, BP is capable of breaking chromosomes and producing micronuclei. This confirms several previously published studies (Heddle *et al.*, 1983).

2AAF and 4AAF

Results with 4AAF show it to be basically negative in the assay system (Tables 8 and 9). In an initial experiment, 4 doses of 4AAF were tested at 4 sample times. Of the 16 combined male and female data sets evaluated, only one was significantly different from the pooled control data set (291 mg/kg × 2 dose at 96 h) (Table 8). Since none of the

Figure 1. Micronucleated polychromatic erythrocyte induction by BP in mice as a function of dose, sample time and route of administration.

Left figure-BP administered by gavage; closed circles (●), 120 mg/kg; open square (□), 150 mg/kg; closed square (■) 188 mg/kg.

Right figure-BP administered by ip; stars (*), 46 mg/kg; closed circles (●), 92 mg/kg; open squares (□), 139 mg/kg; closed squares (●), 185 mg/kg.

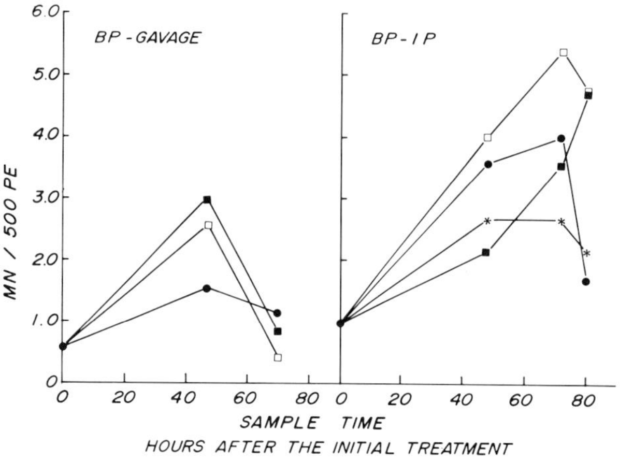

Table 9. *Micronucleus assay results for 4AAF in B6C3F1 mice*

Treatment mode: ip		Solvent: DMSO		Age of mice at time of treatment: 11 wks	
Treatment and dose mg/kg	Sex	47 hour Mean ± SE PE/(NE + PE)	47 hour Mean ± SE MN/500 PE	71 hour Mean ± SE PE/(NE + PE)	71 hour Mean ± SE MN/500 PE
4AAF					
546 × 2	No. of mice treated = 24 (12M + 12F)			No. of survivors = 20 (11M + 9F)	
	M	0.42 ± 0.02	0.17 ± 0.17	0.56 ± 0.08	0.00 ± 0.00
	F	0.50 ± 0.06	0.40 ± 0.40	0.56 ± 0.04	0.00 ± 0.00
	M + F	0.45 ± 0.03	0.27 ± 0.19[a]	0.56 ± 0.05	0.00 ± 0.00[a]
364 × 2	No. of mice treated = 36 (18M + 18F)			No. of survivors = 36	
	M	0.54 ± 0.04	0.00 ± 0.00	0.53 ± 0.04	0.56 ± 0.24
	F	0.55 ± 0.02	0.56 ± 0.24	0.57 ± 0.05	0.75 ± 0.25
	M + F	0.54 ± 0.03	0.38 ± 0.18[a]	0.55 ± 0.03	0.65 ± 0.17[a]
Untreated controls					
0	No. of mice = 4F			No. of survivors = 4F	
	F	No sample taken		0.56 ± 0.03	0.25 ± 0.25
DMSO (solvent control)					
0.1 ml × 2	No. of mice treated = 4F			No. of survivors = 4F	
	F	No sample taken		0.43 ± 0.03	0.50 ± 0.29
DMBA (positive control)					
25 × 2	No. of mice treated = 6 (3M + 3F)			No. of survivors = 6	
	M	0.37 ± 0.05	9.00 ± 1.53		
	F	0.34 ± 0.05	4.00 ± 1.53		
	M + F	0.35 ± 0.03	6.50 ± 1.48[b]		

[a] Not significant.
[b] Significant at $P \leqslant 0.01$.

other data sets was significantly different from the controls. This response was questioned.

As there was no sign of toxicity in the animals during the initial experiment, a confirmation test was conducted with significantly higher doses of 4AAF. During the confirmation test, approximately 17% of the mice treated with the highest dose died. The assay results were still negative at each dose and sample time. This observation supports previous studies that classified 4AAF as negative in the *in vivo* micronucleus assay (Tsuchimoto and Matter, 1981; Salamone *et al.*, 1981).

Results from the 2AAF experiment are reported in Table 10. In that test 2AAF induced significant increases in the MPE level at all sample times for the highest dose. Only the 70 h sample elicited a significant response at the next 2 lower doses, while none of the samples at the lowest dose was significantly different from the pooled control values. These facts (i.e., a dose-related effect at 70 h and all samples positive at the highest dose) strongly suggest that 2AAF is a clastogen. Two reports published in 1981 also concluded that 2AAF was an *in vivo* clastogen (Tsuchimoto and Matter, 1981; Salamone *et al.*, 1981).

4 Conclusions

The *in vivo* micronucleus assay appears to discriminate between the suspect carcinogens, BP and 2AAF, and their supposed non-carcinogenic analogues, PYR and 4AAF, respectively. Both BP and 2AAF induced MPE levels significantly higher than observed for solvent and nontreated animals. BP induced positive assay responses when administered by either ip or gavage, while PYR failed to give a positive response with either route of administration. 2AAF and 4AAF were tested only by ip. Since their close structural analogues were active in the *in vivo* micronucleus assay, the lack of detectable activity for PYR and 4AAF suggests that they pose a lesser hazard in whole animal systems than BP and 2AAF. Either PYR and 4AAF are not genotoxic *in vivo*, or their potencies are so much less than those of BP and 4AAF, respectively, that they are not detectable by current assay procedures.

Table 10. *Micronucleus assay results for 2AAF in B6C3F1 mice*

Treatment mode: ip   Solvent: DMSO   Age of mice at time of treatment: 11–12 wks

Treatment and dose mg/kg	Sex	46 hour Mean ± SE PE/(NE + PE)	46 hour Mean ± SE MN/500 PE	70 hour Mean ± SE PE/(NE + PE)	70 hour Mean ± SE MN/500 PE	80 hour Mean ± SE PE/(NE + PE)	80 hour Mean ± SE MN/500 PE
2AAF							
1540 × 2	No. of mice treated = 24 (12M + 12F)				No. of survivors = 18 (12M + 6F)		
	M	0.34 ± 0.02	9.50 ± 4.56	0.29 ± 0.02	3.50 ± 1.50	0.32 ± 0.00	3.00 ± 0.41[a]
	F	0.33 ± 0.07	2.50 ± 0.96	0.43 ± 0.11	0.00 ± 0.00	Not treated	
	M + F	0.33 ± 0.03	6.00 ± 2.53[b]	0.34 ± 0.04	2.33 ± 1.20[a]		
1110 × 2	No. of mice treated = 24 (12M + 12F)				No. of survivors = 19 (12M + 7F)		
	M	0.33 ± 0.06	1.25 ± 0.63	0.33 ± 0.05	3.50 ± 1.71	0.30 ± 0.05	1.25 ± 0.63[c]
	F	0.35 ± 0.08	1.25 ± 0.48	0.39 ± 0.05	1.67 ± 1.20	Not treated	
	M + F	0.34 ± 0.05	1.25 ± 0.37[c]	0.36 ± 0.03	2.71 ± 1.08[a]		
660 × 2	No. of mice treated = 18 (9M + 9F)				No. of survivors = 18		
	M	0.31 ± 0.06	2.00 ± 1.00	0.40 ± 0.03	3.00 ± 1.53	0.40 ± 0.07	1.00 ± 0.58
	F	0.34 ± 0.05	0.67 ± 0.33	0.42 ± 0.07	0.33 ± 0.33	0.48 ± 0.08	1.00 ± 1.00
	M + F	0.33 ± 0.04	1.33 ± 0.56[c]	0.41 ± 0.03	1.67 ± 0.92[a]	0.44 ± 0.05	1.00 ± 0.52[c]
220 × 2	No. of mice treated = 18 (9M + 9F)				No. of survivors = 16 (8M + 8F)		
	M	0.34 ± 0.02	2.00 ± 1.00	0.44 ± 0.09	0.33 ± 0.33	0.31 ± 0.01	1.00 ± 1.00
	F	0.40 ± 0.05	1.00 ± 0.58	0.48 ± 0.07	0.33 ± 0.33	0.37 ± 0.04	0.00 ± 0.00
	M + F	0.37 ± 0.03	1.50 ± 0.56[c]	0.46 ± 0.05	0.33 ± 0.33[c]	0.34 ± 0.02	0.50 ± 0.29[c]
Untreated controls							
0	No. of mice = 6 (3M + 3F)				No. of survivors = 6		
	M					0.42 ± 0.06	1.00 ± 1.00
	F					0.50 ± 0.07	0.00 ± 0.00
	M + F	No sample taken		No sample taken		0.46 ± 0.04	0.50 ± 0.50
DMSO (solvent control)							
0.1 ml × 2	No. of mice treated = 6 (3M + 3F)				No. of survivors = 6		
	M			0.55 ± 0.13	0.67 ± 0.33		
	F			0.56 ± 0.05	0.33 ± 0.33		
	M + F	No sample taken		0.56 ± 0.06	0.50 ± 0.22	No sample taken	
DMBA (positive control)							
25 × 2	No. of mice treated = 14 (10M + 4F)				No. of survivors = 14		
	M	0.39 ± 0.07	4.00 ± 1.83				
	F	0.33 ± 0.03	3.25 ± 1.44				
	M + F	0.36 ± 0.04	3.63 ± 1.08	No sample taken		No sample taken	
23 × 2	M	0.39 ± 0.05	6.33 ± 3.28[b]	No sample taken		No sample taken	

[a] Significant at $P \leqslant 0.05$.
[b] Significant at $P \leqslant 0.01$.
[c] Not significant.

5 References

Heddle, J.A., E. Stuart and M.F. Salamone (1984) The bone marrow micronucleus test. In B.J. Kilbey, M. Legator, W. Nichols, and C. Ramel (Eds), *Handbook of Mutagenicity Test Procedures*, Elsevier Science Publishers BV, Amsterdam, pp. 441–457.

Heddle, J.A., M. Hite, B. Kirkhart, K. Larsen, J.T. MacGregor, G.W. Newell and M.F. Salamone (1983) The induction of micronuclei as a measure of genotoxicity, *Mutation Res.*, **123**, 61–118.

Kastenbaum, M.A., and K.O. Bowman (1970) Tables for determining the statistical significance of mutation frequencies, *Mutation Res.*, **9**, 527–549.

Kirkhart, B. (1981) Micronucleus tests on 21 compounds. In F.J. de Serres and J. Ashby (Eds), *Progress in Mutation Research*, Vol. 1. Evaluation of Short-Term Tests for Carcinogenicity, Elsevier/North-Holland, Amsterdam, pp. 698–704.

Kliesch, U., I. Roupova and I.-D. Adler (1982) Induction of chromosome damage in mouse bone marrow by benzo(a)pyrene, *Mutation Res.*, **102**, 265–273.

Salamone, M.F., J.A. Heddle and M. Katz (1981) The mutagenic activity of 41 compounds on the *in vivo* micronucleus assay. In F.J. de Serres and J. Ashby (Eds), *Progress in Mutation Research*, Vol. 1. Evaluation of Short-Term Tests for Carcinogenicity, Elsevier/North-Holland, Amsterdam, pp. 686–697.

Salamone, M.F., J.A. Heddle, E. Stuart and M. Katz (1980) Towards an improved micronucleus test: Studies on three model agents, mitomycin C, cyclophosphamide and dimethylbenzanthracene, *Mutation Res.*, **74**, 347–356.

Salamone, M.F., and J.A. Heddle (1983) The bone marrow micronucleus assay: rationale for a revised protocol. In F.J. de Serres (Ed.), *Chemical Mutagens* Vol. 8. Plenum Press, NY, pp. 111–149.

Tsuchimoto, T., and B. Matter (1981) Activity of coded compounds in the micronucleus test. In F.J. de Serres and J. Ashby (Eds), *Progress in Mutation Research*, Vol. 1. Evaluation of Short-Term Tests for Carcinogenicity, Elsevier/North-Holland, Amsterdam, pp. 705–711.

Jeremy A. Styles, Christine E. Phillips, Jennifer Shaw and David Hart
Imperial Chemical Industries PLC, Central Toxicology Laboratories, Alderley Park, Cheshire, UK

30

The clastogenic activity of benzo(a)pyrene, pyrene, 2-acetylaminofluorene and 4-acetylaminofluorene in the mouse bone marrow micronucleus test

1 Summary

The mouse bone marrow micronucleus test was used to compare the clastogenic activity of benzo(a)pyrene (BP), pyrene, 2-acetylaminofluorene (2AAF) and 4-acetylaminofluorene (4AAF) using dose levels for each chemical based on 80% of the $MLD_{50/7}$ or 5000 mg/kg whichever was the lower. The carcinogens BP and 2AAF both gave positive responses 24, 36 and 48 hours after dosing when given by oral or intraperitoneal routes of administration. The test gave the same response when 500 cells per animal were counted compared with counts of 1000 cells per animal. Additional studies showed that the clastogenic effects of BP persisted for at least 96 hours after i.p. dosing but for only 48 hours after oral administration. The effects of 2AAF did not persist beyond 48 hours for either route of administration.

Neither pyrene nor 4AAF gave positive responses at any sampling time.

2 Introduction

Benzo(a)pyrene, pyrene, N-(2-fluorenyl)acetamide (2AAF) and N-(4-fluorenyl)acetamide (4AAF) were tested for clastogenic effects in the mouse micronucleus assay using a modified method of Salamone *et al.* (1981). Cyclophosphamide was used as a positive control to assess the sensitivity of the test system.

The study was divided into two parts:
Part 1 – chemicals administered orally.
Part 2 – chemicals administered intra-peritoneally.

3 Materials and methods

Chemicals

Cyclophosphamide, was supplied by Ward Blenkinsop and Co. Ltd. London, U.K. Benzo(a)pyrene, Pyrene, 2AAF, 4AAF were supplied by Lancaster Synthesis, Morecambe, U.K. The solvent control vehicle/negative control substance was olive oil, which was obtained from Thornton and Ross, Huddersfield, U.K. The compounds were dispersed in olive oil, heated to 60 °C for 18 hours and then sonicated for 5 minutes prior to dosing.

Animals

Male and Female C57BL/6J mice (8–12 weeks old) were used for the study. These animals were supplied by the Animal Breeding Unit, Alderley Park, Macclesfield, Cheshire.

On arrival the mice were housed ten per cage on mobile mouse racks and given food (Porton Combined Diet (PCD) supplied by BP Nutrition Ltd, Stepfield, Witham, Essex, U.K. and water *ad libitum*. After acclimatisation the animals were housed five per cage.

The animal cell was maintained at a mean temperature of 22 °C, a relative humidity mean of 45%, with 22 air changes per hour. Lighting was cycled with 12 hours light and 12 hours dark each 24 hours.

Experimental design

After acclimatisation the mice were distributed on to racks and the animals were identified by cage cards.

Phase 1 Acute Toxicity Determinations: The acute toxicity of each chemical was determined by observation of the animals over a seven day period after a single intragastric dose for Part 1 of the study and a single intraperitoneal dose for Part 2 of the study. (Salamone *et al.* (1981).)

Phase 2 Dosing of Animals for the Micronucleus Test: Part 1 of study – each concentration of test chemical was given as a single intragastric dose at concentrations derived from acute toxicity determinations. High dose levels were 80% of the Median lethal dose (MLD/7) or 5000 mg/kg in the absence of toxicity. Lower doses were based on multiples of the molecular weight of the chemical in mg/kg. Due to toxic effects which occurred in the female mice receiving 3000 mg/kg 2AAF, a repeat experiment was done at 2000 mg/kg (with relevant controls).

Part 2 of study – animals were dosed with a single intraperitoneal injection using doses equivalent to the 80% and 50% of the MLD/7. The lower dose level was used to ensure a test result in the event of excessive cytotoxicity in the bone marrow in the top dose group and to allow the observation of a dose response, should the higher dose give a positive result.

Further studies were carried out to compare the effects of 2AAF and benzo(a)pyrene at 72 and 96 hours after dosing.

Processing of bone marrow

The animals were killed by cervical dislocation at 24, 36 and 48 hours after dosing. Further studies were carried out on benzo(a)pyrene and 2AAF in which animals were killed at 72 and 96 hours after dosing in addition to the earlier sampling times.

Femurs were removed and stripped clean of muscle. Both ends of the femur were removed and a fine tipped

Table 1. *Incidence of Micronuclei/1000 Polychromatic Erythrocytes Following Oral Dosing Combined Mean Values from Individual Experiments Data derived from counting 1000 PCEs per animal*

			Incidence of Micronuclei per 1000 polychromatic erythrocytes at various sampling times after dosing ± SD (no. of group means).				
Group	Compound	Dose	24 hrs	36 hrs	48 hrs	72 hrs	96 hrs
1	Control-olive oil	0.2 ml/10 g	2.6 ± 0.6 (7)	2.4 ± 0.3 (5)	3.2 ± 1.4 (7)	3.3 ± 1.4 (4)	2.8 ± 0.1 (4)
2	Cyclophosphamide	46 mg/kg	19.9 ± 4.8 (7)**	24.2 ± 1 (5)**	11.5 ± 6 (7)**	8.7 ± 2.8 (4)*	4.2 ± 1.8 (4)
3	Benzo(a)pyrene	252 mg/kg	8.8 ± 1.9 (6)**	14.6 ± 3.8 (4)**	11.1 ± 1.9 (6)**	3.5 ± 1.6 (4)	3.8 ± 1.1 (4)
4	Benzo(a)pyrene	5000 mg/kg	13.7 ± 1.2 (6)**	22.3 ± 4.4 (4)**	22.3 ± 3 (6)**	3.6 ± 1.8 (4)	2.9 ± 0.5 (4)
5	Pyrene	202 mg/kg	2.9 ± 0.7 (2)	2.1 ± 2.7 (2)	3.1 ± 0.1 (2)		
6	Pyrene	404 mg/kg	2.7 ± 0.1 (2)	3.0 ± 1.9 (2)	2.6 ± 0 (2)		
7	2-AAF	223 mg/kg	5.0 ± 1.1 (2)**	5.2 ± 1.7 (2)**	5.0 ± 1.4 (2)		
8	2-AAF	2000 mg/kg	3.6 ± 2.1 (1)	19.6 ± 10.2 (1)**	10.6 ± 5.8 (1)**		
		3000 mg/kg	7.0 ± 4.1 (1)**	17.0 ± 10.4 (1)**	10.0 ± 6.0 (1)**		
9	4-AAF	892 mg/kg	2.2 ± 0.6 (2)	2.1 ± 1.3 (2)	3.1 ± 1.5 (2)		
10	4-AAF	5000 mg/kg	2.2 ± 0.3 (2)	2.9 ± 1.2 (2)	2.0 ± 0.6 (2)		

* Statistically significantly different from the control group at the 5% level ('t'-test, one sided).
**Statistically significantly different from the control group at the 1% level ('t'-test, one sided).
Data derived from counting 1000 PCEs per animal.

Table 2. *Incidence of Micronuclei/1000 Polychromatic Erythrocytes Following Intraperitoneal Dosing Combined Mean Values from Individual Experiments Data derived from counting 1000 PCEs per animal*

			Incidence of Micronuclei per 1000 polychromatic erythrocytes at various sampling times after dosing ± SD (no. of group means).				
Group	Compound	Dose	24 hrs	36 hrs	48 hrs	72 hrs	96 hrs
1	Control-olive oil	0.2 ml/10 g	2.7 ± 1.1 (6)	2.2 ± 0.3 (4)	3.3 ± 1.7 (6)	2.6 ± 1.5 (4)	3.8 ± 0.3 (2)
2	Cyclophosphamide	46 mg/kg	23.2 ± 12.9 (6)**	20.1 ± 10.6 (4)*	10.8 ± 5 (6)**	3.3 ± 1.5 (4)	4.4 ± 0.3 (2)
3	Benzo(a)pyrene	3132 mg/kg	8.4 ± 2.5 (4)**	12.7 ± 2.1 (2)**	26.8 ± 4.6 (4)**	23.0 ± 9.0 (2)**	27.9 ± 3.8 (2)**
4	Benzo(a)pyrene	5000 mg/kg	6.7 ± 2.7 (4)**	12.4 ± 0.3 (2)**	23.8 ± 4 (4)**	24.9 ± 3.5 (2)**	33.4 ± 7.6 (2)*
5	Pyrene	340 mg/kg	2.2 ± 0.3 (2)	2.6 ± 0.3 (2)	1.7 ± 1.5 (2)		
6	Pyrene	544 mg/kg	2.8 ± 0.6 (2)	2.2 ± 0.8 (2)	2.4 ± 0.5 (2)		
7	2-AAF	312 mg/kg	10.2 ± 7.1 (2)*	15.3 ± 1.3 (2)**	4.9 ± 2.1 (2)		
8	2-AAF	500 mg/kg	8.4 ± 5.8 (6)*	10.6 ± 1.8 (4)**	12.2 ± 3.8 (6)**	2.1 ± 0.9 (4)	3.9 ± 1.6 (2)
9	4-AAF	750 mg/kg	2.0 ± 1.1 (2)	2.1 ± 0.4 (2)	2.0 ± 0.8 (2)		
10	4-AAF	1200 mg/kg	1.8 ± 0.6 (2)	2.5 ± 1.3 (2)	1.5 ± 0.3 (2)		

* Statistically significantly different from the control group at the 5% level ('t'-test, one sided).
**Statistically significantly different from the control group at the 1% level ('t'-test, one sided).
Data derived from counting 1000 PCEs per animal.

sable hair paint brush (size 100), was dipped into albumin (6% in physiological saline) and introduced into the marrow canal and four streaks of cells were applied to a labelled (coded) glass slide. The slides were allowed to air dry. Marrow suspensions were prepared by injecting 2.0 ml of foetal calf serum through the remaining femur, and collected into a labelled (coded) tube. The paint brush was rinsed in physiological saline before the next animal was sampled. The slides were stained and fixed with Polychrome Methylene Blue (from Miles Laboratory Ltd., Stoke Court, Stoke Poges, Slough, U.K.). The staining procedure was done using the Hema-Tek II Automatic Slide Stainer and Hema-Tek II stain pack. The suspensions of marrow cells in foetal calf serum were used for making smear preparations. These were fixed in methanol before staining with Giemsa stain.

Cytogenetic analysis

The slides from each animal were examined microscopically using a x100 oil immersion objective and x10 eyepieces. For each animal 500 polychromatic erythrocytes were counted and the incidence of micronuclei recorded. A further 500 cells were examined from the same animal and the incidence of micronuclei counted. The polychromatic erythrocyte: normochromatic erythrocyte ratio was also determined for each animal on a sample of 200 erythrocytes per slide. When scoring was completed the slides were decoded and the data arranged for statistical analysis.

Statistical analysis

The micronucleus data were analysed for significant differences from the control group using a one sided Student's 't' test. The incidences of polychromatic erythrocytes to normochromatic erythrocytes were compared using a two sided Student's 't' test.

4 Results and discussion

The accumulated results from all experiments are given in Table 1 (oral dosing) and Table 2 (intraperitoneal dosing). From these it can be seen that cyclophosphamide (CTX), benzo(a)pyrene (BP) and 2AAF all gave positive responses at one or more sampling times, while pyrene and 4AAF were negative at all sampling times.

Analysis of combined individual data (Table 3 oral dosing; Table 4 intra peritoneal dosing) showed that there was more variation between animals than between experiments. The results of counting 500 cells per animal

Table 3. *Incidence of micronuclei/1000 polychromatic erythrocytes following oral dosing. Mean of individual animal results from all experiments. Data derived from counting 1000 PCEs per animal*

			Incidence of Micronuclei per 1000 polychromatic erythrocytes at various sampling times after dosing ± SD (no. of animals).				
Group	Compound	Dose	24 hrs	36 hrs	48 hrs	72 hrs	96 hrs
1	Control-olive oil	0.2 ml/10 g	2.7 ± 2.1 (35)	2.4 ± 1.5 (25)	3.3 ± 2.3 (35)	3.3 ± 2.3 (18)	2.8 ± 1.9 (18)
2	Cyclophosphamide	46 mg/kg	20.6 ± 8.4 (35)**	24.7 ± 8.2 (25)**	12.3 ± 7.9 (35)**	8.8 ± 6.5 (19)**	4.6 ± 2.7 (16)
3	Benzo(a)pyrene	252 mg/kg	8.8 ± 4.9 (30)**	14.5 ± 6 (19)**	11.0 ± 7.2 (29)**	3.7 ± 2.9 (18)	3.8 ± 2.7 (18)
4	Benzo(a)pyrene	5000 mg/kg	13.7 ± 4.4 (30)**	22.3 ± 8.4 (20)**	22.3 ± 9.4 (30)**	3.6 ± 2.5 (20)	2.9 ± 1.9 (20)
5	Pyrene	202 mg/kg	2.9 ± 1.7 (10)	2.1 ± 2.7 (10)	3.1 ± 2.0 (9)		
6	Pyrene	404 mg/kg	2.7 ± 1.6 (10)	3.0 ± 3.0 (10)	2.6 ± 1.4 (10)		
7	2AAF	223 mg/kg	5.0 ± 3.4 (10)*	5.2 ± 2.1 (10)**	5.0 ± 2.5 (10)		
8	2AAF	2000 mg/kg	3.6 ± 2.1 (5)	19.6 ± 10.2 (5)**	10.6 ± 5.8 (5)**		
		3000 mg/kg	7.0 ± 4.1 (5)**	17.0 ± 10.4 (5)**	10.0 ± 6.0 (5)**		
9	4AAF	892 mg/kg	2.2 ± 1.5 (10)	2.1 ± 1.8 (10)	3.1 ± 3.2 (10)		
10	4AAF	5000 mg/kg	2.2 ± 1.7 (10)	2.9 ± 2.0 (10)	2.0 ± 1.3 (10)		

* Statistically different from the control group at the 5% level ('t'-test, one sided).
**Statistically significantly different from the control group at the 1% level ('t'-test, one sided).
Data derived from counting 1000 PCEs per animal.

Table 4. *Incidence of micronuclei/1000 polychromatic erythrocytes following intraperitoneal dosing. Mean of individual animal results from all experiments. Data derived from counting 1000 PCEs per animal*

			Incidence of Micronuclei per 1000 polychromatic erythrocytes at various sampling times after dosing ± SD (no. of animals).				
Group	Compound	Dose	24 hrs	36 hrs	48 hrs	72 hrs	96 hrs
1	Control-olive oil	0.2 ml/10 g	2.7 ± 2.3 (29)	2.0 ± 2.0 (19)	3.3 ± 2.3 (30)	2.6 ± 2.2 (20)	3.8 ± 2.0 (10)
2	Cyclophosphamide	46 mg/kg	23.2 ± 12.0 (30)**	18.3 ± 7.5 (20)**	10.8 ± 7.5 (30)**	3.3 ± 2.2 (20)	4.4 ± 3.3 (10)
3	Benzo(a)pyrene	3132 mg/kg	8.4 ± 6.7 (20)**	12.7 ± 4.8 (10)**	26.8 ± 9.9 (20)**	23 ± 15.8 (10)**	27.9 ± 9.1 (10)**
4	Benzo(a)pyrene	5000 mg/kg	8.5 ± 9.0 (20)**	12.4 ± 5.4 (10)**	23.8 ± 10.1 (20)**	25.2 ± 6.8 (9)**	33.4 ± 1.3 (10)**
5	Pyrene	340 mg/kg	2.2 ± 1.7 (10)	2.6 ± 1.8 (10)	1.7 ± 1.8 (10)		
6	Pyrene	544 mg/kg	2.8 ± 2.2 (10)	2.1 ± 2.1 (9)	2.3 ± 1.4 (9)		
7	2AAF	312 mg/kg	10.2 ± 5.8 (10)**	15.3 ± 7.3 (10)**	4.9 ± 2.8 (10)		
8	2AAF	500 mg/kg	8.4 ± 9.3 (30)**	10.8 ± 7.6 (18)**	11.7 ± 8.7 (29)**	2.2 ± 1.5 (18)	4 ± 1.9 (9)**
9	4AAF	750 mg/kg	2.0 ± 1.8 (10)	2.1 ± 0.9 (10)	2.0 ± 0.8 (10)		
10	4AAF	1200 mg/kg	1.8 ± 1.4 (10)	2.5 ± 2.1 (1)	1.6 ± 1.7 (7)		

* Statistically significantly different from the control group at the 5% level ('t'-test, one sided).
**Statistically significantly different from the control group at the 1% level ('t'-test, one sided).
Data derived from counting 1000 PCEs per animal.

(Tables 3–4) were compared with data from counting 1000 cells from the same animals (Tables 1–2) and it can be seen that in this study the results were quantitatively similar and qualitatively identical. From the control data it was calculated that in terms of experimental design, given the numbers of animals per group (10), the method was capable of detecting a 2.5 fold increase in the incidence of micronuclei ($P<0.5$) or 3 fold increase ($P<.01$) when 1000 cells/animal were counted (10,000 cells per group). If the number of cells counted per animal is reduced to 500 the level of detection is 3.5 fold ($P<.05$) or 5 fold ($P<.01$). Alternatively, to detect a doubling of MN

Table 5. *Incidence of Micronuclei/1000 Polychromatic Erythrocytes Following Oral Dosing Combined Mean Values from Individual Experiments. Data derived from counting 500 PCEs per animal*

			Incidence of Micronuclei per 1000 polychromatic erythrocytes at various sampling times after dosing ± SD (no. of group means).				
Group	Compound	Dose	24 hrs	36 hrs	48 hrs	72 hrs	96 hrs
1	Control-olive oil	0.2 ml/10 g	2.6 ± 0.8 (7)	2.2 ± 1 (5)	3.4 ± 1.2 (7)	3.3 ± 2.1 (4)	3.0 ± 0.3 (4)
2	Cyclophosphamide	46 mg/kg	21.0 ± 4.7 (7)**	23.8 ± 2.2 (5)**	12.8 ± 6.3 (7)**	8.7 ± 3.9 (4)*	4.0 ± 2.2 (4)
3	Benzo(a)pyrene	252 mg/kg	8.9 ± 3.8 (6)**	13.9 ± 5.8 (4)**	11.4 ± 5.5 (6)**	3.6 ± 1.9 (4)	3.6 ± 1.4 (4)
4	Benzo(a)pyrene	5000 mg/kg	14.6 ± 2.3 (6)**	21.6 ± 4.6 (4)**	21.6 ± 6.0 (6)**	3.8 ± 2.0 (4)	2.8 ± 0.7 (4)
5	Pyrene	202 mg/kg	2.8 ± 1.1 (2)	2.0 ± 1.7 (2)	3.6 ± 0.6 (2)		
6	Pyrene	404 mg/kg	2.4 ± 1.1 (2)	3.2 ± 2.3 (2)	2.4 ± 0 (2)		
7	2-AAF	223 mg/kg	5.6 ± 1.7 (2)**	4.4 ± 1.1 (2)*	5.6 ± 1.7 (2)		
8	2-AAF	2000 mg/kg	4.0 (1)	22.4 (1)**	10.4 (1)**		
		3000 mg/kg	7.2 (1)**	14.4 (1)**	8.8 (1)**		
9	4-AAF	892 mg/kg	2.2 ± 1.4 (2)	2.2 ± 1.4 (2)	2.8 ± 2.3 (2)		
10	4-AAF	5000 mg/kg	2.2 ± 0.8 (12)	3.0 ± 2.0 (2)	2.0 ± 0 (2)		

* Statistically significantly different from the control group at the 5% level ('t'-test, one sided).
**Statistically significantly different from the control group at the 1% level ('t'-test, one sided).

Table 6. *Incidence of Micronuclei/1000 Polychromatic Erythrocytes Following Intraperitoneal Dosing. Combined Mean Values from Individual Experiments. Data derived from counting 500 PCEs per animal*

			Incidence of Micronuclei per 1000 polychromatic erythrocytes at various sampling times after dosing ± SD (no. of group means).				
Group	Compound	Dose	24 hrs	36 hrs	48 hrs	72 hrs	96 her
1	Control-olive oil	0.2 ml/10 g	2.4 ± 1.4 (6)	2.1 ± 1.6 (4)	3.3 ± 1.8 (6)	2.5 ± 1.7 (4)	3.8 ± 0.3 (2)
2	Cyclophosphamide	46 mg/kg	23.8 ± 13.1 (6)**	19.7 ± 9.0 (4)**	10.3 ± 5.4 (6)*	3.1 ± 1.9 (4)	4.8 ± 0 (2)
3	Benzo(a)pyrene	3132 mg/kg	8.6 ± 7.0 (4)	12.2 ± 1.4 (2)**	26.7 ± 7.8 (4)**	22.8 ± 9.6 (2)**	28.0 ± 1.7 (2)**
4	Benzo(a)pyrene	5000 mg/kg	6.3 ± 3.6 (4)*	13.0 ± 0.3 (2)**	24.3 ± 5.7 (4)**	25.0 ± 3.6 (2)**	34.8 ± 7.3 (2)
5	Pyrene	340 mg/kg	2.4 ± 0 (2)	2.4 ± 0.6 (2)	1.4 ± 1.4 (2)		
6	Pyrene	544 mg/kg	2.6 ± 0.3 (2)	2.5 ± 0.7 (2)	1.8 ± 0.9 (2)		
7	2-AAF	312 mg/kg	11.6 ± 7.9 (2)*	16.2 ± 1.4 (2)**	4.6 ± 2.0 (2)		
8	2-AAF	500 mg/kg	8.5 ± 8.1 (6)	11.5 ± 7.3 (4)*	12.1 ± 5.6 (6)**	2.1 ± 0.8 (8)	3.4 ± 1.3 (2)
9	4-AAF	750 mg/kg	2.4 ± 1.1 (2)	2.2 ± 0.8 (2)	1.8 ± 0.3 (2)		
10	4-AAF	1200 mg/kg	2.0 ± 0 (2)	2.6 ± 0.8 (2)	1.4 ± 0.1 (2)		

* Statistically significantly different from the control group at the 5% level ('t'-test, one sided).
**Statistically significantly different from the control group at the 1% level ('t'-test, one sided).

Table 7. *Incidence of micronuclei/1000 polychromatic erythrocytes following oral dosing. Mean values from individual animal results from all experiments. Data derived from counting 500 PCEs per animal*

			Incidence of Micronuclei per 1000 polychromatic erythrocytes at various sampling times after dosing ± SD (no. of animals)				
Group	Compound	Dose	24 hrs	36 hrs	48 hrs	72 hrs	96 hrs
1	Control-olive oil	0.2 ml/10 g	2.6 ± 2.1 (35)	2.2 ± 1.8 (25)	3.4 ± 3 (35)	3.1 ± 2.6 (18)	3.0 ± 2.6 (18)
2	Cyclophosphamide	46 mg/kg	21.0 ± 9.4 (35)**	23.8 ± 8.6 (25)**	12.8 ± 8.1 (35)**	11.4 ± 7.6 (19)**	4.4 ± 3.1 (16)
3	Benzo(a)pyrene	252 mg/kg	8.9 ± 5.3 (30)**	13.5 ± 7.0 (19)**	11.2 ± 7.3 (29)**	3.7 ± 2.7 (18)	3.7 ± 3.2 (18)
4	Benzo(a)pyrene	5000 mg/kg	14.2 ± 5.1 (30)**	21.6 ± 9.5 (20)**	21.6 ± 9.3 (30)**	3.8 ± 2.9 (20)	2.8 ± 2.5 (20)
5	Pyrene	202 mg/kg	2.8 ± 2.5 (10)	2.0 ± 3.4 (10)	3.2 ± 1.8 (9)		
6	Pyrene	404 mg/kg	2.4 ± 1.8 (10)	3.2 ± 3.9 (10)	2.4 ± 1.6 (10)		
7	2AAF	223 mg/kg	5.6 ± 4.5 (10)**	4.4 ± 2.5 (10)**	5.6 ± 2.8 (10)**		
8	2AAF	2000 mg/kg	4.0 ± 3.2 (5)	22.4 ± 12.4 (5)**	10.4 ± 5.9 (5)**		
		3000 mg/kg	7.2 ± 5.0 (5)**	14.4 ± 7.5 (5)**	8.8 ± 6.7 (5)**		
9	4AAF	892 mg/kg	2.2 ± 2.0 (10)	2.2 ± 2.6 (10)	2.8 ± 4.0 (10)		
10	4AAF	5000 mg/kg	2.2 ± 2.4 (10)	3.0 ± 3.3 (10)	2.0 ± 1.3 (10)		

* Statistically significantly different from the control group at the 5% level ('t'-test, one sided).
**Statistically significantly different from the control group at the 1% level ('t'-test, one sided).

with 1000 cells per animal the group size would have to be 20 animals ($P < .05$) or 36 animals ($P < .01$).

The experiments incorporating additional sampling times revealed that while 2AAF, BP and CTX all gave positive results up to 48 hours, there were significant differences in the persistence of effects between chemicals. For CTX, a water soluble compound, the oral route appears to be more effective and gave more prolonged responses than i.p. dosing, possibly because absorption and transport to the marrow is more efficient. For compounds with low solubility in aqueous media, intraperitoneal administration results in more prolonged responses than the oral route. This may be due to the formation of a depot of precipitated compound in the peritoneal cavity having a long residence time whereas, when given orally, the precipitated dose will be absorbed poorly and excreted rapidly, and the time available for absorption short in comparison with i.p. dosing.

Table 8. *Incidence of micronuclei/1000 polychromatic erythrocytes following intraperitoneal dosing. Mean of individual animal results from all experiments. Data derived from counting 500 PCEs per animal*

			Incidence of Micronuclei per 1000 polychromatic erythrocytes at various sampling times after dosing ± SD (no. of animals).				
Group	Compound	Dose	24 hrs	36 hrs	48 hrs	72 hrs	96 hrs
1	Control-olive oil	0.2 ml/10 g	2.4 ± 2.5 (29)	2.2 ± 2.7 (19)	3.4 ± 2.6 (30)	2.5 ± 2.4 (20)	3.8 ± 2.4 (10)
2	Cyclophosphamide	46 mg/kg	23.8 ± 13.3 (30)**	19.7 ± 12.8 (20)**	10.3 ± 7.3 (30)**	3.1 ± 2.7 (20)	4.8 ± 4.6 (10)
3	Benzo(a)pyrene	3132 mg/kg	8.6 ± 7.3 (20)**	12.2 ± 3.6 (10)**	26.7 ± 10.1 (20)**	22.8 ± 16.3 (10)**	28.0 ± 8.3 (10)**
4	Benzo(a)pyrene	5000 mg/kg	6.3 ± 7.4 (20)**	13.0 ± 6.9 (10)**	24.3 ± 10.9 (20)**	25.3 ± 8.4 (9)**	34.8 ± 14.7 (10)**
5	Pyrene	340 mg/kg	2.4 ± 2.3 (10)	2.4 ± 2.6 (10)	1.4 ± 1.9 (10)		
6	Pyrene	544 mg/kg	2.6 ± 3.0 (10)	2.4 ± 2.6 (9)	1.8 ± 2.1 (9)		
7	2AAF	312 mg/kg	11.6 ± 6.5 (10)**	16.2 ± 8.7 (10)**	4.6 ± 3.1 (10)		
8	2AAF	500 kg/mg	8.5 ± 9.3 (30)**	10.5 ± 7.2 (18)**	12.3 ± 8.9 (29)**	2.1 ± 2.0 (18)	3.5 ± 2.2 (9)
9	4AAF	750 mg/kg	2.4 ± 2.2 (10)	2.2 ± 1.1 (10)	1.8 ± 1.1 (10)		
10	4AAF	1200 mg/kg	2.0 ± 1.3 (10)	2.8 ± 2.3 (10)	1.4 ± 1.5 (7)		

* Statistically significantly different from the control group at the 5% level ('t'-test, one sided).
**Statistically significantly different from the control group at the 1% level ('t'-test, one sided).

Table 9. *Percentage of polychromatic erythrocytes following oral dosing. Combined group mean values from all experiments*

			Percentage of Polychromatic Erythrocytes ± SD (no. of group means)				
Group	Compound	Dose	24 hrs	36 hrs	48 hrs	72 hrs	96 hrs
1	Control-olive oil	0.2 ml/10 g	27.2 ± 5.9 (7)	27.0 ± 6.0 (5)	26.4 ± 3.9 (7)	23.4 ± 3.9 (7)	27.1 ± 8.1 (4)
2	Cyclophosphamide	46 mg/kg	29.3 ± 6.6 (7)	31.8 ± 9.2 (5)	22.7 ± 4.9 (7)	21.3 ± 6.4 (4)	24.6 ± 6.8 (4)
3	Benzo(a)pyrene	252 mg/kg	24.2 ± 4.7 (6)	31.3 ± 10.5 (4)	15.2 ± 4.1 (6)**	17.0 ± 2.6 (4)	28.5 ± 7.8 (4)
4	Benzo(a)pyrene	5000 mg/kg	26.3 ± 3.7 (6)	30.7 ± 4.4 (4)	15.3 ± 4.4 (6)**	11.1 ± 0.8 (6)*	27.6 ± 8.4 (4)
5	Pyrene	202 mg/kg	30.7 ± 1.0 (2)	42.8 ± 0.3 (2)*	33.7 ± 6.3 (2)		
6	Pyrene	404 mg/kg	36.4 ± 5.9 (2)	28.1 ± 8.0 (2)	28.8 ± 5.1 (2)		
7	2AAF	223 mg/kg	34.2 ± 2.8 (2)	30.6 ± 7.1 (2)	28.1 ± 0.7 (2)		
8	2AAF	3000 mg/kg	30.0 ± 4.5 (2)	23.5 ± 3.0 (2)	25.6 ± 7.6 (2)		
		2000 mg/kg	28.0 (1)	19.0 (1)	12.8 (1)*		
9	4AAF	892 mg/kg	35.1 ± 1.3 (2)	42.6 ± 12.2 (2)	39.9 ± 18.0 (2)		
10	4AAF	5000 mg/kg	28.0 ± 13.6 (2)	49.1 ± 4.9 (2)**	40.7 ± 18.8 (2)		

* Statistically significantly different from the control group at the 5% level ('t'-test, two sided).
**Statistically significantly different from the control group at the 1% level ('t'-test, two sided).
200 erythrocytes counted per animal.

Table 10. *Percentage of polychromatic erythrocytes following intraperitoneal dosing. Combined group mean values from all experiments*

			Percentage of Polychromatic Erythrocytes ± SD (no. of group means)				
Group	Compound	Dose	24 hrs	36 hrs	48 hrs	72 hrs	96 hrs
1	Control-olive oil	0.2 ml/10 g	28.3 ± 4.6 (6)	30.4 ± 9.5 (4)	31.4 ± 8.4 (6)	27.8 ± 5.5 (4)	24.9 ± 7.2 (2)
2	Cyclophosphamide	46 mg/kg	24.5 ± 5.0 (6)	22.8 ± 4.0 (4)	22.5 ± 5.0 (6)*	20.6 ± 2.8 (4)	25.6 ± 7.3 (2)
3	Benzo(a)pyrene	3132 mg/kg	24.5 ± 4.6 (4)	27.5 ± 4.9 (2)	23.0 ± 7.1 (4)	7.8 ± 2.3 (2)**	11.3 ± 8.6 (2)
4	Benzo(a)pyrene	5000 mg/kg	22.8 ± 5.6 (4)	27.3 ± 6.4 (2)	18.5 ± 3.3 (4)**	4.3 ± 0.7 (2)**	11.5 ± 1.5 (2)
5	Pyrene	340 mg/kg	29.2 ± 4.0 (2)	25.9 ± 14.3 (2)	34.7 ± 14.0 (2)		
6	Pyrene	544 mg/kg	25.4 ± 5.9 (2)	27.0 ± 10.2 (2)	32.9 ± 9.1 (2)		
7	2AAF	312 mg/kg	23.8 ± 7.1 (2)	24.1 ± 0.4 (2)	27.5 ± 1.3 (2)		
8	2AAF	500 mg/kg	25.1 ± 5.2 (6)	26.1 ± 6.0 (μ)	24.9 ± 5.9 (6)	21.5 ± 4.3 (4)	24.5 ± 13.0 (2)
9	4AAF	750 mg/kg	30.7 ± 0.7 (2)	25.6 ± 1.1 (2)	29.6 ± 12.1 (2)		
10	4AAF	1200 mg/kg	29.7 ± 1.5 (2)	36.0 ± 13.0 (2)	30.5 ± 7.8 (2)		

* Statistically significantly different from the control group at the 5% level ('t'-test, two sided).
**Statistically significantly different from the control group at the 1% level ('t'-test, two sided).
200 erythrocytes counted per animal.

Table 11. *Summary evaluation of test data*

3 Sampling times (24, 36 and 48 hours), positive at any time. Results from counting 500 cells per animal compared with 1000 cells per animal.

Oral dosing			
Group		500 cells	1000 cells
1	Control-olive oil	–	–
2	Cyclophosphamide 46 mg/kg	+	+
3	Benzo(a)pyrene 252 mg/kg	+	+
4	Benzo(a)pyrene 5000 mg/kg	+	+
5	Pyrene 202 mg/kg	–	–
6	Pyrene 404 mg/kg	–	–
7	2AAF 223 mg/kg	+	+
8	2AAF 3000 mg/kg ♂	+ a	+ a
	2AAF 2000 mg/kg ♀	+	+
9	4AAF 892 mg/kg	–	–
10	4AAF 5000 mg/kg	–	–

a – males only females not analysed.

Intraperitoneal dosing			
Group		500 cells	1000 cells
1	Control-olive oil	–	–
2	Cyclophosphamide 46 mg/kg	+	+
3	Benzo(a)pyrene 3132 mg/kg	+	+
4	Benzo(a)pyrene 5000 mg/kg	+	+
5	Pyrene 340 mg/kg	–	–
6	Pyrene 544 mg/kg	–	–
7	2AAF 312 mg/kg	+	+
8	2AAF 500 mg/kg	+	+
9	4AAF 750 mg/kg	–	–
10	4AAF 1200 mg/kg	–	–

Table 12. *Evaluation of test data: Comparison of test results at five sampling times and two routes of dosing*

				Test result at various sampling times after dosing				
Group	Compound	Route	Dose	24 hrs	36 hrs	48 hrs	72 hrs	96 hrs
1	Control-olive oil	oral	0.2 ml/10 g	–	–	–	–	–
		i.p.	0.2 ml/10 g	–	–	–	–	–
2	Cyclophosphamide	oral	46 mg/kg	+	+	+	+	–
		i.p.	46 mg/kg	+	+	+	+	–
3	Benzo(a)pyrene	oral	252 mg/kg	+	+	+	–	–
		i.p.	3132 mg/kg	+	+	+	+	+
4	Benzo(a)pyrene	oral	5000 mg/kg	+	+	+	–	–
		i.p.	5000 mg/kg	+	+	+	+	+
5	Pyrene	oral	202 mg/kg	–	–	–		
		i.p.	340 mg/kg	–	–	–		
6	Pyrene	oral	404 mg/kg	–	–	–		
		i.p.	544 mg/kg	–	–	–		
7	2AAF	oral	223 mg/kg	+	+	–		
		i.p.	312 mg/kg	+	+	–		
8	2AAF	oral	2000 mg/kg	–	+	+		
			3000 mg/kg	+	+	+		
		i.p.	500 mg/kg	+	+	+	–	–
9	4AAF	oral	892 mg/kg	–	–	–		
		i.p.	750 mg/kg	–	–	–		
10	4AAF	oral	5000 mg/kg	–	–	–		
		i.p.	1200 mg/kg	–	–	–		

The results show that for the detection of clastogenic activity *in vivo* using the MN assay a test chemical must, in general, be given at a high dose. The compound must be subjected to a preliminary toxicity test to establish the MLD/7 and a high (80% MLD/7) and low dose (50% MLD/7) calculated for the micronucleus study. For most chemicals it will be sufficient to read only slides from the high dose group to establish whether or not there is clastogenic activity. For some exceptionally cytotoxic compounds, animals from high dose groups may not give satisfactory preparations for analysis if there has been pronounced disturbance of the cell cycle and inhibition of mitosis. In these instances slides from the lower dose group can be examined. The method is, therefore, not appropriate, in its present form, for the detection of low dose or chronic effects. As noted above, to detect a doubling of the induced incidence of micronuclei compared with controls the numbers of animals in control and test groups would have to be substantially increased and hence the number of slides to be examined increased accordingly. Detection of micronuclei following chronic exposure to a clastogen has been reported in mature circulating erythrocytes by Schlegel and McGregor (1982; 1983).

A measure of the cytotoxic effects of the chemicals is shown in the proportion of polychromatic erythrocytes present in the marrow (Tables 5 and 6). After oral administration there was no effect with CTX but both BP and 2AAF caused a reduction in polychromatic cells at 48 hours and, with BP also at 72 hours.

Both pyrene and 4AAF caused an increase in polychromatics after 36 hours. The effects on polychromatics after intraperitoneal dosing were confined to CTX (48 hours) and BP (48 and 72 hours).

In general, the only consistent relationship between cytotoxicity and clastogenicity was seen with BP, where both were increased after intraperitoneal dosing compared with oral administration. There was no significant cytotoxicity in females given 3000 mg/kg 2AAF although this dose caused non-specific toxicity and resulted in the repeat study of 2000 mg/kg.

The qualitative test responses are summarised in Tables 7 and 8 where it can be seen that at the main sampling times of 24, 48 and 36 hours there were positive responses for CTX, BP and 2AAF by either route of administration when 500 or 1000 cells per animal were counted.

5 References

Salamone, M., J. Heddle and M. Katz (1981) Mutagenic activity of 41 compounds in the *in vivo* micronucleus assay. In *Evaluation of Short-term Tests for Carcinogens* (eds F.J. de Serres and J. Ashby), pp. 686–697, Elsevier, New York.

Schlegel, R. and J.T. McGregor (1982) The persistence of micronuclei in peripheral blood erythrocytes: detection of chronic chromosome breakage in mice. *Mutation Res.*, **104**, 367–369.

Schlegel, R. and J.T. McGregor (1983) A rapid screen for cumulative chromosomal damage in mice. Accumulation of circulating micronucleated erythrocytes. *Mutation Res.*, **113**, 481–487.

Pawan S. Chauhan, Ramesh C. Chaubey
and Krishna Sundaram
Bio-Medical Group, Bhabha Atomic Research Centre, Bombay, India

31

Bone-marrow micronucleus test in Swiss male mice

1 Summary

BP, PYR, 2AAF, and 4AAF were evaluated for clastogenic potential using the bone marrow micronucleus test in Swiss male mice. Animals were exposed to the chemicals, only once, by intraperitoneal injections. Time–response studies followed by dose–response studies were carried out for each of the four chemicals. BP, at 252 mg/kg, provoked a maximum response at 24 h. In the dose–response study carried out at 24 h, no significant effect was observed at 63 and 126 mg/kg BP. At 378 mg/kg, BP significantly increased the frequency of micronucleated erythrocytes ($P < 0.01$) demonstrating the ability of the hydrocarbon to provoke chromosomal damage in the bone marrow of Swiss mice. PYR failed to induce any positive response up to 808 mg/kg. The time–response and dose–response studies demonstrated the ability of 2AAF to induce a moderate positive effect. 4AAF, on the contrary, was negative up to 446 mg/kg. Both the known carcinogens, BP and 2AAF, provoked a positive response in the bone marrow micronucleus assay, while PYR and 4AAF, considered to be noncarcinogenic, were clearly negative.

2 Introduction

The mouse bone marrow micronucleus test (MNT) was developed and introduced in chemical mutagenesis by Schmid (1975) and Heddle (1973). The test has been accepted as an indicator of chromosomal damage *in vivo* with about the same level of sensitivity as that of bone marrow metaphase analysis (Miller, 1973; Kliesch *et al.*, 1982). In our laboratory, the test was introduced by Dr. Schmid in 1974 during a training course on mutagenicity evaluation held at the Research Centre. Studies on mutagenicity evaluation of a number of chemical and physical agents have since been completed in Swiss mice in this laboratory (Chaubey *et al.*, 1978a, b; Chauhan, 1982; Chauhan *et al.*, 1983). The objective of the present investigations was to evaluate the clasogenicity of BP, PYR, 2AAF, and 4AAF in mice by the bone marrow micronucleus assay.

3 Materials and methods

Animals

Adult inbred Swiss male mice (11–13 weeks) bred in the animal house of this Research Centre were used. Animals were maintained under controlled conditions of temperature, humidity, and lighting and had free access to food and water ad libitum. Mice were randomised into various groups and housed individually in aluminium alloy cages with rice husk bedding.

Chemicals
All 4 chemicals (BP, PYR, 2AAF, and 4AAF) were supplied by ICI (UK). Corn oil was purchased from local sources as was dimethylsulphoxide (DMSO, J.T. Baker & Co., USA).

Preparation of solutions
BP and PYR were dissolved in corn oil. The chemicals were dispersed in oil, using a magnetic stirrer and mild warming in water bath, without any difficulty. 2AAF and 4AAF were first dissolved in pure DMSO and diluted with corn oil to produce a 40% DMSO to 60% corn oil solution. The concentrated solutions were kept warm at about 40 °C and needed occasional stirring.

Treatment
Animals were treated intraperitoneally only once. Solutions were prepared immediately before use. The solvent control group received the respective solvent at a dose of 0.1 ml per 10 mg body weight.

No positive control groups were employed with these experiments. The sensitivity of the test strain to radiation and chemical mutagens is well established (Chaubey *et al.*, 1978a, b; Chauhan *et al.*, 1983; Singh *et al.* 1983).

Procedure

Slide preparation Animals were sacrificed by cervical dislocation, both femora were removed and cleaned. Bone marrow was aspirated slowly into foetal calf serum with a 25 g needle and centrifuged at 1000 rpm for 5 min. The supernatant was removed and the sediment thoroughly mixed with a Pasteur capillary pipette. Slides were smeared and allowed to air dry; two slides were prepared from each animal.

Staining Air dried slides were first stained in undiluted May-Gruenwald solution for 3 min and then in diluted May-Gruenwald for 2 min. They were rinsed twice in distilled water, stained with diluted Giemsa solution (1 part Giemsa, 6 parts distilled water) for 10 min, rinsed well in distilled water, cleared in xylene for 5 min, and embedded in Euparal (Chaubey *et al.*, 1982; Chauhan, 1982).

Scoring Coded slides were screened at 1000 times magnification in oil under bright field with green filter for the presence of polychromatic (PE) and normochromatic erythocytes (NE) with or without micronuclei. For each animal, 2000 PEs and corresponding NEs were scored in the time–response study. In the case of the dose–response study, 1000 PEs were scored for each animal.

Statistical evaluation
Data were analysed using a computer programme developed at this Research Centre by B. Singh. Comparisons were made by one tail chi-square test with Yates correction as well as by Fischer's exact test of normal approximation. Differences were considered statistically significant at 95% confidence limits.

4 Results

The data are summarised in Table 1–4. Pooled untreated controls and pooled solvent controls have been included in the tables for comparison with the concurrent controls and the test groups.

BP
In the time–response study there was a significant increase ($P < 0.05$) in the incidence of micronucleated PEs at all the three expression times, but the maximum response (0.74%) was obtained when bone marrow was sampled at 24 h after treatment with BP.

A dose–response study for BP was carried out at 24 h. The solvent control group exhibited a higher response (0.24%) as compared with the untreated control group (0.11%). The mean frequency of micronucleated PE among groups of mice treated with 63 and 126 mg/kg (0.17%) was not significantly higher than the concurrent or pooled solvent control or untreated control groups. However, BP treatment at a dose of 252 mg/kg resulted in an increased incidence of micronucleated PE (0.41%). The PE/PE $\pm$ NE ratio was not drastically affected by BP treatments up to 252 mg/kg. At 378 mg/kg, a marked increase in the frequency of micronucleated PE was observed ($P < 0.01$).

PY
The time–response study with PYR was carried out at 606 mg/kg (3.0 m mol/kg) body weight (Table 2). The mean frequency of micronucleated PE observed in the 24 and 48 h groups was comparable with that of the concurrent and pooled solvent and untreated groups. However, at 36 h, the incidence of micronucleated PE (0.27%) in the PYR treated group was higher ($P < 0.05$) than that of the concurrent solvent control but not that of the concurrent untreated group. The solvent group showed a marginally low frequency as compared with the pooled solvent control groups.

The dose–response study for PYR was carried out at 36 h. There was mortality among mice treated at higher dose levels; one and two mice died in the 606 and 808 mg/kg dose groups, respectively. There were no differences in the mean frequency of micronucleated PE among any of the treatment or control groups (Table 2). PYR failed to induce micronuclei up to 808 mg/kg (4.0 m mol/kg).

2AAF
The time–response study with 2AAF was carried out at 223 mg/kg (1.0 m mol/kg) level. Data shown in Table 3 indicate that at 36 h, the mean frequency of micronucleated PE (0.28%) was significantly higher ($P < 0.01$) than the concurrent or pooled untreated and solvent controls. The frequency at 24 and 48 h expression times were quite low as compared with the frequency observed at 36 h.

The dose–response study for 2AAF was carried out at 36 h using various doses as shown in Table 3. The statistical evaluation showed significant increase in the frequency of micronucleated PE among all the treatment groups as compared with that of the control groups. The dose–response relationship was, not unambiguous, as 111.5 and 223.0 mg/kg exhibited similar response. The

Table 1. *Micronucleus studies with BP in Swiss male mice*

Group (time) (mg/kg)	Polychromatic erythrocytes (PE)	Micronucleated PE		Normochromatic erythrocytes (NE)	Micronucleated NE		PE/(PE + NE)
		No.	% ± SE		No.	% ± SE	
Time–response study: dose 252.0 mg/kg							
Untreated control	6369	6	0.09 ± 0.04	7784	4	0.05 ± 0.02	0.45
Solvent control (48 h)	6014	5	0.08 ± 0.03	8014	5	0.06 ± 0.02	0.43
(24 h)	6034	45	0.74[a] ± 0.05	8386	7	0.08 ± 0.02	0.42
(36 h)	6034	33	0.54[a] ± 0.10	8304	4	0.04 ± 0.03	0.42
(48 h)	6012	16	0.26[a] ± 0.12	7994	7	0.08 ± 0.02	0.43
Dose–response study: expression time 24 h							
Untreated control	5041	6	0.11 ± 0.05	6044	3	0.04 ± 0.01	0.45
Solvent control (24 h)	4535	11	0.24 ± 0.07	5743	9	0.15 ± 0.04	0.44
63.0	5032	9	0.17 ± 0.02	6658	3	0.04 ± 0.01	0.43
126.0	4542	8	0.17 ± 0.07	5968	0	0.00 ± 0.00	0.43
252.0	4543	19	0.41[b] ± 0.14	5910	1	0.01 ± 0.01	0.43
378.0	5041	38	0.75[a] ± 0.11	7474	8	0.11 ± 0.03	0.40
Pooled untreated control	44834	57	0.12 ± 0.01	54384	30	0.05 ± 0.01	0.45
Pooled solvent control	21606	25	0.11 ± 0.02	27798	21	0.07 ± 0.01	0.44

[a] Significantly different from concurrent controls.
[b] Significantly different from pooled controls.

Table 2. *Micronucleus studies with PYR in Swiss male mice*

Group (time) mg/kg	Polychromatic erythrocytes (PE)	Micronucleated PE		Normochromatic erythrocytes (NE)	Micronucleated NE		PE/(PE + NE)
		No.	% ± SE		No.	% ± SE	
Time–response study: dose 606.0 mg/kg							
Untreated control	6225	11	0.17 ± 0.02	7977	5	0.06 ± 0.02	0.44
Solvent control (48 h)	6014	5	0.08 ± 0.03	8014	5	0.06 ± 0.02	0.43
(24 h)	6015	5	0.08 ± 0.06	8131	4	0.05 ± 0.01	0.43
(36 h)	6026	16	0.26[a] ± 0.04	8075	2	0.02 ± 0.01	0.43
(48 h)	6031	9	0.14 ± 0.05	8304	3	0.03 ± 0.00	0.42
Dose–response study: expression time 36 h							
Untreated control	5051	7	0.13 ± 0.04	6138	5	0.08 ± 0.05	0.45
Solvent control (36 h)	5044	4	0.07 ± 0.03	6027	2	0.02 ± 0.01	0.46
202.0	5034	7	0.13 ± 0.04	6402	3	0.04 ± 0.01	0.44
404.0	4531	5	0.11 ± 0.05	5617	1	0.01 ± 0.01	0.45
606.0	4021	2	0.04 ± 0.02	5437	0	0.00 ± 0.00	0.43
808.0	3023	1	0.03 ± 0.03	3898	0	0.00 ± 0.00	0.44
Pooled untreated control	44834	57	0.12 ± 0.01	54384	30	0.05 ± 0.01	0.45
Pooled solvent control	21606	25	0.11 ± 0.02	27798	21	0.07 ± 0.01	0.44

[a] Significantly different from concurrent solvent control.

Table 3. *Micronucleus studies with 2AAF in Swiss male mice*

Group (time) (mg/kg)	Polychromatic erythrocytes (PE)	Micronucleated PE No.	Micronucleated PE % ± SE	Normochromatic erythrocytes (NE)	Micronucleated NE No.	Micronucleated NE % ± SE	PE / PE + NE
Time–response study: dose 223.0 mg/kg							
Untreated control	6028	4	0.06 ± 0.01	7286	5	0.06 ± 0.03	0.45
Solvent control (48 h)	6031	5	0.08 ± 0.04	7968	0	0.00 ± 0.00	0.43
(24 h)	6033	7	0.11 ± 0.09	7862	1	0.01 ± 0.01	0.43
(36 h)	6033	17	0.28[a] ± 0.12	9473	4	0.04 ± 0.02	0.39
(48 h)	6028	6	0.09 ± 0.05	8274	2	0.02 ± 0.02	0.42
Dose–response study: expression time 36 h							
Untreated control	5047	6	0.11 ± 0.02	5994	2	0.03 ± 0.01	0.46
Solvent control (36 h)	5053	5	0.09 ± 0.04	5983	1	0.01 ± 0.01	0.46
55.8	5024	13	0.25[a] ± 0.08	7741	3	0.03 ± 0.04	0.39
111.5	5039	20	0.39[a] ± 0.07	6917	5	0.07 ± 0.03	0.42
223.0	5030	19	0.37[a] ± 0.09	7611	5	0.06 ± 0.02	0.40
334.5	5042	24	0.47[a] − 0.14	6956	4	0.05 ± 0.03	0.42
Pooled untreated control	44834	57	0.12 ± 0.01	54384	30	0.05 ± 0.01	0.45
Pooled solvent control	21191	24	0.11 ± 0.02	26588	8	0.03 ± 0.01	0.44

[a] Significantly different from controls.

Table 4. *Micronucleus studies with 4AAF in Swiss male mice*

Group (time) mg/kg	Polychromatic erythrocytes (PE)	Micronucleated PE No.	Micronucleated PE % ± SE	Normochromatic erythrocytes (NE)	Micronucleated NE No.	Micronucleated NE % ± SE	PE / PE + NE
Time–response study: dose 446.0 mg/kg							
Untreated control	6032	9	0.14 ± 0.02	7118	2	0.02 ± 0.01	0.46
Solvent control (48 h)	5053	7	0.13 ± 0.04	5974	1	0.01 ± 0.01	0.46
(24 h)	10048	24	0.23 ± 0.05	14107	16	0.11 ± 0.02	0.42
(36 h)	4022	9	0.22 ± 0.02	6180	6	0.09 ± 0.03	0.39
(48 h)	8033	13	0.16 ± 0.03	11325	5	0.04 ± 0.01	0.41
Dose–response study: expression time 24 h							
Untreated control	5041	8	0.15 ± 0.04	6043	4	0.06 ± 0.03	0.45
Solvent control (24 h)	5054	7	0.13 ± 0.04	6663	6	0.09 ± 0.04	0.43
223.0	5030	7	0.13 ± 0.02	7691	7	0.09 ± 0.02	0.40
334.5	5039	8	0.15 ± 0.04	7827	5	0.06 ± 0.03	0.39
446.0	5044	8	0.15 ± 0.06	7158	6	0.08 ± 0.02	0.41
Pooled untreated control	44834	57	0.12 ± 0.01	54384	30	0.05 ± 0.01	0.45
Pooled solvent control	21191	24	0.11 ± 0.02	26588	8	0.03 ± 0.01	0.44

frequency increased at the 334.5 mg/kg dose group (0.41%). Doses higher than 334.5 mg/kg could not be tested, as animals looked morbid at this dose.

4AAF

Time–response study for 4AAF was carried out at 446 mg/kg. The frequency of micronucleated PE shown in Table 4 was 0.23, 0.22, and 0.16% at different expression times and was not statistically different from controls. The dose–response study conducted at 24 h failed to demonstrate any increase in the incidence of micronucleated PE. The PE/PE ± NE ratio was markedly affected, suggesting a depression of the proliferative activity of the bone marrow.

5 Discussion

BP was tested at the doses suggested by the coordination committee. At 63 and 126 mg/kg levels, there was no significant increase in the frequency of micronucleated PE. At 252 mg/kg BP, an increase in the frequency of the micronucleated PE during the time–response and dose–response studies was observed. The effect was pronounced and significantly different ($P < 0.01$) from the concurrent solvent and pooled solvent control groups in the time–response study. However, during the dose–response study though, there was a clear-cut increase in the frequency of micronucleated PE (0.41%) at 252 mg/kg dose, and the effect was less pronounced as compared with the time–response study (0.74%). At the same time, the concurrent solvent control group exhibited the highest frequency of micronucleated PE (0.24%) (Table 1) among all the concurrent solvent and untreated control groups. This resulted in the loss of statistical significance of BP-induced micronucleated PE at the 252 mg/kg dose when compared with the solvent control group during the dose–response study. The frequency was higher and significantly different from that of all other control groups. At 378 mg/kg of BP, a significantly high increase was observed in the frequency of micronucleated PE. Thus, BP proved to be clastogenic in Swiss male mice. PYR, tested between 202 and 808 mg/kg, failed to induce any significant response as-compared with the control groups during the dose–response study. The only significance observed at 36 h, at 606 mg/kg, during the time–response study could neither be reproduced during the dose–response study nor did a higher dose of 808 mg/kg induce a significant increase in the frequency of micronucleated PE. PYR exhibited lack of potential to induce an increase in the micronucleated PE in Swiss males. In the case of 2AAF, a significant increase in the frequency of micronucleated PE was observed at 36 h. In addition to the recommended doses between 55.8 and 223 mg/kg, an extra group exposed to 334.5 mg/kg 2AAF was also used. 2AAF provoked a significant increase in the incidence of micronucleated PE at all dose groups. Though the response was not different at 111.5 and 223 mg/kg and the dose–response relationship was not unambigously linear, 2AAF did induce a positive response in this study. The structural analogue, 4AAF, could not be tested at the highest recommended level of 892 mg/kg due to toxicity in Swiss males. The highest dose of 446 mg/kg that could be evaluated failed to induce any change in the frequency of micronucleated PE in the dose–response study. 4AAF was found noneffective up to 446 mg/kg in the present study.

A few micronucleus studies on BP have been published recently in mice. Raj and Katz (1983) reported a frequency of about 2.32% in B6C3F$_1$ females injected twice with 150 mg/kg BP in DMSO at 72 h expression time. Kliesch *et al.* (1982) dissolved BP in corn oil and treated IC3F1 males by oral intubation. These investigators reported BP to be highly clastogenic, exhibiting a linear response up to 250 mg/kg with a frequency of 4.80%. Unlike the other study (Raj and Katz, 1983), the maximum response was observed at 36 h posttreatment. No clear dose–response was observed in 3 studies (Kirkhart, 1981; Salamone *et al.*, 1981; Tsuchimoto and Matter, 1981) when BP was given twice 24 h apart at various dose levels and bone marrow was processed at 6, 24, or 36 h after the second treatment. In the present study, BP did not exhibit a potent clastogenic effect up to 126 mg/kg, though a significant increase in micronucleated PE was observed at relatively higher doses. In a separate study, 50 mg/kg BP given twice a week for 5 or 7 weeks increased the micronucleated NEs in the peripheral blood of Swiss mice, the frequency being 0.70 to 0.90% (Sonalkar *et al.*, 1985).

Corn oil has recently been reported to reduce BP-induced micronuclei in the bone marrow of mice by Raj and Katz (1983). However, Kliesch *et al.* (1982) had used corn oil as solvent for BP and found the carcinogen to be highly clastogenic. In our laboratory, BP dissolved in corn oil induced dominant lethal mutations in Swiss mice (Chauhan *et al.*, these volumes, 1987). In a parallel study, 378 mg/kg BP dissolved in olive oil was evaluated for a time–response study, and 0.90% micronucleated PE were observed at 36 h. The role of corn oil to suppress the mutagenicity of BP in the present micronucleus study need to be ascertained.

All 4 test chemicals require metabolic activation for carcinogenic activity. The levels of polycyclic aromatic hydroxylases are known to vary profoundly among different strains of mice and are genetically determined (Nebert *et al.*, 1972). The detoxifying mechanisms which render the active metabolites inactive may also vary among different strains. It is not unlikely that *Ah* responsive and nonresponsive strains may exhibit large variability in their response to BP. In our laboratory, EMS mutagenicity has been shown to vary among different strains of mice (Chauhan *et al.*, 1983). Such factors may also contribute to variability of response observed among different laboratories in the IPCs study.

6 Conclusions

Micronucleus studies published during the last 12 years from this laboratory have demonstrated that the inbred Swiss male mice used in the present investigations respond adequately to known physical and chemical mutagens e.g., x-rays, gamma rays, EMS, MMS, hycanthone, and lucanthone (Chaubey *et al.*, 1978a, b; Chauhan *et al.*, 1983; Singh *et al.*, 1983).

BP at 252 mg/kg (1 m mol/kg) and above induced a significant increase in the micronucleated PE, demonstrating the clastogenicity of the hydrocarbon in Swiss males. PYR failed to elicit a positive response even up to 808 mg/kg (4.0 m mol/kg).

2AAF induced a positive response in the time–response and in the dose–response studies in the bone marrow micronucleus assay, though the dose–effect relationship was not unambiguous. 4AAF tested up to 446 mg/kg (2.0 m mol/kg) was clearly negative in both the time–response and the dose–response studies. The bone marrow micronucleus assay, provided evidence of clastogenicity for BP and 2AAF and did not respond to PYR and 4AAF.

In view of the different nature of dose–response relationship, it is difficult to make relative assessments of the potency of 2AAF and BP. At 1.50 m mol/kg, BP provoked a more pronounced response.

7 Acknowledgement

The authors wish to thank Shri Balvinder Singh for the computer programme and statistical evaluation of the data. Thanks are also due to Shri H.N. Bhilwade, Shri M.R. Morwekar and Shri B.L. Kulkarni for technical assistance and Mrs. Narayanan, Shri A.T. Nair and Frau Janod for typing this manuscript.

8 References

Chaubey, R.C., P.S. Chauhan and K. Sundaram (1982) The micronucleus test in mammals. In *Book of Articles, Evaluation of Mutagenic and Carcinogenic Potential of Environmental Agents*, EMSI, BARC, Bombay, pp. 128–132.

Chaubey, R.C., K.P. George and K. Sundaram (1978a) X-ray induced micronuclei in the bone marrow erythrocytes of mice. *Int. J. Radiat. Biol.*, **33**, 507–510.

Chaubey, R.C., B.R. Kavi, P.S. Chauhan and K. Sundaram (1978b) The effect of hycanthone and maleic hydrazide on the frequency of micronuclei in the bone marrow erythrocytes of mice. *Mutation Res.*, **57**, 187–191.

Chauhan, P.S. (1982) Whole animal systems in mutagenicity evaluation. I. Chromosomal mutation tests. In *Book of Articles, Evaluation of Mutagenic and Carcinogenic Potential of Environmental Agents*, EMSI, BARC, Bombay, pp. 40–51.

Chauhan, P.S., M. Aravindakshan, R.C. Chaubey and K. Sundaram (1983) Role of rodent dominant lethal mutation test and micronucleus assay in the mutagenicity evaluation of environmental chemicals. In Proceedings of an Indo-FRG Seminar on '*Methodological Approaches for Estimation of Low Dose Effects of Radiations and Chemicals in the Environment*,' BARC, Bombay, pp. 186–203.

Chauhan, P.S., M. Aravindakshan and K. Sundaram (1987) *Dominant lethal mutation test in Swiss male mice...* (this volume).

Heddle, J.A. (1973) A rapid *in vivo* test for chromosomal damage, *Mutation Res.*, **18**, 187–190.

Kirkhart, B. (1981) Micronucleus test on 21 compounds. In F.J. de Serres and J. Ashby (Eds), *Evaluation of Short-Term Tests for Carcinogens*, Elsevier/North Holland, NY, pp. 698–704.

Kliesch, U., I. Roupova and I.D. Adler (1982) Induction of chromosome damage in mouse bone marrow by benzo(a)pyrene, *Mutation Res.*, **102**, 265–273.

Miller, R.C. (1973) The micronucleus test as an *in vivo* cytogenetic method. *Environ. Health Perspect.*, **6**, 167–170.

Nebert, D.W., F.M. Goujon and J.E. Gielen (1972) Aryl hydrocarbon hydroxylase induction by polycyclic hydrocarbons, *Nature* (London), New Biology, 236, 107.

Raj, A.S. and M. Katz (1983) Inhibitory effect of 7, 8 benzoflavone on DMBA and BaP-induced bone marrow micronuclei in mouse. *Mutation Res.*, **110**, 337–342.

Salamone, M.F., J.A. Heddle and M. Katz (1981) Mutagenic activity of 41 compounds in the *in vivo* micronucleus assay. In F.J. de Serres and J. Ashby (Eds), *Evaluation of Short-Term Tests for Carcinogens*, Elsevier/North Holland, NY, pp. 686–697.

Schmid, W. (1975) The micronucleus test, *Mutation Res.*, **31**, 9–15.

Singh, B., R.C. Chaubey, M. Kumar, P.S. Chauhan and K. Sundaram (1983). Evaluation of dose response relationships. Proceedings of an Indo-FRG Seminar on '*Methodological Approaches for Estimation of Low Dose Effects of Radiations and Chemicals in the Environment*', BARC, Bombay, pp. 315–341.

Sonalkar, B., R.C. Chaubey, P.S. Chauhan and K. Sundaram (1985) *Peripheral micronucleated erythrocytes as a simple and rapid measure for chromosomal damage during chronic exposure*, Tenth Annual Conference of EMSI, BARC, Bombay, Fb. 18–21, Book of Extended Summaries and Abstracts, pp. 197–198.

Tsuchimoto, T. and B.E. Matter (1981) Activity of coded compounds in the micronucleus test. In F.J. de Serres and J. Ashby (Eds), *Evaluation of Short Term Tests for Carcinogenicity*, Elsevier/North Holland, NY, pp. 705–711.

Kurt E. Suter
Sandoz Ltd., Preclinical Research, Toxicology Department, Basle, Switzerland

32

Micronucleus incidences induced in bone marrow erythrocytes of pregnant NMRI mice and in liver-derived erythrocytes of their foetuses

1 Summary

Female NMRI mice were treated with the mutagens 2AAF or BP or with the presumed nonmutagens 4AAF or PYR on day 15 of pregnancy. Micronuclei were assessed in bone marrow erythrocytes of the pregnant females as well as in liver-derived blood erythrocytes of their foetuses at 4 time points after treatment. 2AAF and BP were clearly mutagenic in bone marrow erythrocytes as well as in blood erythrocytes. Ambivalent results were obtained for 4AAF and PYR. At the high dose level, there was a tendency for increased micronucleus incidences in maternal bone marrow erythrocytes (4AAF) as well as in foetal blood erythrocytes (4AAF, PYR). However, only one or 2 increases per compound and erythrocyte type were statistically significant. Furthermore, the increases were small and similar to those at the low dose level, where they were not significant. They were, therefore, considered to be of incidental nature. The differences in the results found for the conventional assay and the foetal blood micronucleus assay were of quantitative rather than qualitative nature.

2 Introduction

Although micronuclei can be assessed in any mitotically active tissue, young bone marrow erythrocytes of mice are used for the conventional micronucleus test for practical reasons (von Ledebur and Schmid, 1973; Racine and Matter, 1984). Until recently, it was thought that it is not possible to detect micronuclei in sufficient quantities in blood erythrocytes, because erythrocytes containing these structures would be eliminated in the spleen shortly after entering circulation (von Ledebur and Schmid, 1973). However, it has now been demonstrated that micronuclei can, in fact, be detected in the blood of foetal mice (Cole *et al.*, 1979, King and Wild, 1979, Stoyel and Clark, 1980) and foetal rats (Suter and Kubli, 1982) as well as in the blood of adult mice (MacGregor, 1980). Apparently, the postulated elimination mechanism is not always operative.

The evaluation of micronuclei in foetuses is of particular interest for the following reason. If the treatment period is chosen accordingly, treatment occurs during the phase where the haematopoietic system, as usual for mammals, is temporarily located in the liver. Although the metabolic activity of the fetal liver is low, short-lived mutagenic metabolites may reach the erythrocytes, which are located close to the hepatocytes. Thus, the foetal blood micronucleus assay may be more sensitive than the conventional micronucleus assay for certain indirectly acting mutagens (Cole *et al.*, 1981, 1982; King and Wild, 1979; Stoyel and Clark, 1980).

In the present study, 2 mutagens requiring metabolic activation, 2AAF and BP, and 2 presumed nonmutagens, 4AAF and PYR, were compared for their ability to induce micronuclei in bone marrow erythrocytes of pregnant NMRI mice and in liver derived blood erythrocytes of their foetuses. Procarbazine (PCZ), also an indirect mutagen, served as positive control; the negative control was olive oil.

3 Materials and methods

Pregnant NMRI mice (Ivanovas), 8–12 weeks old, were treated once with one of 4 test compounds: PYR, BP, 4AAF, or 2AAF. PCZ and olive oil served as positive and negative controls, respectively. PYR, BP, 2AAF, and 4AAF were dissolved in olive oil and PCZ in 0.9% saline. Treatment was by gavage on day 15 of pregnancy (day of vaginal plug designated as day 0 of pregnancy). Groups of 5 to 7 animals were killed 24, 30, 48, and 72 h after treatment. For each time point a separate experiment was performed in which the 4 test compounds were investigated at one dose level together with the positive and negative controls.

The high dose experiments were performed at all 4 time points. The doses were 1 m mol/kg (252 mg/kg) for BP, 4 m mol/kg (892 mg/kg) for 4AAF, and 1 m mol/kg (223 mg/kg) for 2AAF. In the case of PYR the first experiment was performed using a dose of 4 m mol/kg (808 mg/kg). As this dose was lethal for 53% of the foetuses, it was reduced to 2 m mol/kg in the following experiments. The low dose experiment was performed only at the 30 h time point, which proved to be optimal for the purpose of this study. The doses were 0.08 m mol/kg (20 mg/kg) for BP, 0.10 m mol/kg (20 mg/kg) for PYR, and 0.23 m mol/kg (50 mg/kg) for 4AAF and 2AAF. PCZ was administered in every experiment at a dose of 0.05 m mol/kg (12.9 mg/kg).

Micronuclei were assessed in bone marrow erythrocytes of each female and in blood erythrocytes of one of her foetuses. Per animal, two bone marrow or blood smears were prepared and stained according to conventional methods (Matter, 1976). 500 polychromatic erythrocytes per slide were evaluated for the presence of micronuclei. All assessments were performed using coded slides.

Statistical evaluation of the data was performed with the Kastenbaum test (Kastenbaum and Bowman, 1970). Values of treatment groups were compared with their concurrent control values as well as with the composite control value calculated from all vehicle controls for foetal blood erythrocytes and maternal bone marrow erythrocytes.

4 Results

Comparison of effects induced by the various chemicals

In the high dose experiments, BP and 2AAF clearly increased the incidences of micronuclei in polychromatic erythrocytes derived from bone marrow of pregnant NMRI mice (Table 1) as well as from the blood of their foetuses (Table 2). In comparison with the negative controls, the incidences were distinctly increased. Statistically significant differences were recorded at 2 or more time intervals after treatment. In bone marrow erythrocytes, the highest incidences were found for both test chemicals

Table 1. *Incidences of micronuclei in bone marrow erythrocytes of pregnant NMRI mice*

Compound	24 h Dose (m mol/kg)	MN[a] (%)	30 h Dose (m mol/kg)	MN (%)	30 h Dose (m mol/kg)	MN (%)	48 h Dose (m mol/kg)	MN (%)	72 h Dose (m mol/kg)	MN (%)
2AAF	1	0.43[b]	1	0.75[b]	0.23	0.18	1	0.32[b]	1	0.20
4AAF	4	0.28	4	0.21	0.23	0.06	4	0.30	4	0.30[b]
BP	1	0.95[b]	1	1.96[b]	0.08	0.34[b]	1	1.18[b]	1	0.48[b]
PYR	4	0.14	2	0.12	0.10	0.22	2	0.10	2	0.22
Olive oil	0	0.10	0	0.17	0	0.13	0	0.10	0	0.10
PCZ	0.05	1.56[b]	0.05	2.02[b]	0.05	1.90[b]	0.05	0.26	0.05	0.10

[a] MN, Micronuclei.
[b] $P \leqslant 0.05$ (Kastenbaum test).

Table 2. *Incidences of micronuclei in liver-derived erythrocytes of foetuses*

Compound	24 h Dose (m mol/kg)	MN[a] (%)	30 h Dose (m mol/kg)	MN (%)	30 h Dose (m mol/kg)	MN (%)	48 h Dose (m mol/kg)	MN (%)	72 h Dose (m mol/kg)	MN (%)
2AAF	1	1.05[b]	1	1.00[b]	0.23	0.72	1	0.58	1	0.58
4AAF	4	0.73[b]	4	0.81	0.23	0.76	4	0.76[b]	4	0.46
BP	1	8.98[b]	1	10.7[b]	0.08	1.12[b]	1	2.78[b]	1	0.97[b]
PYR	4	0.50	2	0.77	0.10	0.74	2	0.83[b]	2	0.42
Olive oil	0	0.44	0	0.57	0	0.55	0	0.42	0	0.52
PCZ	0.05	1.66[b]	0.05	2.83[b]	0.05	2.14[b]	0.05	0.68	0.05	0.42

[a] MN, Micronuclei.
[b] $P \leqslant 0.05$ (Kastenbaum test).

as well as for PCZ at the 30 h time point. The same held true for foetal blood erythrocytes, with the exception of 2AAF, where the incidences at 24 and 30 h were comparable.

In the low dose experiment, BP was still mutagenic, whereas 2AAF caused a negative response.

In contrast to BP and 2AAF, the mutagenic responses to PYR and 4AAF were less clear. In the high dose experiments, PYR showed no mutagenic activity in bone marrow erythrocytes; incidences of micronuclei were, however, marginally elevated in blood erythrocytes of the foetuses 30 and 48 h after treatment. At this latter time point, the difference to the controls was significant. 4AAF-incidences for both bone marrow erythrocytes and foetal blood erythrocytes were slightly but consistently increased. In comparison with the controls, the difference was statistically significant at one time point (72 h) for bone marrow erythrocytes and at 2 time points (24 and 48 h) for foetal blood erythrocytes.

In the low dose experiment, micronucleus incidences for PYR and 4AAF were in the same range as those in the high dose experiment after 30 h. The only exception was the incidence observed in bone marrow erythrocytes for 4AAF, which was lower. In contrast to the high dose experiments, there was no statistically significant difference.

Comparison of effects induced in bone marrow erythrocytes and liver derived fetal erythrocytes.

The micronucleus incidences found in the negative controls were up to 5 times higher in liver derived foetal erythrocytes than in the bone marrow erythrocytes of their mothers.

The micronucleus incidences in BP and 2AAF treated animals were up to 9 and 4 times higher in foetal blood erythrocytes than in maternal bone marrow erythrocytes. The micronucleus incidences for PCZ were comparable in both types of erythrocytes.

5 Discussion and conclusions

From the results of this study, it can be concluded that BP and 2AAF clearly increased incidences of micronuclei in bone marrow erythrocytes of pregnant mice as well as in liver derived blood erythrocytes of their foetuses. PYR marginally increased incidences in foetal erythrocytes as did 4AAF in maternal bone marrow erythrocytes and fetal erythrocytes. However, for these latter two compounds, the increases were small. Furthermore, similar increases were also found in the low dose experiment, except for bone marrow erythrocytes exposed to 4AAF. Considering their extent and the apparent lack of a dose-effect relationship, the increases were regarded as of incidental nature.

In the case of BP, and to a lesser extent also of 2AAF, micronucleus incidences were higher in foetal blood erythrocytes than in maternal bone marrow erythrocytes. It is not clear from the results of this study whether this effect was due to special kinetic or metabolic features of the fetal/placental unit or to a generally greater tendency toward micronucleus formation in foetal erythrocytes. In the negative controls, the micronucleus incidences in foetal blood erythrocytes were also increased in comparison with maternal bone marrow erythrocytes. This fact seems to support the latter possibility. In contrast to the findings with 2AAF, BP and the negative controls, in the case of PCZ micronucleus incidences observed in foetal erythrocytes were similar to those in maternal bone marrow erythrocytes, indicating that micronucleus induction in foetal erythrocytes is a complex process that depends on several different factors.

It is important to stress that the chemicals used in this study caused the same mutagenic responses in both the conventional micronucleus assay and the foetal blood micronucleus assay. Differences in results were of quantitative rather than of qualitative nature.

6 References

Cole, R.J., N.A. Taylor, J. Cole and C.F. Arlett (1979) Transplacental effects of chemical mutagens detected by the micronucleus test, *Nature*, **277**, 317–318.

Cole, R.J., N. Taylor, J. Cole and C.F. Arlett (1981) Short-term tests of transplacentally active carcinogens. I. Micronucleus formation in fetal and maternal mouse erythroblasts, *Mutation Res.*, **80**, 141–157.

Cole, R.J., N. Taylor, J. Cole, L. Henderson and C.F. Arlett (1982) Short-term tests for transplacentally active carcinogens. Sensitivity of the transplacental micronucleus test to diethylnitrosamine, *Mutation Res.*, **104**, 165–171.

Kastenbaum, M.A. and K.O. Bowman (1970) Tables for determining the statistical significance of mutation frequencies, *Mutation Res.*, **9**, 527–549.

King, M.T. and D. Wild (1979) Transplacental mutagenesis: The micronucleus test on fetal mouse blood, *Human Genetics*, **51**, 183–194.

MacGregor, T., C.M. Wehr and H. Gould (1980) Clastogen-induced micronuclei in peripheral blood erythrocytes: The basis of an improved micronucleus test, *Environ. Mutagen.*, **2**, 509–514.

Matter, B.E. (1976) Failure to detect chromosome damage in bone marrow cells of mice and Chinese hamsters exposed *in vivo* to some ergot derivatives, *J. Int. Res.*, **4**, 382–392.

Racine, R.R. and B.E. Matter (1984) The Micronucleus Test as an Indicator of Mutagenic Exposure In A.A. Ansari and F.J. de Serres (Eds), *Single-Cell Mutation Monitoring Systems*, Plenum Publishing Corporation, New York, pp. 217–231.

Stoyel, C.J. and A.M. Clark (1980) The transplacental micronucleus test, *Mutation Res.*, **74**, 393–398.

Suter, K.E. and I. Kubli (1982) The micronucleus test as a model system for transplacentally induced chromosome aberrations, *Teratology*, **26**, 10A.

Von Ledebur, M. and W. Schmid (1973) The micronucleus test. Methodological aspects, *Mutation Res.*, **19**, 109–117.

Ad D. Tates, Nico de Vogel, and Annelies H.M. Rotteveel

Department of Radiation Genetics and Chemical Mutagenesis, State University of Leiden, Leiden, The Netherlands

33

Detection of micronucleated polychromatic erythrocytes in bone marrow of male mice

1 Summary

Male hybrid mice were given single oral exposures of different doses of the four test chemicals BP, PYR, 2AAF, and 4AAF. Bone marrow cells were fixed at 24, 36, or 48 h after treatment. Results for 2AAF and BP were clearly positive and the effects were dose dependent. At equimolar doses BP was about 3 times more clastogenic than 2AAF. 4AAF and PYR were clear negatives.

2 Introduction

As a participating laboratory in the IPCS Collaborative Study on Short-Term Tests for Genotoxicity and Carcinogenicity, we conducted the following three *in vivo* tests for the detection of clastogenic action by the four test chemicals. The 3 *in vivo* tests were (1) the micronucleus test, using bone marrow of mice; (2) the SCE test using bone marrow cells of mice; and (3) a micronucleus test for hepatocytes of rats that has been developed in our laboratory. Results obtained with the latter 2 tests are reported separately. For the first 2 test systems, the same mouse strain was used and for all these systems some of the same doses of the 4 chemicals were used. Thus, the detection capacity of the 3 assay systems can, at least partly, be compared.

3 Materials and methods

Animals

(101 × C3H) F_1 males were used. They were bred in the animal house in our building, housed at 21–23 °C with a humidity of 60–65%, and received drinking water and standard food pellets for mice ad libitum (Hope Farms, Woerden, The Netherlands). Parent strains were obtained one year before the start of the experiments from Dr. Ehling in the Gesellschaft für Strahlen- und Umweltforschung, München, FRG. At the time of treatment animals weighed 25–30 g and were 8 to 12-weeks old.

Chemicals

The 4 tests compounds were received from Lancaster Synthesis, Ltd. (Eastgate, U.K.) and were stored at 4 °C. Olive oil (Ph. Ned.) with 5% Tween 80 was used as the solvent and also as the negative control. Mitomycin C (MMC) (Kyowa, Lamepro B.V.) dissolved in phosphate buffered saline (PBS) served as the positive control. Test compounds and the negative control were administered by cavage (po) in 0.3 ml quantities (1% of body weight).

Animals were not fasted prior to treatment. MMC (1 mg/kg) was given intraperitoneally (ip).

Benzo(a)pyrene (BP; 63, 126, and 252 mg/kg) could be dissolved (with the help of sonication) in olive oil plus Tween 80. Pyrene (PYR; 202, 404, and 808 mg/kg) dissolved easily at the dose of 202 mg/kg. At 404 mg/kg some small crystals remained detectable and at 808 mg/kg PYR remained in suspension despite sonication for 5 min. Consistent with expectation, 2-acetyl aminofluorene (2AAF; 55.8, 111.6 and 223 mg/kg) and 4-acetyl aminofluorene (4AAF; 223, 446, and 892 mg/kg) remained in suspension.

Micronucleus test procedure

Each substance was administered as a single dose. 2AAF, 4AAF, BP, and PYR were first tested at the highest dose and animals were killed 24, 36, and 48 h after oral application (5 animals per dose and fixation time). The fixation time giving the highest frequency of micronucleated polychromatic erythrocytes (MPEs) was generally selected for the testing of the lower doses of each chemical.

Six animals were sacrificed for the negative and positive control, respectively. Control animals were sacrificed only at 36 h after treatment.

Preparation and analysis of slides

Femurs were excised and their proximal ends removed with a pair of scissors. Bone marrow cells were gently aspirated with a 25 g needle filled with 1, 5-ml filtered foetal calf serum. Aspirated bone marrow was centrifuged at 70 times g for 5 min. The supernatant was discarded and the pellet resuspended in a small amount of serum. A smear was made on a slide, cells were fixed for 5 min in absolute methanol, and air dried. One-day old slides were first stained for 3 min in filtered undiluted May-Grünwald (Gurr) and then for 2 min in May-Grünwald diluted with distilled water (1:1). After short rinsing in distilled water, slides were stained for 10 min in a filtered Giemsa solution (1 part in 6 parts distilled water). Slides were then rinsed in tap and distilled water and mounted in Depex.

In each slide, 1000 polychromatic erythrocytes (PE) and 1000 normo-chromatic erythrocytes (NE) were analysed for the presence of micronuclei. In the accompanying tables the frequencies of MPEs and micronucleated normochromatic erythrocytes (MNE) are given as percentages. In addition, we determined the ratio of polychromatic erythrocytes (PEs) among total erythrocytes (PE/PE + NE) by countin PE and NE cells among 1000

Table 1. *Incidence of micronuclei in the bone marrow of (101 × C3H)F_1 male mice*

Chemical	Dose (mg/kg)	Fixation time (h)	MNE/PE %	MPE/PE %	PE/PE + NE
2AAF	55.8	36	0.10 ± 0.07[b,c]	0.60 ± 0.19	28.1 ± 4.0
	111.6		0.18 ± 0.16	0.86 ± 0.26	26.5 ± 6.1
	223		0.18 ± 0.16	1.66 ± 0.34	22.4 ± 4.3
	223	24	0.13 ± 0.10	1.80 ± 0.49	42.5 ± 4.5
		36	0.18 ± 0.16	1.66 ± 0.34	22.4 ± 4.3
		48	0.36 ± 0.11	1.58 ± 0.66	28.0 ± 5.2
4AAF	223	36	0.18 ± 0.13	0.24 ± 0.11	35.5 ± 5.8
	446		0.14 ± 0.09	0.26 ± 0.11	28.2 ± 9.8
	892		0.10 ± 0.12	0.38 ± 0.15	33.3 ± 7.0
	892	24	0.08 ± 0.05	0.30 ± 0.07	39.4 ± 5.3
		36	0.10 ± 0.12	0.38 ± 0.15	33.3 ± 7.0
		48	0.04 ± 0.06	0.16 ± 0.09	41.6 ± 4.9
BP	63	36	0.34 ± 0.09	0.72 ± 0.28	34.6 ± 4.0
	126		0.34 ± 0.24	2.02 ± 0.47	27.8 ± 4.3
	252		0.24 ± 0.13	4.82 ± 0.75	30.8 ± 5.4
	252	24	0.16 ± 0.09	2.10 ± 0.41	46.7 ± 4.7
		36	0.24 ± 0.13	4.82 ± 0.75	30.8 ± 5.4
		48	0.54 ± 0.37	2.14 ± 0.39	25.7 ± 1.9
PYR	202	36	0.12 ± 0.05	0.18 ± 0.08	44.1 ± 4.5
	404		0.14 ± 0.11	0.14 ± 0.17	38.4 ± 7.2
	808		0.16 ± 0.15	0.22 ± 0.23	39.6 ± 7.9
	808	24	0.08 ± 0.08	0.12 ± 0.08	30.0 ± 7.5
		36	0.16 ± 0.15	0.22 ± 0.23	39.6 ± 7.9
		38	0.14 ± 0.09	0.20 ± 0.10	33.8 ± 4.3
OLIVE OIL[a]		36	0.18 ± 0.12	0.23 ± 0.12	32.4 ± 9.0
MMC[a]	1	36	0.52 ± 0.20	4.62 ± 1.19	20.2 ± 7.6

[a] Olive oil and MMC were administered ip, whereas the 4 test chemicals were given by gavage.
[b] Standard deviation.
[c] Usually 5 animals per dose per fixation time were used.

erythrocytes per slide. Data were analysed with Wilcoxon's two-sample test. Two-sided P values were calculated and values below 0.05 were considered to indicate a significant effect of the chemical under test.

4 Results

Primary data were sent to the coordinator for micronucleus assays for blood cells (Dr. D. Wild, Würzburg, FRG). Summary data are given in Table 1. Results for 2AAF were clearly positive at all doses and fixation times tested. The highest frequency of micronucleated polychromatic erythrocytes (MPEs) was observed at 24 h after treatment. The effect of dose was tested 36 h after treatment and was found to be dose dependent. At the highest dose tested (223 mg/kg), there was a 7.8-fold increase of the frequency of MPEs over the control value. In the case of 4AAF no significant effects were detected at any of the doses and fixation times tested. Therefore, this chemical clearly registered negative. BP gave positive results at all doses and fixation times tested. The maximum frequency of MPEs was found at 36 h after treatment and at the maximum dose tested (252 mg/kg). At this dose there was a 21-fold increase of the frequency of MPE over the control. Effects of BP were clearly dose dependent. The clastogenic potential of PYR was tested at doses of 202, 404, and 808 mg/kg and 24, 36 and 48 h after exposure, but the results were completely negative.

5 Conclusions

2AAF is a clear positive even at the lowest dose tested (55.8 mg/kg). It induced a 7.8-fold increase over the control frequency of MPEs at the highest dose tested (223 mg/kg). 4AAF gives a clear negative response. BP is a rather strong clastogen and induced a 21-fold increase over the control frequency of MPEs at the highest dose tested (252 mg/kg). On the other hand, PYR is clearly negative. When tested at equimolar doses, BP (1 m Mol/kg = 252 mg/kg) is a stronger clastogen than 2AAF (1 m Mol/kg = 223 mg/kg).

6 Acknowledgement

Thanks are due to the Ministry of Housing, Physical Planning and Environment for financial support.

James T. MacGregor, Carol M. Wehr and
Philip R. Henika
USDA Western Regional Research Center, Berkeley, California, USA

34

The peripheral blood erythrocyte micronucleus assay in weanling and foetal Swiss mice

1 Summary

The frequency of micronucleated erythrocytes in the peripheral blood of weanling and foetal Swiss mice exposed to benzo(a)pyrene (BP), pyrene (PYR), 2-acetylaminofluorene (2AAF), or 4-acetylaminofluorene (4AAF) was examined. BP induced an approximately linear dose-related increase in the frequency of micronucleated RNA-positive erythrocytes in weanling, foetal, and maternal mice. The frequency of micronucleated cells was increased to 10 to15 times control values at a dose of 1 m mole/kg. Under identical test conditions, PYR failed to increase significantly the frequency of micronucleated cells at doses up to 2 m moles/kg. A very small increase was observed in the pyronin-positive population in foetal mice 48 h after PYR treatment, but the increase was not significant by the criteria for a positive response adopted by the micronucleus test work group. 2AAF increased the frequency of micronucleated polychromatic erythrocytes in weanling male mice to 3.5 to 5.5 times the control level at doses between 1 and 3 m moles/kg but had no significant effect in foetal or maternal mice. 4AAF induced an equivocal increase in the frequency of micronucleated polychromatic erythrocytes in weanling male mice and failed to increase the frequency of micronucleated cells in foetal or maternal mice.

2 Introduction

Chromosomal damage in nucleated erythropoietic cells is causally associated with the occurrence of micronuclei in newly formed bone marrow erythrocytes (Heddle *et al.*, 1983; Schmid, 1976). In the mouse, micronucleated erythrocytes enter the peripheral blood and remain in the circulation with approximately the same lifetime as normal erythrocytes (Schlegel and MacGregor, 1982). The frequency of micronuclei in circulating blood provides a convenient index of in vivo chromosomal damage in this species (MacGregor *et al.*, 1983a). Since active haematopoiesis occurs in the liver of foetal mice, the incidence of micronucleated erythrocytes in foetal blood has proved to be a sensitive index of the genotoxic potential of certain classes of agents that are activated to unstable active intermediates primarily by the liver (Cole *et al.*, 1981, 1982; King and Wild, 1979; Stoyel and Clark, 1980). In this investigation, the frequencies of micronucleated erythrocytes in both weanling and foetal mouse blood were determined following exposures to BP, PYR, 2AAF, and 4AAF.

3 Materials and methods

Chemicals

BP, PYR, 2AAF, and 4AAF were supplied by Dr. John Ashby of Imperial Chemical Industries, Alderley Park, England. Olive oil was obtained from ICN Nutritional Biochemicals, Cleveland, OH (lot #15041), and Napa Valley Olive Oil Manufactory, St. Helena, CA. Cyclophosphamide was obtained from ICN Pharmaceuticals, Plainview, NJ (Lot #27068-A), and methyl methanesulfonate (MMS) from Aldrich Chemical Company, Milwaukee, WI (Lot #7724CK). Wright's and Giemsa powdered blood stains were from Harleco, Gibbstown, NJ; Hoechst 33258 stain was from Calbiochem, La Jolla, CA; and pyronin Y was from Polysciences, Warrington, PA.

Dosing solutions

Test chemicals were administered by gavage in olive oil at a volume of 10 ml per kg body weight. Chemicals were suspended in olive oil, stirred overnight or longer at room temperature, and were suspended with a vortex mixer immediately before making dilutions or dosing. All experiments with BP and PYR used olive oil from ICN Nutritional Biochemicals. 2AAF and 4AAF were tested in weanling mice with olive oil from both ICN Nutritional Biochemicals and from Napa Valley Olive Oil Manufactory. The preliminary rangefinding experiment, using 2 mice per dose point, suggested a possible difference in response to 2AAF with the different oils, but no significant differences were found when the complete experiments were run. The data presented in this report are those using the olive oil obtained from the Napa Valley source, because more dose levels were tested with that vehicle. BP dissolved almost completely at the lowest dose tested but was administered as a suspension at the higher dose levels. PYR, 2AAF, and 4AAF were suspensions at all doses tested.

Negative control animals received an equivalent volume of olive oil. Positive control agents were dissolved in isotonic saline and given intraperitoneally.

Peripheral blood assay in weanling mice

Male weanling (21-day old) Swiss-Webster mice were obtained from Simonsen Laboratories, Gilroy, CA. Mice were housed with pine shavings in plastic shoebox cages and were fed Purina Rodent Chow #5001 (Ralston Purina, St. Louis, MO) and tap water ad libitum. Groups of 8 animals received one dose of test material or control vehicle each day for 4 days. The test doses employed were: BP-63, 126, and 252 mg/kg (0.25, 0.5, and 1.0 m moles/kg); PYR – 101, 202, and 404 mg/kg (0.5, 1.0, and 2.0 m moles/kg); 2AAF – 112, 223, 446, and 624 mg/kg (0.5, 1.0, 2.0, and 2.8 m moles/kg); 4AAF – 223, 446, and 892 mg/kg (1.0, 2.0, and 4.0 m moles/kg). Blood samples were obtained from the ventral tail vessel approximately 20 h after the last dose. Six positive control mice, which received a single dose of 30 mg/kg (0.11 m moles/kg) cyclophosphamide 48 h before sampling, were included in each experiment and were sampled on the same day as the other test groups. Blood smears were prepared, fixed in absolute methanol, and stained with Wright's-Giemsa blood stain as described previously (Schlegel and MacGregor, 1982). Slides were randomized, coded, and scored at 1000X magnification using an oil immersion objective lens. The frequency of micronucleated polychromatic erythrocytes was determined by scoring 1200 polychromatic erythrocytes from each animal. The criteria for identifying micronuclei have been described by Schmid (1976). The percentage of polychromatic erythrocytes among total erythrocytes in each sample was determined by noting the number of polychromatic cells in a single field containing a minimum of 200 erythrocytes.

Foetal blood assay

Timed-mated Swiss-Webster mice, aged 5 to 6 months (32 ± 3 gm), were obtained on day 13 of pregnancy from Simonsen Laboratories, Gilroy, CA. Mice were housed with pine shavings in plastic shoebox cages and were fed Simonsen #7 breeding diet (Simonsen Laboratories) and tap water libitum. Groups of 5 animals per dose/time point received a single dose of test material or control vehicle on day 15 of pregnancy.

The test doses employed were: BP – 63, 126, and 252 mg/kg (0.25, 0.5, and 1.0 m moles/kg); PYR – 202, 404, and 606 mg/kg (1.0, 2.0, and 3.0 m moles/kg); 2AAF – 112, 223, 446, and 624 mg/kg (0.5, 1.0, 2.0 and 2.8 m moles/kg); 4AAF – 223, 446, and 892 mg/kg (1.0, 2.0, and 4.0 m moles/kg). Positive control mice received 50 mg/kg (0.45 m moles/kg) methyl methanesulfonate. Groups of treated animals were sacrificed at 30 and 48 hours after treatment. Two foetuses were removed from each mother. Blood smears were prepared from each foetus by severing the cervical blood vessels, transferring approximately 2 μl of cervical blood to a microscope slide with a 10 μl capillary pipet, spreading the smear, and fixing the air-dried smear for 15 seconds in absolute methanol. Prior to the 48 h sacrifice, samples of maternal blood were obtained and fixed by the same method used for samples from weanling mice. Fixed smears of foetal and maternal blood were stained with Hoechst 33258 and pyronin Y as described previously (MacGregor *et al.*, 1983b), randomized and coded, and scored at 1000X magnification with epifluorescent illumination.

Two classes of RNA-containing (pyronin-positive) erythrocytes (PPEs) were examined for the presence of micronuclei: those in which uniformly distributed RNA resulted in a cell uniformly stained with pyronin Y (uniformly stained pyronin-positive erythrocytes, UEs) and those in which aggregated RNA resulted in a cell stippled with pyronin-positive, RNA-containing bodies (stippled pyronin-positive erythrocytes, SEs). The appearance of these 2 classes of RNA-positive erythrocytes is shown in Figure 1. The uniformly stained PPEs constitute the youngest class of a nucleate erythrocytes, while the stippled PPEs are intermediate in age between the uniformly stained PPEs and the older RNA-negative erythrocytes (Cole *et al.*, 1981; vide infra). The percentages of uniformly stained and stippled PPEs among total mature erythrocytes were determined based on the numbers of these cell types among 100 to 200 erythrocytes from each foetus. The frequencies of micronucleated UEs and micronucleated SEs were determined by scoring 500 erythrocytes of each type per foetus. The criteria for identifying micronuclei were those of Schmid (1976), with the ad-

Figure 1. Foetal mouse blood stained with Hoechst 33258 and pyronin Y.
Left: Green (546 nm) excitation showing uniformly stained pyronin-positive erythrocytes (solid arrow), stippled pyronin-positive erythrocytes (dashed arrow), and a micronucleus (M) in a stippled erythrocyte. A leukocyte nucleus (L) is also shown. The pyronin-positive features fluoresce orange.
Right: Ultraviolet (360 nm) excitation of the same field showing Hoechst-positive emission (blue) from the chromatin-containing micronucleus (M) and leukocyte nucleus (L).

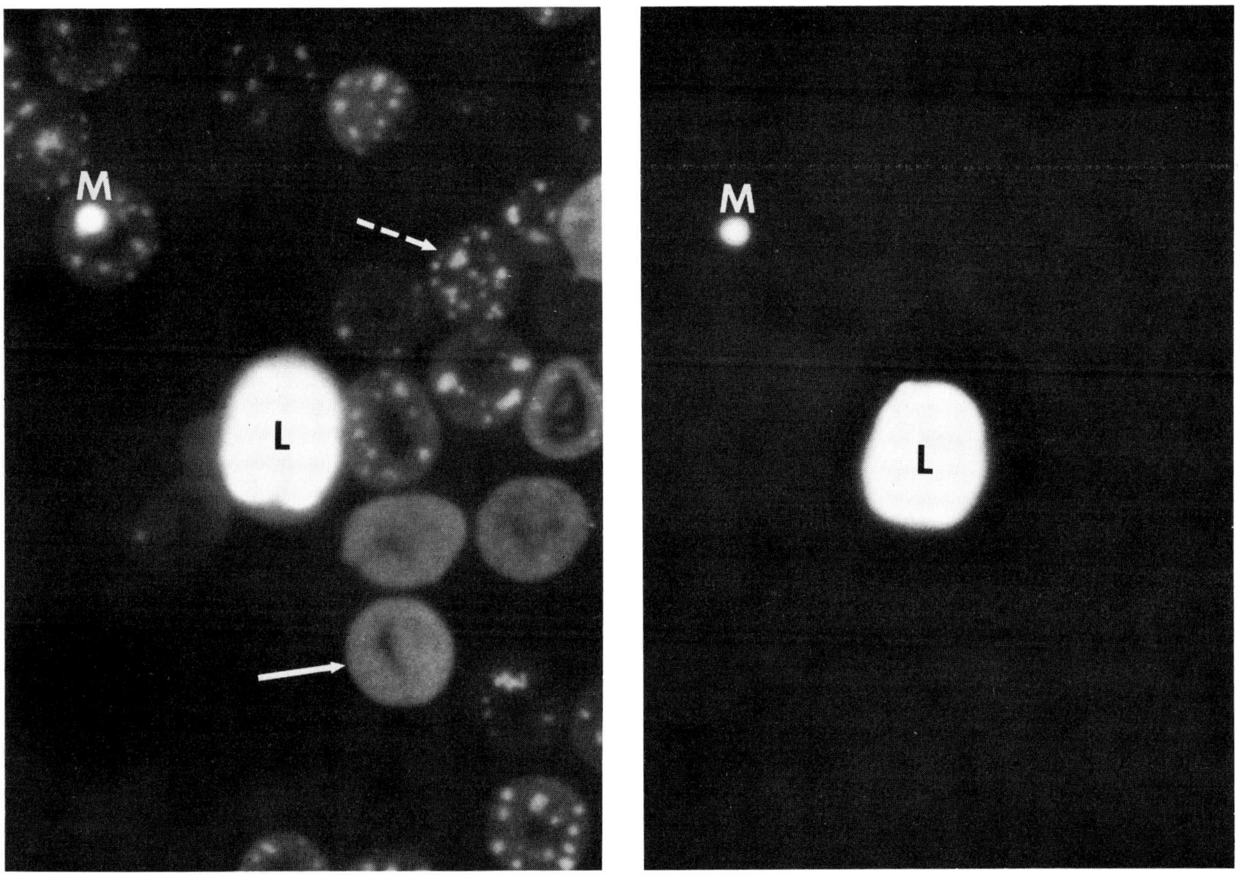

Figure 2. Frequencies of micronucleated polychromatic erythrocytes in the peripheral blood of weanling male mice exposed to benzo(a)pyrene, pyrene, 2-acetylaminofluorene, or 4-acetylaminofluorene.

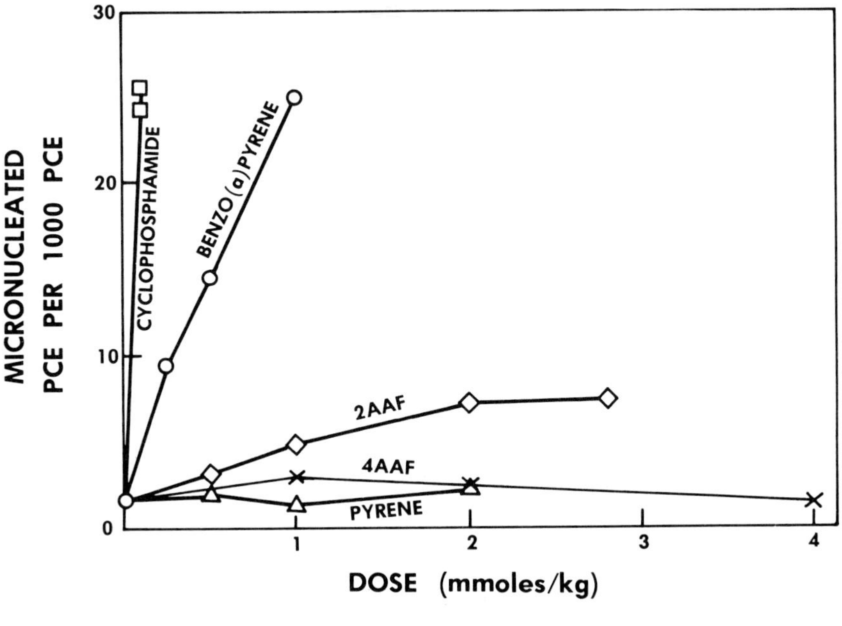

ditional requirement that the characteristic fluorescence emissions of DNA be present, [blue with UV (360 nm) excitation and orange with green (546 nm) excitation (see Figure 1)] (MacGregor *et al.*, 1983b).

Dose selection
The doses administered were based on rangefinding experiments using 2 animals per dose group. The upper dose for each compound was chosen to be either that expected to cause a measurable alteration in RNA-positive erythrocyte frequency, in micronucleated erythrocyte frequency, or in test animal survival, or to be the maximum practical dose which could be given.

4 Results

Effect of BP and PYR
BP induced a statistically significant, dose-related in-

Table 1. *Frequencies of micronucleated polychromatic erythrocytes (MPCE) in weanling male mice exposed to benzo(a)pyrene or pyrene*

Chemical, Dose (mg/kg)		Per cent PCE[a] ± SEM	MPCE/1200 PCE	Mean MPCE/ 1000 PCE[b]
Olive oil, 10 ml/kg		3.0 ± 0.40	4, 3, 1, 2 0, 0, 3, 2	1.6
BP,	63	1.6[†] ± 0.32	9, 12, 15, 12, 12, 3, 7, 21	9.5**
	126	1.1[††] ± 0.26	20, 7, 10, 19, 10, 36, 30, 7	14.5**
	252[c]	0.9[††] ± 0.21	27, 22[d], 18, 17, 19[d], 28, 18[d]	23.7**
PYR,	101	4.1 ± 0.58	2, 1, 2, 1, 0, 3, 7, 3	2.0
	202	3.3 ± 0.70	0, 0, 1, 1, 3, 3, 3, 1	1.3
	404[e]	2.1 ± 0.52	5, 1, 0, 3, 4	2.2
CP[f],	30	0.5[††] ± 0.13	14, 28, 15 76, 34, 17	25.6**

[a] Polychromatic erythrocytes (PCE). Significance tested by Duncan's Multiple-Range Test (Bliss, 1967). Values differing from the control group are indicated: †, $p \leqslant .05$; ††, $p \leqslant .01$.
[b] Significance tested by the binomial comparison of Kastenbaum and Bowman (1970). Values differing from the control group are indicated: *, $p \leqslant .05$; **, $p \leqslant .01$.
[c] One animal not scored due to severe PCE suppression.
[d] Number per 500 cells.
[e] Three animals died prior to sampling.
[f] Cyclophosphamide (CP).

Table 2. *Frequencies of micronucleated polychromatic erythrocytes (MPCE) in weanling male mice exposed to 2-acetylaminofluorene or 4-acetylaminofluorene*

Chemical, Dose (mg/kg)		Per cent PCE[a] ± SEM	MPCE/1200 PCE	Mean MPCE/ 1000 PCE[b]
Olive oil, 10 ml/kg		3.2 ± 0.42	1, 0, 2, 0, 1, 3, 3, 3	1.4
2-AAF,	112[c]	3.0 ± 0.53	1, 1, 6, 4, 7, 2, 5	3.1*
	223[c]	2.2 ± 0.51	10,9,5,5, 3, 4, 5	4.9**
	446	2.6 ± 0.25	11, 4, 8, 5, 4, 16, 7, 14	7.2**
	624[c]	1.9[†] ± 0.35	8, 10, 7, 16, 4, 4, 14	7.5**
4-AAF,	223	2.8 ± 0.28	2, 3, 3, 4, 4, 6, 2, 5	3.0**
	446	2.5 ± 0.49	2, 1, 4, 3, 6, 1, 3, 1	2.2
	892[c]	3.1 ± 0.44	3, 1, 3, 2, 1, 0, 2	1.4
CP[d],	30	1.2[††] ± 0.25	32, 27, 29, 32, 17, 39	24.4**

[a–b] See footnotes to Table 1.
[c] One animal died prior to sampling.
[d] Cyclophosphamide (CP).

crease in the frequency of micronucleated RNA-positive erythrocytes in both weanling and foetal mice at doses between 63 and 252 mg/kg (Figures 2, 3; Tables 1, 3, and 4). In the weanling mice, the response was almost linear with dose, with an induced frequency of approximately 25 micronucleated polychromatic erythrocytes per thousand polychromatic erythrocytes at 252 mg/kg (Figure 2, Table 1). In the foetal mice, a similar response was observed, with a maximum observed frequency of slightly more than 30 micronucleated uniformly stained pyronin-positive erythrocytes (UEs) per thousand UEs at 252 mg/kg (Figure 3, Tables 3 and 4).

PYR failed to induce statistically significant increases in the frequency of micronucleated erythrocytes in the weanling mice at doses up to 404 mg/kg (Figure 2 and Table 1). In the foetal mice, PYR caused an apparent weak elevation in the frequency of micronucleated uniform pyronin-positive erythrocytes at doses of 202 and 404 mg/kg at the 48 h sampling time (Table 4). This slight effect could have been due to a real effect in foetal liver erythropoietic cells, which was not detected in adult bone marrow erythropoietic cells, but was more likely an artifact of the lower than usual control value that was observed in the uniform pyronin-positive erythrocyte

Table 3. *Frequencies of micronucleated pyronin-positive erythrocytes in foetal mice 30 hours after exposure to benzo(a)pyrene or pyrene*

Chemical, Dose (mg/kg)		Per cent UE[a,b] ± SEM	Per cent SE[a,b] ± SEM	Micronucleated UE/UEs Scored[c]	Micronucleated SE/SEs Scored[c]	Mean MUE/ 1000 UE[d]	Mean MSE/ 1000 SE[d]
Olive oil, 10 ml/kg		54 ± 3.1	38 ± 2.0	3/507, 2/503; 0/532, 3/506; 3/540, 3/503; 2/525, 1/505; 0/507, 4/505	0/518, 0/500; 1/531, 3/519; 2/501, 0/515; 1/502, 2/502; 1/513, 1/500	4.1	2.2
MMS,	50	52 ± 1.8	39 ± 1.9	16/594, 8/513; 20/559, 22/535; 24/516, 19/503; 8/510, 9/507; 43/542, 19/529	5/519, 2/506; 4/520, 7/500; 5/508, 3/512; 2/513, 2/510; 3/504, 0/500	35.4**	6.5**
BP,	63	52 ± 1.2	40 ± 1.2	0/532,0/510;3/511,2/520;3/528, 5/533; 5/513, 7/506; 0/527, 3/510	0/518, 2/512; 3/526, 2/517; 2/508, 2/507; 3/512, 0/500; 1/519; 0/513	5.4	2.9
	126[e]	56 ± 3.3	38 ± 2.9	3/518, 0/516; 2/520, 4/524; 2/500, 3/501; 8/529, 4/503	1/528, 2/516; 1/503, 0/514; 1/500, 0/565; 0/504, 1/504	6.3	1.5
	252[f]	53 ± 3.6	38 ± 3.8	17/505, 10/519; 2/521, 2/515; 4/507, 6/512; 6/510, 9/520	2/510, 0/504; 1/518, 1/502; 1/525, 1/565; 2/520, 1/537	13.6**	2.2
PYR,	202[f]	48 ± 1.7	42 ± 1.4	0/522, 3/516; 0/526, 3/526; 1/513, 1/549;3/500,2/539	2/506, 0/522; 2/520, 2/501; 0/501, 2/530;0/514,2/500	3.1	2.4
	404	51 ± 1.7	43 ± 1.4	2/512, 2/529; 1/500, 0/508; 2/509 2/508, 0/513, 1/500; 1/506, 6/524	0/509, 1/528; 2/519, 0/501; 2/529, 2/510;2/501,3/503;1/510,7/504	3.3	3.9
	606[g]	52 ± 3.6	42 ± 3.4	2/500, 1/513; 2/515, 1/531	2/500, 2/504; 2/533, 0/506	2.9	2.9

[a] UE, uniformly pyronin-positive erythrocytes; SE, stippled pyronin-positive erythrocytes.
[b] Significance tested by Duncan's Multiple-Range Test (Bliss, 1967). Values differing from the control group are indicated: †, $p \leqslant .05$; ††, $p \leqslant .01$.
[c] Values between semicolons are individual foetuses from a single mother.
[d] Micronucleated UE (MUE); micronucleated SE (MSE). Significance tested by the binomial comparison of Kastenbaum and Bowman (1970). Values differing from the control group are indicated: *, $p \leqslant .05$; **, $p \leqslant .01$.
[e] All foetuses dead in one mother.
[f] One mother died prior to sampling.
[g] All foetuses dead in two mothers; one mother aborted.

Figure 3. Frequencies of micronucleated RNA-positive erythrocytes in the peripheral blood of foetal mice 48 h after transplacental exposure to benzo(a)pyrene or pyrene. Solid lines denote uniformly stained pyronin-positive erythrocytes (UE). Dashed lines denote stippled pyronin-positive erythrocytes (SE).

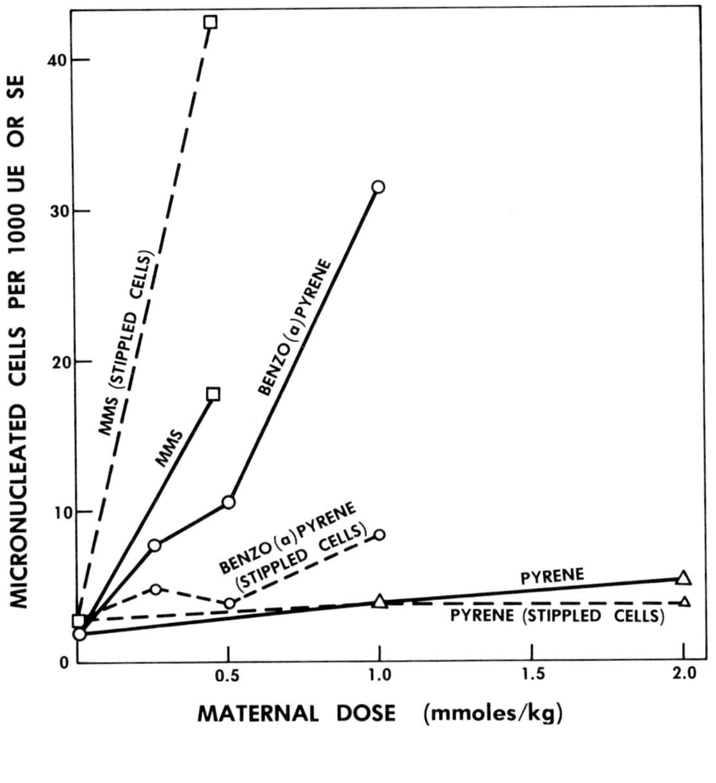

Table 4. *Frequencies of micronucleated pyronin-positive erythrocytes in foetal mice 48 hours after exposure to benzo(a)pyrene or pyrene*

Chemical,	Dose (mg/kg)	Per cent UE[a,b] ± SEM	Per cent SE[a,b] ± SEM	Micronucleated UE/UEs Scored[c]	Micronucleated SE/SEs Scored[c]	Mean MUE/ 1000 UE[d]	Mean MSE/ 1000 SE[d]
Olive oil, 10 ml/kg		34 ± 2.2	41 ± 2.0	0/500, 0/537; 2/505, 2/502; 1/507, 0/501; 1/540, 1/528; 2/535, 1/501	3/515, 3/523; 2/501, 1/501; 0/505, 1/502;1/505,2/524;1/508,1/546	1.9	2.9
MMS,	50[e]	30 ± 3.5	43 ± 2.6	18/541, 16/520; 7/519, 9/507; 10/538, 3/542; 7/517, 5/518	23/560, 28/547; 25/513, 35/522; 11/503, 16/516; 20/501, 20/503	17.8**	42.7**
BP,	63[f]	31 ± 2.3	42 ± 2.5	4/508, 12/508; 5/526, 3/511; 0/500, 5/526; 1/551, 2/547	2/542, 5/516; 2/563, 2/504; 2/521, 3/500; 2/535, 2/519	7.7**	4.8
	126[g]	30 ± 4.1	41 ± 2.8	5/510, 5/511; 14/522, 4/501; 2/573 3/508	1/508, 1/517; 4/548, 5/530; 0/532, 1/512	10.6**	3.8
	252[g]	24 ± 4.7	48 ± 3.4	9/503, 11/527; 33/523, 32/519; 7/515, 6/528	4/535, 3/506; 5/506, 2/501; 9/537, 3/547	31.5**	8.3**
PYR,	202	30 ± 3.2	43 ± 1.4	1/504, 2/521; 1/536, 4/514; 3/504, 2/520; 0/512, 1/500; 4/515, 2/516	0/511, 2/544; 2/548, 2/562; 4/565, 0/515; 2/513, 1/518; 3/508, 3/544	3.9*	3.6
	404[e]	37 ± 1.8	46 ± 1.8	2/539, 2/513; 4/524, 3/535; 2/511, 5/532; 1/517, 3/546	0/532, 2/576; 3/524, 2/537; 2/523, 3/503; 2/553, 2/540	5.2**	3.7
	606[h]	—	—	—	—	—	—

[a–d]See footnotes to Table 3.
[e]One mother littered prior to sampling.
[f]One mother died prior to sampling.
[g]All foetuses resorbed in one mother; one mother littered prior to sampling.
[h]Two mothers died; all foetuses resorbed in 2 mothers; one mother littered prior to sampling.

Table 5. *Frequencies of micronucleated pyronin-positive erythrocytes in foetal mice 30 hours after exposure to 2-acetylaminofluorene or 4-acetylaminofluorene*

Chemical,	Dose (mg/kg)	Per cent UE[a,b] ± SEM	Per cent SE[a,b] ± SEM	Micronucleated UE/UEs Scored[c]	Micronucleated SE/SEs Scored[c]	Mean MUE/ 1000 UE[d]	Mean MSE/ 1000 SE[d]
Olive oil, 10 ml/kg		54 ± 2.4	39 ± 2.0	1/517, 3/500; 1/535, 1/514; 2/527, 4/512; 3/514, 3/531; 1/520, 2/505	0/502, 0/503; 1/515, 2/508; 2/509, 2/510; 1/505, 1/531; 2/502, 2/510	4.1	2.6
MMS,	50	50 ± 2.8	44† ± 1.3	14/510, 16/526; 16/512, 9/517; 7/508, 9/512; 32/512, 30/519; 15/526, 12/532	3/500, 2/565; 3/510, 1/526; 2/518, 0/526; 2/501, 3/511; 1/500, 2/506	30.9**	3.7
2-AAF,	112	51 ± 1.6	43 ± 1.6	3/523, 0/505; 2/510, 3/515; 2/502, 3/500; 1/508, 1/529; 3/515, 2/566	0/520, 1/511; 3/508, 0/517; 0/508, 1/527; 1/502, 1/542; 0/522, 2/501	3.9	1.7
	223	51 ± 1.6	43 ± 1.6	2/502, 1/515; 2/514, 2/525; 3/518, 1/527; 1/523, 2/509; 2/516, 3/503	3/511, 2/566; 1/551, 2/508; 1/500, 2/508; 1/547, 2/520; 0/506,1/500	3.7	2.9
	446	50 ± 2.3	43 ± 1.6	0/504, 3/500; 3/500, 5/505; 1/502, 3/153; 1/514, 1/500; 2/525, 3/518	1/500, 0/520; 2/508, 3/506; 3/506, 2/501; 2/507, 0/518; 0/518, 5/501	4.3	3.5
	624	48 ± 1.8	44† ± 1.0	0/500, 3/502; 4/502, 1/501; 3/511, 1/509; 3/510, 0/506; 0/517, 1/532	0/512, 0/527; 0/511, 2/503; 0/527, 1/510; 0/510, 1/510; 0/508, 1/519	3.1	1.0
4-AAF,	223	52 ± 1.5	41 ± 1.0	3/502, 2/509; 2/508, 2/512; 1/507, 2/518; 2/516, 4/514; 2/544, 0/505	2/514, 1/502; 1/518, 2/521, 1/530, 3/523; 2/502, 2/530; 1/565, 1/500	3.9	3.1
	446	45†† ± 0.9	46†† ± 0.8	1/524, 4/503; 2/519, 1/520; 1/504, 2/528; 4/504, 2/510; 0/512, 3/528	1/508, 2/513; 1/501, 2/502; 0/511, 3/516; 0/506, 0/536; 1/518, 3/527	3.9	2.5
	892	47† ± 0.5	46†† ± 0.6	1/513, 1/510; 2/500, 2/525; 0/508, 3/519; 0/513, 3/505; 1/510, 1/508	0/504, 0/526; 0/505, 1/515; 2/521, 2/513; 0/513, 0/510; 0/512, 0/501	2.7	1.0

[a–d]See footnotes to Table 3.

Table 6. *Frequencies of micronucleated pyronin-positive erythrocytes in foetal mice 48 hours after exposure to 2-acetylaminofluorene 4-acetylaminofluorene*

Chemical,	Dose (mg/kg)	Per cent UE[a,b] ± SEM	Per cent SE[a,b] ± SEM	Micronucleated UE/UEs Scored[c]	Micronucleated SE/SEs Scored[c]	Mean MUE/ 1000 UE[d]	Mean MSE/ 1000 SE[d]
Olive oil, 10 ml/kg[e]		47 ± 1.8	46 ± 1.5	1/547, 5/500; 2/512, 4/514; 2/507, 1/508; 3/500, 1/511	3/562, 3/518; 1/527, 3/510; 3/515, 1/520; 0/506, 1/518	4.6	3.6
MMS,	50	45 ± 1.7	46 ± 1.1	1/520, 4/515; 4/515, 2/547; 3/503, 2/508; 2/518, 3/528; 7/564, 4/508	21/509, 7/530; 12/508, 18/518; 4/539, 14/510; 8/512, 7/501; 23/520, 19/500	6.1	25.8*
2-AAF,	112[e]	45 ± 2.6	45 ± 2.6	0/500, 4/519; 3/500, 2/514; 3/519, 4/500;0/507,3/508	0/513, 1/507; 3/515, 1/527; 2/507, 3/500;2/523,0/511	4.7	2.9
	223[f]	57†† ± 3.7	30†† ± 2.5	0/523, 2/500; 4/529, 1/535; 3/517, 3/521; 4/500, 1/501	2/503,1/548;3/500,2/526;2/507, 3/516; 1/502, 1/500	4.4	3.7
	446[f]	41 ± 2.8	40 ± 2.7	1/531, 6/521; 3/519, 2/542; 2/501, 3/515; 1/500, 2/515	2/502, 3/501; 1/529, 2/508; 1/500, 3/527; 1/517, 2/536	4.8	3.6
	624	44 ± 1.2	45 ± 0.9	1/506, 1/515; 2/530, 3/507; 1/515, 1/505; 1/518, 3/509; 2/538, 5/543	0/520, 1/500; 3/556, 2/541; 2/532, 1/533; 1/504, 3/529; 1/507, 1/574	3.9	2.9
4-AAF,	223[g]	38 ± 4.7	42 ± 1.4	2/504, 3/502; 0/514, 2/519	1/514, 0/515; 3/523, 3/503	3.4	3.4
	446[h]	41 ± 1.2	49 ± 1.1	4/510, 5/518; 1/513, 6/500	4/552, 2/510; 4/528, 1/552	7.8	5.1
	892	47 ± 1.5	43 ± 1.8	0/517, 4/501; 6/543, 3/510; 2/545, 2/534; 2/510, 2/510; 3/500, 3/500	4/502, 2/504; 2/564, 3/500; 0/534, 0/515; 0/524, 2/515; 3/500, 2/518	5.2	3.5

[a–d]See footnotes to Table 3.
[e]All foetuses from one mother were resorbed.
[f]One mother died prior to sampling.
[g]All foetuses from 3 mothers were resorbed.
[h]Two mothers died prior to sampling; all foetuses from one mother were resorbed.

Table 7. *Frequencies of micronucleated pyronin-positive erythrocytes in maternal blood 48 hours after treatment with benzo(a)pyrene, pyrene, 2-acetylaminofluorene, or 4-acetylaminofluorene*

Chemical, Dose (mg/kg)		Per cent UE[a,b] ± SEM	MUE/1000 UE[c]	Mean MUE/1000 UE[c]
Olive oil, 10 ml/kg		4.6 ± 1.1	0, 1, 2, 3, 2	1.6
MMS,	50[d]	2.9 ± 0.50	22, 37, 25, 24	27**
BP,	63[e]	2.8 ± 0.17	3, 1, 1, 1	1.5
	126	4.6 ± 1.1	19, 1, 17, 4, 1	8.4**
	252[f]	2.7 ± 0.77	34, 24, 9, 5	18.**
PYR,	202	5.1 ± 1.2	1, 2, 2, 1, 6	2.4
	404	4.7 ± 0.91	1, 2, 1, 0, 2	1.2
	606[g]	5.1 ± 2.8	2, 1	1.5
Olive oil, 10 ml/kg		6.3 ± 0.87	2, 2, 2, 1, 3	2.0
MMS,	50	4.2 ± 0.60	16, 23, 9, 10, 25	16.5**
2-AAF,	112	6.0 ± 0.97	0, 1, 3, 1, 0	1.0
	223[e]	5.0 ± 0.46	1, 2, 3, 3	2.2
	446[e]	4.0 ± 0.55	2, 2, 1, 0	1.2
	624	4.8 ± 0.71	2, 0, 2, 1, 3	1.6
4-AAF,	223	5.4 ± 0.67	2, 2, 0, 3, 1	1.6
	446[h]	5.0 ± 0.82	1, 2, 1	1.3
	892	6.2 ± 0.71	3, 2, 3, 2, 4	2.8

[a–b] See footnotes to Table 3.
[c] Micronucleated UE (MUE). Significance tested by the binomial comparison of Kastenbaum and Bowman (1970). Values which differ from the control group are indicated: *, $p \leqslant .05$; **, $p \leqslant .01$.
[d] One animal littered prior to sampling
[e] One animal died prior to sampling.
[f] One animal not scored due to severe PPE suppression.
[g] Two animals died and one littered prior to sampling.
[h] Two animals died prior to sampling.

Figure 4. Frequencies of micronucleated uniform pyronin-positive erythrocytes in the peripheral blood of 17-day pregnant mice 48 h after exposure to benzo(a)pyrene, pyrene, 2-acetylaminofluorene, or 4-acetylaminofluorene.

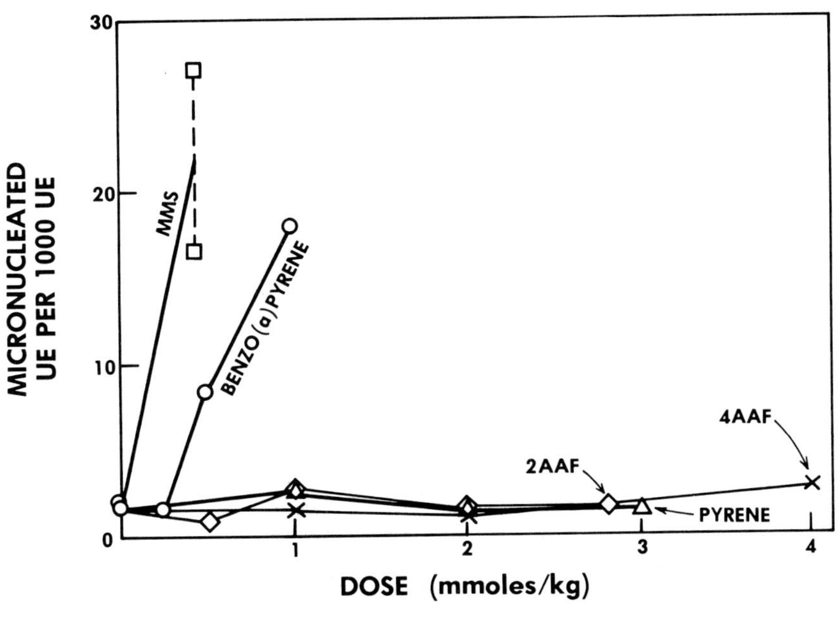

population in this particular experiment (compare with control values in Tables 3, 5, and 6). This response was not statistically significant when compared with the pooled control values from Tables 3–6. No increase was observed in the stippled erythrocytes at either sampling time (Tables 3 and 4).

An increased frequency of micronucleated pyronin-positive erythrocytes was also observed in the maternal blood sample taken 48 h after treatment of the pregnant females with BP (Figure 4). Since the time of the peak response in the maternal blood is not known, the observed values may be an underestimate of the maternal bone marrow response. Nevertheless, it is noteworthy that the response observed in maternal blood was similar to that in the weanling male blood (cf. Figures 2 and 4). PYR did not elevate the frequency of micronucleated pyroninpositive erythrocytes in maternal blood at doses up to 606 mg/kg.

Effect of 2AAF and 4AAF

2AAF induced a statistically significant, dose-related increase in the frequency of micronucleated polychromatic erythrocytes in the weanling male mice at doses between 223 and 624 mg/kg (Figure 2, Table 2). Over the same dose range, no significant increases in micronucleated cells were observed among either uniform or stippled pyronin-positive erythrocytes in the foetal mice (Tables 5, 6) or in the maternal blood sample obtained 48 h after exposure (Table 7). The effect of 2AAF in the weanling mice, although highly significant statistically, was considerably weaker than that observed after BP exposure (Figure 2).

Exposure to 4AAF did not significantly increase the frequency of micronucleated RNA-positive erythrocytes in foetal or maternal mice (Tables 5, 6, and 7), but an equivocal response was observed in the weanling mice (Table 2). In the weanling mice, the frequency of micronucleated polychromatic erythrocytes observed after exposure to 223 mg/kg was significantly higher ($p \cong 0.01$) than the control, but significant increases were not observed after exposure to doses of 446 or 892 mg/kg. In a replicate experiment (not shown), one dose point (892 mg/kg) was also significantly elevated relative to the control value ($p \cong 0.01$). Thus, it is likely that 4AAF induces a very weak response in the weanling mice which is at the borderline of detectability with the experimental design employed.

5 Discussion

BP was a potent inducer of micronuclei in weanling, foetal, and maternal mice. In each case, the response was almost a linear function of dose, and the groups receiving 252 mg/kg BP had 10- to 15-fold higher frequencies of micronucleated cells than the concurrent control groups. PYR was negative, or at most very weakly active, under identical test conditions. Since approximately a 2-fold increase over the average control value would have been statistically significant by the binomial comparison employed, the activity of BP in the assays performed was at least 5 to 8 times greater than that of PYR.

In the foetal mouse, the time between dosing and the peak incidence of micronuclei in the RNA-positive erythrocyte populations scored was considerably longer for BP than for the positive control agent, methyl methanesulfonate. A relatively long period between dosing with BP and the peak incidence of micronuclei in the foetal mouse, compared with other known genotoxins, has been noted previously by Cole *et al.* (1981). In the present experiments, the frequency of micronucleated cells among the uniform pyronin-positive erythrocytes was markedly elevated and the frequency in the older stippled population had already increased significantly at 30 h after dosing with methyl methanesulfonate (Figure 5). At 48 h after dosing, the frequency had declined in the uniformly stained population and was markedly elevated in the stippled cells. After treating with BP, the micronucleated cells had not yet reached their maximum

Figure 5. Frequencies of micronucleated uniformly stained pyronin-positive erythrocytes (UE) and stippled erythrocytes (SE) in the peripheral blood of foetal mice at 30 and 48 h after transplacental exposure to benzo(a)pyrene or methyl methanesulfonate.

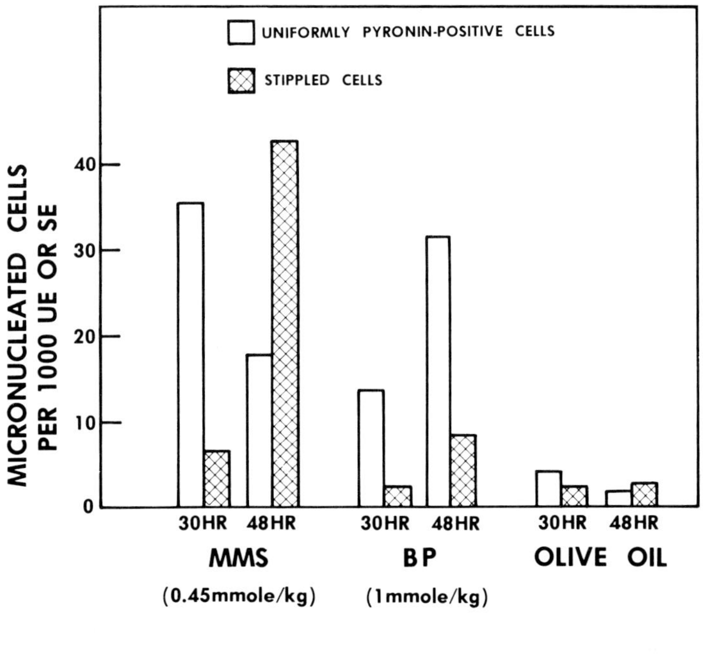

frequency in the uniformly stained population at 30 h and did not begin to enter the stippled population until 48 h after dosing (Figure 5). These data clearly demonstrate the advantage of scoring separately the uniformly stained and stippled pyronin-positive erythrocyte subpopulations of RNA-positive erythrocytes in experiments in which a single dose of test agent is given. Since the micronucleated cells appear first in the uniformly stained population and then move into the stippled population, fewer samples are required to ensure that the peak frequency is observed if these subpopulations are scored separately in each sample.

2AAF increased the frequency of micronucleated polychromatic erythrocytes in the weanling males but failed to alter the micronucleus frequency in foetal blood. This difference in response between the foetal and weanling male mice presumably is due to a difference in the levels of metabolically active intermediate reaching the target erythropoietic cell population. In the foetal animals, erythropoiesis occurs mainly in the liver, while in weanling and adult mice, the bone marrow is the principal site of erythropoiesis (Cole *et al.*, 1982). It has been shown previously that micronuclei may be induced in foetal erythrocytes by compounds which fail to induce micronuclei in the adult mouse (Cole *et al.*, 1982; Stoyel and Clark, 1980). These differences are thought to be due to higher levels of the metabolically activated proximate mutagen reaching the erythropoietic tissue in the foetal mouse (Cole *et al.*, 1982; Stoyel and Clark, 1980). In the present case, the weanling males were more sensitive than the foetal mice.

2AAF increased the frequency of micronucleated polychromatic erythrocytes approximately 3.5- to 5.5-fold in the weanling male mice. The response did not appear to be linearly related to dose, although the statistical error in the observed values makes the actual shape of the dose–response curve uncertain. Little increase in the response was obtained as the dose was increased beyond 446 mg/kg (2 m moles/kg) (Figure 2). The very poor solubility of both 2AAF and 4AAF in the olive oil vehicle employed suggests that this may be due to a limitation in the absorption rate as the dose is increased. Because the increases in the frequency of micronucleated cells induced by 2AAF were not large, the activity of 2AAF vs 4AAF cannot be determined precisely. However, the data show that 2AAF has at least twice the activity of 4AAF.

The results obtained with the 4 chemicals tested are consistent with previously published data on their activities in adult and foetal mouse erythropoietic tissues. BP has been tested in several laboratories and has been reported to increase the frequency of micronucleated erythrocytes in adult bone marrow (Heddle *et al.*, 1983) and in foetal liver (Cole *et al.*, 1981). Cole *et al.* (1981) compared the induction of micronuclei in erythrocytes from foetal liver and maternal bone marrow at the times of maximum response in each tissue following ip administration of BP in corn oil, and reported a greater effect in maternal bone marrow than in foetal liver. The liver erythrocyte response reported by these workers was somewhat weaker than that we obtained in foetal blood erythrocytes. PYR has been consistently reported not to induce micronucleated polychromatic erythrocytes in adult bone marrow (Heddle *et al.*, 1983), but has apparently not been tested in foetal erythrocytes.

2AAF and 4AAF have previously been tested in the adult bone marrow erythrocyte micronucleus test as part of an earlier International Collaborative Program for the evaluation of short-term tests for carcinogens (Salamone *et al.*, 1981; Tsuchimoto and Matter, 1981). Both laboratories which tested these agents found 2AAF, but not 4AAF, to increase the frequency of micronucleated polychromatic erythrocytes following ip administration in dimethylsulphoxide. In agreement with the present findings in peripheral blood, these investigators found 2AAF to induce maximum frequencies in bone marrow 4 to 5 times the control values at doses between 280 and 1120 mg/kg.

6 References

Bliss, C.I. (1967) *Statistics in Biology*, Vol. 1, McGraw-Hill, N.Y., pp. 253–257.

Cole, R.J., N. Taylor, J. Cole and C.F. Arlett (1981) Short-term tests for transplacentally active carcinogens. I. Micronucleus formation in fetal and maternal mouse erythroblasts, *Mutation Res.*, **80**, 141–157.

Cole, R.J., N. Taylor, J. Cole, L. Henderson and C.F. Arlett (1982) Short-term tests for transplacentally active carcinogens. Sensitivity of the transplacental micronucleus test to diethylnitrosamine, *Mutation Res.*, **104**, 165–171.

Heddle, J.A., M. Hite, B. Kirkhart, K. Mavournin, J.T. MacGregor, G.W. Newell and M.F. Salamone (1983) The induction of micronuclei as a measure of genotoxicity, *Mutation Res.*, **123**, 61–118.

Kastenbaum, M.A. and K.O. Bowman (1970) Tables for determining the statistical significance of mutation frequencies, *Mutation Res.*, **9**, 527–549.

King, M.-T. and D. Wild (1979) Transplacental mutagenesis: The micronucleus test on fetal mouse blood, *Human Genet.*, **51**, 183–194.

MacGregor, J.T., R. Schlegel, W.N. Choy and C.M. Wehr (1983a) Micronuclei in circulating erythrocytes: A rapid screen for chromosomal damage during routine toxicity testing in mice. In A.W. Hayes, R.C. Schnell and T.S. Miya (Eds), *Developments in The Science and Practice of Toxicology*, Elsevier, Amsterdam, pp. 555–558.

MacGregor, J.T., C.M. Wehr and R.G. Langlois (1983b) A simple fluorescent staining procedure for micronuclei and RNA in erythrocytes using Hoechst 33258 and pyronin Y, *Mutation Res.*, **120**, 269–275.

Salamone, M.F., J.A. Heddle and M. Katz (1981) Mutagenic activity of 41 compounds in the *in vivo* micronucleus assay. In F.J. de Serres and J. Ashby (Eds), *Evaluation of Short-Term Tests for Carcinogens*, Vol. 1, *Prog. Mutation Research*, Elsevier, Amsterdam, pp. 686–697.

Schlegel, R. and J.T. MacGregor (1982) The persistance of micronuclei in peripheral blood erythrocytes: Detection of chronic chromosome breakage in mice, *Mutation Res.*, **104**, 367–369.

Schmid, W. (1976) The micronucleus test for cytogenetic analysis. In A. Hollaender (Ed.), *Chemical Mutagens*, Plenum Press, New York, Vol. 4, pp. 31–53.

Stoyel, C.J. and A.M. Clark (1980) The transplacental micronucleus test, *Mutation Res.*, **74**, 393–398.

Tsuchimoto, T. and B.E. Matter (1981) Activity of coded compounds in the micronucleus test. In F.J. de Serres and J. Ashby (Eds), *Evaluation of Short-Term Tests for Carcinogens*, Vol. I, *Prog. Mutation Research*, Elsevier, Amsterdam, pp. 705–711.

Brita Beije and John Ashby[1]
Genetic Toxicology Laboratory
Environmental Toxicology Unit, Wallenberg Laboratory
Stockholm University, Sweden
[1]*Central Toxicology Laboratory*
Imperial Chemical Industries PLC
Alderley Park, Cheshire, UK

35

Micronucleus test in bone marrow of rats exposed to 2-acetylaminofluorene and 4-acetylaminofluorene

1 Summary

Male Alderley Park rats were dosed orally with 2-acetylaminofluorene (2AAF) or 4-acetylaminofluorene (4AAF). The dose levels were between 50 and 3000 mg/kg for 2AAF and between 1000 and 3000 mg/kg for 4AAF. No significant increase in micronucleated polychromatic erythrocytes were observed after 4AAF treatment, whereas 2AAF caused a weak response at dose levels between 500 and 3000 mg/kg.

2 Materials and methods

Chemicals

Cyclophosphamide (CP) was dissolved in water, whereas 2AAF and 4AAF were dissolved or suspended by sonication in corn oil. The concentration of each dosing solution was made such that 10 ml/kg body weight would be the volume administered. For the higher dose levels of 2AAF, it was finely ground and maintained as a suspension in corn oil for 4 h at 70 °C. Before dosing, the mixture was cooled. All 2AAF and 4AAF suspensions were sonicated and well shaken prior to dosing by gavage.

Animals and dosing

Male Alderley Park (AP) rats (9–13 weeks old, 190–250 g) were maintained as described earlier (Ashby *et al.*, 1985). They were dosed by gavage at 9 pm (12 and 36 h exposure) or at 9 am (24 h exposure). The rats were weighed before administration of the dosing solution (10 ml/kg body weight). Control animals were treated with corn oil or water. CP was used as the positive control. As the micronucleus assay was performed concomitantly with unscheduled DNA synthesis (UDS) in the liver, some of the 2AAF and 4AAF experiments were done without a CP control being performed the same day.

Bone marrow micronucleus assay

The assay was conducted as originally described by Schmid (1975). A single femur was aspirated with foetal calf serum (fcs) and the suspension centrifuged at 1000 g for 5 min. The pellet was resuspended in fcs by gentle agitation with a thin glass rod. After a further centrifugation, the pellet was rinsed with fcs and suspended in an equal volume fcs. Smears were prepared and stained with a Haematech automatic staining machine (Ames stain pack). The bone marrow micronuclei were scored under a light microscope; 1000 polychromatic erythrocytes (PE) were assessed per animal.

3 Results

All the bone marrow data are summarized in Table 1. The corn oil controls and water controls are presented separately. However, as the exposure time had no effect on the vehicle controls, the 24 h and 36 h values have been pooled in Table 1. The positive control, CP, caused a significant increase in micronucleated polychromatic erythrocytes (MPE) at both exposure times. 4AAF had no effect, whereas a small increase in MPE was observed after 36 h exposure at high levels of 2AAF (0.5–3.0 g/kg).

As mentioned above, micronucleus assays were performed concomitantly with the rat liver UDS assay. In some cases, it was possible to use the same animal for both genetic endpoints. The individual animal data from these experiments are shown in Table 2. CP had no effect on liver UDS in these rats but caused a marked increase in MPEs. 2AAF, on the other hand, caused a strong induction in liver UDS activity, and a larger number of animals was needed to detect a weak increase in MPEs. Finally, 4AAF did not affect either endpoint in the rats investigated. Details regarding the rat liver UDS assay are given in the IPCS report on "Unscheduled DNA Synthesis in Rat Liver After *in Vivo* Treatment with 2AAF and 4AAF" by Beije and Ashby (this volume).

To emphasize the difference in effect between CP

Table 1. *Incidence of MPEs in the bone marrow of rats treated with CP, 2AAF or 4AAF at the dose levels and exposure periods shown*

Compound	Exposure time (h)	Dose (mg/kg)	No. of animals	MPE/1000 PEs (mean ± SD)[a,b]	PE/PE + NE[c] (mean ± SD)
Corn oil	24/36	10 ml/kg	28	1.5 ± 1.2	0.47 ± 0.06; $n = 12$
Water		10 ml/kg	6	0.83 ± 0.41	0.52 ± 0.08; $n = 5$
CP	24	10	5	4.8 ± 2.8***	0.45; $n = 2$
		15	6	9.8 ± 3.9***	0.44 ± 0.07; $n = 4$
	36	10	11	14.6 ± 6.23***	—
		15	6	15.5 ± 6.12***	0.45; $n = 2$
2AAF	24	2000	6	2.0 ± 1.41	—
	36	50	5	1.6 ± 1.14	—
		100	5	2.6 ± 2.51	—
		250	8	2.13 ± 0.99	—
		500	19	3.5 ± 2.29***	0.40 ± 0.05; $n = 5$
		1000	10	3.0 ± 1.41**	—
		2000	10	4.8 ± 2.04***	0.46 ± 0.06; $n = 10$
		3000	18	4.0 ± 3.45***	0.40 ± 0.05; $n = 8$
4AAF	36	1000	5	1.8 ± 1.48	—
		2000	5	1.8 ± 1.64	0.43 ± 0.10; $n = 5$
		3000	11	2.45 ± 2.07	0.47 ± 0.06; $n = 11$

[a] MPE, micronucleated polychromatic erythrocyte; PE, polychromatic erythrocyte.
[b] Significance levels 0.01 (**) and 0.001 (***) by a one-sided Student's t-test.
[c] NE, normochromatic erythrocyte.

Table 2. *Bone marrow data and liver UDS obtained concurrently from the same animal at the time and dose levels shown*

Compound	Exposure Time (h)	Dose (mg/kg)	Liver UDS (NG)[a]	Bone marrow (MPE/1000 PE)[b]
Corn oil	36	10 ml/kg	−1.8	0
CP	36	10	−2.5	9
			−4.1	18
			−5.4	22
			−9.20	13
		15	−7.0	13
			−5.6	25
2AAF	36	50	25.17	0
		500	41.63	1
4AAF	36	2000	−2.46	0
			−1.92	0
			−3.52	3
			−2.84	3
			−4.51	3

[a] UDS, unscheduled DNA synthesis; NG, net nuclear grains.
[b] MPE, micronucleated polychromatic erythrocytes.
PE, polychromatic erythrocytes.

and 2AAF and 4AAF, a figure illustrating micronucleus induction after 36 h exposure, has been included (Figure 1). In this figure, three more CP doses have been added, and a maximum response in the rat at 10–15 mg CP/kg is indicated (from Ashby and Beije, 1985).

4 Discussion

In the present study, no statistically significant effect on bone marrow micronuclei was observed after 4AAF treatment. There was a slight tendency towards an induction of MPEs at the highest 4AAF dose used (3000 mg/kg). With 2AAF, both positive and negative results have been reported earlier (Heddle *et al.*, 1983). However, only fairly high dose levels of 2AAF, such as 1000 mg 2AAF/kg body weight, have been shown to induce micronuclei in the bone marrow (Friedman and Staub, 1977). Our results support these earlier findings, with a weak effect being observed at dose levels from 500 mg/kg. No induction of MPE was noticed at the lower dose levels used.

5 Acknowledgement

The financial support to B. Beije from the Swedish Work Environment Protection Fund is acknowledged.

6 References

Ashby, J., and B. Beije (1985) Concomitant observations of UDS in the liver and micronuclei in the bone marrow of rats exposed to cyclophosphamide or 2-acetylamino-fluorene. *Mutation Res.*, **150**, 383–392.

Ashby, J., P.A. Lefevre, B. Burlinson and M.G. Penman (1985) An assessment of the *in vivo* hepatocyte DNA repair assay. *Mutation Res.*, **156**, 1–18.

Beije, B., and J. Ashby, *Unscheduled DNA synthesis in rat liver after in vivo treatment with 2AAF and 4AAF*. This volume.

Friedman, M.A., and J. Staub (1977) Induction of micronuclei in mouse and hamster bone-marrow by chemical carcinogens. *Mutation Res.*, **43**, 255–262.

Heddle, J.A., M. Hite, B. Kirkhart, K. Mavournin, J.T. MacGregor, G.W. Newell and M.F. Salamone (1983) The induction of micronuclei as a measure of genotoxicity. A report of the U.S. Environmental Protection Agency Gene-Tox Program. *Mutation Res.*, **123**, 61–118.

Schmid, W. (1975) The micronucleus test. *Mutation Res.*, **31**, 9–15.

Figure 1. The effect of cyclophosphamide (CP), 2AAF, and 4AAF on the induction of micronuclei in rat bone marrow. Each data point represents the mean ± S.D. of the number of animals indicated. Probability, using Student's test; $**p < 0.01$; $***p < 0.001$.

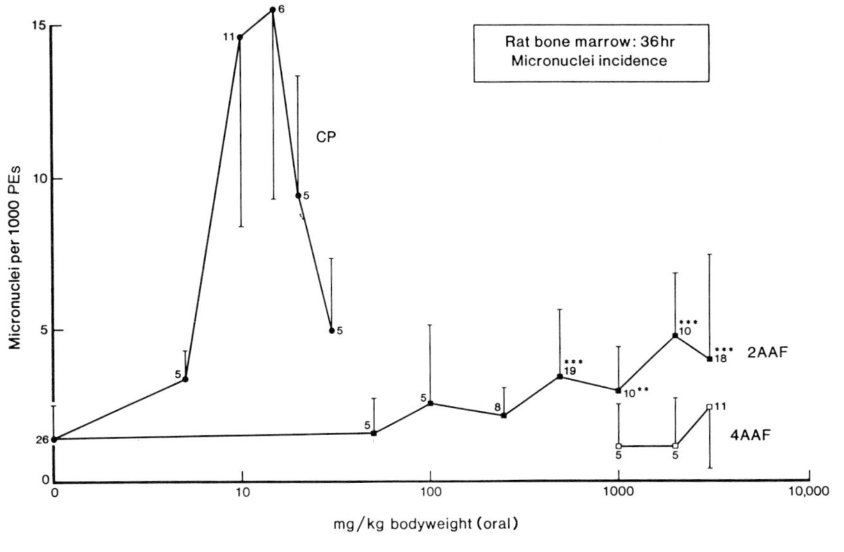

Paul A. Watkins
Imperial Chemical Industries PLC, Pharmaceuticals Division, Mereside, Cheshire, UK

36

Micronucleus test in the Alderley Park Wistar-derived albino rat

1 Summary

On the basis of provided toxicity data, benzo(a)pyrene (BP), pyrene (PYR), 4-acetylamino-fluorene (2AAF), and 2-acetylaminofluorene (4AAF) were administered to Alderley Park Wistar derived female rats. The compounds were formulated as ball-milled suspensions in Kraft Wesson corn oil. The corn oil and cyclophosphamide were used as the vehicle (negative) and reference (positive) controls, respectively. Corn oil, BP, and PYR were administered as a single dose by gavage; 4AAF, 2AAF, and cyclophosphamide were given as a single intraperitoneal injection. Samples of bone marrow were examined for the incidence of micronuclei in the polychromatic erythrocytes 24 h after dosing. The bone marrow was also used for metaphase analysis.

BP (200 mg/kg) and 4AAF (500 mg/kg) gave a statistically significant increase (2.4-fold and 2.9-fold, respectively) in the incidence of micronuclei when compared with the negative control group. The incidence of micronuclei with BP (100 mg/kg) and 2AAF (150 mg/kg) was increased approximately 2-fold but did not achieve statistical significance. Cyclophosphamide, as expected, caused a statistically significant increase in the incidence of micronuclei.

2 Introduction

The scoring of the incidence of micronuclei in bone marrow cells (the micronucleus test) is an alternative procedure to classical analysis for the detection of agents that cause chromosomal damage (Heddle *et al.*, 1983).

Until now, the rat has not been the species of choice for the micronucleus test because of problems of mast cell granules that are indistinguishable from micronuclei in the bone marrow smears. A method of preparing bone marrow cells was used that overcomes this problem and allows examination of bone marrow from an animal for both chromosome damage and induction of micronuclei [i.e. the "cytonucleus test" (Albanese and Watkins, 1983; Albanese, 1986, 1985; Watkins, 1985)]. Benzo(a)pyrene (BP), pyrene (PYR), 2-acetylaminofluorene (2AAF) and 4-acetylaminofluorene (4AAF) have been examined in the micronucleus test as part of the IPCS collaborative study. Cytogenetic damage and the induction of micronuclei were compared in the same animals.

3 Materials and methods

Test compounds

BP, PYR, 4AAF, and 2AAF were formulated as suspensions in corn oil (Kraft Wesson) by ball-milling overnight. The control (negative) article was the vehicle corn oil. The reference (positive) control cyclophosphamide (CP) was supplied by Koch Light Laboratories, Colnbrook, England. CP was dissolved in physiological saline immediately prior to dosing. The concentration of each formulation was calculated to give a constant dose volume of 5 ml/kg body weight. Doses were chosen on the basis of information supplied by the cytogenetic studies group coordinator. The doses were BP, 100 and 200 mg/kg orally; PYR, 250 and 500 mg/kg orally; 2AAF, 150 and 300 mg/kg intraperitoneally (ip); and 4AAF 500 and 1000 mg/kg ip. The highest dose of 4AAF was later reduced in steps to 375 mg/kg ip. Cyclophosphamide was dosed at 20 mg/kg ip corn oil at 5 ml/kg orally.

The stability of each test article formulation was not measured. The suspensions were stored at 4 °C over the period of dosing and resuspended by vigorous shaking prior to dosing.

Animals

110 female Alderley Park Wistar derived albino rats weighing 100–120 g at the start of the test. The animals were acclimatized in the animal room for 7 days prior to the start of the test.

The rats were housed up to 4 per cage with food and water freely available. Each rat was individually numbered by an ear mark.

Study design

For convenience of handling each test article group and the positive control group was divided into 5 replicates and consisted of 10 rats. The vehicle control group was divided into 5 replicates and consisted of 20 rats. Each replicate was sampled 24 h after dosing with a 24 h interval between replicates. Mortality and atypical signs were monitored daily.

Necropsy and bone marrow slide preparation

One and a half hours before necropsy, vinblastine sulphate was administered (5 mg/kg ip) to arrest dividing cells at metaphase for the cytogenetic component of the study (Albanese, this volume).

Table 1. *IPCS micronucleus test in Alderley Park Wistar-derived albino rat. Total and percentage incidence of micronucleus per group and statistical significant difference to Group* I *given in mg/kg*

Compound	Corn oil	BP	BP	PYR	PYR	4AAF	4AAF	2AAF	2AAF	CP
Dose	5	100	200	250	500	500	375	150	300	20
Group	I	II	III	IV	V	VI	VII	VIII	IX	X
Replicate	1 0	1	5	1	1	4	Dead[b]	0	0	4
1	0 2	4	6	1	0	2	Dead[b]	1	0	4
	2 0	2	1	0	0	5	0[a,c]	1	0	6
2	1 2	5	1	2	0	1	Dead[b]	3	1	10
	1 0	0	1	0	0	Poor[a] Slide	0	2	0	5
3	0 0	0	1	1	0	2	0	0	1	Poor[a] Slide
	0 1	0	1	1	1	0	2	2	0	6
4	0 0	2	1	1	0	0	1	1	Poor[a] Slide	8
	3 0	0	0	2	4	Dead	1	3	Dead	7
5	1 0	0	0	0	1	Dead	1	0	Dead	5
Total incidence	14/20000	14/10000	17/10000	9/10000	7/10000	14/7000	5/6000	13/10000	2/7000	55/9000
Frequency (%)	0.07	0.04	0.17	0.09	0.07	0.2	0.083	0.13	0.029	0.61
Significant difference from Group I	—	(NS)	($p - < 0.05$)	(NS)	(NS)	($p - < 0.05$)	(NS)	(NS)	(NS)	($p < 0.001$)

Figures in parenthesis indicate statistical significance (students t test) compared with negative control (Group I).
NS = Not significant.
[a] = Not included in the analysis.
[b] = Dosed 4AAF 1000 mg/kg.
[c] = Dosed 4AAF 750 mg/kg.

The rats were killed by carbon dioxide inhalation and one femur removed. The bone marrow was flushed with 3 ml of foetal calf serum (Flow Laboratories Ltd). The resulting cell suspension was centrifuged at 1000 rpm for 5 minutes. The supernatant was discarded and the cells resuspended in the remaining drops of supernatant. One drop of this cell suspension was then spotted onto a microscope slide and a film made by drawing another microscope slide over it. The remaining cell suspension was processed further for cytogenetic metaphase analysis (Albanese 1986). The film was allowed to air dry and stained, using an automatic staining machine (Ames Haemotec) applying Wright's stain to the cells.

Evaluation of slides for micronuclei and toxicity
The slides were coded and assessed blind. The number of micronuclei per 1000 PCEs was counted and recorded. As the first 200 PCEs were counted, the number of normocytes seen in the same field of view was also noted as a measure of cell toxicity.

4 Results

Animal observations
Deaths occurred in Replicate 1 of group VII (4AAF 1000 mg/kg). This dose was then lowered to 750 mg/kg. A death at this new dose in Replicate 2 resulted in a further reduction to 375 mg/kg. In Groups VI (4AAF, 500 mg/kg) and IX (2AAF, 300 mg/kg), both rats from each group in Replicate 5 were killed in extremis.

Incidence of micronuclei
The incidence of micronuclei for each animal per replicate and group with the percentage frequency and results of the statistical analysis is shown in Table 1. The one-sided statistical significances given in Table 1 are based on fitting a Poisson distribution to the number of micronuclei per animal in each of the groups using the GLIM programme and a logarithmic-link function (Baker and Nelder, 1978). The mean of each dosed group was compared with the control by dividing the parameter estimate by its standard error and referring the result to a Student's t distribution. The data exhibits extra-Poisson variability and the standard errors were inflated by the estimated scale parameter. The analysis should be regarded as approximate, since further analysis indicated that differences between the groups varied significantly from replicate to replicate (e.g., Groups II, III and VI are high relative to Group I in the first 2 replicates but not in the latter replicates). This makes the interpretation difficult, and the statistical significances should be regarded as a rough guide only. The groups that show an approximate 2-fold or more increase in the percentage frequency of micronuclei are BP 100 mg/kg (2.0), BP 200 mg/kg (2.4 statistically significant), 4AAF 500 mg/kg (2.9 statistically significant), 2AAF 150 mg/kg (1.9), and CP 20 mg/kg (8.7 statistically significant).

The normocyte counts for each slide are not shown but these and visual appraisal of the slides showed that the compounds did not affect the production of PCEs by the bone marrow.

5 Discussion

The statistical analysis of the data was confounded by the unexpected toxicity of 4AAF and 2AAF. The dose level of 4AAF was adjusted twice downward to a final dose of 375 mg/kg. Also with these compounds, 2 more data points were lost because of slides that were unreadable. However, BP at 200 mg/kg and 4AAF at 500 mg/kg did show a statistically significant increase in the incidence of micronuclei when compared with the vehicle control group. As already discussed the statistical analysis should only be regarded as a rough guide. A more simplistic approach of looking for an increase greater than twice the vehicle control group incidence of micronuclei showed that BP gave a dose-related positive result and PYR was consistently negative. 4AAF was positive at the highest dose level (500 mg/kg) only, whilst 2AAF at most, only showed a 1.9-fold increase in the incidence of micronuclei at the lower dose level (150 mg/kg). Three data points were lost due to deaths or technical reasons.

The constraints of time set by the collaborative programme prevented the investigation of more doses and time points. The results of the simultaneous metaphase analysis of the same bone marrow are detailed elsewhere in this volume (Albanese, 1986). These show BP to give a dose-related positive result. The increased incidence of chromosomally aberrant cells in the 2AAF (9-fold increase) and 4AAF (2-fold increase) was not considered to be biologically significant, since it was not dose related (2AAF) or it was within the historical control range for this laboratory (4AAF).

Although a few problems were encountered with this study this result and others (Watkins, 1985) show that it is possible to do the micronucleus test in the rat and simultaneously do metaphase analysis on the same bone marrow. Improvements to the staining methods and slide preparation since this study have resolved the technical problems encountered.

6 Conclusion

Statistical analysis of the results shows an inter-replicate variation that makes interpretation difficult. The results indicate that BP at 100 and 200 mg/kg, 4AAF at 500 mg/kg and 2AAF at 150 mg/kg have increased the incidence of micronuclei in Alderley Park Wistar derived albino female rats under the conditions of the test.

7 Acknowledgements

The author would like to thank Mr. S. Moore and Miss M. Osman for their excellent technical assistance and also Dr. R. Ferguson for his valued statistical analysis of the results.

8 References

Albanese, R. (1986) *In vivo cytogenetic assay in the Alderley Park Wistar derived albino rat* (this volume).

Albanese, R. (1985) In Parry, J.M. and Artlett, C.F. (Eds), 4-dimethylaminoazobenzene (DAB) and 4-cyanodimethylaniline (CDA) – An *in vivo* cytogenetic study in the Alderley Park (Wistar) rat, Parry, *Comparative Genetic Toxicology: The Second UKEMS Collaborative Study.* MacMillan, London, pp. 475–481.

Albanese, R. and P.A. Watkins (1983) Is the "Cyto-

Nucleus Test" a possibility? *Environ. Mutagenesis,* **5(6)**, 945.

Baker, R.J. and J.A. Nelder (1978) The GLIM system, release 3 manual. *Numerical Algorithms Group,* Oxford.

Heddle, J.A., M. Hite, B. Kirkhart, K. Mavournin, J.T. MacGregor, G.W. Newell and M.F. Salamone (1983) The induction of micronuclei as a measure of genotoxicity. A report of the U.S. Environmental Protection Agency Gene-Tox Program. *Mutation Res.*, **123**, 61–118.

Watkins, P.A., (1985) In Parry, J.M. and Arlett, C.F. (Eds), 4-dimethylazobenzene and 4-cyanodimethylaniline: Micronucleus test in the Alderley Park Wistar derived rat, *Comparative Genetic Toxicology: The Second UKEMS Collaborative Study.* MacMillan, London, pp. 453–459.

PART III
In vivo assays: bone marrow

C
Sister chromatid exchanges

Raymond R. Tice (Workgroup Chairman)[1]
Medical Department
Brookhaven National Laboratory Upton, New York, USA

37

Summary report on the performance of the in vivo sister chromatid exchange assays

1 Summary

Nine laboratories participated in a comparative analysis of the ability of BP/PYR and 2AAF/4AAF to induce sister chromatid exchanges (SCEs) *in vivo*, using mice, rats and Chinese hamsters. Bone marrow, intestinal epithelium and peripheral blood lymphocytes were used as target tissues, following either gavage or intraperitoneal (ip) administration of the test chemical. Based on a statistical analysis of the SCE data by ANOVA and using an α of 0.05, BP induced a highly significant increase in SCEs in every tissue sampled in each species and among all strains of mice by both routes. Treatment with PYR induced a marginally positive SCE response in 5 of 15 experiments (considering male and female responses separately). By both routes, treatment with 2AAF induced a highly significant increase in SCE frequency in bone marrow cells of all mouse strains, in bone marrow cells of rats, and in bone marrow and intestinal epithelial cells of Chinese hamsters. Treatment with 4AAF induced a marginally positive SCE response in 6 of 15 experiments (considering male and female responses separately). The 2 carcinogens, BP and 2AAF, were clearly active regardless of species, strain, sex, tissue and route of administration and more potent than either of the two putative non-carcinogenic structural analogues, PYR and 4AAF, respectively.

2 Introduction

As part of the IPCs Collaborative Study on Short-Term *In Vivo* Tests for Genotoxicity and Carcinogenicity, nine laboratories participated in an evaluation of the *in vivo* induction of sister chromatid exchanges (SCEs) by 3,4-benzo(a)pyrene (BP), pyrene (PYR), 2-acetamino-fluorene (2AAF) and 4-acetylaminofluorene (4AAF). SCE formation involves the breakage and rejoining of the two DNA duplexes (one in each sister chromatid) present in an eukaryote chromosome after the completion of DNA replication (for reviews, see Wolff, 1982; Sandberg, 1982; Tice and Hollaender, 1984a, b). SCEs are elicited primarily in response to DNA synthetic activity on a damaged template (e.g., Wolff *et al.*, 1974; Shafer, 1982), the magni-

[1] Workgroup participants: James W. Allen, Sheila M. Galloway (absent, but data considered), Andrew D. Kligerman, Steven B. Neal (absent, but data considered by Gregory S. Probst), Fabrizio Paliti, Annalisa Zunino (absent, but data considered), Ad D. Tates, Demetra Vlachos, and with statistical assistance of Brian G. Ormiston

tude of the response being proportional to the level of DNA damage (e.g., Perry and Evans, 1975; Reynolds *et al.*, 1979; Abe and Sasaki, 1982; Littlefield, 1982; Takehisa, 1982). In the absence of cell lethality, the exact number of SCEs elicited by any DNA-damaging treatment depends on: (1) the nature (e.g., Perry and Evans, 1975; Littlefield, 1982; Takehisa, 1982) and the genomic distribution (e.g., Carrano and Wolff, 1975; Schubert and Rieger, 1981) of the pertinent DNA lesion(s); (2) the efficiency of the appropriate DNA repair system in removing the lesion(s) (e.g., deWeerd-Kastelein *et al.*, 1977; Sasaki, 1982; Shafer, 1982; Tice and Schvartzman, 1982); (3) the temporal relationship between the formation of the primary DNA damage and the onset of scheduled DNA synthesis in the presence of bromodeoxyuridine (BrdUrd) (e.g., Sasaki, 1982; Shafer, 1982); and (4) the detection procedure (Natarajan *et al.*, 1983). The incorporation into DNA of BrdUrd or some other heavy analogue of thymidine for at least the first of two consecutive cell cycles is necessary for the visualisation of SCEs in second generation metaphase cells (e.g., Latt, 1974; Schvartzman and Tice, 1982).

The usefulness of assessing SCEs is based on their sensitivity as an indicator of genotoxic damage and on their ease of ascertainment. Depending on the methodology used (Perry and Evans, 1975; Natarajan *et al.*, 1983), the frequency of SCEs is increased significantly by many genotoxic agents following treatments at doses too low to induce a significant increase in chromosomal aberrations. Compared with chromosomal aberration analysis, SCE analysis is less subjective and requires the scoring of fewer cells for the same statistical power of test. However, the observation that SCEs are induced by chemicals [e.g., antimetabolites (Guglielmi *et al.*, 1982; Takehisa, 1982)] or conditions [e.g., psychological stress (Fischman *et al.*, 1985)] that are believed not to induce direct DNA damage, suggests that small but statistically significant increases in SCE frequency may not always indicate direct genotoxic activity. Rather, the positive responses may result from altered homeostasis, may or may not have a threshold and may or may not be mechanistically related to phenomena such as mutations and/or cancer. This limitation is not unique to SCE analysis; for example, chromosomal aberrations are also induced by antimetabolites (Kihlman, 1977) and by changes in osmolality (Galloway *et al.*, 1985).

For *in vivo* SCE analysis, several independent approaches have been developed to ensure adequate incorporation of BrdUrd for at least one round of DNA replication. The most commonly used techniques include multiple intraperitoneal (ip) injections of BrdUrd (Latt *et al.*, 1977), subcutaneous implantation of a BrdUrd Tablet (Allen *et al.*, 1978) coated with agar (King *et al.*, 1982) or partially coated with paraffin (McFee *et al.*, 1983), or continuous intravenous infusion of a BrdUrd isotonic solution (Schneider *et al.*, 1978). A variety of tissues has been used to examine the ability of chemicals to induce SCEs *in vivo*. Tissue analysed include: bone marrow (Latt *et al.*, 1977; Schneider *et al.*, 1978), alveolar macrophages (Conner *et al.*, 1979), regenerating liver (Conner *et al.*, 1979), fetal tissues (Kram, 1982) and germ cells (Allen, 1982).

To assess the induction of SCEs *in vivo*, several different protocols have been utilized. The protocols differ in the time of test agent treatment and the sample time, relative to the onset of BrdUrd administration. The difference in the 2 treatment time approaches used most often is determined by whether the cells to be scored were exposed to the chemical prior to or after the first cycle of BrdUrd incorporation. Treatment of animals just after the onset of exposure to BrdUrd increases the length of time available for the formation of potentially reactive metabolite(s), increases the opportunity for a persistent DNA lesion to elicit a SCE response in 2 cell cycles, lessens the opportunity for a confounding interaction between the test chemical and/or its reactive intermediate(s) and BrdUrd-substituted DNA, and, for many laboratories, is more practical. Treatment of animals 8 to 10 h after the onset of BrdUrd exposure (i.e., just prior to the second cell cycle for many somatic cell types) reduces bias in SCE yields due to selective toxicity and enables a direct comparison of the induction of SCEs and chromosomal aberrations within the same cell. Experiments to assess the impact of treatment time on SCE yields, using a limited number of genotoxic chemicals demonstrate responses that differed quantitatively but not qualitatively between early and late treatment times (Tice *et al.*, 1987). While the lack of concordance in the direction of the quantitatively different response (increased, decreased, and remaining unchanged) suggested that no single treatment time can a priori be assumed to provide the maximal sensitivity, earlier treatment time resulted in an appreciably better SCE response for slowly metabolized chemicals (e.g. dimethylbenzanthracene).

The second protocol variable is sample time. SCE analysis depends upon obtaining metaphase cells in the second generation after BrdUrd administration. *In vitro*, the yield of induced SCEs may be affected by sample time (Carrano *et al.*, 1978). However, *in vivo* bone marrow studies have not detected appreciable differences in SCE yield with sample time (McFee *et al.*, 1984). For most *in vivo* somatic cell experiments, including those using bone marrow, tissue samples can be analyzed generally between 18 and 30 hours after the beginning of the BrdUrd exposure, with the maximal number of second generation metaphase cells appearing between 22 and 26 h. Because of concern about treatment-induced cell cycle delay and the possibility that the more heavily damaged cells cycle more slowly, sample times greater than 22–26 have been used also.

Of great importance in any protocol design is the procedure used in the selection of the dose(s) to be tested. Each investigator in the SCE study group was provided with the available information on the LD_{50} for each chemical and with the testing regimens suggested by other coordinators. Each investigator was asked to assess to the fullest extent possible and/or by his/her standard testing protocol, the ability of the 4 chemicals to induce SCEs. In an attempt to compare directly the results obtained by different laboratories, each investigator was requested also to assess the ability of a fixed dose of mitomycin C, a direct acting genotoxicant, to induce SCEs.

Protocol information for each investigator can be

Table 1. *Individual laboratory protocols for the SCE assay*[a]

Species	Strain	Sex	Age (mo.)	Animals/ group	Tissue	Cells/ tissue	BrdUrd[b] Wt	Tablet[b] coat	Treat. time[c]	Sample time[c]	Stat. Anal.[d]	Investi- gator
Chinese hamster		F	3–4	3	bm, ie	25	20–30	A	5	26	Dunnett's test	Neal
Mouse	B6C3F1	M	2.5–3.5	4	bm	25	50	P	0.5	24	ANOVA	Tice
	(C3H × 101)F_1	M	3–4	5	bm	25	25	A	10	24, 30	ANOVA	Tates
	C57B1/6	M, F	4–5	4	bm	25	50	P	0.5	24	Dunnett's test	Allen
		M	3–4	4	pbl	25	NA	NA	NA	24[de]	Dunnett's test	Kligerman
	CD-1	M, F	1–2	4	bm	20	50	P[f]	1	24–30	ANOVA, trend test	Galloway
		M, F	2.5–3.5	3	bm	25	15	A	8	24	Student's t test	Palitti
		M	2–3	4	bm	25	12–13	A	1	25	Student's t test Mann-Whitney U test Correlation analysis	Zunino
Rat	S-D	M, F	1.5–2.0	5	bm	25	5 × 50	P	2	26	ANOVA	Vlachos

[a] Abbreviations: M = male; F = female; bm = bone marrow; ie = intestinal epithelium; pbl = peripheral blood lymphocytes; NA = not applicable; A = agar; P = paraffin; S-D = Sprague-Dawley.
[b] BrdUrd tablet weight (wt) in mg and type of coating (coat).
[c] Time in hours after BrdUrd tablet implantation.
[d] Statistical tests used by each investigator to analyze the data. For information on whether the data were transformed prior to analysis, see individual reports.
[e] Time at which blood was drawn for culturing.
[f] Animals exposed to BP were implanted with uncoated tablets.

Table 2. *Data receipt form: IPCS Collaborative study (CSSTT) IN VIVO SCE assay*

	Compound:	Route:
	Solvent:	Time of Admin:
	Dose (mg/kg):	Injected Volume:
Data Reporting Sheet	Species:	Strain:
Name of Participant:	Age:	Weight:
Laboratory:	BrdUrd Route:	BrdUrd Dose:
Date of Assay:	Colc. Durat:	Exp. Durat:

	Animal I.D./Sex/Slide I.D.											
Cell number/	No. SCE	No. Chr/	No. SCE	No. Chr/	No. SCE	No. Chr/	No. SCE	No. Chr/	No. SCE	No. Chr/	No. SCE	No. Chr
1.												
2.												
3.												
4.												
5												
6.												
7.												
8.												
9.												
10.												
11.												
12.												
13.												
14.												
15.												
16.												
17.												
18.												
19.												
20.												
21.												
22.												
23.												
24.												
25.												
$\bar{X}$												
S.D.												
%I												
%II												
%III												
No. Cells												

found in Table 1. Investigators who assessed SCEs *in vivo* directly used agar- or paraffin-coated BrdUrd tablets in most studies. Treatment times were within one or at 5 (Neal), 8 (Palitti) or 10 (Tates) h following BrdUrd tablet implantation. Sample times were about 24 post BrdUrd tablet implantation; 2 investigators (Galloway, Tates) sampled up to 30 post BrdUrd tablet implantation. Chinese hamsters (Neal), rats (Vlachos) and mice were used in these studies. Although bone marrow cells were analysed most often, intestinal epithelial cells (Neal) and peripheral blood lymphocytes (Kligerman) were analysed also. The results of each investigator's study and the inferences that can be made from interlaboratory comparisons follow.

3 Methods

Data receipt and validation

All data were received from each investigator on a common data entry form (Table 2). Data on the experimental protocol, chemical, dose, animals, and SCEs (on a per cell basis), along with cell kinetic data for the target tissue, were listed on these forms. All data were reviewed for completeness of information and then forwarded to Drs. F.J. de Serres and B. Margolin at NIEHS, Research Triangle Park (RTP), NC, USA, for computer input. Prior to statistical analysis, all SCE data, on a per cell basis, were verified twice, once at RTP and once at Brookhaven National Laboratory, and data entry errors were corrected.

Statistical analysis

An α level of 0.05 was used in all statistical evaluations. Mean SCE frequencies from individual animals, considered either as raw variates or determined from square-root transformed $(\text{SCE/cell} + 0.5)^{1/2}$ individual SCE/cell data (Whorton *et al.*, 1984), were analysed by a one-way Analysis of Variance (ANOVA). Data on animals for which less than 20 cells were scored were omitted from any statistical analysis, as were dose groups that contained only one animal. In laboratories where both male and female animals were used, a two-way ANOVA was conducted to determine if a sex difference in SCE response existed. If a significant sex difference response was obtained, male and female data were analysed separately; otherwise, male and female data were combined for analysis by a one-way ANOVA. To estimate potency for chemicals that induced an increase in SCE frequency, a dose that would have induced an absolute increase of 5 SCE/cell (the SCE-5 dose) above the concurrent solvent control frequency was calculated by inverse prediction, using linear regression analysis. For nonlinear SCE data, data at higher doses were omitted from the analysis if those data were not elevated significantly over the next lower dose. The linear regression statistics were used to estimate the SCE-5 dose and, where feasible, the asymmetric 95% confidence limits for that dose (Sokal and Rohlf, 1981).

Interlaboratory Mitomycin C (MMC) Comparison

As an additional method for evaluating inter-laboratory variability, each investigator was requested to conduct a positive control experiment with MMC using the same protocol used to assess the SCE inducing capability of the 4 test chemicals, with the following exceptions: the dose of MMC was to be 0.5 mg/kg and the route was to be ip. A common lot of MMC was not used in these studies. MMC-induced responses among animals within each laboratory were analysed with the same statistical approaches that were used to analyse the test chemical SCE data.

4 Results

Intra- and Interlaboratory Control Animal SCE Data Frequency Comparisons

For each laboratory, the control group mean SCE per cell frequency and the 95% confidence limits by species, tissue and route are presented in Table 3 and Fig. 1. Within a laboratory, control data were combined across subsets of control animals (e.g. male, female; solvent 1, solvent 2) if a significant difference was not found by a two-way ANOVA of nontransformed data. No significant difference in SCE frequencies were detected between untreated and corn-oil treated animals (Allen, $P = 0.2025$,), between corn oil and corn oil plus DMSO-treated animals (Neal, $P = 0.6187$,), or between corn oil and carboxymethylcellulose-treated animals (Zunio, $P = 0.2256$,). Upon analyses of male and female control data, significant differences in SCE frequencies were not found in 2 studies of mice (Allen, $P = 0.1025$; Palitti, $P = 0.7741$), but were found in another study of mice (Galloway, $P = 0.0021$) and in the single rat study (Vlachos, $P = 0.0003$). No significant difference in control SCE frequency was found between bone marrow and intestinal epithelium (Neal, $P = 0.2047$) or between bone marrow cells sampled at 24 and at 30 h post BrdUrd tablet implantation (Tates, $P = 0.6144$,). However, a significant difference in SCE frequency was found between mice implanted with uncoated BrdUrd tablets and those implanted with paraffin-coated BrdUrd tablets (Galloway, $P = 0.0012$).

After combining, where appropriate, control data within a laboratory, interlaboratory comparisons were conducted. Significant differences in control animal SCE frequencies occurred among laboratories in which bone marrow tissue was analyzed ($P < 0.0001$), between species ($P < 0.0001$) and, not unexpectedly, between studies analysing tissues *in vivo* directly and the one study (Kligerman) in which SCE levels were analyzed *in vitro* using an *in vivo* exposed tissue (i.e. peripheral blood lymphocytes) ($P < 0.0001$).

The Induction of SCEs by BP

In Table 4, data on the ability of BP to induce SCEs are presented along with each investigator's conclusion and the conclusion of the present analysis based on a one-way ANOVA of raw or square-root transformed SCE per cell data. There was complete agreement beetween each investigator's conclusion and the results of the present analysis. Use of nontransformed *vs.* square-root transformed SCE per cell data did not alter the conclusion nor, to any appreciable extent, the calculated P values. Treatment with BP induced a significant increase in SCE frequency in all species, in all strains of mice, in both sexes,

Table 3. *Control SCE data*[a]

Species	Strain	Sex	Tissue	Number animals	Solvent	Route	SCE frequency[b]	Investigator
Chinese hamster		F	bm, ie	32	CO + CO/DMSO	ip/gav	2.83 ± 0.11	Neal
Mouse	B6C3F1	M	bm	12	CO	ip	5.03 ± 0.55	Tice
	(C3H × 101)F_1	M	bm	12	OO	gav	5.58 ± 0.26	Tates[c]
	C57B1/6	M + F	bm	67	None + CO	gav	3.91 ± 0.17	Allen
		M	pb1	20	CO	gav	10.13 ± 0.62	Kligerman
	CD-1	M[d]	bm	8	CO	ip	5.50 ± 0.45	Galloway
		F[d]	bm	8	CO	ip	5.55 ± 0.28	Galloway
		M[e]	bm	4	CO	ip	5.61 ± 0.90	Galloway
		F[e]	bm	4	CO	ip	6.96 ± 1.67	Galloway
		M + F	bm	23	CO	gav	3.60 ± 0.37	Palitti
		M	bm	24	CO + CMC	ip/gav	2.90 ± 0.13	Zunino
Rat	S-D	M	bm	10	CO	gav	2.60 ± 0.59	Vlachos
		F	bm	10	CO	gav	4.27 ± 0.59	Vlachos

[a] Abbreviations: F = female; M = male; CMC = carboxymethyl cellulose; CO = corn oil; DMSO = dimethylsulphoxide; OO = olive oil; bm = bone marrow; ie = intestinal epithelium; pbl = peripheral blood lymphocytes; ip = intraperitoneal; gav = gavage; S-D = Sprague-Dawley.
[b] Group men SCE/cell frequency ± 95% confidence limits.
[c] Data from 24 and 30 h sample times combined.
[d] Data on control animals implanted with paraffin-coated BrdUrd tablets.
[e] Data on control animals implanted with noncoated BrdUrd tablets.

Figure 1. SCE frequencies for control animals by investigator. Each point represents the mean SCE/cell frequency while the range bars give the 95% confidence limits. The numbers in parentheses indicate the number of control animals analysed. Abbreviations: IE = intestinal epithelium; PBL = peripheral blood lymphocytes; F = female; M = male. *Control animals implanted with paraffin-coated BrdUrd pellets. **Control animals implanted with uncoated BrdUrd pellets.

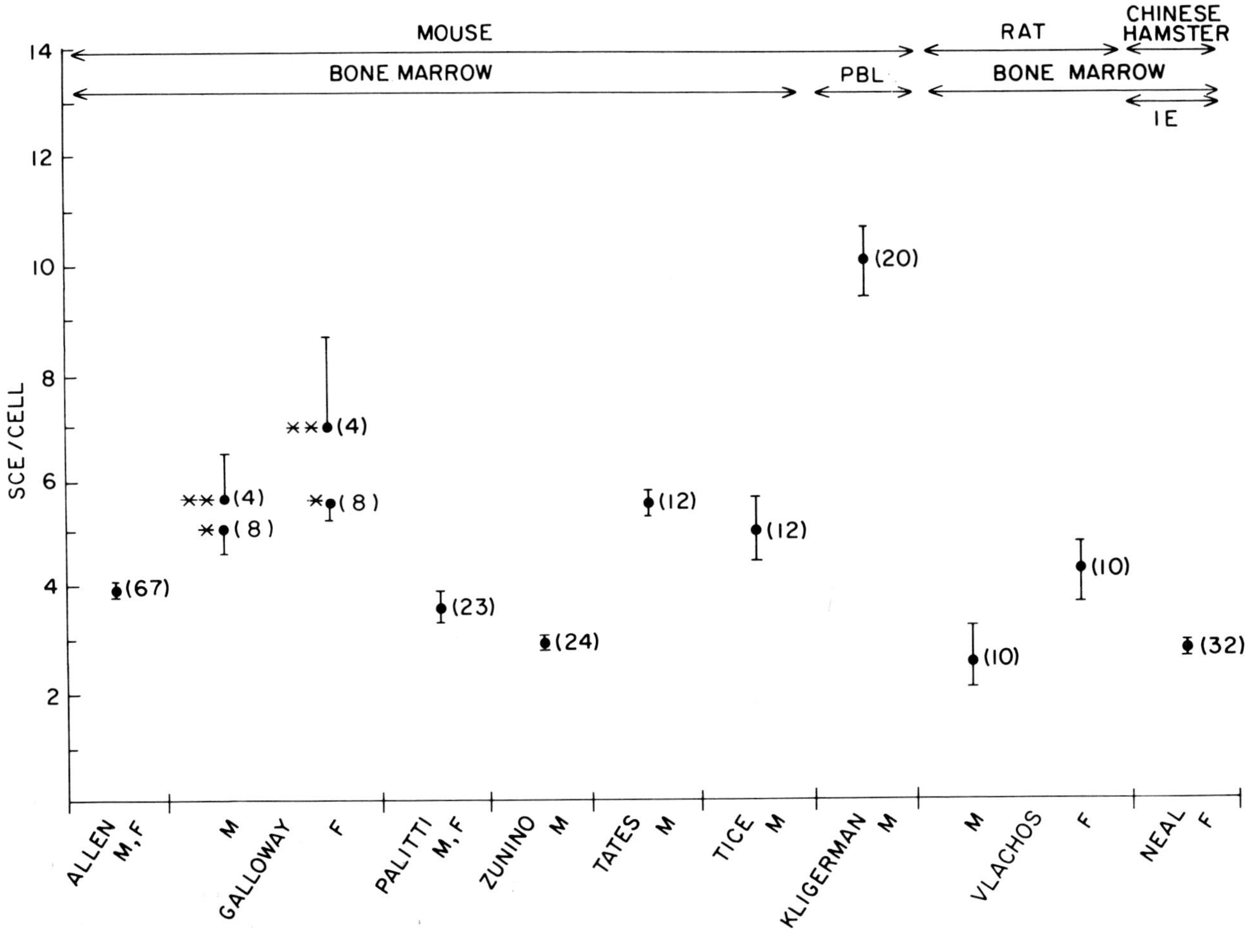

in every tissue assayed and when administered by either ip injection or gavage. Significant differences between male and female animals (Table 5) were found for 2 of 3 studies on mice (Allen, Galloway) and in the single rat study (Vlachos). In the third mouse study in which male and female animals were used (Palitti), no sex difference in response was detected. In the only study (Neal) in which the same tissue (bone marrow) was examined using the same experimental protocol except for route of administration, treatment by ip injection was more effective in inducing SCEs than was treatment by gavage ($P = 0.0001$). In the studies in which more than one tissue was analysed for SCE induction following administration of BP by the same route, the magnitude of the SCE response did not differ significantly between peripheral blood lymphocytes (Kligerman) and bone marrow cells (Allen) in male C57B1/6 mice treated by gavage (~ 10 SCE per cell over control levels), nor between intestinal epithelial cells and bone marrow cells of gavage-treated female Chinese hamsters (Neal, $P = 0.2214$). Also, BP-induced SCE frequencies in bone marrow cells collected at 24 h were not significantly different from those collected at 30 h after BrdUrd tablet implantation (Tates, $P = 0.1294$).

The shapes of the dose response for SCE induction, regardless of species, strain, tissue, sex, route of exposure, or solvent, were strikingly similar (Figure 2). In all cases, except for Chinese hamsters administered BP by ip injection, the dose response curve saturated at higher doses after a sharp initial rise to a maximal SCE level ranging from 3 to 10 SCE per cell above control frequencies. In Chinese hamsters, the dose response curve for SCE induction by BP in bone marrow cells was linear following ip treatment but appeared to saturate in both bone marrow cells and intestinal epithelial cells following treatment by gavage.

To compare the relative potency of BP among the different studies, a SCE-5 dose for each study was calculated (Table 6). In Figure 3, these SCE-5 doses are ordered by species and route. For mice the calculated SCE-5 doses ranged from a low of 27.3 mg/kg to a high of 121.3 mg/kg. The SCE-5 doses for Sprague-Dawley rats and Chinese hamsters were higher than those determined for mice. Where male and female animal SCE-5 doses could be compared, only in rats was there an appreciable sex difference. Male rats appeared more sensitive than females (237.1 vs 391.5 mg/kg, respectively). Where tissue comparisons were possible, the SCE-5 doses for BP

Table 5. *Male vs female responses to BP*[a]

Investigator	Species	Strain	Route	P Value[b]
Allen	Mouse	C57B1/6	gav	0.0038[c]
Galloway	Mouse	CD-1	ip	0.0100[c]
Palitti	Mouse	CD-1	gav	0.2949
Vlachos	Rat	S-D	gav	0.0001[c]

[a] See Table 4 for explanation of abbreviations.
[b] P values determined by two-way ANOVA, the sex by treatment interaction for the S-D rat data was significant.
[c] Significantly different at an α of 0.05.

Table 4. *Induction of SCEs by BP*[a]

Species	Strain	Sex	Tissue	Solvent	Route	Doses (mg/kg)	SCE response[b] x	SCE response[b] $\sqrt{x}$	Investigator	Investigator conclusion
Chinese hamster		F	bm	CO	ip	62,5,125,250, 500	0.0001[d]	0.0001[d]	Neal	+
		F	bm	CO	gav	125,250,500, 1000	0.0464[d]	0.0318[d]	Neal	+
		F	ie	CO	gav	125,250,500, 1000	0.0067[d]	0.0063[d]	Neal	+
Mouse	B6C3F1	M	bm	CO	ip	62,5,125,250	0.0001[d]	0.0001[d]	Tice	+
	(C3H × 101)F_1	M	bm	OO	gav	63,126,252	0.0001[d]	0.0001[d]	Tates[c]	+
	C57B1/6	M	bm	CO	gav	63,126,252, 504,1008	0.0001[d]	0.0001[d]	Allen	+
		F	bm	CO	gav	63,126,252, 504,1008	0.0001[d]	0.0001[d]	Allen	+
		M	pbl	CO	gav	63,252,504	0.0001[d]	0.0001[d]	Kligerman	+
	CD-1	M	bm	CO	ip	25,50,100	0.0001[d]	0.0001[d]	Galloway	+
		F	bm	CO	ip	25,50,100, 200	0.0003[d]	0.0001[d]	Galloway	+
		M, F	bm	CO	gav	63,126,252	0.0001[d]	0.0001[d]	Palitti	+
		M	bm	CO	gav	63,126,252	0.0001[d]	0.0001[d]	Zunino	+
Rat	S-D	M	bm	CO	gav	62.5,125, 250,500	0.0001[d]	0.0001[d]	Vlachos	+
		F	bm	CO	gav	62.5,125, 250,500	0.0042[d]	0.0044[d]	Vlachos	+

[a] Abbreviations: F = female; M = male; CMC = carboxymethyl cellulose; CO = corn oil; DMSO = dimethylsulphoxide; NaCl = 0.9% sodium chloride; OO = olive oil; bm = bone marrow; ie = intestinal epithelium; pbl = peripheral blood lymphocytes; ip = intraperitoneal; gav = gavage; + = positive response; + w = weakly positive response; − = negative response.
[b] P value for SCE response analyzed by one-way ANOVA using raw SCE per cell means (X) or square root transformed data ($\sqrt{x + 0.5}$) for cells analyzed within each animal.
[c] Data from 24 and 30 h sample times combined.
[d] Significantly different at $\alpha = 0.05$.

Figure 2. SCE dose response curves for BP. A. Administration of BP by ip injection. B. Administration of BP by gavage. Symbols: ○ = male; □ = female; △ = combined male and female; ○, □ = mouse; ●, ■ = rat; ■ = Chinese hamster;——— = bone marrow;— — = intestinal epithelium;— · — = peripheral blood.

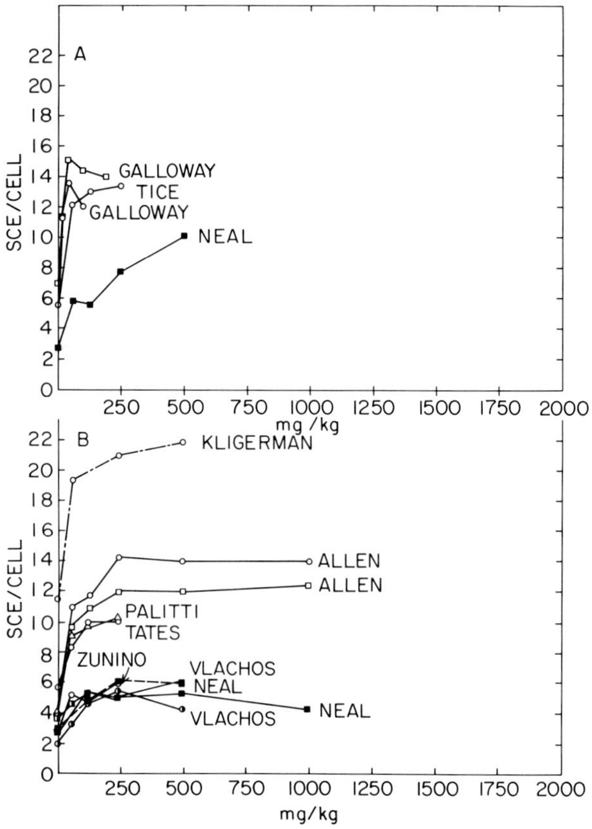

administered by gavage did not differ significantly between bone marrow and peripheral blood lymphocytes of male C57B1/6 mice (44.9 mg/kg vs 40.1 mg/kg, respectively) but appeared to differ between bone marrow and intestinal epithelial cells of Chinese hamsters (241.6 mg/kg vs 401.9 mg/kg, respectively). Finally, the SCE-5 doses for bone marrow cells of Chinese hamsters were similar regardless of whether BP was administered by ip injection (287.9 mg/kg) or by gavage (241.6 mg/kg).

Induction of SCEs by PYR

The results of the PYR studies conducted by the nine laboratories are summarized in Table 7. Treatment with PYR induced a positive SCE response in 5 of 15 trials: following treatment by ip injection, in male and female CD-1 mice (Galloway) and, following treatment by gavage, in male CD-1 mice (Zunino), female C57B1/6 mice (Allen), and male Sprague Dawley rats (Vlachos). Use of square-root transformed data did not alter significantly the P value calculated using raw variates. Differences between male and female animals were found in the 3 mouse studies (Allen, Galloway, Palitti) and in the single rat study (Vlachos)(Table 8). While the ANOVA P value was significant for both male and female rats, only the male data exhibited a positive increase. With one exception (Zunino), there was concordance between the conclusion of the investigator and that of the present analysis. This discordance in conclusions arises from differences in statistical approaches. Zunino obtained a significant difference for PYR using the Mann-Whitney U Test and comparing all PYR-treated animals against

Figure 3. SCE-5 doses for BP by investigator. Each point represents the SCE-5 dose. The range bars indicate the asymetric 95% confidence limits. Male and female animals were combined if no significant difference (at an alpha of 0.05) were found. Range bars not presented where the analysis was considered inappropriate. Abbreviations: IE = intestinal epithelium; PBL = peripheral blood lymphocytes: IP = intraperitoneal; GAV = gavage, F = female; M = male.

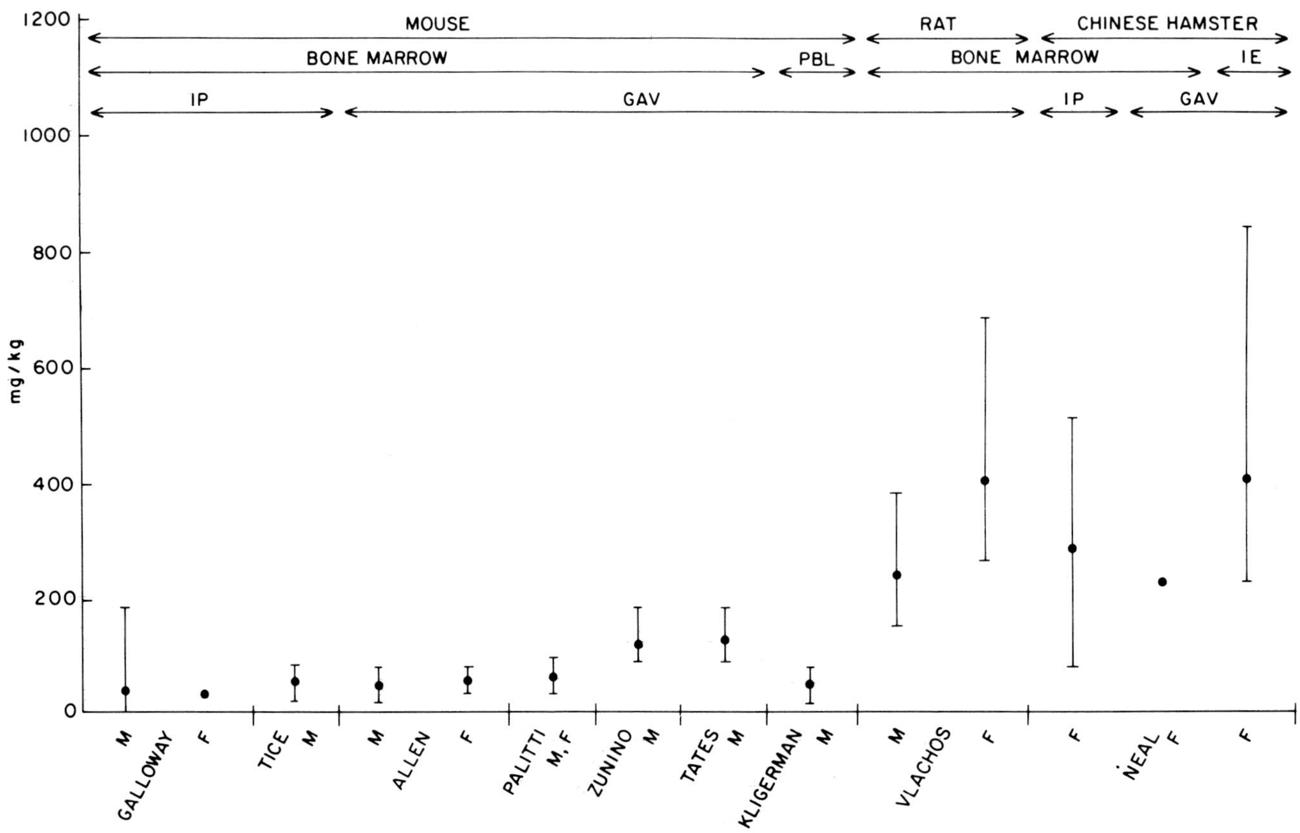

Table 6. *SCE-5 doses for BP*[a]

Species	Strain	Sex	Tissue	Solvent	Route	Doses used[b] (mg/kg)	SCE-5 (mg/kg)	L95%CL[c] (mg/kg)	U95%CL[d] (mg/kg)	Investi-gator
Chinese hamster		F	bm	CO	ip	62.5,125,250 500	287.9	87.6	504.2	Neal
		F	bm	CO	gav	125	241.6	nc	nc	Neal
		F	bm	CO	gav	125,250	401.9	236.0	839.8	Neal
Mouse	B6C3F1	M	bm	CO	ip	62.5	48.2	20.9	78.5	Tice
	(C3H × 101)F_1	M	bm	OO	gav	63	121.3	81.6	173.3	Tates[e]
	C57B1/6	M	bm	CO	gav	62.5	44.9	17.2	75.9	Allen
		F	bm	CO	gav	62.5	54.4	34.0	77.2	Allen
		M	pbl	CO	gav	63	40.1	12.2	69.6	Kligerman
	CD-1	M	bm	CO	ip	25,50	27.3	nc	nc	Galloway
		F	bm	CO	ip	25,50	29.7	10.9	49.2	Galloway
		M,F	bm	CO	gav	63	66.7	37.8	99.8	Palliti
		M	bm	CO	gav	63	117.3	84.8	171.3	Zunino
Rat	S-D	M	bm	CO	gav	125	237.1	152.1	381.6	Vlachos
		F	bm	CO	gav	125	391.5	263.9	681.5	Vlachos

[a] Abbreviations: nc = not possible to calculate, see Table 4 for explanation of additional abbreviations.
[b] Doses used in the calculation of the SCE-5 dose.
[c] Lower 95% confidence limit.
[d] Upper 95% confidence limit.
[e] Data from 24 and 30 h sample times combined.

Table 7. *Induction of SCEs by PYR*[a]

Species	Strain	Sex	Tissue	Solvent	Route	Doses (mg/kg)	SCE response[b] x	$\sqrt{x}$	Investi-gator	Investigator Conclusion
Chinese hamster		F	bm	CO/DMSO	ip	62.5,125,250, 500	0.8842	0.8835	Neal	–
		F	bm	CO/DMSO	gav	125,250,500 1000	0.3457	0.3542	Neal	–
		F	ie	CO/DMSO	gav	125,250,500, 1000	0.8511	0.8456	Neal	–
Mouse	B6C3F1	M	bm	CO	ip	500,1000,2000	0.8010	0.8143	Tice	–
	(C3H × 101)F_1	M	bm	OO	gav	202,404,808	0.3573	0.3562	Tates[c]	–
	C57B1/6	M	bm	CO	gav	202,404,808, 1212	0.4581	0.4578	Allen	–
		F	bm	CO	gav	202,404,808	0.0400[d]	0.0397[d]	Allen	+ W
		M	pbl	CO	gav	202,404,808	0.3741	0.3834	Kligerman	–
	CD-1	M	bm	CO	ip	100,400,800, 1000	0.0434[d]	0.0433[d]	Galloway	+ W
		F	bm	CO	ip	100,400,800, 1000	0.0162[d]	0.0166[d]	Galloway	+ W
		M	bm	CO	gav	200,400,800,	0.9916	0.9915	Palitti	–
		F	bm	CO	gav	200,400,800,	0.4573	0.4555	Palitti	–
		M	bm	CO	gav	202,404,808,	0.0267[d]	0.0270[d]	Zunino	–
Rat	S-D	M	bm	CO	gav	125,250,500, 1000	0.0101[d]	0.0094[d]	Vlachos	+
		F	bm	CO	gav	125,250,500, 1000	0.0045[d]	0.0035[d]	Vlachos	–

[a] See Table 4 for explanation of abbreviations and footnotes.
[b] Although a significant treatment effect was found, the response was for a decline in SCEs.

Table 8. *Male vs female animal response to PYR*[a]

Investigator	Species	Strain	Route	P Value[b]
Allen	Mouse	C57B1/6	gav	0.0473[c]
Galloway	Mouse	CD-1	ip	0.0046[c]
Palitti	Mouse	CD-1	gav	0.0187[c]
Vlachos	Rat	S-D	gav	0.0001[c]

[a] See Table 4 for explanation of abbreviations.
[b] P values determined by two-way ANOVA, the sex by treatment interaction for the S-D rat data was significant.
[c] Significantly different at an α of 0.05.

Figure 4. SCE dose response curves for PYR. A. Administration of PYR by ip injection. B. Administration of PYR by gavage. Symbols: ○ = male; □ = female; △ = combined male and female; ○, □ = mouse; ◐, ◧ = rat; ■ = Chinese hamster; ——— = bone marrow; --- = intestinal epithelium; —·— = peripheral blood lymphocytes.

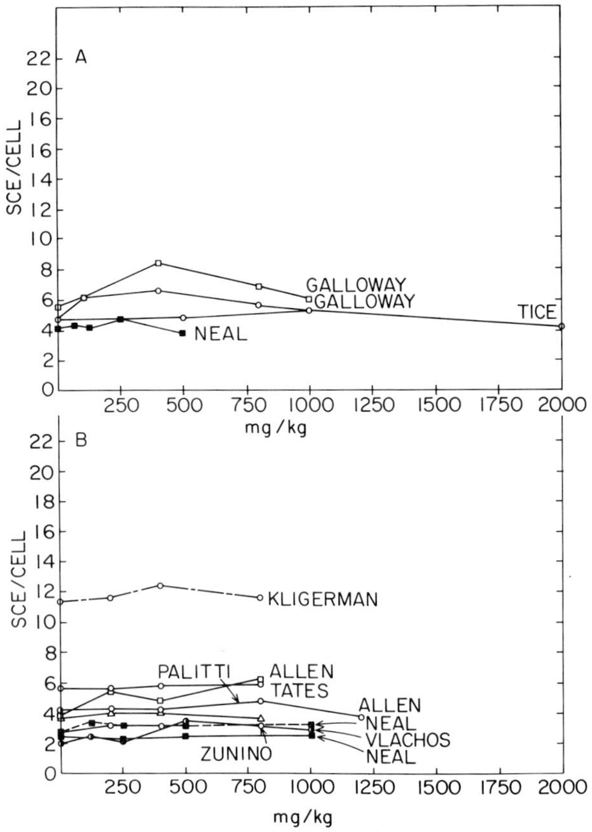

Figure 6. SCE dose response curves for 2AAF. A. Administration of 2AAF by ip injection. B. Administration of 2AAF by gavage. Symbols: ○ = male; □ = female; △ = combined male and female; ○, □ = mouse; ◐, ◧ = rat; ■ = Chinese hamster, ——— = bone marrow; --- = intestinal epithelium; —·— = peripheral blood lymphocytes.

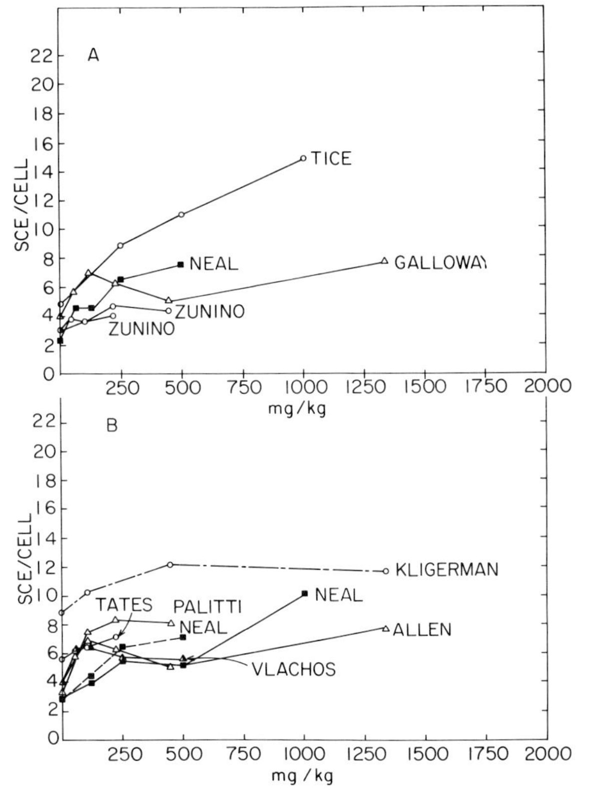

Figure 5. SCE-5 doses for PYR by investigator. Each point represents the SCE-5 dose. The range bars indicate the asymmetric 95% confidence limits. Range bars are not presented where the analysis was considered inappropriate. Abbreviations: IP = intraperitoneal; GAV = gavage; F = female; M = male. The SCE-5 doses calculated for BP under similar experimental conditions are provided for comparative purposes.

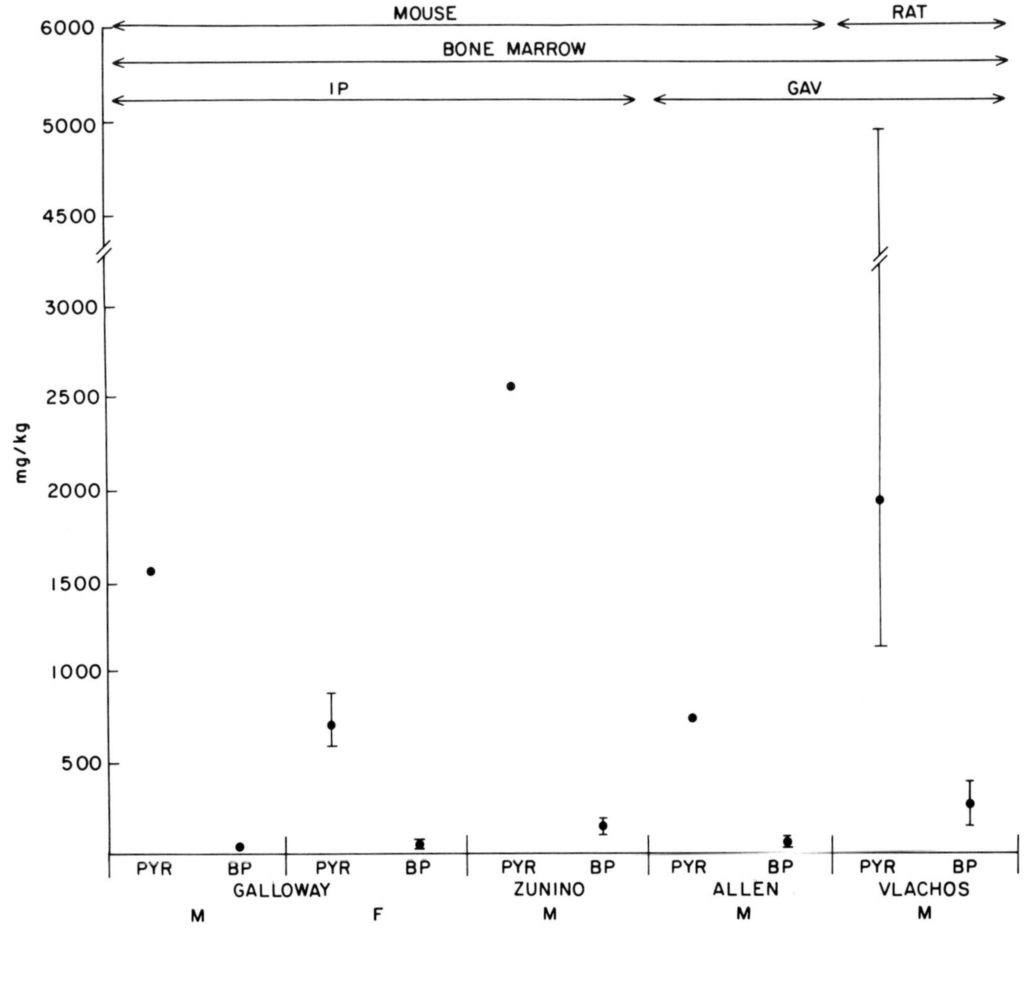

combined control animal data. However, because of nonsignificant P value following an analysis for a linear trend in the dose-response curve and finding also no dose which produced an elevation in SCE frequency of 2.2 standard deviations above the combined control group data, Zunino concluded that PYR was ineffective in inducing SCEs.

The SCE dose-response curves are presented in Figure 4. Where feasible, an SCE-5 dose and the 95% confidence limits were calculated for those experiments in which a positive response was detected (Table 9). These values are presented graphically in Figure 5, along with the BP SCE-5 doses calculated for the same investigator. Dividing the SCE-5 doses for PYR by the SCE-5 doses for BP provides one measure of the relative potency of BP when compared to that of PYR (Table 10). The relative potency for BP over PYR ranged from a low of 8.1 in rats to a high of 42.3 in male CD-1 mice.

Induction of SCEs by 2AAF

Treatment with 2AAF induced a significant increase in SCEs (Table 11) in male and female animals, in each strain of mice, in every species, by both routes and in almost every tissue. The single exception was for peripheral blood lymphocytes of C57B1/6 mice following treatment with 2AAF by gavage (Kligerman). It should be noted that by

Table 9. *SCE-5 doses for PYR*[a]

Species	Strain	Sex	Tissue	Solvent	Route	Doses used[b] (mg/kg)	SCE-5 (mg/kg)	L95%CL[c] (mg/kg)	U95%CL[d] (mg/kg)	Investigator
Mouse	C57B1/6	F	bm	CO	gav	202	727.1	nc	nc	Allen
	CD-1	M	bm	CO	ip	100,400	1154	nc	nc	Galloway
		F	bm	CO	ip	100,400	732.5	613.5	880.2	Galloway
		M	bm	CO	gav	223	2569	nc	nc	Zunino
Rat	S-D	M	bm	CO	gav	125,250,500	1918	1158	4595	Vlachos

[a] See Table 4 for explanations of abbreviations and Table 6 for explanations of footnotes.

Table 10. *Relative potency of BP vs PYR in bone marrow*[a]

Species	Strain	Sex	Solvent	Route	SCE-5 dose PYR/BP	Investigator
Mouse	C57B1/6	F	CO	gav	13.4	Allen
	CD-1	M	CO	ip	42.3	Galloway
		F	CO	ip	24.7	Galloway
		M	CO	gav	21.9	Zunino
Rat	S-D	M	CO	gav	8.1	Vlachos

[a] See Table 4 for explanation of abbreviations.

Table 11. *Induction of SCEs by 2AAF*[a]

Species	Strain	Sex	Tissue	Solvent	Route	Doses (mg/kg)	SCE response[b] x	SCE response[b] $\sqrt{x}$	Investigator	Investigator conclusion
Chinese hamster		F	bm	CO/DMSO	ip	62.5,125,250, 500	0.0011[d]	0.0004[d]	Neal	+
		F	bm	CO/DMSO	gav	125,250,500, 1000	0.0001[d]	0.0001[d]	Neal	+
		F	ie	CO/DMSO	gav	125,250,500, 1000	0.0004[d]	0.0005[d]	Neal	+
Mouse	B6C3F1	M	bm	CO	ip	250,500,1000	0.0001[d]	0.0001[d]	Tice	+
	(C3H × 101)F_1	M	bm	CO	gav	55.8,111.5, 223,446,892[e]	0.0001[d]	0.0001[d]	Tates[e]	+
	C57B1/6	M,F	bm	CO	gav	55.8,111.5, 223,446,1338	0.0001[d]	0.0001[d]	Allen	+
		M	pbl	CO	gav	111.5,446,1338	0.0674	0.0624	Kligerman	+
	CD-1	M,F	bm	CO	ip	250,500,750, 1000	0.0009[d]	0.0002[d]	Galloway	+
		M,F	bm	CO/DMSO	gav	110,220,440	0.0001[d]	0.0001[d]	Palitti	+
		M	bm	CMC/NaCl	ip	55.8,111.5, 223,446[b]	0.0296[d]	0.0233[d]	Zunino	+
Rat	S-D	M,F	bm	CO	gav	62,5,125, 250,500	0.0029*	0.0017*	Vlachos	+

[a] See Table 4 for explanation of abbreviations and footnotes.
[e] Data combined from two separate experiments with 2AAF.

the investigator's statistical criteria (i.e. Dunnett's test), 2AAF was positive in this experiment. The conclusion of the common analysis was not affected by transformation of the SCE per cell data. No differences between male and female animals in SCE response were found in any study (Table 12). The dose–response curves are presented in Figure 6.

After omitting the nonlinear portions of each dose–response curve from the analysis, a SCE-5 dose for each study was calculated (Table 12 and Figure 7). The SCE-5 doses and their corresponding 95% confidence limits are more variable among strains and between species and tissues than those calculated for BP. This variability probably reflects the relatively lower potency of 2AAF in inducing SCEs when compared with BP, a greater range in sensitivity, and the greater complexity of the observed SCE dose–response curves.

When the magnitude of the SCE response in different tissues was compared, 2AAF administered by gavage induced in mice a greater response in bone marrow cells (Allen, 5.5 SCEs per cell above control levels) when compared to peripheral blood lymphocytes (Kligerman, 3.5 SCEs per cell above control levels) and in Chinese hamsters (Neal) a greater response in intestinal epithelial cells (4.3) than in bone marrow cells (2.1). Similarly, when the effect of exposure route on the induction of SCEs by 2AAF in bone marrow cells of Chinese hamsters was analysed, administration by ip injection induced more SCEs (4.5 SCE/cell) than did administration by gavage (2.3 SCE/cell; $p = 0.0004$). There was not a significant difference between samples times (24 vs 30 h) in the SCE response to 2AAF (Tates, $P = 0.9746$). Finally, while mice had the highest response to BP, rats responded slightly higher to the induction of SCEs by 2AAF than did either mice or Chinese hamsters. These differences among tissues, routes, and species are supported by the calculated SCE-5 doses (Table 13).

Table 12. *Male vs female responses to 2AAF*[a]

Investigator	Species	Strain	Route	P Value[b]
Allen	Mouse	C57B1/6	gav	0.8091
Galloway	Mouse	CD-1	ip	0.1848
Palitti	Mouse	CD-1	gav	0.2822
Vlachos	Rat	S-D	gav	0.3994

[a] See Table 4 for explanation of abbreviations.
[b] P values determined by two-way ANOVA, all sex by treatment interactions were nonsignificant.

Figure 7. SCE-5 doses for 2AAF by investigator. Each point represents the SCE-5 dose. The range bars indicate the asymmetric 95% confidence limits. Male and female animals were combined if no significant difference (at an alpha of 0.05) were found. Abbreviations: IE = intestinal epithelium; PBL = peripheral blood lymphocytes; IP = intraperitoneal; GAV = gavage; F = female; M = male.

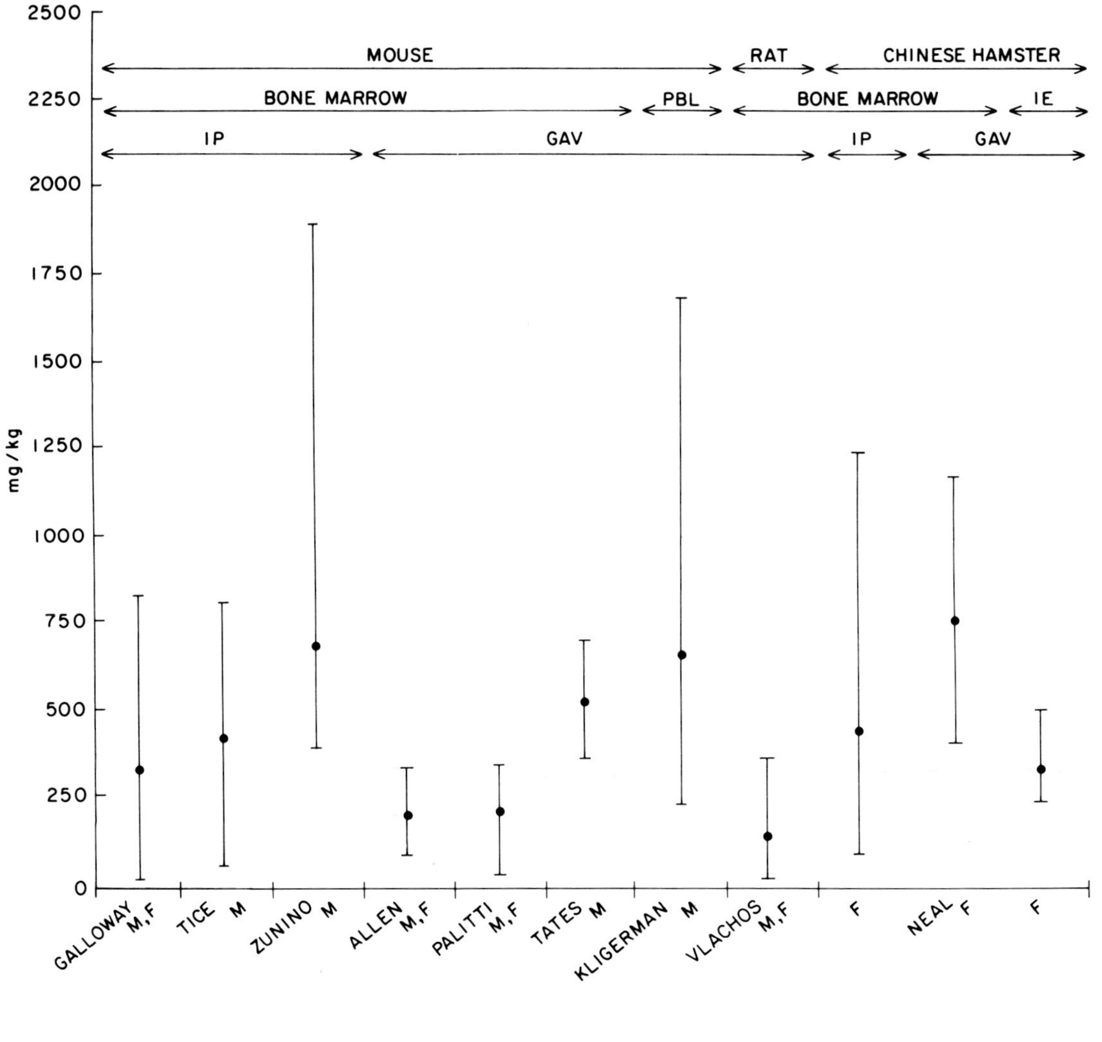

Table 13. *SCE-5 doses for 2AAF*[a]

Species	Strain	Sex	Tissue	Solvent	Route	Doses used[b] (mg/kg)	SCE-5 (mg/kg)	L95%CL[c] (mg/kg)	U95%CL[d] (mg/kg)	Investigator
Chinese hamster		F	bm	CO/DMSO	ip	62.5,125,250, 500	496.5	82.9	1243	Neal
		F	bm	CO/DMSO	gav	125,250,500, 1000	753.9	418.2	1157	Neal
		F	ie	CO/DMSO	gav	125,250,500	332.0	224.0	494.8	Neal
Mouse	B6C3F1	M	bm	CO	ip	250,500,1000	431.5	53.4	809.0	Tice
	(C3H × 101)F_1	M	bm	OO	gav	55.8,111.6,223, 446,892[f]	521.8	355.7	692.8	Tates[e]
	C57B1/6	M,F	bm	CO	gav	55.8,111.5	182.8	86.3	324.8	Allen
		M	pbl	CO	gav	111.5,446	661.4	235.5	1686	Kligerman
	CD-1	M,F	bm	CO	ip	250	326.4	21.1	836.7	Galloway
		M,F	bm	CO/DMSO	gav	110,220,440	200.3	35.3	402.5	Palitti
		M	bm	CMC/NaCl	ip	110,111.5,223	697.4	404.1	1882	Zunino
Rat	S-D	M,F	bm	CO	gav	67.5,125,250, 500	132.9	38.4	387.7	Vlachos

[a] See Table 4 for explanation of abbreviations and Table 6 for explanation of footnotes.
[f] Data obtained from two experiments with 2AAF.

Table 14. *Induction of SCEs by 4AAF*[a]

Species	Strain	Sex	Tissue	Solvent	Route	Doses (mg/kg)	SCE response[b] x	SCE response[b] $\sqrt{x}$	Investigator	Investigator conclusion
Chinese hamster		F	bm	CO/DMSO	ip	62.5,125,250, 500	0.3519	0.3366	Neal	–
		F	bm	CO/DMSO	gav	125,250,500, 1000	0.2928	0.2898	Neal	–
		F	ie	CO/DMSO	gav	125,250,500, 1000	0.2992	0.2942	Neal	–
Mouse	B6C3F1	M	bm	CO	ip	500,1000,2000	0.0386[d]	0.0382[d]	Tice	+ W
	(C3H1 × 101)F_1	M	bm	OO	gav	223,446,892	0.0020[d]	0.0022[d]	Tates[e]	+ W
	C57B1/6	M,F	bm	CO	gav	223,446,892, 1338	0.0029[d]	0.0030[d]	Allen	+ W
		M	pbl	CO	gav	111.5,446, 1338	0.1019	0.1054	Kligerman	–
	CD-1	M	bm	CO	ip	250,500,1000	0.5425	0.5412	Galloway	–
		F	bm	CO	ip	250,500,1000	0.0395[d]	0.0333[d]	Galloway	+ W
		M,F	bm	CO/DMSO	gav	223,446,892	0.9746	0.9724	Palitti	–
		M	bm	CMC/NaCl	ip	223,446,892	0.2509	0.2606	Zunino	–
Rat	S-D	M	bm	CO	gav	125,250,500, 1000	0.0169[d]	0.0133[d]	Vlachos	+
		F	bm	CO	gav	125,250,500, 1000	0.1731	0.1764	Vlachos	–

[a] See Table 4 for explanation of abbreviations and footnotes.

Table 15. *Male vs female responses to 4AAF*[a]

Investigator	Species	Strain	Route	P Value[b]
Allen	Mouse	C57B1/6	gav	0.1879
Galloway	Mouse	CD-1	ip	0.0003[c]
Palitti	Mouse	CD-1	gav	0.2135
Vlachos	Rat	S-D	gav	0.0006[c]

[a] See Table 4 for explanation of abbreviations.
[b] P values determined by two-way ANOVA, all sex by treatment interactions were nonsignificant.
[c] Significantly different at an α of 0.05.

Table 16. *SCE-5 doses for 4AAF*[a]

Species	Strain	Sex	Tissue	Solvent	Route	Doses used[b] (mg/kg)	SCE-5 (mg/kg)	L95%CL[c] (mg/kg)	U95%CL[d] (mg/kg)	Investigator
Mouse	B6C3F1	M	bm	CO	ip	500,1000	2528	1331	7804	Tice
	(C3H × 101)F_1	M	bm	OO	gav	223,446,892	3390	2189	5904	Tates
	C57B1/6	M,F	bm	CO	gav	223,446	2334	1467	4855	Allen
	CD-1	F	bm	CO	ip	250,500,1000	2012	824	7376	Galloway
Rat	S-D	M	bm	CO	gav	125	448	251	1398	Vlachos

[a] See Table 4 for explanation of abbreviations and Table 6 for explanation of footnotes.

Induction of SCEs by 4AAF

Data from each investigator on the ability of 4AAF to induce SCEs are presented in Table 14. In 6 experiments, treatment with 4AAF gave a positive SCE response: following ip administration in female but not male CD-1 mice (Galloway) and in male B6C3F1 mice (Tice), and following treatment by gavage in male but not female Sprague-Dawley rats (Vlachos), in male and female C57B1/6 mice (Allen), and in male (C3H × 101)F1 mice (Tates). Sex differences in response are analysed in Table 15. There was concordance between the conclusion of the individual investigator and that of the present analysis. Sample time (24 vs 30 h) had no effect on the SCE response in (C3H × 101) F1 mice treated by gavage (Tates, $P = 0.3995$). The individual dose–response curves for SCE induction by 4AAF are presented in Figure 8.

The SCE-5 doses for 4AAF in mice (Table 16, Figure 9) were significantly different from the SCE-5 doses calculated for 2AAF in the corresponding experiments (Table 13). The relative potency with 2AAF compared to 4AAF in the mouse studies ranged from a minimum of 5.9 to a maximum of 12.8 (Table 17). In the

Table 17. *Relative potency of 2AAF vs 4AAF in bone marrow*[a]

Species	Strain	Sex	Solvent	Route	SCE-5 dose 4AAF/2AAF	Investigator
Mouse	B6C3F1	M	CO	ip	5.9	Tice
	(C3H × 101)F_1	M	OO	gav	6.5	Tates
	C57B1/6	M,F	CO	gav	12.8	Allen
	CD-1	F	CO	ip	6.2	Galloway
Rat	S-D	M	CO	gav	3.4	Vlachos

[a] See Table 4 for explanation of abbreviations.

Table 18. *SCE induction by MMC (0.5 mg/kg)*[a]

Species	Strain	Sex	No. Anim. per group	Tissue	Treatment[b] time	Sample[b] Time	SCE frequency[c] Control	Treated	P value[d]	Investi-gator
Chinese hamster		F	4	bm	5	26	2.72 ± 0.54	5.05 ± 1.47	0.0001[g]	Neal
Mouse	B6C3F1	M	8	bm	0.5	24	5.26 ± 0.43	10.40 ± 0.76	0.0001[g]	Tice[e]
	(C3H × 101)F_1	M	5	bm	10	30	5.65 ± 0.50	8.49 ± 0.53	0.0001[g]	Tates
	C57B1/6	M	4	bm	0.5	24	4.82 ± 1.27	12.55 ± 0.59	0.0001[g]	Allen[e]
		M	5	pbl	0.5	24[f]	9.02 ± 1.27	14.00 ± 2.58	0.0007[g]	Kligerman[e]
	CD-1	M	4	bm	1	24	5.23 ± 1.53	10.61 ± 1.07	0.0001[g]	Galloway
		F	4	bm	1	24	5.49 ± 0.94	12.26 ± 1.23	0.0001[g]	Galloway
		M		bm	8	24	3.07 ± 0.16	6.23 ± 1.62	0.0017[g]	Palitti
		M	4	bm	1	25	3.16 ± 0.97	6.10 ± 0.79	0.0002[g]	Zunino
Rat	S-D	M	4	bm	2	25	2.96 ± 0.86	2.71 ± 0.30	0.4161	Vlachos

[a] See Table 4 for explanation of abbreviations.
[b] Time in hours after BrdUrd tablet implantation.
[c] Group mean SCE per cell frequency ± 95% confidence limits.
[d] P value calculated from one-tailed Student's test.
[e] Data combined from 2 experiments.
[f] Time at which blood sample obtained.
[g] Significantly different at $\alpha = 0.005$.

Table 19. *SCE-5 doses for MMC*[a]

Species	Strain	Sex	Tissue	SCE-5 (mg/kg)	L95%CL[b] (mg/kg)	U95%CL[c] (mg/kg)	Investi-gator
Chinese hamster		F	bm	1.07	0.64	2.10	Neal
Mouse	B6C3F1	M	bm	0.49	0.32	0.66	Tice
	(C3H × 101)F_1	M	bm	0.88	0.67	1.14	Tates
	C57B1/6	M	bm	0.32	0.23	0.42	Allen
		M	pbl	0.50	0.13	1.00	Kligerman
	CD-1	M	bm	0.46	0.11	0.95	Galloway
		F	bm	0.37	0.14	0.63	Galloway
		M	bm	0.79	0.44	1.46	Palitti
		M	bm	0.85	0.58	1.26	Zunino
Rat	S-D	M	bm	—	—	—	Vlachos

[a] See Table 4 for explanation of abbreviations and Table 6 for explanation of footnotes. Route of treatment was by intraperitoneal injection.

male rat, a comparatively low SCE-5 dose of 448 mg/kg was calculated. When the SCE-5 doses for 2AAF and 4AAF in the rat were compared, the potency for 2AAF was 3.4 times that of 4AAF.

MMC positive control studies

The ability of MMC (0.5 mg/kg) administered ip to induce SCEs was examined by each investigator, using that laboratory's protocol (e.g. species, strain, treatment time, sample time). The protocols and resulting data are presented in Table 18. MMC at this dose induced a significant increase in SCEs in all study animals except rats. The SCE-5 dose for each positive study is presented in Table 19 and, ranked according to magnitude of the calculated dose, in Figure 10. There was generally good agreement among the laboratories conducting mouse studies with the SCE-5 doses ranging from 0.32 mg/kg to 0.88 mg/kg.

5 Discussion

Interlaboratory variability in control data

Within a laboratory, control animal SCE frequencies were independent of the choice of solvent, the sample time, the tissue analysed following administration of BrdUrd *in vivo* and, in two laboratories (Allen, Palitti) of the sex of the animals. However, SCE frequencies in bone marrow cells differed significantly between mice implanted with coated BrdUrd tablets and uncoated BrdUrd tablets (Galloway), between male and female mice in one laboratory (Galloway) and between male and female Sprague-Dawley rats (Vlachos). Where a significant sex difference among control animal SCE frequencies was observed, SCE levels were higher in female animals (i.e., ~.7 SCE per cell in mice; ~1.7 SCE per cell in rats). The difference in bone marrow SCE levels between males and females was probably due to the use of younger animals (< 2 mo of age), with a corresponding greater weight difference between the sexes. Because the weight of the BrdUrd tablets remained constant in each laboratory, female animals, and especially female rats, experienced a greater level of BrdUrd in their bone marrow. Since BrdUrd induces SCEs (Latt *et al.*, 1981), the increased levels would have resulted in a higher frequency of SCEs. In Galloway's laboratory, the use of paraffin-coated BrdUrd tablets lowered significantly the SCE frequency among control animals. This observation suggests that coating BrdUrd tablets with paraffin (or presumably with agar) both improves the *in vivo* availability of BrdUrd to the proliferating cells (King *et al.*, 1982; McFee *et al.*, 1983) and minimises the frequency of BrdUrd-induced SCEs.

Figure 8. SCE dose response curves for 4AAF. A. Administration of 4AAF by ip injection. B. Administration of 4AAF by gavage. Symbols: ○ = male; □ = female; △ = combined male and female; ○, □ = mouse; ◑, ◨ = rat; ■ = Chinese hamster; ——— = bone marrow; — — — = intestinal epithelium; —·— = peripheral blood lymphocytes.

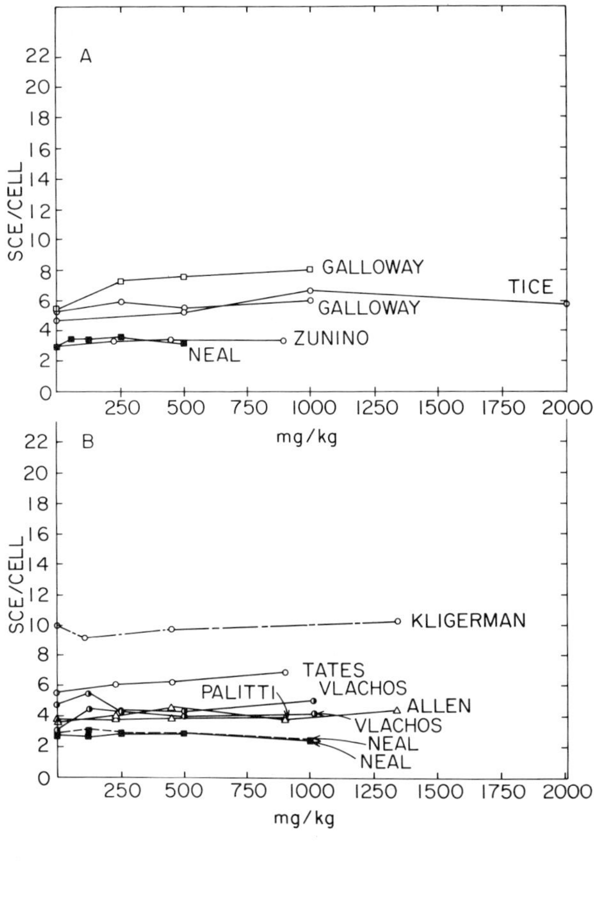

In comparisons among laboratories, control animal bone marrow SCE frequencies differed significantly among mice used in different laboratories, among mice, rats, and Chinese hamsters, and from peripheral blood lymphocytes mitogenically stimulated to proliferate *in vitro*. The result of the last comparison was not surprising considering the difference in cell type and in BrdUrd exposure conditions. The significant differences in control animal SCE frequencies among mouse studies and among mouse, rat and Chinese hamsters might be accounted for by differences in BrdUrd metabolism and utilization, animal age and weight, weight of the implanted BrdUrd tablet, BrdUrd tablet coating method, and SCE scoring method. The data available are too limited to permit for accurate inferences as to the source(s) of interlaboratory variability (see positive control studies below).

Induction of SCEs by the Four Test Chemicals

Several statistical approaches were used by the investigators to analyse SCE data. With one exception (Zunino), parametric tests were used only. Zunino, in addition to analysing the data by parametric tests, used a Mann-Whitney U test to compare the SCE frequencies in all animals treated with one of the 4 test chemicals against SCE frequencies in the pooled controls. All analyses were based on SCE per cell data, either as raw variates or following some type of transformation to normalize the inter-group variances, and used individual animals as the unit of variability. Where the SCE data for each dose was compared against the control values, α was generally conserved by taking into account the number of camparisons being made. The SCE frequency for each animals was based on 20–25 cells and 3 to 5 animals comprised each exposure group.

In the statistical analysis conducted to evaluate all laboratory data by a uniform approach, animal SCE per cell frequencies based either on the raw variates or on square-root transformed ($\sqrt{SCE/cell + 0.5}$) variates (Whorton *et al.*, 1984) were used in an analysis of variance. While square-root transformation resulted in greater

Figure 9. SCE-5 doses for 4AAF by investigator. Each point represents the SCE-5 dose. The range bars indicate the asymmetric 95% confidence limits. Male and female animals were combined if no significant difference (at an alpha of 0.05) were found. Abbreviations: IP = intraperitoneal; GAV = gavage; F = female; M = male. The SCE-5 doses calculated for 2AAF under similar experimental conditions are provided for comparative purposes.

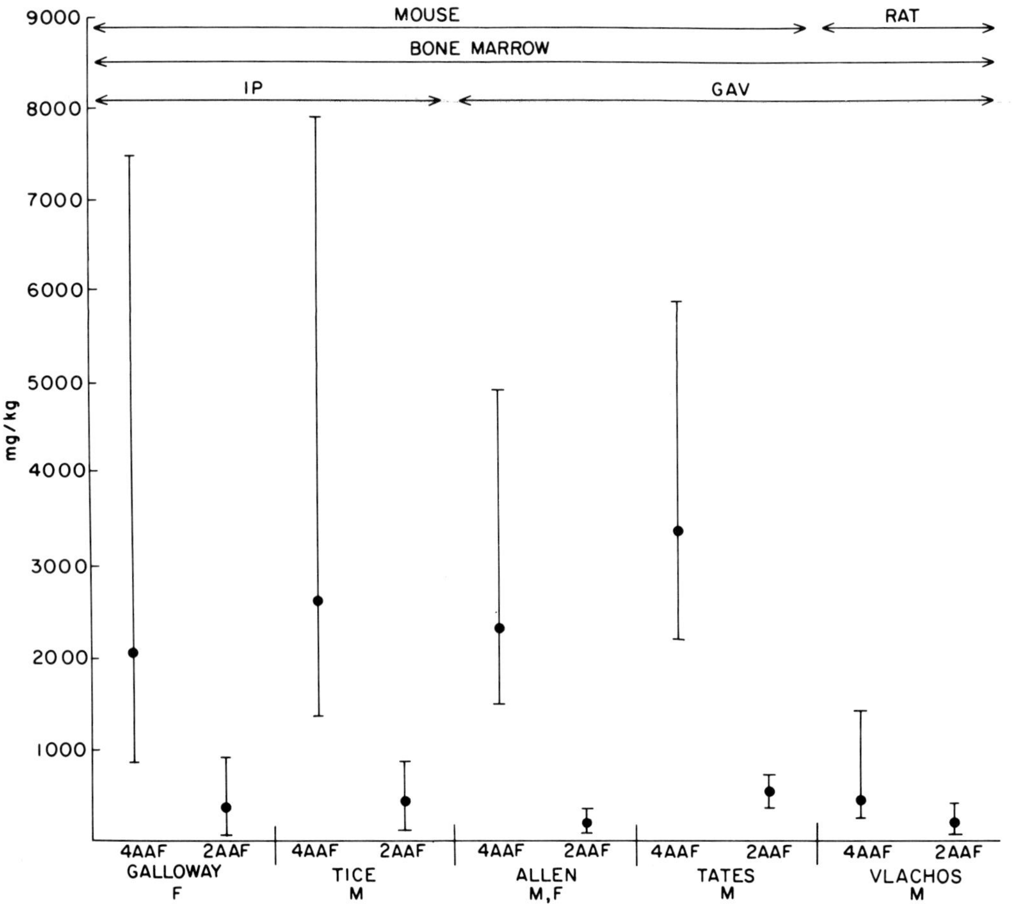

Figure 10. SCE-5 doses for MMC by investigator in increasing quantity of MMC needed. Each point represents the SCE-5 dose. The range bars indicate the asymmetric 95% confidence limits. Abbreviations: PBL = peripheral blood lymphocytes; F = female; M = male. Route of administration was by ip injection.

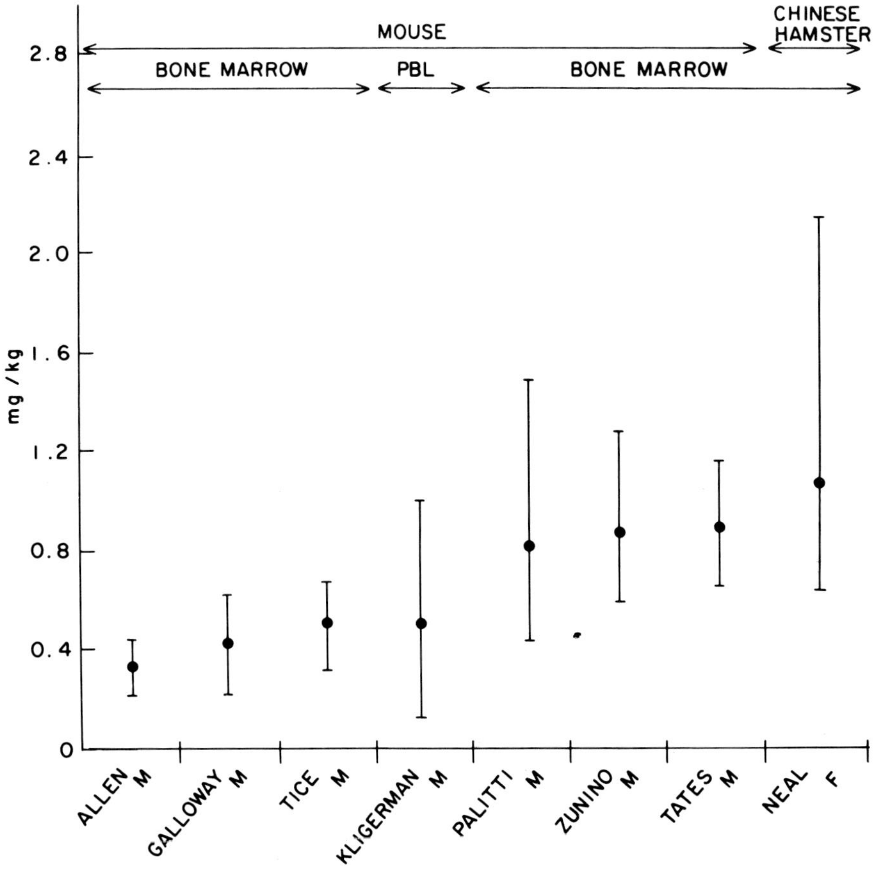

homoscedascity, in no case did the transformation result in an appreciable difference in the resulting P values.

IP administration or treatment by gavage with the 2 established carcinogens BP and 2AAF induced, with one exception, a highly significant increase in SCEs in all tissues, in both sexes, in all strains of mice, and in all 3 species. The single questionable result occurred with 2AAF in peripheral blood lymphocytes of C57B1/6 mice following treatment by gavage. The lack of a significant induction of SCEs in peripheral blood lymphocytes depended on the statistical approach used; the original investigator (Kligerman) called the response in these cells positive. The weak response obtained in these cells could be attributed to the route of administration since, in a previous study in which 2AAF was administered ip, Kligerman *et al.* (1985) obtained a strong positive SCE response under otherwise identical experimental conditions.

Although both BP and 2AAF induced strong positive SCE responses, the shapes of the dose–response curves, the magnitudes of the effects, and species and sex sensitivity differed. For BP, the SCE response in mice, rats, and in Chinese hamsters following administration by gavage was curvilinear, and appeared to saturate at low doses (i.e. < 50 mg/kg). This type of *in vivo* SCE dose–response is common for polycyclic aromatic hydrocarbons (Lockard *et al.*, 1982; Paika *et al.*, 1981). The shape and magnitude of the SCE response is strongly dependent upon the duration between treatment time and sample time; the shorter the time the more curvilinear the shape of the dose–response curve and the lower the absolute increase in SCE frequency (Tice *et al.*, 1987). For 2AAF, the dose–response curves were more complex. Following ip injection, a linear dose–response curve was induced in male B6C3F1 mice while in male and female CD-1 mice (Galloway), the SCE response saturated with increasing dose. Following treatment by gavage, the SCE response in C57B1/6 mice and Sprague-Dawley rats was complex, increasing and decreasing with increasing dose, while in Chinese hamsters (Neal), in male CD-1 mice (Palitti) and in male (C3H $\times$ 101)F1 mice (Tates), the SCE response curves were more linear. These differences resulted from specie/strain differences in 2AAF metabolism, coupled with differences due to the route of exposure and time of treatment.

Sex differences in response to BP were observed for 2 mouse studies, with males exhibiting a higher response in C57B1/6 mice (Allen) and a lower response in CD-1 mice (Galloway). However, the calculated SCE-5 doses were not appreciably different between the sexes, suggesting a similar initial rate of metabolism. The lack of a sex difference in the third mouse study (Palitti) was probably due to the use of a later treatment time (i.e. $+ 8$ h). This later treatment time may limit the resolution of male-female differences in PAH metabolism and/or in the bioavailability of the reactive metabolite(s) of BP. Only in Sprague-Dawley rats was there an appreciable sex difference in the SCE-5 dose, males being $\sim 40\%$ more responsive than females. Sex differences in the SCE response among mice and rats following treatment with 2AAF were not observed.

There were no striking differences in sensitivity to BP among strains of mice (as determined by comparing their SCE-5 doses), independent of choice of solvent or route of exposure (see Figure 3). For 2AAF, greater variability among strains of mice occurred. Mice appeared to be more sensitive to BP than either Sprague-Dawley rats or female Chinese hamster: for 2AAF, Sprague-Dawley rats were the most sensitive. Under most experimental conditions, BP was more potent than 2AAF in inducing SCEs (Table 20). The exceptions were bone marrow cells of Sprague-Dawley rats and intestinal epithelial cells of Chinese hamsters, both following treatment by gavage.

Although the route of exposure did not appear to affect the magnitude of the SCE-5 doses for BP among the various strains of mice, administration by ip injection was more effective than administration by gavage in inducing SCEs in bone marrow cells of Chinese hamsters (Figure 3). Furthermore, in Chinese hamsters the bone marrow SCE response curve for BP administered ip did not saturate over the dose range tested, as it did in mice similarly treated, or in Chinese hamsters administered BP orally. A similar route dependent difference in the magnitude of the SCE response in bone marrow cells of Chinese hamsters was observed for 2AAF.

Following treatment with BP by gavage, the intestinal epithelial cells and the bone marrow cells of Chinese hamsters exhibited the same SCE response. This similarity in SCE yields presumably reflected genotoxic damage induced by circulating metabolites. However, for 2AAF, the intestinal epithelial cells of gavage-treated Chinese hamsters exhibited a higher SCE response than did bone marrow cells sampled from the same animals. The differential response for 2AAF presumably reflected preferential metabolism in the intestinal tissue. Consistent with these tissue-activity differences between BP and 2AAF in Chinese hamsters, the 2AAF SCE-5 dose for peripheral blood lymphocytes in male C57B1/6 mice following treatment by gavage was ~ 3.5 times higher than that for bone marrow cells. This significant difference could have resulted from differences in bioavailability of the reactive metabolite(s) of 2AAF, differences in cell type sensitivity, and/or differences in the extent of DNA repair of 2AAF-induced lesions prior to sampling. The sampling time chosen for the peripheral blood lymphocyte assay (i.e. 24 h) could have provided more time for repair of 2AAF-induced DNA lesions than would have occurred in the bone marrow assay. In contrast, the SCE-5 doses for BP in peripheral blood lymphocytes and bone marrow cells of gavage-treated male C57B1/6 mice were strikingly similar. This degree of concordance could have resulted from a comparable level of genotoxic metabolites of BP in the blood and bone marrow, with little DNA repair of the SCE-inducing lesions in either cell type.

In vitro, both PYR and 4AAF have demonstrable genotoxic activity (see Ashby, these proceedings). PYR, in the presence of rat liver S9, was shown to induce SCEs in two studies (Perry and Thomson, 1981; Evans and Mitchell 1981). *In vivo*, treatment with PYR and 4AAF, even at very high doses, failed to induce a positive SCE response in the majority of the experimental protocols. However, for PYR and 4AAF, 5 and 6, respectively, of 15 experimental trials (considering the use of male and female animals as separate trials) were positive for SCE induction at an α level of 0.05. For PYR, 3 gavage studies (male Sprague-Dawley rats, male CD-1 mice and female C57B1/6 mice) and one ip study (male and female CD-1 mice) were positive. The negative SCE response in another study of CD-1 mice treated with PYR by gavage could be attributed to administration of the test chemical 8 hours after BrdUrd tablet implantation compared with ~ 1 hour in the positive studies. For 4AAF, the 6 positive SCE responses were observed in female CD-1 mice and in male B6C3F1 mice administered 4AAF ip and in male and female C57B1/6 mice, in male (101 × C3H)F1 mice, and in male Sprague-Dawley rats administered 4AAF by gavage.

It may be informative to compare the shapes of the dose response curves, and the apparent strain, sex and species specificity for the two non carcinogenic analogues to that observed for the carcinogens. For BP and for PYR, the dose response curves appeared to saturate with increasing dose. However, because the PYR-induced SCE response was so small, the apparent saturation in SCE yields may not have biological relevance. CD-1 mice, the mouse strain apparently most responsive to PYR, did not appear to be any more sensitive to BP than any other strain. Male Sprague-Dawley rats appeared slightly more sensitive than female Sprague-Dawley rats to BP; also male but not female rats were sensitive to PYR. Male and female C57B1/6 mice exhibited increased SCE levels after treatment with BP, but only females exhibited an increased SCE response to PYR. Rats were less sensitive than mice to BP but not always less sensitive than mice to PYR. Thus, while the SCE responses to BP and PYR had some features in common, significant differences were present. Based on calculated SCE-5 doses, the potency of PYR, where a positive SCE response was obtained, was less in male rats (i.e. ~ 8 times) and much less in mice (i.e. ~ 21–42 times) than that of BP. *In vitro*, in Chinese hamster ovary (CHO) cells in the presence of rat liver S9, PYR was some 6 times less active than BP at a dose of 100 μg/ml (Perry and Thomson, 1981).

Differences and similarities in the SCE response to 2AAF and 4AAF were readily apparent also. For 2AAF, sex differences in SCE response were not observed, but for 4AAF, male rats and not female rats exhibited a significant SCE response and in one mouse study (Galloway), SCEs were induced in female but not male mice. In another mouse study (Allen), male and female mice were affected to a similar extent. The shapes of the dose–response curves for 2AAF and 4AAF administered by gavage were similar, especially in the rat study. Sprague-Dawley rats were more sensitive to the SCE-inducing capability of both 2AAF and 4AAF. Although relatively sensitive to 2AAF when administered by gavage, the intestinal epithelium of Chinese hamsters did not exhibit a positive response to 4AAF. Finally, all the positive responses to 4AAF were significantly less (~ 3 to 13 times, based on SCE-5 doses) than that observed for 2AAF.

Where it was possible to compare the relative potency of PYR and 4AAF (Table 20), PYR was 2.7 and 3.2 times more active than 4AAF in mice but only 0.2 times as active as 4AAF in male Sprague-Dawley rats.

It was interesting to compare the dose ranges over which each investigator examined the ability of the carcinogen/noncarcinogen pair to induce SCEs. It was possible that one or more of the negative responses for PYR or 4AAF would have been positive if higher doses had been tested. For example, 4AAF was positive for SCE induction at a dose of 1000 mg/kg when administered ip to male B6C3F1 mice (Tice). However, the highest dose tested in two laboratories in which a negative response was obtained following a similar protocol was 832 mg/kg (Zunino) and 500 mg/kg (Neal).

It is necessary to question the biological relevance of a statistically positive SCE response obtained at doses that are probably not compatible with the long-term survival of the test animals. It is likely that multiple mechanisms are involved in the formation of SCEs, and that one of these mechanisms may not reflect direct DNA damage, but rather stress, either at the cell (Galloway *et al.*, 1985) or the animal level (Fischman *et al.*, 1985). The small but statistically significant increases in SCEs observed for chemicals like PYR and 4AAF at high doses may not be related mechanistically to the larger SCE increases observed at low doses of BP and 2AAF. If the positive responses to PYR/4AAF resulted from cellular/animal stress than the responses may have an effective threshold. Thus, attempts to compare simply the relative potencies of these 2 pairs of chemicals may be unrealistic. Yet, the lack of a positive SCE response (or other response indicative of genotoxic damage) under a variety of experimental conditions certainly gives credence to a conclusion of nongenotoxicity, while the magnitude and shape of the dose–response curve for a positive response, even under high-dose conditions, provides data which can be manipulated to assign risk under worst-case conditions. The possibility for false positive results is not limited to the SCE assay. Chromosomal aberration and presumably micronucleus induction can result when conditions (e.g. osmolality, pH) are altered to such an extent that normal cellular growth is affected (e.g. Galloway *et al.*, 1985).

The SCE-5 dose, or some other arbitrary measure of potency, is useful in that it allows for a comparison of activities between experiments, chemicals, laboratories and endpoints, regardless of the dose range tested or the base-line frequency. However, the methods by which the SCE-5 doses and the 95% confidence limits are calculated include assumptions that require further evaluation. The calculated SCE-5 dose assumes an additive response and linearity for the critical portion of the dose–response curve. Further, the method by which the linear portion of the dose–response curve is estimated is somewhat arbitrary. Consequently, further efforts are required to test the validity of this approach for examining the potency of SCE inducing agents and, by extrapolation, other genotoxic endpoints.

MMC studies

A study to assess the SCE-inducing capability of MMC at a fixed dose was included to help evaluate laboratory differences in SCE ascertainment. MMC is a potent, direct acting SCE-inducer whose ability to induce SCEs is not altered significantly by strain, species, route, time of administration or sample time (Tice *et al.*, 1987; Tice,

Table 20. *Relative potency for BP vs 2AAF and PYR vs 4AAF*[a]

Species	Strain	Sex	Tissue	BP vs 2AAF: Solvent BP	BP vs 2AAF: Solvent 2AAF	BP vs 2AAF: Route BP	BP vs 2AAF: Route 2AAF	SCE-5 dose[b] 2AAF/BP	PYR vs 4AAF: Solvent PYR	PYR vs 4AAF: Solvent 4AAF	PYR vs 4AAF: Route PYR	PYR vs 4AAF: Route 4AAF	SCE-5 dose[b] 4AAF/PYR	Investigator
Chinese hamster		F	bm	CO	CO/DMSO	ip	ip	2.0						Neal
		F	bm	CO	CO/DMSO	gav	gav	3.1						Neal
		F	ie	CO	CO/DMSO	gav	gav	0.9						Neal
Mouse	B6C3F1	M	bm	CO	CO	ip	ip	10.2						Tice
	(C3H × 101)F_1	M	bm	OO	OO	gav	gav	4.9						Tates
	C57B1/6	M	bm	CO	CO	gav	gav	4.6						Allen
		F	bm	CO	CO	gav	gav	3.8	CO	CO	gav	gav	3.2	Allen
		M	pbl	CO	CO	gav	gav	18.7						Kligerman
	CD-1	M	bm	CO	CO	ip	ip	12.0						Galloway
		F	bm	CO	CO	ip	ip	11.0	CO	CO	ip	ip	2.7	Galloway
		M,F	bm	CO	CO/DMSO	gav	gav	3.4						Palitti
		M	bm	CO	CMC/NaCl	gav	ip	6.7						Zunino
Rat	S-D	M	bm	CO	CO	gav	gav	0.6	CO	CO	gav	gav	0.2	Vlachos
		F	bm	CO	CO	gav	gav	0.4						Vlachos

[a] See Table 4 for explanation of abbreviations.
[b] SCE-5 doses compared using moles/kg.

unpublished data; McFee *et al.*, 1984). A relatively low level of MMC (0.5 mg/kg) was selected for comparative purposes to avoid differences in scoring accuracy at high SCE levels. MMC was administered by ip injection to mice, rats and Chinese hamsters using different treatment times. Only the study involving rats (Vlachos) failed to obtain a positive response for MMC at this dose. Among the remaining 9 studies, there was a good degree of concordance among laboratories in the magnitude of the SCE response, as indicated by the calculated SCE-5 doses. The doses ranged from 0.32 mg/kg to 1.07 mg/kg, with a median dose of 50 mg/kg. Only the 2 extreme values were significantly different from one another. This degree of concordance is especially surprising in that a common lot or source of MMC was not used in these studies and differences in genotoxic activity for MMC depending on lot and source have been reported (Proctor and Gaulden, 1985). The observation that the 9 investigators were strikingly similar in their ability to obtain a positive SCE response using a direct-acting genotoxic chemical supports the sex, strain, and species differences observed for the 4 test chemicals.

6 Conclusion

The two carcinogens, BP and 2AAF, were strongly active in the *in vivo* SCE test, regardless of sex, strain, species, route of exposure and whether bone marrow cells or intestinal epithelial cells were sampled. The two purported noncarcinogens, PYR and 4AAF, were weakly active in a limited number of assays (5 and 6 of 15, respectively). The SCE responses for PYR and 4AAF were far more dependent upon sex, strain, species, route of exposure, and tissue than those observed for BP and 2AAF. Where positive responses were obtained for PYR and 4AAF, the calculated SCE-5 dose approach indicates that BP is ~8 to 9 times more active than PYR, and 2AAF is ~3 to 13 times more active than 4AAF in inducing SCEs.

7 Acknowledgements

The authors would like to thank Ms. L. Wasson and Ms. G. Callister for typing this contribution. RRT was supported by the National Toxicology Program under National Institute of Health Sciences Interagency Agreement No. Y01-ES-20098-05 and by the U.S. Department of Energy under Contract DE-AC02-76CH00016. Accordingly, the U.S. Government retains a nonexclusive, royalty-free license to publish or reproduce the published form of this contribution, or allow others to do so, for U.S. Government purposes.

8 References

Abe, S. and M. Sasaki (1982) SCE as an index of mutagenesis and/or carcinogenesis. In A.A. Sandberg (Ed.), Sister *Chromatid Exchange,* Alan R. Liss, Inc., New York, pp. 461–514.

Allen, J.W. (1982) SCE and meiotic crossover exchange in germ cells. In A.A. Sandberg (Ed.), *Sister Chromatid Exchange,* Alan R. Liss, Enc., New York, pp. 297–314.

Allen, J.W., C.F. Shuler and S.A. Latt (1978) Bromodeoxyuridine tablet methodology for *in vivo* studies of DNA synthesis, *Somatic Cell Genet.*, **4**, 393–405.

Carrano, A.V., L.H. Thompson, P.A. Lindl and J.L. Minkler (1978) Sister chromatid exchanges as an indicator of mutagenesis, *Nature* (London), **271**, 551–553.

Carrano, A.V. and S. Wolff (1975) Distribution of sister chromatid exchanges in the euchromatin and heterochromatin of the Indian muntjac, *Chromosoma,* **53**, 361–369.

Conner, M.K., Y. Alarie and R.L. Dombroske (1979) Sister chromatid exchange in murine alveolar macrophages, regenerating liver and bone marrow cells – a simultaneous multicellular *in vivo* assay, *Chromosoma,* **74**, 51–55.

DeWeer-Kastelein, E.A., W. Keijzer, G. Rainaldi and D. Bootsma (1977) Induction of sister chromatid exchanges in xeroderma pigmentosum cells after exposure to ultraviolet light, *Mutation Res.*, **45**, 253–261.

Fischman, H.K., D.D. Kelly and S.D. Rainer (1985) Behavioral stress and sister chromatid exchanges, *Environ. Mutagen.*, **7**, (Suppl 3), 51.

Galloway, S.M., C.L. Bean, M.A. Armstrong, D. Deasy, A. Kraynak and M.O. Bradley (1985) False positive *in vitro* chromosome aberration tests with non-mutagens at high concentrations and osmolalities, *Environ. Mutagen,* **7**, (Suppl. 3), 48–49.

Guglielmi, G.E., T.F. Vogt and R.R. Tice (1982) Induction of sister chromatid exchanges and inhibition of cellular proliferation *in vitro*, I. Caffeine, *Environ. Mutagen.*, **4**, 191–200.

Kihlman, B.A. (1977) *Caffeine and Chromosomes*, Elsevier Scientific Publishing Company, Amsterdam, 504 pp.

King, M.-T., D. Wild, E. Gocke and K. Eckhardt (1982) 5-Bromodeoxuridine tablets with improved depot effect for analysis *in vivo* of sister chromatid exchanges in bone marrow and spermatogonial cells, *Mutation Res.*, **97**, 117–129.

Kligerman, A.D., G.L. Erexson and J.L. Wilmer (1985) Induction of sister chromatid exchange (SCE) and cell cycle inhibition in mouse peripheral blood lymphocytes exposed to mutagenic carcinogens *in vivo, Mutation Res.*, **157**, 181–187.

Kram, D. (1982) In utero analysis of sister chromatid exchange. In A.A. Sandberg (Ed.), *Sister Chromatid Exchange*, Alan R. Liss, Inc., New York, pp. 205–216.

Latt, S.A. (1974) Sister chromatid exchanges, indices of human chromosome damage and repair: Detection by fluorescence and induction by mitomycin C, *Proc. Natl. Acad. Sci.* USA, **71**, 3162–3166.

Latt, S.A., J.W. Allen, W.E. Rogers and L.A. Juergens (1977) *In vitro* and *in vivo* analysis of sister chromatid exchange formation. In B.J. Kilbey, M. Legator, W. Nichols and C. Ramel (Eds), *Handbook of Mutagenicity Test Procedures*, Elsevier Scientific Publishing Co., New York, pp. 275–291.

Littlefield, L.G. (1982) Effects of DNA-damaging agents on SCE. In A.A. Sandberg (Ed.), *Sister Chromatid Exchange*, Alan R. Liss, New York, pp. 355–394.

Lockard, J.M., J.W. Prater, C.J. Viau, H.G. Enoch and P.S. Sabharwal (1982) Comparative study of the genotoxic properties of Eastern and Western U.S. shale oils, crude petroleum, and coal-derived oil, *Mutation Res.*, **102**, 221–235.

McFee, A.F., K.W. Lowe and J.R. San Sebastian (1983) Improved sister chromatid differentiation using paraffin-coated bromodeoxyuridine tablets in mice, *Mutation Res.*, **119**, 83–88.

McFee, A.F., K.W. Lowe and M.N. Sherrell (1984) Influence of time on the *in vivo* induction of sister chromatid exchanges and chromosomal aberrations, *Environ. Mutagen.*, **6**, 469.

Paika, I.J., M.T. Beauchesne, M. Randall, R.R. Schreck and S.A. Latt (1981) *In vivo* SCE analysis of 20 coded compounds. In F.J. de Serres and J. Ashby (Eds), Evaluation of Short-Term Tests for Carcinogens, *Prog.*

Mutation Research I, Elsevier Scientific Publishing Co., New York, pp. 673–681.

Perry, P.E. and H.J. Evans (1975) Cytological detection of mutagen-carcinogen exposure by sister chromatid exchange, *Nature* (London), **258**, 121–124.

Perry, P.E. and E.J. Thomson (1981) Evaluation of the sister chromatid exchange method in mammalian cells as a screening system for carcinogens. In F.J. de Serres and J. Ashby (Eds), *Evaluation of Short-Term Tests for Carcinogens, Prog. Mutation Res. I*, Elsevier Scientific Publishing Co., New York, pp. 560–569.

Proctor, B.L. and M.E. Gaulden (1985) Interactions of mitomycin C with culture medium and fetal calf serum components as determined by mass spectrometry Environ. *Mutagen.*, **7**, (Suppl. 3), 49.

Reynolds, R.J., A.T. Natarajan and P.H.M. Lohman (1979) Micrococcus luteus uv endonuclease-sensitive sites and sister chromatid exchanges in Chinese hamster ovary cells, *Mutation Res.*, **64**, 353–356.

Sandberg, A.A. (Ed.) (1982) *Sister Chromatid Exchange, Progress and Topics in Cytogenetics*, Vol. 2, Alan R. Liss, Inc., New York, 707 pp.

Sasaki, M. (1982) Sister chromatid exchange as a reflection of cellular DNA repair. In A.A. Sandberg (Ed.), *Sister Chromatid Exchange*, Alan R. Liss, Inc., New York, pp. 135–161.

Schneider, E.L., R.R. Tice and D. Kram (1978) Bromodeoxyuridine differential chromatid staining technique: A new approach to examining sister chromatid exchange and cell replication kinetics. In D.M. Prescott (Ed.), *Methods in Cell Biology*, Vol. 20, Academic Press, New York, pp. 379–409.

Schubert, I. and R. Rieger (1981) Sister chromatid exchanges and heterochromatin, *Human Genetics*, **57**, 119–130.

Schvartzman, J.B. and R.R. Tice (1982) 5-Bromodeoxyuridine and its role in the production of sister chromatid exchanges. In A.A. Sandberg (Ed.), *Sister Chromatid Exchange*, Alan R. Liss, New York, pp. 123–134.

Shafer, D.A. (1982) Alternative replication bypass mechanisms for sister chromatid exchange formation. In A.A. Sandberg (Ed.), *Sister Chromatid Exchange*, Alan R. Liss, Inc., New York, pp. 67–98.

Sokal, R.R. and F.J. Rohlf (1981) *Biometry*, Third Edition, W. Freeman and Co., San Francisco.

Takehisa, S. (1982) Induction of sister chromatid exchanges by chemical agents. In S. Wolff (Ed.), *Sister Chromatid Exchange*, John Wiley and Sons, Inc., New York, pp. 87–147.

Tice, R.R. and A. Hollaender (Eds) (1984a) *Sister Chromatid Exchange, Part A, The Nature of SCE*, Plenum Publishing Corp., New York, 492 pp.

Tice, R.R. and A. Hollaender (Eds) (1984b) *Sister Chromatid Exchange, Part B, Genetic Toxicology and Human Studies*, Plenum Publishing Corp., New York, 540 pp.

Tice, R.R. and J.B. Schvartzman (1982) Sister chromatid exchange: A measure of DNA lesion persistence. In A.A. Sandberg (Ed.), *Sister Chromatid Exchange*, Alan R. Liss, New York, pp. 33–45.

Tice, R.R., J.L. Ivett and A.F. McFee (1987) The effect of agent treatment time on the induction of sister chromatid exchanges in mouse bone marrow cells *in vivo, Mutation Res.*, **182**, 15–29.

Whorton, E.B., Jr. R.R. Tice and D.G. Stetka (1984) Statistical design, analysis, and inference issues in studies using sister chromatid exchange. In R.R. Tice and A. Hollaender (Eds), *Sister Chromatid Exchange, Part A, The Nature of SCE*, Plenum Publishing Corp., New York, pp. 431–440.

Wolff, S. (Ed.) (1982) *Sister Chromatid Exchange*, John Wiley and Sons, Inc., New York, 306 pp.

Wolff, S., J. Bodycote and R.B. Painter (1974) Sister chromatid exchanges induced in Chinese hamster cells by UV irradiation at different stages of the cell cycle: The necessity for cells to pass through S, *Mutation Res.*, **25**, 73–81.

Raymond R. Tice, Carol A. Luke, Valerie M. Miller, Brian G. Ormiston, Anita C. Wilton and Michael D. Shelby[1]

Brookhaven National Laboratory, Medical Department, Upton, New York, USA, and [1]National Institute of Environmental Health Sciences, Cellular and Genetic Toxicology Branch, Research Triangle Park, North Carolina USA

38

A comparative analysis of the induction of sister chromatid exchanges in the bone marrow cells of B6C3F1 mice

1 Summary

The ability of two established carcinogens – benzo(a)pyrene and 2-acetylaminofluorene – to induce sister chromatid exchanges (SCEs) and to inhibit cellular proliferation in bone marrow cells of male B6C3F1 mice was compared with that of two, reportedly noncarcinogenic, structural analogues – pyrene and 4-acetylaminofluorene, respectively. Based on an analysis of variance using an α of 0.05, treatment with BP and 2AAF induced a highly significant increase ($P < 0.0001$) in SCE frequency, 4AAF induced a marginally significant increase ($P = 0.0424$) in SCE frequency, while PYR was clearly negative for SCE induction ($P = 0.8905$). To compare SCE-inducing potency, SCE-5 values (i.e., the dose of a chemical calculated to induce an increase of 5 SCEs/cell above concurrent control values) for BP, 2AAF and 4AAF were compared. BP was nine times more active than 2AAF, while 2AAF was six times more active than 4AAF in inducing SCEs. With regard to bone marrow proliferation kinetics, 4AAF induced a highly significant increase ($P = 0.0002$) and 2AAF a marginally significant ($P = 0.0445$) increase in cell cycle duration; treatment with BP ($P = 0.5924$) or PYR ($P = 0.2979$) did not significantly affect bone marrow cell cycle duration.

2 Introduction

Benzo(a)pyrene (BP) and 2-acetylaminofluorene (2AAF) are widely recognized as animal carcinogens (Kriek, 1974; NCI, 1951–1978; IARC 1982, 1983) while pyrene (PYR) and 4-acetylaminofluorene (4AAF), structural analogues of BP and 2AAF, respectively, have not demonstrated carcinogenic acitivity (Hicks and Ashby, this volume). Both carcinogens are active in a variety of short-term *in vivo* and, with metabolic activation, *in vitro* assays, including cytogenetic, DNA damage and repair, cell transformation, and prokaryotic/eukaryotic mutational tests (IARC, 1983; Ashby, 1981). In several of these *in vitro* assays, PYR and 4AAF were found also to induce a positive genotoxic response (Ashby, 1981). Because of these *in vitro* findings, collaborative studies focusing on different biological endpoints were undertaken to compare, as extensively as possible, the *in vivo* genotoxicity of these 4 chemicals. In this report, we compare the abilities of 2AAF, 4AAF, BP, and PYR to induce sister chromatid exchanges (SCEs), a sensitive indicator of DNA damage (see Sandberg, 1982; Wolff, 1982; Tice and Hollaender, 1984a,b), *in vivo* in mouse bone marrow cells and, as a secondary measure of toxicity, to inhibit bone marrow cellular proliferation (Tice and Ivett, 1985).

2 Materials and methods

The experimental design and methods used in this study follow the U.S. National Toxicology Program standard protocol for *in vivo* mouse bone marrow cytogenetics evaluation (McFee *et al.*, 1987). This methodology is presented in part in the following sections.

Animals

Male B6C3F1 mice, received from Frederick Cancer Research Center as weanlings, were used throughout these studies. Until selected for experimentation, mice were housed at a population density no greater than one animal per 90 sq cm under controlled environmental conditions [i.e., 25 °C; 50% relative humidity; 20 air exchanges/h and a 12 h photoperiod (light 0700–1900 h)]. All mice were bedded on corn cob litter (Bed O'Cobs, J.R. Nielsen and Sons, Inc.) with food (Purina Laboratory Rodent Chow 5001) and water available ad libitum. Within each experiment, animals used were of the same date of birth (between 10 to 16 weeks) and weighed within a 2 g range of a mean weight of 25 to 30 g. For maximum tolerated dose determinations, generally 3 animals per exposure group were used; for SCE analysis, 4 to 7 animals per dose were used, depending on the expected rate of lethality, to ensure that 4 animals per exposure group could be analysed.

Chemicals

BP (WHO 12), PYR (WHO 11), 2AAF (WHO 14), and 4AAF (WHO 13) were obtained from Lancaster Synthesis Ltd. (UK). Mitomycin C (MMC; Lot No. 12F-0600), 5-bromodeoxyuridine (BrdUrd; Lot No. 73F-0548), and reagent grade corn oil (Lot No. 131-076) were obtained from Sigma; colchicine from Eli Lilly; Hoechst 33258 from American Hoechst; and Giemsa from Harleco. The 4 test chemicals and MMC were stored at 4 °C. Within one-half hour of administration, MMC was dissolved in phosphate buffered saline (PBS; pH 7.4), and the 4 test chemicals were dissolved or suspended in corn oil. When necessary (e.g., 2AAF, 4AAF, PYR), sonication (5 min, 37°C) and/or syringing through an 18 g needle were used to prepare a homogenous suspension prior to injection. All injections were intraperitoneal (ip) at a volume of 0.4 ml.

3 Experimental protocol

Each animal was anaesthetised lightly with Metofane (Pitman-Moore, Inc.), a small incision (~ 6 mm) made on the right lateral aspect of the abdomen, a 50 mg paraffin-coated BrdUrd tablet implanted subcutaneously, and the incision closed with wound clips (Clay-Adams). The BrdUrd tablets were made on a Parr pellet press using a die (0.178 in diam) and 55 ft lbs of pressure (Allen *et al.*, 1979), and coated (~ 70%) with paraffin (McFee *et al.*, 1983). Thirty minutes after BrdUrd tablet implantation (0900–1100 h), each animal was injected ip with the test chemical or the appropriate solvent. Twenty-one and one-half hours later (22 h after BrdUrd tablet implantation), each surviving animal was injected ip with colchicine (2 mg/kg). Two hours later (24 h after BrdUrd tablet implantation), the animals were killed by cervical dislocation, and the bone marrow from both femurs removed by flushing with PBS. The aspirated bone marrow was pelleted by centrifugation, incubated in 0.075 M KCl (25 min; 37 °C), repelleted, fixed with ice cold 3:1 methanol:glacial acetic acid, and stored at 4° to −5 °C until used for slide preparation. The MMC positive control experiments, which used a dose of 0.5 mg/kg, were conducted using the same protocol.

Slide preparation, staining, and analysis

Each fixed bone marrow sample was washed twice in fixative and coded flame-dried slides were prepared. All slides were differentially stained for SCE and cell kinetic analysis using a modified (Tice *et al.*, 1978) fluorescence-plus-Giemsa technique (Goto *et al.*, 1975). Slides were scored without knowledge of the treatment. Slides for BP and 2AAF were scored concurrently, as were the slides for PYR and 4AAF. In each of these two sets of slides, similarly coded MMC positive control and MMC solvent control slides were placed randomly. To examine individual animal bone marrow proliferation kinetics, 100 randomly selected metaphase cells were scored for replicative history (i.e., the number of S phases completed since the time of BrdUrd tablet implantation) (Tice *et al.*, 1976; Tice and Ivett, 1985). For SCE analysis, 25 second generation metaphase cells were selected for scoring based on good chromatid differential staining, lack of overlapping chromosomes, and chromosome number (40 ± 3 chromosomes). In each SCE study, 2 individuals scored one-half of each animals's data (i.e., 12 or 13 cells).

Maximum tolerated dose (MTD) determinations

For BP and 2AAF, the test doses used in these studies were the same as those used in previous SCE studies on the same chemical (Ivett *et al.*, 1984). For PYR and 4AAF, preliminary studies were conducted to determine the optimal dose range to be tested for each chemical. The purpose of the MTD determination was to identify a range of doses that would evaluate, as completely as possible, the ability of a test chemical to induce SCEs in mouse bone marrow cells *in vivo*. Thus, selection of the MTD was made on the basis of one or more of the following criteria: over a 24 h period and, when necessary, including the maximum dose possible based on obtaining a homogeneous injection suspension, (1) animal lethality was apparent and/or, (2) significant delay in bone marrow cellular proliferation occurred to the extent that too few metaphase cells scorable for SCEs was observed (Ivett *et al.*, 1984; McFee *et al.*, 1987). In the absence of available lethality data, groups of animals (generally 3 per group) were injected ip following the standard protocol for SCE analysis at doses of 10, 100, and 1000 mg/kg (Ivett *et al.*, 1984). Where lethality information was available, doses around the LD_{50} were tested. Cell proliferation kinetics were determined in bone marrow material from all survivors. Based on the lethality and/or bone marrow kinetic response, higher, intermediate or lower doses were tested in the same manner until a MTD was determined. For SCE analysis, the MTD, half of the MTD and a quarter of the MTD were tested.

Statistical analysis
For cell proliferation kinetic analysis, the proportion of cells among the first, second, and third generations in each animal was transformed into an average generation time (AGT):

AGT = BrdUrd exposure duration/Replicative Index,

where the Replicative Index (RI) = 1 times the frequency of first generation metaphase cells + 2 times the frequency of second generation metaphase cells + 3 times the frequency of third generation metaphase cells... (Ivett and Tice, 1982). AGT data were analysed by one-way ANOVA, using an α of 0.05, to determine whether the treatment had any effect on bone marrow cell kinetics. Individual animal data were omitted from the analysis if differential staining quality was considered inadequate.

SCE data were square-root transformed (Whorton *et al.*, 1984) and analyzed by a one-way ANOVA using individual animal frequencies. If a significant P value was obtained (at $\alpha = 0.05$), the minimally effective dose tested was determined by comparing the square-root transformed data from each treatment dose against the solvent control group data using Student's t test (including a Bonferroni correction for multiple comparisons). For chemicals inducing a positive increase in SCE frequency, a dose which would have induced an absolute increase of 5 SCE/cell (the SCE-5 dose) above the concurrent solvent control frequency was calculated by inverse prediction using linear regression (Guglielmi *et al.*, 1982). For non-linear SCE data, data at higher doses were omitted from the analysis if the data at that dose was not apparently different from the next lower dose. The linearly calculated slope associated regression statistics were then used to estimate the SCE-5 dose and the asymmetric 95% confidence limits for that dose (Sokol and Rohlf, 1969).

4 Results

Maximum tolerated dose (MTD) determinations
To allow an examination of experiment-to-experiment variability for BP and 2AAF, MTD values based on a previous study with these two chemicals (Ivett *et al.*, 1984) were used. Thus, an MTD of 250 mg/kg for BP, based on

Table 1. *Maximum tolerated dose determinations*[a]

Chemical	Dose (mg/kg)	AGT (h)	Chemical	Experiment	Dose (mg/kg)	AGT (h)
PYR[b]	0	11.4	4AAF[b]	I	0	12.4
	500	11.1			250	12.2
	1000	10.6			500	12.6
	1500	10.9			1000	12.4
				II	0	11.4
					1500	13.6
					2000	12.3

[a] Three animals were treated and survived for 24 h at each dose.
[b] MTD of 2000 mg/kg selected based on suspension characteristics.

Table 2. *SCE and AGT data for BP*

Dose (mg/kg)	Number treated	Number survivors	SCE Frequency[a]	N	P[b] value	AGT(h)[c]	N	P[d] value
0	5	5	5.61 ± 0.49	4	—	12.3 ± 0.2	4	—
62.5	6	6	12.13 ± 0.57	4	<0.0001[f]	12.4 ± 0.1	6	NC
125.0	6	6	12.94 ± 1.38	4	<0.0001[f]	12.5 ± 0.3	6	NC
250.0	6	6	13.33 ± 0.55	4	<0.0001[f]	12.7 ± 0.1	6	NC
P value[e]			< 0.0001[g]			0.5924		
SCE-5 = 48.2 mg/kg		95% confidence limits	lower = 20.9 mg/kg upper = 78.5 mg/kg					
MMC positive control data								
0	10	10	4.97 ± 0.17	4		10.8 ± 0.5	10	
0.5	10	10	9.82 ± 0.38	4		12.7 ± 0.5	6	
P value[e]			<0.0001[g]			0.0184[g]		

[a] Group mean SCE per cell frequency ± the standard error of the mean among N animals.
[b] P values for pairwise comparison between control and treatment group data using square-root transformed SCE/cell data and pooled group variances in Student's t test.
[c] Group mean AGT (h) values ± the standard error of the mean among N animals.
[d] P values for pairwise comparisons between control and treatment group data using separate group variances in Student's t test.
[e] P value obtained from ANOVA analysis using square-root transformed SCE/cell data or untransformed AGT data.
[f] Significantly different at $\alpha = 0.0167$ (.05/3 using Boneferroni correction for multiple comparisons).
[g] Significantly different at $\alpha = 0.05$.
NC = not calculated.

lethality, and of 1000 mg/kg for 2AAF, based on cell cycle delay, were used in this study. Treatment with PYR, at doses of 500, 1000, and 1500 mg/kg, resulted in no animal lethality nor in an apparent delay in bone marrow cellular proliferation (Table 1). A dose of 2000 mg/kg was selected as the MTD because of difficulty in obtaining a homogeneous suspension at higher doses. Exposure to 4AAF also did not cause any animal lethality nor did it inhibit bone marrow proliferation when tested at doses of 250, 500, and 1000 mg/kg. In a second study, conducted at 1500 and 2000 mg/kg, treatment with 4AAF again induced no animal lethality nor inhibited cell proliferation kinetics. Because of difficulty in achieving a homogenous suspension at higher doses, an MTD of 2000 mg/kg was selected (Table 1).

SCE and average generation time (AGT) evaluation
BP. Data on the SCE and AGT response to treatment with BP are presented in Table 2. Treatment with BP at doses of 62.5, 125, and 250 mg/kg induced a highly significant increase ($P < 0.0001$) in SCE frequency, as did treatment with the positive control, MMC, at a dose of

Table 3. *SCE and AGT data for PYR*

Dose (mg/kg)	Number treated	Number survivors	SCE Frequency[a]	N	P[b] value	AGT(h)[c]	N	P[d] value
0	5	5	4.68 ± 0.28	4	—	12.4 ± 0.4	4	—
500	5	5	4.74 ± 0.34	4	NC	13.7 ± 1.1	5	NC
1000	5	5	5.04 ± 0.46	4	NC	12.9 ± 0.6	5	NC
2000	5	4	4.80 ± 0.52	3	NC	14.8 ± 0.9	3	NC
P value[e]			0.8905			0.2979		
SCE-5	NC							
MMC positive control data								
0	10	10	5.54 ± 0.27	4		10.6 ± 0.5	10	
0.5	10	10	11.24 ± 0.38	4		14.4 ± 0.5	10	
P value[e]			< 0.0001[g]			< 0.0001[g]		

See Table 2 for explanation of footnotes

Table 4. *SCE and AGT data for 2AAF*

Dose (mg/kg)	Number treated	Number survivors	SCE Frequency[a]	N	P[b] value	AGT(h)[c]	N	P[d] value
0	6	6	4.78 ± 0.37	4	—	12.5 ± 0.3	6	—
250	7	7	8.78 ± 0.40	4	< 0.0016[f]	12.7 ± 0.2	6	NC
500	5	5	10.97 ± 1.26	4	< 0.0001[f]	13.3 ± 0.7	5	NC
1000	6	6	14.81 ± 0.72	4	< 0.0001[f]	14.1 ± 0.4	6	NC
P value[e]			< 0.0001[g]			0.0445[g]		
SCE-5 = 431.5 mg/kg		95% confidence limits		upper = 53.4 mg/kg lower = 801.0 mg/kg				
MMC positive control data								
0	10	10	4.97 ± 0.17	4		10.8 ± 0.5	10	
0.5	10	10	9.82 ± 0.38	4		12.7 ± 0.5	6	
P value[e]			0.0001[g]			0.0184[g]		

See Table 2 for explanation of footnotes

Table 5. *SCE and AGT data for 4AAF*

Dose (mg/kg)	Number treated	Number survivors	SCE Frequency[a]	N	P[b] value	AGT(h)[c]	N	P[d] value
0	5	5	4.68 ± 0.28	4	—	12.4 ± 0.4	6	—
500	5	5	5.13 ± 0.30	4	0.4250	12.3 ± 0.2	5	0.9209
1000	5	5	6.73 ± 0.68	4	0.0067[f]	19.1 ± 1.1	4	0.0056[f]
2000	7	7	5.47 ± 0.40	4	0.2066	14.4 ± 1.00	4	0.1208
P value[e]			0.0424[g]			0.0002[g]		
SCE-5 = 2527.6 mg/kg		95% confidence limits		lower = 1330.7 mg/kg upper = 7804.5 mg/kg				
MMC positive control data								
0	10	10	5.54 ± 0.27	4		10.6 ± 0.5	10	
0.5	10	10	11.28 ± 0.38	4		14.4 ± 0.5	10	
P value[e]			< 0.0001[g]			< 0.0001[g]		

See Table 2 for explanation of footnotes.

0.5 mg/kg ($P = 0.0001$). The magnitude of the SCE response for BP was not significantly different among the three test doses, indicating a saturation in response. Pairwise comparison between each dose and the solvent control group of animals determined that the lowest effective dose tested (at $\alpha = 0.05$) was 62.5 mg/kg. The calculated SCE-5 value for BP (omitting the group data for doses at 125 and 250 mg/kg) was 48.2 mg/kg. Over the dose range tested, no effect on cellular proliferation kinetics, as measured by AGT, was observed ($P = 0.5924$).

PYR At doses of 500, 1000, and 2000 mg/kg, treatment with PYR did not induce a significant increase in SCEs ($P = 0.8905$) nor did it effect the AGT ($P = 0.2979$) (Table 3). One animal treated with 2000 mg/kg died within the 24 hour experimental period. MMC at a dose of 0.5 mg/kg induced a significant increase in SCEs ($P < 0.0001$).

2AAF At dose of 250, 500, and 1000 mg/kg (see Table 4), exposure to 2AAF induced a highly significant ($P < 0.0001$) linear increase ($r = 0.922$) in SCE frequencies. Pairwise comparisons between each treated and the control group indicated that 250 mg/kg was the lowest effective dose tested. The calculated SCE-5 value for 2AAF (using all dose-response data) was 431.5 mg/kg. A marginally significant delay in bone marrow cell cycle kinetics was observed ($P = 0.0445$). For MMC analysis, see the BP results.

4AAF As shown in Table 5, Treatment with 4AAF at doses of 500, 1000, and 2000 mg/kg induced a weak but significant increase in SCE frequency ($P = 0.0424$). A maximal response of $\sim 44\%$ (~ 2 SCE/cell above control level) occurred at the intermediate dose of 1000 mg/kg. The rate of bone marrow cellular proliferation was significantly affected ($P = 0.0002$) by treatment with 4AAF, with the greatest increase in cell cycle duration ($\sim + 6.7$ h) occurring at 1000 mg/kg. The SCE-5 value for 4AAF (omitting the data at 2000 mg/kg) was calculated to be 2527.6 mg/kg. For MMC analysis, see the PYR results.

5 Discussion

Exposure to the 2, well-established carcinogens 2AAF and BP induced a highly significant SCE response at all doses tested, but with different dose response curves. SCE frequency increased linearly with increasing dose for 2AAF, while the SCE response at all three doses of BP appeared to have saturated. This linear increase in SCE frequency observed for 2AAF occurred at doses which exceeded the solubility for 2AAF in corn oil, suggesting that for ip administration, using doses only which are completely dissolved in the solvent is not necessary. The most likely explanations for the apparent saturation in the SCE response for BP is either that the levels of mixed functions oxidases capable of metabolizing BP into a SCE-inducing reactive intermediate(s) have been exceeded at all 3 doses, or that saturation in the effective dose administered to the animal occurred due to technical reasons. Since the induction of chromosomally aberrant cells by BP over the same dose range increased with increasing dose (Luke *et al.*, this volume), the latter explanation is not likely. The difference in dose–response for BP-induced SCEs and BP-induced chromosomally aberrant cells suggests also the involvement of different metabolites of BP in the induction of these 2 genotoxic endpoints.

When the potency of these 2 carcinogenic chemicals is compared using the calculated SCE-5 values, BP (SCE-5 = 48.2 mg/kg) is ~ 9 times more potent than 2AAF (SCE-5 = 431.5 mg/kg). In calculating the SCE-5 value for BP, however, it must be remembered that the linear portion of the SCE dose response curve might not have been attained at the lowest dose tested, potentially making the calculated SCE-5 value an overestimation of the true value. Treatment with 2AAF but not BP significantly altered the rate of cell proliferation kinetics in the bone marrow, while exposure to neither carcinogen at these doses caused lethality among the treated animals.

Treatment with PYR, a purported noncarcinogenic structural analogue of BP, to the highest dose feasible considering the administration technique, was negative for SCE induction. Treatment with 4AAF, a noncarcinogenic structural analogue of 2AAF, induced a marginally significant increase in SCE frequency. Exposure to 4AAF induced also an elongation of cell cycle duration (~ 6.5 hr) but no animal lethality. Exposure to PYR had no effect on AGT, but one animal among the group treated with the highest dose (i.e. 2000 mg/kg) died.

Because of the weak but positive response for 4AAF, a SCE-5 value for this compound was calculated. When this value (2527.6 mg/kg) is compared with the SCE-5 value calculated for 2AAF (i.e. 431.5 mg/kg), the carcinogen 2AAF was some six times more potent in inducing SCEs than the purported noncarcinogen 4AAF. Due to the lack of a positive trend in SCE data for PYR, a SCE-5 value could not be calculated. The difference in potency between 2AAF and 4AAF suggests that if a mechanistic relationship exists between the induction of SCEs and carcinogenic activity and if the SCEs induced by the low doses of 2AAF and the high doses of 4AAF arise through a common process, then to achieve the same tumor yield at least six times more chemical must be used or for the same power of test, six times more animals per dose would be necessary. This proportionately assumes, of course, a linear relationship between dose and tumor yields, an assumption which may not be supportable scientifically.

6 Conclusion

When administered ip, BP and 2AAF were effective inducers of SCEs in bone marrow cells of male B6C3F1 mice. Up to the highest dose feasible within the constraints of the test system, PYR was completely ineffective in inducing SCEs while 4AAF induced a slight but significant increase in SCEs at an intermediate dose.

7 Acknowledgments

The authors would like to thank Ms. L. Wasson and Ms. G. Callister for typing this contribution. The work in this contribution was supported by the National Toxicology Program under National Institute of Health Sciences Interagency Agreement No. YOI-ES-20098-05 and by the U.S. Department of Energy under Contract DE-AC02-76CH00016. According, the U.S. Government retains a nonexclusive, royalty-free license to publish or reproduce the published form of this contribution, or allow others to do so, for U.S. Government purposes.

8 References

Allen, J.W., C.F. Shuler, R.W. Mendes, and S.A. Latt (1979) A simplified technique for *in vivo* analysis of sister chromatid exchanges using 5-bromodeoxyuridine tablets, *Cytogent. Cell Genet.*, **18**, 231–237.

Ashby, J. (1981) Overview of study and test chemical activities. In F.J. de Serres and J. Ashby (Eds.), *Evalution of Short-Term Tests for Carcinogens, Report of the International Collaborate Program*, Elsevier/North-Holland, Amsterdam, pp. 112–171.

Hicks, M. and Ashby, J. (1986) This volume.

Goto, K., T. Akematasa, H. Shimazu, and T. Sugiyama (1975) Simple differential Giemsa staining of sister chromatids after treatment with photosensitive dyes and exposure to light and the mechanism of staining, *Chromosoma*, **53**, 223–230.

Guglielmi, G.E., T.F. Vogt, and R.R. Tice (1982) Induction of sister chromatid exchanges and inhibition of cellular proliferation *in vitro I.* Caffeine. *Environ. Mutagen.*, **4**, 191–200.

IARC (1982) *Evaluation of the carcinogenic Risk of Chemicals to Humans*, Suppl. 4, Chemicals, Industrial Processes, and Industries Associated with Cancer in Humans, World Health Organization International Agency for Research on Cancer, Lyon, France.

IARC (1983) *Evaluation of the Carcinogenic Risk of Chemicals to Humans*, Vol. 32, Polynuclear Aromatic Compounds, Part I, Chemical, Environmental, and Experimental Data, World Health Organization International Agency for Research in Cancer, Lyon, France.

Ivett, J.L., and R.R. Tice (1982) Average Generation Time: A new method of analysis and quantitation of cellular proliferation kinetics, *Environ. Mutagen.*, **4**, 358.

Ivett, J.L., C.A. Luke, and R.R. Tice (1984) Range finding studies for determining the optimal doses and termination times for evaluating chromosomal aberrations and sister chromatid exchanges *in vivo*, *Environ. Mutagen.*, **6**, 468.

Kriek, E. (1974) Carcinogenesis by aromatic amines, *Biochim. Biophys. Acta*, **355**, 177–203.

Luke, C.A., R.R. Tice, V.M. Miller, B.G. Ormiston, A.C. Wilton, and M.D. Shelby (1986) A comparative analysis of the induction of chromosomal aberrations in the bone marrow cells of B6C3F1 mice, This volume.

McFee, A.F., K.W. Lowe, and J.R. San Sebastion (1983) Improved sister chromatid differentiation using paraffincoated bromodeoxyuridine tablets in mice, *Mutation Res.*, **119**, 83–88.

McFee, A.F., R.R. Tice, and M.D. Shelby (1986) Interlaboratory standardization of procedures for *in vivo* mouse bone marrow cytogenetics assay, In preparation.

NCI (1951–1978), *Survey of Compounds Which Have Been Tested for Carcinogenic Activity*, US Dept. of Health and Human Services, Washington, DC.

Sandberg, A.A. (Ed.) (1982) Sister Chromatid Exchange, Program and Topics in *Cytogenetics*, Vol. 2, Alan R. Liss, NY.

Sokal, R.R. and F.J. Rohlf (1969) *Biometry*, Second Edition, W.H. Freemand and Co., San Francisco, CA.

Tice, R.R., E.L. Schneider, and J.M. Rary (1976) The utilization of bromodeoxyuridine incorporation into DNA for the analysis of cellular kinetics, *Exp. Cell Res.*, **102**, 232–236.

Tice, R.R., M.A. Bender, J.L. Ivett, and R.T. Drew (1978) Cytogenetic effects of inhaled ozone, *Mutation Res.*, **58**, 293–304.

Tice, R.R., and A. Hollaender (Eds), (1984a) *Sister Chromatid Exchanges, Book A. The Nature of Sister Chromatid Exchanges*, Plenum Publishing Corp., NY.

Tice, R.R. and A. Hollaender (Eds) (1984b) *Sister Chromatid Exchanges, Book B. Genetic Toxicology and Human Studies*, Plenum Publishing Corp., NY.

Tice, R.R., and J.L. Ivett (1985) Cytogenetic analysis of bone marrow damage. In R.D. Irons (Ed.), *Toxicology of the Blood and Bone Marrow*, Raven Press, NY, pp. 119–140.

Whorton, E.B., Jr., R.R. Tice, and D.G. Stetka (1984) Statistical design, analysis, and inference issues in studies using sister chromatid exchange. In R.R. Tice and A. Hollaender (Eds), *Sister Chromatid Exchange, Book A: The Nature of Sister Chromatid Exchanges*, Plenum Publishing Corp., NY, pp. 431–440.

Wolff, S. (ed.) (1982) *Sister Chromatid Exchange*, John Wiley and Sons, NY.

James W. Allen, Lorraine C. Backer, James Campbell [1], Yousuf Sharief[1], and Barbara Westbrook-Collins[1]

US Environmental Protection Agency
Genetic Toxicology Division
Research Triangle Park, North Carolina, USA
[1]*Environmental Health Research & Testing, Inc.*
Research Triangle Park, North Carolina, USA

39

Analysis of sister chromatid exchange in C57BL/6 mouse bone marrow cells following in vivo exposure to carcinogen/noncarcinogen chemical pairs

1 Summary

In vivo SCE analyses in mice have revealed clear differences in SCE induction potencies between the members of two carcinogen/noncarcinogen chemical pairs: 2AAF/4AAF and BP/PYR. 2AAF and BP were moderately efficient inducers of SCE. BP was the most active of the four chemicals. A clear dose–response pattern was not observed for 2AAF. 4AAF and PYR caused minimal increases in SCE frequencies at one or two doses. The strength of the SCE induction response did not always correspond to the degree of cell-cycle inhibition. It was concluded that *in vivo* SCE analyses could separate carcinogens from noncarcinogens on the basis of their relative abilities to cause this form of genetic damage.

2 Introduction

Male and female C57B1/6 mice were exposed by gavage to the presumed carcinogen/noncarcinogen chemical pairs: 2-acetylaminofluorene (2AAF)/4-acetylaminofluorene (4AAF) and benzo(a)pyrene (BP)/pyrene (PYR). Bone marrow cells were analysed for sister chromated exchange (SCE) induction and cell-cycle kinetics.

3 Material and methods

Animals

Male and female mice of strain C57B1/6 were obtained from Jackson Laboratory (Bar Harbor, ME) and acclimated for several weeks prior to experimental use. They were maintained at the U.S. EPA animal facility in laminar flow rooms with 15 cycles per h of biocleaned air at 60–68% relative humidity, 20–21 °C temperature, and a 12 h light-dark cycle (07.00–19.00h). The animals were housed 5 per cage and fed Lab Chow (Purina) and water ad libitum. At the time of the study, the mice were 18 to 19 weeks old, males weighing 29–33 grams and females weighing 22–26 grams.

Chemicals

The test chemicals 2AAF, 4AAF, BP and PYR were received from Lancaster Synthesis, Ltd., UK, and stored unopened at −20 °C. 5-Bromo-2-deoxyuridine (BrdUrd) and cyclophosphamide (CP) were purchased from Sigma Chemical Co. (St. Louis, MO) and dissolved in 0.85% saline. All the test chemicals were suspended in corn oil (Fisher Scientific Co., Raleigh, NC) and were partially dissolved by shaking in a vortex mixer. None of the chemicals dissolved completely, even at the lowest dose. The interlaboratory control, mitomycin C (MMC; Sigma Chem. Co.), was dissolved in 0.85% saline.

Experimental procedures
BrdUrd pellet methodology (Allen *et al.*, 1978; McFee *et al.*, 1983) was used to label mouse bone marrow cells *in vivo* with BrdUrd. Fifty-milligram BrdUrd pellets were pressed with a Parr pellet press and a 0.178-in.-diameter punch and die (Parr Instrument Co., Moline, IL). The pellets were coated with embedding paraffin (Fisher Scientific Co., Raleigh, NC) to cover approximately 85% of the surface area. They were implanted subcutaneously in the lateral abdominal region of mice anaesthetised with Metofane (Pitman-Moore, Washington Crossing, NJ).

Thirty minutes after BrdUrd treatment, the test chemical was administered by gavage (1 ml/100 g body weight) using a 20 g stainless steel animal-feeding needle (Popper and Sons, New Hyde Park, NY). The interlaboratory control, MMC, was administered in a 0.2–0.3 ml volume by intraperitoneal (ip) injection. Twenty-one hours later, 2 mg/kg of colchicine (Sigma Chemical Co., St. Louis, MO) was injected ip. Animals were sacrificed by cervical dislocation after 2–3 h, and marrow cells were harvested from the femurs. The cells were treated with 0.075 *M* KCl hypotonic solution for 15 min and fixed 3 times with methanol: acetic acid (3:1). Slides were prepared and differentially stained (FPG) in accordance with standard cytogenetic methodology (Latt *et al.*, 1981).

SCE frequencies were analysed from coded slides. Twenty-five well-differentiated second division cells with 40 ± 2 chromosomes were assessed from each animal. Cell replication indices were calculated from 100 metaphases per animal revealing first (M_1), second (M_2), or third (M_3) division stain patterns. A replicative index (RI) = (1 times M_1) + (2 times M_2) + (3 times M_3) frequency (Schneider and Lewis, 1981) was calculated for each animal.

Statistical analysis
The raw data for SCE frequencies and cell-cycle kinetics were first analysed using an analysis of variance (Sokal and Rohlf, 1969). Effects due to chemical exposure, the sex of the animal, the interaction between the two, and the individual animal variation defined by sex and chemical dose were examined using a nested model. A two-way analysis of variance of the means of each group as defined by sex and dose was also conducted to confirm the conclusions of the first analysis. A Dunnett's multiple comparisons of means test (Dunnett, 1955) was used to compare each treatment group with the controls (corn oil). The test was done on the data pooled over males and females if there was no significant effect due to sex and no significant interaction between sex and dose in the analysis of variance. When either of these terms was statistically significant, Dunnett's test was applied to the data for males and females separately.

4 Results

2AAF
SCE data for mice exposed to 2AAF are presented in Table 1. Analysis of variance (ANOVA) showed a statistically significant increase in SCEs due to the treatment ($P < 0.001$). There was no statistically significant effect due to sex or to the interaction between sex and dose ($P > 0.05$ for both). The combined male and female data are shown in Figure 1. Although three doses (111.5, 223, and 1338 mg/kg) caused statistically significant increases in SCE levels as demonstrated using Dunnett's test ($P < 0.05$), there was no clear dose-dependent response.

Figure 1. Comparative SCE-inducing potential of 2AAF and 4AAF in male and female mice combined. Symbols represent mean values for each dose group ± standard error of the mean.

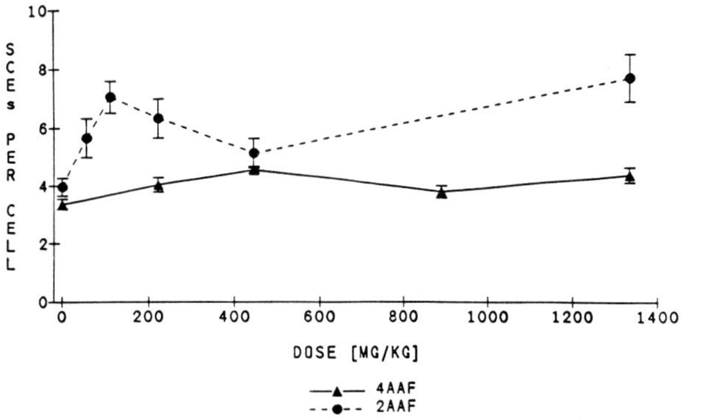

Table 1. *Effect of 2AAF and 4AAF on SCE frequencies: Means/dose group*

Treatment dose (mg/kg)	Male Mean SCE/cell ± SD(n)	Female Mean SCE/cell ± SD(n)
2AAF		
0.0	4.4 ± 0.39(4)	3.6 ± 1.11(4)
55.8	5.3 ± 2.41(4)	6.0 ± 1.46(4)
115.5	7.8 ± 1.98(3)	6.5 ± 0.72(4)
223.0	5.4 ± 1.80(4)	7.3 ± 1.67(4)
446.0	4.6 ± 1.05(4)	5.7 ± 1.71(4)
1338.0	9.0 ± 2.34(4)	6.4 ± 1.51(4)
4AAF		
0.0	3.3 ± 0.43(4)	3.4 ± 0.67(4)
223.0	4.5 ± 0.56(4)	3.5 ± 0.44(4)
446.0	4.7 ± 0.37(4)	4.4 ± 0.25(4)
892.0	3.5 ± 0.69(4)	4.0 ± 0.47(4)
1338.0	4.6 ± 1.10(3)	4.2 ± 0.27(4)

The male and female results for RIs are graphed in Figure 2. RI data could not be combined because the ANOVA showed a significant effect due to sex and treatment ($P < 0.001$ for both) as well as a significant interaction between the two ($P < 0.001$). When analysed separately by the Dunnett's test, the data for female mice showed a statistically significant ($P < 0.05$) decrease in RI only at the 223 mg/kg dose, while the data for the males showed a statistically significant ($P < 0.05$) decrease at all doses. This difference may be attributed to the unusually high RI mean of 2.5 for this particular group of corn oil control males. The slides were reanalysed, and the original findings of very high individual and group RIs were confirmed. If male control values more closely resembling those of the other male control groups in this study were used, there would not be a significant effect on RI by 2AAF exposure.

4AAF

SCE data from 4AAF-exposed mice are also presented in Table 1. There was a statistically significant ($0.001 < P < 0.01$) increase in SCE frequency because of the 4AAF treatment, with no significant ($P > 0.10$) difference in response due to sex and no significant ($P > 0.010$) interaction between dose and sex. A Dunnett's comparison done on the males and females combined detected statistically significant ($P < 0.05$) increases in SCE frequency at the 223, 446, and 1338 mg/kg doses.

The male and female results for RIs are graphed in Figure 3. 4AAF exposure caused no statistically significant ($P > 0.10$) effect on RIs in this study.

BP

SCE frequencies for BP-treated mice are shown in Table 2. One male and one female died within 24 h of exposure to the highest dose of BP. An ANOVA showed a statistically significant ($P < 0.001$) treatment effect (Figure 4). There was also a statistically significant ($0.001 < P < 0.01$) effect due to the sex of the animal, but no significant ($P > 0.10$) interaction between dose and sex. Analysed separately using Dunnett's test, both males and females showed statistically significant ($P < 0.05$) increases in SCE values at all doses.

The actual values for SCE frequencies were greater in males than in females at each dose. Males showed an increase in SCE frequency of up to 14.3 per cell following exposure to 252 mg/kg, with a plateau of about 14 SCEs per cell at the higher doses. The SCE frequency for females showed a continuing increase over all doses, with a gentler plateauing effect.

The male and female results for RIs are graphed in Figure 5. There was a statistically significant ($P < 0.001$)

Figure 2. The effect of 2AAF on replicative indices in male and female mice. Symbols represent mean values for each dose group ± standard error of the mean.

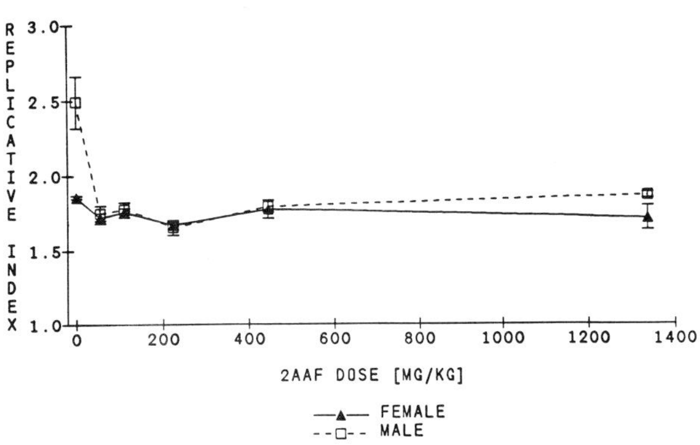

Figure 3. The effect of 4AAF on replicative indices in male and female mice. Symbols represent mean values for each dose group ± standard error of the mean.

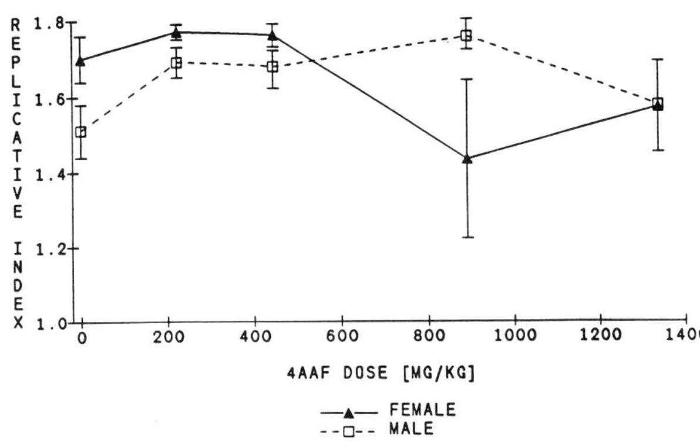

Table 2. *Effect of BP and PYR on SCE frequencies: Mean/dose group*

Treatment dose (mg/kg)	Male Mean SCE/cell ± SD(n)	Female Mean SCE/cell ± SD(n)
BP		
0.0	4.2 ± 0.76(4)	3.8 ± 0.78(4)
63.0	11.0 ± 2.21(4)	9.5 ± 1.21(4)
126.0	11.7 ± 1.58(4)	10.8 ± 1.31(4)
252.0	15.3 ± 1.07(3)	11.9 ± 0.64(4)
504.0	14.0 ± 1.31(4)	12.1 ± 1.92(4)
1008.0	14.1 ± 3.77(3)	12.6 ± 1.35(3)
PYR		
0.0	4.0 ± 0.81(4)	3.3 ± 0.70(3)
202.0	4.3 ± 0.46(4)	5.2 ± 1.46(3)
404.0	4.2 ± 0.37(4)	4.8 ± 0.83(4)
808.0	4.8 ± 1.14(3)	6.2 ± 1.41(2)
1212.0	3.8 ± 0.23(3)	(a)

[a] No analysis due to animal death or insufficient number of second division cells.

decrease in RI due to BP exposure and no effect due to the sex of the animal ($P < 0.10$). Because the test for an effect due to the interaction of sex and dose was marginally nonsignificant ($P = 0.072$), the male and female data were analysed separately in the Dunnett's test. For the females, the 252 and 1008 mg/kg doses caused statistically significant ($P < 0.05$) decreases in RI. In the males, all doses except the lowest (63 mg/kg) caused statistically significant ($P < 0.05$) decreases in replicative index.

PYR

SCE frequencies for PYR are listed in Table 2. The analysis of this experiment was complicated by the toxicity of the compound; 3 females died after exposure to the highest dose of 1212 mg/kg PYR.

Figure 4. Comparative SCE-inducing potential of BP and PYR in male and female mice. Symbols represent mean values for each dose group ± standard error of the mean.

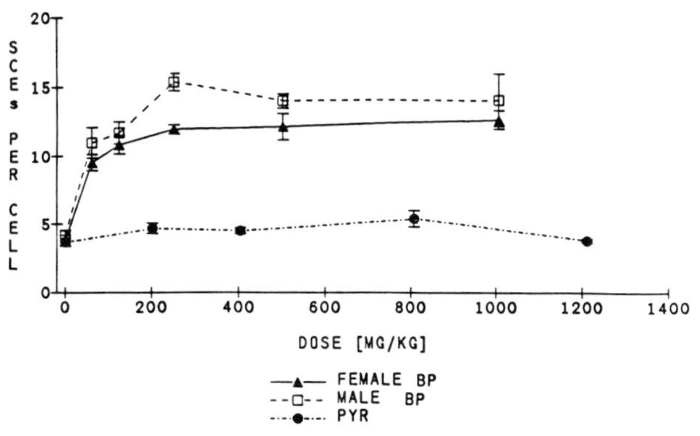

Figure 5. The effect of BP on replicative indices in male and female mice. Symbols represent mean values for each dose group ± standard error of the mean.

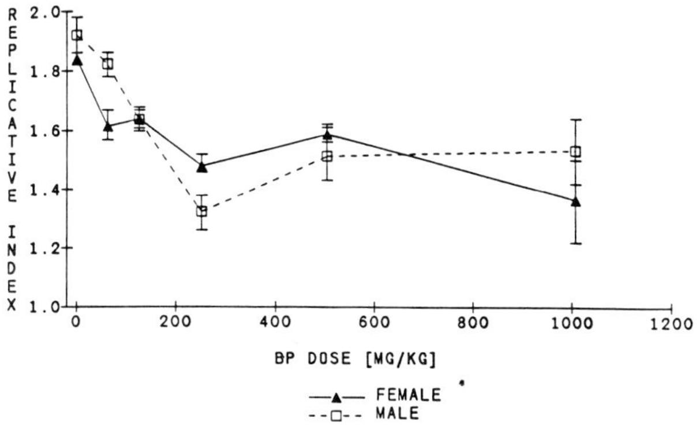

Figure 6. The effect of PYR on replicative indices in male and female mice. Symbols represent mean values for each dose group ± standard error of the mean.

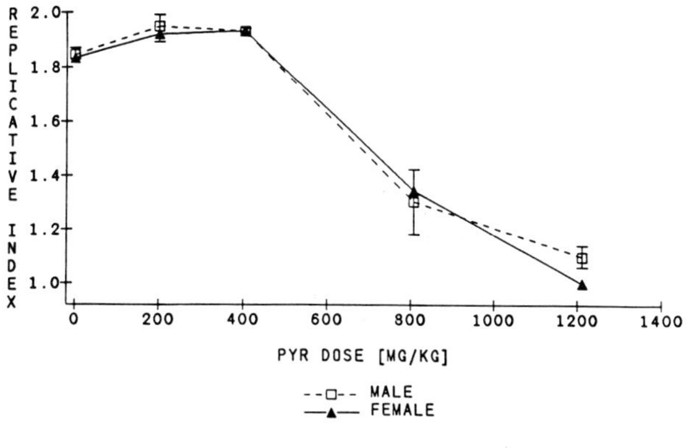

An ANOVA showed a statistically significant ($0.01 < P < 0.05$) increase in SCE frequency due to PYR exposure with no significant ($P > 0.10$) effect due to sex or the interaction between sex and dose ($P > 0.10$). A Dunnett's test done on the data pooled over both sexes showed a statistically significant ($P < 0.05$) increase in SCEs for the 808 mg/kg dose only. The response of the females at the 808 mg/kg dose, the only dose statistically significant in Dunnett's tests done on males and females separately, is the primary reason for the overall positive response of SCE frequency to PYR exposure.

The male and female results for RIs are graphed in Figure 6. There was a statistically significant ($P < 0.001$) decrease in RI after exposure to PYR, and no significant ($P > 0.10$) effects due to either the sex of the animal or the interaction between sex and dose. The Dunnett's test done on the data from males and females separately showed significant ($P < 0.05$) decreases in RI at only the 2 highest doses and only for males. Since there was a single RI value of 1.0 for the female group exposed to the highest dose of PYR, the Dunnett's comparison was not computed.

MMC

Results from the interlaboratory control experiments with mitomycin C (MMC) are presented in Appendix I. An increase of nearly 3 times the baseline SCE frequency was achieved at the dose of 0.5 mg/kg. The exposed animals did not reveal significantly diminished RI values from those of the control animals.

5 Discussion

All of the chemicals tested in the present study showed some activity for SCE induction. However, the carcinogens were clearly more potent than the presumed noncarcinogens for producing this effect. Differences in activity were greater between BP and PYR than between 2AAF and 4AAF. The relation between cytogenetic damage and effects upon cell cycling was inconsistent.

2AAF caused moderate increases in SCE frequencies over a broad dose range without sex specificity or clear dose dependence. The highest increments of approximately 2 times baseline values were in the general range of those reported in other *in vivo* studies (Takehisa and Wolff, 1978; Schreck *et al.*, 1979; Neal and Probst, 1983; Parodi *et al.*, 1983). Our results indicate a considerable amount of animal-to-animal variation in SCE response to 2AAF (data not shown). This variation is consistent with observations in rabbit lymphocytes (Takehisa and Wolff, 1978), and it may have obscured small dose-related changes in response. Some cell-cycle delay consistently occurred, although, as with SCE results, the effects were not necessarily greater at higher doses than at lower doses. Extreme cellular toxicity or animal death did not occur at any dose. These results may reflect the interplay of exposure by gavage with complex metabolism (Miller, 1978) and specific DNA damage repair capacities (Lieberman *et al.*, 1976).

4AAF, administered over the same dose range as 2AAF, showed relatively little cytogenetic activity. A weak SCE induction response was observed at 2 doses, and there was no apparent effect upon cell cycling. These results are in accord with the general view of this chemical as being unremarkable for causing genetic damage. 4AAF also has been reported to be negative for the induction of micronuclei (Salamone *et al.*, 1981), unscheduled DNA synthesis (Martin and McDermid, 1981), liver DNA repair (Williams, 1977), and gene mutations (Clive *et al.*, 1979) in mammalian cells.

The curvilinear SCE induction pattern produced by BP, an increasing response at low doses followed by a plateau effect at higher doses, was similar to that which has been produced in other *in vivo* mouse bone marrow (Bayer and Bauknecht, 1977; Schreck and Latt, 1980) and lymphocyte (Kligerman *et al.*, 1985) SCE assays using this chemical. The other studies used the ip route of administration. Thus, the gavage exposure mode used here probably was not a critical determinant of the dose–response pattern obtained. More likely, complexities in the metabolism of this chemical may have led to a saturation of the activating enzyme systems, thereby placing an upper limit on the amount of active metabolites that may reach the marrow cells. That male mice showed a significantly greater SCE induction response than females is interesting in that males of other mouse strains have also shown greater tumour induction responses to polycyclic aromatic hydrocarbon exposure (Roe, 1975). A possible explanation for some relation between these two responses is unclear, although sex-specific differences in levels of AHH activities may be implicated in these results. However, quantitative strain-specific variations in AHH levels have not correlated with differences in SCE induction (Schreck and Latt, 1980).

PYR was weakly positive for SCE induction at 1 dose due to a slightly elevated response in females. This minimal response is in agreement with the findings from other *in vivo* (Paika *et al.*, 1981; Sirianni and Huang, 1978) and *in vitro* (Popescu *et al.*, 1977; Evans and Mitchell, 1981) rodent cell studies in which little or no SCE induction resulted from exposure to this compound. Despite the relative inactivity of PYR for SCE induction, its ability at high doses to slow cell cycling was as great, or greater, than that of BP. These contrasting responses for the BP/PYR chemical pair indicate that SCE and cell-cycling endpoints are not necessarily related.

6 Conclusions

Our results have indicated that the carcinogens 2AAF and BP are more efficient inducers of SCE than their presumed noncarcinogenic analogues. The sample size used was sufficient to allow this discrimination. However, additional animals per dose group may restore sensitivity occasionally lost because of excessive inter-animal variability or death. SCE induction did not provide an absolute distinction; rather, it defined relative strengths for inducing this specific form of genetic damage. Whether or not the minimal SCE-inducing potentials of 4AAF and PYR reflect comparable potentials as weak carcinogens cannot be concluded.

7 Acknowledgements

The authors would like to thank A. Petro, D. Harrison and E. Crosson for excellent technical assistance; A. Stead for help with statistical analyses; and S. Milton and C. Kuehnhoff for dedicated typing of the manuscript. Thanks also are due to A. Kligerman and G. Erexson for their help with data analyses.

8 References

Allen, J.W., C.F. Shuler and S.A. Latt (1978) Bromodeoxyuridine tablet methodology for *in vivo* studies of DNA synthesis, *Somatic Cell Genet.*, **4**, 393–405.

Bayer, U. and T. Bauknecht (1977) The dose-dependence of sister chromatid exchanges induced by 3 hydrocarbons, in the *in vivo* bone marrow test with Chinese hamsters, *Experientia*, **15**, 25.

Clive, D., K.O. Johnson, J.F.S. Spector, A.G. Batson, and M.M.M. Brown (1979) Validation and characterization of the L5178Y/TK$^{+/-}$ mouse lymphoma mutagen assay system, *Mutation Res.*, **59**, 61–108.

Dunnett, C.W. (1955) A multiple comparisons procedure for comparing several treatments with a control, *J. Am. Stat. Assoc.*, **50**, 1096–1121.

Evans, E.L. and A.D. Mitchell (1981) Effects of 20 coded chemicals on sister chromatid exchange frequencies in cultured Chinese hamster cells. In, F.J. de Serres and J. Ashby (Eds), Evaluation of Short-Term Tests for Carcinogens, *Progress in Mutation Research*, Vol. 1, Elsevier/North-Holland, NY, pp. 528–550.

Kligerman, A.D., G.L. Erexson and J.L. Wilmer (1985) Induction of sister-chromatid exchange (SCE) and cell-cycle inhibition in mouse peripheral blood B lymphocytes exposed to mutagenic carcinogens *in vivo*, *Mutation Res.*, **157**, 181–187.

Latt, S.A., J. Allen, S.E. Bloom, A. Carrano, E. Falke, D. Kram, E. Schneider, R. Schreck, R. Tice, B. Whitfield and S. Wolff (1981) Sister chromatid exchanges: A report of the Gene-Tox program, *Mutation Res.*, **87**, 17–62.

Lieberman, M.W., D.E. Amacher, J.A. Elliott and S.L. Huang (1976) Removal of bound acetylaminofluorene from the DNA of mammalian cells in culture. In, P.N. Magee *et al.* (Eds), *Fundamentals in Cancer Prevention*, Univ. Park Press, Baltimore, MD, pp. 335–348.

Martin, C.N. and A.C. McDermid (1981) Testing of 42 coded compounds for their ability to induce unscheduled DNA repair synthesis in Hela Cells. In, F.J. de Serres and J. Ashby (Eds), *Evaluation of Short-Term Tests for Carcinogens, Report of the International Collaborative Program Progress in Mutation Research*, Vol. 1, Elsevier/North-Holland, NY, pp. 531–537.

McFee, A.F., K.W. Lowe and J.R. San Sebastian (1983) Improved sister-chromatid differentiation using paraffin coated bromodeoxyuridine tablets in mice, *Mutation Res.*, **119**, 83–88.

Miller, E.C. (1978) Some current perspectives on chemical carcinogenesis in humans and experimental animals, *Cancer Res.*, **38**, 1479–1496.

Neal, S.B. and G.S. Probst (1983) Chemically-induced sister-chromatid exchange *in vivo* in bone marrow of Chinese hamsters, *Mutation Res.*, **113**, 33–43.

Paika, I.J., M.T. Beauchesne, M. Randall, R.R. Schreck and S.A. Latt (1981) *In vivo* analysis of 20 coded compounds. In F.J. de Serres and J. Ashby (Eds), Evaluation of Short-Term Tests for Carcinogens, *Progress in Mutation Research*, Vol. 1, Elsevier/North-Holland, NY, pp. 673–681.

Parodi, S., A. Zunino, L. Ottaggio, M. de Ferrari and L. Santi (1983) Lack of correlation between the capacity of inducing sister-chromatid exchanges *in vivo* and carcinogenic potential, for 16 aromatic amines and azo derivatives, *Mutation Res.*, **108**, 225–238.

Popescu, N.C., D. Turnbull and J.A. DiPaolo (1977) Sister chromatid exchange and chromosome aberration analysis with the use of several carcinogens and

noncarcinogens: Brief communication, *J. Natl. Cancer Inst.*, **59**, 289–293.
Roe, F.J.C. (1975) Neonatal induction of hepatic and other tumours. In W.H. Butler and P.M. Newberne (Eds), *Mouse Hepatic Neoplasia*, Elsevier, Amsterdam, 133 pp.
Salamone, M.F., J.A. Heddle and M. Katz (1981) Mutagenic activity of 41 compounds in the *in vivo* micronucleus assay. In F.J. de Serres and J. Ashby (Eds), Evaluation of Short-Term Tests for Carcinogens, Report of the International Collaborative Program, *Progress in Mutation Research*, Vol. 1, Elsevier/North-Holland, NY, pp. 686–690.
Schneider, E.L. and J. Lewis (1981) Aging and sister chromatid exchange. VIII. Effect of aging environment on sister chromatid exchange induction and cell cycle kinetics in Ehrlich ascites tumor cells. A brief note, *Mech. Ageing Dev.*, **17**, 327–330.
Schreck, R.R. and S.A. Latt (1980) Comparison of benzo(a)pyrene metabolism and sister chromatid exchange induction in mice, *Nature* (London), **288**, 407–408.
Schreck, R.R., I.J. Paika and S.A. Latt (1979) *In vivo* induction of sister-chromatid exchanges in liver and marrow cells by drugs requiring metabolic activation, *Mutation Res.*, **64**, 315–328.
Sirianni, S.R. and C.C. Huang (1978) Sister chromatid exchange induced by promutagens/carcinogens in Chinese hamster cells cultured in diffusion chambers in mice, *Proc. Soc. Exp. Biol. Med.*, **158**, 269–274.
Sokal, R.R. and F.J. Rohlf (1969) *Biometry, the principles and practice of statistics in biological research*, W.H. Freeman and Company, San Francisco, CA, 776 pp.
Takehisa, S. and S. Wolff (1978) Sister-chromatid exchanges induced in rabbit lymphocytes by 2-aminofluorene and 2-acetylaminofluorene after *in vitro* and *in vivo* metabolic activation, *Mutation Res.*, **58**, 321–329.
Williams, G.M. (1977) Detection of chemical carcinogens by unscheduled DNA synthesis in rat liver primary cell cultures, *Cancer Res.*, **37**, 1845–1851.

Appendix 1. *Effect of MMC on SCE frequencies and replicative indices*

		Male	
Treatment (mg/kg)	Animal	SCE/cell mean ± SD	RI
Solvent control	1	5.9 ± 1.72	1.9
	2	4.6 ± 1.71	1.8
	3	4.7 ± 1.68	1.9
	4	4.2 ± 1.80	1.9
Group mean		4.85 ± 0.73	1.88 ± 0.35
MMC 0.5 mg/kg	1	12.5 ± 3.42	1.8
	2	12.5 ± 3.36	1.8
	3	12.4 ± 2.58	1.7
	4	13.0 ± 3.46	1.9
Group mean		12.52 ± 0.34	1.80 ± 0.08

Ad D. Tates, Nico de Vogel, Isaac Neuteboom and Annelies H.M. Rotteveel
Department of Radiation Genetics and Chemical Mutagenesis, State University of Leiden, The Netherlands

40

Detection of sister chromated exchanges in bone marrow of male mice

1 Summary

Ten hours after subcutaneous implantation of BrdUrd tablets, male mice received oral doses of 2AAF, 4AAF, BP, or PYR. They were sacrificed 24 or 30 h after tablet implantation. Bone marrow cells were then analysed for induction of SCEs. The effect of 2AAF was negative of 55.8 and 111.6 mg/kg, but a positive dose-dependent effect was observed at 223, 446, and 892 mg/kg. 4AAF was a weak positive at 892 mg/kg. BP registered positive at all doses tested (63, 126, and 252 mg/kg), whereas PYR was clearly negative (202, 404, and 808 mg/kg).

2 Introduction

As participants in the IPCS Collaborative Study on Short-Term Tests for Genotoxicity and Carcinogenicity, we conducted three *in vivo* tests for the detection of cytogenetic damage induced by the 4 test compounds: (1) the SCE test using bone marrow cells of mice; (2) the micronucleus assay using bone marrow cells of mice; and (3) a micronucleus test using hepatocytes of rats. Results obtained with the latter 2 tests are reported separately. For the first 2 test systems, the same mouse strain was used, and for all these test systems some of the same doses of the 4 chemicals were used. Thus, the detection capacity of the 3 assay systems can, at least partly, be compared.

When carrying out *in vivo* studies on induction of SCEs, 2 different administration procedures for chemicals have been employed routinely. In some studies, the potential clastogen is administered at or immediately after the beginning of BrdUrd administration to minimize the likelihood of an interaction between the clastogen and BrdUrd-substituted DNA. In other studies the clastogen is administered 8–10 h after the start of the BrdUrd treatment to minimize the possibility of cytotoxicity having a selective effect on frequencies of SCEs. Each of these 2 procedures has its advantages and disadvantages. In the present study, the latter procedure has been followed.

3 Materials and methods

Animals

(101 × C3H) Fl male mice were used. They were bred in the Gesellschaft für Strahlen- und Umweltforschung, München, FRG, and sent to us when 4 to 8-weeks old. At Leiden, they were housed at 21–23 °C and a humidity of

60–65% and received drinking water and standard food pellets for mice (Hope Farms, Woerden, The Netherlands). Animals were used for treatment when 12 to 15-weeks old. Weight of the animals varied from 25–30 g.

Chemicals

The 4 test compounds were received from Lancaster Synthesis Ltd. (Eastgate, U.K.). They were stored at 4 °C. Olive oil (Ph. Ned.) was used as the solvent and also as the negative control. Mitomycin C (MMC) (Kyowa, Lampro, B.V.) dissolved in phosphate-buffered saline (PBS) served as the positive control. Test compounds and the negative control were administered orally in 0.3 ml quantities (1% of body weight). Animals were not fasted prior to treatment. MMC (0.5 and 1 mg/kg) was given intraperitoneally (ip). Benzo(a)pyrene (BP; 63, 126, and 252 mg/kg) could be dissolved (with the help of sonication) in olive oil. Pyrene (PYR; 202, 404, and 808 mg/kg) dissolved easily at the concentration of 202 mg/kg. At 404 mg/kg, some small crystals remained detectable and at 808 mg/kg, PYR remained in suspension despite sonication for 5 min. Consistent with expectation, 2-acetylaminofluorene (2AAF; 55.8, 111.6, 223, 446, and 892 mg/kg) and 4-acetylaminofluorene (4AAF; 223, 446, and 892 mg/kg) remained in suspension.

Preparation of BrdUrd tablets and tablet implantation

Tablets (7 mm in diameter) were manufactured in the pharmacy of the University Hospital, Leiden. Each tablet consisted of a mixture of 25 mg BrdUrd (Sigma) and 50 mg granulatum lactosii (granulatum placebo). The tablet was coated with agar (coating was carried out with 4.5% agar in water; the agar layer was allowed to dry in air).

Tablets were implanted as follows: mice were anaesthetized with ether, attached to a cork plate, ventral side up, a selected region of the abdomen shaved, a small (1 cm) cutaneous incision made, the BrdUrd tablet implanted into a pocket made by the circular movement of scissors with blunt ends, and the opening closed with a wound clip.

Treatment protocol

Protocol A:

Tablet implantation	Application of test chemical	Colchicine	Sacrifice of animal 24 h after tablet implantation
↓	↓	↓	↓
X----------	X----------	X--------	X----------
10.00 h	20.00 h	8.00 h	10.00 h

Protocol B:

Tablet implantation	Application of test chemical	Colchicine	Sacrifice of animal 30 h after tablet implantation
↓	↓	↓	↓
X----------	X----------	X--------	X----------
10.00 h	20.00 h	14.00 h	16.00 h

For each test chemical, both protocols were used and 5 animals were used per protocol per concentration of the chemical. Five or six animals were used for the negative and positive controls.

Isolation of bone marrow cells

Each mouse was injected ip with 0.3 ml of a 0.06% colchicine (Sigma) solution. After 2 h, the mice were killed by cervical dislocation, both femurs were excised. The proximal end of each femur was removed carefully with a pair of scissors until a small opening to the marrow canal became visible. The bone marrow was gently aspirated in a centrifuge tube (1 tube/femur) with 1.5 ml PBS. The aspirated bone marrow was then centrifuged at 90 times g (700 rpm) for 6 min, the supernatant removed completely, the pellet suspended in 1.5 ml 0.05 *M* KCl and clots removed, if present. The centrifuge tube was transferred for 12 min to a waterbath (37 °C), and centrifuged at 90 times g for 6 min. The supernatant was removed up to 1 cm above pellet, the pellet resuspended, and centrifuged again for about 3 min.

The supernatant was removed completely and the pellet carefully left intact. One ml of freshly prepared fixative (ethanol:acetic acid = 3:1) was added slowly and allowed to stand for 5 min at room temperature. The fixative was then aspirated and new fixative added (leaving the pellet undisturbed) twice. The fixative was subsequently removed and the pellet suspended vigorously. More fixative was added and allowed to stand for 15 min. Cell suspension was then centrifuged at 90 g for 6 min and allowed to stand overnight.

Slide preparation, staining and analysis

After renewal of fixative, the cell suspension was taken up in a drawn-out Pasteur pipette. A few drops of suspension were placed on a dry clean slide. When slides were dry, they were stained for 15 min with a Hoechst solution (0.05 mg Hoechst 33258 in 100 ml PBS). The Hoechst solution was then replaced by 4 drops of staining solution and slides were sealed with a cover slip and rubber cement. Slides were exposed overnight to a TL lamp (Philips TL 40W/33 or Sylvania FR40 T12/CW). Cover slips were then removed and slides were treated for 1.5 h with a 5 times SSC solution at pH 7.0, at 65 °C. Slides were rinsed 3 times in distilled water, stained for 15 min at room temperature in a 5% Giemsa solution, rinsed in distilled water, and mounted in Depex.

For each animal, 25 cells were scored for number of chromosomes and SCEs. Only cells with a minimum of 38 chromosomes were taken into consideration. In addition, 100 dividing cells were examined for the percentage of 1st, 2nd, and 3rd division cells.

Statistical analysis

Most of the data were analysed using a one-way analysis of variance (ANOVA), followed by the Scheffé method for contrasts (Armitage, 1971). The mean frequency of SCEs/animal was used for the calculations. Student's two-sample test was used to compare frequencies of SCEs at 24 h with those at 30 h. Changes in cell cycle progression have not been analysed statistically. Average cell generation time was determined as follows: time of sacrifice (24 h or 30 h) divided by 1 times frequency MI cells + 2 times frequency MII cells + 3 times frequency MIII cells.

Table 1. *SCE induction in (101 × C3H)F1 male mice*

Chemical	concentration (mg/kg)	Time of sacrifice (h)	SCEs/cell ± S.D.[a]	P-value[b] (one-sided)	% M I	% M II	% M III
Olive oil		24	5.7 ± 0.5 (6)		15.2 ± 8.0	84.2 ± 8.8	0.7 ± 1.0 (12.9)[e]
		30	5.5 ± 0.4 (6)		10.7 ± 10.0	67.7 ± 18.1	21.7 ± 15.5 (14.2)
			5.6 ± 0.4[c]				
MMC	1	24	21.9 ± 1.9 (6)		46.7 ± 11.1	53.3 ± 11.1	0.0 (15.7)
		30	23.6 ± 2.7 (6)		19.0 ± 8.2	79.0 ± 7.8	2.0 ± 1.7 (16.4)
			22.8 ± 2.4				
	0.5	30	8.5 ± 0.4 (5)[d]		16.0 ± 14.1	74.8 ± 16.7	9.2 ± 4.8 (15.5)
2AAF	55.8 I[f]	24	6.3 ± 0.8 (5)	n.s.	30.8 ± 13.9	69.2 ± 13.9	0.0 (14.2)
		30	6.0 ± 0.6 (4)	n.s.	55.3 ± 13.2	44.3 ± 12.5	0.5 ± 1.0 (20.6)
			6.2 ± 0.7				
	111.6 I	24	6.5 ± 0.8 (5)	n.s.	33.0 ± 5.8	67.0 ± 5.8	0.0 (14.4)
		30	6.4 ± 0.7 (5)	n.s.	34.4 ± 13.6	62.8 ± 8.8	2.8 ± 5.2 (17.8)
			6.5 ± 0.7				
	223 I	24	6.8 ± 1.0 (5)	0.03	46.4 ± 14.1	53.6 ± 14.1	0.0 (15.6)
	I	30	7.4 ± 1.2 (5)	0.0007	50.2 ± 15.3	46.4 ± 17.1	3.4 ± 3.9 (19.6)
			7.1 ± 1.1				
	223 II[f]	30	7.5 ± 0.7 (5)[d]	0.0003	16.0 ± 1.6	72.8 ± 2.3	11.2 ± 2.4 (15.4)
	446 II	30	10.3 ± 0.6 (5)	0.0001	24.8 ± 11.3	73.0 ± 10.1	2.2 ± 1.8 (16.9)
	892 II	30	14.3 ± 1.2 (5)	0.0001	18.4 ± 7.4	80.2 ± 7.3	1.4 ± 0.9 (14.9)
4AAF	223	24	6.5 ± 0.7 (5)	0.035	13.4 ± 6.4	86.2 ± 6.3	0.4 ± 0.6 (12.8)
		30	5.8 ± 0.8 (5)	n.s.	18.2 ± 12.2	73.6 ± 5.1	8.2 ± 9.2 (14.4)
			6.1 ± 0.8				
	446	24	5.9 ± 0.7 (5)	n.s.	8.2 ± 3.9	91.8 ± 3.9	0.0 (12.5)
		30	6.6 ± 0.8 (5)	n.s.	8.0 ± 5.8	87.2 ± 6.1	4.8 ± 9.6 (15.2)
			6.3 ± 0.8				
	892	24	6.5 ± 0.6 (5)	0.048	13.2 ± 4.9	86.4 ± 5.2	0.4 ± 0.6 (12.8)
		30	7.4 ± 1.0 (5)	0.0008	15.8 ± 4.3	82.2 ± 5.5	2.0 ± 1.4 (16.1)
			6.9 ± 0.9				
BP	63	24	7.9 ± 0.9 (5)	0.0001	39.4 ± 10.3	59.0 ± 9.5	1.6 ± 1.1 (14.8)
		30	8.5 ± 1.1 (5)	0.0001	16.0 ± 6.0	76.8 ± 11.4	7.2 ± 6.6 (15.7)
			8.2 ± 1.0				
	126	24	9.2 ± 1.1 (5)	0.0001	18.4 ± 4.8	74.2 ± 9.1	7.4 ± 5.2 (12.7)
		30	10.5 ± 0.6 (5)	0.0001	13.6 ± 3.9	78.4 ± 3.6	8.0 ± 1.4 (15.4)
			9.9 ± 1.1				
	252	24	10.1 ± 1.0 (4)	0.0001	26.0 ± 7.5	72.5 ± 7.6	1.5 ± 1.3 (13.7)
		30	10.0 ± 1.2 (5)	0.0001	18.2 ± 8.6	73.4 ± 11.6	8.4 ± 8.9 (15.8)
			10.0 ± 1.0				
PYR	202	24	5.7 ± 0.2 (4)	n.s.	34.0 ± 5.5	64.8 ± 5.4	1.3 ± 0.5 (14.3)
		30	5.6 ± 0.3 (4)	n.s.	28.8 ± 7.1	67.8 ± 4.8	3.5 ± 2.4 (17.2)
			5.6 ± 0.2				
	404	24	5.9 ± 0.6 (5)	n.s.	22.2 ± 2.4	76.4 ± 2.7	1.4 ± 0.6 (13.4)
		30	5.7 ± 0.3 (4)	n.s.	38.3 ± 16.7	58.8 ± 17.1	3.0 ± 1.2 (18.2)
			5.8 ± 0.5				
	808	24	5.9 ± 0.5 (4)	n.s.	26.3 ± 5.7	70.5 ± 4.0	3.3 ± 3.4 (13.5)
		30	5.8 ± 0.2 (4)	n.s.	30.3 ± 11.4	68.8 ± 10.6	1.0 ± 1.4 (17.6)
			5.9 ± 0.4				

[a] S.D. standard deviation; number of animals is in parentheses.
[b] P-values for comparisons of frequencies of SCEs in treated animals sacrificed at 24 or 30 h with such frequencies measured in negative control animals (olive oil) sacrificed at 24 and 30 h.
[c] Pooled data for 24 and 30 h.
[d] In this experiment hypotonic treatment was 15 min with a 1% Na-citrate solution.
[e] Average cell generation time (in hours).
[f] Experiment I and II.

4 Results

Summary data and results of the statistical analysis are given in Tables 1 and 2.

Effects of 2AAF

In the first experiment, a significant effect was detected at the highest dose tested (223 mg/kg) irrespective of the time of sacrifice. According to toxicity data provided by Salamone (1981) this dose is about one-tenth of the LD50/7 (2,220 mg/kg) for ip injection. Although the frequencies of SCEs increased slightly with dose, they were not significantly different. Data on the percentages of MI, MII, and MIII cells and the average cell generation time in control and treated animals indicate that 2AAF causes a slight slowing down of the cell cycle progression.

In the second experiment, the effect of 223 mg/kg was retested along with doses of 446 and 892 mg/kg. With respect to the frequency of SECs per cell, the results of the 2 tests at 223 mg/kg are in excellent agreement, but there are differences in the percentages of MI, MII, and MIII cells. In the second experiment, here was no indication for a slowing down of the cell cycle progression at any of the 3 doses tested. We cannot offer an explanation for this difference between the 2AAF experiments. The same strain of mice and batch of 2AAF were used in both experiments, but there was a time difference of one year between the two experiments. The SCE-inducing effect of 2AAF in the second experiment was clearly significant and dose dependent. At all 3 doses, the effect of 2AAF was decidedly larger than that at equivalent doses of 4AAF.

Effects of 4AAF

A weak positive effect was found at the lowest dose tested (223 mg/kg) and only in animals sacrificed at 24 h. Further positive effects were detected at the highest dose tested (892 mg/kg). At this dose level, the effect was marginal when animals were sacrificed at 24 h but significant when sacrificed at 30 h. There is an indication for a dose-dependent effect of 4AAF, because its positive effect was significantly stronger at 892 mg/kg than at 223 mg/kg. It is strange, however, that at the intermediate dose of 446 mg/kg, the effect was negative. In the present experiment, the strongest effect of 4AAF was found well above the LD50/7 dose (364 mg/kg) for ip administration of the chemical (Salamone, 1981). Treatment with 4AAF had little or no effect on the cell-cycle progression, and there was a slight, but probably insignificant, increase of the average cell generation with dose.

Effect of BP

At all dose levels tested, and irrespective of the time of sacrifice, treatments with BP resulted in a significant increase in the frequency of SCEs. When sacrificed at 24 h, the effect of BP was significantly stronger at 252 mg/kg than at 63 mg/kg. At the sacrifice time of 30 h, the effects at 126 and 252 mg/kg were significantly more pronounced than at 63 mg/kg. It seems that the effect of BP is dose dependent with a tendency for a flattening of the curve at the highest dose. The lowest effective dose tested is about one-fourth of the LD50/7 (232 mg/kg) for ip treatment (Salamone, 1981). BP had a slight depressing effect on the cell cycle progression but an effect of dose was not detectable.

Effects of PYR

The effect of PYR was negative at all dose levels tested. The highest dose used was above the LD50/7 of 514 mg/kg for ip administration (Salamone, 1981). There was a depressing effect of PYR on the cell cycle progression that seemed to be independent of dose but fairly persistent.

5 Discussion

When carrying out cytogenetic experiments, it is customary to test the effect of potential clastogenic agents at more than one fixation time. In the present set of experiments, 2 fixation times were generally used (24 and 30 h). When significant effects of the test chemicals were found, the frequencies of SCEs at 30 h tended to be a little higher than those at 24 h but never significantly so. In

Table 2. *Further statistical analysis of data*[a]

1. Comparison between frequencies of SCEs at 24 and 30 h:
   No significant differences were recorded
2. Comparison between frequencies of SCEs after treatment with different dose levels of 2AAF and 4AAF:
   Frequencies of SCEs after treatment with 446 and 892 mg/kg 2AAF were significantly higher than those recorded after treatment with 446 and 892 mg/kg 4AAF
3. Comparison between frequencies of SCEs after treatment with different dose levels of BP and PYR:
   For all possible comparisons the frequencies of SCEs after treatment with BP were always significantly higher (6 times P 0.01; 12 times P 0.001) than those recorded after treatment with PYR.
4. Comparison between frequencies of SCEs at different dose levels of a particular chemical:

2AAF	II[b]	30 h	223 mg/kg	versus	2AAF	II	30 h	446 mg/kg:	P = 0.0006
2AAF	II	30 h	223 mg/kg	versus	2AAF	II	30 h	892 mg/kg:	P = 0.0001
2AAF	II	30 h	446 mg/kg	versus	2AAF	II	30 h	892 mg/kg:	P = 0.0007
4AAF		30 h	223 mg/kg	versus	4AAF		30 h	892 mg/kg:	P = 0.012
BP		24 h	63 mg/kg	versus	BP		24 h	252 mg/kg:	P = 0.003
BP		30 h	63 mg/kg	versus	BP		30 h	126 mg/kg:	P = 0.005
BP		30 h	63 mg/kg	versus	BP		30 h	252 mg/kg:	P = 0.042

[a] Only significant differences have been listed.
[b] Data from second experiment.

none of the experimental series was there a significant difference between results obtained at 24 and 30 h (Table 2). Therefore, it is permissible to pool the data. This procedure essentially does not change the outcome of the experiments, with the exception of the effect of 4AAF at 223 mg/kg, which is then no longer significant.

2AAF versus 4AAF

Combined data of the two experiments with 2AAF clearly indicate a positive and dose-dependent effect. The lowest effective dose tested was 223 mg/kg. At 446 mg/kg, the effect was significantly higher than at 223 mg/kg, and the effect of 892 mg/kg was significantly higher than that of 446 mg/kg. In the case of 4AAF, the effects are much less pronounced. At the dose of 223 mg/kg, the effect is at the borderline of significance for the fixation time of 24 h, and the effect disappears when data for the two fixation times are pooled. The effect at 446 mg/kg and the slight difference between 223 and 446 mg/kg are not significantly different. With the statistical test used, the effect of 892 mg/kg was significantly higher than that of 223 mg/kg, but at face value the difference seems to be nonexistent or very slight. In our opnion, 4AAF must be classified in the SCE bone marrow test as negative or at best weakly positive.

BP versus PYR

BP clearly registered as positive at the 3 dose levels tested and the effect was dose dependent. In view of the significant effect at the lowest dose, it is possible that even lower doses of BP would result in increased frequencies of SCEs.

PYR was negative at all dose levels tested. Because the effect of PYR on cell cycle progression was not very pronounced, there remains a small possibility that PYR may react positively at higher dose levels. The present results with the SCE test are generally in good agreement with those obtained with the micronucleus test in our laboratory. In the latter experiments, the same mouse strain and doses (with exception of 2AAF) of the 4 chemicals were used. In both the SCE and micronucleus test, the effects of BP were positive at all doses and fixation times used, whereas the effects of PYR were always negative. Effects of 2AAF were more readily detectable with the micronucleus assay (lowest effective dose (LED) 55.8 mg/kg) that with the SCE test (LED 223 mg/kg). 4AAF was negative in the micronucleus assay and a very weak positive in the SCE test.

6 Conclusions

2AAF registered as clearly positive with a lowest effective dose tested of 223 mg/kg. Effects were clearly dose dependent. In the SCE test, the effect of 4AAF was very weak. BP significantly increased frequencies of SCEs at all dose levels and in a dose dependent manner. Results with PYR were entirely negative.

7 Acknowledgement

Thanks are due to the Ministry of Housing, Physical Planning and Environment for financial support.

8 References

Armitage, P. (1971) *Statistical methods in medical research*, Chapter 7.3, Blackwell, Oxford, pp. 202–207.

Salmone, M.F. (1981) Toxicity of 41 carcinogens and noncarcinogens. In: de Serres, F.J., and J. Ashby (Eds), Evaluation of Short-Term Tests for Carcinogens. *Progress in Mutation Research*, Vol. I, Elsevier, N.Y., p. 682.

Sheila M. Galloway, Christian L. Bean, Michael J. Armstrong, Leonard Oppenheimer and Timothy Schofield
Merck Sharp & Dohme Research Laboratories, West Point, Pennsylvania, USA

41

In vivo sister chromatid exchanges in mouse bone marrow: test of four compounds in male and female CD-1 mice

1 Summary

Our objective was to investigate the effect on sister chromatid exchanges (SCEs) in mouse bone marrow cells of 2 pairs of compounds. Each pair comprised a carcinogen and a closely related analogue of greatly reduced or undetectable carcinogenic potency. They were benzo(a)pyrene (BP) and pyrene (PYR), and 2-acetylaminofluorene (2AAF) and 4-acetylaminofluorene (4AAF).

We treated CD-1 mice by intraperitoneal injection of compounds suspended in corn oil an hour after subcutaneous implantation of wax-coated tablets of 5-bromodeoxyuridine. The following day bone marrow cells were fixed, stained, and scored for SCEs and cell cycle kinetics.

BP and 2AAF induced convincing increases in SCEs in both males and females. The SCE response to both compounds showed a plateau. There was no consistent sex difference for BP, although females were higher responders in one of 2 trials. There was a slightly greater response to 2AAF in males compared with females. PYR induced a very weak but statistically significant increase in SCEs that was slightly more marked in females. PYR was also more toxic to females than to males. 4AAF did not produce significant increases in SCEs in males, but there was a weak response in females. There was no marked increase in cell cycle length in any of the tests, even at toxic doses of PYR.

2 Introduction

Sister chromatid exchanges (SCEs) are a sensitive indicator of exposure to many carcinogens and mutagens. The adaptation of the SCE test to mouse bone marrow provides a simple cytogenetic test system, and the validity of SCE induction *in vivo* as a test for potential carcinogens is under investigation. As part of a program to compare a series of short-term *in vivo* tests, we tested 2 carcinogens and 2 compounds with greatly reduced or undetectable carcinogenicity.

3 Materials and methods

Mice

CD-1 weanling mice [CD-1 Crl (ICR) Br] were obtained from Charles River and quarantined for at least one week before use, in environmentally controlled rooms. Animals were housed 8–12 of one sex per cage and given Purina

Certified Rodent Chow and tap water ad libitum. The bedding was "Beta-Chip" heat-treated hardwood (Northeastern Products Corp.). For the SCE tests animals were 5 weeks old. We used 4 animals of each sex at each dose level, randomly selected and identified by unique numbered metal ear tags.

Compounds
Compounds were received on 2 occasions from ICI, UK.

Vehicle
Mazola corn oil was used to suspend the compounds. The oil was purchased fresh at the beginning of the study and the same bottle was used throughout (i.e., it was about 11 weeks old at last use).

Storage
Because we did not have any information on the stability of 4AAF and 2AAF in corn oil, these compounds were prepared fresh before each test. The suspensions of BP and PYR were stored in the dark and used more than once.

Bromodeoxyuridine (BrdUrd)
This chemical was administered as tablets implanted subcutaneously (Allen *et al.*, 1977). BrdUrd tablets (50 mg BrdUrd with unknown "filler") were purchased from Boehringer-Mannheim (FRG). Except for the first SCE test of BP and several range-finding tests, the tablets were coated thinly with paraffin embedding wax, leaving a small opening in the coating (McFee *et al.*, 1983).

Treatment regimen
Tablets were implanted subcutaneously on day 1, without the use of anaesthetic. About one h later, the animals were treated intraperitoneally (ip) with test compounds suspended in corn oil. The dosing volume was 0.1 ml per 10 g body weight. On day 2, animals were treated ip with 2 mg/kg colchicine 2.5 to 3 hours before sacrifice. In tests where 2 compounds were used and controls were shared, all control and test animals were treated on the same day and as close together in time as possible.

Bone marrow preparations
Animals were sacrificed by cervical dislocation and the femurs removed. To obtain bone marrow we crushed the femurs in Hanks balanced salt solution, shook the collection tubes vigorously to dislodge the marrow into suspension, and after letting bone fragments settle out for about a minute, decanted the suspension. Cells were treated with 5 ml hypotonic KCl for about 20 min (0.5% KCl at 37 °C), then fixed by adding about 0.5 ml of fixative to each tube and by allowing the tubes to stand for about 30 min before washing 3 times in freshly prepared fixative (3:1 v/v methanol:glacial acetic acid). Slides were prepared by air drying techniques and stained by a modification (Goto *et al.*, 1978) of the Fluorescence-plus-Giemsa technique (Perry and Wolff, 1974).

Staining
After staining in Hoechst 33258 at 0.5 ug/ml in pH6.8 phosphate buffer (M/15 Sorensen's buffer), slides were rinsed, mounted in the same buffer, and exposed to "black light" from two 15 W tubes on a hotplate at about 60 °C, for 10–30 min. Finally, slides were stained with Giemsa.

Scoring
The first experiment with BP was scored by one reader without coding the slides. The other tests were scored by one to three readers (slides were distributed to avoid any effect of reader differences and were read without knowledge of the dose group of the animals). Twenty M2 cells were scored per animal and cell cycle kinetics were determined for 100 metaphase cells.

Calculation of average generation time (AGT)
The AGT was calculated by dividing the length of treatment with BrdUrd by the replicative index (RI). The RI was obtained as follows:

$$RI = (M1 \times 1) + (M1+ \times 1.5) + (M2 \times 2) + (M2+ \times 2.5) + (M3 \times 3)$$

where M1, M2, and M3 are the percentages of metaphases that have undergone one, two, or three complete cell cycles in BrdUrd, and M1+ and M2+ have undergone partial S phases in BrdUrd.

Statistical analysis

Experimental unit
The individual animal was used as the experimental unit. The variability between and within animals was compared for each of the 5 studies (i.e., BPI, BPII, PYR, 2AAF, and 4AAF). In 4 of the 5 studies, there was a statistically significant non-zero animal-to-animal component. The fifth study was borderline. This variability is not surprising in an outbred strain of mice (CD-1). If the cell rather than the animal is used as the unit, the result will be to increase the false positive rate.

Chromosome number
We chose to evaluate SCEs by themselves, ignoring chromosome number. Only cells with close to the expected value of 40 chromosomes were scored. Although there were some statistically significant differences, these were small and there were no systematic effects across studies (e.g., there was no indication that chromosome number decreased with increasing doses).

Transformation
All analyses were performed both on the untransformed SCE value and by use of a square root transformation. Such a transformation is appropriate for Poisson distributed data and has been found useful for *in vitro* SCE assays. However, there is no indication from the present data that a Poisson distribution applies. Similar interpretations resulted from analysing the untransformed SCEs or the respective square roots.

High frequency cells
These cells were examined for 2 reasons. One or more high frequency cells (HFC) within an animal might distort the measure of "average response" for that animal, and there was interest in studying the occurrence of HFC alone. It is

possible that a carcinogen might not affect the average SCE response greatly, but might result in an increase in the numbers of cells with high SCE values. The square root transformation will reduce the impact of HFC on the resulting measure of central tendency. We also evaluated trimmed means and medians for SCE values per animal along with arithmetic means. The trimmed mean is the arithmetic mean of the 18 cells remaining after dropping the high and low cell (largest and lowest SCE value). The median is the middle value (i.e., the arithmetic mean of the tenth and eleventh largest SCE value for an animal). Both the trimmed mean and median would be less affected than the arithmetic mean by the occurrence of one or more HFC. Although there were minor differences for some of the analyses, the mean SCE, trimmed mean, or median in general led to the same interpretation of results; results shown here are for the arithmetic mean.

We also analysed the proportions of HFC. A HFC was defined as approximately the upper fifth percentile in the control group, ignoring sex effects. Because 20 cells were scored per animal, an approximation of the upper fifth percentile was calculated by obtaining the maximum SCE/cell for each of the 8 animals in the control group (4 males, 4 females) and then taking the median of these 8 values. The resulting percentiles calculated for the control groups using this algorithm were as follows: BP I, 19; BP II/PYR, 10; 2AAF/4AAF, 11.

The resulting estimated average proportions for the various control groups (3.7% to 4.4%) is close to the desired 5% result. To analyse the proportions of HFC we then calculated for the treated animals the proportion of cells with SCE values greater than or equal to the 95th percentile and applied an arcsine transformation to these data (i.e., arcsine $\sqrt{P}$, where P is the proportion of HFC for a given animal). An arcsine transformation would be appropriate if the proportion of HFC followed a binominal distribution. The transformed values were analysed in the same way as the other SCE data.

Two-way ANOVA (analysis of variance)

For each of the response variables, a two-way ANOVA was performed. The response variables were as follows: (i) mean SCE per animal (average over 20 cells), (ii) trimmed mean SCE, (iii) median SCE, (iv) mean SCE following a square root transformation, (v) proportion of HFC (arcsine transformation applied to these proportions). The two factors in the ANOVA were sex and dose (including a zero dose control group).

A test for sex by dose interaction was examined to see whether any differences between sexes at each dose (including the control) were consistent across the different doses. All doses were compared with each other using the least squares means (appropriate for unbalanced data). This is the same as doing 2 sample t tests for every dose combination but using the error term from the ANOVA analysis. In interpreting these results one should note that these are two-sided tests with no adjustment for the multiplicity of testing (e.g., for 5 doses there are $\binom{5}{2} = 10$ two-dose comparisons). This will result in a tendency to achieve statistical significance more often than expected under a 5% error rate.

Trend test

The trend test (Tukey *et al*., 1985) examined whether there was a linear increase in response (one-sided test) with increase in arithmetic dose (the control was taken as a zero dose). Alternatives to the arithmetic scaling would be unlikely to alter results dramatically with the doses employed in these studies. Results at the higher doses in the present studies may be complicated by lack of solubility of test compound. Downturns in response at higher doses and a non-significant trend result may not be a reliable indication that there is no adverse effect. We also examined trend statistics eliminating the higher doses one at a time. When interpreting these trend results, one should be aware of the greater than 5% chance of false positives due to multiple testing.

4 Results

BP

Dose selection The doses chosen were 25, 50, 100, 200 and 400 mg/kg. BP was a suspension at the doses tested and became unmanageable for ip injection at higher doses. The sacrifice time was about 24.5 hours after implantation of the BrdUrd tablet, but half the 400 mg/kg group were sacrificed later, at about 28 h, to allow for cell cycle delay.

Toxicity No overt signs of toxicity were noted.

BrdUrd administration Tablets were wax coated in the second, but not in the first, SCE test of BP.

SCE The results are shown in Table 1. There was a clear, apparently dose-related increase in SCEs in both sexes, reaching a plateau at more than twice the control values. In the first trial, SCEs in females reached higher values than in males, but when the slightly higher control level in females was taken into account, there was little sex difference. (The sex-by-dose interaction was not statistically significant.) For example, the absolute increases at 50 mg/kg were 8.1 and 8.2 SCE per cell in males and females, respectively. Also, while one female at 50 mg/kg had a mean SCE/cell of 19.0, considerably higher than the highest male level of 16.4, the males had cells with very high SCE levels (e.g., up to 44 SCE). The slightly higher SCE response in females was consistent across doses and was thus statistically significant ($P < 0.05$). One factor that may contribute to the slightly higher control levels in females may be the lower average weight of the female mice, so that the dose of BrdUrd is relatively higher in females. However, no simple relationship between SCE/cell and animal weight is apparent immediately in these small groups of animals (Table 1a).

The SCE test was repeated to confirm the results but using BrdUrd tablets coated with wax to slow down dissolution. The results for this test (trial II) are shown in Table 1. Only 2 animals per sex per dose were used except in the controls (4 males and 4 females). The lowest 3 doses were re-tested (25, 50, and 100 mg/kg of BP). Once again, clearly positive results were obtained. The plateau in mean SCE level was apparent in the males in this test, but not clear in the females. However, no statistically significant sex difference was found by ANOVA. The trial II results did show a significant sex-by-dose interaction, mainly due to the elevated response in the low-dose males (16.3 SCE/cell). AGT values did not show any significant cell cycle delay, although there were slightly longer AGTs at 25 and 50 mg/kg BP (13.6 hours) than in controls (13.0 hours).

PYR

Dose selection Doses were selected to overlap with the BP range but to reach higher levels, because, based on published information, less marked toxicity and genetic effects were expected.

Toxicity In a preliminary range-finding test, one male and one female per dose were treated with PYR at 0, 500, 1000, 1500 and 2000 mg/kg. In this test the animals were younger and smaller than usual, with mean weights of 11.3 g (males) and 17.2 g (females). The animals all looked unhealthy, and one male at 1500 mg/kg was moribund on day 2 before sacrifice. However, the bone marrow preparations showed no sign of reduced mitotic index or delay of cell cycle progression.

For the second test we used animals with a mean weight of 23.9 g, and PYR doses of 100, 500, and 1000 mg/kg. A top dose of 1000 mg/kg was selected, because PYR was difficult to inject at higher doses and the dose range was selected to overlap with that for BP. Four animals of each sex were used for each dose; 2 of the 4 females at 1000 mg/kg died before sacrifice, and a third was moribund at harvest. This test was not used for SCE analysis due to poor quality of differentiation on the slides. It was clear from the poor staining and the appearance of cells with darkly stained Y chromosomes that the BrdUrd was not incorporated throughout two S phases and that the tablets were dissolving too quickly.

Table 1. *Sister chromatid exchange in mouse bone marrow after treatment with BP*

	Trial I[a]						Trial II[b]					
	Mean SCE/cell (Range)						Mean SCE/cell (Range)					
Dose (mg/kg)	Males	ANOVA[c]	Females		Total		Males		Females		Total	
0	5.6 (0–16)	A	7.0 (0–30)	A	6.3	A	4.8 (1–9)	A	5.6 (1–13)	A	5.2	A
25	11.4*** (3–29)	B	11.4** (2–29)	B	11.4***	B	16.3*** (5–32)	B	9.6* (3–24)	B	12.9***	B
50	13.7*** (5–44)	C	15.1*** (5–35)	C	14.4***	C	11.3*** (4–22)	C	11.9** (5–25)	B	11.6***	B
100	12.1*** (3–25)	BC	14.6*** (4–40)	C	13.3***	C	13.2*** (7–27)	BC	13.8*** (5–26)	B	13.5***	B
200	NR	—	13.9*** (3–28)	—[d]		—	—	—	—	—	—	—

NR = Not scored

[a] Uncoated BrdUrd tablet, 4 animals/sex/dose.

[b] Wax-coated BrdUrd tablet, 2 animals/sex/dose except controls; 4/sex/dose.

[c] Results of ANOVA: Treatment groups with no letters in common are statistically significantly different from each other ($P < 0.05$).

[d] ANOVA not performed because only one sex was scored.

*Result of sequential trend test: one-sided test result of arithmetic trend from the control group up to and including the dose indicated. *$0.01 < P \leqslant 0.05$; **$0.001 < P \leqslant 0.01$; ***$P \leqslant 0.001$.

Table 1a. *Weights and SCE frequencies in control animals*

	Males					Females				
	1	2	3	4	Mean	1	2	3	4	Mean
SCE/cell	6.2	5.5	4.9	5.9	5.6	7.8	8.0	6.4	5.8	7.0
Weight (g)	26	23	28	25	25.6	23	24	19	20	22.1

Table 2. *Sister chromatid exchange in mouse bone marrow after treatment with PYR*

	Mean SCE/cell (Range)					
(mg/kg)	Males	ANOVA[a]	Females		Total	
0	4.8 (1–10)	A	5.6 (1–13)	A	5.2	A
100	6.1 (1–12)	BC	6.2 (2–12)	A	6.2*	B
400	6.6** (1–27)	C	8.4** (3–17)	B	7.5***	C
800	5.7 NS (2–11)	AC	6.8* (2–15)	A	6.2*	B
1000	5.2 NS (2–12)	AB	6.0 NS[b] (2–14)	A	5.6 NS	AB

[a] Results of ANOVA: Treatment groups with no letters in common are statistically significantly different from each other ($P < 0.05$).

[b] Mean for 2 surviving animals.

*Results of sequential trend test: one-sided test result of arithmetic trend from the control group up to and including the dose indicated. *$0.01 < P \leqslant 0.05$; **$0.001 < P \leqslant 0.01$; ***$P \leqslant 0.001$; NS $P > 0.05$.

For all subsequent experiments, tablets were coated over most of the surface with paraffin wax.

SCE In the second SCE test (Table 2), we used doses of 100, 400, 800 and 1000 mg/kg of PYR. Once again, two of the four females at 1000 mg/kg died the day after treatment.

A weak positive effect was found ($0.01 < P \leqslant 0.05$). There was a parabolic response pattern, with increases at 100 and 400 mg/kg but reduced SCEs at higher doses. This general pattern was seen in both sexes and for all variables analyzed, and possibly may be due to a lack of solubility of PYR at the higher doses tested.

There was a statistically significant sex difference ($P \leqslant 0.01$) but no sex-by-dose interaction. Because of the parabolic shape the trend over all doses was not statistically significant, although a trend through the 400 mg/kg dose was significant ($P \leqslant 0.001$). Because the estimate of biological/experimental variability for the PYR study is reduced by a factor of 3 or more compared with that in the BP studies, smaller effects more easily can achieve statistical significance.

2AAF

Toxicity No deaths occurred in a range-finding test of doses 100, 250, 500, 750 and 1000 mg/kg of 2AAF in males weighing an average of 26.6 g. Two animals were treated at each dose level; the compound was extremely difficult to administer ip at the higher doses, due to lack of solubility. Although AGTs were not estimated in this range-finding because of the poor differentiation (due to use of uncoated BrdUrd tablets), there was an apparent reduction in M2 cells with dose so that the sacrifice time for the subsequent SCE test was extended for the higher dose groups.

SCE The results are shown in Table 3 and are clearly positive ($P < 0.01$). The doses used in the SCE test were 250, 500, 750 and 1000 mg/kg. There was a similar response pattern to that seen for BP (i.e., a statistically significant increase at the lowest dose tested, followed by a plateau). There was no apparent difference in control values between males and females in this test. There appeared to be slightly greater SCE induction by 2AAF in males, with a maximum twice the control value at 750 mg/kg. The sex difference was not statistically significant, nor was there a sex-by-dose interaction. At 500 mg/kg, the 2 males sacrificed later (at 30 h compared with 26 h) appeared to have slightly higher SCEs than the 26 h males. In contrast, the SCE values in the female mice sacrificed at 30 h at the same dose level appeared to be lower than in the 26 h females, so that sacrifice time did not consistently affect the results.

Once again there was no clear increase in AGT in these animals; at 1000 mg/kg, in males the AGT of 14.4 h appeared higher than the 12.4 h mean for controls, but the animal variability is large and this difference is unlikely to be significant.

4AAF

Dose selection Doses were selected to overlap with the 2AAF dose range and were limited by our ability to suspend 4AAF in manageable form for ip injection.

Toxicity In a preliminary range-finding test of doses from 500 to 2000 mg/kg, all the animals of both sexes were dead or moribund the day after treatment. These animals were younger and smaller than those used for SCE tests (males 11.3 g; females 17.3 g), so the test was repeated in males with an average weight of 26.6 g. Two males per dose were treated with 4AAF at 100, 250, 500, 1000, and 1500 mg/kg. All survived, but one male at 1500 mg/kg was moribund at harvest. In the SCE test, the top dose used was 1000 mg/kg and no deaths occurred.

SCE The results are shown in Table 3. All animals were killed 26 h after implantation of the coated BrdUrd tablet. There was no apparent increase in SCE in the males. In females there were slight increases in SCEs at all 3 dose levels; these ranged from 29 to 43% above the mean control level: the effect was statistically significant ($P < 0.05$) in the ANOVA and also resulted in a weak positive trend (Table 3). As with PYR compared with BP, the biological/experimental variability for 4AAF was much less than for 2AAF, so that statistical significance could be achieved more readily. The range of SCEs per cell also increased, since cells with up to 24 SCEs appeared in treated but not in control animals. Overall, this is weak

Table 3. *Sister chromatid exchange in mouse bone marrow after treatment with 2AAF of 4AAF*

Dose (mg/kg)	2AAF[a] SCE/cell (Range) Males	ANOVA[b]	Females		Total		4AAF SCE/cell (Range) Males		Females		Total	
0	5.3 (0–13)	A	5.5 (2–13)	A	5.4	A	5.3 (0–13)	A	5.5 (2–13)	A	5.4	A
250	10.5** (3–34)	B	8.1 NS (1–19)	AB	9.3***	B	5.9 NS (0–17)	A	7.2* (1–21)	AB	6.6*	B
500	9.1* (1–24)	B	9.5** (3–24)	B	9.3***	B	5.5 NS (0–15)	A	7.6* (1–18)	B	6.6*	B
750	11.0** (2–25)	B	9.5** (3–16)	B	10.2***	B	ND		ND		ND	
1000	10.5** (4–24)	B	8.9* (3–29)	B*	9.7***	B	6.0 NS (1–20)	A	8.0** (2–24)	B	7.0**	B

[a] Fixation at 30 h for 2 mice at 500 mg/kg and all 4 mice at each of 750 and 1000 mg/kg doses.
[b] Result of ANOVA: Treatment groups with no letters in common are statistically significantly different from each other ($P < 0.05$).
*Results of sequential trend test; one-sided test result of arithmetic trend from the control group up to and including the dose indicated. *$0.01 < P \leqslant 0.05$; **$0.001 < P \leqslant 0.01$; ***$P \leqslant 0.001$; NS $P > 0.05$.
ND = Not done

evidence for an SCE response. There was a slight increase in AGT in both sexes, (14.7 hours at 1000 mg/kg compared with 12.0 hours in controls).

5 Discussion

BP and PYR

The data summarised in Tables 1 and 2 show a clearly positive result for BP but a very much weaker response for PYR up to the maximum dose testable (limited by lethality and by the difficulty of administering the compound).

The increase in SCEs in mice treated with BP reached a plateau; such a plateau has been repeatedly observed (e.g., *in vitro*: Galloway *et al.*, 1985; *in vivo*: Paika *et al.*, 1981) and may result from a combination of limiting solubility and saturation of metabolic capacity of the animal.

In the first of the 2 tests of BP, the SCE levels were slightly higher in females than in males ($P < 0.05$), but control levels were also higher in females. The results of the second trial did not support the suggestion that the sexes differed in their SCE response to BP. This second trial was the only example of a statistically significant sex-by-dose interaction among the five studies and was caused by an unexpectedly high response for the two low-dose males, so that for routine analysis, results usually may be combined across sexes.

Data were analysed using mean SCE levels and proportions of high frequency cells (HFC). Results of HFC analyses were consistent with the mean SCE analysis (i.e., a significant effect observed at the low dose, 25 mg/kg, and a rather flat response over the range 25 to 100 mg/kg). The cut point (95th percentile) used in trial I was 19, while that used in trial II was 10. This discrepancy was a result of the higher SCE levels observed in the trial I controls and led to a statistically much stronger response in the trial II treated animals. For example, at 50 mg/kg of BP, 20 and 64% of the cells were HFC cells in trial I and II, respectively.

Note that for an arcsine transformation the binominal HFC proportion variability would be expected to be $1/4n$; since $n = 20$, $1/4n$ equals 1/80, or 0.0125. The observed error mean squares are at least 50% higher than this, indicating extra binominal variability that is probably the added animal-to-animal variance component.

The reason for this animal-to-animal variability is not known, but could involve the rate of administration of BrdUrd. At sacrifice, it was found that wax-coated tablets did not dissolve to the same extent in all animals. It is possible that if a tablet dissolved very rapidly, the high dose of BrdUrd in the first cycle might induce higher SCE values than would a steadier exposure to BrdUrd during two S phases. This possibility might explain the greater variability and wide range in SCE/cell in the first test with BP (uncoated tablets were used and probably dissolved within the first 8 to 12 h, giving a relatively large dose of BrdUrd during the first S phase).

For PYR a weak, parabolic response was detectable statistically, but the biological significance of small effects of this type is not known. The pattern of maximal response at 400 mg/kg followed by a decrease at higher doses was seen for both sexes and was detected by statistical analysis either of mean SCE or of proportions of HFC.

The overall variability in the PYR study was less than in the BP data by a factor of more than 3; therefore, it is easier to achieve statistical significance in the PYR study.

The trend analyses (Table 2) indicate no overall trend (over all doses of PYR), but a significant trend was observed through the 400 mg/kg dose.

It is interesting that the AGT was not increased in animals treated with BP at doses that induced large increase in SCEs, nor was there any marked increase in cell cycle length in PYR-treated animals, up to lethal doses of PYR.

2AAF and 4AAF

The results are summarised in Table 3. The carcinogen 2AAF was positive in the SCE test, while 4AAF (which, if carcinogenic, is much weaker than 2AAF) did not induce SCEs in males although there was a possible weak effect in females. There was some indication that the induction of SCEs by 2AAF might be greater in males than in females, but the variability among males was wide, and it is not possible to conclude that there is a real sex difference without repeating the assay.

In common with the BP vs PYR comparison, variability in the 2AAF test was much greater than in the 4AAF data (error mean squares of 4.98 and 0.87, respectively). Thus, statistical significance for small increases was easier to achieve for the "noncarcinogens" PYR and 4AAF.

6 Conclusions

In testing 2 pairs of compounds, each of which comprises a carcinogen and a compound that is closely related structurally but of greatly reduced or undetectable carcinogenic potency, this SCE test discriminated between the carcinogen and "noncarcinogen" in both cases. There was a dose-related increase in SCEs for BP and for 2AAF, with maximum levels at or approaching a level double the background control values. In contrast, very slight increases were found for both PYR and 4AAF. Since the "noncarcinogens" did produce statistically significant increases, it will be important to determine criteria for positive decisions for *in vivo* SCE tests.

7 References

Allen, J.W., C.F. Shuler, R.W. Mendes and S.A. Latt (1977) A simplified technique for *in vivo* analysis of sister-chromatid exchanges using 5-bromodeoxyuridine tablets, *Cytogenet. Cell Genet.*, **18**, 231–237.

Galloway, S.M., A.D. Bloom, M. Resnick, B.H. Margolin, F. Nakamura, P. Archer and E. Zeiger (1985) Development of a standard protocol for *in vitro* cytogenetic testing with Chinese hamster ovary cells: Comparison of results for 22 compounds in two laboratories, *Environ. Mutagen.*, **7**, 1–51.

Goto, K., S. Maeda, Y. Kano and T. Sugiyama (1978) Factors involved in differential Giemsa-staining of sister chromatids, *Chromosoma*, **66**, 351–359.

McFee, A.F., K.W. Lowe and J.R. San Sebastian (1983) Improved sister-chromatid differentiation using paraffin-coated bromodeoxyuridine tablets in mice, *Mutation Res.*, **119**, 83–88.

Paika, I.J., M.T. Beauchesne, M. Randall, R.R. Schreck and S.A. Latt (1981) *In vivo* SCE analysis of 20 coded compounds. In F.J. de Serres and J. Ashby (Eds), Evaluation of Short-Term Tests for Carcinogens, *Prog. Mutation Res.*, **1**, 673–681.

Perry, P., and S. Wolff (1974) New Giemsa method for the differential staining of sister chromatids, *Nature* (London), **251**, 156–158.

Tukey, J.W., J.L. Ciminera and J.F. Heyse (1985) Testing the statistical certainty of a response to increasing doses of a drug, *Biometrics*, **41**, 295–301.

Fabrizio Palitti[1,4], Mario Fiore[1], Renata Cozzi[1], Ruggero Ricordy[1], Caterina Tanzarella[1], Pasquale Mosesso[2], and Nicola Loprieno[3]

[1]Centro di Studio di Genetica Evoluzionistica del CNR, c/o Dipartimento di Genetica e Biologia Molecolare, Città Universitaria, Roma, Italy
[2]Life Science Research, Roma Toxicology Centre, Pomezia, Italy
[3]Istituto di Biochimica, Biofisica e Genetica, Università di Pisa, Italy
[4]Dipartimento di Biologia Animale, Università di Catania, Italy

42

Tests with Swiss albino mice to detect the induction of sister chromatid exchanges in bone marrow cells

1 Summary

The ability of BP, PYR, 4AAF and 2AAF to induce SCEs in bone marrow cells of mice treated orally was determined. BP and 2AAF gave positive results. PYR and 4AAF gave negative results. In this type of investigation, a good correlation was found with carcinogenic and non-carcinogenic activity.

2 Introduction

As a contribution to the international Program of Chemical Safety, we assayed 4 chemicals for the induction of sister chromatid exchanges (SCEs) in bone marrow cells of mice. The chemicals tested were BP, PYR, 2AAF and 4AAF. BP and 2AAF are well known mutagenic and carcinogenic compounds; PYR and 4AAF are mutagenic compounds in *in vitro* tests and are apparently inactive in *in vivo* rodent bioassays for cancer.

Agar-coated BrdUrd tablets (King *et al.*, 1982) were used to obtain differential staining of the sister chromatids. To minimize the possibility of cytotoxicity having a selective effect on cell proliferation the test compounds were administered orally 8 h after the start of BrdUrd treatment.

3 Materials and methods

Animals

Male and female Swiss mice (Crl:CD-1(CR)BR) obtained from Charles River Italia Spa (Calco, Como, Italy) were used. Mice were housed under controlled environmental conditions (20 °C, 60% relative humidity and an 8 hrs photo period). All mice were bedded on sawdust with standard food pellets (Italiana Mangimi) and water available ad libitum. Animals were treated when they were 12 to 15 weeks old and weighed 28–35 g.

Preparation of BrdUrd tablets and their implantation

Each tablet consisted of a mixture of 15 mg of BrdUrd (Fluka) and 50 mg of corn starch (CMRS). Tablets (6 mm in diameter) were manufactured by an RIIC tablet press and then coated with 4.5% Difco bacto agar according to the procedure of King *et al.* (1982). Tablets were implanted subcutaneously into the abdomen of aetherized mice. After implantation, the incision was closed with 2 autoclip sutures.

Test chemicals
The test chemicals were received through the IPCS organisation. PYR and BP were dissolved in corn oil (Sigma), 2AAF and 4AAF in corn oil containing 10% dimethylsulphoxide (DMSO). The negative controls consisted of solvent vehicle treated mice. The positive control used was mitomycin C [MMC, Kyowa, 0.5 mg/kg body weight (b.w.) dissolved in phosphate-buffered saline (PBS)].

All test compounds were administered orally in 0.6 ml quantities. MMC was given by intraperitoneal (ip) injection. PYR (200, 400 and 800 mg/kg b w) and BP (63, 126 and 252 mg/kg) dissolved only at the lowest concentration and suspensions were obtained at the other two doses; 2AAF (110, 220 and 440 mg/kg) and 4AAF (223, 446 and 892 mg/kg) remained in suspension.

Assay procedure
BrdUrd-tablets were implanted in an appropriate number of mice. Eight hours post implantation, the animals were treated with test compounds and were sacrificed 16 h later (24 h after the BrdUrd table implantation).

Colchicine (3.2 μg/g) was injected ip 2 h before sacrifice.

Harvesting and slide preparations
Air-dried preparations of bone marrow cells were made according to the technique developed by Ford and Hamerton (1956). The slides were stained using the fluorescence-plus-Giemsa technique developed by Perry and Wolff (1974).

Scoring of the preparations
At least 2 slides per mouse were coded and 25 well-spread metaphases with 40 chromosomes were analysed for SCEs. Cell cycle progression were evaluated by scoring 100 cells and recording the number of metaphases in first (M_1), second (M_2) and third (M_3) mitosis.

Analysis of the results
The number of SCEs per cell was utilized for statistical analysis. All data were transformed according to the

Table 1. *Effect of the test compounds on the frequency of SCEs in bone marrow cells of female and male Swiss mice*

Chemical and concentration (mg/kg)		Sex	SCE/cell ± S.D.[a] (no. of animals)	P-value[b]	%M_1	%M_2	%M_3
Corn oil:		M	4.1 ± 2.0(8)		21.6	76	2.4
		F	3.6 ± 1.9(3)		16.7	81.4	1.7
Corn oil + 10% DMSO:		M	3.0 ± 1.8(3)		40.6	59.3	
		F	3.2 ± 1.6(6)		18	79.6	2.4
2AAF	110	M	6.2 ± 3.3(3)	0.00230	38.3	61.6	
		F	5.8 ± 2.3(3)	0.00052	15.3	84.3	0.3
	220	M	7.9 ± 3.8(3)	0.00012	45.6	54	1
		F	6.6 ± 2.9(3)	0.00077	24.6	75	0.3
	440	M	8.9 ± 3.2(3)	0.00002	43.3	56.6	
		F	8.1 ± 3.3(3)	0.00616	26.6	73	0.3
4AAF	223	M	2.9 ± 1.9(3)	n.s.[c]	14.6	79	6.3
		F	4.5 ± 2.0(3)	n.s.	17.6	79.6	2.6
	446	M	3.5 ± 2.0(3)	n.s.	14.6	81	4.3
		F	3.9 ± 2.0(3)	n.s.	10.3	87.6	2
	892	M	4.1 ± 2.3(3)	n.s.	14	85	1
		F	3.9 ± 2.0(3)	n.s.	11.6	83.3	5
BP	63	M	9.2 ± 4.2(4)	0.00000	25.2	72.5	2.2
		F	8.8 ± 3.3(3)	0.00000	25.6	72.6	1.6
	126	M	9.7 ± 3.9(4)	0.00000	20.7	78.7	0.5
		F	9.4 ± 2.9(3)	0.00000	14.6	83	2.3
	252	M	10.3 ± 4.0(4)	0.00000	39.3	70.5	
		F	10.0 ± 3.5(3)	0.00000	22	77.6	0.3
PYR	200	M	3.5 ± 1.8(4)	n.s.	19	79	2
		F	4.2 ± 2.1(3)	n.s.	24	76	
	400	M	3.5 ± 1.8(4)	n.s.	37.7	62.2	
		F	4.6 ± 2.2(3)	n.s.	17	81.3	1.6
	800	M	3.4 ± 1.8(4)	n.s.	38.2	61,5	0.2
		F	3.7 ± 1.8(3)	n.s.	28.6	71.3	
PBS		M	3.0 ± 1.5(4)		11	78.5	10.5
MMC	0.5	M	6.2 ± 3.0(4)	0.00000	19.5	75.5	5

[a]S.D. = standard deviation. $M_1 - M_2 - M_3$ = 1st, 2nd and 3rd mitosis.
[b]P-values for comparison of frequencies of SCEs in treated animals with such frequencies in solvent treated animals.
[c]n.s. = not significant.

equation $y = log\ (y+1)$ and subjected to a nested analysis of variance (within animals and within groups). The estimated error variance obtained was used to perform T tests between the different groups. Our criterion for a positive result is not a dose-dependent effect but a statistically significant difference, using an α at 0.05 in at least 2 consecutive doses or in the highest but not toxic dose.

4 Results

The detailed analysis of the SCE incidence in bone marrow cells of treated animals is shown in Table 1.

Positive controls
MMC increased the frequency of SCEs in comparison with its relevant control.

BP
This compound gave a positive response in the assay in both sexes. At all dose levels tested, there was a significant increase in the frequency of SCEs but a dose effect was not clearly detectable. Cell cycle progression was only slightly depressed with increasing dose.

PYR
This compound gave a negative response in both sexes in the assay. Cell cycle progression was affected with increasing dose.

2AAF
This compound gave a positive response in both sexes in the assay. At all doses tested there was a significant increase in the frequency of SCEs that was only slightly dose-dependent. Cell cycle progression was not clearly depressed.

4AAF
This compound gave a negative response in both sexes in the assay. Cell cycle progression was not depressed.

5 Discussion

In our protocol, the test compounds were administered 8 h after BrdUrd tablet implantation. This procedure has advantages and disadvantages. It makes the study more directly comparable with the chromosomal aberration study but enhances the opportunity for an interaction between BrdUrd-substituted DNA and the compound, and does not give an opportunity for persistent DNA lesions or may not provide adequate time for the formation of reactive metabolites to elicit SCEs (Natarajan *et al.*, 1983). It could be that by using another protocol the data would be qualitatively or quantitatively different.

For all tested compounds there were problems of solubility and this may be one of the explanations for the absence of a clear dose effect.

BP versus PYR
BP clearly increased the frequency of SCEs at the 3 tested doses while PYR was negative. The fact that cell cycle progression was affected by PYR might suggest that the compound reached the bone marrow cells but had no effect on the yield of SCEs.

2AAF versus 4AAF
2AAF clearly increased the frequency of SCEs in a dose-dependent manner, while 4AAF was negative. The fact that 4AAF had no effect on cell cycle progression would suggest that it did not reach the bone marrow cells; higher doses could be tested if the problem of solubility could be overcome.

6 Conclusions

BP and 2AAF gave a positive result, based on a marked statistically significant increase in SCEs seen with all 3 tested doses, while PYR and 4AAF gave a negative result at all tested doses.

These findings would have been more significant if a parallel study of the metabolism and the kinetics of the test compounds would have been performed.

It can be concluded that, in this type of assay, a good correlation was found with carcinogenic and noncarcinogenic activity.

7 Acknowledgements

This research was supported by a grant from Progetto Finalizzato Oncologia and Progetto Finalizzato del CNR Medicina Preventiva No. 72.02256.56.

8 References

Ford, C.E. and J.L. Hamerton (1956) A colchicine hypotonic citrate squash technique sequence of mammalian chromosomes, *Stain Technol.*, **31**, 247–251.

King, M.T., D. Wild, E. Gocke and K. Eckhardt (1982) 5-Bromodeoxy-uridine tablets with improved depot effect for analysis *in vivo* of sister chromatid exchanges in bone marrow and spermatogonial cells, *Mutation Res.*, **97**, 117–129.

Natarajan, A.T., A.D. Tates, M. Meijers, I. Neuteboom and N. de Vogel (1983) Induction of sister chromatid exchanges (SCEs) and chromosomal aberrations by mitomycin C and methyl methane sulfonate in Chinese hamster ovary cells: An evaluation of methodology for detection of SCEs and of persistent DNA lesions towards the frequencies of observed SCEs, *Mutation Res.*, **121**, 211–223.

Perry, S. and S. Wolff (1974) New Giemsa method for the differential staining of sister chromatid, *Nature* (London), **251**, 156–158.

Annalisa Zunino, Patrizia Russo, Mauro Pala, Marcella De Ferrari, Silvio Parodi and Leonardo Santi
Istituto Nazionale per la Ricerca sul Cancro, Department of Clinical and Experimental Oncology, University of Genoa, Genoa, Italy

43

Sister-chromatid exchange induction in bone marrow cells of Swiss mice

1 Summary

Benzo(a)pyrene (BP), pyrene (PYR). 2-acetylaminofluorene (2AAF) and 4-acetylaminofluorene (4AAF) were tested for their ability to induce sister chromatid exchange (SCE) in mouse bone marrow cells after treatment of adult Swiss male mice. BP induced a 109% increase, 2AAF induced a 51% increase, 4AAF induced a 14% increase, and PYR induced an 11% increase at the most effective doses. The statistical significance of the results obtained was much higher for BP and 2AAF than for 4AAF and PYR.

2 Introduction

Results of several studies indicate BP and 2AAF to be carcinogenic in rodents (NCI 1972–73, NCI 1978). Recently it has been shown that both BP and 2AAF are initiators in the rat liver model (Tsuda *et al.*, 1980).

The respective analogues, PYR and 4AAF, of the two above listed compounds, were found very weak or negative in carcinogenicity studies (NCI 1972–73, NCI 1978).

It was interesting to investigate if a concordance with the carcinogenicity data could be found using bone marrow as the target organ and sister chromatid exchanges as the endpoint.

3 Materials and methods

Chemicals

4AAF, 2AAF, PYR, and BP were obtained from Lancaster synthesis Ltd. (U.K.). Other chemicals were purchased: carboxymethylcellulose and corn oil from SIGMA, St. Louis (USA), 5-bromo-2′-deoxyuridine, BrdUrd and colchicine from SERVA. Heidelberg (GFR), Hoechst 33258 from HOECHST, Hannover (GFR), Giemsa stain from E. Merck, Darmstadt (GFR), Bacto agar from DIFCO, Detroit (USA), Mitomycin C from Kyowa, Tokyo (Japan).

Animals

Swiss male mice [CD-1 Charles River, Calco (Italy)] 2 to 3 months old were used in all experiments. Animals were housed under controlled climate conditions (i.e., 25 °C, 60% relative humidity for an 8 h photoperiod). They were provided animal chow (4RF21 standard diet, "Italiana Mangimi") and water *ad libitum.*

Experimental procedure
BP and PYR were dissolved in corn oil. Complete solution of the chemicals was never achieved. The partial suspension was administered by gavage (0.1 ml/10 g mouse). 2AAF and 4AAF were suspended in a 0.9% NaCl solution containing 1% carboxymethylcellulose as a suspending agent. The suspension was administered ip (0.1 ml/10 g mouse). In both cases, controls received the same amount of vehicle. The dosages administered are reported in Tables 1–3. MMC, used as positive control, was administered ip in PBS (0.16 ml/10 g mouse).

One agar coated BrdUrd tablet (12–13 mg) was implanted subcutaneously (King *et al.*, 1982) between 9 and 12 am. One h later, the compound under test or the vehicle was administered. Twenty-two hours after treatment, colchicine (5 mg/kg) was administered ip. Animals were killed by cervical dislocation at 25 h. Using a saline solution, the bone marrow cells were removed from the femurs. Hypotonic treatment was carried out with 0.075 M KCl for 15 min at 37°C. The cells were fixed with methanol: acetic acid (3:1). The slides were prepared and stained as previously described (Parodi *et al.*, 1983).

Twenty-five well spread metaphases of the second cell cycle after BrdUrd exposure were scored for chromosome number and SCEs.

Table 1. *Number of SCEs induced at different dosages by PYR and BP in Swiss mice*

Chemicals	Dose (mg/kg) and route	SCEs	Mean ± S.D.
Corn oil	—p.o	3.00	2.83 ± 0.23
		3.04	
		2.88	
		2.40	
		2.76	
		2.92	
PYR	202p.o.	3.12	3.22 ± 0.14
		3.32	
		3.08	
		3.36	
	404p.o	3.48	3.34 ± 0.2
		3.20	
	808p.o	2.84	3.17 ± 0.34
		3.52	
		3.16	
BP	63p.o	5.16	5.09 ± 0.73
		4.32	
		5.78	
	126p.o	4.40	4.7 ± 0.42
		5.00	
	252p.o	6.40	5.97 ± 0.36
		6.12	
		5.64	
		5.72	

Table 2. *Number of SCEs induced at different doses by 4AAF and 2AAF in Swiss mice*

Chemicals	Dose (mg/kg) and route	SCEs	Mean ± S.D.
CMC[a]	—i.p	2.80	2.89 ± 0.24
		2.72	
		3.24	
		2.80	
4AAF	223 ip	3.24	3.28 ± 0.06
		3.32	
	446ip	4.12	3.54 ± 0.63
		3.64	
		3.76	
		2.64	
	892ip	2.80	3.26 ± 0.38
		3.16	
		3.72	
		3.36	
2AAF	55.8ip	4.72	3.38 ± 0.87
		3.84	
		4.12	
		2.64	
	(111.5)	3.80	3.74 ± 0.61
		3.04	
		3.60	
		4.52	
	223ip	4.08	4.14 ± 0.11
		4.04	
		4.28	
		4.16	

[a]CMC carboxymethylcellulose.

Statistical analysis
To determine statistical significance of the data at an α of 0.05 three different statistical tests were used and their conclusions summed. First, a Student's t test was used to determine the number of treatment group doses for each chemical significantly different from the combined control data (Snedecor and Cochran, 1967). Second, a Mann-Whitney U test was performed after comparing all treated animals with the combined control group (Siegel, 1956). Third a positive linear dose response trend was evaluated by examining the correlation between SCE and dose, again using combined controls (Snedecor and Cochran, 1967).

4 Results

In Tables 1–3, the detailed results obtained from treating adult Swiss male mice with the 4 compounds under investigation are shown.

Table 3. *Number of SCEs induced by 2AAF (Experiment 2) and mytomycin C in Swiss mice*

Chemicals	Dose (mg/kg) and route	SCEs	Mean ± S.D.
CMC[a]	—ip	3.44	3.16 ± 0.61
		2.92	
		3.84	
		2.44	
2AAF	223ip	5.56	4.76 ± 0.61
		4.64	
		4.08	
		4.76	
	446ip	3.29	4.56 ± 0.9
		4.4	
		4.04	
		5.88	
MMC	0.5ip	6.36	6.1 ± 0.49
		5.88	
		5.52	
		6.64	

[a]CMC carboxymethylcellulose.

For PYR and 4AAF, the highest dose used was in the range of a LD_{50} ip (Adler, personal communication), for 2AAF, the highest dose used was in the range of a LD_{50} by the oral route. The highest dose of BP was in the range of a LD_{50} ip (Salamone, 1981). However, by the oral route, twice the highest used dose was found non toxic (Adler, personal communication). Some animals were found dead 24 hours after treatment by gavage with a dose of 808 mg/kg PYR. The data reported in Tables 1–3 were submitted to statistical analysis. Different statistical parameters were considered.

(a) We evaluated whether the mean of the results obtained at a given dose fell within the 2.5% right tail of a Student's t distribution calculated from the 14 control experiments (it was implicitly assumed that our results could be treated with parametric statistics),

(b) because usually we deal with small samples, for which the use of parametric statistics could be arbitrary, we used the non parametric Mann-Whitney U test (Siegel, 1956),

(c) we evaluated the correlation between doses of the compounds tested and the number of SCEs induced.

In the right column of Table 4 is a cumulative statistical evaluation of the results, which seems to suggest the following conclusions.

(1) A very small (10–15%) increase in SCEs frequency was induced with LD_{50} levels of either PYR or 4AAF.

(2) At the most effective dose tested a 51% increase was observed for 2AAF and 109% increase was observed for BP.

(3) The cumulative statistical analysis of the data suggests a rather clear cut difference between 2AAF and BP and 4AAF and PYR.

In arbitrary units, in the case of 2AAF and BP a level of statistical significance of 6 was reached, while only a level of one was reached in the case of 4AAF and PYR.

5 Discussion

Only results obtained *in vivo* are discussed, as related to the induction of SCEs in bone marrow cells. *In vitro* data can differ for important metabolic reasons making comparisons difficult to analyse. We were able to find information on BP, PYR and 2AAF. For BP, we surveyed five publications (Paika *et al.*, 1981, Schreck and Latt, 1980, Bayer and Bauknecht, 1977, Roszinsky-Kocher *et al.*, 1979. Raffetto *et al.*, 1981) and no selection of data was performed. Results from the above studies appear to be proportional to the basal number of SCEs in controls.

As a consequence, if we analyse the BP data in terms of increment percentage over controls, we observe that the responses are homogeneous. In the 5 quoted papers, increments are between 100 and 170% for dose levels of about 100 mg/kg. These doses are already in a plateau region for SCE induction. In the present investigation, we have found a maximum increment of 109% at 252 mg/kg.

For PYR we were able to find only one publication (Paika *et al.*, 1981). The results reported in that work are practically superimposable with our results. The response was essentially negative, with, at most, a 10% increase around a dose of 300 mg/kg.

For 2AAF we were able to find 3 papers related to SCE induction in bone marrow cells *in vivo* (Neal and Probst, 1983; Schreck *et al.*, 1979, Parodi *et al.*, 1983). The slope of the dose–response curve was already flattening above 25 mg/kg (Schreck *et al.*, 1979). In a plateau region, increments between 110 and 170% were reported for 2AAF. In the present investigation, a smaller increment was found (about 51% for the most active dosage). This increment is lower than results reported previously (Parodi *et al.*, 1983). There were 2 obvious differences between the present and the previous experiment. The BrdUrd tablet used in the last experiment contained only 12 mg instead of the 50 used before. The chemical sources were different in the 2 experiments; the chemical used in the first experiment came from Sigma and was 96% pure.

We were unable to found any published work on *in vivo* treatments with 4AAF. We observed an increment of 15% without any dose dependence. The global statistical evaluation, reported in Table 4, was essentially negative for 4AAF.

As a positive control we have used the dose of 0.5 mg/kg MMC, given ip. The increment observed in

Table 4. *Statistical treatment of SCE data*

Chemicals	Results of statistical tests: Response > (control mean + 2.2 S.D.)[a]	Mann-Whitney[b]	Correlation coefficient[c]	Cumulative statistical evaluation[d]
PYR	−	+	− (0.30)	1
BP	+ +	+ +	+ + (0.89)	6
4AAF	−	+	− (0.36)	1
2AAF	+ +	+ +	+ + (0.70)	6

[a] Control mean ± S.D. from 14 data was 2.94 ± 0.38; − = no dosage with a mean > 2.2 S.D.; ± = one dosage with a mean > 2.2 S.D.; + = two dosages with a mean > 2.2 S.D.; + + = three dosages with a mean > 2.2 S.D.

[b] For statistical analysis the data of all three dosages were cumulated together; − = $p > 0.05$; + = $p < 0.05$; + + = $p < 0.002$;

[c] The correlation coefficient is referred to the relationship between drug concentration and SCEs frequency. The individual data in Tables 1–3 were used. Values of r are given in parenthesis; − = $p > 0.05$; + = $p < 0.05$; + + = $p < 0.01$.

[d] ± ≡ 0.5; + ≡ 1; + + ≡ 2.

respect to the corresponding controls performed in the third experimental group (Table 3) was of about 93%. We reviewed 3 papers on MMC (Kram *et al.*, 1979, Nakanishi and Schneider, 1979; Palitti *et al.*, 1982).

In these publications the increment that could be extrapolated for a dose of 0.5 mg/kg covered a range from 160 to 250%. We were unable to pick up a specific experimental difference that could explain why the response obtained in our experimental conditions was somewhat lower than the response reported by previous authors. The response to MMC observed in our experimental conditions could be utilised as a useful reference point for a normalization of the results obtained with the 4 compounds investigated.

6 Conclusions

From our results (Table 4) and the results reported in the literature, it seems that the bone marrow in vivo SCE test is capable of detecting correctly the 2 active carcinogens with respect to their 2 structural analogues. The response observed was never very strong, the maximum effect was about twice control level for BP, while for 2AAF we found a positive but unusually weak response.

7 Acknowledgements

This research was supported by grants N. 83.00901.96 and N. 83.02850.56 from Consiglio Nazionale delle Ricerche.

8 References

Bayer, U. and T. Bauknecht (1977) The dose-dependence of sister chromatid exchanges induced by 3 hydrocarbons in the *in vivo* bone marrow test with Chinese Hamster, *Experientia*, **33**, 25.

King, M.T., D. Wild, A. Gocke and K. Eckhardt (1982) 5-Bromodeoxyuridine tablets with improved depot effect for analysis *in vivo* of sister-chromatid exchanges in bone marrow and spermatogonial cells, *Mutation Res.*, **97**, 117–129.

Kram, D., E.L. Schneider, G.C. Senula dn Y. Nakanishi (1979) Spontaneous and Mytomicin-c induced sister-chromatid exchanges. Comparison of *in vivo* and *in vitro* systems, *Mutation Res.*, **60**, 339–347.

Nakanishi, Y., and E.L. Schneider (1979) *In vivo* sister chromatid exchange: a sensitive measure of DNA damage, *Mutation Res.*, **60**, 329–337.

National Cancer Institute (1972–1973) *Survey of compounds which have been tested for carcinogenic activity*, Government Printing Office, Washington D.C.

National Cancer Institute (1978) *Survey of the compounds which have been tested for carcinogenic activity*, Government Printing Office, Washington D.C.

Neal, S.B. and G.S. Probst (1983) Chemically induced sister-chromatid exchange *in vivo* in bone marrow of Chinese hamsters. An evaluation of 24 compounds, *Mutation Res.*, **113**, 33–43.

Paika, I.J., M.T. Beauchesne, M. Randall, R.R. Schreck and S.A. Latt (1981) *In vivo* SCE analysis of 20 coded compounds. In F.J. de Serres and J. Ashby (Eds), Evaluation of Short Term Tests for Carcinogens. Report of the International Collaborative Program, *Progress in Mutation Research*, Vol. 1, Elsevier/North Holland N.Y. pp. 673–681.

Palitti, F., C. Tanzarella, R. Cozzi, R. Ricordi, E. Vitagliano and M. Fiore (1982) Comparison of the frequencies of SCEs induced by chemical mutagens in bone marrow, spleen and spermatogonial cells of mice, *Mutation Res.*, **103**, 191–195.

Parodi, S., A. Zunino, L. Ottaggio, M. De Ferrari and L. Santi (1983) Lack of correlation between the capability of inducing sister-chromatid exchanges *in vivo* and carcinogenic potency, for 16 aromatic amines and azoderivatives, *Mutation Res.*, **108**, 225–238.

Raffetto, G., F. Valerio, R. Puntoni, P. Faggin and A. Zunino (1981) Possibilita' offerte dallo studio degli scambi tra i cromatidi gemelli (SCE) per una valutazione del rischio oncogeno in ambiente di lavoro, *Annali dell'Istituto Superiore di Sanita'*, **17**, 603–606.

Roszinsky-Köcher, G., A. Basler and G. Röhrborn (1979) Mutagenicity of polycyclic hydrocarbons. V. Induction of sister-chromatid exchanges *in vivo*, *Mutation Res.*, **66**, 65–67.

Salamone, M.F. (1981) Toxicity of 41 carcinogens and carcinogenic analogs. In F.J. de Serres and J. Ashby (Eds), Evaluation of Short-Term Tests for Carcinogens. Report of the International Collaborative Program. *Progress in Mutation Research*, Vol. 1, Elsevier/North Holland, N.Y. pp. 682–685.

Schreck. R.R., I.J. Paika and S.A. Latt (1979) *In vivo* induction of sister chromatid exchanges in liver and marrow cells by drugs requiring metabolic activation, *Mutation Res.*, **64**, 315–328.

Schreck, R.R. and S.A. Latt (1980) Comparison of benzo(a)pyrene metabolism and sister chromatid exchange induction in mice, *Nature*, **288**, 407–408.

Siegel. S. (1956) Non parametric statistic for the behavioural Sciences, McGraw-Hill Book Company Inc., N.Y. pp. 116–127.

Snedecor, G.W. and W.G. Cochran (1967) *Statistical methods*, The Iowa State University Press, Ames, Iowa, U.S.A. p. 59 and p. 173.

Tsuda, H., G. Lee and E. Farber (1980) Induction of resistant hepatocytes as a new principle for a possible short term *in vivo* test for carcinogens, *Cancer Res.*, **40**, 1157–1164.

Demetra A. Vlachos, Lauren B. Rickard, G. Jay Graepel and Nancy C. Chromey
Haskell Laboratory for Toxicology and Industrial Medicine, E.I. du Pont de Nemours & Co., Inc., Newark, Delaware, USA

44

In vivo sister chromatid exchange evaluation in Sprague-Dawley rats

1 Summary

Sister chromatid exchange (SCE) induction in bone marrow cells was evaluated following oral administration of benzo(a)pyrene (BP), pyrene (PYR), 2-acetylamino-fluorene (2AAF), or 4-acetylaminofluorene (4AAF) to male and female Sprague-Dawley rats. All 4 test compounds induced significant increases in SCEs in males; BP was the most potent inducer, followed by 2AAF, 4AAF and PYR. The only compound clearly positive in females was BP. Combined analysis of the data for both sexes, however, showed significant SCE increases with 2AAF and 4AAF. Interaction between the response curves for males and females precluded combining these data for BP or PYR. A single intraperitoneal (ip) dose of 0.5 mg/kg mitomycin C (MMC) when administered to male rats failed to increase the SCE frequency.

2 Introduction

Haskell Laboratory participated in the IPCS Collaborative Study on *In Vivo* Short-term Tests. Using the standard protocol for this laboratory, the carcinogens BP and 2AAF and their putative noncarcinogenic analogs PYR and 4AAF were evaluated for their ability to induce SCEs in male and female rat bone marrow cells. Single acute doses of these compounds were administered by gavage at multiple levels.

Bone marrow SCE induction following a single ip dose (0.5 mg/kg) of MMC was also studied in male rats. This was done to provide a basis for comparing this laboratory's findings with those of other SCE investigators participating in the collaborative study.

3 Materials and methods

Animals

Seven- to eight-week-old male (200–268 g) and female (150–177 g) Sprague-Dawley rats (Crl:CD®(SD) BR, Charles River, Kingston, NY) were used. Animals were housed individually in standard wire mesh cages at 22–24 °C, 35–60% relative humidity, with a 12-h light/dark cycle. Purina Certified Rodent Chow® (Purina, St. Louis, MO) and water were provided ad libitum.

Chemicals

Sources were as follows: BP, PYR, 2AAF, and 4AAF, Imperial Chemical Industries, Ltd., (UK); MMC, cyclophosphamide (CP) and colchicine, Sigma, (St. Louis,

MO); 5-bromodeoxyuridine (BrdUrd) tablets, Boehringer Mannheim Biochemicals, (Indianapolis, IN); Metofane® (methoxyflurane), Pitman-Moore, Inc. (Washington Crossing, NJ).

BP, PYR, 2AAF, and 4AAF treatments

Five 50 mg BrdUrd tablets were implanted subcutaneously (Allen *et al.*, 1978) between the shoulder blades of Metofane®-anaesthetised rats 2 hours prior to dosing. The tablets used were half coated with household paraffin (McFee, 1983). The incisions were closed with 1–3 wound clips.

All dosing solutions/suspensions were prepared immediately prior to administration. BP, PYR, 2AAF, and 4AAF were suspended in corn oil (Mazola®) using a Polytron® homogeniser (Brinkman Instruments, Westbury, NY). Appropriate dilutions were made and administered by gavage in a dosing volume of 8 mL/kg (2AAF and BP dose groups) or 10 mL/kg body weight (4AAF and PYR dose groups). The vehicle control groups received 10 mL corn oil/kg. The positive control, CP, was dissolved in distilled deionized water and administered by gavage at 10 mg/kg, using a dosing volume of 4 mL/kg.

BP and 2AAF were each tested at 62.5, 125, 250, and 500 mg/kg body weight and PYR and 4AAF at 125, 250, 500 and 1000 mg/kg. Each dose group and control group consisted of 5 males and 5 females. BP and PYR were evaluated concurrently using common vehicle and positive control groups; 2AAF and 4AAF were similarly tested. Males and females were studied in separate experiments.

The doses were selected on the basis of preliminary toxicity studies, which indicated that rats survived at least one day after oral administration of 500 mg/kg BP or 2AAF. This dose was chosen as the maximum for the 2 carcinogens and was in the range of doses used by other IPCS investigators. The maximum test level for PYR and 4AAF was set at 1000 mg/kg, as these compounds were expected to have lower, if any, genotoxicity.

MMC treatment

One hour after BrdUrd tablet implantation, 4 male rats were injected ip with 0.5 mg MMC/kg body weight and 4 received the solvent, phosphate-buffered saline (PBS, with Ca^{++} and Mg^{++}). The dosing volume for both groups was 16 mL/kg. This procedure was used at the request of R.R. Tice, SCE workgroup coordinator. (Other investigators were instructed to use the same dose, route, and vehicle for their evaluation of SCE induction by MMC.)

Preparation and staining of metaphase cells

Animals were killed 24 h after dosing by CO_2 asphyxiation. Two hours prior, the animals were injected ip with 1 mg colchicine/kg to arrest metaphase cells. The bone marrow was flushed from both femurs with Hanks' balanced salt solution (37 °C), treated with 0.075 *M* KCl (37 °C) for 15 min, and fixed twice in cold methanol: glacial acetic acid (3:1). Following overnight refrigeration in fixative, the cell suspensions were fixed a third time and dropped onto microslides and flame-dried.

A modification of the fluorescence-plus-Giemsa (FPG) technique (Perry and Wolff, 1974; Goto *et al.*, 1978) was used to differentially stain sister chromatids. The slides were stained for 10 min with a solution of 8×10^{-5} 10^{-5} *M* Hoecsht 33258 in pH 6.8 phosphate buffer. After rinsing in distilled water, the slides were placed in a shallow pan of buffer on a 62–64 °C slide warmer and exposed to two 15 W UV fluorescent "black" lightbulbs (type F15T8) for 2 hours at a distance of approximately 8 cm. Slides were rinsed in distilled water, stained for 20 min with 5% Giemsa (Fisher) in pH 6.8 phosphate buffer, and rinsed again. After drying, the slides were

Table 1. *Sister chromatid exchange induction by BP and PYR in rat bone marrow cells*

	Males		Females	
Dose (mg/kg)	SCEs/cell/rat[a]	$\%M_{II}^{b}$	SCEs/cell/rat[a]	$\%M_{II}^{b}$
Corn oil	2.0 ± 0.26	96.8 ± 0.49	3.7 ± 0.19	94.0 ± 2.97
BP[c]				
62.5	3.2 ± 0.21*	83.6 ± 9.15	4.7 ± 0.21	88.4 ± 4.40
125.0	4.6 ± 0.50***	94.4 ± 1.60	5.4 ± 0.29**	85.2 ± 6.18
250.0	5.3 ± 0.35***	98.4 ± 1.17	5.0 ± 0.53*	89.6 ± 4.26
500.0	4.1 ± 0.60**	95.2 ± 1.96	6.0 ± 0.38**[d]	90.0 ± 4.40
PYR[e]				
125.0	2.5 ± 0.23	96.4 ± 1.60	3.6 ± 0.16	94.4 ± 2.71
250.0	2.2 ± 0.23	98.8 ± 0.80	2.6 ± 0.22	89.2 ± 2.42
500.0	3.4 ± 0.37**	97.6 ± 1.60	3.5 ± 0.27[d]	96.0 ± 3.37
1000.0	2.9 ± 0.19*	100.0 ± 0.00	3.6 ± 0.13	79.6 ± 8.38
CP				
10.0	15.5 ± 0.82***	85.2 ± 7.30	16.2 ± 0.76***	88.8 ± 6.53

[a]Mean ± SEM. Except where noted otherwise, N = 5.
[b]Mean % second metaphase cells ± SEM; remaining cells were first metaphases. (No third metaphases found.)
[c]Data for both sexes could not be combined due to interaction between response curves, $p < 0.05$.
[d]N = 4.
[e]Data for both sexes could not be combined due to interaction between response curves, $p < 0.01$.
*$p = 0.05$; **$p < 0.01$; ***$p < 0.001$.

soaked in xylene for 5 min and coverslipped using Permount®.

Slide scoring

Where possible, 25 second-division metaphase cells per animal were analysed for SCEs using 1000X magnification. Only cells having 42 ± 2 centromeres were scored. Fifty metaphases per animal were evaluated to determine the percentages of first-, second-, and third-division cells.

Statistical analysis

SCE data for BP, PYR, 2AAF, and 4AAF were analysed using a two-way (dose, sex) analysis of variance (ANOVA). Animals yielding fewer than 20 scorable second-division cells were excluded from the calculations. Individual animal mean SCE frequencies were transformed using the formula $(\text{SCEs/cells/rat})^{0.25}$. This power transformation, which was calculated using the data from this study, provided data that satisfied the assumptions of the ANOVA (i.e., homogeneity of variance and normally distributed random errors) (Snee and Irr, 1981). Each treatment group was compared with the concurrent vehicle control by Dunnett's test (Dunnett, 1964). The data were also evaluated for linear and quadratic dose–response relationships.

Dose effects were examined for each sex independently. In the absence of significant interaction between dose and sex, the data for males and females were combined and re-analysed. Any observed differences in response between males and females were assumed to be due to sex and not to experiment-to-experiment variation. Statistical significance was judged at the 5% level.

Figure 1. Dose–Response Relationships for BP- and PYR-Treated Rats. ●, BP; ○, PYR.

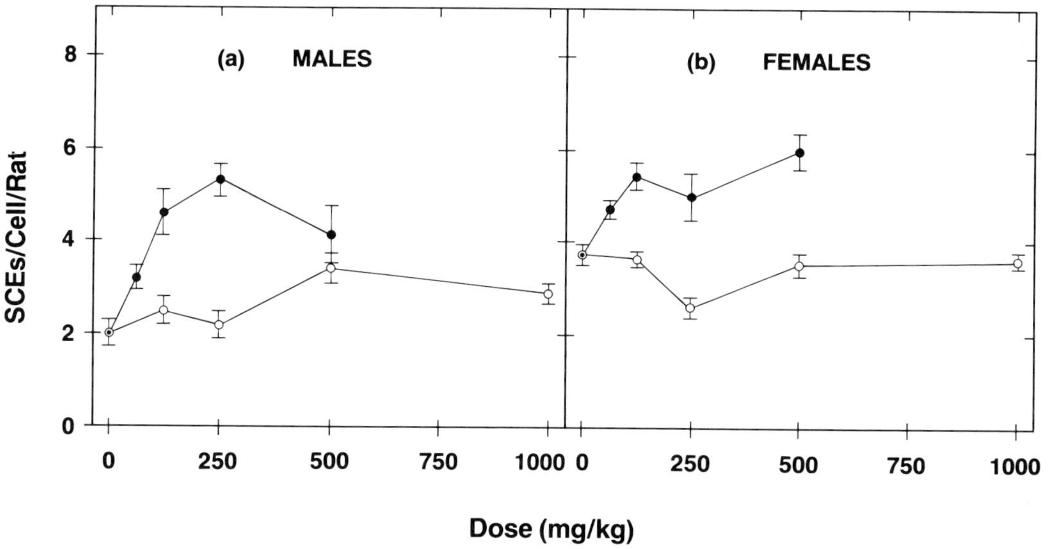

Table 2. *Sister chromatid exchange induction by 2AAF and 4AAF in rat bone marrow cells*

	Males		Females		Combined
Dose (mg/kg)	SCEs/cell/rat[a]	$\%M_{II}^{b}$	SCEs/cell/rat[a]	$\%M_{II}^{b}$	SCEs/cell/rat[c]
Corn oil	3.2 ± 0.22	96.8 ± 1.02	4.8 ± 0.36	96.4 ± 1.83	4.0 ± 0.33
2AAF					
62.5	5.6 ± 0.91*	98.4 ± 0.40	7.1 ± 0.92	84.4 ± 8.20	6.4 ± 0.66**
125.0	6.3 ± 0.59**	99.2 ± 0.49	6.5 ± 0.29	96.0 ± 1.89	6.4 ± 0.31***
250.0	5.8 ± 0.46**	91.6 ± 3.37	5.5 ± 0.50	96.4 ± 1.17	5.7 ± 0.33*
500.0	6.2 ± 0.70**	96.4 ± 0.75	4.9 ± 0.57	95.6 ± 2.04	5.6 ± 0.48*
4AAF					
125.0	4.6 ± 0.37**	99.2 ± 0.49	5.5 ± 0.35	95.2 ± 1.02	5.1 ± 0.28**
250.0	4.3 ± 0.19*	97.5 ± 0.75	4.4 ± 0.41	97.2 ± 1.02	4.4 ± 0.21
500.0	4.1 ± 0.21	98.0 ± 0.89	4.4 ± 0.27[d]	98.5 ± 0.96	4.2 ± 0.17
1000.0	4.3 ± 0.32*[d]	90.0 ± 7.54	5.1 ± 0.32	95.6 ± 2.22	4.8 ± 0.26
CP					
10.0	19.9 ± 0.97***	96.8 ± 1.36	15.7 ± 1.26***[d]	94.0 ± 3.37	18.0 ± 1.02***

[a] Mean ± SEM. Except where noted otherwise, N = 5.
[b] Mean % second metaphase cells ± SEM; remaining cells were first metaphases. (No third metaphases found.)
[c] Mean ± SEM.
[d] N = 4.
* $p < 0.05$; ** $p < 0.01$; *** $p < 0.001$.

4 Results

BP and PYR

The untransformed summary data for BP and PYR are shown in Table 1; the indicated results of the statistical analysis are based on the transformed data. For both BP and PYR, the data for males and females could not be combined due to interaction ($p < 0.05$) between dose–response curves. Significant increases in SCEs were seen at all BP doses tested in male rats (62.5–500 mg/kg) and at the 3 highest doses in females (125–500 mg/kg). The overall dose response in males was significantly quadratic, but appeared linear at the lower doses (Figure 1a). A linear response was obtained for females (Figure 1b).

In males treated with PYR, significant increases in SCEs were induced at 500 and 1000 mg/kg. The dose response was quadratic overall but appeared linear below 500 mg/kg (Figure 1a). PYR-treated females showed no increases in SCEs (Figure 1b).

2AAF and 4AAF

The data for 2AAF and 4AAF are given in Table 2. Significant increases in SCEs were seen in 2AAF-treated male rats at all doses (62.5–1000 mg/kg). The dose response was linear up to 125 mg/kg but appeared to plateau at higher doses, giving an overall quadratic relationship (Figure 2a). SCE levels in 2AAF-treated females were elevated at the 2 lower doses (62.5 and 125 mg/kg), but the increases were not statistically significant, and no dose response was identified (Figure 2b). The combined analysis of 2AAF-treated males and females (Table 2), however, showed significant SCE induction at all doses and a quadratic dose response, with a linear component (not shown).

SCEs were significantly increased in males treated with 125, 250, or 1000 mg 4AAF/kg. No significant increases were seen in females; however, the combined analysis of both sexes yielded a positive effect at 125 mg/kg. No positive dose–response relationships were obtained with 4AAF treatment (Figures 2a and 2b).

Corn oil controls, CP, and MMC

A comparison among all corn oil-treated rats showed significantly higher baseline SCE frequencies in females as compared with males ($p < 0.001$). CP, the positive control, gave significant SCE increases under all conditions (Tables 1 and 2). No increases in SCE frequency were seen at the single dose level of MMC evaluated in males (0.5 mg/kg) (Table 3).

Cell kinetics

A comparison of the relative numbers of first- and second-division metaphases in treated vs control animals showed no evidence of cell-cycle delay in any of the experiments (Tables 1, 2, and 3). A statistical evaluation of these data, however, was not done. No third-division metaphases were found during the course of this study.

5 Discussion

All 4 test compounds were judged positive in males, while only BP was positive in females, according to the statistical methods applied. It is worth noting that for the 2AAF-treated females, the 2 lowest doses (62.5 and 125 mg/kg) would have been positive if Student's t test, a less conservative test, had been used instead of Dunnett's test, which corrects for multiple comparisons. According to the combined analysis of males and females (using Dunnett's test), 2AAF and 4AAF were positive.

BP was the most potent of the 4 compounds, inducing a 165% increase in SCE frequency above control

Table 3. *SCE induction by mitomycin C in male rat bone marrow cells*

Dose (mg/kg)	SCEs/cell/rat[a]	$\%M_{II}^{b}$
0 (PBS)	3.0 ± 0.27	83.5 ± 4.65
0.5	2.7 ± 0.11	83.5 ± 6.70

[a] Mean ± SEM, N = 4.
[b] Mean % second metaphase cells ± SEM; remaining cells were first metaphases. (No third metaphases found.)

Figure 2. Dose–Response Relationships for 2AAF- and 4AAF-Treated Rats. ●, 2AAF; ○, 4AAF.

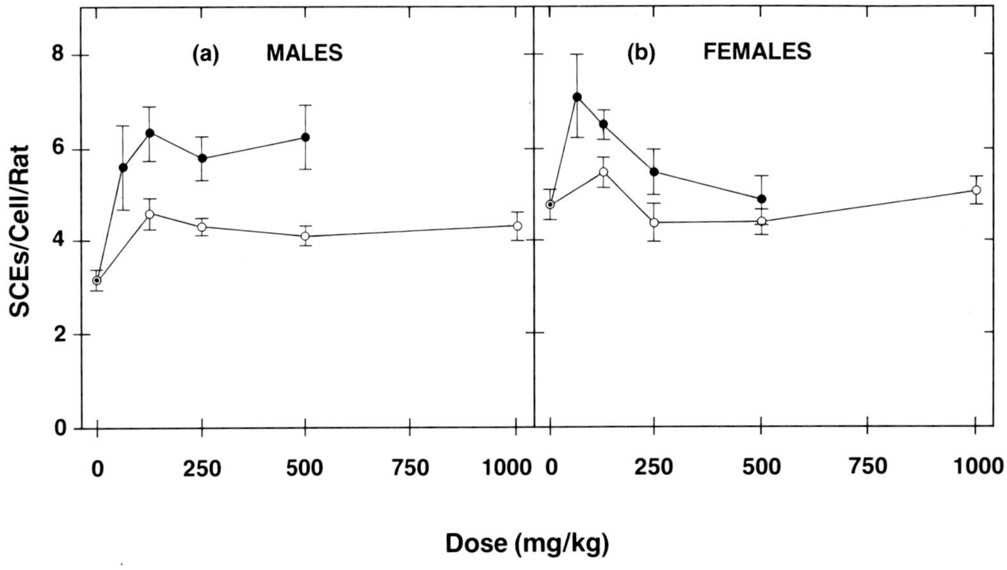

values at 250 mg/kg, the maximally effective dose in males and a 62% increase in females at 500 mg/kg. By contrast, the frequency was increased (by 70% at 500 mg/kg) only in males by PYR.

With 2AAF and 4AAF, SCE induction peaked at the lower doses. The maximum SCE increase above control levels in 2AAF-treated rats was approximately 100% in males at 125 mg/kg, 50% in females at 62.5 mg/kg, and 60% in the combined analysis of both sexes at 125 mg/kg. The analogue 4AAF was less potent, inducing increases of 44% in males, 14% in females, and 26% overall at 125 mg/kg.

The increased baseline SCE frequencies seen in female rats may be attributed to the higher relative dose of BrdUrd which the females received compared with that of the males. (All animals, regardless of weight, received 250 mg of BrdUrd; at the time of tablet implantation, the females weighed an average of about 30% less than the males.)

While female rats had a higher baseline SCE frequency, males were more sensitive to SCE induction by the 2 carcinogens and their analogues, and especially to CP, the positive control compound. The average SCE increase observed with 10 mg CP/kg was approximately 600% above control levels in males and less than 300% above female control frequencies.

The lack of increased SCE response with MMC was unexpected, as the dose tested previously has been shown to be positive in rodents (R.R. Tice, personal communication). Unfortunately, time did not permit repeating the MMC study or establishing the potency of the particular lot of MMC used.

6 Conclusions

When administered orally in corn oil, BP induced significant increases in SCEs in male and female Sprague-Dawley rats, while PYR was less potent and induced SCEs only in males. Combined analyses of SCE data for males and females could not be done for BP or PYR due to significant interaction between the response curves for each sex. 2AAF and 4AAF were judged positive in males but not females, with more activity shown by 2AAF. The combined analysis of males and females showed significant SCE induction by both 2AAF and 4AAF. While female rats showed a higher baseline SCE frequency, male rats appeared more sensitive to the effects of the 4 test compounds.

7 Acknowledgements

The authors wish to thank Dr. Awni M. Sarrif for constructive comments and acknowledge Ms. Lynn S. Ford for expert technical assistance.

8 References

Allen, J.W., C.F. Shuler and S.A. Latt (1978) BrdUrd tablet methodology for *in vivo* studies of DNA synthesis, *Somatic Cell Genet.*, **4**, 393–405.

Dunnett, C.W. (1964) New tables for multiple comparisons with a control, *Biometrics*, **20**, 482–491.

Goto, K., S. Maeda, Y. Kano, and T. Sugiyama (1978) Factors involved in differential Giemsa-staining of sister chromatids, *Chromosoma*, **66**, 351–359.

McFee, A.F., K.W. Lowe, and J.R. San Sebastian (1983) Delivery of BrdUrd for mouse bone marrow assay is improved by partial coating of tablets, *Environ. Mutagen.*, **5**, 454 (Abstract).

Perry, P., and S. Wolff (1974) New Giemsa method for the differential staining of sister chromatids, *Nature* (London), **251**, 156–158.

Snee, R.D., and J.D. Irr (1981) Design of a statistical method for the analysis of mutagenesis at the HGPRT locus of cultured Chinese hamster ovary cells, *Mutation Res.*, **85**, 77–93.

Andrew D. Kligerman, Gregory L. Erexson, Barbara Westbrook-Collins and James W. Allen[1]

Environmental Health Research and Testing, Inc., Research Triangle Park, North Carolina, (USA) and [1]U.S. Environmental Protection Agency, Genetic Toxicology Division, Research Triangle Park, North Carolina, USA

45

Analysis of sister chromatid exchange in C57BL/6 mouse peripheral blood lymphocytes following in vivo exposure to carcinogen/ noncarcinogen pairs

1 Summary

The mouse peripheral blood lymphocyte (PBL) culture system was modified to include density gradient separation of mononuclear cells to increase mitotic yield and to lessen dependence of culture success on animal white blood cell counts. This system was used to evaluate the sister chromatid exchange (SCE)-inducing potential of the carcinogen/noncarcinogen pairs: 2-acetylaminofluorene (2AAF)/4-acetylaminofluorene (4AAF) and benzo(a)pyrene (BP)/pyrene (PYR) following *in vivo* exposure by gavage. 2AAF and BP caused dose-related increases in SCE frequency, but 4AAF and PYR failed to cause a significant increase in SCE frequency at any of the doses examined. Only PYR caused a clear dose-related depression of cell replication kinetics as measured by the replicative index.

2 Introduction

As part of the IPCS collaborative study on *in vivo* tests for genotoxicity, we have developed a modified peripheral blood lymphocyte (PBL) culture method and used it to study SCE and cell replication kinetics in PBLs of male mice exposed by gavage to either 2AAF, 4AAF, BP, or PYR. The system is a modification of the one developed by Erexson *et al.* (1983) and makes use of phytohaemagglutinin-stimulated PBLs (T-lymphocytes) for analysis of genotoxic effects. The PBLs are exposed *in vivo* at the G_0 stage of the cell cycle; 24 h later they are stimulated to divide *in vitro* for two cell cycles in the presence of 5-bromodeoxyuridine (BrdUrd). The results from this study are to be used to help evaluate the sensitivity and selectivity of the mouse PBL assay compared with other mammalian *in vivo* cytogenetic assays. Furthermore, it is of interest to determine how responses to *in vivo* chemical exposure in G_0 cells compare with those of rapidly dividing cells.

3 Materials and methods

Animals

Male C57BL/6 mice, 8 weeks of age, were purchased from Jackson Laboratory (Bar Harbor, ME) and housed in the animal facility in laminar-flow rooms with 15 cycles/h of biocleaned air at 60–68% relative humidity. Room temperature was maintained at 20° to 21 °C with a 12 h light-dark cycle. Animals were fed rodent lab chow (Purina) and water ad libitum. Animals used in the study were between 19 and 27 weeks of age and weighed between 30 and 35 g. During an experiment with any one chemical, the animals were all the same age. For the replicate mitomycin C (MMC) experiment (interlaboratory positive control) and for those involving BP and PYR, the animals were housed 1 to a cage at least 1 week prior to gavage. In all other experiments, the animals were housed 5 to a cage.

Chemical treatment

BP, PYR, 2AAF, and 4AAF were supplied by Imperial Chemical Industries (Cheshire, UK). MMC (lot #111F–0324) was purchased from Sigma Chemical Co. (St. Louis, MO). MMC was dissolved in 0.85% saline; BP, PYR, 2AAF, and 4AAF were suspended in corn oil (Fisher Scientific, Raleigh, NC). The animals received BP, PYR, 2AAF, 4AAF, or corn oil by gavage (0.10 ml/10 g body weight); those receiving MMC or saline were injected intraperitoneally. Blood was removed by cardiac puncture 23.5 to 25 h after treatment.

Blood removal and culture procedure

Mice were anaesthetised with methoxyflurane (Pitman-Moore, Washington Crossing, NJ) and swabbed over the thorax with 70% ethanol. Blood (0.8–1.5 ml) was removed by cardiac puncture using a 3-ml heparinized syringe fitted with a 25-gauge, 5/8-in. needle. The blood samples were placed in 6-ml Falcon tubes and diluted 1:1 with phosphate-buffered saline (PBS) (PH 7.4). The diluted blood was layered over 1.5 ml of Lymphocyte Separation Medium® (Litton Bionetics, Kensington, MD) and centrifuged at 400 times g for 40 min. The lymphocyte-rich layer was removed and washed once with PBS and twice with PBS supplemented with 2% heat-inactivated foetal bovine serum (HI-FBS) (GIBCO, Grand Island, NY). After the final wash, the cells were resuspended in complete medium consisting of RPMI 1640 plus 25 mM Hepes buffer with L-glutamine (GIBCO), 20% HI-FBS (GIBCO), 6% mouse serum (Cedarlane Laboratories, Hornby, Ontario, Canada), 100 units penicillin/ml, and 100 μg streptomycin (GIBCO)/ml, and an additional 292 μg L-glutamine (GIBCO)/ml. Phytohaemagglutinin (12 μg/ml, 14 mitogenic units, lot #0951; or 10 μg/ml, 17 mitogenic units, lot #840020, HA-16, Burroughs-Wellcome, Greenville, NC) was added to the medium at the time of culture initiation to stimulate blastogenesis.

The cells were counted using a Coulter® counter model ZBI (Coulter Electronics, Hialeah, FL), and when possible, 2 replicate cultures containing 500,000 mononuclear cells in 1 ml of complete medium were established for each animal. The cultures were incubated at 37 °C in a humidified 5% CO_2 atmosphere. After 21 h, the medium was changed and complete medium (as described above) without mouse serum but with 5 μM BrdUrd (Sigma) was added. The cultures were incubated in the dark for an additional 25 h, 0.5 μg colcemid (GIBCO)/ml was added, and the cultures were harvested by centrifugation (700 times g, 7 min) 4 h later (total culture time: 50 h). Hypotonic treatment, fixation, and slide preparation were as described previously (Wilmer *et al.*, 1983). For animals for which more than one culture were prepared, the cells from each culture were pooled prior to slide preparation. Differential staining was achieved using the fluorescence-plus-Giemsa techniques (Wolff and Perry, 1974; Goto *et al.*, 1978).

Data collection and statistical analysis

Slides were coded so that the scorer had no knowledge of the treatment dose, and a single scorer read all the slides for an individual experiment (1 chemical). A total of 25 second division cells containing 40 ± 2 chromosomes were scored for SCE from each animal, and a total of 100 consecutive metaphases were analysed to determine cell cycle kinetics. The replicative index (RI) (Schneider and Lewis, 1981) was calculated according to the formula:

$$RI = \text{frequency of } M_1 + (2 \times \text{frequency } M_2) + (3 \times \text{frequency } M_3).$$

A one-way analysis of variance (Snedecor and Cochran, 1967) and Dunnett's test for comparing treatment to control groups (Dunnett, 1955) were used to determine the statistical significance of the data. Tests involving the SCE data were one-tailed ($\alpha = 0.05$), and analyses of RI were two-tailed ($\alpha = 0.05$).

4 Results

2AAF caused a small dose-related increase in the SCE frequency. A statistically significant increase was evident at doses of 446 and 1338 mg/kg, but little difference was apparent among the 3 doses examined (Table 1, Figure 1). 4AAF did not cause a statistically significant increase in SCE frequency at any of the doses examined (Table 1, Figure 1). Neither compound produced a significant alteration in cell cycle kinetics. BP

Figure 1. Induction of SCEs (mean ± S.D.) in mouse PBLs after *in vivo* exposure to either 2AAF (●——●) or 4AAF (▲– – –▲).

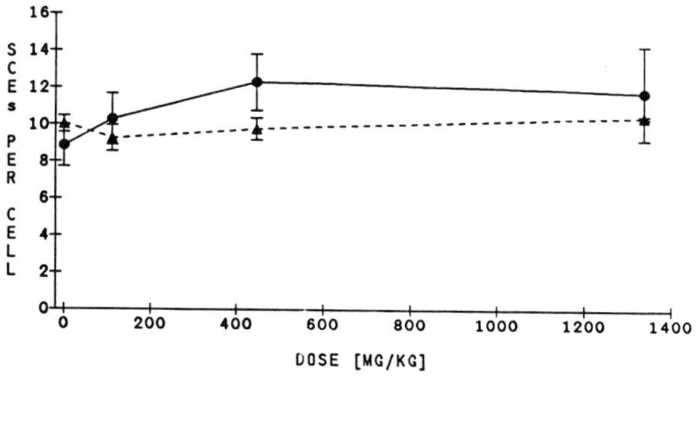

Figure 2. Induction of SCE (mean ± S.D.) in mouse PBLs after *in vivo* exposure to either BP (●——●) or PYR (▲– – –▲).

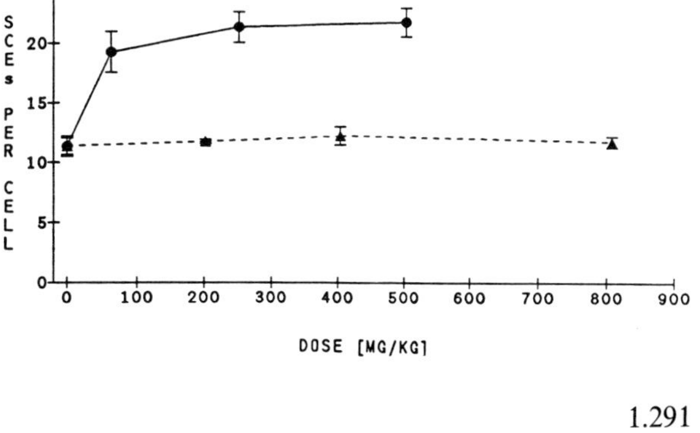

caused a significant dose-related increase in the SCE frequency at all doses examined (Table 1). The shape of the dose–response curve indicates an apparent plateau in the SCE frequency at doses higher than 63 mg/kg (Figure 2). BP did not produce a consistent dose-related slowing of cell cycle kinetics; only at a dose of 63 mg/kg was the RI significantly less than that of the control. PYR did not increase the SCE frequency at any of the doses tested; however, there was a significant slowing of cell cycling at 808 mg/kg (Table 1).

MMC, which served as an interlaboratory standard, induced a significant increase in SCE frequency in each replicate experiment (Table 1). In the first replicate, MMC (0.5 mg/kg) induced 6.1 SCEs/metaphase as opposed to 4.4 SCEs/metaphase in the second replicate. The pooled data for the RIs for the 2 experiments with MMC show significant slowing in cell replication kinetics.

5 Discussion

The results obtained with the mouse PBL culture system generally verify the published results from other short-term cytogenetic and mutation assays. Takehisa and Wolff (1978), Schreck *et al.* (1979), Neal and Probst (1983), and Parodi *et al.* (1983) found that *in vivo* exposure to the carcinogen 2AAF increased the frequency of SCE in various small mammals. The weak response obtained in the isolated PBL culture system compared with our previously published results with 2AAF using mouse whole blood culture (Kligerman *et al.*, 1985) is most likely due to the poor uptake of 2AAF administered by gavage compared with ip injection used in our previous study.

The failure of 4AAF to induce SCEs is in agreement with its non- or very weak carcinogenic potential (Flaks *et al.*, 1982). However, 4AAF has shown some weak genotoxic responses in *Salmonella typhimurium* (McCann *et al.*, 1975; Simmon, 1979) in the *in vitro* rat liver DNA repair assay (Williams, 1977) and the mouse lymphoma mutation assay (Clive *et al.*, 1979).

The carcinogen BP was the most potent SCE inducer of the 4 chemicals tested. It produced a curvilinear dose response very similar to the one obtained using the mouse whole blood culture system (Kligerman *et al.*, 1985). Similar dose–response relations have been obtained by others using bone marrow SCE assays (Bayer and Bauknecht, 1977; Schreck and Latt, 1980; Schreck *et al.*, 1982; Paika *et al.*, 1981). The observed biphasic nature of the slope of the dose–response curve at higher doses may be due either to saturation of the metabolic capabilities of the mouse or to difficulty in delivery of a suspension of a polycyclic aromatic hydrocarbon to the target cells.

PYR is considered to be noncarcinogenic and is generally negative in most short-term tests for genotoxicity (IARC, 1983). The results of the present study are in agreement with most of those of the previous studies. No increase in the SCE frequency was detected in mouse PBLs at the highest dose examined (808 mg/kg), which caused a significant slowing of cell replication kinetics. In agreement with these findings, Paika *et al.* (1981) found no increase in the SCE frequencies in either liver or bone marrow, and Sirianni and Huang (1978) detected no elevation in the SCE frequency in Chinese hamster cells placed in diffusion chambers in PYR-treated mice. Similarly, Tong *et al.* (1981) and Cunningham and Ringrose (1983) found no increase in SCE frequency in metabolically competent rat liver epithelial cells in culture. However, Popescu *et al.* (1977) and Evans and

Table 1. *Effect of test chemicals on SCE induction and cell replication kinetics of mouse peripheral blood lymphocytes*

Chemical	Dose (mg/kg)	Number of animals	SCEs/ metaphase ($x \pm$ s.d.)[a]	Replicative index ($x \pm$ s.d.)[a]
2AAF	0	4	8.87 ± 1.13	2.03 ± 0.36
	111.5	4	10.27 ± 1.38	1.97 ± 0.09
	446	4	12.24 ± 1.53*	1.79 ± 0.13
	1338	4	11.70 ± 2.52*	1.84 ± 0.20
4AAF	0	4	10.02 ± 0.44	1.69 ± 0.22
	111.5	3	9.25 ± 0.70	1.72 ± 0.36
	446	4	9.73 ± 0.60	1.89 ± 0.23
	1338	3	10.40 ± 0.08	1.85 ± 0.46
BP	0	4	11.38 ± 0.71	1.30 ± 0.05
	63	4	19.23 ± 1.72*	1.17 ± 0.07*
	252	4	21.31 ± 1.28*	1.22 ± 0.09
	504	4	21.69 ± 1.19*	1.25 ± 0.03
PYR	0	4	11.37 ± 0.86	2.08 ± 0.13
	202	4	11.73 ± 0.17	1.83 ± 0.12
	404	4	12.20 ± 0.77	1.98 ± 0.05
	808	3	11.68 ± 0.45	1.62 ± 0.27*
MMC[b]	0	2	9.68 ± 0.11	1.63 ± 0.01
	0.5	2	15.82 ± 0.03*	1.35 ± 0.20
MMC[c]	0	2	8.36 ± 0.40	1.61 ± 0.09
	0.5	3	12.79 ± 1.03*	1.38 ± 0.09

[a] Mean ± standard deviation among animals; * indicates significantly different from control ($p < 0.05$) by Dunnett's test.
[b] Replicate #1.
[c] Replicate #2.

Mitchell (1981) reported slight increases in the SCE frequencies in Chinese hamster cells treated with PYR *in vitro* with added metabolic activation.

During the course of this study, it was observed that the use of isolated PBLs permitted the analysis of SCEs in animals with low leukocyte counts (< 3,000,000/ml) without resulting culture failure, which would have occurred if whole blood inoculums were used from animals having low leukocyte counts. However, problems were still encountered in instances where the leukocyte counts were so low that even the minimal number of cells (500,000/culture) was impossible to obtain. In addition, it was found that housing the male mice 1 per cage for at least 1 week prior to the start of an experiment prevented aggression, thus reducing the stress on the animals. This lack of stress probably accounted for the resultant large increase in the number of peripheral blood leukocytes recovered (Kligerman, unpublished) (cf. Keller *et al.*, 1981).

6 Acknowledgements

The authors would like to thank M. Bryant, D. Harrison, and J. Campbell for excellent technical assistance and A. Stead and J. Tabor for help with statistical analysis and computer programming, respectively.

7 References

Bayer, U. and T. Bauknecht (1977) The dose-dependence of sister chromatid exchanges induced by 3 hydrocarbons, in the *in vivo* bone marrow test with Chinese hamsters, *Experientia*, **15**, 25.

Clive, D., K.O. Johnson, J.F.S. Spector, A.G. Batson and M.M.M. Brown (1979) Validation and characterization of the L518Y/TK$^{+/-}$ mouse lymphoma mutagen assay system, *Mutation Res.*, **59**, 61–108.

Cunningham, M.L. and P.S. Ringrose (1983) Benzo(a)pyrene and aniline increase sister chromatid exchanges in cultured rat liver fibroblasts without addition of activating enzymes, *Toxicol. Lett.*, **16**, 235–239.

Dunnett, C.W. (1955) A multiple comparisons procedure for comparing several treatments with a control, *J. Am. Stat. Assoc.*, **50**, 1096–1121.

Erexson, G.L., J.L. Wilmer and A.D. Kligerman (1983) Analysis of sister-chromatid exchange and cell-cycle kinetics in mouse T- and B-lymphocytes from peripheral blood cultures, *Mutation Res.*, **109**, 271–281.

Evans, E.L. and A.D. Mitchell (1981) Effects of 20 coded chemicals on sister-chromatid exchange frequencies in cultured Chinese hamster cells. In F.J. de Serres and J. Ashby (Eds), Evaluation of Short-Term Tests for Carcinogens, *Progress in Mutation Research*, Vol. 1, Elsevier/North-Holland, NY, pp. 528–550.

Flaks, B, Y. Yosi and A. Flaks, (1982) Effects of prolonged oral treatment with N-fluoren-4-ylacetohydroxamic acid on rat liver fine structure, *Carcinogenesis*, **3**, 103–109.

Goto, K., S. Maeda, Y. Kano and T. Sugiyama (1978) Factors involved in differential Giemsa-staining of sister chromatids, *Chromosoma*, **66**, 351–359.

IARC (1983) Polynuclear Aromatic Compounds, Part 1, *Chemical, Environmental and Experimental Data*, Vol. 32, IARC, Lyon, pp. 431–445.

Keller, S.E., J.M. Weiss, S.J. Schleifer, N.E. Miller and M. Stein, (1981) Suppression of immunity by stress: Effect of a graded series of stressors on lymphocyte stimulation in the rat, *Science*, **213**, 1397–1400.

Kligerman, A.D., G.L. Erexson and J.L. Wilmer (1985) Induction of sister-chromatid exchange (SCE) and cell-cycle inhibition in mouse peripheral blood B-lymphocytes exposed to mutagenic carcinogens *in vivo*, *Mutation Res.* **157**, 181–187.

McCann, J., E. Choi, E. Yamasaki and B.N. Ames (1975) Detection of carcinogens as mutagens in the *Salmonella*/microsome test: Assay of 300 chemicals, *Proc. Natl. Acad. Sci.*, USA, **72**, 5135–5139.

Neal, S.B. and G.S. Probst (1983) Chemically-induced sister-chromatid exchange *in vivo* in bone marrow of Chinese hamsters, *Mutation Res.*, **113**, 33–43.

Paika, I.J., M.T. Beauchesne, M. Randall, R.R. Schreck and S.A. Latt (1981) *in vivo* analysis of 20 coded compounds, In F.J. de Serres and J. Ashby, (Eds), Evaluation of Short-Term Tests for Carcinogens, *Progress in Mutation Research*, Vol. 1, Elsevier/North-Holland, NY, pp. 673–681.

Parodi, S., A. Zunino, L. Ottaggio, M. de Ferrari and L. Santi (1983) Lack of correlation between the capacity of inducing sister-chromatid exchanges *in vivo* and carcinogenic potential, for 16 aromatic amines and azo derivatives, *Mutation Res.*, **108**, 225–238.

Popescu, N.C., D. Turnbill and J.A. DiPaolo (1977) Sister-chromatid exchange and chromosome aberration analysis with the use of several carcinogens and noncarcinogens: Brief communication, *J. Natl. Cancer Inst.*, **59**, 289–293.

Schneider, E.L. and J. Lewis (1981) Aging and sister-chromatid exchange. VIII. Effect of aging environment on sister-chromatid exchange induction and cell cycle kinetics in Ehrlich ascites tumor cells. A brief note, *Mech. Ageing Dev.*, **17**, 327–330.

Schreck, R.R. and S.A. Latt (1980) Comparison of benzo(a)pyrene metabolism and sister chromatid exchange induction in mice, *Nature* (London), **288**, 407–408.

Schreck, R.R., I.J. Paika and S.A. Latt (1979) *In vivo* induction of sister-chromatid exchanges in liver and marrow cells by drugs requiring metabolic activation, *Mutation Res.*, **64**, 315–328.

Schreck, R.R., I.J. Paika and S.A. Latt (1982) Differences in murine procarcinogen activation enzymes are not accompanied by parallel differences in procarcinogen-induced sister-chromatid exchanges, *Mutation Res.*, **94**, 143–153.

Simmon, V.F. (1979) *In vitro* mutagenicity assays of chemical carcinogens and related compounds with *Salmonella typhimurium*, *J. Natl. Cancer Inst.*, **62**, 893–899.

Sirianni, S.R. and C.C. Huang (1978) Sister-chromatid exchange induced by promutagens/carcinogens in Chinese hamster cells cultured in diffusion chambers in mice, *Proc. Soc. Exp. Biol. Med.*, **158**, 269–274.

Snedecor, G.W. and W.G. Cochran (1967) Statistical Methods, The Iowa State University Press, *Ames*, 593 pp.

Takehisa, S. and S. Wolff (1978) Sister-chromatid exchanges induced in rabbit lymphocytes by 2-aminofluorene and 2-acetylaminofluorene after *in vitro* and *in vivo* metabolic activation, *Mutation Res.*, **58**, 321–329.

Tong, C., S. Ved Brat and G.M. Williams (1981) Sister-chromatid exchange induction by polycyclic aromatic hydrocarbons in an intact cell system of adult rat-liver epithelial cells, *Mutation Res.*, **91**, 467–473.

Williams, G.M. (1977) Detection of chemical carcinogens by unscheduled DNA synthesis in rat liver primary cell cultures, *Cancer Res.*, **37**, 1845–1851.

Wilmer, J.L., G.L. Erexson and A.D. Kligerman (1983) Implications of an elevated sister-chromatid exchange frequency in rat lymphocytes cultured in the absence of erythrocytes, *Mutation Res.*, **109**, 231–248.

Wolff, S. and P. Perry (1974) Differential Giemsa staining of sister-chromatids and the study of sister-chromatid exchanges without autoradiography, *Chromosoma*, **48**, 341–353.

Steven B. Neal and Gregory S. Probst
Lilly Research Laboratories, Toxicology Division, Division of Eli Lilly and Company, Greenfield, Indiana, USA

46

Discrimination of structurally similar carcinogen/noncarcinogen analogues by analysis of sister chromatid exchange in vivo in bone marrow and intestinal epithelium of Chinese hamsters

1 Summary

Two carcinogens, 2-acetylaminofluorene (2AAF) and benzo(a)pyrene (BP), and their respective noncarcinogenic analogues, 4-acetylaminofluorene (4AAF) and pyrene (PYR), have been tested in a variety of *in vitro* assays for genotoxicity; in some studies the noncarcinogenic analogues rendered positive responses (Ashby, 1981). These compounds were evaluated *in vivo* for the induction of sister chromatid exchange (SCE) to determine if this system could discriminate their carcinogenic activity. Each compound was administered to Chinese hamsters by the oral route and SCE frequencies in bone marrow and intestinal epithelium were compared in the same animal. The effects in bone marrow were also evaluated following intraperitoneal (ip) dosing. Both 2AAF and BP induced a dose-dependent increase in SCE in each target tissue, irrespective of the route of administration. This response was most pronounced in bone marrow, especially following ip exposure. The oral administration of BP produced a "saturation effect" in each tissue, as the magnitude of the SCE response was not significantly increased at doses in excess of 100 mg/kg. Maximum doses of 1000 mg/kg by gavage or 500 mg/kg ip of 4AAF or PYR did not induce SCE in bone marrow or intestinal epithelium. It was concluded that these structurally similar carcinogen/noncarcinogen pairs were discriminated *in vivo* by an analysis of SCE. The effect was evident in 2 tissues, and the magnitude of the response was dependent on the route of exposure.

2 Introduction

Sister chromatid exchanges (SCEs) represent the reciprocal interchange of DNA between homologous chromatids observable in metaphase chromosomes. Although neither the molecular mechanism nor the biological significance of SCE formation is thoroughly understood, independent studies have established that most mutagens/carcinogens are inducers of SCEs (Carrano *et al.*, 1978; Kato, 1977; Neal and Probst, 1983; Perry *et. al.*, 1975).

The advent of the bromodeoxyuridine tablet method (Allen *et al.*, 1977) for *in vivo* sister chromatid differentiation has provided a convenient approach for the identification of SCEs induced *in vivo*. Owing to the simplicity of this technique, the readily scorable chromosomal endpoint, and excellent sensitivity to genotoxins (Wolff, 1979), this system presents an especially attractive test for chemically induced chromosomal damage. Chinese hamsters are desirable for studies of this type due to their small size, low chromosome number, chromosome size and morphology, and low background SCE frequency.

The *in vivo* application of this test provides: (1) the advantage of multiple sites and pathways of metabolic activation as well as the opportunity for metabolic detoxification (2) multiple routes of administration to reflect relevant routes of exposure and (3) systemic distribution of genotoxins to multiple target tissues. Consequently, this system may reflect the behaviour of potential genotoxins in the intact target animal. Furthermore, the use of bone marrow cells provides a target removed from the major sites of metabolism and necessitates the systemic distribution of active metabolites to induce SCEs. Positive findings in this *in vivo* system are significant and complement results obtained in concurrent *in vitro* tests for genotoxins.

This report describes experiments in which Chinese hamsters were treated parenterally and orally with 4 suspect genotoxins to evaluate the *in vivo* induction of SCEs in bone marrow and intestinal epithelium. To confirm the responsiveness of the system, the promutagen cyclophosphamide (CP) was also tested and served as the positive control. Animals that served as negative controls were treated with corn oil or corn oil-DMSO.

3 Materials and methods

Animals

Three- to four-month-old female Chinese hamsters (*Cricetulus griseus*) were selected from a breeding colony established in the Toxicology Division, Eli Lilly and Company (Calland *et al*., 1985). Each animal, after reaching adulthood, was allowed to acclimate to the animal facilities for at least 2 weeks prior to use. Animals were housed individually in shoe-box cages containing hardwood chip bedding and excelsior nesting material. The animal quarters were maintained at $24 \pm 3\,^{\circ}C$ with 40% minimum relative humidity and a reverse lighting cycle (dark: 1300 to 2400). Purina laboratory rodent chow and tap water were provided ad libitum.

Three animals were used for each treatment group, 2 animals served as solvent-treated controls, and one animal served as the positive control and received cyclophosphamide. Historical data gathered in this laboratory (Neal and Probst, 1983) have indicated that 2 animals are sufficient as negative controls and that a single animal is sufficient as a positive control.

Chemicals

Bromodeoxyuridine (BrdUrd) was obtained from Calbiochem (LaJolla, CA), cyclophosphamide (CP) and mitomycin C (MMC) from Sigma (St. Louis, MO), and vinblastine sulfate (Velban®) from Eli Lilly and Company (Indianapolis, IN).

2AAF, 4AAF, BP and PYR were obtained from Lancaster Synthesis Ltd. (UK) and were used as supplied without further purification or attempt to identify impurities. The test chemicals were administered on an individual mg/kg body weight basis by the ip or oral route as solutions or suspensions in phosphate buffered saline or corn oil. In a few cases, compounds were initially dissolved in DMSO and then diluted in corn oil. In these cases the final concentration of DMSO never exceeded 3%, and the resulting maximum dose of DMSO was 0.3 mg/kg. The corn oil-DMSO combination did not induce SCEs in significantly greater number than those observed in corn oil treated controls.

Justification of doses

Upper treatment levels of the 4 test chemicals were selected using rodent LD_{50} values (based upon values provided to the investigators) and consideration of chemical solubility.

Preparation of BrdUrd Tablets: Tablets of pure BrdUrd were prepared using a Stokes-Eureka tablet press (Model 511–1) fitted with a 4 mm concave punch.

Agar-coating technique

Pure tablets of BrdUrd were coated with agar according to a modified procedure of King *et al.* (1982). A 4.5% Difco Bacto-Agar® solution was prepared in Milli-Q® purified distilled water, autoclaved, and placed in a 55 °C water bath. Aliquots of 1.5 ml of the 4.5% Bacto-Agar were dispensed into each well of a Linbro® multi-well culture plate (24-well multiplate, well capacity 3.5 mm, area per well $2.0\,cm^2$).

The agar solution in each well was allowed to cool until it was semisolidified, then a BrdUrd tablet was placed into each well. Immediately, an additional 1.5 ml of 55 °C agar was overlayed onto the tablet. The agar was allowed to solidify at room temperature (cooling at 4 °C resulted in condensation between the two agar layers).

Agar-coated BrdUrd (AC-BrdUrd) tablets were removed by punching each cylinder with a 8 mm glass tube. Each AC-BrdUrd tablet was trimmed to a thickness of 4 mm, placed in a petri plate lined with Saran Wrap®, and dehydrated in a desiccator.

In vivo assay for sister chromatid exchange

Bone Marrow: To obtain sister chromatid differentiation (SCD) in bone marrow cells, a 20–30 mg AC-BrdUrd tablet was subcutaneously implanted into the abdomen of each animal. Compounds were administered by either the ip or oral routes to animals 5 h after tablet implantation. Eighteen h after compound administration mitotic arrest was established by ip administration of 1 mg/kg Velban® and 2 h later the animals were sacrificed. Bone marrow was flushed from both femurs with 75 mM KCl (37 °C) for 20–25 minutes, then centrifused for 5 min at 176 times g. The supernatant was discarded and the cell pellet was resuspended by hand agitation in 0.1 ml of residual supernatant. The cells were fixed by adding 5 ml of methanol-acetic acid (3:1, v/v) and subsequently washed with 3 changes of this fixative by centrifugation for 5 min at 176 times g. Metaphase preparations were made by standard flame-drying techniques.

Intestinal epithelium

Sister chromatid differentiation was achieved in intestinal epithelial cells of the ileum by the subcutaneous (sc) implantation of a 20–30 mg AC-BrdUrd tablet into a 30–35 g female Chinese hamster. Five hours after tablet implantation, the compound was administered by the oral route in a volume not exceeding 10 ml/kg. Mitotic arrest was established by ip administration of 1 mg/kg Valban® 23 h after tablet implantation.

Intestinal epithelial cells were isolated from the ileum by dissecting the ileum free from the animal. Gastrointestinal contents were flushed with a syringe containing 1% sodium citrate (37 °C). The intestine was split longitudinally, cut into small pieces, and placed in a petri plate. Small sections of intestine were transferred to a conical centrifuge tube containing 10 ml of 1% sodium citrate (37 °C). Total hypotonic time was 30 min at room temperature, followed by centrifugation for 3 min at 176 times g. The supernatant was discarded, and the packed tissue was resuspended by hand agitation. Tissue was fixed by the addition of cold methanol-glacial acetic acid (3:1, v/v), transferred to capped vials, and stored at 4 °C.

To obtain metaphases, a few pieces of fixed intestine were removed and placed into centrifuge tubes containing freshly prepared 3:1 methanol-acetic acid. Following centrifugation for 10 min at 176 times g, the supernatant was discarded and a sufficient volume of 60% glacial acetic acid (GAA) was added to obtain a cell suspension. A 4 min treatment with 60% GAA was usually sufficient to obtain a cell suspension. The cell suspension was flooded onto a 65 °C prewarmed slide and then was immediately withdrawn. A siliconized Pasteur pipette was used for this procedure.

Sister chromatid differentiation
Slides were dried overnight at room temperature and stained by a modification of the fluorescence-plus-Giemsa (FPG) technique of Perry and Wolff (1974). Briefly, slides were stained for 12 min in an aqueous solution (1 μg/ml) of Hoechst 33258, then rinsed for 4 min in distilled water. The slides were coverslipped with phosphate buffered saline, and the coverslip edges were sealed with lacquer. Following exposure to ultraviolet (UV) light ($\lambda = 366$ nm, General Electric No. F40BL) for one hour at a distance of 3 cm, the coverslips were removed. Final processing included a 20 min incubation in 10XSSC at 60 °C, a distilled water rinse, and staining in a 3% aqueous solution of Giemsa.

Criteria for a positive response and cytotoxicity
For each animal, 25 well-spread representative metaphases from each tissue, containing the diploid number of chromosomes and demonstrating typical second division sister chromatid differentiation, were scored for SCEs. A positive response was judged for compounds that induced a dose-related increase in SCE frequency in which at least 2 doses were statistically different from controls as determined by Dunnett's t test (Dunnett, 1964) at a probability level < 0.05. As a measure of cytotoxicity, a total of 100 metaphase figures per animal were scored for characteristic first, second, or third division staining. An increase in the number of first division metaphase figures with a corresponding decrease in second or third division metaphase figures was taken as evidence of perturbation of the cell cycle and was indicative of a cytotoxic effect.

4 Results

Influence of BrdUrd on SCE Frequency: It has been reported that the BrdUrd required for sister chromatid differentiation (SCD) can influence the frequency of SCEs (Kram *et al.*, 1979; Tice *et al.*, 1976; Wilmer and Soares, 1980). Acceptable SCD was obtained in bone marrow and intestinal epithelium with a 20–30 mg AC-BrdUrd tablet, and the baseline responses for SCE in those tissues were similar (Table 1). The baseline SCE frequency observed in bone marrow of vehicle control animals was similar to baseline frequencies reported by other investigators using Chinese hamsters (Basler *et al.*, 1979; Bayer, 1978; Roszinsky-Kocher *et al.*, 1979).

Consistency of SCE frequency in vivo
Experiment-to-experiment variation and reproducibility of the *in vivo* SCE assay are important factors to consider when evaluating marginal SCE responses due to chemical treatment. Historical results from this laboratory, summarized in Figures 1 and 2, show a consistent and reproducible baseline SCE frequency in vehicle control animals, as well as a consistent response for CP-treated animals.

Chemically induced SCE
The *in vivo* induction of SCEs was measured in bone marrow and intestinal epithelial cells of Chinese hamsters treated ip or orally with 2AAF, 4AAF, BP, or PYR.

2AAF
2AAF was dissolved in corn oil/DMSO (final DMSO concentration 1.8%) and administered to animals parenterally at doses of 500, 250, 125, and 62.5 mg/kg. Quantification of SCEs in bone marrow revealed a dose-related, statistically significant increase in SCE formation in all animals treated with 2AAF (Table 2). No deaths or cytotoxicity were observed in any animal treated parenterally with 2AAF.

2AAF was also administered orally as a suspension in corn oil to another group of animals at doses of 1000, 500, 250, and 125 mg/kg. One of 3 animals receiving

Table 1. *Baseline response for SCE in bone marrow and intestinal epithelium of Chinese hamsters*

SCE/Metaphase[a] Bone Marrow	Intestine
AC-BrdUrd (20–30 mg)	AC-BrdUrd (20–30 mg)
2.9 ± 1.3	3.9 ± 1.0
3.2 ± 1.1	4.2 ± 2.1
3.2 ± 1.3	3.4 ± 1.4
2.9 ± 1.4	3.2 ± 1.0
3.6 ± 1.9	3.5 ± 1.3
3.0 ± 1.5	3.3 ± 1.7

[a] Counts represent mean ± S.D. of 25 second division metaphases from each of 2 animals per experiment. Bone marrow and intestinal epithelium were obtained from the same animal.

1000 mg/kg 2AAF was found dead 16 h after treatment. SCEs in bone marrow as well as intestinal epithelium were evaluated 19 h after chemical treatment, and a dose-related, statistically significant increase in SCE formation was noted in bone marrow at doses of 1000, 500, and 250 mg/kg (Table 3) and in intestinal epithelium at doses of 500, 250, and 125 mg/kg (Table 4). Cytotoxicity, characterized by an increase in the number of first division staining cells and a decrease in second or third division metaphases, was noted in bone marrow of animals treated orally with 1000 and 500 mg/kg 2AAF. No mitotic figures were observed in intestinal epithelium of surviving animals receiving 1000 mg/kg 2AAF indicative of a cytotoxic effect due to chemical treatment.

4AAF

4AAF was administered ip as a suspension in corn oil at doses of 500, 250, 125, and 62.5 mg/kg (Table 2) and orally

Figure 1. Historical vehicle control response for SCE in bone marrow of Chinese hamsters: 2/81–5/84. Each symbol represents mean of two animals from which 50 metaphases were evaluated.

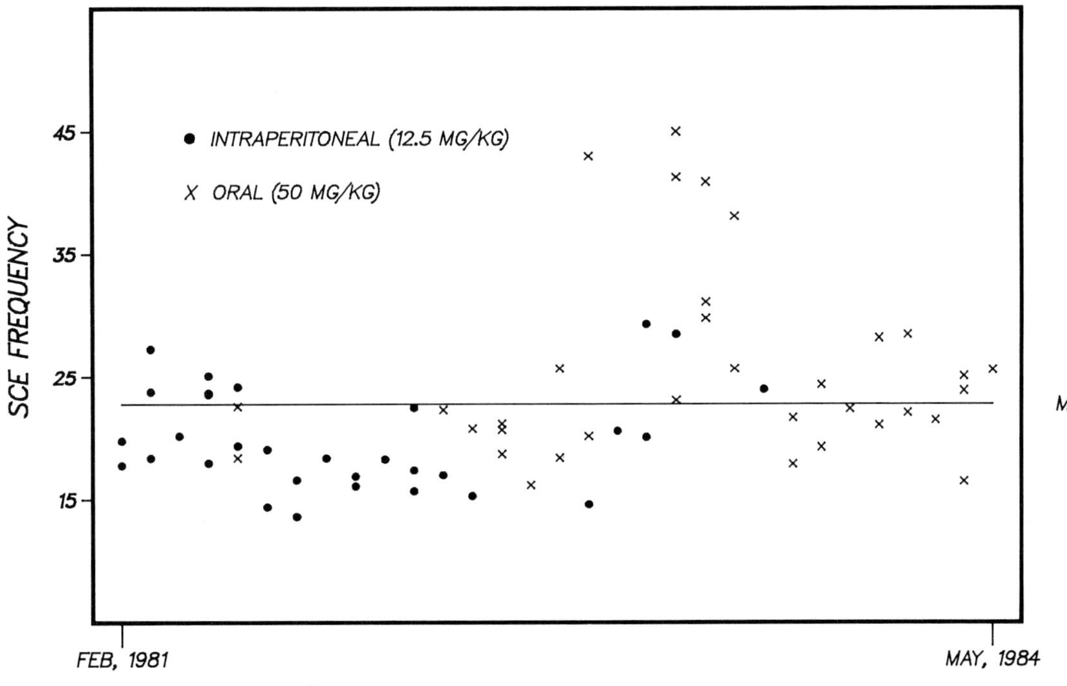

Figure 2. Historical cyclophosphamide response for SCE in bone marrow of Chinese hamsters: 2/81–5/84. Each symbol represents mean of a single animal from which 25 metaphases were evaluated.

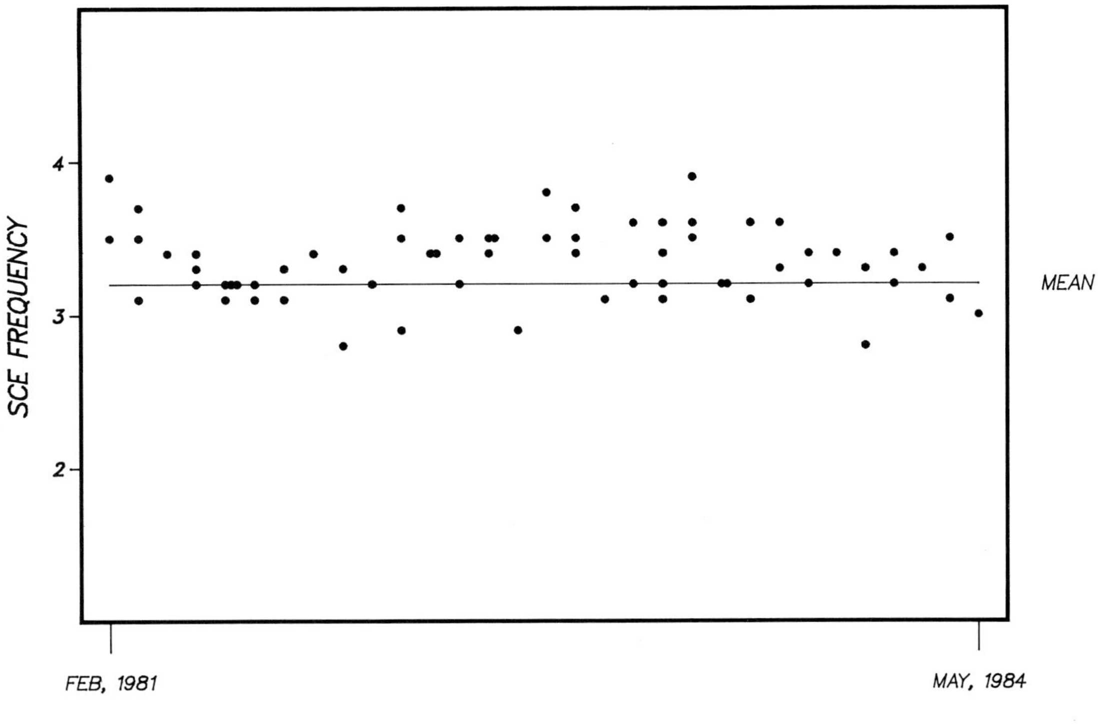

Table 2. *Summary values for the in vivo induction of sister chromatid exchange in bone marrow of Chinese halsters treated intraperitoneally with 2AAF, 4AAF, BP, or PYR*

Chemical	Dose (mg/kg)	Metaphases scored	SCE/Metaphase (Mean ± S.D.)	Cell cycle (% M_1)
2AAF	500	75[a]	7.5 ± 3.5[d]	21
	250	75[a]	6.5 ± 3.8[d]	24
	125	75[a]	6.9 ± 3.7[d]	30
	62.5	75[a]	4.6 ± 2.8[d]	29
Corn oil/DMSO	10 ml/kg	50[b]	3.0 ± 1.3	22
CP	12.5	25[c]	13.5 ± 5.5[d]	41[e]
4AAF	500	75[a]	3.3 ± 1.6	17
	250	75[a]	3.6 ± 1.6	16
	125	75[a]	3.4 ± 1.7	15
	62.5	75[a]	3.4 ± 1.7	20
Corn oil	10 ml/kg	50[b]	2.9 ± 1.2	20
CP	12.5	25[c]	11.2 ± 4.8[d]	33[e]
BP	500	75[a]	10.1 ± 4.2[d]	20
	250	75[a]	7.4 ± 3.7[d]	19
	125	75[a]	5.6 ± 2.8[d]	20
	62.5	75[a]	5.8 ± 2.9[d]	18
Corn oil	10 ml/kg	50[b]	2.7 ± 1.3	22
CP	12.5	25[c]	20.8 ± 7.5[d]	49[e]
PYR	500	75[a]	3.0 ± 1.5	12
	250	75[a]	3.0 ± 1.5	13
	125	75[a]	2.9 ± 1.6	17
	62.5	75[a]	2.9 ± 1.8	18
Corn oil/DMSO	10 ml/kg	50[b]	3.1 ± 1.4	16
CP	12.5	25[c]	24.7 ± 8.3[d]	35[e]

[a] Three animals per group.
[b] Negative control, 2 animals per group.
[c] Positive control (cyclophosphamide) one animals per group.
[d] Significantly different from vehicle control as determined by Dunnett's t test ($p < 0.05$).
[e] Cytotoxic response. Historical % M_1 (vehicle controls) 21 ± 5.

Table 3. *Summary values for the in vivo induction of sister chromatid exchange in bone marrow of Chinese hamster treated orally with 2AAF, 4AAF, BP, or PYR*

Chemical	Dose (mg/kg)	Metaphases scored	SCE/Metaphase (Mean ± S.D.)	Cell cycle (% M_1)
2AAF	1000	50[a]	10.3 ± 3.9[e]	47[f]
	500	75[b]	5.2 ± 2.3[e]	45[f]
	250	75[b]	5.4 ± 2.8[e]	23
	125	75[b]	3.9 ± 1.4	17
Corn oil	10 ml/kg	50[c]	3.1 ± 1.3	17
CP	50	25[d]	16.1 ± 6.6[e]	45[f]
4AAF	1000	75[b]	2.5 ± 1.2	20
	500	75[b]	2.9 ± 1.3	16
	250	75[b]	3.0 ± 1.6	16
	125	75[b]	2.7 ± 1.7	18
Corn oil	10 ml/kg	50[c]	2.7 ± 1.3	17
CP	50	25[d]	21.4 ± 8.2[e]	76[f]
BP	1000	75[b]	4.4 ± 1.8[e]	35[f]
	500	50[a]	5.3 ± 2.4[e]	16
	250	50[a]	5.0 ± 2.4[e]	20
	125	75[b]	5.3 ± 3.3[e]	11
Corn oil	10 ml/kg	50[c]	2.7 ± 1.3	18
CP	50	25[d]	20.0 ± 10.3[e]	72[f]
PYR	1000	75[b]	2.6 ± 1.2	15
	500	75[b]	2.5 ± 1.5	25
	250	75[b]	2.3 ± 1.4	18
	125	75[b]	2.7 ± 1.4	25
Corn oil	10 ml/kg	50[c]	2.5 ± 1.5	28
CP	50	25[d]	22.4 ± 9.9[e]	69[f]

[a] Results from two animals, one animal died.
[b] Three animals per group.
[c] Negative control, 2 animals per group.
[d] Positive control, one animal per group.
[e] Significantly different from vehicle control as determined by Dunnett's t test ($P < 0.05$).
[f] Cytotoxic response. Historical % M_1 (vehicle controls) 21 ± 5.

as a suspension in corn oil at doses of 1000, 500, 250, and 125 mg/kg (Tables 3 and 4). No increase in SCE frequency as compared with the frequency in the vehicle treated control groups was noted in bone marrow or intestinal epithelium, nor was there any evidence of cytotoxicity.

BP

Parenteral administration of BP suspended in corn oil produced dose-dependent, statistically significant increases in SCE formation at doses of 500, 250, 125, and 62.5 mg/kg (Table 2). No cytotoxicity was noted in bone marrow of any animal receiving BP parenterally.

Additional groups of animals were treated orally with suspensions of BP in corn oil at doses of 1000, 500, 250, or 125 mg/kg (Tables 3 and 4). One animal receiving 500 mg/kg and one animal receiving 250 mg/kg BP were found dead 16 h after dosing, apparently due to toxicity of chemical treatment. Quantification of SCEs in bone marrow revealed a statistically significant increase in SCE formation at treatments of 1000, 500, 250, and 125 mg/kg and in intestinal epithelium at doses of 500, 250, and 125 mg/kg. An increase in the number of first division metaphase figures of bone marrow was noted in 2 of 3 animals treated with 1000 mg/kg BP. Cell cycle analysis of

Table 4. *Summary values for the in vivo induction of sister chromatid exchange in intestinal epithelium of Chinese hamsters treated orally with 2AAF, 4AAF, BP, or PYR*

Chemical	Dose (mg/kg)	Metaphases scored	SCE/Metaphase (Mean ± S.D.)	Cell cycle (% M_1)
2AAF	1000	0[a]	—	—
	500	75[c]	7.1 ± 2.7[g]	13
	250	75[c]	6.5 ± 3.2[g]	9
	125	75[c]	4.5 ± 2.2[g]	8
Corn oil	10 ml/kg	50[d]	2.8 ± 1.4	15
CP	50	25[e]	6.7 ± 3.1[g]	32[h]
4AAF	1000	75[c]	2.6 ± 1.0	5
	500	75[c]	2.9 ± 1.3	10
	250	75[c]	3.1 ± 1.9	11
	125	75[c]	3.1 ± 1.3	9
Corn oil	10 ml/kg	50[d]	2.9 ± 1.4	8
CP	50	25[e]	7.1 ± 2.2[g]	20[h]
BP	1000	0[b]	—	100[h]
	500	50[f]	5.8 ± 2.6[g]	7
	250	50[f]	6.3 ± 2.7[g]	4
	125	75[c]	4.8 ± 2.6	5
Corn oil	10 ml/kg	50[d]	3.2 ± 1.2	3
CP	50	25[e]	6.5 ± 2.4[g]	29[h]
PYR	1000	75[c]	3.2 ± 1.5	4
	500	75[c]	3.1 ± 1.5	6
	250	75[c]	3.1 ± 1.4	5
	125	75[c]	3.2 ± 1.6	5
Corn oil	10 ml/kg	50[d]	2.9 ± 1.2	3
CP	50	25[e]	5.6 ± 2.3[g]	10

[a] No observable mitotic figures.
[b] No observable second division metaphases.
[c] Three animals per group.
[d] Negative control, 2 animals per group.
[e] Positive control, one animal per group.
[f] Results from two animals, one animal died.
[g] Significantly different from vehicle control as determined by Dunnett's t test ($p < 0.05$).
[h] Cytotoxic response. Historical % M_1 (vehicle controls) 9 ± 5.

Table 5. *Summary values for in vivo induction of sister chromatid exchange in bone marrow of Chinese hamsters treated intraperitoneally with mitomycin C*

Chemical treatment	Dose (mg/kg)	Total metaphases scored[a]	SCE/Metaphase (Mean ± S.D.)	Cell cycle (% M_1)
Mitomycin C	0.5	100	5.1 ± 2.3[c]	23
Control (Phosphate buffered saline)	10 ml/kg	100	2.7 ± 1.3	18

[a] Four animals per treatment group.
[b] Historical % M_1 (vehicle controls) 21 ± 5.
[c] Significantly different from control as determined by Dunnett's t test ($p < 0.05$).

intestinal epithelium metaphases revealed that 100% of metaphases from animals treated with 1000 mg/kg BP were characteristically first division staining.

PYR

PYR was administered parenterally as a suspension in corn oil/DMSO (final DMSO concentration 1.2%) to animals at doses of 500, 250, 125, and 62.5 mg/kg (Table 2) and orally as a suspension in corn oil at doses of 1000, 500, 250, and 125 mg/kg (Tables 3 and 4). No induction of SCEs or cytotoxicity was noted in any of the treatment groups.

MMC

MMC was administered ip as a solution in phosphate buffered saline to animals at a dose of 0.5 mg/kg. A statistically significant increase in SCE formation was noted in all animals treated with MMC (Table 5). No cell cycle delay was noted.

5 Discussion

The potential genotoxicity of 2AAF, 4AAF, BP, and PYR was evaluated *in vivo* in Chinese hamsters by measuring SCE frequencies in metaphases of bone marrow and intestinal epithelium. This *in vivo* test is sensitive to both direct-acting and activation-dependent genotoxins (Neal and Probst, 1983, 1984; Wolff, 1979). Findings in this system, coupled with results from appropriate *in vitro* tests, provide a relevant measure of potential genotoxicity of chemical agents.

In these studies, compounds were administered ip and SCE formation in bone marrow cells was measured. Additional studies were conducted after oral administration of compounds, and both bone marrow and intestinal epithelium were evaluated for SCEs. Epithelium of the small intestine represents a mitotically active tissue, having a cell cycle similar to that of bone marrow. Therefore, both tissues can be obtained from the same animal. Epithelium of the small intestine also serves as a proximal target at the site of absorption and enteric biotransformation for compounds administered orally, while bone marrow may be evaluated as a secondary target for compounds that are absorbed, metabolized, and systemically distributed.

A dose-related positive response for chemically induced SCE was found with 2AAF and BP after ip administration. These results are especially meaningful in that the target tissue bone marrow is removed from the major tissues that function *in vivo* both for activation and detoxification of xenobiotics. A positive response for the induction of SCEs was also noted in bone marrow as well as intestinal epithelium after oral administration of 2AAF and BP. The SCE response obtained for 2AAF in bone marrow was lower for the oral route than that observed for ip exposure (Tables 2 and 3). Likewise, with BP a similar response in bone marrow was noted; however, a saturation effect was observed in that no substantial increase in SCE formation was seen with BP treatment in excess of 125 mg/kg. This saturation effect was also observed in intestinal epithelium after oral treatment with BP, but the magnitude of SCE response was greater than that obtained in bone marrow (Table 4).

Two analogues, 4AAF and PYR, which on the basis of *in vitro* tests for genotoxicity were classified as potentially genotoxic (Ashby, 1981), failed to induce SCE *in vivo* (Tables 2, 3 and 4). The *in vivo* SCE studies reported here support the conclusion that 4AAF and PYR have no genotoxic potential *in vivo*.

6 Conclusion

The identification of SCE *in vivo* in Chinese hamster bone marrow and intestinal epithelium allows a comparison of the response of 2 target tissues to chemical treatment. These studies indicate that 2 structurally similar carcinogen/noncarcinogen analogues can accurately be discriminated in 2 tissues, irrespective of the route of administration. However, difference in the magnitude of the SCE response were noted with the 2 carcinogens, 2AAF and BP. Factors contributing to this effect may include: (1) route of administration; (2) differences in absorption, metabolism, and distribution; and (3) differential target tissue sensitivity.

7 References

Allen, J.W., C.F. Shuler, R.W. Mendes and S.A. Latt (1977), A simplified technique for *in vivo* analysis of sister chromatid exchanges using 5-bromodeoxyuridine tablets, *Cytogenet. Cell Genet.*, **18**, 231–237.

Allen, J.W., C.F. Shuler and S.A. Latt (1978), Bromodeoxyuridine tablet methodology for *in vivo* studies of DNA synthesis, *Som. Cell Genet.*, **4**, 393–405.

Ashby, J. (1981), Overview of Study and Test Chemical Activities. In F.J. de Serres and J. Ashby (Eds), Evaluation of Short-Term Tests for Carcinogens, *Report of the International Collaborative Program*, Elsevier/North-Holland, NY, pp. 112–171.

Basler, A., U. Bachmann, G. Roszinsky-Köcher and G. Röhrborn (1979), Effects of caffeine on sister-chromatid exchanges (SCE) *in vivo*, *Mutation.*, **59**, 209–214.

Bayer, U., (1978), The *in vivo* induction of sister chromatid exchanges in the bone marrow of the Chinese hamsters, II. N-Nitrosodiethylamine (DEN) and N-Isopropyl-α(2-methyl-hydrazino)-p-toluamide (Natulan), Two carcinogenic compounds with specific mutagenicity problems, *Mutation Res*, **56**, 305–309.

Calland, C.J., S.R. Wightman, and S.B. Neal (1985), Establishment of a Chinese Hamster Colony, *Lab Animal Sci.* **36**, 183–185.

Carrano, A.V., L.H. Thompson, P.A. Lindl and J.L. Minkler (1978), Sister chromatid exchange as an indicator of mutagenesis, *Nature* (London), **271**, 551–553.

Dunnett, C.W. (1964), New tables for multiple comparison with a control, *Biometrics*, **20**, 482.

Kato, H. (1977), Spontaneous and induced sister chromatid exchanges as revealed by the BUdR-labeling method, *Int. Rev. Cytol.*, **49**, 55–97.

King, M.T., D. Wild, E. Gocke and K. Eckhardt (1982), 5-Bromodeoxyuridine tablets with improved depot effect for analysis *in vivo* of sister chromatid exchanges in bone marrow and spermatogonial cells. *Mutation Res.*, **97**, 117–129.

Kram, E., G. Schneider, G.C. Senula and Y. Nakanishi (1979), Spontaneous and mitomycin-C induced sister chromatid exchange: comparison of *in vivo* and *in vitro* systems, *Mutation Res.*, **60**, 339–347.

Neal, S.B. and G.S. Probst (1983), Chemically induced sister chromatid exchange *in vivo* in bone marrow of Chinese hamsters, An evaluation of 24 compounds, *Mutation. Res.*, **113**, 33–43.

Neal, S.B. and G.S. Probst (1984), Assessment of sister chromatid exchange in spermatogonia and intestinal epithelium in Chinese hamsters. In R.R. Tice and A.

Hollaneder (Eds), *Sister Chromatid Exchanges*, Plenum, NY, pp. 613–628.
Perry, P., and S. Wolff (1974), New Giemsa method for the differential staining of sister chromatids, *Nature* (London), **251**, 156–158.
Perry, P., and J. Evans (1975), Cytological detection of mutagencarcinogen exposure by sister chromatid exchange, *Nature* (London), **258**, 121–125.
Roszinsky-Köcher, G., A. Basler and G.R. Röhrborn (1976), Mutagenicity of polycyclic hydrocarbons V. Induction of sister-chromatid exchanges *in vivo*, *Mutation. Res.*, **66**, 65–67.
Tice, R., J. Chaillet and E.L. Schneider (1976), Demonstration of spontaneous sister chromatid exchanges *in vivo*, *Exp. Cell Res.*, **102**, 426–429.
Wilmer, J.L., and E.R. Soares (1980), Sister chromatid exchange *in vivo* in mice: I. The influence of increasing doses of bromodeoxyuridine, *Environ. Mutagan.*, **2**, 35–42.
Wolff, S. (1979), Sister chromatid exchange: The most sensitive mammalian system for determining the effects of mutagenic compounds. In K. Berg (Ed.), *Genetic Damage in Man Caused by Environmental Agents*, Academic Press, Inc., NY, pp. 229–258.

PART IV

In vivo assays: liver

Byron E. Butterworth
Chemical Industry Institute of Technology Research Triangle Park, North Carolina, USA

47

Introduction: liver-specific assays

The assays that comprised the Liver Specific Assay Group were: 1) the focal hepatic proliferative lesion (FHPL) (usually gamma-glutamyl transpeptidase (GGT)) protocols to detect initiation and promotion; 2) cytogenetic endpoints in liver cells; 3) DNA repair as evidence by unscheduled DNA synthesis (UDS); and 4) DNA strand breaks. The summary reports of discussion groups dealing with these four assay categories precede each group of investigators' reports.

This summary presents a personal perspective of the performance and value of these assays. Both liver DNA adduct studies and liver-specific host-mediated assays were added later in the study, but were placed in the Miscellaneous category under Dr. McGregor to even out the administrative responsibilities. A comparison and discussion of all the test systems and endpoints in this tissue can be found in the report of the Liver Specific Assay Discussion Group headed by Dr. Hunt (Chapter 48).

In order to compare results from one laboratory to the next it is important that the same batch of chemical, species of animal, doses, route of administration, and time points be used. Although some investigators had begun their studies before I was appointed coordinator, I was able to suggest uniform guidelines that most researchers followed and proved to facilitate comparisons among laboratories. The guidelines for testing in the liver-specific group were different from those of other groups in that modest doses of chemicals were used and the i.p., route of administration was avoided. In taking this conservative approach to testing, we tried to avoid disturbing the normal patterns of uptake, distribution, detoxification, and excretion that occur in the whole animal. In general, all chemicals were administered by gavage in corn oil at 10 ml/kg body weight. 2AAF was given at 10 to 50 mg/kg, 4AAF at 50 to 200 mg/kg and BP and PYR at 100 to 500 mg/kg.

It is difficult to use *in vitro* cell culture genotoxicity assays to predict carcinogenic potency, organ specificity, and mechanisms in the intact animal. If a chemical produces only kidney tumors in the male rat, a positive response in the Ames test does not help explain why other tissues, females, and other species remained unaffected. If a compound produces only mouse liver tumors and is not genotoxic in cell culture models, one is left to wonder as to the mechanism of action in the animal. As more systems become available in the whole animal, more meaningful information is obtained. One can make valuable comparisons between different endpoints when the same tissue from the same treated animal are examined. The fact that so many assays were represented in this IPCS study is a testament to just how much progress has been made in techniques to measure genotoxic activity in the animal. In general, the liver specific assays yielded a cohesive story as to the activity of the test chemicals in this target organ. These points are addressed in detail in Chapter 48. Observations of particular interest will be briefly highlighted here.

As a point of comparison, the amount of binding to rat and mouse liver DNA was determined by Dr. Randerath's postlabeling technique, and it is clear that the greatest binding was by 2AAF to rat liver DNA. The initiation (FHPL) assays were consistent with this observation. Administration of a single dose of 2AAF to rats with no promotion regimen resulted in no observable FHPL after 6 weeks. If a 70-day promotion regimen was used, then 2AAF was positive and BP showed a marginal response. If a 6-month promotion protocol was employed, then both 2AAF and BP were positive with a marginal response for 4AAF. Thus, increased promotion results in increased sensitivity. In an assay in which the compounds were administered in the diet to assess for promotional activity, 2AAF was positive with some response by 4AAF and BP as well. While these initiation and promotion studies provide valuable information, they are so time-consuming that they cannot be viewed as routine short-term tests.

The micronucleus assay in hepatocytes was clearly positive with 2AAF but did not detect other chemicals which bound to the DNA to a lesser extent. Although this assay is still in the developmental stages it was judged as a promising new system. Clastogenicity and sister chromatid exchange assays were performed in liver epitheloid cells. The only positive response was with BP, but it is difficult to compare these results with those from assays that employ hepatocytes. It was felt that more work was required to characterize this system.

The largest group in the liver specific section comprised those performing the *in vivo* DNA repair assay. Agreement among the seven laboratories was excellent. Minor variations in technique and experimental protocol did not significantly affect the final results. 2AAF consistently yielded a strong, dose-related response in the range from 10 to 50 mg/kg. 4AAF was weakly positive in three laboratories and negative in the other four. Both BP and PYR were negative. These results are consistent with the binding studies and suggest that the sensitivity of the DNA repair assay is determined by the number of DNA adducts formed.

Interestingly, 2AAF was negative in the DNA repair assay in mouse hepatocytes both *in vivo* and *in vitro*. This was unexpected because 2AAF is a mouse liver carcinogen. However, the DNA binding was about an order of magnitude lower in the mouse liver compared to the rat liver. Further, liver tumors appear early in rats and late in mice in response to 2AAF. These observations suggest that there may be fundamental differences between the rat and mouse liver in the metabolic activation or mechanism of action of 2AAF.

The autoradiographic DNA repair assay can also provide information on the stimulation of hepatocytes into the DNA-synthesis (S-phase) of the cell cycle. While the relationship of semi-conservative DNA synthesis to carcinogenesis remains to be established, it appears that there is a correlation between the ability to induce hepatocyte turnover and the induction of hepatocellular carcinomas, even for chemicals which appear to have little or no genotoxic activity. One might also predict that agents which produce cell division may have promoting or co-carcinogenic activity. All investigators who examined 4AAF found it to be a potent inducer of S-phase in rat hepatocytes. Thus, one might predict that, in rats, 4AAF would: 1) produce hyperplasia in the liver; 2) would act in place of a partial hepatectomy in initiation-promotion protocols; and 3) would induce liver tumors late in a long-term carcinogenicity bioassay.

Two laboratories that conducted the alkaline elution assay found no activity in rat hepatocytes with any of the chemicals. This was unexpected, particularly since some of the very same hepatocyte preparations exhibited DNA repair with 2AAF. The alkaline elution assay, however, did appear to be very sensitive in detecting methylating agents such as dimethylnitrosamine. In contrast, there was one laboratory which reported that either DNA breaks or cross links were induced by all of the compounds except PYR. More effort will be required with the alkaline elution assay to understand the insensitivity to 2AAF and the interlaboratory variations in results.

In comparing the results of the liver-specific assays with the IPCS study as a whole and in considering all factors, it would appear that the combination of the bone marrow micronucleus assay and the *in vivo* hepatocyte DNA repair assay would be a reasonable *in vivo* screen for genotoxic and potential carcinogenic activity.

1 Acknowledgements

I would like to thank all the investigators in the liver-specific assays section. The amount of time and money donated to this study was substantial. Without exception, everyone was most helpful and conducted their studies in the highest professional manner. The information gained greatly contributes to the progress of the field of genetic toxicology.

John M. Hunt (Workgroup Chairman)[1]
Department of Pathology and Laboratory Medicine University of Texas Medical School, Houston, Texas, USA

48

Summary report on liver-specific short-term tests

The results of the liver specific short-term tests have been summarized in two tables. In Table 1, consensus results are presented for each type of assay system used in the testing. The criteria used by individual investigators to decide if a particular assay gave a postive or negative response with the test chemicals varied depending on the nature of the assay endpoint. The reports of the individual investigators should be consulted to ascertain these criteria. The mg/kg doses of test chemicals used in the assays are also presented in Table 1 according to the following convention: for positive responses the dose is the lowest tested that gave a positive response; for negative responses, the dose is the highest tested.

Table 2 represents the fraction of laboratories reporting positive results for the test chemicals using the diverse liver specific assay systems. Table 2 is intended to indicate the number of laboratories performing a given assay, the agreement among different laboratories performing the same assay and the overall fraction of positive responses obtained with each of the 4 test chemicals. Detailed discussions of the correlation with long-term carcinogenesis bioassay data (Table 3) can be found in the individual investigator reports for each assay. Footnote a of Table 1 keys the assays to the investigators.

In this summary section, the relative responses of the individual liver specific assay systems will be considered with reference to long term carcinogenesis bioassay data for the 4 test chemicals (Table 3). A detailed discussion of the bioassay data is presented in Appendix 3. Adequate bioassay data indicate that 2AAF is carcinogenic in rats and mice for liver and for general (nonhepatic) sites. However, 2AAF is a more potent hepatocarcinogen in rats than in mice, and time to tumour is also much shorter in rats than in mice. BP has been shown to be carcinogenic in rats, but only for nonhepatic sites, usually the site of administration. In mice, BP has only been shown to induce cancers at nonhepatic sites. Adequate data are lacking on the carcinogenicity of BP for mouse

[1] Workgroup participants: Diana Anderson, Peter Bannasch, David H. Blakey, Byron D. Butterworth, Riccardo Crebelli, Gary M. Decad, David J. Doolittle, Charles N. Edwards, Howard P. Glauert, Makoto Hayashi, Vladimir I. Ivanov, James T. MacGregor, Douglas McGregor, Dennis A. Pagano, Silvio Parodi, Gregory S. Probst, Kurt Randerath, Christopher Richardson, Margaret Richold, Ad D. Tates, Raymond R. Tice, and G. Ashley Wickramaratne

Table 1. *Summary of liver specific assay results*

Assay[a]	Test Chemicals[b]			
	2AAF	4AAF	BP	PYR
A. DNA repair (UDS)				
1. Rat	++++	–[c]	–	–
	5	2000	500	500
2. Mouse	–	–	–	
	50	200	200	NT[d]
B. S-Phase DNA synthesis				
1. Rat				
24 h	–	++++	–	–
	200	50	500	500
48 h	+	–	+	–
	50	200	100	500
Weanling rats 4h	NT	NT	+	NT
			100	
2. Mouse	–	+	–	NT
24 h	200	200	200	
48 h	+	–	NT	NT
	200	200		
C. DNA strand breaks				
1. Alkaline elution				
Rat	–	–	–	–
	700	700	250	500
2. Crosslinks				
Rat	+	+	+	–
	20	20	52	136
3. Unwinding				
Mouse	+	–	–	–
	50	50	100	100
Plus repair inhibitor	+	–	+	–
	10	50	100	100
D. DNA adducts				
1. Rat	++++	+	++	–
	150	150	150	450
2. Mouse	+++	–	++	–
	150	150	150	150
E. Micronucleus (rat)	+	–	–	–
	7	890	250	800
F. Clastogenesis (rat)	–	–	–	–
	50	50	100	100
G. SCE (rat)	–	–	+	–
	100	100	100	100
H. Cell ploidy change	+	–	+	–
(rat and mouse)	500	1000	5000	404
I. Enzyme altered liver foci				
Initiation with one dose (rat)				
1. No promotion (8 wk)	–	–		
	500	500	NT	NT
2. Promote for 70 days	+	–	+	–
	10	300	100	380
3. Promote for 6 months	+	+	+	–
	10	250	100	500
Chemical in diet as promoter[e]				
4. Six wk	+	+	+	–
	12	12	12	12
5. 14 days (transplantation)	+	–	–	–
	12	36	36	36
J. Host-mediated assay mutagenesis (mouse)				
1. Salmonella TA100	–	–	–	–
	250	250	250	250
2. Yeast, uninduced, 4 h	–	+	+	+
	2500	2500	625	625
3. Yeast, uninduced, 16 h	+	–		
	2500	2500	NT	NT
4. Yeast, induced, 4 h	–	–		
	2500	2500	NT	NT
K. Foetal blood micronucleus (mouse)				
1. Strain SW	–	–	+	–
	625	893	63	404
2. Strain NMRI	+	–	+	–
	223	893	252	404

liver. Long-term carcinogenesis bioassay data for 4AAF or PYR in either rats or mice is more limited and this limits the strength of the conclusions drawn as to the predictive value of the short-term tests evaluated in this study. The data presented underscore the important question of whether liver specific assays may be useful in detecting carcinogens that act at sites other than liver.

The DNA repair or unscheduled DNA synthesis (UDS), assay (Table 1,A) yielded excellent agreement among laboratories, although marginal responses were obtained with 4AAF (Table 1, footnote c). Correlation with the bioassay data for the rat is what might be

Note to Table 1

[a] Investigators performing the assays listed in Table 1 are as follows: A: Ashby, Beije, Burlinson, Butterworth, Doolittle, Kornbrust, Mirsalis, Salocks. B: Beije, Burlinson, Butterworth, Doolittle, Kornbrust, Mirsalis, Salocks. C.1: Bermudez, Elliott, Russo; C.1: Russo; C.3: Birnboim. D: Randerath. E: Tates. F: Meyer. G: Meyer. H: Styles. I.1: Bannasch; I.2: Pereira; I.3. Glauert; I.4: Ito; I.5 Hunt. J: Bronzetti, Forster, Pagano; K: MacGregor, Suter.

[b] For each of the 4 test chemicals, a positive (+) or negative (−) assay consensus result in reported. Numbers are lowest tested dose producing a (+) response or highest tested dose producing a (−) response. Doses are in mg/kg. To give a feeling for potency, each (+) represents an approximate order of magnitude compared to the other chemicals and historical controls. This semi-quantitation should be considered meaningful only within a particular assay.

[c] Three laboratories reported a marginal response at 1000 mg/kg.

[d] NT is not tested.

[e] Doses are daily mg/kg intake of test chemicals in the diet.

Table 2. *Fraction of laboratories reporting positive results in liver-specific assays*

Assay	2AAF	4AAF	BP	PYR
A. DNA repair (UDS) (rat)	7/7	0/7[a]	0/3	0/2
(mouse)	0/2	0/2	0/1	NT
B. S-Phase DNA synthesis				
Rat 24 h	0/3	6/6	0/1	0/2
48 h	2/2	0/1	1/1	0/1
Weanling rat 4 h	NT	NT	1/1	NT
Mouse 24 h	0/1	2/2	0/1	NT
48 h	1/1	0/1	NT	NT
C. DNA strand breaks				
1. Alkaline elution (rat)	1/3	1/3	0/3	0/3
2. Crosslinks (rat)	1/1	1/1	1/1	0/1
3. Unwinding (mouse)				
− inhibitor	1/1	0/1	0/1	0/1
+ inhibitor	1/1	0/1	1/1	0/1
D. DNA adducts (rat)	1/1	1/1	1/1	0/1
(mouse)	1/1	0/1	1/1	0/1
E. Micronucleus (rat)	1/1	0/1	0/1	0/1
F. Clastogenesis (rat)	0/1	0/1	0/1	0/1
G. SCE (rat)	0/1	0/1	1/1	0/1
H. Cell ploidy change (rat)	1/1	0/1	1/1	0/1
(mouse)	1/1	0/1	1/1	0/1
I. Enzyme altered liver foci (rat)				
Initiation				
1. No promotion (8 weeks)	0/1	0/1	NT	NT
2. 70 days promotion	1/1	0/1	1/1	0/1
3. Six month promotion	1/1	1/1	1/1	0/1
Promotion				
4. Diet for 6 wk	1/1	1/1	1/1	0/1
5. Diet for 2 wk (transplant)	1/1	0/1	0/1	0/1
J. Host mediated assay (mouse)	1/4	1/4	1/2	1/2
K. Foetal blood micronucleus (mouse)	1/2	0/2	2/2	0/2
Overall fraction of (+) Results	25/40	14/43	15/29	1/26
(only rat hepatocyte DNA endpoints)	16/19	3/19	6/14	0/13
Overall percentage of (+) Results	63%	33%	52%	4%
(only rat hepatocyte DNA endpoints)	84%	16%	43%	0%

[a] 3 laboratories reported a marginal response

predicted with 2AAF and BP. The failure of any investigators to obtain a positive response with 2AAF in UDS assays with the mouse is striking and difficult to explain, because this chemical is a mouse hepatocarcinogen. This discrepancy, however, suggests a quantitative difference in the metabolic activation of 2AAF in these two species. The DNA adduct assay (Table 1, D) in the rat indicates over an order of magnitude less binding of 4AAF, BP, and PYR compared with 2AAF and an alteration of mouse liver DNA by 2AAF much lower than the level for rat liver DNA. This difference in extent of DNA adduct formation could account for the apparently all-or-none UDS assay result difference between rat and mouse with 2AAF. Additional work is needed to test this possibility. The very low minimum dose (5 mg/kg) of 2AAF producing a positive response in the rat UDS assay is notable and similar in magnitude to the 7 mg/kg dose effective over a narrow dose range in the rat liver micronucleus assay (Table 1, E). Positive responses were reported for other assays at 10 mg/kg of 2AAF (Table 1).

Detection of *in vitro* S-phase DNA synthesis induced by the test chemicals (Table 1, B) represents a novel observation. This assay has an autoradiographically scored endpoint analogous to, but clearly distinguishable from, that of UDS. The time of scoring the S-phase DNA assay endpoint is critical (Table 1, B). A regenerative response following hepatotoxicity peaks at about 48 h following a single treatment. A mitogenic response leading to hyperlasia usually peaks at about 24 h. 4AAF was only much more potent than 2AAF in the S-phase assay, but the peak of activity came at 24 h, suggesting a mitogenic effect. One striking observation with the S-phase assay was its uniformly positive response with 4AAF in the rat, with a weaker response in the mouse. For 4AAF, this result contrasted with the moderately low percentage of positive responses (33%) (Table 2) in the short-term assays. Thus, the S-phase assay may provide unique information about the mechanism of action of 4AAF as a weak genotoxin in this study.

Three assay systems were used to detect DNA single-strand breaks (SSB) (Table 1, C). An unresolved disagreement among laboratories using the alkaline elution method (Table 1, C.1) was noted in that 2 laboratories found no activity with any chemicals, while one found activity with both 2AAF and 4AAF. The nondetection of 2AAF and the variation from laboratory to laboratory indicated that alkaline elution is of limited predictive utility. The DNA-protein crosslinking assay, done in only one laboratory, produced qualitatively different results from the other two assays done in the same laboratory (Table 1, C.2). A fluorimetric assay (Table 1, C.3) which detects SSB by quantitating the extent of alkali-induced DNA unwinding in solution produced different results for 2AAF and BP depending on the treatment with a DNA repair inhibitor, adenosine arabinoside, of mice prior to isolation of liver cell DNA. Use of the inhibitor permitted clear differentiation between 2AAF and 4AAF and between BP and PYR.

Table 3. *Long-term carcinogenesis bioassay data for test chemicals in rat and mouse*[a]

Species	Test Chemicals 2AAF	4AAF	BP	PYR
Rat	L+[b]	Li	L−	Li
	G+	Gi	G+	Gi
Mouse	L+	Li	Li	Li
	G+	Gi	G+	Gi

[a] See chapter by Hicks (this volume) for supporting literature citations.
[b] Abbreviations used: L, liver carcinogenesis; G, general carcinogenesis (i.e., carcinogenesis at nonhepatic sites); +, positive cancer endpoint in adequate study; −, negative cancer endpoint in adequate study; i, inadequate cancer endpoint data available.

Formation of DNA adducts was assessed by a sensitive ^{32}P-postlabeling method (Table 1, D). This method permits quantitation of the exposure of cells to test chemicals which can themselves form covalent adducts with DNA nucleotides or which induce the formation of DNA adducts with as yet unknown endogenous reactants. The significance of DNA adducts is as yet unknown with regard to the bioassay endpoint of cancer. However, the quantitation obtained with the ^{32}P-postlabeling assay makes possible tentative correlations of levels of adduct formation and the positive or negative responses of the other short-term assay systems using the 4 test chemicals. Figure 1 graphically illustrates the magnitude of the adduct formation response for each chemical. If several other assays are ranked by sensitivity (Figure 1), their response fits well with what might be predicted by the adduct data. This illustration that the data from different systems correlate in a quantitative manner reveals the strength of these whole animal assays relative to short-term *in vitro* assays. The postlabeling technique yields results that appear to correlate well with carcinogenic potency.

The 3 rat cytogenetic assays in the study – the micronucleus test, chromosomal aberration (clastogenesis), and sister chromatid exchange (SCE) – differed in their responses to the 4 test chemicals (Table 1, E, F, G). A dose-dependent positive response with 2AAF was noted in the rat liver micronucleus test, whereas negative results were obtained with the 3 other test chemicals. Interestingly, this was the same profile of activity observed with the UDS assay and the liver cell transplantation assay in rats (Table 1, A.1 and I.5, respectively). The micronucleus assay endpoint was scored microscopically and the micronuclei scored were observed in hepatocytes. In contrast, both the clastogenesis and SCE assays scored endpoints in liver-derived cells that underwent mitosis in cell culture *in vitro*. It is likely that the dividing cells were not hepatocytes because adult rat hepatocytes have a very limited capacity for mitosis in culture. The clastogenesis assay was negative for all 4 test chemicals, and a positive result for SCE was observed only with BP. Thus, further work will be necessary to identify the liver-derived cells which may be responding in these latter 2 cytogenetic assays.

A cytofluorimetric method was employed to assess the effect of high doses of the 4 test chemicals on the relative liver cell ploidy in rats and mice (Table 1, H). An increase in the ratio of diploid to tetraploid liver cell number was induced by 2AAF and BP but not by 4AAF or PYR. The assay used a flow cytometer to detect cell ploidy in intact liver-derived cells. Further characterisation of these cells will be helpful in interpreting the assay endpoint and making correlations between the ploidy results and those from other short-term tests.

The endpoint scored in the rat enzyme-altered liver foci assays (Table 1, I) was the appearance of liver foci, also known as areas or colonies, identifiable histochemically as having elevated levels of the marker enzyme gamma-glutamyltranspeptidase (GGT). This marker enzyme has gained wide use in identifying early stages of rat liver carcinogenesis, although the significance of the altered phenotype is not presently known. The appearance of GGT positive foci in rat liver often precedes the appearance of liver carcinomas and is thus correlated with hepatocarcinogenicity. Liver foci assays have been devised to test for initiating activity (Table 1, I. 1,2,3) or promoting activity (Table 1, I. 4,5). However, the extreme complexity of the assay protocols makes it difficult to conclude with absolute certainty which effect of a test chemical, initiation or promotion, is producing the endpoint. With this caveat in mind, the assays have been tabulated in separate groups in Tables 1 and 2. With the exception of the uniformly negative response of the 1-dose 8-week assay (Table 1, I.1), all assay systems gave a positive response with 2AAF. The foci assays gave varied responses with 4AAF and BP and were uniformly negative with PYR. It is notable that in 3 foci assays (Table 1, I.2,3,4), BP produced a positive response, despite the fact that BP has nor been shown to be a liver carcinogen in the rat (Table 3). The use of partial hepatectomy in these protocols, may allow the detection of chemicals which are carcinogenic in tissues other than liver. As the time of promotion was increased, more positive endpoints were observed (Table 1, I.2, 3, 4, 5). Promotion activity yielded an additional perspective that could not be provided by *in vitro* short-term tests.

Host-mediated mutagenesis assays in mouse liver, employing either Salmonella TA100 or yeast, can be considered liver specific only in that the liver was the organ excised, homogenized and cultured to search for microbial revertants. A limitation of this approach is the entrapment of blood in this highly vascular organ and the question of the exact site in the intact animal in which the microorganisms acquire their revertant phenotypes. The tests using TA100 were all negative whereas an unusual patterm of positive responses was observed using yeast in uninduced mouse liver (Table 1, J.2, 3). The intrasanguinous host-mediated mutagenesis assay, as well as the urine mutagenesis assays in which all 4 test chemicals gave positive responses with TA100 (results not included in Table 1), may represent methodologies unusually sensitive to the special chemical properties of the test chemicals. For example, solubility, hydrophobicity, dissociation constant of chemical-plasma protein complexes, and metabolic fates could make the urine assay and host-mediated assay sensitive only to certain types of potential carcinogens. This caution should, of course, be applied to all assay in the present study.

Results of a mouse foetal blood cell micronucleus assay have been included in (Table 1, K.1, 2) because hemopoiesis occurs in the liver during foetal development.

Figure 1. Comparison of the responses of assays with altered hepatocyte DNA as the endpoint. The extent of DNA adduct formation is shown by the bars. The response of each assay for each chemical is denoted by a + (positive), − (negative), or w + (marginal response). No symbol indicates that the test was not done.

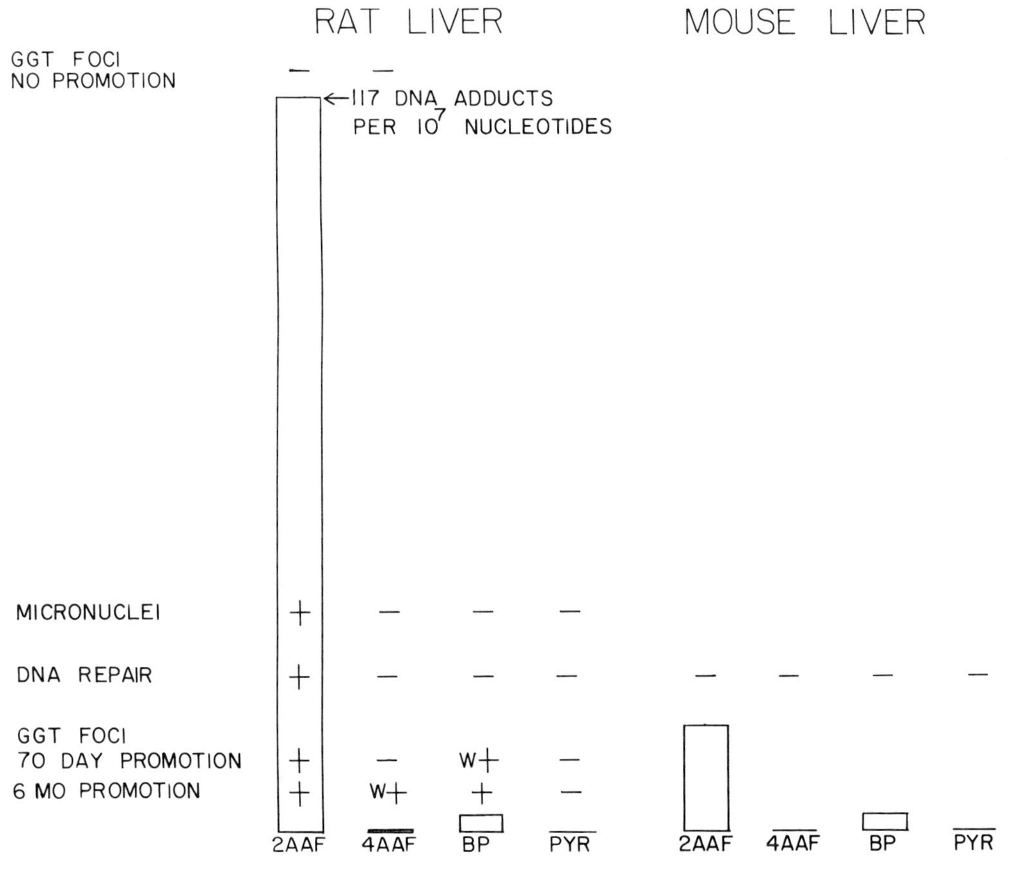

Response of this assay to BP in 2 mouse strains was positive, whereas 2AAF gave a positive response in only one of the 2 strains. The result with 2AAF indicates the need to choose strains carefully as new assays are developed.

The overall fractions and percentages of positive responses obtained with each of the 4 test chemicals in the liver specific assay systems are presented at the end of Table 2. Sixty-three per cent of the assays were positive with 2AAF, an established carcinogen for rat and mouse liver (Table 3). Out of 26 assays, PYR produced a response in only one (4%). The percentages of assays positive for 4AAF and BP were 33% and 52%, respectively. It is evident that with whole animal assays, one cannot use the simple plus-minus evaluation commonly employed with *in vitro* short-term tests but must examine the responses on a quantitative, case-by-case basis. In so doing, a much clearer understanding of potency, mechanism of action, and organ specificity may be obtained. In preparing Table 2, no selection of assay systems was done, and all results were tallied with equal weight. As has been stressed at the beginning of this summary chapter, the individual investigator reports should be consulted to determine, for example, the criteria used to differentiate a positive from a negative response, the extent to which variation in endpoint sampling times may affect responses, and the extent of quantitation of potency possible for a particular assay. When adequate data (Table 3) become available for 4AAF, BP, and PYR in mice and rats, the outcome of the short-term *in vivo* assays can be better assessed as predictive or not predictive of carcinogenicity.

In conclusion, the liver specific assays evaluated in this study represent highly diverse methodologies for detecting exposure and response of liver cells to potential carcinogens. Some assays emerged as potentially valuable tools for predicting the carcinogenicity of chemical agents. These whole animal assay systems should provide information as to the organ and species specificity and mechanism of action of the test chemicals not obtainable by *in vitro* short-term tests. As research continues to improve these methodologies and as newer assay systems are developed, it should be possible to assemble a set of assays which can serve as accurate short-term tests for predicting carcinogenic potency in specific target organs.

PART IV
In vivo assays: liver

A
Initiation and promotion

Howard P. Glauert (Workgroup Chairman)[1]
McArdle Laboratory for Cancer Research, University of Wisconsin, Madison, Wisconsin, USA

49

Summary report on the performance of the liver initiation and promotion assays

1 Introduction

Foci of altered hepatocytes have been shown to precede the development of hepatocellular tumors induced by chemicals in a number of laboratories (Bannasch *et al.*, 1980; Emmelot and Scherer, 1980; Farber, 1980; Pitot and Sirica, 1980; Williams, 1980). A close statistical correlation between the appearance of the later development of tumors has been reported by different groups (Emmelot and Scherer, 1980; Kunz *et al.*, 1983).

In initiation – promotion studies, "initiation" and "promotion" are defined operationally. Initiation can be accomplished by the short-term administration of a chemical, which begins the process of carcinogenesis. The subsequent process that facilitates the development of initiated cells into carcinomas is termed promotion. The detection of promotion has been improved over the last 20 years. An unequivocal differentiation between the biologically different processes of initiation and promotion during chemical carcinogenesis is not possible with this approach (Pitot and Sirica, 1980). However, the carcinogenic potential of a chemical may be detected within a few weeks or months by the assay protocols used in this study. Many carcinogens with target tissues other than liver can also be detected by these assays; it is possible, however, that some may escape detection (Pereira and Stoner, 1985).

The assay used by Bannasch has been previously applied to N-nitrosomorpholine (Bannasch *et al.*, 1979) and should be considered a research tool. The assays of Pitot and Glauert (Pitot *et al.*, 1978; Goldsworthy and Pitot, 1985; Pitot *et al.*, 1985) and of Pereira *et al.* (1982) have been successfully used with many test chemicals but cannot be considered established assays at present. The validity of the protocol for the detection of promoters employed by Ito *et al.* (1982, 1983; Tsuda *et al.*, 1984a, b) has been examined by testing over 30 compounds. The liver cell transplantation assay of Hunt represents a new assay and, without further characterization, is probably only useful as a research technique (Hunt *et al.*, 1982, 1985). The characteristics of these assays are summarized in Table 1.

All the assays used, even those reported in the literature previously, suffer from the limitation that they may not be able to unequivocally differentiate initiation

[1] Workgroup participants: Peter Bannasch, John M. Hunt, Nobuyuki Ito, and Michael A. Pereira

Table 1. *Characteristics of liver initiation/promotion assays*

	Bannasch	Pereira	Glauert	Ito	Hunt
Endpoint	GGT^+, $ATPase^-$, $G6Pase^-$, $G6PDH^-$, & glycogen storage foci[a]	GGT^+foci	GGT^+foci	GGT^+foci	GGT^+colonies
Rat strain	Fischer 344	Fischer 344	Charles River	Fischer 344	Fischer 344
Sex	Male	Male	Female	Male	Male
Method of promotion (for "initiation tests")	None	0.05% PB in water	0.05% PB in diet	NA[b]	NA
Method of initiation (for "promotion tests")	NA	NA	NA	DEN, ip (200 mg/kg)	Solt-Farber protocol (Solt *et al.*, 1977)
Duration of experiment	2 h–8 wk	11 wks	17 or 25 wks	8 wks	14 days
Additional treatment	None	PH at time of "initiation"	PH at time of "initiation"	PH 1 week after starting "promotion"	0.02% test chemical fed for 1 wk before cells' transplantation, PH at time of donor liver cell transplantation
Positive control	None	DEN (10 mg/kg)	DEN (10 mg/kg)	None	None
Diet	Altromin cereal-based diet	Purina lab chow (cereal-based diet)	NIH-07 cereal-based diet	Oriental M. cereal-based diet	Dyets No. 101 purified diet
Definition of positive results		Nonparametric Kruskal-Wallis test for difference among dose levels of the test substance followed by a multiple comparison test based on the Kruskal-Wallis test	Means for test chemicals were compared to control mean using a one-sided nonparametric test. A confidence level of 95% or greater was considered to be significant. A positive dose–response (derived from linear regression) was also necessary to declare a chemical as positive	Means for test chemicals for areas of foci (mm^2/cm^2) were compared to control means using Students t-test. A confidence level of 95% or greater was considered to be significant	Range of donor liver colony detection in host rat liver > 10-fold above background levels

[a] Abbreviation: GGT, gamma-glutamyltranspeptidase; ATPase, canalicular ATPase, G6Pase, glucose-6-phosphatase; G6PDH, glucose-6-phosphate dehydrogenase; PB, phenobarbital; DEN, diethylnitrosamine; PH, partial hepatectomy.

[b] NA = not applicable

from promotion (Pitot and Sirica, 1980). Nonetheless, *in vivo* assays that pay careful attention to the sequence in which a test chemical and the other chemicals are administered are useful in ascertaining the initiating and promoting properties of a test substance.

2 Protocol factors of importance

A factor of major importance in initiation-promotion studies is the performance of a partial hepatectomy (PH), which has been shown to make the liver considerably more sensitive to the carcinogenic effects of chemicals (Craddock, 1978; Cayama *et al.*, 1978). It is especially important to recognize the crucial relationship between the time of PH and the administration of the test chemical, as the optimum time varies from one chemical to another (Stewart, 1981; Tsuda *et al.*, 1980; Mirsalis *et al.*, 1982; Pereira *et al.*, 1983). Therefore, it is important to test an unknown substance at various time intervals in relation to the PH.

The use of phenobarbital, which has no or only weak hepatocarcinogenicity, to promote the appearance of foci or tumors is especially helpful to determine if a chemical is a carcinogen. Phenobarbital is administered for several weeks or months after a single application of the test chemical. The use of phenobarbital greatly increases the sensitivity of the assay.

3 Data base for the assay

Control data for the rat liver focus assay have been collected in a number of laboratories. The appearance of "spontaneous" foci varies among different strains and laboratories, and also among different experiments conducted in the same laboratory with animals of the same strain. Thus, proper concurrent controls are indispensable for each experiment.

With respect to the specific assays, data have been obtained over a number of years. The assays as performed by Pereira and Herren-Freund, Pitot and Glauert, and Ito and Tsuda have proven to be highly reproducible. Dose–response studies performed at different times for the positive control diethylnitrosamine (DEN) gave quantitatively similar results (Goldsworthy and Pitot, 1985; Pereira *et al.*, 1982). Previous studies of 2AAF, BP, and PYR also gave results similar to the results reported in the present study (Herren *et al.*, 1982; Tsuda *et al.*, 1980). Thus, the assays appear to be reproducible. The control data for each laboratory are as follows;

a. Pereira and Herren-Freund: an acceptable level for a group of vehicle-treated (control) animals is less than one focus/cm^2. These values have not been previously exceeded by the vehicle controls (water, corn oil, dimethylsulphoxide, and tricaprylin).
b. Glauert: 21 rats were used as controls in the first experiment and 11 were used in the second. This number is probably sufficient since the number of foci per liver induced in the 2 experiments was very close (289 $\pm$ 64 foci per liver in the first experiment and 284 $\pm$ 61 in the second). Similar control values have been seen in other experiments from this laboratory. Defining an absolute range of acceptable control values, however, would be difficult because of slight variations between experiments.
c. Ito: Control data obtained from the protocol of Tsuda *et al.* (1984a, 1984b), in which rats (more than 20 per group) are subjected to DEN injection followed by an 8-week latency period and the performance of a PH during the latency period, has ranged from 5 to 10 gamma-glutamyl-transpetidase (GGT)-positive foci/cm^2 and 0.2 to 1.0 mm^2/cm^2 in over 300 animals.
d. Bannasch: The assay as used by Bannasch *et al.* (1979) gave positive results for N-nitrosomorpholine within 4 weeks after a single application of the LD50 of this compound.
e. Hunt: The background levels of GGT-positive colony formation in the assay of Hunt have been shown previously to be in the range of 0–4 colonies per standard liver section (less than 2 colonies/cm^2) (Hunt *et al.*, 1982).

4 Criteria for evaluation

Responses should be classified as negative, positive, or inconclusive. The use of words such as clearly, weak, or questionable should be avoided. A response in the assays should be judged as either positive or negative as determined by the appropriate statistical procedure. A study could be judged to be inconclusive if the experiment was terminated prematurely, if the dose was too low, or if not enough animals survived for an adequate evaluation.

Relative sensitivity between assays

In the 3 assays in which a single dose of the test compound was administered, a wide range of sensitivity is seen. In the study by Bannasch, neither 2AAF nor 4AAF induced an effect within 8 weeks, when applied at a dose of 500 mg/kg without any further treatment. Glauert and Pereira, however, both found 2AAF to produce an effect at 10 mg/kg. The increased sensitivity in these assays may be related to the performance of PH on the rats before 2AAF administration, to the administration of phenobarbital, and to the much longer time between administration of 2AAF and sacrifice of the rats. Both Pereira and Glauert found BP to be positive and PYR to be negative, although Glauert saw a greater response with BP after longer treatment with phenobarbital. With 4AAF, Pereira saw no effect at 300 mg/kg, whereas Glauert found it to be positive at 250 mg/kg. The withdrawal of phenobarbital for 7 days before sacrifice in Pereira's study and the longer promotion period in Glauert's study may have contributed to this result.

In the 2 assays examining promoting activity, that of Ito *et al.* appears to be more sensitive: it gave positive results for BP and 4AAF, whereas Hunt's assay did not. Both Ito and Hunt, however, found 2AAF to be positive at a similar dose. It should be noted that Ito's method required only 8 weeks to obtain sufficient results.

Group size

The groups used in the studies reported here were of sufficient size. Increasing the size of the groups would not have greatly increased the statistical power of the assays.

Artifacts that might have resulted in false positive results

"False positive" responses may be due to cellular changes

that are phenotypically similar to those induced by unequivocal carcinogens but are to a large extent reversible after withdrawal of the respective compounds. An example of a false positive response is the induction of GGT activity in the periportal area (zone 1) of the liver by phenobarbital and other chemicals (Pereira *et al.*, 1982). The zonal induction of GGT might be mistaken for the occurrence of GGT-positive foci. But once the possibility of this false positive response is recognized, it can easily be distinguished from the induction of GGT-positive foci. In the present study, induction of GGT in the periportal region was not observed.

Interpretation of negative results

Negative results may be caused by several factors including the absence of phenotypic expression of the initial carcinogenic event. Therefore, negative results in these assays cannot be taken to disprove a carcinogenic potential of the compound tested. But negative results in a validated bioassay do indicate a lower probability that the chemical is a carcinogen.

Use of several endpoints

The foci of altered hepatocytes induced by known carcinogens show a number of cytomorphological and cytochemical changes, such as changes in glycogen content or cytoplasmic basophilia, decrease or increase in the activity of many enzymes, or deficiency of iron storage in an artificially siderotic liver. None of these "markers" alone can be considered to be absolutely reliable in the detection of the foci. Thus, the evaluation of more than one marker is recommended. Simple histology in hematoxylin and eosin stained tissue sections appears to be a reliable method. GGT, which has been frequently used up to the present, has been shown to be an inadequate marker under certain experimental conditions (Rao *et al.*, 1982; Shirai *et al.*, 1985; Ito *et al.*, 1984; Pereira *et al.*, 1984; Glauert *et al.*, 1986).

5 Quantitative differences between carcinogen–noncarcinogen pairs

In the study by Glauert, quantitative differences between the carcinogen–noncarcinogen pairs were evident. At the highest dose tested (100 mg/kg), BP induced a 13-fold increase in the number of foci per liver and a 5-fold increase in the focal volume, whereas PYR did not alter either parameter at the highest dose tested (500 mg/kg). Both 2AAF and 4AAF increased the number and volume of GGT-positive foci, but 2AAF was much more potent. At the 50 mg/kg dose, 2AAF increased both the number of foci per liver and the focal volume as a percent of liver volume 14-fold; at the same dose, 4AAF increased both the number and volume of foci about 3-fold (which was not statistically significant). Therefore, 2AAF appears to be about 5 times more potent than 4AAF. At the highest dose tested (250 mg/kg), 4AAF increased the number of foci per liver 5-fold and the focal volume 4-fold. It is difficult to compare the response at this level with the response to 2AAF or 4AAF at 50 mg/kg; at the 250 mg/kg dose, 4AAF would not completely dissolve in corn oil and had to be intubated as a slurry. Therefore, it is highly questionable whether the dose of 4AAF was completely absorbed in the gut, making comparisons with lower doses difficult. In addition, the high dose of 4AAF was tested in a separate experiment.

In the study by Pereira, both 2AAF and BP were positive, and both 4AAF and PYR were negative. Therefore, only the minimum quantitative difference between the carcinogen-noncarcinogen pairs can be estimated. For 2AAF/4AAF, the minimum positive dose for 2AAF tested was 10 mg/kg. Thus, 2AAF was at least 30 fold more active than 4AAF. The values for BP/PYR were 20 mg/kg for BP and 380 mg/kg for PYR, so that BP was at least 19-fold more active than PYR.

In the study by Hunt, at doses of 12 mg/kg/day, taking 2AAF data as the positive control (12–14 for number of colonies/cm^2 liver), the relative responses of 4AAF, BP and PYR ranged from 0.0 to 0.41 colonies/cm^2. When doses of these latter 3 test chemicals were increased to 36 mg/kg/day, the relative responses still ranged from 1–3 colonies/cm^2 of liver (i.e. control or background levels of colonies).

In the assay of Ito *et al.*, all of the compounds were fed in the diet at the concentration of 0.02% (approximately 12 mg/kg/day). If relative potencies are compared, 2AAF is 50 times more potent than 4AAF and 25 times more potent than BP.

6 Evaluation of new assays

The liver cell transplantation experiment of Hunt is an attempt to apply a selection procedure (Laishes and Farber, 1978) to the short-term detection of carcinogens. Lack of carcinogenicity cannot be reasonably inferred from negative results obtained using this liver cell transplantation procedure. The positive control test chemical, 2AAF, may exert its selective effect by facilitating proliferation of transplanted donor liver cells by physiologically altering the biochemical environment of the host rat liver. Whatever the exact nature of this assay, the negative test chemicals 4AAF, BP, and PYR apparently do not result in the appearance of GGT-positive donor liver colonies in host livers.

Table 2. *Summary and comparison of the data presented*

Investigator	BP	PYR	2AAF	4AAF
Bannasch[a]	ND[b]	ND	– (500)[c]	– (500)
Glauert[d]	+ (100)	– (500)	+ (10)	+ (250)
Pereira[d]	+ (20)	– (380)	+ (10)	– (300)
Ito[e]	+ (12)	– (12)	+ (12)	+ (12)
Hunt[f]	– (36)	– (36)	+ (12)	– (36)

[a] Single dose without further treatment.
[b] ND = not done.
[c] The dose of compound (in mg/kg/day) is in parentheses; For positive results, the lowest effective dose is shown; for negative results, the maximum dose tested is shown.
[d] Single dose followed by promotion by phenobarbital.
[e] Initiation by DEN followed by feeding of test compound.
[f] Proliferation of transplanted altered cell population by feeding of test compound to recipient rat.

7 Performance of assays

The work group appraised the overall performance of the assays for initiation/promotion (Table 2). The chemicals were evaluated as follows:

BP

In the assays of Pereira and Glauert, BP was judged to be positive for initiating activity (Bannasch did not test BP). However, Pereira found that foci were not induced if the PH was not performed at or near the time of carcinogen administration.

In the assays of Ito *et al.*, BP was found to be positive. Therefore, it appears that BP has promoting activity. BP was negative in Hunt's assay.

PYR

Glauert and Pereira both found PYR to be negative for initiating activity (Bannasch did not test it); Hunt and Ito both found PYR to be negative in their assays. Therefore, it appears that PYR does not have initiating or promoting activities.

2AAF

In the assays of Glauert and Pereira, 2AAF was found to be positive; in the assay of Bannasch, it was found to be negative. Pereira noted that the performance of a PH was necessary to induce foci in his experiments. (All of Glauert's rats received PH, none of Bannasch's did.) It is possible that Bannasch did not see a positive result because he did not perform a PH or treat his rats with phenobarbital during the latency period. It was concluded, therefore, that 2AAF has initiating activity.

In the assay of Ito *et al.*, 2AAF exerted promoting activity. In Hunt's assay, 2AAF was also positive.

4AAF

4AAF was found to be negative in the assays by Bannasch and Pereira, but positive in the assay by Glauert. It is possible that Glauert and Pereira had different results because of slight differences in their protocols: Glauert promoted with phenobarbital for 4 or 6 months whereas Pereira used a 70-day promotion period; Pereira withdrew his animals from phenobarbital 7 days before sacrifice whereas Glauert did not; Glauert used female rats whereas Pereira used male rats; and Glauert used outbred rats whereas Pereira used inbred Fischer rats. Two of these factors possibly make Glauert and Pitot's assay more sensitive: a longer promotion time (up to 3 or 4 months) increases the number of GGT-positive foci (Goldsworthy *et al.*, 1984); and the withdrawal of phenobarbital was found to lead to a decrease in the number of GGT-positive foci within 10 days by Glauert *et al.* (1986), although Herren and Pereira (1983) did not observe a decrease. Therefore, it was concluded that 4AAF appears to have initiating activity.

4AAF was found to be positive in the assay of Ito *et al.* therefore, it appears to have promoting activity. In Hunt's assay, 4AAF was negative.

8 Conclusions

In summary, 2AAF, 4AAF, and BP were found to have both initiating and promoting activities. PYR was not found to have either initiating or promoting activities.

9 Acknowledgement

The authors wish to thank Dr. Hiroyuki Tsuda for his participation in the writing of this manuscript.

10 References

Bannasch, P., D. Mayer and H.J. Hacker (1980) Hepatocellular glycogenosis and hepatocarcinogenesis, *Biochim. Biophys. Acta.*, **605**, 217–245.

Bannasch, P., D. Mayer and R. Krech (1979) Neoplastische und praeneoplastische Veranderungen bei Ratten nach einmaliger oraler Applikation von N-Nitrosomorpholin, *J. Cancer Res. Clin. Oncol.*, **94**, 233–248.

Cayama, E., H. Tsuda, D.S.R. Sarma and E. Farber (1978) Initiation of chemical carcinogenesis requires cell proliferation, *Nature* (London), **275**, 60–62.

Craddock. V.M. (1978) Cell proliferation and induction of liver cancer. In (H. Remmer, H.M. Bolt, P. Bannasch and H. Popper, Eds), *Primary Liver Tumors*, MTP Press, Lancaster, U.K., pp. 377–384.

Emmelot, P. and E. Scherer (1980) The first relevant cell stage in rat liver carcinogenesis. A quantitative approach, *Biochem. Biophys. Acta.*, **605**, 247–304.

Farber, E. (1980) The sequential analysis of liver cancer induction, *Biochem. Biophys. Acta.*, **605**, 149–166.

Glauert, H.P., M. Schwarz and H.C. Pitot (1986) The phenotypic stability of altered hepatic foci: effect of the short-term withdrawal of phenobarbital and of the long-term feeding of purified diets after the withdrawal of phenobarbital. *Carcinogenesis*, **7**, 117–121.

Goldsworthy, T.L. and H.C. Pitot (1985) An approach to the development of a short-term whole-animal bioassay to distinguish initiating agents (incomplete carcinogens), promoting agents, complete carcinogens, and non-carcinogens in rat liver, *J. Toxicol. Environ. Health*, **16**, 389–402.

Goldsworthy, T., H.A. Campbell and H.C. Pitot (1984) The natural history and dose-response characteristics of enzyme-altered foci in rat liver following phenobarbital and diethylnitrosamine administration, *Carcinogenesis*, **5**, 67–71.

Herren, S.L. and M.A. Pereira (1983) Tumor promotion in rat liver, *Environ. Health Perspect.*, **50**, 123–129.

Herren, S.L., M.A. Pereira, A.L. Britt and M.K. Khoury (1982) Initiation/promotion assay for chemical carcinogens in rat liver. *Toxicol. Lett.*, **12**, 143–150.

Hunt, J.M., M.T. Buckley, B.A. Laishes and H.A. Dunsford (1985) Immunological approaches to the purification of putative premalignant hepatocytes from genotypic mosaic rat livers, *Cancer Res.*, **45**, 2226–2233.

Hunt, J.M., M.T. Buckley, P.A. Onnink, P.B. Rolfe and B.A. Laishes (1982) Liver cell membrane alloantigens as cellular markers in genotypic mosaic rat livers undergoing chemically induced hepatocarcinogenesis, *Cancer Res.*, **42**, 227–236.

Ito, N., M.A. Moore and P. Bannasch (1984) Modification of the development of N-nitrosomorpholine-induced hepatic lesions by 2-acetylaminofluorene, phenobarbital and 4, 4′- diaminodiphenylmethane: a sequential histological and histochemical analysis, *Carcinogenesis*, **5**, 335–342.

Ito, N., H. Tsuda, R. Hasegawa and K. Imaida (1982) Sequential observation of pathomorphologic alterations in preneoplastic lesions during the promoting stage of hepatocarcinogenesis and the development of short-term test system for hepatopromoters and hepatocarcinogens. *Toxicol. Pathol.*, **10**, 37–49.

Ito, N., H. Tsuda, R. Hasegawa and K. Imaida (1983) Comparison of the promoting effects of various agents in induction of preneoplastic lesions in rat liver. *Environ. Health Perspect.*, **50**, 131–138.

Kunz, H.W., H.A. Tennekes, R.E. Port, M. Schwarz, D. Lorke and G. Schaude (1983) Quantitative aspects of chemical carcinogenesis and tumor promotion in liver, *Environ. Health Prespect.*, **50**, 113–122.

Laishes, B.A. and E. Farber (1978) Transfer of viable putative preneoplastic hepatocytes to the livers of syngeneic host rats, *J. Natl. Cancer Inst.*, **61**, 507–512.

Mirsalis, J.C., C.K. Tyson and B.E. Butterworth (1982) Detection of genotoxic carcinogens in the *in vivo–in vitro* hepatocyte DNA repair assay, *Environ. Mutag.*, **4**, 553–562.

Pereira, M.A. and G.D. Stoner (1985) Comparison of rat liver foci assay and strain A mouse lung tumor assay to detect carcinogens: a review, *Fund. Applied Toxicol.*, **5**, 688–699.

Pereira, M.A., S.L. Herren, A.L. Britt and M.K. Khoury (1982) Initiation/promotion bioassay in rat liver: use of gammaglutamyltranspeptidase-positive foci to indicate carcinogenic activity, *Toxicol. Pathol.*, **10**, 11–18.

Pereira, M.A., S.L. Herren-Freund, A.L. Britt and M.K. Khoury (1983) Effect of strain, sex, route of administration, and partial hepatectomy on the induction by chemical carcinogens of gamma-glutamyltranspeptidase foci in rat liver, *Cancer Lett.*, **20**, 207–214.

Pereira, M.A., S.L. Herren-Freund, A.L. Britt and M.K. Khoury (1984) Effect of coadministration of phenobarbital sodium on N-nitrosodiethylamine-induced gamma-glutamyltranspeptidase-positive foci and hepatocellular carcinoma in rats, *J. Natl. Cancer Inst.*, **72**, 741–744.

Pitot, H.C. and A.E. Sirica (1980) The stages of initiation and promotion in hepatocarcinogenesis, *Biochim. Biophys. Acta.*, **605**, 191–215.

Pitot, H.C., L. Barsness, T. Goldsworthy and T. Kitagawa (1978) Biochemical characterization of stages of hepatocarcinogenesis after a single dose of diethylnitrosamine, *Nature* (London), **271**, 456–458.

Pitot, H.C., H.P. Glauert and M. Hanigan (1985) The significance of biochemical markers in the characterization of putative initiated cell populations in rodent liver, *Cancer Lett.*, **29**, 1–14.

Rao, M.S., N.D. Lalwani, D.G. Scarpelli and J.K. Reddy (1982) The absence of gamma-glutamyl transpeptidase activity in putative preneoplastic lesions and in hepatocellular carcinomas induced in rats by the hypolipidemic peroxisome proliferator Wy-14, 643, *Carcinogenesis*, **3**, 1231–1233.

Shirai, T., K. Imaida, M. Ohshima, S. Fukushima, M. Lee, C.M. King and N. Ito (1985) Different responses to phenobarbital promotion in the development of gamma-glutamyltranspeptidase-positive foci in the liver of rats initiated with diethylnitrosamine, N-hydroxy-2-acetylaminofluorene and aflatoxin B_1, *Gann*, **76**, 16–19.

Solt, D.B., A. Medline and E. Farber (1977) Rapid emergence of carcinogen-induced hyperplastic lesions in a new model for the sequential analysis of liver carcinogenesis, *Am. J. Pathol.*, **88**, 595–618.

Stewart, B.W. (1981) Generation and persistance of carcinogen-induced repair intermediates in rat liver DNA *in vivo*, *Cancer Res.*, **41**, 3238–3243.

Tsuda, H., R. Hasegawa, K. Imaida, T. Masui, M.A. Moore and N. Ito (1984a) Modifying potential of 31 chemicals on the short-term development of gamma-glutamyl transpeptidase-positive foci in diethylnitrosamine-initiated rat liver, *Gann*, **75**, 876–883.

Tsuda, H., G. Lee and E. Farber (1980) Induction of resistant hepatocytes as a new principle for a short-term *in vivo* test for carcinogens, *Cancer Res.*, **40**, 1157–1164.

Tsuda, H., T. Masui, K. Imaida, S. Fukushima and N. Ito (1984b) Promotive effect of primary and secondary bile acids on the induction of gamma-glutamyl transpeptidase-positive liver cell foci as a possible endogenous factor for hepatocarcinogenesis in rats, *Gann*, **75**, 871–875.

Williams, G.M. (1980) The pathogenesis of rat liver cancer caused by chemical carcinogens, *Biochem. Biophys. Acta.*, **605**, 167–189.

Peter Bannasch, Hiroo Katayama[1], Heide Zerban and Hans Jörg Hacker
Institut für Experimentelle Pathologie, Deutsches Krebsforschungszentrum Heidelberg, W. Germany

50

Early cytomorphological and cytochemical changes induced in liver by single large doses of 2-acetylaminofluorene and 4-acetylaminofluorene

1 Summary

Groups of 20 male Fischer 344 rats were treated with a single high dose (200 or 500 mg/kg bw) of either 2AAF or 4AAF. Liver tissue was taken from 4 animals of each group at 2 h, 12 h, 24 h, 14 days and 8 weeks and investigated by histological and enzyme histochemical methods. Only unspecific toxic liver lesions such as zonal loss of glycogen, disorganization of the rough endoplasmic reticulum, and zonal decrease or increase in the activity of some enzymes were observed. Foci of altered hepatocytes as described earlier after administration of many hepatocarcinogens were not found under the experimental conditions chosen. Although it is conceivable that such focal lesions indicate a carcinogenic potential, the negative result of the present experiments does not allow any final conclusion as to whether the compounds tested are carcinogenic or not.

2 Introduction

Persistent foci of altered hepatocytes that show characteristic changes in their morphology and biochemistry have been observed prior to the development of hepatic tumours in various experimental models, especially in the rat liver treated with chemical carcinogens (Bannasch *et al.*, 1982). On the basis of conventional light and electron microscopical investigations, the foci were classified as clear or acidophilic cell foci storing glycogen in excess; mixed cell foci; and basophilic foci, poor in glycogen but rich in ribosomes (Bannasch, 1968; Daoust and Calamai, 1971; Squire and Levitt, 1975; Stewart *et al.*, 1980). The persistent foci have been shown to represent distinct stages in the process of hepatocarcinogenesis (Bannasch, 1968). Comparative enzyme histochemical studies revealed that the activity of many enzymes, including key enzymes of the carbohydrate metabolism, is decreased or increased in the focal lesions (Bannasch *et al.*, 1980; Emmelot and Scherer, 1980; Farber, 1980; Pitot and Sirica, 1980; Williams, 1980). As a rule, the foci were induced by long-term administration of hepatocarcinogens. Some potent carcinogens produced the foci even when they were applied in single large doses with (Scherer *et al.*, 1972) or without (Bannasch *et al.*, 1979) prior partial hepatectomy. However, the foci developed only after lag periods of some weeks or months in these cases. Recently,

[1]Visiting Research Scientist from Saitama Medical School, Saitama, Japan

a hitherto not clearly defined type of focal liver lesion has been described as "tigroid cell focus" in rats treated with a single dose (5 mg/kg bw) of aflatoxin B_1 (Bannasch *et al.*, 1984). The tigroid cell foci started to develop 8 weeks after administration of the carcinogen and progressed to expansively growing nodules but not to hepatocarcinomas within 104 weeks under the conditions of the experiment.

The observations outlined prompted us to investigate whether or not the induction of persistent focal liver lesions might serve as an indicator of the carcinogenic potential of chemicals in a short-term experiment without any additional manipulation such as partial hepatectomy or administration of a so-called "promoter". The chemicals studied were the known hepatocarcinogen 2-acetylaminofluorene (2AAF) and its reported noncarcinogenic analogue 4-acetylaminofluorene (4AAF). Comparative electron microscopical investigations of the early changes induced by these compounds in rat liver have been published by Flaks (Flaks, 1972; Flaks and Basley, 1979). However, no enzyme histochemical methods were used in these experiments.

3 Materials and methods

Chemicals

The chemicals tested, 2AAF and 4AAF, were provided by the Pharmaceutical Division of Imperial Chemical Industries Ltd. (Macclesfield, UK).

Animal experiments

We used a total of 100 male Fischer 344 rats (purchased from Charles River) that were about 8 weeks old and had a weight of 230 ± 20 g at the beginning of the experiment. The animals were kept under constant conditions (room temperature, 22 °C; humidity, 55%; light:dark cycle, 12 h: 12 h), and had free access to the commercial laboratory chow Altromin® (Altromin Spezialfutterfabrik, Lage, FRG) and to tap water. The rats were divided into 5 groups of 20 animals each. Four groups of 20 animals received either 2AAF or 4AAF at 200 or 500 mg/kg bw, respectively, as a single dose by gavage. Both the 2AAF and the 4AAF were suspended in sesame oil at 50 mg/ml for the low dose and at 100 mg/ml for the high dose experiment. The 20 control animals received sesame oil alone by gavage. Four animals of each group including controls were sacrificed at 2 h, 12 h, 24 h, 14 days, and 8 weeks after administration of the chemicals. The liver was removed under light ether anesthesia and further processed for conventional light microscopy and cytochemistry.

Light microscopy

Two adjacent slices of the left median and right liver lobe were rapidly cut with a razor blade. One slice was fixed in Carnoy's fluid and automatically processed for paraffin sections. The other portion was immediately frozen in isopentane at −150 °C and stored at −80 °C in air tight vessels until used.

Serial paraffin sections 2 μm thick were stained with hematoxylin and eosin (H & E) or the periodic acid Schiff (PAS) reaction using orange G as a counterstain. The frozen tissue was handled as described by Hacker *et al.* (1982) and stained with toluidine blue, methyl green-pyronine, or the PAS reaction using toluidine blue as a counterstain.

Enzyme histochemistry

Correlative histochemical investigations in serial sections (7 to 14 μm) of frozen liver tissue were performed as described (Hacker *et al.*, 1982). The activity of the following enzymes was investigated: glucose-6-phosphatase (G6Pase; after Benner *et al.*, 1979), glycogen phosphorylase (PHO; after Hacker, 1978), glucose-6-phosphate dehydrogenase (G6PDH; after Meijer and de Vries, 1974), adenosine triphosphatase (ATPase; after Wachstein and Meisel, 1957), and γ-glutamyltranspeptidase (γ-GT; after Lojda *et al.*, 1976).

4 Results

2AAF

Clearly demarcated hepatic foci composed of clear, acidophilic, basophilic, or mixed cell populations as described after many hepatocarcinogens were not found in any group. At 2, 12, and 24 hours, nearly all animals treated with 2AAF showed a zonal loss of hepatocellular glycogen stores, predominantly in zones 1 and 3 of the functional liver acinus as defined by Rappaport (1963). A similar but much less pronounced change was also seen in a few control animals. In the liver of the treated rats, the loss of glycogen usually was accompanied by a "chromatolysis," a disaggregation of the basophilic (or pyroninophilic) bodies which correspond to the rough endoplasmic reticulum, resulting in a finely granular or diffuse cytoplasmic basophilia (or pyroninophilia). Both the zonal loss of glycogen and the chromatolysis did not show considerable differences in its intensity in the 2 dose groups but were most pronounced at 12 and 24 h. At the latter time point, single necrotic hepatocytes were sometimes observed within the zone of glycogen loss and chromatolysis. However, neither a zonal loss of glycogen nor a chromatolysis or necrotic cells were observed 14 days and 8 weeks after administration of 2-AAF.

Whereas the glycogen content had clearly decreased in the zones described, the liver parenchyma of animals killed at 24 h or later was frequently exceptionally rich in glycogen in other areas, especially in zone 2 of the functional acinus. Great variations in the animals of all groups, including controls, were evident, and a clearcut dose-dependence could not be established for this phenomenon. An increase in acidophilic cytoplasmic components (most probably corresponding to a proliferation of the smooth endoplasmic reticulum) was observed in acinus central areas of the liver lobule in two animals of the low dose group at 12 h and in all animals in both groups at 24 h. This cellular alteration was never found in the controls and also not in the animals treated with 2AAF at 2 h, 14 days, or 8 weeks.

Clearcut enzyme-altered foci were not observed at any time. The zonal loss of glycogen was usually associated with a loss in the activity of the ATPase, with variable alterations (decrease or increase) in G6PDH activity and sometimes also with a slight increase in the activity of the G6Pase. All of these changes had disappeared at 8 weeks. The only finding that persisted from 12 h up to 8 weeks was a very strong activity of PHO in the areas of the liver parenchyma, which were exceptionally rich in glycogen. Of particular interest is that the animals treated with 2AAF showed an increased number of γ-GT-positive hepatocytes in zone 1 of the liver acinus at 8 weeks.

4AAF

With respect to the changes in the glycogen content and the concomitant alterations in the activity of enzymes, 4AAF had virtually the same effect on the liver parenchyma as did 2AAF. However, one remarkable difference was that γ-GT-positive hepatocytes were not encountered after treatment with 4AAF at any time. Another difference is that 4AAF only very rarely produced necrosis of single hepatocytes and led to an increase in acidophilic cytoplasmic components of acinuscentral hepatocytes exclusively in animals of the high dose group at 24 h.

5 Discussion

The results reported show that a single high dose of either 2AAF or 4AAF leads to a number of cellular changes in the liver parenchyma. Most of these changes are similar after treatment with the 2 compounds and have been known for a long time as unspecific-toxic liver lesions. Thus, a zonal loss of glycogen and a chromatolysis (which has been shown to be the light microscopic equivalent of a disorganization of the rough endoplasmic reticulum) may appear as a consequence of many liver intoxications and prolonged starvation (Bannasch, 1968; Svoboda and Higginson, 1968). In stop experiments with N-nitrosomorpholine, we established many years ago that these alterations frequently appear early under the influence of carcinogens but are reversible after withdrawal of the respective compounds (Bannasch, 1968; Bannasch *et al.*, 1972). In investigations with 2AAF, 4AAF, and other chemicals, Flaks has repeatedly postulated that the disorganization of the rough endoplasmic reticulum is a persistent lesion that might be specific for hepatocarcinogens (Flaks, 1972; Flaks and Basley, 1979; Flaks *et al.*, 1982). However, the present findings prove once again that these changes do not persist after withdrawal of the 2AAF or 4AAF. The fact that Flaks (1972) did not observe a disorganization of the rough endoplasmic reticulum in ultrastructural investigations of the 4AAF-treated rat liver might be explained by the zonal appearance of these alterations, which may be missed when the samples for electron microscopy are taken at random.

The transient increase in acidophilic cytoplasmic material occurring predominantly in acinuscentral hepatocytes after treatment with both 2AAF and 4AAF is most probably due to a proliferation of the smooth endoplasmic reticulum as demonstrated earlier by electron microscopy (Flaks, 1972; Flaks and Basley, 1979). Flaks suggested that this change might be related to the induction of drug-metabolising enzymes by the 2 compounds. At all events, this proliferation of the smooth endoplasmic reticulum appears not to be comparable to the persistent changes that are characteristic for the acidophilic cells of the putative focal preneoplastic liver lesions.

Unequivocal focal liver lesions which might be considered as preneoplastic changes did not develop under the experimental conditions chosen. This result certainly does not rule out the possibility that such lesions might appear after administration of higher single doses of 2AAF or 4AAF or after longer lag periods. Using N-nitrosomorpholine as a carcinogen, we observed in rat liver persistent foci composed of clear or acidophilic glycogen storage cells as early as 4 weeks after application of a very high single dose (320 mg/kg bw), which corresponded to the LD_{50} of this compound (Bannasch *et al.*, 1979). After a lag period beyond 55 weeks 4/11 animals developed neoplastic hepatic nodules and 1/11 animals a hepatocellular carcinoma. The few cells positive for γ-GT seen 8 weeks after application of 2AAF might reflect incipient development of focal preneoplastic lesions. It is evident from our observations, that it is difficult to establish a reliable *in vivo* short-term test for the evaluation of the carcinogenic potential of chemicals using the foci of altered hepatocytes as an indicator. If persistent foci develop after single or repeated doses of chemicals, it is conceivable that these compounds have a carcinogenic potential. However, the negative result does not allow any final conclusion regarding the carcinogenicity of the substances tested.

6 Acknowledgements

The authors would like to express their appreciation to D. Greulich, G. Schmidt, M. Frank and E. Schneider for excellent technical assistance and to A. Lampe for typing the manuscript.

7 References

Bannasch, P. (1968) The cytoplasm of hepatocytes during carcinogenesis. *Rec. Res. Cancer Res.*, **19**, 1–100.

Bannasch, P., J. Papenburg and W. Ross (1972) Cytomorphologische und morphometrische Studien der Hepatocarcinogenese. I. Reversible und irreversible Veränderungen am Cytoplasma der Leberparenchymzellen bei Nitrosomorpholin-vergifteten Ratten. *Z. Krebsforsch.*, **77**, 108–133.

Bannasch, P., D. Mayer and R. Krech (1979) Neoplastische und präneoplastische Veränderungen bei Ratten nach einmaliger oraler Applikation von N-Nitrosomorpholin. *J. Cancer Res. Clin. Oncol.*, **94**, 233–248.

Bannasch, P., D. Mayer and H.J. Hacker (1980) Hepatocellular glycogenesis and hepatocarcinogenesis *Biochim. Biophys. Acta*, **605**, 217–245.

Bannasch, P., M.A. Moore, F. Klimek and H. Zerban (1982) Biological markers of preneoplastic foci and neoplastic nodules in rodent liver. *Toxicologic Pathology*, **10**, 19–34.

Bannasch, P., U. Benner and H.J. Hacker (1984) Tigroid cell foci and their possible relation to hepatic tumors. In C.F.A. van Bezooijen (Ed.), *Pharmacological, Morphological and Physiological Aspects of Liver Aging*, EURAGE Publ., Rijswijk, pp. 195–204.

Benner, U., H.J. Hacker and P. Bannnasch (1979) Electron microscopical demonstration of glucose-6-phosphatase in native cryostat sections fixed with glutaraldehyde through semipermeable membranes. *Histochemistry*, **65**, 41–47.

Daoust, R. and R. Calamai (1971) Hyperbasophilic foci as sites of neoplastic transformation in hepatic parenchyma. *Cancer Res.*, **31**, 1290–1296.

Emmelot, P., and Scherer, E. (1980) The first relevant cell stage in rat liver carcinogenesis. A quantitative approach. *Biochim. Biophys. Acta*, **605**, 247–304.

Farber, E. (1980) The sequential analysis of liver cancer induction. *Biochim. Biophys. Acta*, **605**, 149–166.

Flaks, B. (1972) Early changes in the fine structure of rat hepatocytes, induced by the non-carcinogen 4-acetyl-aminofluorene. *Chem.-Biol. Interact.*, **5**, 127–137.

Flaks, B. and W.A. Basley (1979) Acute fine structural changes in rat hepatocytes induced by a single large dose of 2-acetylaminofluorene. *Virchows Arch. B. Cell Pathol.*, **29**, 309–320.

Flaks, B., Y. Yost and A. Flaks (1982) Effects of prolonged oral treatment with N-fluoren-4-ylacetohydroxamic acid on rat liver fine structure. *Carcinogenesis*, **3**, 103–109.

Hacker, H.J. (1978) Histochemical demonstration of glycogen phosphorylase (EC 2.4.1.1) through the use of semipermeable membranes. *Histochemistry*, **58**, 289–296.

Hacker, H.J., M.A. Moore, D. Mayer and P. Bannasch (1982) Correlative histochemistry of some enzymes of carbohydrate metabolism in preneoplastic and neoplastic lesions in the rat liver. *Carcinogenesis*, **3**, 1265–1272.

Lojda, Z., R. Gossrau and T.H. Schiebler (1976) Enzyme histochemical methods. Springer Verlag, Berlin-Heidelberg-N.Y.

Meijer, A.E.F.H. and G.P. de Vries (1974) Semipermeable membranes for improving the histochemical demonstration of enzyme activities in tissue sections. IV. Glucose-6-phosphate dehydrogenase and 6-phosphogluconate dehydrogenase (decarboxylating). *Histochemistry*, **40**, 349–359.

Pitot, H.C. and A.E. Sirica (1980) The stages of initiation and promotion in hepatocarcinogenesis. *Biochim. Biophys. Acta*, **605**, 191–215.

Rappaport, A.M. (1963) Acinar units and the pathophysiology of the liver. In CH. Rouiller (Ed.), *The Liver*, **1**, Academic Press, New York-London, pp. 265–328.

Scherer, E., M. Hoffmann, P. Emmelot and Friedrich-Freska, H. (1972) Quantitative study on foci of altered liver cells induced in the rat by a single dose of diethylnitrosamine and partial hepatectomy. *J. Natl. Cancer Inst.*, **49**, 93–106.

Squire, R.A. and M.H. Levitt (1975) Report of a workshop on classification of specific hepatocellular lesions in rats. *Cancer Res.*, **35**, 3214–3223.

Stewart, H.L., G. Williams, C.H. Keysser, L.S. Lombard and R.J. Montali, (1980) Histologic typing of liver tumors of the rat. *J. Natl. Cancer Inst.*, **65**, 179–206.

Svoboda, D. and J. Higginson (1968) A comparison of ultrastructural changes in rat liver due to chemical carcinogens, *Cancer Res.*, **28**, 1703–1733.

Wachstein, M. and E. Meisel (1951) Histochemistry of hepatic phosphatase at a physiological pH with special reference to the demonstration of bile canaliculi. *Am. J. Clin. Pathol.*, **27**, 13–23.

Williams, G.M. (1980) The pathogenesis of rat liver cancer caused by chemical carcinogens. *Biochim. Biophys. Acta*, **605**, 167–189.

Michael A. Pereira and Sydna L. Herren-Freund

U.S. Environmental Protection Agency Health Effects Research Laboratory 26 W. St. Clair Street Cincinnati, Ohio 45268

51

Liver initiation assay: the rat liver foci bioassay of carcinogens and noncarcinogens[1]

1 Summary

2-Acetylaminofluorene, 4-acetylaminofluorene, benzo(a)pyrene, and pyrene were evaluated in the rat liver foci bioassay for their ability to initiate gamma-glutamyltranspeptidase-positive foci. Under the conditions used, 2-acetylaminofluorene and benzo(a)pyrene were active in the assay, while 4-acetylaminofluorene and pyrene were not active.

2 Introduction

This report contains the results of our participation in the International Collaborative Study to evaluate why some genotoxic chemicals were inactive in *in vivo* rodent carcinogenesis bioassays. 4-Acetylaminofluorene (4AAF) and pyrene (PYR), two genotoxic chemicals that have been inactive in *in vivo* rodent carcinogenesis bioassays, and their genotoxic and carcinogenic analogues, 2-acetylaminofluorene (2AAF) and benzo(a)pyrene (BP), were evaluated in the rat liver foci bioassay for tumour initiating activity.

The rat liver foci assay has been proposed for use to demonstrate the carcinogenic potential of chemicals, especially those determined to be genotoxic (Bull and Pereira, 1982; Weisburger and Williams, 1983). The substance under test is administered to rats that have received a partial hepatectomy so that the first peak of regenerative DNA synthesis occurs during the period of maximum binding of the test substance to cellular DNA. Thus, an attempt is made to maximize the extent of fixation in the daughter strands of the alteration crucial for the initiation of carcinogenesis. The fact that the time of maximum DNA binding for most test substances is not known has resulted in the need to evaluate the activity of a test substance by administering it at different times after the partial hepatectomy in this bioassay. After a period of 1–2 weeks for the liver to recuperate from the toxicity of the test substance, the animals received 500 ppm sodium phenobarbital in their drinking water to stimulate initiated cells to grow into clones of altered hepatocytes. These clones or foci of initiated cells can be recognized by alteration in their phenotype, such as an increase in gamma-glutamyltranspeptidase (GGT) activity (Pugh and Goldfarb, 1978; Pitot and Sirica, 1980), a decrease in adenosine triphosphate (Scherer and Emmelot, 1976), a decrease in glucose 6-phosphatase activity (Tatematsu *et al.* 1977) or a decrease in their ability to accumulate iron (Williams, 1980). It is believed that the presence of these altered-foci demonstrate the occurrence of initiation of carcinogenesis and can predict the eventual appearance of tumors (Farber, 1980; Pitot and Sirica, 1980; Pereira, 1985). Therefore, the ability of a substance to initiate the formation of altered-foci has become the basis of a short-term *in vivo* bioassay in rat liver for the determination of whether a genotoxic chemical is capable of initiating carcinogenesis.

3 Material and methods

Animals
Male Fischer-344 rats were purchased from Charles River Co. (Portage, MI). The animals received rodent laboratory chow (Ralston Purina Co., St. Louis, MO) and drinking water ad libitum.

Figure 1. Experimental protocols for the rat liver foci bioassay in which the chemicals under test were administered either 6 or 12 hours after a two-thirds partial hepatectomy.

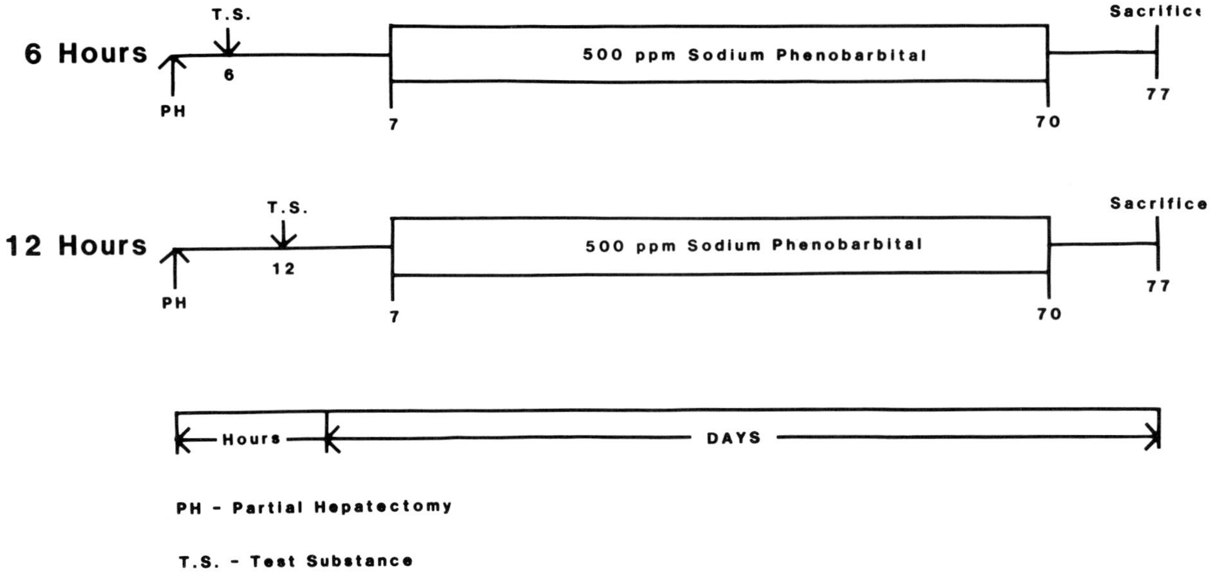

Table 1. *Rat liver foci bioassay*

		GGT-Foci/cm^3			
Chemical[a]		Partial hepatectomy		Sham operation	
mmole/kg		6 hours	12 hours	6 hours	12 hours
2-AAF					
	0.045	679 ± 221 (6)[b]	370 ± 90.5 (9)	ND	ND
	0.225	698 ± 210 (6)	1208 ± 246 (6)	ND	ND
	0.450	1030 ± 249 (7)	588 ± 196 (9)	ND	ND
	1.35	424 ± 233 (4)	1013 ± 288 (9)	80.6 ± 54.4 (10)	74.0 ± 74.0 (9)
4AAF					
	0.225	38.2 ± 29.3 (9)	20.9 ± 20.9 (8)	ND	ND
	1.35	21.7 ± 21.7 (4)	110 ± 110 (4)	0 ± 0 (10)	105 ± 75.6 (7)
BP					
	0.08	39.8 ± 39.8 (6)	196 ± 125 (4)	ND	ND
	0.40	254 ± 120 (7)	513 ± 322 (5)	ND	ND
	1.20	341 ± 134 (6)	538 ± 105 (5)	0 ± 0 (8)	22.8 ± 22.8 (6)
PYR					
	0.50	0 ± 0 (7)	90.4 ± 25.4 (5)	ND	ND
	1.50	0 ± 0 (8)	71.0 ± 71.0 (5)	0 ± 0 (8)	29.6 ± 20.8 (8)
Corn oil					
	2 ml/kg	189 ± 130 (5)	0 ± 0 (7)	78.5 ± 61.8 (6)	0 ± 0
DEN		12 hours	18 hours	12 hours	18 hours
	0.030	1470 ± 448 (6)	1120 ± 287 (6)	ND	ND
	0.060	1280 ± 646 (6)	1800 ± 348 (4)	751 ± 413 (6)	491 ± 285 (4)

[a]The chemicals were administered either 6 or 12 h after a 2/3 partial hepatectomy or a sham operation except for DEN, which was administered either 12 or 18 h after the operation.
[b]Results are means ± S.E. for the number of animals in parentheses.
[c]ND, not determined.

Materials

Diethylnitrosamine (DEN) was purchased from Eastman Kodak Co. (Rochester, NY); N-gamma-L-glutamyl-4-methoxy-2-naphthylamide from Bachem (Torrence, CA); and sodium phenobarbital from J.T. Baker (Philipsburg, NJ). 2AAF, 4AAF, BP and PYR were supplied by the International Programme on Chemical Safety/World Health Organization.

Experimental design

The protocol of the rat liver foci bioassay is presented in Figure 1. The rats, weighing 200–225 gm at the start of the study, received while under ether anesthesia either a 2/3 partial hepatectomy or a sham operation. The sham operation consisted of externalizing and then returning the liver to the peritoneal cavity. The test chemical was dissolved in corn oil on the day of treatment and administered orally by stomach tube either 6 or 12 h after the operation. Each animal received 2 ml/kg corn oil. Some animals received DEN which is used in our laboratory as a positive control to ensure that the sensitivity of the assay has not changed. Seven days after the operation, the animals started to receive 500 ppm sodium phenobarbital in their drinking water for the next 70 days. Seventy-seven days after the operation, the sodium phenobarbital was removed and the animals killed 7 days later by decapitation. Four tissue blocks, 2 each from the remaining 2 lobes were taken and quickly frozen on dry ice in Optimum Cutting Temperature Compound (Fisher Chemical Co. Pittsburgh, Pa). Sections 8 μ thick were obtained and stained for GGT activity by the procedure of Rutenburg *et al.* (1969) and counter-stained with haematoxylin.

Three sections totalling 2–4 cm^2 were evaluated from each animal for the presence of GGT-foci. The area of each focus and section were obtained by projection of the image onto a digitizing board (Elographics, Oak Ridge, TN), connected to a WANG Model 2200 VP computer (Lowell, MA). The number of foci/cm^3 was calculated as described by Campbell *et al.* (1982).

4 Results

The results of the rat liver foci bioassay are presented in Table 1. 2AAF, BP, and DEN were active in the bioassay, while 4AAF (1.35 m mole/kg) and PYR (1.50 m mole/kg) were inactive. The minimum tested dose of 2AAF (0.045 m mole/kg), BP (0.08 m mole/kg), and DEN (0.03 m mole/kg) were positive in the assay, so that the assay should be capable of detecting even lower doses of these chemicals.

The rat liver foci assay was more sensitive to BP administered 12 rather than 6 h after a partial hepatectomy. No consistent difference in the sensitivity of the assay was observed between 2AAF administered at either 6 or 12 h after partial hepatectomy or between DEN administered at either 12 or 18 h. 2AAF and BP were not active in the assay when administered to sham-operated animals, while DEN was active in sham-operated animals, although at a lower level of activity.

5 Discussion

Initiation of chemical carcinogenesis requires the simultaneous occurrence of DNA alteration/binding and DNA replication. The liver has a very low level of DNA replication and cellular proliferation, so that initiation in the liver requires the induction of DNA replication. The induction of DNA replication can result from the mitogenic action of a chemical or from the regenerative response to the toxicity of a chemical. In the rat liver foci bioassay, a 2/3 partial hepatectomy is performed, and the resulting regenerative growth of the liver is used to fix the alteration in the DNA into the daughter strands. Thus, initiation in the rat liver foci bioassay requires that the test substance binds/alters DNA in the crucial manner for initiation to occur, while no longer requiring that it stimulate DNA replication because of the regenerative response to the partial hepatectomy. This results in the rat liver foci bioassay being sensitive to chemicals such as BP which in intact animals are not carcinogenic toward the liver (Craddock, 1976; Tsuda *et al.*, 1980; Herren *et al.*, 1982).

The apparent requirement for a substance to be active in the rat liver foci bioassay is that it binds DNA in such a manner as to be genotoxic and, thereby, initiate carcinogenesis. In the International Program to Evaluate Short-Term Tests for Carcinogens, 4AAF and PYR demonstrated genotoxic activity in bacterial mutation and eukaryotic systems (Ashby, 1981). Hence, it appears that these 2 chemicals, which were not active in prior long-term rodent carcinogenesis bioassays, are capable of producing genotoxic alterations in DNA. 4AAF was less than 1/30 as active as 2AAF in the rat liver foci bioassay since 1.35 m mole/kg 4AAF was inactive while 0.045 m mole/kg 2AAF was active in the assay. PYR was less than 1/20 as active as BP in the assay.

The inactivity of 4AAF and PYR in our study could have resulted either from their being noncarcinogens or from their possessing less activity than their carcinogenic analogues. If the latter case in true, they may not have been tested in our study at a dose sufficient to demonstrate their activity in the rat liver foci bioassay. In the bacterial mutation assays for carcinogens, 2AAF was approximately 30–100 times more active than 4AAF (Bridges *et al.*, 1981). In the rat liver foci bioassay, 4AAF was not tested at a dose greater than 30–100 times the 2AAF dose. As a result, if 4AAF has the same relative activity as 2AAF in the rat liver foci bioassay as in the bacterial mutation assays, then 4AAF would have been tested in the liver foci bioassay at doses below the sensitivity of the assay. Thus, we are unable to conclude that 4AAF as well as PYR are noncarcinogens instead of weak carcinogens that possess activity below the sensitivity of the bioassay as performed in our study.

The potency of a chemical in the rat liver foci bioassay is dependent on the extent of crucial alterations produced in the DNA. It is expected that a relationship (possibly linear) should exist between the response in the rat liver foci bioassay and the amount of genotoxic adduct formed in DNA, which is crucial for initiation of carcinogenesis. It is recognized that the rate of repair of the genotoxic adduct and possibly other pharmacokinetic parameters would alter this relationship. It would be very productive to compare in rat liver the extent of DNA adduct formation for 4AAF and 2AAF and for PYR and BP. If equal potency in the rat liver foci bioassay is assumed for the adducts of the noncarcinogenic and of the carcinogenic analogue, one could estimate whether the extent of DNA alteration induced by the noncarcinogenic analogue was sufficient for activity in the assay. If sufficient DNA binding occurs with the noncarcinogenic analogue, then another property would have to be lacking, thereby, resulting, in its inability to initiate carcinogenesis even though it is genotoxic.

6 References

Ashby, J. (1981) Overview of Study and Test Chemical Activities. In F.J. de Serres and J. Ashby (Eds), Evaluation of Short-term Test for Carcinogens, Elsevier/North-Holland, New York, pp. 112–172.

Bridges, B.A., A. Bonin, B.J. Dean, R.C. Garner, D. Gatehouse, S. Hubbard, D. Ichinotsubo, F. Lorenzo, D. MacDonald, D. MacGregor, G. Martire, T. Matsushima, G. Mohn, M. Nagao, M. Richold, I. Rowland, V. Simmon, T. Skopek, R. Turman, S. Venitt and E. Zeiger (1981), Report on the Performance of Bacterial Mutation Assays. In F.J. de Serres and J. Ashby (Eds), *Evaluation of Short-Term Tests for Carcinogens*, Elsevier/North Holland, New York, pp. 49–67.

Bull, R.J. and M.A. Pereira (1982) Development of a short-term testing matrix for estimating relative carcinogenic risk, *J. Am. Coll. Toxicol.*, **1**, 1–15.

Campbell, H.A., H.C. Pitot, V.R. Potter and B.A. Laishes (1982) Application of quantitative stereology to the evaluation of enzyme-altered foci in rat liver, *Cancer Res.*, **42**, 465–472.

Craddock, V.M. (1976) Cell Proliferation and Experimental Liver Cancer. In M.M. Cameron, D.A. Linsell and G.P. Warwick (Eds), *Liver Cell Cancer*, Elsevier, Amsterdam, pp. 153–201.

Farber, E. (1980) The sequential analysis of liver cancer induction, *Biochim. Biophys. Acta*, **605**, 149–166.

Herren, S.L., M.A. Pereira, A.L. Britt, and M.M. Khoury (1982) Initiation/promotion assay for chemical carcinogens in rat liver, *Toxicol. Lett.*, **12**, 143–150.

Pereira, M.A. (1985) Rat liver foci assay. In H.A. Milman and E.K. Weisburger (Eds), *Handbook of Carcinogen Testing*, Noyes Publications, Park Ridge, pp. 152–178.

Pitot, H.C. and A.E. Sirica (1980) The stages of initiation and promotion in hepatocarcinogenesis, *Biochim. Biophys. Acta*, **605**, 191–215.

Pugh, T.O. and S. Goldfarb (1978) Quantitative histochemical and autoradiographic studies of hepatocarcinogenesis in rats fed 2-acetylaminofluorene followed by phenobarbital, *Cancer Res.*, **38**, 4450–4457.

Rutenburg, A.M., H. Kim, J.W. Fischbein, J.S. Hanker, H.L. Wasserkrug and A.M. Seligman (1969) Histochemical and ultrastructural demonstration of γ-glutamyl transpeptidase activity, *J. Histochem. Cytochem.*, **17**, 517–526.

Scherer, E. and P. Emmelot (1976) Kinetics of induction and growth of enzyme-deficient islands involved in hepatocarcinogenesis, *Cancer Res.*, **36**, 2544–2554.

Tatematsu, M., T. Shirai, H. Tsuda, Y. Miyata, Y. Shinohara and N. Ito (1977) Rapid production of hyperplastic liver nodules in rats treated with carcinogenic chemicals: A new approach for an *in vivo* short-term screening test for hepatocarcinogens, *Gann*, **68**, 499–507.

Tsuda, H., G. Lee and E. Farber (1980) Induction of resistant hepatocytes as a new principle for a possible short-term *in vivo* test for carcinogens. *Cancer Res.*, **40**, 1157–1164.

Weisburger, J.H. and G.M. Williams (1983) The distinct health risk analyses required for genotoxic carcinogens and promoting agents, *Environ. Health Perspect.*, **50**, 233–245.

Williams, G.M. (1980) The pathogenesis of rat liver cancer caused by chemical carcinogens, *Biochim. Biophys. Acta*, **605**, 167–189.

[1] Disclaimer: The research described in this article has been reviewed by the Health Effects Research Laboratory and approved for publication. Mention of trade names or commercial products does not constitute endorsement or recommendation for use.

Howard P. Glauert and Henry C. Pitot
McArdle Laboratory For Cancer Research, University of Wisconsin, Madison, Wisconsin, USA

52

Ability of the test chemicals to initiate gamma-glutamyltranspeptidase-positive foci in female rats

1 Summary

BP, PYR, 2AAF, and 4AAF were tested for their ability to initiate gamma-glutamyltranspeptidase (GGT)-positive foci in rat liver. The test chemicals were administered intragastrically to 200-gm female CD rats either immediately after (BP, PYR) or 12 h after (2AAF, 4AAF) partial hepatectomy. Rats received 0.05% phenobarbital in the diet as a promoter for 4 or 6 months. BP, 2AAF, and 4AAF significantly increased the number and volume of GGT-positive foci, whereas PYR did not affect either parameter. This study confirms the initiating ability of the complete carcinogens BP and 2AAF and the noncarcinogenicity of PYR. This study suggests that 4AAF may act as a pure initiator, since its carcinogenicity has never been demonstrated.

2 Introduction

The identification of potential carcinogens in the environment is critical for the potential prevention of cancer in the human population. The exposure of humans to new chemicals released into the environment requires that some mechanism exist for determining a chemical's carcinogenicity. Traditionally, the carcinogenicity of a chemical has been determined by administering that chemical to an experimental animal for the lifespan of the animal. This method, however, is expensive and time-consuming, allowing the testing of only a relatively few chemicals per year. Furthermore, these studies provide little information as to the mechanism by which a chemical may induce cancer.

For these reasons, short-term carcinogenicity tests have been developed to assess the carcinogenic potential of chemicals. These tests are quicker, less expensive, allow the testing of a large number of chemicals, and often provide information as to the mechanism of a chemical's carcinogenicity. Many of these assays use cultured cells from rodents or humans (i.e. *in vitro* tests). Several noncarcinogens, however, are positive in a wide variety of *in vitro* assays (de Serres and Ashby, 1981).

The ability of the liver to metabolize a wide variety of xenobiotics as well as the appearance of preneoplastic focal lesions during the carcinogenic process makes this tissue ideal for screening chemical carcinogens (Pitot and Sirica, 1980). Several model systems have been developed to study initiation and promotion in liver (Goldsworthy *et al.*, 1986). Therefore, it is possible to study not only the ability of chemicals to produce tumors or preneoplastic lesions but also to determine the stage of carcinogenesis in which an agent acts. In addition, one can detect chemicals that are incomplete carcinogens (pure initiators) or tumor promoters. Ordinarily, these agents would exhibit weakly positive or negative results in existing long-term animal bioassays for carcinogenicity.

The purpose of this study was to determine the ability of the carcinogens, benzo(a)pyrene (BP) and 2-acetylaminofluorene (2AAF), and their noncarcinogenic analogues, pyrene (PYR) and 4-acetylaminofluorene (4AAF), to initiate gamma-glutamyltranspeptidase (GGT)-positive foci in rat liver using the protocol of Pitot *et al.* (1978). PYR and 4-AAF, although never shown to be carcinogenic, are active in a number of *in vitro* assays (de Serres and Ashby, 1981; Althaus *et al.*, 1982; International Agency for Research in Cancer, 1983). The test chemicals were administered shortly after 70% partial hepatectomy (PH). Rats were then fed 0.05% phenobarbital (PB) in the diet as a promoter. We found that BP, 2AAF, and 4AAF initiated GGT-positive foci, whereas PYR did not.

3 Materials and methods

Chemicals

BP, PYR, 2AAF, and 4AAF were from Lancaster Synthesis Ltd., Morecambe, England; diethylnitrosamine (DEN) was obtained from Eastman Organic Chemicals, Rochester, NY; PB was from Sigma Chemical Co., St. Louis, MO.

Animals and diets

Female CD rats (180 to 200 g) were obtained from Charles River. Rats were housed in groups of 2 or 3 in plastic cages containing wood shavings for litter. Rats were given food and water ad libitum. Rats were fed the NIH-07 cereal-based diet (American Institute of Nutrition, 1977), which was obtained from either Zeigler Brother, Garners, PA, or Teklad Test Diets, Madison, WI. All rats were allowed to adjust for one week before an experiment was started.

4 Experimental protocol

The rats were subjected to a 70% partial hepatectomy (Higgins and Anderson, 1931). At various times afterward, the rats were given a single intragastric dose of one of the test chemicals: for BP and PYR, rats were intubated immediately after partial hepatectomy; for 2AAF and 4AAF, rats were intubated 12 h after partial hepatectomy; for DEN, rats were intubated 24 h after partial hepatectomy. Controls received an equivalent dose of corn oil 12 h after partial hepatectomy. The test chemicals were first dissolved in acetone (2AAF and 4AAF) or benzene (BP and PYR). Corn oil was added to the solution and the solvent was boiled off. DEN was dissolved in water and administered immediately thereafter. For control rats, acetone or benzene was added to corn oil and then boiled off.

One week after the intubations, rats were fed 0.05% PB in the diet for 4 or 6 months. Rats were then killed by decapitation and frozen liver sections were prepared as previously described (Goldsworthy *et al.*, 1984) and subsequently stored at −70 °C. Additional liver pieces were fixed in formalin and processed for histological analysis. Frozen liver sections were prepared and stained for GGT as described previously (Goldsworthy *et al.*, 1984). The number of GGT-positive foci per liver and the volume of liver occupied by the foci were determined as described previously (Campbell *et al.*, 1982).

Statistical analysis

Data for the treatment groups were compared with the negative control group data for the number of foci/cm^2, foci/cm^3, foci/liver, and focal volume as a per cent of liver volume. Because the data were not normally distributed and because treatment variances were heterogeneous, a nonparametric, one-sided multiple comparison test was used (Hollander and Wolfe, 1973). Foci diameters were analyzed using a nonparametric, two-sided multiple comparison test. A confidence level of 95% or greater was considered significant. For all 4 chemicals, the values at

Table 1. *Effect of the test chemicals on the initiation of GGT-positive foci in the first experiment*

Treatment (mg/kg)		Foci/cm^2	Foci/cm^3	Foci diameter (μm)	Focal volume (% of liver)	Foci/liver
Control	(21)[a]	0.6 ± 0.1[b]	23 ± 5	376 ± 78	0.09 ± 0.03	289 ± 64
4AAF						
10	(12)	0.7 ± 0.1	30 ± 6	272 ± 27	0.10 ± 0.04	432 ± 102
50	(11)	1.5 ± 0.5	61 ± 15	249 ± 30	0.30 ± 0.17	826 ± 248
2AAF						
10	(12)	2.7 ± 0.6[c]	125 ± 26[c]	292 ± 59	0.25 ± 0.07	1478 ± 287[c]
50	(11)	8.3 ± 2.9[c]	268 ± 82[c]	297 ± 28	1.22 ± 0.57[c]	3947 ± 1376[c]
PYR						
20	(14)	0.7 ± 0.1	27 ± 5	323 ± 43	0.10 ± 0.04	359 ± 66
100	(12)	0.9 ± 0.3	31 ± 9	341 ± 48	0.22 ± 0.17	410 ± 135
BP						
20	(7)	0.6 ± 0.1	15 ± 4	459 ± 74	0.23 ± 0.16	210 ± 53
100	(8)	6.6 ± 2.0[c]	306 ± 102[c]	254 ± 24	0.44 ± 0.13[c]	3859 ± 1356[c]
DEN						
10	(6)	18.8 ± 3.5[c]	740 ± 171[c]	289 ± 36	1.32 ± 0.46[c]	9388 ± 2149[c]

[a]Number of rats per group is in parentheses
[b]Values are means ± SEM
[c]Values significantly greater than control values ($p < 0.05$)

each dose and values for the negative control were subjected to linear regression for the focal volume and the number of foci per liver (Gill, 1978). To check for dose response, we tested the slope of the derived line to determine if it was greater than zero. A confidence level of 95% or greater was considered to be a significant indication of a positive response. For a chemical to be considered positive, the number of foci per liver and the focal volume had to be greater than the control for at least one dose *and* a significant dose response had to exist for these two parameters. Differences in the incidence of neoplastic nodules were analyzed using X^2 statistics (Gill, 1978). Differences in rat weights and rat liver weights were analyzed using one-way analysis of variance (Gill, 1978).

5 Results

The two-stage hepatocarcinogenesis system in rats (Pitot *et al.*, 1978, 1980; Goldsworthy *et al.*, 1984) was used to evaluate the initiating ability of the test chemicals. Rats received an intragastric dose of the test chemicals at various times after PH. The timing of the intubation was dependent on the metabolism of the chemical: the more fat-soluble chemicals take longer to induce DNA damage in the liver (Stewart, 1981; Mirsalis *et al.*, 1982) and were therefore intubated sooner after PH. All rats were fed 0.05% PB in the diet as a promoting agent.

In the first experiment, rats received 2 doses of each of the test chemicals: 2AAF and 4AAF were intubated at doses of 10 and 50 mg/kg, while BP and PYR were intubated at doses at 20 and 100 mg/kg. DEN, the positive control, was intubated at a dose of 10 mg/kg. All rats were fed PB for 6 months.

The effect of the test chemicals on the number and volume of GGT-positive foci is shown in Table 1. Both BP and 2AAF induced significant increases in the number of foci per liver and in the focal volume, both in a dose-related manner. We therefore classified BP and 2AAF as positive. PYR did not induce significant increases in the number of foci per liver or in the focal volume at either dose, nor was a dose response seen for either parameter. 4AAF did not induce significant increases in the number of foci per liver or in the focal volume. The highest dose, however, did increase the number of foci per liver at the 90% confidence level. In addition, the dose response for both the focal volume and number of foci per liver was significant.

The incidence of tumors is shown in Table 2. Neoplastic nodules were seen in one or more rats in all groups, but no hepatocarcinomas were seen. None of the treatments induced a significant increase in the incidence of neoplastic nodules.

Table 2. *Incidence of tumors in the first experiment*

Treatment[a] (mg/kg)	Rat with Neoplastic Nodules[b] (%)
Control	19 (4/21)
4AAF	
10	17 (2/12)
50	18 (2/11)
2AAF	
10	17 (2/12)
50	45 (5/11)
PYR	
20	7 (1/14)
100	8 (1/12)
BP	
20	43 (3/7)
100	13 (1/8)
DEN	
10	33 (2/6)

[a] All rats were fed 0.05% PB during the 6-month promotion period.
[b] The number of rats with neoplastic nodules over the total number of rats per treatment is in parentheses.

Because PYR and 4AAF did not induce significant increases in the number and volume of GGT-positive foci, we decided to retest these 2 chemicals at higher doses. In the second experiment, PYR was intubated at a dose of 500 mg/kg and 4AAF at 250 mg/kg. At this dose, 4AAF would not completely dissolve in corn oil and had to be intubated as a slurry. DEN was again used as a positive control at a dose of 10 mg/kg. Rats were fed PB for 4 months.

The results of the second experiment are shown in Table 3. The 500 mg/kg dose of PYR had no effect on the number of foci per liver or on the focal volume; we therefore classified it as negative. The 250 mg/kg dose of 4AAF induced significant increases in the number of foci per liver and in the focal volume. Furthermore, in comparing the two experiments, a dose response appears to exist: the 50 mg/kg dose in the first experiment induced a 3-fold increase in focal volume and number; the 250 mg/kg dose in the second experiment induced a 4- to 5-fold increase in the focal volume and number. Also, a

Table 3. *Effect of the test chemicals on the initiation of GGT-positive foci in the second experiment*

Treatment (mg/kg)		Foci/cm^2	Foci/cm^3	Foci diameter (μm)	Focal volume (% of liver)	Foci/liver
Control	(11)[a]	0.5 ± 0.1[b]	24 ± 5	240 ± 33	0.018 ± 0.005	284 ± 61
PYR						
500	(12)	0.5 ± 0.1	29 ± 6	188 ± 18	0.032 ± 0.016	334 ± 78
4AAF						
250	(10)	2.1 ± 0.5[c]	128 ± 30[c]	179 ± 15	0.071 ± 0.017[c]	1419 ± 342[c]
DEN						
10	(3)	24.7 ± 3.4[c]	1115 ± 136[c]	227 ± 12	1.597 ± 0.400[c]	11,740 ± 1882[c]

[a] Number of rats per groups is in parentheses.
[b] Values are means ± SEM.
[c] Values significantly greater than control values ($p < 0.05$).

significant dose response was seen in the first experiment. We therefore classified 4AAF as positive. Only two rats developed neoplastic nodules in the second experiment: one receiving 4AAF and one receiving DEN.

In both experiments, the treatments did not affect the rat weights or the rat liver weights (data not shown).

6 Discussion

In this study we examined the ability of the test chemicals to act as initiators in a two-stage hepatocarcinogenesis model in the rat using PB as the promoting agent (Pitot *et al.*, 1978; Goldsworthy *et al.*, 1984). The ability to induce GGT-positive foci was examined. When PB is used as the promoting agent, GGT is the most efficient marker, since 80–90% of PB-promoted foci exhibit this marker (Goldsworthy and Pitot, 1985; Peraino *et al.*, 1984).

We found that 3 of the 4 test compounds initiated GGT-positive foci: BP, 2AAF, and 4AAF. The method discriminates satisfactorily between the carcinogen, BP, and the noncarcinogen, PYR. At the highest dose tested (100 mg/kg, BP induced a 13-fold increase in the number of foci per liver and a 5-fold increase in the focal volume, whereas PYR did not alter either parameter at the highest dose tested (500 mg/kg). This study confirms the ability of BP to initiate liver foci (Kitagawa *et al.*, 1980; Peraino *et al.*, 1984). Although 2AAF and 4AAF both significantly increased the number and volume of GGT-positive foci, 2AAF was much more potent. At the 50 mg/kg dose, 2AAF increased both the number of foci per liver and the focal volume 14-fold; at the same dose, 4AAF increased both the number and volume of foci about 3-fold (which was not statistically significant). Therefore, 2AAF appears to be about 5 times more potent than 4AAF. At the highest dose tested (250 mg/kg), 4AAF increased the number of foci per liver 5-fold and the focal volume 4-fold. It is difficult to compare the response at this level with the response to 2AAF or 4AAF at 50 mg/kg; at the 250 mg/kg dose, 4AAF would not completely dissolve in corn oil and had to be intubated as a slurry. Therefore, it is highly questionable whether the dose of 4AAF was completely absorbed in the gut, making comparisons with lower doses difficult.

The positive result seen with 4AAF is of interest in view of 2 earlier studies examining its carcinogenicity after chronic administration. Weisburger *et al.* (1952) found that feeding 0.25% 4AAF in the diet did not induce tumors after 13 months; Schinz *et al.* (1955) found that after feeding 0.1% 4AAF in the diet for 116 days only one colon tumor was seen after 900 days. If 4AAF is a weak complete carcinogen, it is likely that it must be administered for a longer period of time than in these two studies. 4AAF might also be a pure initiator (incomplete carcinogen), in which case long-term treatment with a promoter would be necessary to induce tumors.

7 Conclusions

BP, 2AAF, and 4AAF are initiators of PB-promoted GGT-positive foci in female CD rats; PYR is not.

8 Acknowledgements

We thank Dr. Harold Campbell for his assistance in the use of the computer to quantitate foci; Chen-Ching Lai for verifying the purity of 2AAF and 4AAF; Susan Moran and Wendy Kennan for excellent technical assistance; Jane Weeks, Mary Folz-Erbs, Susan Carlson, and Patty Collard for preparing the histological slides; and Kristen Luick for typing the manuscript. This work was supported by DHHS PHS National Research Service Award 5 T32 ES07015 from the National Institute of Environmental Health Sciences (H.P.G. was an N.I.E.H.S. Postdoctoral Trainee); grants CA-07175 and CA-22484 from the National Cancer Institute; and contract N01-ES-3-5024 from the National Toxicology Program.

9 References

Althaus, F.R., S.D. Lawrence, G.L. Sattler, D.G. Longfellow and H.C. Pitot (1982). Chemical quantification of unscheduled DNA synthesis in cultured hepatocytes as an assay for the rapid screening of potential chemical carcinogens, *Cancer Res.*, **42**, 3010–3015.

American Institute of Nutrition (1977). Report of the American Institute of Nutrition ad hoc committee on standards for nutritional studies, *J. Nutr.*, **107**, 1340–1348.

Campbell, H.A., H.C. Pitot, V.R. Potter and B.A. Laishes (1982). Application of quantitative stereology to the evaluation of enzyme-altered foci in rat liver, *Cancer Res.*, **42**, 465–472.

de Serres, F.J. and J. Ashby (1981). *Evaluation of Short-Term Tests for Carcinogens: Report of the International Collaborative Program*, Elsevier/North Holland, New York.

Gill, J.L. (1978) *Design and Analysis of Experiments in the Animal and Medical Sciences*, Vol. I, Iowa State University Press, Ames, IA.

Goldsworthy, T.L. and H.C. Pitot (1985). The quantitative analysis and stability of histochemical markers of altered hepatic foci in rat liver following initiation by diethylnitrosamine administation and promotion with phenobarbital, *Carcinogenesis*, **6**, 1261–1269.

Goldsworthy, T., H.A. Campbell and H.C. Pitot (1984). The natural history and dose–response characteristics of enzyme-altered foci in rat liver following phenobarbital and diethylnitrosamine administration, *Carcinogenesis*, **5**, 67–71.

Goldsworthy, T., M.H. Hanigan and H.C. Pitot (1986). Models of hepatocarcinogenesis in the rat: contrasts and comparisons, *CRC Critical Rev. Toxicol.*, **17**, 61–89.

Higgins, G.M. and R.M. Anderson (1931). Experimental pathology of the liver. I. Restoration of the liver of the white rat following partial surgical removal, *Arch. Pathol.*, **12**, 186–202.

Hollander, M. and D.A. Wolfe (1973). *Nonparametric Statistical Methods*, John Wiley and Sons, New York.

International Agency for Research in Cancer (1983). *Pyrene, IARC monographs on the Evaluation of the Carcinogenic Risk of Chemicals to Humans*, **32**, 431–445.

Kitagawa, T., T. Hirakawa, T. Ishikawa, N. Nemoto and S. Takayama (1980). Induction of hepatocellular carcinoma in rat liver by initial treatment with benzo(a)pyrene after partial hepatectomy and promotion by phenobarbital, *Toxicol. Lett.*, **6**, 167–171.

Mirsalis, J.C., C.K. Tyson and B.E. Butterworth (1982). Detection of genotoxic carcinogens in the *in vivo – in vitro* hepatocyte DNA repair assay, *Environ. Mutag.*, **4**, 553–562.

Peraino, C., E.R. Staffeldt, B.A. Carnes, V.A. Ludeman, J.A. Blomquist and S.D. Vesselinovitch (1984). Characterization of histochemically detectable altered hepatocyte foci and their relationship to hepatic tumorigenesis in rats treated once with

diethylnitrosamine or benzo(a)pyrene within one day after birth, *Cancer Res.*, **44**, 3340–3347.
Pitot, H.C., and A.E. Sirica (1980). The stages of initiation and promotion in hepatocarcinogenesis, *Biochim. Biophys. Acta*, **605**, 191–215.
Pitot, H.C., L. Barsness, T. Goldsworthy and T. Kitagawa (1978). Biochemical characterizations of stages of hepatocarcinogenesis after a single dose of diethylnitrosamine, *Nature*, **271**, 456–458.
Pitot, H.C., T. Goldsworthy, H. Campbell and A. Poland (1980). Quantitative evaluation of the promotion by 2, 3, 7, 8-tetrachloride-p-dioxin of hepatocarcinogenesis from diethylnitrosamine, *Cancer Res.*, **40**, 3616–3620.
Schinz, H.R., H. Fritz-Niggli, T.W. Campbell and H. Schmid (1955). Krebsbildung durch Aminofluorene und verwandte Korper, *Oncologia*, **8**, 233–245.
Stewart, B.W. (1981). Generation and persistance of carcinogen-induced repair intermediates in rat liver DNA *in vivo, Cancer Res.*, **41**, 3238–3243.
Weisburger, J.H., E.K. Weisburger and H.P. Morris (1952). Analogs of the carcinogen 2-acetylaminofluorene: the isomeric 4-acetylaminofluorene, *J. Am. Chem. Soc.*, **74**, 4540–5443.

Nobuyuki Ito, Hiroyuki Tsuda and Tsuneo Masui
First Department of Pathology, Nagoya City University Medical School, Japan

53

In vivo promotion assay of four compounds on enhancement of the development of γ-glutamyl transpeptidase-positive liver cell foci in male F344 rats

1 Summary

The promoting effect of 4 test compounds 2-acetylaminofluorene (2AAF), 4-acetylaminofluorene (4AAF), benzo(a)pyrene (BP) and pyrene (PYR) on liver carcinogenesis was tested in male F344 rats initially treated with diethylnitrosamine (DEN). Two weeks after a single dose of DEN (200 mg/kg, intraperitoneally), rats were given one of the test compounds for 6 weeks. At 3 weeks following administration of DEN, the rats were subjected to partial hepatectomy. Results were judged by comparing quantitative values, number and area (mm^2) per unit area (cm^2), of histochemical γ-glutamyl transpeptidase-positive (γ-GT^+) liver cell foci of experimental groups as opposed to the control groups given DEN alone. Of the compounds tested, 2AAF exerted an obvious promoting effect by enhancing a significantly increased number ($28.57 \pm 14.81/cm^2$) and area ($61.60 \pm 11.89\ mm^2/cm^2$) of γ-GT^+ foci over those in the control group ($9.65 \pm 3.55/cm^2$ and $0.53 \pm 0.20\ mm^2/cm^2$, respectively). 4AAF and BP also showed a promoting effect, by less potent, as assessed the significantly increased area (1.25 ± 0.72 and $2.19 \pm 1.03\ mm^2/cm^2$, respectively) of foci. PYR did not cause either increase in number or area of the foci showing negative in this promotion assay system.

2 Introduction

Since the application of the 2-stage concept of chemical carcinogenesis (Belenblum and Shubik, 1947), separate initiation and promotion steps have been evaluated in organs other than the skin, and an important role for promoting agents in the development of tumours has been indicated (Cohen *et al.*, 1979; Hicks *et al.*, 1975; Kitagawa and Sugano, 1978; Narisawa *et al.*, 1974; Peraino *et al.*, 1977; Witchi, 1981). Recent results indicate that short-term assay systems offer a useful approach for the detection of promoters of liver carcinogenesis (Cameron *et al.*, 1982; Ito *et al.*, 1980; Tatematsu *et al.*, 1983). These short-term assay systems are characterized by early detection of preneoplastic lesions in the liver such as hyperplastic nodules or γ-GT^+ foci. γ-Glutamyl transpeptidase has found approval as a marker enzyme for altered islands and is particularly useful, since it stains as positive in these preneoplastic lesions (Ogawa *et al.*, 1980; Solt *et al.*, 1983; Tamano *et al.*, 1981; Tsuda *et al.*, 1980).

3 Materials and methods

Animals

Male, 6-week-old, F344 rats (Charles River Japan, Inc., Atsugi, Japan) were housed in plastic cages in an air-conditioned room at 24 °C with a 12 hr-12 hr light-dark cycle. All animals were fasted for 18 hours prior to sacrifice.

Chemicals and diet

DEN was purchased from Tokyo Chemical Industry Co., Ltd., Tokyo, Japan; the 4 test compounds were supplied by IPCS. They were mixed into Oriental M powdered basal diet (Oriental Yeast Co., Tokyo, Japan) at a concentration of 0.02%.

4 Experimental protocol

A total of 225 rats were divided into 3 groups. Group 1 was given a single intraperitoneal (ip) injection of 200 mg/kg of DEN dissolved in 0.9% NaCl to initiate hepatocarcinogenesis (Solt and Farber, 1976; Solt *et al.*, 1983). After 2 weeks on basal diet, rats received one of the test compounds in their diet at a concentration of 0.02% for 6 weeks as shown in Figure 1. Animals were subjected to two-thirds partial hepatectomy at week 3 after injection of DEN. They were killed for examination at week 8. Group 2 was given DEN and partial hepatectomy at week 3 without feeding of test compound. Group 3 was injected with 0.9% NaCl instead of DEN solution; subjected to feeding of test compounds and partial hepatectomy similarly to Group 1 (Figure 1).

Figure 1

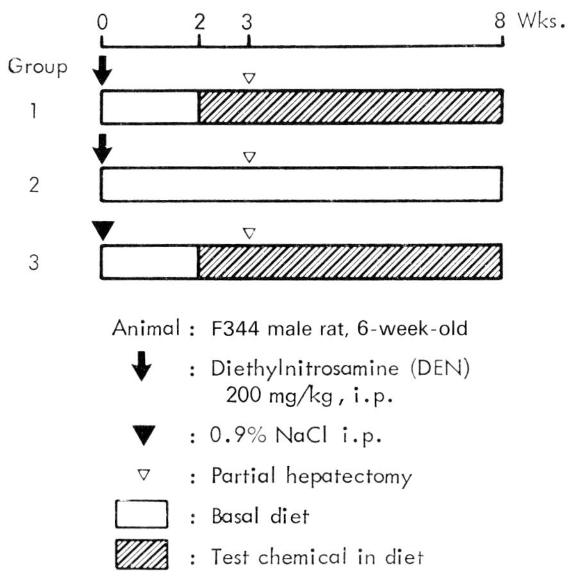

EXPERIMENTAL DESIGN

Body weight and food consumption

The body weight of each animal was recorded at the start of the experiment and then weekly throughout the study period. The quantity of food consumed by each cage of rats during the 2-day period was recorded and group mean daily consumption per rat was calculated. At weekly intervals, the intake of the test material was calculated from the group mean food intake and body weight data and dietary inclusion levels.

Histological examination and quantitation of induced lesions

At the time of sacrifice, the liver was excised and sections of 1- to 2-mm thickness were cut with a razor blade. Four slices, one each from the right posterior and caudate lobes and 2 from the right anterior lobes, were fixed in ice-cold acetone and processed for paraffin sectioning for subsequent histological examination of γ-GT$^+$ foci as described previously (Ogawa *et al.*, 1980; Tamano *et al.*, 1981; Tsuda *et al.*, 1980). The numbers and areas of γ-GT$^+$ foci were measured by a color video image processor (VIP-21C, Olympus-Ikegami Tsushin Co., Tokyo, Japan). The measurement was carried out without reference to the treatment group of the slide (blind method). Statistical analysis was performed with Student's t test.

Since most hyperplastic nodules show positive for γ-glutamyl transpeptidase activity, γ-GT$^+$ foci in this experiment included hyperplastic nodules.

5 Results

Body and liver weights (Table 1)

Animals given 2AAF in Group 1 had significantly de-

Table 1. *Body and liver weights and food and chemical intakes*

Group	Treatment	No.	Body weight (bw)	Liver weight	Liver/bw (%)	Intake/day/kg Food (g)	Intake/day/kg Test chem. (mg)
1	DEN→2AAF	19	230.37 ± 14.37***[a]	12.60 ± 2.11***	5.45 ± 0.73***	57.61	11.52
	DEN→4AAF	25	267.78 ± 21.17*	7.30 ± 0.89	2.72 ± 0.22**	58.81	11.76
	DEN→BP	24	273.20 ± 17.25	7.88 ± 1.10**	2.88 ± 0.26***	63.10	12.61
	DEN→PYR	19	276.92 ± 29.17	7.92 ± 1.28*	2.85 ± 0.23***	60.23	12.04
2	DEN (−)	23	279.90 ± 14.09	7.14 ± 0.56	2.55 ± 0.14	67.14	0
3	(−) 2AAF	17	234.75 ± 28.43	7.77 ± 1.29	3.30 ± 0.28	55.95	11.19
	4AAF	24	280.43 ± 14.33	7.96 ± 1.04	2.83 ± 0.27	60.95	12.19
	BP	24	279.33 ± 16.59	8.02 ± 1.33	2.86 ± 0.33	58.33	11.67
	PYR	26	271.13 ± 20.51	7.36 ± 1.40	2.70 ± 0.33	60.48	12.10

[a]Significantly different from Group 2 at $*P < 0.05$, $**P < 0.01$, $***P < 0.001$.

creased body weight compared with Group 2. Treatment with 2AAF without DEN also caused decrease of body weight (Group 3). Animals given 4AAF showed less pronouced but significant decrease of body weight.

Remarkable increase of liver weight was observed in animals given 2AAF after DEN due to induced hyperplastic nodules. Groups given BP or PYR after DEN showed a slight but statistically significant decrease of liver weight. All animals in Group 1 showed significantly increased values of liver/body weight ratio (%).

Food and test compound intake

All animals given test compounds (Groups 1 and 3) showed decreased food consumption as compared with Group 2. Test compound intake did not differ between the groups, with or without DEN.

Quantitative values of γ-GT$^+$ foci (Table 2)

Of the 4 compounds tested in this system, 2AAF induced the highest values of γ-GT$^+$ foci with regard to both number ($28.57 \pm 14.81/cm^2$) and area ($61.60 \pm 11.89\ mm^2/cm^2$), with a significant difference compared with Group 2 given DEN alone. Although the number of foci was not altered, 4AAF was responsible for a significantly increased area of the foci ($1.25 \pm 0.72\ mm^2/cm^2$). BP induced the smallest number of foci among Group 1 and showed a significant difference compared with Group 2 ($7.58 \pm 2.87/cm^2$). In contrast, the area was significantly increased from that of Group 1 (2.19 ± 1.03 mm^2/cm^2). The size of foci was obviously larger than other groups in Group 1. PYR did not induce increased values in both number and area with significant difference.

In Group 3, only 2AAF induced an obvious increase in number and area of foci by itself ($16.51 \pm 7.33/cm^2$ and 3.97 ± 2.17, respectively).

6 Discussion

In agreement with previous investigations, 2AAF used as a positive control compound exerted the most potent promoting activity, inducing the highest values of γ-GT$^+$ foci. Since 2AAF is a hepatocarcinogen, it should theoretically possess potential for both initiation of liver cells and promotion. In this experiment, the maximal initiating activity of 2AAF in Group 3 could be estimated, by the number of foci, at no more than $16.51 \pm 7.33/cm^2$. On the other hand, the number of foci after initiation with DEN in Group 2 is shown as 9.65 ± 3.55. Therefore, the sum total of the two values probable represents the summation of cells initiated by both DEN and 2AAF, and it is noteworthy that the value is close to that induced by DEN followed by 2AAF (28.57 ± 14.81). In contrast, the area was far greater than the simple sum total of these in Groups 2 and 3 (approximately 14-fold). Therefore, it could be concluded that the promoting activity of 2AAF is more potent than its initiation potential under the present conditions. This is consistent with our previous results (Ito *et al.*, 1980, 1982, 1983; Tsuda *et al.*, 1980, 1984a, 1984b). The promoting effect of 4AAF is indicated because of significantly increased area of foci.

In the group given BP after DEN, the number of foci was unexpectedly smaller than in any other groups with statistical difference as compared with Group 2. On the other hand, the area of foci showed clear increase. This apparent paradoxical result may be due to the fusiform appearance of large foci causing a decreased number of foci. Accordingly, it is concluded that 2AAF is a far more potent promoting agent than BP and 4AAF. PYR was negative for promotion.

7 Conclusion

The results of this investigation of the influence of chemicals on the induction of γ-GT$^+$ foci in an *in vivo* short-term assay system indicate that while 2AAF exerts an obvious promoting effect, 4AAF and BP show only enhancing activity and PYR in lacking in promoting potential.

8 References

Belenblum, I. and P.A. Shubik (1947) A new quantitative approach to the study of the stages of chemical carcinogenesis in the mouse's skin, *Br. J. Cancer*, **1**, 383–391.

Cameron, R.G., K. Imaida, H. Tsuda and N. Ito (1982) Promotive effects of steroids and bile acids on hepatocarcinogenesis initiated by diethylnitrosamine, *Cancer Res.*, **42**, 2426–2428.

Cohen, S.M., M. Arai, J.B. Jacobs and G.H. Friedell (1979) Promoting effect of saccharin and DL-tryptophan in urinary bladder carcinogenesis, *Cancer Res.*, **39**, 1207–1217.

Hicks, R.M., J. Wakefield and J. Chowaniec (1975) Evaluation of a new model to detect bladder

Table 2. *Quantitative values of γ-GT$^+$ foci in the liver*

Group	Treatment	No. of rats	γ-GT$^+$ foci No./cm^2	γ-GT$^+$ foci Area (mm^2)/cm^2
1	DEN→2AAF	19	28.57 ± 14.81***[a]	61.60 ± 11.89***
	DEN→4AAF	25	9.24 ± 2.14	1.25 ± 0.72***
	DEN→BP	24	7.58 ± 2.87*	2.19 ± 1.03***
	DEN→PYR	19	10.20 ± 2.24	0.61 ± 0.29
2	DEN (−)	23	9.65 ± 3.55	0.53 ± 0.20
3	(−) 2AAF	17	16.51 ± 7.33	3.97 ± 2.17
	4AAF	24	0.01 ± 0.05	0.00 ± 0.00
	BP	24	0.00 ± 0.00	0.00 ± 0.00
	PYR	26	0.02 ± 0.07	0.00 ± 0.01

[a]Significantly different from Group 2 at *$P < 0.05$, **$P < 0.01$, ***$P < 0.001$.

carcinogens or co-carcinogens: Results obtained with saccharin, cyclamate, and cyclophosphamide, *Chem.-Biol. Interact.*, **11**, 225–233.

Ito, N., M. Tatematsu, K. Nakanishi, R. Hasegawa, T. Takano, K. Imaida and T. Ogiso (1980) The effects of various chemicals on the development of hyperplastic liver nodules in hepatectomized rats treated with N-nitrosodiethylamine or N-2-fluorenylacetamide, *Gann*, **71**, 832–842.

Ito, N., H. Tsuda, R. Hasegawa and K. Imaida (1982) Sequential observation of pathomorphologic alterations in preneoplastic lesions during the promoting stage of hepatocarcinogenesis and the development of short-term test system for hepatopromoters and hepatocarcinogens, *Toxicol. Pathol.*, **10**, 37–49.

Ito, N., H. Tsuda, R. Hasegawa and K. Imaida (1983) Comparison of the promoting effects of various agents in induction of preneoplastic lesions in rat liver, *Environ. Health Perspect.*, **50**, 131–138.

Kitagawa, T. and H. Sugano (1978) Enhancing effect of phenobarbital on the development of enzyme-altered islands and hepato-cellular carcinomas initiated by 3′-methyl-4-diethylaminoazobenzene or diethylnitrosamine, *Gann*, **69**, 679–687.

Narisawa, T., N.E. Magadia, J.H. Weisburger and E.L. Wynder (1974) Promoting effects of bile acids on colon carcinogenesis after intrarectal instillation of N-methyl-N′-nitro-N-nitrosoguanidine in rats, *J. Natl. Cancer Inst.*, **53**, 1093–1097.

Ogawa, K., D.B. Solt and E. Farber (1980) Phenotypic diversity as an early property of putative preneoplastic hepatocyte populations in liver carcinogenesis, *Cancer Res.*, **40**, 725–733.

Peraino, C., R.J.M. Fry, E. Staffeldt and J.P. Christopher (1977) Enhancing effects of phenobarbitone and butylated hydroxytoluene on 2-acetylaminofluorene-induced hepatic tumorigenesis in the rat, *Fd. Cosmet. Toxicol.*, **15**, 93–96.

Solt, D.B., E. Cayama, H. Tsuda, K. Enomoto, G. Lee and E. Farber (1983) Promotion of liver cancer development by brief exposure to dietary 2-acetylaminofluorene plus partial hepatectomy or carbon tetrachloride, *Cancer Res.*, **43**, 188–191.

Solt, D. and E. Farber (1976) New Principle for the analysis of chemical carcinogenesis, *Nature* (London), **263**, 701–703.

Tamano, S., H. Tsuda, M. Tatematsu, R. Hasegawa, K. Imaida and N. Ito (1981) Induction of γ-glutamyl transpeptidase positive foci in rat liver by pyrolysis products of amino acids, *Gann*, **72**, 747–753.

Tatematsu, M., R. Hasegawa, K. Imaida, H. Tsuda and N. Ito (1983) Survey of various chemicals for initiating and promoting activities in a short-term *in vivo* system based on generation of hyperplastic liver nodules in rats, *Carcinogenesis*, **4**, 381–386.

Tsuda, H., R. Hasegawa, K. Imaida, T. Masui, M.A. Moore and N. Ito (1984a) Modifying potential of thirty-one chemicals on the short-term development of γ-glutamyl transpeptidase-positive foci in diethylnitrosamine-initiated rat liver, *Gann*, **75**, 876–883.

Tsuda, H., G. Lee and E. Farber (1980) Induction of resistant hepatocytes as a new principle for a possible short-term *in vivo* test for carcinogens, *Cancer Res.*, **40**, 1157–1164.

Tsuda, H., T. Masui, K. Imaida, S. Fukushima and N. Ito (1984b) Promotive effect of primary and secondary bile acids on the induction of γ-glutamyl transpeptidase-positive liver cell foci as a possible endogeneous factor for hepatocarcinogenesis in rats, *Gann*, **75**, 871–875.

Witchi, H. (1981) Enhancement of tumor formation in mouse lung by dietary butylated hydroxytoluene, *Toxicology*, **21**, 95–104.

John M. Hunt
Department of Pathology and Laboratory Medicine University of Texas Medical School, Houston, Texas, USA

54

A liver cell transplantation assay for promoting activity in rats

1 Summary

A rat liver cell transplantation assay was used to compare the strongly selective regimen of short-term dietary 2-acetylaminofluorene (2AAF) plus two-thirds partial hepatectomy (PH) (2AAF/PH regimen) with regimens using the other potential liver tumour promoters 4-acetylaminofluorene (4AAF), benzo(a)pyrene (BP) and pyrene (PYR) in place of 2AAF, in their capacity to stimulate gammaglutamyltranspeptidase-positive (GT^+) liver colony formation in host rats bearing intravenously transplanted carcinogen-altered donor liver cells.

Male F344 donor rats received 200 mg/kg diethylnitrosamine (DEN) intraperitoneally followed two weeks later by two weeks of dietary 2AAF (0.02%) plus PH (2AAF/PH) (DEN/2AAF/PH regimen). As a positive control, $(WF \times F344)F_1$ male host rats received only the 2AAF/PH regimen and supported donor cell colony growth. However, donor rat liver cells failed to yield above background levels of GT^+ liver colonies in host rat livers when the host rats received the test chemicals 4AAF, BP or PYR plus PH instead of the 2AAF/PH regimen. The dietary concentration of all test chemicals was 0.02% (w/w). Indirect immunofluorescent staining of the genotypic mosaic host rat livers confirmed that no donor origin colonies were present in host rats unless 2AAF was used. Increasing the test chemicals' dietary concentrations to 0.06% (w/w) failed to yield colony production in syngeneic F344 host rat livers above background levels. Strictly speaking, the assay discriminates between 2AAF and 4AAF, but not between BP and PYR. Although the strongly selective 2AAF/PH host rat regimen appears to act as a very strong promoter for premalignant donor liver cells, the regimen may depend for its efficacy on chemical properties unique to 2AAF.

2 Introduction

Tumour promotion in rodent liver carcinogenesis is classically demonstrated by treatment with promoting chemicals of animals that have received a prior initiating carcinogenic stimulus (Pitot and Sirica, 1980; Peraino *et al.*, 1983). In such experimental systems it is difficult to rule out synergism of initiating and promoting agent in the promotion process, particularly since the carcinogens used as initiators may possess promoting activity (Miller, 1978).

In this report, a liver cell transplantation assay has been used in which donor-origin liver cell colonies were scored histochemically in livers of host rats that had received donor liver cells intravenously and which were treated with one of four possible promotion regimens. This assay is based on the liver cell transplantation technique originally developed by Laishes and Farber (1978) in which premalignant liver cells generated in donor rats were shown to be capable of colonizing host rat livers only if the host rats received a strongly selective regimen consisting of dietary 2AAF and two-thirds partial hepatectomy (PH). The use of F344 donor rats and (WF × F344)F_1 hybrid host rats permits immunological identification of donor-origin liver cell colonies in the host rat livers independently of any histochemically detected enzyme marker such as gamma-glutamyltranspeptidase (GT) (Hunt *et al.*, 1982). Substitution in the host rat selective regimen (Figure 1) of 4AAF PYR, or BP in place of 2AAF resulted in a background level of donor-origin colony production in host rat livers. Thus, the promoting activity of 2AAF coupled with PH appears not to be shared with any of the other 3 chemicals tested.

3 Materials and methods

Donor rats

(Figure 1) F344 male rats (Harlan) (8–10 wks) were given 200 mg/kg DEN (Alfa) and 2 weeks later they were treated with a selective regimen consisting of 2-weeks feeding of diet containing 0.02% (w/w) 2AAF (Aldrich) with a two-thirds PH surgically performed one week after starting the dietary 2AAF. Eight days after discontinuation of dietary 2AAF, the livers were dissociated by in situ perfusion with 0.05% (w/v) Type I collagenase (Sigma). Donor liver cell suspensions were prepared in phosphate buffered saline

Figure 1. Schematic outline of the experimental regimens for donor and host rats.

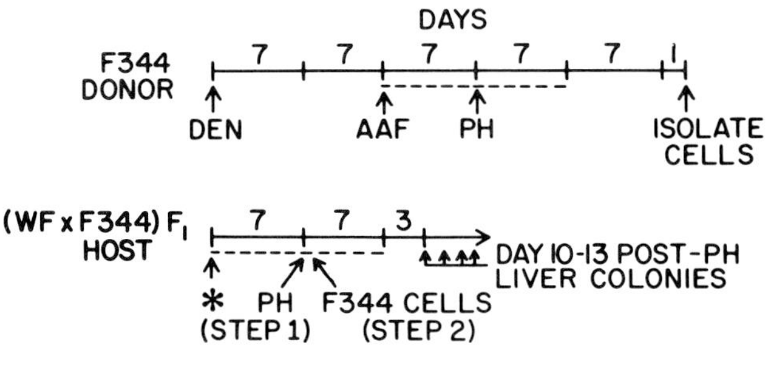

and the number of viable (trypan blue-excluding) and histochemically GT^+ liver cells determined. Donor liver cell suspensions were diluted with phosphate buffered saline to give the desired concentration, usually 1×10^6 viable liver cells per milliliter, and the standard injection volume per host rat was always 1.0 ml. These donor liver cell preparations have previously been shown to contain cells that colonize livers of (WF × F344)F_1 host rats treated with dietary 2AAF and PH (Hunt *et al.*, 1982).

Host rats

(Figure 1)(WF × F344)F_1 male host rats (8–10 weeks) were bred in a colony at the University of Texas Medical School using WF female and F344 male rats (Harlan). Test chemicals were a gift of Dr. John Ashby. The host rats received Diet 101 (Dyets) containing 0.02% (w/w) of an individual test chemical for 2 weeks, with administration beginning at the time indicated by the asterisk in Figure 1. The host rat experimental regimens consisted of 14 days of dietary administration of test chemical (0.02% or 0.06%, w/w) beginning at the point indicated by the asterisk (*), plus PH (Step 1) immediately prior to donor liver cell transplantation (Step 2) after 7 days of test chemical diet feeding. Host rats received control diet for 3 days before sampling at the standard time of 10 days post-PH. Liver sections were sampled and analysed for GT^+ histochemistry and by indirect immunofluorescence using F344 anti-WF alloantiserum. Diets were prepared in polyethylene bags by evenly spreading an acetone solution (2AAF or PYR) or slurry (4AAF or BP) of test chemical in the appropriate amount over a flattened pre-weighed portion of powdered diet. After allowing the acetone to evaporate in a chemical fume hood the bags were closed securely and mixed by hand tumbling in a second bag in the hood for 10 minutes. This method of mixing previously had been shown to adequately mix a dye at 0.02% into 1.5 kg of diet. Diets were administered ad libitum to host rats from aluminium cups (Hazleton) in solid bottom plastic cages covered with fiberglass bonnets. Feeding was done in a plexiglass fan-ventilated isolation cabinet in the Biohazard Animal Facility at the University of Texas Medical School. Unused contaminated diet, animal carcasses and unused tissue were sequested for commercial incineration at a minimum temperature of 1800° F.

Table 1. *Liver colony formation in host rats receiving 2AAF*

Host rat	Kill wt, g	Liver wt, g (% bw)	No. GT^+ liver colonies/cm^2
T1–2[a]	178	3.12 (1.75)	73.9
T1–3	187	4.10 (2.19)	11.9
T1–4	206	5.16 (2.50)	19.6
T1–5	244	6.16 (2.52)	19.2
T1–6	239	5.73 (2.40)	39.2
T6–4[b]	294	6.0 (2.07)	15.7

[a] Expt. T1: Donor rat = F344 male; donor liver cells were 82% viable (trypan blue exclusion) and 20% GT^+ histochemically. Cell Dose = 1×10^6 viable donor liver cells per host rat.

[b] Expt. T6: Donor rat = F344 male; donor liver cells were 72% viable (trypan blue exclusion) and 7.6% GT^+. Cell dose = 1×10^6 viable donor liver cells per host rat.

Liver cell transplantation and liver colony assay
Host rats received an intravenous transplantation of donor liver cells via a mesenteric vein of the hepatic portal system immediately after the two-thirds PH performed one week after initiating the test chemical diet (Figure 1). Host rats received test chemical diets for 1 more week and were placed on Diet 101 alone for 3 days prior to being killed with diethylether over-anesthesia. Livers of the host rats were excised, weighed, and median sections of the lobes frozen on dry ice for preparation of cryostat sections. Six or 16 micrometer cryostat sections were fixed in acetone and either stained histochemically to detect colonies of GT^+ liver cells, or by indirect immunofluorescence as described by Hunt *et al.* (1982) to detect donor-origin liver colonies independently of their GT^+ phenotype. The area of the cryostat liver sections was determined by Ms. Pratibha Desai using a Hewlett-Packard 9874A digitizing tablet and HP9845 computer, using a program kindly provided by Dr. Ruth Bulger. Data are expressed as numbers of colonies per cm^2 liver tissue sectioned.

Table 2. *Liver colony formation in host rats receiving 4AAF*

Host rat	Kill wt, g	Liver wt, g (% bw)	No. (GT^+) liver colonies/cm^2
T2–1[a]	171	3.33 (1.95)	0
T2–2	150	2.79 (1.86)	0
T2–3	151	4.91 (3.25)	0
T2–4	262	9.24 (3.52)	0.41
T2–5	259	9.27 (3.58)	0.21
T3–1[b]	176	5.03 (2.86)	0
T3–2	187	5.32 (2.84)	0
T3–3	153	4.35 (2.84)	0.28

[a] Expt. T2: Donor rat = F344 male; donor liver cells were 76% viable (trypan blue exclusion) and 12% GT^+ histochemically. Cell dose = 1×10^6 viable donor liver cells per host rat.
[b] Expt. T3: Donor rat = F344 male; donor liver cells were 70% viable (trypan blue exclusion) and 10% GT^+ histochemically. Cell dose = 1×10^6 (T3–1 and T3–2) or 5×10^6 (T3–3) viable donor liver cells per host rat.

Table 3. *Liver colony formation in host rats receiving BP*

Host rat	Kill wt, g	Liver wt, g (% bw)	No. (GT^+) liver colonies/cm^2
T4–1[a]	212	8.30 (3.91)	0
T4–2	217	8.83 (4.07)	0
T4–3	191	7.70 (4.03)	0
T4–4	198	7.60 (3.83)	0
T4–5	223	8.51 (3.82)	0
T4–6	200	8.80 (4.40)	0
T5–1[b]	212	7.44 (3.51)	0.25
T5–2	202	7.74 (3.03)	0

[a] Expt. T4: Donor rat = F344 male; donor liver cells were 72% viable (trypan blue exclusion) and 56% GT^+ histochemically. Cell dose = 1×10^6 viable donor liver cells per host rat except for T 4–5 and T 4–6, which received phosphate buffered saline only (no cells).
[b] Expt. T5: Donor rat = F344 male; donor liver cells were 67% viable (trypan blue exclusion) and 12% GT^+ histochemically. Cell dose = 1×10^6 viable donor liver cells per host rat

Table 4. *Liver colony formation in host rats receiving PYR*

Host rat	Kill wt, g	Liver wt, g (% bw)	No. (GT^+) liver colonies/cm^2
T6–1[a]	315	9.03 (2.87)	0
T6–2	334	9.54 (2.86)	0
T6–3	326	10.48 (3.22)	0

[a] Expt. T6: Donor rat = F344 male; donor liver cells were 72% viable (trypan blue exclusion) and 7.6% GT^+ histochemically. Cell dose = 1×10^6 viable donor liver cells per host rat.

4 Results

Table 1 presents the liver colony data typically observed when host rats receive the strongly selective 2AAF/PH regimen (asterisk = 2AAF in Figure 1). In this assay, the colonies were scored histochemically by virtue of their elevated level of GT. The donor liver cell origin of the majority of such GT^+ colonies had been previously demonstrated by indirect immunofluorescence using F344 anti-WF alloantiserum to stain the $(WF \times F344)F_1$ host liver brightly and leave the GT^+ colonies unstained (Hunt *et al.*, 1982). The GT^+ colony transections in cryostat sections ranged in diameter from 111 μm to 815 μm, and the mean transection diameter was $345 \pm 132\,\mu$m (mean $\pm$ SD; $n = 209$ colonies in experiment T1). These data represent the positive control using 2AAF in the selection regimen.

The results presented in Tables 2, 3 and 4 show that incorporation of test chemicals 4AAF (Table 2) BP (Table 3) or PYR (Table 4) resulted in extremely low levels of GT^+ colony production in host rats receiving these chemicals in conjunction with PH (Figure 1). The level of colony production observed in these sets of experimental animals is virtually the same as the background level of colony production observed when 2AAF/PH host rats receive saline injections instead of donor rat liver cells (Laishes and Farber, 1978, Hunt *et al.*, 1982). Staining by indirect immunofluorescence (Hunt *et al.*, 1982) of representative host rat liver cryostat sections from all experiments using test chemicals other than 2AAF revealed no instances of donor origin colonies in these host rat livers.

The testing data presented in Tables 1–4 were obtained using test chemicals at dietary concentrations of 0.02% (w/w). An additional experiment was done in which syngeneic F344 host rats received 0.06% dietary test chemicals with PH (Figure 1) and were transplanted with 1.0×10^6 viable F344 donor liver cells. The numbers of GT^+ liver colonies/cm^2 host rat liver tissue observed with the test chemicals at 0.06% dietary concentration were (mean $\pm$ SD; $n = 3$ host rat for each test chemical): 2.6 $\pm$ 2.7 for 4AAF; 0.9 ± 0.9 for BP; and 1.3 ± 0.6 for PYR. In contrast, the positive control (0.02% 2AAF plus PH) host rats supported colony formation similar to that reported in Table 1 (74 ± 29 GT^+ colonies/cm^2). The GT^+ liver colonies observed in the host rats that received 0.06% dietary test chemicals were small, transections ranging from 40 to 200 μm in diameter.

The fractional liver weights expressed as per cent of body weight (bw) at time of host rat liver sampling (% bw) are tabulated for each of the 4 test chemicals in Tables 1–4. A small difference in liver % bw was observed between host rats receiving 2AAF as compared with the liver % bw of host rats receiving the other 3 test chemicals which did not affect colony production in the assay: the average ($\pm$ SD) liver % bw for the 4 test chemicals were: 2.24 $\pm$ 0.30 for 2AAF; 2.84 ± 0.65 for 4AAF; 3.82 ± 0.41 for BP; and 2.98 ± 0.20 for PYR (data extracted from % bw values in Tables 1–4.

5 Discussion

A liver cell transplantation assay for tumor promotion is attractive conceptually, because it offers the potential for experimentally isolating initiating from promoting effects of test chemicals. The use of 2AAF in conjunction with PH in such an assay (Laishes and Farber, 1978; Hunt *et al.*, 1982) appears to be a requirement for donor colony production. The relatively crude liver % body weight data presented in Tables 1–4 suggest that the effectiveness of 2AAF correlates with its capacity to inhibit normal host rat liver cell division in response to the growth stimulus of PH. Thus, this mitoinhibitory activity of 2AAF (Kitagawa, 1971) may be the essential activity responsible for the success of the 2AAF/PH regimen in permitting donor liver cell of colonization of host rat livers, culminating ultimately in donor-origin hepatocellular carcinomas in these host rats (Hunt *et al.*, 1985). In contrast, studies with small numbers of animals have tentatively shown that DEN/2AAF/PH donor liver cells intravenously transplanted into livers of host rats fail to produce hepatocellular carcinomas at high incidence when the host rats are maintained 2 years on 0.05% dietary phenobarbital or dichlorodiphenyltrichloroethane (Hunt *et al.*, unpublished). Taken together, these results suggest that the regimen of brief dietary 2AAF exposure and partial hepatectomy acts as an extremely strong promoter of hepatocarcinogenesis. A similar conclusion has been drawn by Solt *et al.* (1983) for a different experimental system that does not utilize liver cell transplantation.

Strictly speaking, the results presented in Tables 1 and 2 demonstrate discrimination between 2AAF and 4AAF by the liver cell transplantation assay. Since 4AAF is not an hepatocarcinogen in rats (Morris *et al.*, 1960) this result is not unexpected. Little is known about the interaction of 4AAF with liver cells and with liver cell DNA in contrast to a relatively large body of such information for the carcinogen 2AAF (Arcos and Argus, 1974; Miller, 1978). Further *in vivo* testing using the liver cell transplantation assay might be profitably undertaken using other structural analogues of 2AAF screened for carcinogenicity and mutagenicity by Morris *et al.* (1960) and Poirier and de Serres (1979), respectively, to elucidate the chemical requirement for the strong promoting activity of 2AAF in this assay.

The liver cell transplantation assay failed to discriminate between the carcinogen BP (IARC, 1973) and the non-carcinogen PYR (Tatematsu *et al.*, 1983). The failure of the BP to exert 2AAF-like promoting effect on transplanted donor liver cells in the host rat livers may reflect the great difference in binding to rat liver cell DNA of these two carcinogens. Lutz *et al.* (1978) report a plateau level of ca. 0.6 f mol BP/ug DNA in contrast to plateau levels of 300–400 f mol 2AAF adduct/μg DNA (Poirier *et al.*, 1984). Even with an *in vitro* hamster embryo cell system, BP binding did not exceed 18 f mol BP/μg DNA (Sebti *et al.*, 1982). BP is hepatocarcinogenic in rats only under such extreme conditions as injection of neonates (Peraino *et al.*, 1981). BP can act as an inducer of enzyme-altered GT^+ foci, but BP-induced foci are detectable in livers of treated rats only if appropriate promoting conditions are used (Deml *et al.*, 1983; Tatematsu *et al.*, 1983).

6 Conclusions

The rat liver cell transplantation assay for promoting activity reported here appears to discriminate between 2AAF and 4AAF, but not between BP and PYR. A more conservative conclusion is that 2AAF cannot be replaced in the selective 2AAF/PH regimen applied to host rats by any of the 3 other chemicals tested in this study. Although the strongly selective 2AAF/PH regimen appears to act as a very strong promoter for GT^+ donor liver cells, the regimen's success may depend upon chemical properties unique to 2AAF.

7 Acknowledgements

This investigation was supported by US PHS Grant Number R01-CA37150-01, awarded by the National Cancer Institute, DHHS. Ms. Pratibha Desai provided technical assistance. Word processing assistance was provided by Ms. Annie Rose and Ms. Jackie Sanders Fagan.

8 References

Arcos, J.C. and M.F. Argus (1974) *Chemical Induction of Cancer*, Vol. II B, Structural Bases and Biological Mechanisms, Academic Press, New York, NY, pp. 253–269.

Deml, E., D. Oesterle and F.J. Wiebel (1983) Benzo(a)pyrene Initiates Enzyme-Altered Islands in the Liver of Adult Rats Following Single Pretreatment and Promotion with Polychlorinated Biphenyls, *Cancer Lett.*, **19**, 301–304.

Hunt, J.M., M.T. Buckley, B.A. Laishes and H.A. Dunsford (1985) Immunological Approaches to the Purification of Putative Premalignant Hepatocytes from Genotypic Mosaic Rat Livers, *Cancer Res.*, **45**, 2226–2233.

Hunt, J.M., M.T. Buckley, P.A. Onnink, P.B. Rolfe and B.A. Laishes (1982) Liver cell Membrane Alloantigens as Cellular Markers in Genotypic Mosaic Rat Livers Undergoing Chemically Induced Hepatocarcinogenesis, *Cancer Res.*, **42**, 227–236.

IARC (1973) *IARC Monographs on the Evaluation of Carcinogenic Risk of the Chemical to Man*, Vol. 3, Certain Polycyclic Aromatic Hydrocarbons and Heterocyclic Compounds, International Agency for Research on Cancer, Lyon, France, pp. 91–136.

Kitagawa, T. (1971) Responsiveness of Hyperplastic Lesions and Hepatomas to Partial Hepatectomy, *Gann*, **62**, 217–224.

Laishes, B.A., and E. Farber (1978) Transfer of Viable Putative Preneoplastic Hepatocytes to the Livers of Syngeneic Host Rats, *J. Natl Cancer Inst.*, **61**, 507–512.

Lutz, W.K., A. Viviani and C. Schlatter (1978) Nonlinear Dose Response Relationship for the Binding of the Carcinogen Benzo(a)pyrene to Rat Liver DNA *In Vivo*, *Cancer Res.*, **38**, 575–578.

Miller, E.C. (1978) Some Current Perspectives on Chemical Carcinogenesis in Humans and Experimental Animals, *Cancer Res.*, **38**, 1479–1496.

Morris, H.P., C.A. Velat, B.P. Wagner, M. Dahlgard and F.E. Ray (1960) Studies of Carcinogenicity in the Rat of Derivatives of Aromatic Amines Related to N-2-Fluorenylacetamide, *J. Natl. Cancer Inst.*, **24**, 149–180.

Peraino, C., R.L. Richards, and F.J. Stevens (1983) Multistage Hepatocarcinogenesis. In T.J. Slaga (Ed.), *Mechanisms of Tumor Promotion*, Vol. 1 Tumor Promotion in Internal Organs, CRC Press, Boca Raton, FL, pp. 1–53.

Peraino, C., E.F. Staffeldt and V.A. Ludeman (1981) Early appearance of Histochemically Altered Hepatocyte Foci and Liver Tumors in Female Rats Treated with Carcinogens One Day After Birth, *Carcinogenesis*, **2**, 463–465.

Pitot, H.C., and A.E. Sirica (1980) The Stages of Initiation and Promotion in Hepatocarcinogenesis, *Biochim. Biophys. Acta*, **605**, 191–215.

Poirier, L.A., and F.J. de Serres (1979) Initial National Cancer Institute Studies on Mutagenesis as a Prescreen for Chemical Carcinogens: An Appraisal, *J. Natl. Cancer Inst.*, **62**, 919–926.

Poirier, M.C., J.M. Hunt, B. True, B.A. Laishes, J.F. Young and F.A. Beland (1984) DNA Adduct Formation, Removal and Persistence in Rat Liver During One Month of Feeding 2-Acetylaminofluorene, *Carcinogenesis*, **5**, 1591–1596.

Sebti, S.M., Baird, B.B. Knowles and L. Diamond (1982) Benzo(a)pyrene-DNA Adduct Formation in Target Cells in a Cell-mediated Mutation Assay, *Carcinogenesis*, **3**, 1317–1320.

Solt, D.B., E. Cayama, H. Tsuda, K. Enomoto, G. Lee, and E. Farber (1983) Promotion of Liver Cancer Development by Brief Exposure to Dietary 2-Acetylaminofluorene Plus Partial Hepatectomy or Carbon Tetrachloride, *Cancer Res.*, **43**, 188–191.

Tatematsu, M., R. Hasegawa, K. Imaida, H. Tsuda and N. Ito (1983) Survey of Various Chemicals for Initiating and Promoting Activities in a Short-term *In Vivo* System Based on Generation of Hyperplastic Liver Nodules in Rats, *Carcinogenesis*, **4**, 381–386.

PART IV
In vivo assays: liver

B
Unscheduled DNA synthesis and S-phase hepatocytes

Jon C. Mirsalis (Workgroup Chairman)[1]
Cellular and Genetic Toxicology Department, SRI International, Menlo Park, California, USA

55

Summary report on the performance of the in vivo DNA repair assays

1 Introduction

Measurement of chemically induced DNA repair as unscheduled DNA synthesis (UDS) is currently accepted as an indicator of genotoxicity and potential carcinogenicity of chemicals in selected target organs. The most widely used UDS assays employ rodent hepatocyte cultures following either *in vitro* (Williams, 1977) or *in vivo* (Mirsalis and Butterworth, 1980) treatment. Both systems have specific strengths and weaknesses for the screening of potential genotoxic agents (Mitchell and Mirsalis, 1984).

In vivo assays have been reported since the early 1970s (Stich and Kieser, 1974), but the currently accepted *in vivo–in vitro* methodology has been in use for only a relatively short period of time (Mirsalis and Butterworth, 1980). Several laboratories have adopted these procedures, including the seven that participated in the present study, and a relatively large historical data base is now available in both rats and mice. Since 1980 over 100 compounds have been examined in the *in vivo* DNA repair assay. The number of treated animals is large with over 4000 rats and over 1500 mice examined. These efforts have also generated an historical vehicle control data base in excess of 350 rats and 100 mice as of 1985.

In addition to measurement of UDS, autoradiographic techniques allow simultaneous measurement of cells undergoing DNA replication. Cells in S-phase are easily recognized by the dense labeling pattern over the nuclei. This offers a valuable means to assess the effects of chemicals on cell proliferation within the liver.

The data base for S-phase synthesis is somewhat less extensive than that for UDS, with over 40 compounds having been examined for hepatocellular proliferation. There are over 100 vehicle control rats and over 60 vehicle control mice in the S-phase data base as of 1985.

In vivo UDS assays have been reported in several tissues other than the liver, including lymphocytes (Skinner *et al.*, 1980), tracheal epithelium (Doolittle and Butterworth, 1984), pancreas (Steinmetz and Mirsalis, 1984), spermatocytes (Working and Butterworth, 1984), stomach (Furihata and Matsushima, 1982; Furihata *et al.*, 1984a, 1984b, 1985), kidney (Tyson and Mirsalis, 1985), nasal epithelium (Bermudez and Allen, 1984), and alveolar macrophages (Cheng *et al.*, 1983). Most of these

[1] Workgroup participants: John Ashby, Brita Beije, Byron D. Butterworth, Gary M. Decad, David J. Doolittle, Chie Furihata, Douglas Kornbrust, Gregory S. Probst, and Charles B. Salocks

assays employ techniques similar to the liver assay and data are expressed in a similar format, but these organ systems will not be considered in depth in this report.

In the present collaborative study, seven different laboratories examined 2AAF and 4AAF for induction of UDS and the percent of cells in S-phase in hepatocytes of male rats. In addition, two laboratories tested these compounds as well as BP in mice. One laboratory examined all four compounds for induction of UDS and total DNA synthesis in rat forestomach. Summaries of the results are shown in Tables 1 and 2. Data from each laboratory are presented in the individual reports. In the following summary, we have attempted to answer several important questions concerning the performance and evaluation of this assay including: (1) Did different methodologies affect quantitative and qualitative results between laboratories? (2) Are there certain basic guidelines that should be followed by all laboratories performing this assay? (3) What are the criteria for establishing a positive or negative response? (4) What are

Table 1. *In vivo hepatocyte DNA repair (UDS)*

	Rat				Mouse			
DNA Repair	2AAF	4AAF	BP	PYR	2AAF	4AAF	BP	PYR
Ashby	+[a] 5	w+ 1000 mg/kg						
Burlinson			– 100	(weanling rats)				
Butterworth	+ 10	– 200	– 200	– 200	– 50	– 200	– 200	
Mirsalis	+ 10	w+ 200	– 500	– 500	– 200	– 200		
Salocks	+ 10	w+ 500						
Beije	+ 50	– 2000						
Kornbrust	+ 10	– 200		(UDS also observed in pups when 2AAF administered to lactating mother)				
Doolittle	+ 50	– 500						
Furihata Forestomach UDS	– 800	– 1000	w+ 1000	– 800				

[a] + is positive, with the number below being the lowest dose which gave a positive response.
– is negative, with the number below being the highest dose which gave a negative response.
w+ is a weak positive or marginal response.

Table 2. *Induction of S-phase*

	Rat				Mouse			
S-Phase	2AAF	4AAF	BP	PYR	2AAF	4AAF	BP	PYR
Burlinson			+[a] 100 mg/kg	(weanling rats)				
Butterworth		+[b] 200	w+[c] 100	– 200		w+ 200	– 200	
Mirsalis	+[c] 50	+ 200	– 500	– 500	w+[c] 200	+ 200		
Salocks S-phase	– 50	+ 200						
Beije S-phase	w+[d] 500	+[d] 2000						
Kornbrust S-phase	– 50	+ 200						
Doolittle S-phase	– 200	+ 50						

See footnotes to Table 1.
[a] Measurement done at 4 h post treatment.
[b] Unless otherwise noted all measurements were done at 24 h post treatment.
[c] Measurements done at 48 h post treatment.
[d] Measurement done at 36 h post treatment.

the strengths and limitations of this assay when used as part of a battery for the screening of potential carcinogens?

2 Comparison of methodologies

Induction of UDS was tested in mice, weanling rats, and two strains of adult rats, with the majority of data being derived from studies using adult male rats. The assay does not appear to be influenced greatly by variations in methodology, because there was excellent agreement in the observations and conclusions of the participating investigators. This consistency is an extremely important attribute for an assay that may have widespread applications as a testing protocol.

Prior to dosing, animals were either allowed free access to food or were fasted for 12 h. Some investigators felt that fasting the animals made dosing with large volumes (in this case, up to 1.0 ml/100 g body weight) an easier task. None of the chemicals were very soluble in corn oil, and there were differences in the degree of solubilization achieved. This variable may affect the kinetics of uptake and metabolism of a compound and thereby influence the time at which maximal induction of UDS is achieved.

Due to feeding patterns that affect absorption of compounds from the gastrointestinal tract, or to diurnal cycles that affect hepatocellular DNA repair capacity, the time of day chosen for administration of test compounds or for isolation of hepatocytes could also influence results. There is evidence that maintaining the normal feeding pattern and light-dark cycle is important in reducing the variability when measuring the induction of S-phase. For the most part, cell isolations were performed in the morning so that culturing and incubation with ^{3}H-thymidine could be completed before the end of the day. Therefore, animals were routinely dosed at various intervals prior to hepatocyte isolations.

Two investigators used biopsy perfusion techniques to obtain hepatocyte cultures (Byard *et al.*, 1983), whereas a two-step *in situ* perfusion method (Mirsalis and Butterworth, 1980) was employed by all other investigators in this study. Procedures for incubation with ^{3}H-thymidine were similar in all laboratories except for one in which ethanol was present at a 0.7% concentration (Salocks and Decad). Incubation of the hepatocytes with ^{3}H-thymidine was followed by overnight incubation with unlabeled thymidine in all laboratories except two (Doolittle *et al.*; Kornbrust and Dietz) in which thymidine was excluded from the overnight incubation medium. These methodological discrepancies did not have a significant effect on the results obtained by the investigators.

Cell fixation procedures varied considerably. Different investigators used one, two or three exposures to methanol or ethanol: acetic acid (3:1) for 10 to 30 minutes. The duration of autoradiogram exposure ranged from 1 to 2 weeks. Neither of these variables appeared to have a qualitative impact on the responses. Several investigators noted that cooling the developing chemicals to 15 °C produced smaller, more easily quantifiable grains.

Autoradiographic grains were quantified in one laboratory by hand-counting, and nuclear-sized areas of the cytoplasm were estimated visually (Beije and Ashby). This procedure appeared to increase the variation of the data but did not result in conclusions that were at variance with other laboratories in the group. Both the "area" mode and the "count" mode of the colony counters were utilized for automatic grain counting and similar results were achieved with either method. Raw grain counts were corrected for cytoplasmic background grains either by subtracting the average number of grains present in one to three equal-sized cytoplasmic areas or by subtracting the highest of the cytoplasmic counts. The latter method is obviously more conservative. Although differences in the procedures chosen for determination of net nuclear grains produced minor quantitative alterations in the magnitude of the response, qualitative interpretation of results was not affected.

Only one laboratory examined UDS induction in forestomach epidermis; therefore, methodology comparisons were not possible. Techniques for the stomach UDS assay were substantially different from those of the liver assay in that incubation of ^{3}H-thymidine in the presence of 10 mM hydroxyurea (to inhibit S-phase) and liquid scintillation counting were used instead of autoradiography. Total DNA synthesis (TDS) measures total ^{3}H-thymidine incorporation in the absence of hydroxyurea and is similar to measurement of S-phase synthesis in liver. Ornithine decarboxylase activity, a marker enzyme of tumor promoting activity, was induced simultaneously with TDS by the putative glandular stomach tumor promoter NaCl and 5 glandular stomach carcinogens in the rat glandular stomach mucosa (Furihata *et al.*, 1984a, 1985). Thus, induction of TDS may be a marker of tumor promoting activity.

3 Results

2AAF was tested by seven laboratories using the *in vivo–in vitro* rat hepatocyte DNA repair protocol and was a potent inducer of UDS in male rats in all cases in the dose range from (5 to 50 mg/kg). The maximum response was observed 12 h after dosing. An increase in the number of cells in S-phase was also observed following 2AAF exposure, but only at post-treatment times greater than 36 h indicating that 2AAF is both genotoxic and an inducer of hepatic cell proliferation. 2AAF was also examined in male B6C3F1 mice by two laboratories. While 2AAF failed to induce UDS in either lab, it was a weak inducer of S-phase in mice. However, only low doses (20 mg/kg) were examined. This observation indicates dramatic and important species differences in the metabolic activation of 2AAF.

Administration of 4AAF to rats produced clearly negative results in 4 laboratories and borderline responses in three laboratories. The latter responses were judged to be weakly positive, based on average net grains per nucleus (NG) values in the 0–5 range. Although statistical evaluations were not performed, it was evident that the responses were qualitatively different from the vehicle controls. However, it was concluded that further data analysis or experimentation would be required to establish these responses as unequivocally positive. In contrast to the UDS data, a dramatic increase in the percentage of S-phase cells at 24 h after treatment of rats with 4AAF was reported by all six laboratories that measured this endpoint. The hepatotoxicity of 4AAF was evaluated in two laboratories by examining serum enzyme levels 24 h after dosing. Neither laboratory reported any indications of hepatotoxicity in either the rat or mouse suggesting that 4AAF may be acting as a mitogen. As with the rat, 4AAF failed to induce UDS in mouse liver. 4AAF did induce an increase in the percentage of cells in S-phase in the mouse liver, although the magnitude of the response was much less than that observed in rats. Thus, just as with 2AAF, the metabolic activation of 4AAF may be species specific. The observed activity of 4AAF suggests that the compound may have promoting or carcinogenic activity.

2AAF and 4AAF were also tested by two laboratories in the *in vitro* hepatocyte UDS assay. Both compounds increased UDS in rat hepatocytes. As was observed in the whole animal, neither 2AAF nor 4AAF produced an increase in UDS following *in vitro* exposure in mouse hepatocytes. 4AAF appeared to stimulate a slight increase in S-phase synthesis *in vitro* in mouse hepatocytes.

Unlike 2AAF, BP is not a hepatocarcinogen, but tends to produce tumors at the site of application. BP and PYR were tested in rats by two laboratories and in mice by one laboratory. Both compounds failed to induce UDS in rats or mice treated *in vivo* by oral gavage. In contrast, BP did induce UDS in both rat and mouse hepatocytes treated *in vitro*.

BP and PYR were also found to be inactive in weanling rats treated *in vivo*. However, BP did increase the incidence of S-phase cells in hepatocytes of weanling rats exposed to the compound *in vivo*. In adult rats treated with BP, a slight increase in the incidence of S-phase was noted by one investigator. In mice, BP administration had no effect on the number of S-phase cells. Likewise, PYR had no effect on the incidence of S-phase when administered to rats.

In rat forestomach BP produced equivocal results while 2AAF, 4AAF and PYR all failed to induce UDS. 2AAF did induce an increase in total DNA synthesis, suggesting a possible stimulation of cell proliferation in the stomach.

In summary, consistency between the seven laboratories was very good. 2AAF was clearly a strong inducer of UDS and a weak inducer of S-phase in rats, but did not induce UDS in mice at the levels examined. These findings suggest that 2AAF will be a more potent carcinogen in rats than in mice and that there may be fundamental differences in the metabolic activation and the mechanism of action of this compound in these two species. Despite the potent UDS response to 2AAF in rat liver, no effect on UDS was observed in rat stomach, which is consistent with the strict tissue specific carcinogenicity of this chemical. 4AAF was marginal in inducing UDS, but was a very potent S-phase inducer, particularly in the rat. Both BP and PYR failed to elicit DNA repair in rat hepatocytes, although a weak S-phase response was observed in rats treated with BP. The fact that BP induces UDS in hepatocytes *in vitro* suggests that factors operative in the whole animal such as the amount of BP distributed to the liver, the rate of formation of genotoxic metabolites, and/or the rate of detoxification, yield an insufficient number of DNA adducts to produce a detectable UDS response.

4 Recommendation for conducting the *in vivo–in vitro* UDS assays

The basic protocol for conducting the assay has been described (Mirsalis *et al.*, 1982b). While minor technical differences were noted between laboratories, all investigators agreed upon certain basic guidelines for the conduct of *in vivo–in vitro* UDS/S-phase assays.

An appropriate experimental design should include at least 3 animals per group and 3 slides per animal. A minimum of 20 cells per slide should be scored for UDS and at least 1000 cells per slide for S-phase. The unit of measurement for UDS is the net grains per nucleus (NG) and should be presented as the average for at least 3 animals with animal to animal variation. The number of treatment groups should include at least 2 doses of test compound, a vehicle control, and a positive control. Treatment levels should be determined on a compound by compound basis. Treatments equal to one-half the LD_{50} are considered a sensible upper treatment level and generally range between 10 and 500 mg/kg. An assay should consist of measurement of both UDS and the percentage of cells in S-phase. For UDS, one time point between 2 and 5 hours and one between 12 and 20 hours are recommended. For evaluating replicative synthesis (S-phase), both 24 and 48 h timepoints should be examined.

Some investigators recommended fasting of animals prior to treatment with test compounds. Other participants cautioned that fasting might affect the homoeostasis of the animal, which might indirectly influence compound distribution or DNA repair capacity in target cells. It is particularly important not to alter feeding patterns or the light–dark cycle when conducting S-phase studies. Fasting for more than 24 hours caused severe inhibition of total DNA synthesis and UDS in rat forestomach epidermis and in rat glandular stomach mucosa. However, test chemicals may not be absorbed or may be degraded by interaction with dietary constituents if the stomach is full. One solution to the problem is to house rats in separate cages and allow them a limited amount of food overnight to reduce the stomach contents (e.g., 4 g of commercial pellet diet per 200 g rat).

It was agreed that hepatocyte populations for the vehicle control should be at least 65% viable, as assessed by trypan blue exclusion. Cultures of over 90% viability are not uncommon. The viability of cultures prepared from treated animals may be below 65%. Responses obtained by the participating investigators for vehicle controls were invariably less than zero average NG with 0–15% of the cells in repair (nuclei containing $\geqslant$ 5 NG). All of the many historical control values from this group of investigators were consistent with these observations. Vehicle controls outside this range are, therefore, highly questionable. Responses for positive control compounds should be at least 5 NG with over 50% of the cells in repair and should also be within acceptable deviations from the laboratory's historical data. Treatment with 50 mg/kg 2AAF may be expected to yield over 20 NG with over 90% of the hepatocytes in repair.

Criteria for classifying weak responses in the *in vivo*

UDS assay as either positive or negative are still in the process of being defined. Substantial progress towards this goal was made during the course of study due to consideration of the points raised by Margolin and Risko (Chapter 3). The following discussion is therefore a compromise adopted between the currently perceived extremes of biological and statistical significance of data. Compounds that reproducibly induce greater than +5 NG as an animal mean are considered unequivocally positive. However, now that more historical data are available, it is clear that the +5 NG count chosen originally is a very conservative criterion for a positive response. Compounds that induce NG values between 0 and +5 NG require individual consideration. In such cases both a positive dose–response relationship as well as an increase in the percentage of cells in repair are required for the definition of a positive response. Further consideration is required to decide how to classify agents which induce dose-related increases in NG values, but for which the effects lie at or below zero NG. The investigators present at the meeting adopted the view that NG values greater than zero are necessary to define activity in the assay, but this leads to the definition of some problem areas which require further data and comparative studies to resolve. The first is that no use is currently made of control values with this assay. Individual controls are run with each experiment in order to confirm the technical integrity of that assay, but the control value obtained is not used in the interpretation of that specific test, but rather is added to the historical control database. However, this historical control pool itself is not used in a defined statistical test when evaluating individual data sets. Thus, a laboratory with a control mean of −8 NG applies the same assay response criteria as one whose mean is −2 NG. This situation is currently tolerated because the negative NG value for controls is subject to individual data acquisition techniques, but it can lead to conflicts between the statistical assessment of data and the conclusion regarding activity drawn by individual investigators (see Chapter 3). One laboratory, whose mean NG count for 93 controls is -4.2 ± 1.5 NG, is considering the use of three standard deviations above −4.2 as the positive test criterion. This would yield a positive conclusion for NG values greater than 0.3 NG with a p-value of less than 0.001. The percentage of cells in repair for the same control data set is $2 \pm 3\%$, and use of three standard deviations above 2 would yield a positive conclusion for values greater than 11% in repair, with a p-value again of less than 0.001. That laboratory believes that in the comparison of small data sets to the entire control data base, that choosing this higher level of significance, instead of the traditionally employed value of $p < 0.05$, yields positive calls that are more in keeping with the actual biological significance of this assay.

It should be noted that the control NG values for this assay vary among laboratories, but several of the sources of variability cannot be adequately controlled, e.g. the concentration and specific activity of the ^{3}H-thymidine used, the type and extent of dilution of the photographic emulsion, and the period and temperature of autoradiographic exposure. Such variables indicate that each laboratory should acquire an historical control database, even though at present this is not actively used for the assessment of data. Nevertheless, if such a control database existed, and if it was shown that day to day variability did not exceed animal to animal variability, then it could be employed to define a more rational basis for data interpretation (see Chapter 3).

For the present, it was agreed that in order to classify a compound as capable of inducing nuclear DNA repair, then it must be shown that a greater number of silver grains lie over the nucleus of the isolated hepatocytes than in an equal area of their cytoplasm (i.e. > 0 NG). Finally, it is concluded that further studies relating to the role of mitochondrial DNA synthesis or chemically-induced modifications of thymidine uptake mechanisms, etc., will be required to enable data to be interpreted when NG values of about zero are observed, irrespective of the statistical significance of such values.

It has been proposed that an analysis of variance is a suitable statistical test to determine the potential of a compound to increase replicative (S-phase) synthesis if the data for both control and chemical treatments is normally distributed and have equal variance. The data may need to be transformed (e.g., to a natural logarithm) to satisfy these assumptions. In general, control values will be in the range from less than 0.1% to 0.3% of the cells in S-phase. A positive response will generally be in the range from 1% to 10% of the cells in S-phase. At the present time no standard statistical models for analysis of UDS or S-phase are uniformly accepted.

5 Assessment of *in vivo* UDS assays

As a short-term test for chemicals which damage DNA, the *in vivo* UDS assay has several important advantages. Obviously, since the chemical is administered to an intact animal, *in vivo* absorption and distribution as well as extrahepatic metabolism, which may be involved in the overall response, are considered in this assay. Use of this assay may therefore provide the investigator with a more realistic appraisal of a chemical's ability to damage DNA *in vivo* under various exposure conditions. For chemicals such as some nitroaromatic carcinogens in which gut flora are obligatory in the metabolic activation, it is imperative to employ the *in vivo* protocol in order to obtain the mechanistically relevant positive response (Mirsalis *et al.*, 1982a; Doolittle *et al.*, 1983).

The assay is of relatively short duration. Results are available in 2–3 weeks, whereas, in the stomach assay results are obtained in 2 days. The grain counting procedure is tedious, but scoring can be automated. The assay is fairly expensive, with the collagenase being one of the most costly items. Animal usage is low relative to other assays such as the micronucleus test. The schedules for fasting and dosing of animals can be demanding.

One can also use the assay to qualitatively determine the ability of a chemical to induce DNA replication (S-phase) as well as DNA repair thereby providing a "mitogenic" endpoint along with a "genetic" endpoint. The role of chemically induced cell proliferation in carcinogenesis is not completely understood. There is evidence that increased cell turnover may increase the spontaneous mutational frequency and may result in the conversion of some pre-mutagenic lesions to actual mutations before they can be repaired. Moreover, forced proliferation alone may play a major role in producing an increase in liver tumors, especially in mice, and even in the apparent absence of genotoxicity (Butterworth *et al.*, 1984; Mirsalis *et al.*, 1985). Increased cell turnover can result from (1) chemically induced hepatotoxicity from agents such as carbon tetrachloride leading to necrosis and subsequent regeneration of tissue (Mirsalis *et al.*, 1982); or (2) a direct mitogenic effect usually resulting in hyperplasia from agents such as di(2-ethylhexyl)phthalate (DEHP) (Butterworth *et al.*, 1984). Elucidation of the mechanism involved requires the examination of parameters such as liver-to-body weight ratios, histopathology, and the appearance of liver enzymes in serum. Regardless of the mechanism involved, increased hepatocellular proliferation may be an important factor in the hepatocarcinogenicity of chemicals. The quantification of this phenomenon provided by the *in vivo* UDS assay is therefore an important feature of this system.

One strong point of the UDS assay is that it has thus far produced no examples of a false positive response. Positive responses obtained in the assay are, therefore, strongly indicative of carcinogenic potential. The assay may also be used to examine potential initiating (and possibly promoting) activities in specific target organs. At the present time, correlation to carcinogenicity is very good for both stomach and liver carcinogens.

For both the *in vivo* and *in vitro* assays, the use of a genetic endpoint (DNA repair) precludes the detection of carcinogens which act by mechanisms other than formation of repairable DNA adducts. In addition, the *in vivo* assay suffers from the limitation that test chemicals must be distributed to the target cells in sufficient quantities to elicit a detectable amount of DNA damage, particularly since weak responses tend to be obscured by cytoplasmic (background) grains. Thus, chemicals which are hepatocarcinogenic, such as 2AAF, are readily detected in the *in vivo* hepatocyte assay, whereas BP, which produces tumors in tissues other than liver was inactive in the assay. The possibility also exists that test chemicals may inhibit hepatocellular DNA repair mechanisms, either directly, or indirectly, by perturbing cellular homeostasis, which could also preclude detection of genotoxicity (see Kornbrust and Dietz). Hence, there are several mechanisms by which carcinogenic compounds may fail to elicit positive responses in the UDS assay, which obviously confounds interpretation of negative data. Uncertainty over the nature of negative responses may be lessened somewhat by extending the testing protocol to additional doses, timepoints or route of administration, or by determining responses in more than one tissue or species. The profound tissue (liver vs. stomach) and species (rat vs. mouse) specificity observed for the UDS responses to 2AAF attest to the value of this approach. In addition, evidence that a chemical is well distributed to the target cells (e.g., induction of S-phase by 4AAF and/or pharmacokinetic data) greatly strengthens the confidence in negative UDS responses.

In summary, results of the collaborative study confirmed the previously recognized reliability of the *in vivo* hepatocyte UDS assay for detecting genotoxic hepatocarcinogens. The observed tissue and species specificity (e.g., for 2AAF) may compromise the usefulness of the assay as a quick screen for genotoxicity, but at the same time provides important information about *in vivo* targets of test compounds. The observed consistency in results between different laboratories was indicative of good reproducibility in performing the assay. The lack of sensitivity to detect some carcinogens remains the primary issue of concern with this assay.

6 References

Bermudez, E. and P.F. Allen (1984) The assessment of DNA damage and repair in rat nasal epithelial cells, *Carcinogenesis*, **5**, 1453–1458.

Butterworth, B.E., E. Bermudez, T. Smith-Oliver, L. Earl, R. Cattley, J. Martin, S. Strom, R. Jirtle and G. Michalopoulos (1984) Lack of genotoxic activity of di(2-ethylhexyl)phthalate (DEHP) in rodent and human hepatocytes, *Carcinogenesis*, **5**, 1329–1335.

Byard, J.L., J.A. Reese and S.A. Knadel (1983) Isolation and culture of hepatocytes from liver biopsies. In *Isolation, Characterization and Use of Hepatocytes*, R.A. Harris and N.W. Cornell (Eds), Elsevier, New York.

Cheng, M., A.D. Kligerman and B.E. Butterworth (1983) Chemical induction of unscheduled DNA synthesis in rat alveolar macrophages. The Pharmacologist 25, 190.

Doolittle, D.J., J.M. Sherrill and B.E. Butterworth (1983) Influence of intestinal bacteria, sex of the animal and position of the nitrogroup on the hepatic genotoxicity of nitrotoluene isomers *in vivo*. *Cancer Res.* **43**, 2836–2842.

Doolittle, D.J. and B.E. Butterworth (1984) Assessment of chemical-induced DNA repair in rat tracheal epithelial cells, *Carcinogenesis*, **5**, 773–779.

Furihata, C. and T. Matsushima (1982) Unscheduled DNA synthesis in rat stomach. Short-term assay of potential stomach carcinogens. In Indicators of Genotoxic Exposure, Banbury Report No. 13, Bridges, B., Butterworth, B.E., and Weinstein, I.B. (Eds), Cold Spring Harbor Laboratory, New York, NY pp. 123–135.

Furihata, C., Y. Sato, M. Hosaka, T. Matsushima, F. Furukawa, and M. Takahashi, M. (1984a) NaCl induced ornithine decarboxylase and DNA synthesis in rat stomach mucosa, *Biochem. Biophys. Res. Comm.*, **121**, 1027–1032.

Furihata, C., Y. Yamawaki, S. Jin, H. Moriya, K. Kodama, T. Matsushima, T. Ishikawa, S. Takayama, and M. Nakadate (1984b) Induction of unscheduled DNA synthesis in rat stomach mucosa by glandular stomach carcinogens. *J. Natl. Cancer Inst.* **72**, 1327–1334.

Furihata, C., Y. Sato, T. Matsushima and M. Tatematsu (1985) Induction of ornithine decarboxylase and DNA synthesis in rat stomach mucosa by methylglyoxal. *Carcinogenesis*, **6**, 91–94.

Mirsalis, J.C. and B.E. Butterworth (1980) Detection of unscheduled DNA synthesis in hepatocytes isolated from rats treated with genotoxic agents: an *in vivo–in vitro* assay for potential carcinogens and mutagens, *Carcinogenesis*, **1**, 621–625.

Mirsalis, J.C., T.E. Hamm, Jr., Sherrill, J.M. and B.E. Butterworth (1982a). Role of gut flora in the genotoxicity of dinitrotoluene. Nature **295**, 322–323.

Mirsalis, J.C., C.K. Tyson and B.E. Butterworth (1982b) Detection of genotoxic carcinogens in the *in vivo–in vitro* hepatocyte DNA repair assay, *Environ. Mutagen.*, **4**, 553–562.

Mirsalis, J., K. Tyson, E.N. Loh, K.L. Steinmetz, J.P. Bakke, C.M. Hamilton, D.K. Spak and J.W. Spalding (1985). Induction of hepatic cell proliferation and unscheduled DNA synthesis in mouse hepatocytes following *in vivo* treatment, *Carcinogenesis*, **6**, 1521–1524.

Mitchell, A.D., and J.C. Mirsalis (1984) Unscheduled DNA synthesis as an indicator of genotoxic exposure. In

Single Cell Mutation Monitoring Systems: Methodologies and Applications, A.A. Ansari, and F. de Serres (Eds), Plenum Publishing Corp., New York.
Skinner, M.J., B. DeCastro, and J.F. Eyre (1980) Detection of unscheduled DNA synthesis in rat lymphocytes treated *in vivo* with cyclophosphamide and triethylenemelamine, *Environ. Mutagen.*, **2**, 277–278.
Steinmetz, K.L., and J.C. Mirsalis (1984) Induction of unscheduled DNA synthesis in primary cultures of rat pancreatic cells following *in vivo* and *in vitro* treatment with genotoxic agents, *Environ. Mutagen.*, **6**, 331–330.
Stich, H.F., and D. Kieser (1974) Use of DNA repair synthesis in detecting organotropic actions of chemical carcinogens. *Proc. Soc. Exp. Biol. Med.*, **145**, 1339–1342.
Tyson, C.K., and J.C. Mirsalis (1985) Measurement of unscheduled DNA synthesis in rat kidney cells following *in vivo* treatment with genotoxic agents, *Environ. Mutagen.*, **7**, 889–899.
Williams, G.M. (1977) Detection of chemical carcinogens by unscheduled DNA synthesis in rat liver primary cell cultures, *Cancer Res.* **37**, 1845–1851.
Working, P. and B.E. Butterworth (1984) An assay to detect chemically induced DNA repair in rat spermatocytes, *Environ. Mutagen.*, **6**, 273–286.

Byron E. Butterworth and Tracey Smith-Oliver
Chemical Industry Institute of Toxicology Research Triangle Park, North Carolina, USA

56

A comparison of the response of carcinogen/noncarcinogen pairs in the rat and mouse hepatocyte DNA repair assays in vivo and in vitro

1 Summary

2-Acetylaminofluorene (2AAF)/4-acetylaminofluorene (4AAF) and benzo(a)pyrene (BP)/pyrene (PYR) are purported to represent two carcinogen/noncarcinogen pairs of structurally similar compounds. These agents were examined for their ability to induce DNA repair *in vivo* or *in vitro* in hepatocytes of Fischer-344 (F344) rats and $B6C3F_1$ mice. For the *in vivo* studies, compounds were administered in corn oil by gavage. Hepatocytes were isolated and incubated with ^{3}H-thymidine, and unscheduled DNA synthesis (UDS) was measured by quantitative autoradiography. For the *in vitro* studies, chemically induced UDS was assessed in treated primary hepatocyte cultures. 2AAF induced DNA repair in rat hepatocytes both *in vivo* and *in vitro*, consistent with the observation that 2AAF is a genotoxic hepatocarcinogen. 2AAF was negative in mouse hepatocytes both *in vivo* and *in vitro*, indicating species differences in the metabolic activation of 2AAF. This observation suggests a reduced carcinogenic potential for 2AAF in the mouse liver relative to the rat liver, and possibly even a different mechanism of action of carcinogenicity in these two species. 4AAF induced DNA repair in rat hepatocytes in culture but did not induce a UDS response in the whole animal. In rats, 4AAF did induce a dramatic increase in the percentage of cells in S-phase at 24 hours following treatment. This pattern of activity indicates that while 4AAF may not be a genotoxic carcinogen it could well possess promoting or cocarcinogenic activity. Unlike carbon tetrachloride, no liver toxicity was observed with 4AAF, suggesting that the increase in S-phase was a mitogenic response rather than regenerative synthesis caused by hepatotoxicity. No UDS response was seen with 4AAF in mouse hepatocytes *in vivo* or *in vitro*. There was only a weak induction of cells in S-phase in the mouse by 4AAF suggesting that a metabolite of 4AAF may be responsible for this effect that is produced in the rat but not in the mouse. BP induced DNA repair *in vitro* but not *in vivo* in both the mouse and rat. This is consistent with the observation that BP is not a liver carcinogen in feeding studies and is more likely to produce tumors at the site of administration. PYR was negative in all systems examined consistent with the observation that it is not a carcinogen. These data illustrate the value of comparison of genotoxic activity in the whole animal with cell culture models to permit a more meaningful understanding of mechanism of action and more realistic predictions of risk.

2 Introduction

2-Acetylaminofluorene (2AAF)/4-acetylaminofluorene (4AAF) and benzo(a)pyrene (BP)/pyrene (PYR) are examples of pairs of structurally similar compounds which exhibit dramatic differences in carcinogenic potency. They were chosen because cell culture assays for genotoxicity did not reflect the observed differences in carcinogenic activity in the whole animal (Ashby and de Serres, 1981). The assays employed in this study measure chemically induced DNA repair in hepatocytes from the treated animal (Mirsalis and Butterworth, 1980). In addition, stimulation of cells in S-phase can also be assessed in this assay as an indication of potential promoting or cocarcinogenic activity (Doolittle *et al.*, 1983). The *in vivo* results were compared with those obtained by treating primary hepatocyte cultures directly (Williams, 1977). The results between rats and mice were also contrasted. The information obtained clearly demonstrated that whole animal assays such as these provide a far better understanding of differences in activity relating to species, target organ, and cell culture vs the intact animal.

3 Materials and methods

CDF (F-344)/CrIBR male rats weighing 180–220 g and male B6C3F1 mice weighing 28–33 g were obtained from Charles River Breeding Laboratories (Kingston, NY). Upon arrival, all animals were housed in hanging stainless steel mesh cages (Hazelton Systems inc., Vienna, VA) in a mass air displacement room (Bioclean, Hazleton Systems Inc., Vienna, VA). Animals were given food (NIH-07) open formula pelleted chow (Ziegler Bros., Gardner, PA) analyzed for nutrients and contaminants (Lancaster Labs, York, PA) ad libitum. Filter-purified tap water was given ad libitum using an automatic watering system. Randomly selected animals necropsied on arrival were free of known rodent pathogens (standard screening necropsy, Microbiological Associates, Bethesda, MD). Sentinel animals, bled at weekly intervals, were free of virus titers using the standard rat murine viral antibody screening procedure (Microbiological Associates, Bethesda, MD). Lights were on an automatic timer set for 12 h of light and 12 h of darkness. Animal room temperature was 70 ± 2 °F and humidity was 50 ± 5%.

All test chemicals were of the highest purity prepared especially for this study and provided to the investigators by the IPCS. Dimethylnitrosamine (DMN)

Table 1. *DNA repair in VIVO in rat hepatocytes*

Compound	Dose (mg/kg)	Time (hours)	N	Net nuclear grains (NG) ± SE	Per cent of cells with 5 NG ± SE	Per cent of cells in S-phase ± SD	Score[a] UDS	Score[a] S-phase
2AAF	10	12	3[b]	18.1 ± 5.1	84 ± 10		+	
2AAF	50	12	3	43.6 ± 4.3	100 ± 0		+	
4AAF	50	2	2	− 5.4 ± 0.2	1 ± 0		−	
4AAF	50	12	2	− 2.9 ± 0.1	4 ± 1		−	
4AAF	50	24	2	− 3.9 ± 0.4	0 ± 0	0.3 ± 0.1	−	−
4AAF	200	2	1	− 6.4	1		−	
4AAF	200	12	2	− 2.9 ± 1.3	3 ± 2		−	
4AAF	200	24	2	− 4.6 ± 0.1	3 ± 0	10.1 ± 3.0	−	+
BP	100	2	1	− 5.9	1		−	
BP	100	12	2	− 3.8 ± 0.6	2 ± 1		−	
BP	100	24	2	− 5.4 ± 0.1	0 ± 0	0.5	−	
BP	100	48	2	− 8.5 ± 1.6	0 ± 0	1.5	−	w +
BP	200	12	1	− 8.2	0		−	
BP	200	24	1	− 6.3	1		−	
BP	200	48	1	− 8.5	0		−	
PYR	100	12	2	− 5.3 ± 0.9	1 ± 0		−	
PYR	100	24	2	− 5.8 ± 0.9	0 ± 0		−	
PYR	100	48	2	− 7.1 ± 2.0	0 ± 0		−	
PYR	200	12	1	− 4.4	0		−	
PYR	200	24	1	− 5.4	0	0.2	−	−
PYR	200	48	1	− 5.3	1	0.1	−	−
Corn oil	10 ml/kg	2	1	− 5.7	0		−	
Corn oil	10 ml/kg	12	6	− 5.2 ± 0.5	1 ± 0		−	
Corn oil	10 ml/kg	24	3	− 6.5 ± 2.3	0 ± 0	0.2 ± 0.0	−	
Corn oil	10 ml/kg	48	1	− 5.7	0 ± 0		−	
Average of 8 controls from various time points						0.2 ± 0.2		
Control	Media	*In vitro*	2	− 9.6 ± 0.1	1 ± 0		−	
DMSO	1%	*In vitro*	6	− 8.4 ± 1.2	5 ± 1		−	
2AAF	0.01 mM	*In vitro*	8	73.2 ± 8.7	100 ± 0		+	
DMN	1.0 mM	*In vitro*	1	42.6 ± 5.6	97 ± 5		+	

[a] + positive; − negative; w + weak positive or marginal.
[b] Number of animals.

(Aldrich, Milwaukee, WI, > 99% pure) was used as a positive control.

The *in vitro* hepatocyte assays were performed as described by Williams (1977). Mouse hepatocytes were prepared as described by Smith-Oliver and Butterworth (1986). Medium containing the compound was prepared from a stock solution in dimethyl sulfoxide (DMSO). The maximum DMSO concentration was 1%. Medium containing the DMN was prepared from a freshly prepared stock solution in medium. After an attachment period of 1.5 h the hepatocyte cultures were incubated for 18 h in the presence of the test compound and 10μCi/ml ^{3}H-thymidine. Slides were dipped in NTB-2 photographic emulsion (Kodak, Rochester, NY) and exposed for 12 to 14 days at − 20 C.

The *in vivo–in vitro* rat hepatocyte DNA repair assay was performed as described by Mirsalis *et al.*, 1982. Compounds were given by gavage in solution or suspension in corn oil (10 ml/kg body weight). DMN was given by gavage as a freshly prepared water solution (0.2 ml/kg). Hepatocyte cultures were prepared at the times indicated after treatment. After an attachment period of 1.5 h, the cultures were incubated with 10 μCi/ml ^{3}H-thymidine for 4 h followed by 14 h in medium containing no radioactivity and supplemented with 0.25 mM thymidine. Autoradiography was performed as described for the *in vitro* rat hepatocytes with a 12 to 14 day exposure period.

The number of silver grains over the nucleus minus the grains over an equal sized area in the cytoplasm was defined as net grains per nucleus (NG) and was quantitated with an automatic grain counter as described previously (Mirsalis *et al.*, 1982). A cell with greater than 5 NG was considered in repair for both the rat and mouse cells. In most cases 50 cells were counted per slide for 3 slides (150 cells) for each data point for the *in vitro* studies. 50 cells were counted per slide, for 3 slides, for 3 animals (450 cells) for each data point for the *in vivo* experiments. Near the end of these studies it was determined that 25 cells per slide was sufficient to obtain an accurate count. For a few of the experiments the number of cells counted was 25 per slide. A cell in S-phase is easily recognized by the very dense pattern of silver grains over the nucleus. The percentage of cells in S-phase were scored for selected *in vivo* animals. 1000 cells were scored per slide for 3 slides (3000 cells). Data are presented as the value ± SE (animal to animal variation) unless otherwise noted.

The serum alanine aminotransferase assays were performed by the method of Bergmeyer *et al.* (1978) with an Abbott VP.

Table 2. *Serum alanine aminotransferase activities in treated animals*

Compound	Dose (mg/kg)	Time	Unit/liter ± SD
Rats			
2AAF	50	24	89 ± 13
4AAF	200	24	54 ± 1
CCl_4	250	24	2200 ± 1500
Corn oil	10 ml/kg	24	47 ± 1
Mice			
2AAF	50	24	67 ± 26
4AAF	200	24	56 ± 19
CCl_4	50	24	> 8000
Corn oil	10 ml/kg	24	67 ± 22

4 Results

2AAF induced DNA repair in the rat *in vivo* in a dose dependent manner (Table 1). 4AAF did not induce DNA repair but did produce a strong S-phase response at 24 h posttreatment. BP did not induce DNA repair in the rat *in vivo* but did produce a small S-phase response. PYR was negative for both DNA repair and S-phase induction (Table 1). In every experiment a portion of control cells were treated with 2AAF or DMN *in vitro* to verify that a positive response could be detected (Table 1). 4AAF did not produce liver toxicity as evidenced by no increase in the serum concentration of alanine animotransferase (Table 2).

In the *in vitro* rat hepatocyte DNA repair assay

Table 3. *DNA repair in vitro in rat hepatocytes*

Chemical	Dose (mM)	NG ± SE[a]	%IR ± SD	Score[b]
2AAF	0.001	56.3 ± 7.6	100 ± 0	+
2AAF	0.01	59.6 ± 3.9	100 ± 0	+
2AAF	0.1	100.0 ± 5.6	100 ± 0	+
4AAF	0.001	− 4.4 ± 0.6	1 ± 2	−
4AAF	0.01	1.2 ± 1.5	22 ± 16	w +
4AAF	0.1	21.4 ± 2.0	98 ± 2	+
BP	0.001	− 4.0 ± 0.8	0 ± 0	−
BP	0.01	0.1 ± 0.8	17 ± 5	w +
BP	0.1	11.8 ± 1.3	88 ± 8	+
PYR	0.001	− 3.7 ± 0.3	0 ± 0	−
PYR	0.01	− 3.8 ± 1.1	2 ± 5	−
PYR	0.1	− 3.9 ± 1.0	0 ± 0	−
DMN	1.0	40.2 ± 1.1	100 ± 0	+
DMN	10.0	62.0 ± 1.6	100 ± 0	+
Control	Media	− 4.7 ± 0.4	0 ± 0	−
DMSO	1%	− 3.3 ± 0.9	0 ± 0	−

[a] 3 slides examined from cells from one animal.
[b] See Table 1.

2AAF, 4AAF, and BP were positive in a dose related fashion while PYR was negative (Table 3).

In the mouse *in vivo* hepatocyte DNA repair assay neither 2AAF nor 4AAF produced a response (Table 4). 4AAF, however, did produce a weak S-phase response. BP was negative for both UDS and S-phase. DMN does produce a strong DNA repair response in the mouse (Table 4).

Both 2AAF and 4AAF were negative in the mouse *in vitro* hepatocyte DNA repair assay (Table 5). BP produced a UDS response in the primary mouse hepatocyte cultures, while PYR did not. DMN was a potent inducer of UDS in the mouse hepatocytes (Table 5).

5 Discussion

2AAF induced DNA repair in rat hepatocytes both *in vivo* and *in vitro*, consistent with the observation that 2AAF is a genotoxic hepatocarcinogen. 2AAF, however, was without activity in the mouse *in vivo* and *in vitro* DNA repair assays. These data indicate dramatic and significant species differences in the metabolic activation of 2AAF to the ultimate genotoxic form and suggest that 2AAF will be a more potent carcinogen in the rat than in the mouse. The binding data presented in this study confirm that there is far greater binding of 2AAF to rat DNA than to mouse DNA (Randerath, this volume). It is probable that there was not sufficient binding of 2AAF to mouse liver DNA to produce a measurable DNA repair response. 2AAF is a liver carcinogen in the mouse (Staffa and Mehlman, 1979); however, the time to tumor is much longer than in the rat (Hicks, this volume). This suggests that the principal mechanism of action of 2AAF may be direct genotoxicity in the rat and forced cell proliferation in the mouse.

4AAF induced DNA repair in treated cultures of rat hepatocytes but was without this activity in the treated animal. 4AAF did induce a dramatic increase in the percentage of cells in S-phase 24 hours following treat-

Table 4. *DNA repair in vivo in mouse hepatocytes*

Compound	Dose (mg/kg)	Time (hours)	N	Net Nuclear grains (NG) ± SE	Per cent of cells with 5 NG ± SE	Per cent of cells in S-phase ± SD	Score[a] UDS	Score[a] S-phase
2AAF	50	12	2b	− 2.6 ± 0.9	0 ± 0		−	
4AAF	200	24	2	− 3.9 ± 0.7	1 ± 1	1.0 ± 0.4	−	w +
BP	200	12	2	− 2.1 ± 0.5	0 ± 0		−	
BP	200	24	2	− 2.3 ± 1.0	0 ± 0	0.4 ± 0.5	−	−
DMN	10	2	2	22.5 ± 10.3	85 ± 11		+	
Corn oil	10 ml/kg	12	1	− 5.5	0 ± 0	0.2	−	
Corn oil	10 ml/kg	24	1	− 2.1	3 ± 5	0.7	−	
Average of the 12 and 24 hr controls						0.5 ± 0.4		
DMN	1.0 mM	*In vitro*	2	54.2 ± 12.7	100 ± 0		+	
Control	Media	*In vitro*	2	− 6.6 ± 0.6	1 ± 1		−	

[a] See Table 1.
[b] Number of animals.

Table 5. *DNA repair in vitro in mouse hepatocytes*

Chemical	Dose (mM)	NG ± SE[a]	% of cells > 5 NG ± SD	Score[b]
2AAF	0.001	− 8.8 ± 4.2	4 ± 7	−
2AAF	0.01	− 4.0 ± 4.0	4 ± 4	−
2AAF	0.1	− 2.7 ± 0.3	7 ± 2	−
4AAF	0.001	− 7.2 ± 2.2	1 ± 2	−
4AAF	0.01	− 10.2 ± 1.9	1 ± 2	−
4AAF	0.1	− 7.8 ± 3.0	4 ± 4	−
BP	0.001	− 7.0 ± 1.0	0 ± 0	−
BP	0.01	22.2 ± 5.6	90 ± 6	+
BP	0.1	24.7 ± 8.3	83 ± 20	+
PYR	0.001	− 9.2 ± 1.3	0 ± 0	−
PYR	0.01	− 8.9 ± 1.3	0 ± 0	−
PYR	0.1	− 7.9 ± 1.6	3 ± 2	−
DMN	10.0	56.8 ± 5.0	100 ± 0	+
DMN	1.0	49.6 ± 3.6	100 ± 0	+
Control	Media	− 8.6 ± 0.7	0 ± 0	−
DMSO	1%	− 8.5 ± 0.9	0 ± 0	−

[a] 3 slides examined for cells from one animal.
[b] See Table 1.

ment of the rats confirming that the compound did reach the liver. It has been proposed that there are three useful classifications of *in vivo* hepatotoxic agents (Butterworth *et al.*, 1985). Type 1 are agents that are genotoxic in the liver as evidenced by induced DNA repair, identification of DNA adducts, etc. The second are agents such as di(2-ethylhexyl)phthalate that show no primary genotoxicity but induce hyperplasia and yield a maximum S-phase response following a single dose at about 24 hours (Butterworth *et al.*, 1984). The third are agents such as carbon tetrachloride that also exhibit no primary genotoxic activity but are hepatotoxic and yield liver specific enzymes in the serum. Hepatocyte proliferation represents regenerative growth and is seen at about 48 hours after treatment. 4AAF appears to be a type 2 agent according to the classification of Butterworth *et al.* (1985). There are indications that forced cell proliferation may result in promotional effects (Argyris and Slaga, 1981). Further, cell turnover may increase the rate of spontaneous mutations (Stott *et al.*, 1981) and can have the effect of converting premutagenic lesions into mutations before they can be repaired. For example, the fact that technical grade dinitrotoluene is both genotoxic and forces cell proliferation contributes to its strong potency as a liver carcinogen (Mirsalis and Butterworth, 1982). The induction of mouse liver tumors may be particularly sensitive to agents that force cell proliferation (Doull *et al.*, 1983). There appears to be a class of nongenotoxic agents which produce mouse liver tumors but do not produce tumors at other sites or in other species (Ward *et al.*, 1979). The profile of activity of 4AAF suggests that while the compound may not be a genotoxic carcinogen it could well possess promoting or cocarcinogenic activity. 4AAF is without UDS activity in mouse hepatocytes *in vitro* or *in vivo*, corresponding to the species differences observed for 2AAF. In general, the chemically induced S-phase response in mouse hepatocytes is much stronger than that seen in the rat (Butterworth *et al.*, 1984; Smith-Oliver and Butterworth, 1986). Yet, 4AAF induced only a weak S-phase response in mouse hepatocytes compared with the strong response observed in the rat. This suggests that a metabolite of 4AAF is responsible for the S-phase induction with far greater production in the rat than in the mouse.

BP induced DNA repair in primary cultures of rat and mouse hepatocytes but not in hepatocytes in the intact animal of either species. BP does not produce liver tumors in feeding studies (IARC, 1973). Studies confirm that there is far less binding of BP to liver DNA than of 2AAF when the compounds are administered to the animal (Randerath, this volume). Feeding, skin painting, and inhalation studies clearly indicate that tumor induction by BP is related to the route of administration and predominantly affects those cells that come in direct contact with the compound. The potent genotoxicant 1, 6-dinitropyrene (1, 6-DNP) exhibits an identical pattern of activity to BP, being active *in vitro* but not *in vivo* (Butterworth *et al.*, 1983). It is possible that there is limited distribution of polycyclic aromatic hydrocarbons such that there is not enough delivered to the liver to produce a measureable DNA response. If BP administration to rats was preceded by a two thirds partial hepatectomy and followed by feeding a diet containing phenobarbital for a year then some liver tumors were observed in addition to the tumors at the site of BP administration (Kitagawa *et al.*, 1980).

No genotoxic activity was observed for PYR in any of the systems employed, consistent with its observed lack of carcinogenic activity in animal models.

6 Acknowledgments

We thank Douglas Neptun for performing the serum enzyme analysis.

7 References

Argyris, T.S. and T.J. Slaga (1981) Promotion of carcinomas by repeated abrasion in initiated skin of mice, *Cancer Res.*, **41**, 5193–5195.

Ashby, J. and F.J. de Serres (1981) Evaluation of Short-Term Tests for Carcinogens – Report of the International Collaborative Program. *Progress in Mutation Research*, Volume 1, Elsevier/North-Holland.

Bergmeyer, H.U., P. Scheibe and A.W. Wahlefeld (1978) Optimization of methods for aspartate aminotransferase and alanine aminotransferase, *Clin. Chem.*, **24**, 58–73.

Butterworth, B.E., L.L. Earle, S. Strom, R. Jirtle and G. Michalopoulos (1983) Induction of DNA repair in human and rat hepatocytes by 1, 6-dinitropyrene, *Mutat. Res.*, **122**, 73–80.

Butterworth, B.E., E. Bermudez, T. Smith-Oliver, L. Earle, R. Cattley, J. Martin, J Popp, S. Strom, R. Jirtle and G. Michalopoulos (1984) Lack of genotoxic activity of di(2-ethylhexyl)phthalate (DEHP) in rat and human hepatocytes, *Carcinogenesis*, **5**, 1329–1335.

Butterworth, B.E., D. Loury and T. Smith-Oliver (1985) The value of measurement of both genotoxicity and forced cell proliferation in assessing the potential carcinogenicity of chemicals. *Proceedings of the Fourth International Conference on Environmental Mutagens*, p. 214.

Doolittle, D.J., J.M. Sherrill and B.E. Butterworth (1983) Influence of intestinal bacteria, sex of the animal, and position of the nitrogroup on the hepatic genotoxicity of nitrotoluene isomers *in vivo, Cancer Res.*, **43**, 2836–2842.

Doull, J., B.A. Bridges, R. Kroes, L. Golberg, I.C. Munro, O.E. Paynter, E., H.C. Pitot, R. Squire, G.M. Williams and W.J. Darby (1983) *The relevance of mouse liver hepatoma to human carcinogenic risk.* Report of the International Expert Advisory Committee to the Nutrition Foundation. The Nutrition Foundation, 888 Seventeenth Street, N.W., Washington, D.C. 20006.

IARC Monographs on the Evaluation of Carcinogenic Risk of the Chemical to Man (1973) Volume 3. International Agency for Research on Cancer, Lyon, France, pp. 91–136.

Kitagawa, T., T. Hirakawa, T. Ishikawa, N. Nemoto and S. Takayama (1980) Induction of hepatocellular carcinoma in rat liver by initial treatment with benzo(a)pyrene after partial hepatectomy and promotion by phenobarbital, *Toxicol. Lett.*, **6**, 167–171.

Mirsalis, J.C. and B.E. Butterworth (1980) Detection of unscheduled DNA synthesis in hepatocytes isolated from rats treated with genotoxic agents: an *in vivo–in vitro* assay for potential carcinogens and mutagens, *Carcinogenesis*, **1**, 621–625.

Mirsalis, J.C., K.C. Tyson and B.E. Butterworth (1982) Detection of genotoxic carcinogens in the *in vivo–in vitro* hepatocyte DNA repair assay, *Environ. Mutagen.*, **4**, 553–562.

Staffa, J.A. and M.A. Mehlman, (Eds), (1979) *Innovations in Cancer Risk Assessment* (ED01 Study). Pathotox Publishers, Park Forest South, IL, 246 pp.

Smith-Oliver, T. and B.E. Butterworth, (1986) Correlation of the carcinogenic potential of di(2-ethylhexyl)phthalate

(DEHP) with induced hyperplasia rather than with genotoxic activity, *Mutat. Res.* (in press).
Stott, W.T., R.H. Reitz, A.M. Schumann and P.G. Watanabe (1981) Genetic and nongenetic events in neoplasia, *Food and Cosmet. Toxicol.*, **19**, 567–576.
Ward, J.M., R.A. Griesemer and E.K. Weisburger (1979) The mouse tumor as an endpoint in carcinogenesis testing, *Toxicol. Appl. Pharmacol.*, **51**, 389–397.
Williams, G.M. (1977) Detection of chemical carcinogens by unscheduled DNA synthesis in rat liver primary cell cultures, *Cancer Res.*, **37**, 1845–1851.

John Ashby, Brian Burlinson and
Paul A. Lefevre
Imperial Chemical Industries PLC, Central Toxicology Laboratories, Alderley Park, Cheshire, UK

57

Activity of 2-acetylaminofluorene and 4-acetylaminofluorene in an in vitro and an in vivo/in vitro assay for unscheduled DNA synthesis in the rat liver

1 Summary

2AAF gave a clear positive response in the *in vivo/in vitro* liver UDS assay at dose levels between 5-100 mg/kg. In contrast, 4AAF gave only a very weak positive at the elevated dose level of 1000 mg/kg. The mean net grain count (NG) observed for 4AAF (2.18 ± 2.4 based on 13 rats derived from 3 separate experiments) is below our positive test criterion (3NG). Due to the reproducibility of the response, we have concluded weak positive activity for this agent. All determinations were made 12 h after oral gavage of the test agent in corn oil to either Alderly Park or Fischer 344 rats.

2AAF appears to be about 500 times as active as 4AAF, i.e., about 2 mg/kg of 2AAF gives a similar UDS response to 1000 mg/kg of 4AAF at the same timepoint.

Additional UDS data on 2AAF and 4AAF are described in the paper by Beije and Ashby (submitted to work group as a separate document, experiments conducted over the same period in the same laboratory). A similar profile of activities was observed for each chemical in the *in vitro* hepatocyte UDS assay, with 2AAF giving a strong positive response and 4AAF being weakly active at elevated dose levels.

2 Materials and methods

Chemicals
From the central stock

Table 1. *Activity observed for 2AAF and 4AAF in an in vivo/in vitro assay for UDS in rat liver*

						Individual animal data		Cumulative data				
Cmpd	Strain	Length of exp. (hrs)	Dose mg/kg	No. of cells	No. of slides	Net grains (NG)	% repair	No. of animals	(NG) ± SD	% repair ± SD		(NG) for cells in repair (N-C ≥ 5) ± SD
2AAF	AP	12	5	150	3	3.83	41				8.3	
				150	3	2.41	25	3	4.4 ± 2.3	44 ± 20.8	7.2	8.5 ± 1.4
				100	2	6.98	66				10.0	
			1	150	3	0.82	12				6.5	
				125	3	0.14	4	3	0.28 ± 0.58	6.7 ± 4.6	6.6	7.4 ± 1.4
				100	2	0.35	4				9	
			0.1	150	3	−0.42	2				6	
				150	3	−0.33	0	3	−0.22 ± 0.23	1.7 ± 1.5	—	5.9 ± 0.2
				100	2	0.04	3				5.7	

Table 1 (*Cont.*)

Cmpd	Strain	Length of exp. (hrs)	Dose mg/kg	No. of cells	No. of slides	Individual animal data: Net grains (NG)	Individual animal data: % repair	Cumulative data: No. of animals	(NG) ± SD	% repair ± SD		(NG) for cells in repair (N-C ≥ 5) ± SD
Corn Oil	AP	12	10 ml/kg	150	3	−1.4	0	1	−1.4	0		
2AAF	AP	12	5	100	2	10.31	70	3	11.56 ± 2.75	70.9 ± 11.3	14.11	15.6 ± 1.9
				150	3	14.72	38				17.8	
				50	1	9.66	60				14.9	
Corn Oil	AP	12	—	150	3	−2.23	9.3	1	−2.23	9.33	6.7	6.71
2AAF	F344	12	100	100	3	43.13	100	2	41.03 ± 2.96	100 ± 0	43.1	41.03 ± 2.96
				50	2	38.94	100				38.9	
Corn Oil				100	2	−2.16	6	1	−2.16	6	5.7	5.67
2AAF	F344	12	100	150	3	50.22	98	1	50.22	98	51.2	51.22
Corn Oil				100	3	−2.21	2	1	−2.21	2	5.7	5.67
4AAF	F344	12	1000	150	3	6.06	58	4	5.08 ± 2.22	47.7 ± 15.5	9.7	9.33 ± 1.96
				150	3	3.45	36				8.7	
				125	3	7.74	64				11.8	
				100	2	3.07	33				7.1	
6BT	F344	12	40	150	3	31.31	95	2	35.36 ± 5.73	97.67 ± 3.30	32.8	36.12 ± 4.65
				150	3	39.41	100				39.4	
Corn Oil	F344	12	—	100	2	−2.67	0	1	−2.67	0	—	0
4AAF	AP	12	1000	100	2	−2.35	4	1	−2.35	0.44	8.8	8.75
6BT	AP	12	50	150	3	53.71	99.33	2	34.45 ± 27.23	89.66 ± 13.67	54.1	36.53 ± 24.79
				150	3	15.19	80				19	
2AAF	AP	12	100	150	3	57.81	98	4	59.4 ± 10.2	99.5 ± 1	58.9	59.66 ± 10.12
				150	3	65.89	100				65.9	
				150	3	45.67	100				45.7	
				150	3	68.15	100				68.2	
4AF	AP	12	1000	75	3	1.79	25	2	1.81 ± 0.03	26 ± 0.95	9.3	9.75 ± 0.73
				150	3	1.83	27				8.2	
Corn Oil	AP	12	—	150	3	−3.35	1	1	−3.35	1.33	5.5	5.5
2AAF	F344	12	100	150	3	32.19	92	3	25.64 ± 8.66	93.53 ± 3.21	34.8	27.22 ± 9.5
				75	3	28.92	97				29.7	
				150	3	15.82	91				17.2	
2AAF	F344	12	15	150	3	8.41	61	2	11.54 ± 4.43	70.35 ± 13.67	12.6	15.37 ± 3.95
				150	3	14.67	80				18.2	
4AAF	F344	12	1000	150	3	0.12	14	3	0.18 ± 0.13	15.11 ± 5.09	6.6	6.39 ± 0.29
				150	3	0.09	21				6.5	
				150	3	0.33	11				6.1	
4AAF	F344	12	100	150	3	0.31	15	3	−0.12 ± 0.59	9.77 ± 4.68	10.6	7.9 ± 2.43
				150	3	−0.79	5				6	
				150	3	0.13	9				7.1	
Corn Oil	F344	12	—	150	3	−3.36	1	1	−3.36	1.33	5.5	5.50
2AAF	F344	12	100	125	3	10.12	73	3	13.8 ± 2.7	85 ± 10.4	13.9	15.8 ± 1.6
				125	3	15.14	90				16.5	
				150	3	15.67	92				16.8	
4AAF	F344	12	1000	150	3	−0.17	16	6	1.26 ± 0.99	21.5 ± 3.6	7.8	8.5 ± 1.5
				150	3	1.9	24				7.9	
				150	3	1.9	23				8.1	
				150	3	1.4	23				8.8	
				150	3	2.3	25				11.3	
				150	3	0.25	18				7	
Corn Oil	F344	12	10 ml/kg	100	2	−1.98	1	1	−1.98	1	5	
4AAF	F344	12	1000	Cumulative for 3 experiments				13	2.18 ± 2.4	28 ± 16.2		8.3 ± 1.8
Historical controls	AP			1225	25			40	−2.12 ± 1.09	2.96 ± 3.72		6.34 ± 1.83
Historical controls	F344			500	10			4	−2.6 ± 0.48	2.00 ± 2.5		5.67 ± 0

See legend to Table 2. 6BT = 6-p-dimethylaminophenylazobenzthinzole (Ashby *et al.*, 1985).

3 Methods

In vivo/in vitro assay. This was conducted exactly as described in Ashby *et al.* (1985). Dosing was by gavage in corn oil, perfusions were conducted 12 h later. *In vitro* assay. Only a single experiment was conducted, as described earlier for 2AAF (Ashby *et al.*, 1985).

4 Results

The *in vivo* data are shown in Table 1, presented in the order of their conduct (double lines separate different experiments). 2AAF gave a positive response, 4AAF gave a very weak positive response. The *in vitro* data are shown in Table 2. In this experiment the 2AAF response was unusually strong (cf. other examples in Ashby *et al.*, 1985). Also, the control value was slightly elevated. 4AAF gave a dose-related increase in UDS and both compounds were clearly positive in the assay. Both compounds initiated UDS *in vitro*.

5 References

Ashby, J., P.A. Lefevre, B. Burlinson and M.P. Penman (1985). An assessment of the *in vivo* rat liver UDS assay, *Mutation Res.*, **156**, 1–18.

Table 2. *In vitro hepatocyte UDS response for 2AAF and 4AAF*

Cmpd	Dose M	No. of cells	No. of slides	(NG) (a)	% repair	SE cell	SD slide	(NG) for cells in repair (N-C $\geqslant$ 5) $\pm$ SD (b)
DMSO	—	75	3	1.5	37	1.1	5.0	9.8
2AAF	10^{-7}	75	3	136.4	100	6.4	16.5	136.4
4AAF	10^{-7}	75	3	4.6	55	1.1	2.3	11.1
	10^{-6}	75	3	4.2	31	0.9	2.2	11.8
	10^{-5}	75	3	14.2	80	1.2	5.1	17.9
	10^{-4}	75	3	17.6	89	1.4	2.4	19.9
	10^{-3}	75	3	25.6	96	1.6	1.7	26.7

[a] NG represents the net grain count

[b] The NG count is derived from the nuclear count minus the cytoplasmic count (N-C). When this number is 5 or greater the cell is considered to be undertaking DNA repair. The extent of this repair is shown by this column.

Jon C. Mirsalis, C. Kim Tyson, Erica K.N. Loh, James P. Bakke, Marilyn J. Ramsey, Carol M. Hamilton and Karen L. Steinmetz

Cellular and Genetic Toxicology Department, SRI International, Menlo Park, California, USA

58

An evaluation of the ability of benzo(a)pyrene, pyrene, 2- and 4-acetylaminofluorene to induce unscheduled DNA synthesis and cell proliferation in the livers of male rats and mice treated in vivo

1 Summary

The *in vivo*/*in vitro* hepatocyte DNA repair assay was used to examine the induction of unscheduled DNA synthesis (UDS) or hepatic cell proliferation by 2-acetylaminofluorene (2AAF), 4-acetylaminofluorene (4AAF), benzo(a)pyrene (BP) or pyrene (PYR) in male B6C3F1 mice or male Fischer-344 rats. BP and PYR showed no elevations in UDS or cell proliferation in rats following *in vivo* treatment. BP did induce UDS in isolated hepatocytes from male Cynomolgus monkey and adult human following *in vitro* treatment. 2AAF showed a significant elevation in both UDS and cell proliferation in rats, but in mice no UDS was induced and an equivocal cell proliferation response was observed. In the *in vitro* UDS assay, 2AAF induced UDS in hepatocytes from rat, male Syrian Golden hamster, and adult male human, but not in mouse or monkey. 4AAF induced an equivocal UDS response in rat and no increase in UDS in mouse. 4AAF induced significant increases in cell proliferation in both rat and mouse. These results indicate that in rat, BP and PYR are clearly non-genotoxic in the liver which is consistent with the non-hepatocarcinogenicity of these compounds. 2AAF is a potent genotoxic agent in rat liver, but not in mouse. 4AAF is, at best, a weak genotoxicant in rat, but the significant increase 4AAF induces in hepatic cell proliferation suggests that it may be a potent hepatic tumor promotor in both rats and mice.

2 Introduction

The measurement of chemically induced DNA repair as unscheduled DNA synthesis (UDS) has been shown to be an excellent predictor of genotoxic and carcinogenic potential. A number of *in vitro* assays for measurement of UDS have been described. Some of these assays employ a variety of cell lines (Trosko and Yager, 1974; San and Stich, 1975; Martin *et al*., 1978), including primary cultures of rat hepatocytes (Williams, 1977). This latter assay offers the advantage of using a target cell that is metabolically competent. Still, these systems often do not reflect the true genotoxicity present in the whole animal, and even the *in vitro* hepatocyte DNA repair assay has been shown to be unresponsive to some classes of compounds, such as nitroaromatics (Bermudez *et al*., 1979; Probst *et al*., 1981).

The *in vivo*/*in vitro* hepatocyte DNA repair assay (Mirsalis and Butterworth, 1980) involves treatment of animals rather than cells in culture and has been shown to be responsive to a wide variety of genotoxic agents, including nitroaromatics (Mirsalis *et al.*, 1982a). In an *in vivo* system, factors related to uptake, metabolic activation, distribution, detoxification, and elimination are inherently accounted for. This assay has been demonstrated to be useful in the study of sex differences, effects of chronic exposure (Mirsalis and Butterworth, 1982), species differences (Mirsalis *et al.*, 1985) and the role of activation of compounds by intestinal microflora (Mirsalis *et al.*, 1982a). In particular, this assay is extremely useful for the detection and study of genotoxic hepatocarcinogens (Mirsalis *et al.*, 1982b).

This assay can also be used for the measurement of hepatic cell proliferation, determined autoradiographically as S-phase synthesis (Mirsalis *et al.*, 1982b). Induction of S-phase synthesis may be an important mechanism in the hepatocarcinogenicity of some nongenotoxic chemicals, especially in the mouse (Mirsalis *et al.*, 1985).

This assay was used to examine the ability of benzo(a)pyrene (BP), 2-acetylaminofluorene (2AAF), pyrene (PYR), and 4-acetylaminofluorene (4AAF) to induce UDS or cell proliferation in the liver from male rats or mice.

3 Methods

Treatment of animals

Fischer-344 male rats (200 to 275 g) were obtained from Hilltop Animal Supply (Chatsworth, CA). B6C3F1 mice (B6C3F1/CrlBR, 19 to 35 g) were obtained from Charles River Breeding Labs (Kingston, NY). Animals were maintained on Purina Rodent Chow #5001 (Ralston Purina Co., St. Louis, MO) and untreated tap water ad libitum. Animals were identified by ear punch and were treated by gavage with the chemical suspended in corn oil. A total volume of 0.3 to 0.5 ml of test chemical solution was administered per 100 g of body weight. Animals were treated from 2–48 h prior to sacrifice.

Chemicals

BP, PYR, 2AAF, and 4AAF were supplied by the International Program on Chemical Safety (IPCS). A negative control (corn oil) was used in each assay. Because 2AAF is a known inducer of UDS in this assay, no additional positive control was used.

Hepatocyte cultures

Primary hepatocyte cultures were obtained from the livers of treated rats and mice according to the procedure of Mirsalis *et al.* (1982b). The cells were inoculated into Linbro 6-well culture dishes (containing coverslips) in Williams' Medium E supplemented with 2 mM l-glutamine, 50 μg/ml gentamicin (WEI), and 10% fetal bovine serum. Cultures were incubated at 37 °C in a 5% CO_2:95% air atmosphere. After 1.5 to 2 h, the cultures were washed to remove nonviable cells (those not attached to the coverslips). All washes and subsequent culturing were done using serum-free medium.

Cultures were incubated in WEI containing 10 μCi/ml ^{3}H-thymidine (specific activity, approximately 80 Ci/m mole) for 4 h, followed by 14 to 16 h in WEI containing 0.25 mM unlabeled thymidine at 37 °C. Cultures were washed, swelled in a hypotonic solution, fixed, and washed with water. The coverslips were mounted to slides, dipped in Kodak NTB-2 emulsion (diluted 1:1 with water), and exposed at −20 °C for 12 to 14 days prior to development. Cells were stained with 1% methyl-green Pyronin Y.

Measurement of UDS

Quantitative autoradiographic grain counting was accomplished using an ARTEK Model 880 or 980 colony

Table 1. *Induction of unscheduled DNA synthesis and hepatic cell proliferation in hepatocytes from male Fischer-344 rats treated in vivo*

Treatment	Dose (mg/kg)	Time (h)	(n)[a]	N.G.[b] ± S.E.[c]	% IR[d] ± S.E.	% S[e] ± S.E.
Control/corn oil	0[f]	2–12[g]	4	−6.2 ± 1.0	1 ± 0	—
		48	6	—	—	0.13 ± 0.06
2AAF	10	12	3	12.8 ± 4.7	74 ± 12	—
		48	3	−0.9 ± 0.6	15 ± 2	0.07 ± 0.03
	50	12	5	22.5 ± 3.9	87 ± 7	—
		48	3	3.0 ± 0.6	34 ± 4	1.15 ± 0.55
4AAF	200	2	3	−5.9 ± 0.2	1 ± 0	—
		12	3	−1.1 ± 2.1	19 ± 8	—
		24	3	−1.0 ± 0.7	21 ± 5	5.26 ± 1.03
		48	3	−4.1 ± 0.1	2 ± 0	0.17 ± 0.03
BP	500	12	3	−3.7 ± 0.5	2 ± 1	—
		24	3	−4.9 ± 0.4	2 ± 1	0.10 ± 0.03
PYR	500	12	3	−4.4 ± 0.2	1 ± 1	—
		24	3	−2.9 ± 1.1	5 ± 2	0.37 ± 0.49

[a] (n) is the number of animals treated.
[b] N.G. is the net grains/nucleus.
[c] S.E. represents variation between animals.
[d] % IR is the percent of cells with ⩾ 5 NG.
[e] % S is the percent of cells in S-phase.
[f] Controls received 3 ml/kg corn oil.
[g] Controls pooled from 2 and 12 h treatments.

counter interfaced with a Zeiss Universal microscope via an ARTEK TV camera. Data were fed directly into a VAX 11/782 computer via an ARTEK BCD-RS232 Omni-Interface. Fifty morphologically unaltered cells on a randomly selected area of the slide were counted. The highest count from two nuclear-size areas over the most heavily labeled cytoplasmic areas adjacent to the nucleus was subtracted from the nuclear count to give the net grains/nucleus (NG). The percentage of cells in repair, which indicates the extent of the response throughout the liver, was calculated as those cells exhibiting at least 5 NG. All collection of data and pooling of slides or animals was done using programs in the VAX 11/782 computer. A minimum of 3 slides were scored for each of 3 animals, for a minimum total sample of 3 animals, 9 slides, and 450 cells/dose/time point.

Measurement of hepatic cell proliferation

Cells in S-phase are easily distinguished from nonreplicating cells in autoradiographic preparations. Slides were scored manually for S-phase cells. Approximately 2000 cells were counted from a randomly selected area of each slide. Replicating and nonreplicating cells were counted separately using a two-position hand counter.

Figure 1. The distribution of net grains/nucleus in 47 male F-344 rats treated with 50 mg/kg 2AAF 12 h prior to sacrifice. The frequency represents the percentage of animals within the specified range of N.G. The crosshatched bars represent the N.G. of the 5 rats tested in this study.

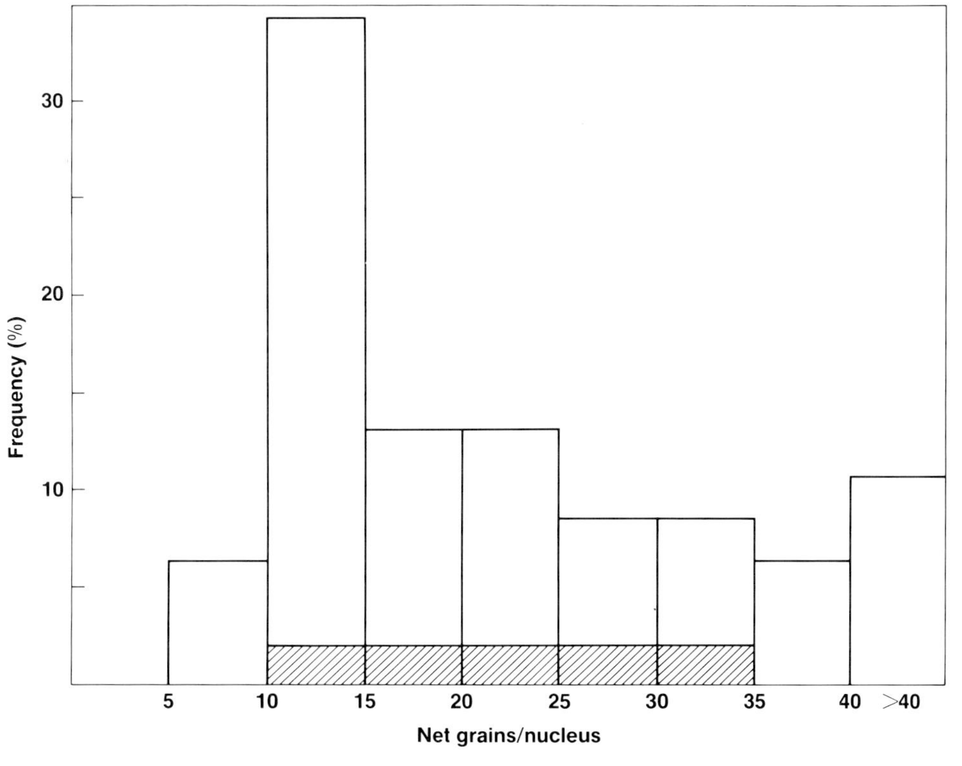

Table 2. *Induction of unscheduled DNA synthesis and hepatic cell proliferation in hepatocytes from male B6C3F1 mice treated in vivo*

Treatment	Dose (mg/kg)	Time (h)	(n)[a]	N.G.[b] ± S.E.[c]	%IR[d] ± S.E.	%S[e] ± S.E.
Control/corn oil	0[f]	2–12[g]	8	−9.5 ± 1.0	1 ± 1	—
		48	7	—	—	0.23 ± 0.06
2AAF	50	12	6	−8.6 ± 1.0	2 ± 1	—
	200	12	3	−4.5 ± 0.7	3 ± 1	—
		24	3	−11.4 ± 1.9	0	0.17 ± 0.06
		48	3	−8.8 ± 0.8	1 ± 0	0.45 ± 0.06
4AAF	200	12	3	−10.4 ± 0.5	0	—
		24	3	−12.0 ± 1.0	1 ± 0	1.31 ± 0.74
		48	3	−9.4 ± 1.6	0	0.07 ± 0.02

[a] (n) is the number of animals treated.
[b] N.G. is the net grains/nucleus.
[c] S.E. represents variation between animals.
[d] % IR is the percent of cells with ⩾ 5 NG.
[e] % S is the percent of cells in S-phase.
[f] Controls received 3 ml/kg corn oil.
[g] Controls pooled from 2 and 12 h treatment.

Table 3. *Induction of unscheduled DNA synthesis in primary cultures of hepatocytes following in vitro treatment with genotoxic agents*

Chemical	Concentration[a] (μg/ml)	Rat NG[b]·SE[c]	%IR[d]	Mouse NG ± SE	%IR	Hamster NG ± SE	%IR	Monkey NG ± SE	%IR	Human NG ± SE	%IR
Control/medium	—	−6.2 ± 0.9	0	−10.0 ± 0.8	0	−3.5 ± 0.6	2	−4.5 ± 0.7	1	−5.7 ± 1.3	8
Control/DMSO	1%	−8.1 ± 1.3	1	−8.8 ± 1.6	0			−3.3 ± 1.7	1	−7.6 ± 1.3	3
2AAF	0.05	11.3 ± 12.5	31	−5.4 ± 1.1	0	−5.3 ± 0.6	11				
	0.1			−7.6 ± 1.3	1	53.2 ± 21.5	97	−5.5 ± 0.6	1	−3.3 ± 4.2	22
	0.5	25.0 ± 12.6	75			36.5 ± 12.8	91	−4.9 ± 0.8	0	12.2 ± 7.6	66
	1			−3.9 ± 2.6	15	e		−4.2 ± 2.3	1	16.7 ± 2.8	88
	5	47.3 ± 4.5	99			e		−6.7 ± 0.8	0	7.8 ± 4.3	51
	10			−8.3 ± 2.5	10	toxic		toxic		19.3 ± 2.6	90
BP	0.1							−3.9 ± 0.9	0	1.1 ± 0.2	34
	0.5							−2.5 ± 0.4	6	22.0 ± 3.1	94
	1							0.4 ± 1.3	20	11.9 ± 3.5	78
	5							4.0 ± 1.7	33	13.0 ± 5.4	77
	10							20.3 ± 2.8	94	21.0 ± 4.4	88
	20							17.2 ± 3.0	100	toxic	

[a] Each concentration represents the mean of 3 cultures, 30 cells scored/culture.
[b] NG is the net grains/nucleus.
[c] SE represents the variation between cultures.
[d] %IR is the percentage of cells with ⩾5 NG.
[e] Grains over the nucleus too dense to count accurately.

In vitro UDS assay

In vitro UDS experiments were performed using 2AAF and BP to examine interspecies differences in liver cells from Fischer-344 rats, B6C3F1 mice, Syrian Golden hamsters, Cynomolgus monkeys, and humans. Isolated hepatocyte cultures were obtained from rat, mouse, and hamster by in situ liver perfusion as described above. Human and monkey liver cell cultures were obtained by biopsy perfusion (Green *et al.*, 1983). Following a 1.5 to 2 h attachment period as described above, cultures were exposed to selected concentrations of 2AAF or BP diluted in DMSO in WEI containing 10 μCi/ml ^{3}H-thymidine for 17 to 21 h at 37 °C. Cultures were fixed, and autoradiography and slide scoring was performed as for *in vivo* experiments, except that autoradiograms were exposed for 7 days prior to development.

4 Results

The *in vivo/in vitro* hepatocyte DNA repair assay was used to examine 2AAF and 4AAF for their ability to induce UDS or hepatic cell proliferation in male rat or mouse liver. In addition, BP and PYR were examined for their ability to induce UDS or S-phase synthesis in rat liver.

2AAF induced a significant elevation in UDS in rats, with the peak response observed 12 h posttreatment (Table 1). The distribution of responses between animals was consistent with the large historical data base available for this compound in our laboratory (Figure 1). In contrast, 2AAF failed to induce UDS in mouse liver (Table 2). 4AAF induced very slight increases in UDS at both 12- and 24-h in rat liver, but no UDS was observed in mouse liver (Table 2). Both BP and PYR failed to induce any increase in UDS in rat liver.

2AAF induced a significant elevation in cell proliferation 48 h after treatment in rats but only a very slight increase in mice. 4AAF induced a significant elevation in the number of cells in S-phase 24 h after treatment in rats and mice. BP and PYR failed to induce an increase in S-phase synthesis in rats; however, only the 24-h time point was examined for this parameter.

2AAF was also a significant inducer of UDS in isolated rat, hamster, and human hepatocytes treated *in vitro* but did not induce UDS in mouse or monkey hepatocytes (Table 3). BP induced UDS *in vitro* in hepatocytes from human and monkey.

5 Discussion

The *in vivo/in vitro* hepatocyte DNA repair assay is widely used for the examination of potential genotoxic and cell proliferative effects of chemicals on rodent liver. This assay was used to examine the effects of 2AAF, 4AAF, BP and PYR in rat and mouse liver.

The strong positive UDS response induced by 2AAF *in vivo* in rat liver suggests that this compound is a potent genotoxic agent in rat liver. 2AAF also induced a significant increase in UDS in rat, hamster and human hepatocytes treated *in vitro*, but not in mouse or monkey hepatocytes. The genotoxicity of 2AAF, combined with the potent induction of S-phase synthesis, would indicate that this compound should be a potent rat hepatocarcinogen and, in fact, 2AAF induces a very high incidence of hepatocellular carcinomas in rats (Tisdel *et al.*, 1984). The negative UDS response observed in mouse liver indicates that an entirely different mechanism of metabolic activation is operating in mouse liver. Pharmacokinetic differences could result in different levels of distribution of 2AAF to mouse liver; however, the negative response observed in isolated mouse hepatocytes is strong evidence that these cells are incapable of metabolising 2AAF to a genotoxic form. 2AAF did induce a very slight increase in S-phase synthesis in mouse liver, indicating that a nongenotoxic metabolite may result in increased cell proliferation in the liver. This increased cell proliferation alone may account for the weak carcinogenicity observed for 2AAF in mouse liver (Tisdel *et al.*, 1984).

4AAF induced a very weak positive UDS response in the rat, although there was considerable variation between animals, and the average net grain count was less than zero. Therefore, this response may be considered to be of limited biological significance. The very significant increase in cell proliferation induced by 4AAF indicates that this compound is effective at stimulating S-phase synthesis in rat liver. A weaker, but still significant, elevation was observed in mouse liver. Compounds that induce increased hepatic cell proliferation often produce an increase in liver tumours, especially in B6C3F1 mice (Mirsalis *et al.*, 1985). Unfortunately, no conclusive bioassay data are available for 4AAF, so correlations to hepatocarcinogenicity are impossible.

The failure of BP to induce UDS in rat liver *in vivo* suggests that this assay may be highly tissue-specific in the detection of genotoxic agents. BP is a potent multi-site carcinogen in many tissues other than the liver; however, numerous carcinogenicity studies have failed to induce liver tumours with this compound in any rodent species. BP does induce UDS in hepatocytes treated *in vitro*, suggesting that the activation and/or detoxification observed *in vitro* is different than that observed *in vivo*. PYR has not been adequately tested in a lifetime bioassay, although preliminary studies have failed to demonstrate any carcinogenicity for this compound.

Measurement of UDS and cell proliferation in the liver following *in vivo* treatment provides needed information on liver-specific effects of chemicals in the whole animal. In addition, species and sex differences are easily examined in these whole animal systems. Therefore, use of the endpoints should provide a valuable addition to the growing battery of *in vivo* genotoxicity tests used for the screening of potential chemical carcinogens.

6 References

Bermudez, E., D. Tillery and B.E. Butterworth (1979) The effect of 2,4-diaminotoluene and isomers of dinitrotoluene on unscheduled DNA synthesis in primary rat hepatocytes, *Environ. Mutagen.*, **1**, 391–398.

Green, C.E., J.E. Dabbs and C.A. Tyson (1983) Functional integrity of isolated rat hepatocytes prepared by whole liver vs. biopsy perfusion, *Anal. Biochem.*, **129**, 269–276.

Martin, C.N., A.C. McDermid and R.C. Garner (1978) Testing of known carcinogens and noncarcinogens for their ability to induce unscheduled DNA synthesis in HeLa cells, *Cancer Res.*, **38**, 2621–2627.

Mirsalis, J.C. and B.E. Butterworth (1980) Detection of unscheduled DNA synthesis in hepatocytes isolated from rats treated with genotoxic agents: An *in vivo–in vitro* assay for potential carcinogens and mutagens, *Carcinogenesis*, **1**, 621–625.

Mirsalis, J.C. and B.E. Butterworth (1982) Induction of unscheduled DNA synthesis in rat hepatocytes following *in vivo* treatment with dinitrotolune, *Carcinogenesis*, **3**, 241–245.

Mirsalis, J.C., T.E. Hamm, Jr., J.M. Sherrill and B.E. Butterworth (1982a) The role of gut flora in the genotoxicity of dinitrotoluene, *Nature*, **295**, 322–323.

Mirsalis, J.C., C.K. Tyson and B.E. Butterworth (1982b) Detection of genotoxic hepatocarcinogens in the *in vivo–in vitro* hepatocyte DNA repair assay, *Environ. Mutagen.*, **4**, 553–562.

Mirsalis, J.C, C.K. Tyson, E.N. Loh, K.L. Steinmetz, J.P. Bakke, C.M. Hamilton, D.K. Spak and J.W. Spalding (1985) Induction of hepatic cell proliferation and unscheduled DNA synthesis in mouse hepatocytes following *in vivo* treatment, *Carcinogenesis*, **6**, 1521–1524.

Probst, G.S., R.E. McMahon, L.E. Hill, C.Z. Thompson, J.K. Epp and S.B. Neal (1981) Chemically induced unscheduled DNA synthesis in primary rat hepatocyte cultures: A comparison with bacterial mutagenicity using 218 compounds, *Environ. Mutagen.*, **3**, 11–32.

San, R.H.C. and H.F. Stich (1975) DNA repair synthesis of cultured human cells as a rapid bioassay for chemical carcinogens, *Int. J. Cancer*, **16**, 284–291.

Tisdel, M.O., R.H. Weltman and R.J. Ryzia (1984) Two-year feeding studies with 2-acetylaminofluorene: comparative histopathology in Sprague-Dawley rats and CD-1 mice, *The Toxicologist*, **4**, 36. (Abstr.)

Trosko, J.E., and J.D. Yager (1974) A sensitive method to measure physical and chemical carcinogen-induced "unscheduled DNA synthesis" in rapidly dividing eukaryotic cells, *Exp. Cell Res.*, **88**, 47–55.

Williams, G.M. (1977) Detection of chemical carcinogens by unscheduled DNA synthesis in rat liver primary cell cultures, *Cancer Res.*, **37**, 1845–1851.

Brita Beije and John Ashby[1]

Genetic Toxicology Laboratory, Environmental Toxicology Unit, Wallenberg Laboratory, Stockholm University, Sweden
[1]Central Toxicology Laboratory, Imperial Chemical Industry PLC, Alderley Park, Cheshire, UK

59

Unscheduled DNA synthesis in rat liver after in vivo treatment with 2-acetylaminofluorene and 4-acetylaminofluorene

1 Summary

2-Acetylaminofluorene (2AAF) or 4-acetylaminofluorene (4AAF) was administered orally to male Alderley Park rats, and 4–36 h later hepatocytes were isolated and processed for autoradiographic measurement of unscheduled DNA synthesis (UDS). 2AAF induced UDS in the liver at dose levels between 50–500 mg/kg body weight. At 50 mg/kg, the induction of UDS seemed to have reached a plateau, with only minor increases in the activity at higher dose levels. 4AAF did not induce UDS at the dose level used, instead there was a dramatic increase in the percentage of hepatocytes in S-phase.

2 Materials and methods

Chemicals
2AAF and 4AAF were dissolved or suspended by sonication in corn oil to concentrations that would allow the administration of 10 ml/kg body weight. For the higher dose levels of 2AAF, the compound was finely ground and maintained as a suspension in corn oil for 4 h at 70 °C. Before dosing the mixture was cooled with concomitant sonication and finally shaken before dosing by gavage.

Animals and dosing
Male Alderley Park rats (9–13 weeks; 190–250 g) were maintained as described earlier (Ashby *et al.*, 1985, Beije and Ashby, 1985). Dosing was by gavage at 9 pm (12- and 36 h exposure), 9 am (24 h exposure), or at 7 am (4 h exposure). The controls received 10 ml corn oil/kg body weight.

Liver UDS assay
The hepatocytes were isolated and processed exactly as described previously (Ashby *et al.*, 1985, Beije and Ashby, 1985). Three slides per animal were read by eye, and 50 cells of normal morphology were assessed from each slide. Eight to twelve rats were used in each experiment. In some cases, the femur was removed from a rat during the liver perfusion. This was done to obtain bone marrow micronucleus data. If the bone marrow micronucleus test was done concomitantly with the liver UDS assay, the exposure time was extended up to 36 h.

3 Results

2AAF induced liver UDS at all dose levels and exposure periods used, whereas 4AAF had no effect on

Table 1. *UDS and S-phase incidence in hepatocytes from AP rats treated in vivo with 2AAF or 4AAF*

						Individual animal data					Cumulative data				% S phases	
						Net grains (NG)										
Cmpd	Strain of rat	Length of exp. (hrs)	Dose mg/kg	No. of cells	No. of slides	Mean	SE/Cells	SD/slides	% repair	No. of animals	(NG ± SD)	%repair ± SD	(NG) for cells in repair (NG ⩾ 5) ± SD	S phases/ 1000 cells	Mean per rat	mean ± SD
Corn Oil	AP	12	10 ml/kg	150	3	− 2.81	0.29	0.59	2.67					—		
				150	3	− 3.22	0.33	1.30	1.33					—		
				100	2	− 2.53	0.37	—	1.0	3	− 2.85 ± 0.35	1.67 ± 0.88	5.75 ± 0.66	0, 1 —	0.05	
		24	10 ml/kg	150	3	− 2.56	0.29	0.30	2.0					2, 0, 0	0.067	
				150	3	− 2.72	0.40	0.77	2.0	2	− 2.64	2.0	7.5	2, 2, 2	0.2	0.13
		36	10 ml/kg	125	3	− 2.45	0.66	0.49	0					0, 2, 0	0.067	
				150	3	− 1.76*	0.32	0.32	5.33					1, 0, 0	0.033	0.05 ± 0.02
				150	3	− 4.59	0.31	0.98	0.67					0, 1, 0	0.033	
				150	3	− 2.37	0.34	0.64	5.33	4	− 2.79 ± 1.24	2.83 ± 2.9	5.96 ± 0.89	1, 0, 1	0.067	
2AAF	AP	5	100	150	3	9.6	0.68	1.95	71.33					—		
				75	2	8.89	1.34	—	64.0	2	9.25	67.67	14.12	—		
		12	50	100	2	40.51	2.05	7.54	100							
				150	3	37.81	1.83	11.54	97.33							
				150	3	45.74	1.94	1.46	98.67							
				100	2	45.35	2.54	2.81	96.00							
				150	3	44.19	2.17	14.52	96.67							
				100	2	48.24	2.70	23.65	100							
				150	3	43.67	2.18	7.21	99.33							
				100	2	40.92	2.15	11.17	98							
				100	2	31.76	2.27	8.63	94	9	42.02 ± 4.97	97.78 ± 2.00	42.93 ± 4.67			
			100	100	3	40.47	1.83	3.33	100					—		
				100	3	34.96	1.93	6.08	100					—		
				100	2	64.41	2.39	—	100					0, 1 —	0.05	
				150	3	57.74	1.83	12.18	100					2, 1, 0	0.1	0.12 ± 0.08
				150	3	62.79	1.79	13.15	100	5	52.07 ± 13.48	100	52.07 ± 13.48	2, 2, 2	0.2	
		24	50	150	3	23.11	1.18	4.33	94					0, 1, 0	0.033	
				100	2	18.08	1.19	2.63	96					1, 1, —	0.1	
				150	3	17.33	0.90	2.09	89.33					2, 0, 0	0.067	0.06 ± 0.03
				150	3	18.41	0.93	4.19	92					0, 1, 0	0.033	
				150	3	19.43	0.98	2.86	93.33	5	19.27 ± 2.27	92.93 ± 2.48	20.61 ± 2.26	0, 0, 2	0.067	
			500	150	3	21.61	1.14	7.16	96					0, 1, 0	0.033	
				150	3	21.25	1.46	6.26	88.67	2	21.43	92.43	23.18	0, 0, 0	0	0.02

2AAF	AP	36	50	150	3	31.35	1.63	0.35	93.33					0, 1, 1	0.067	
				175	3	21.10	0.98	2.60	94.86					0, 1, 1	0.067	
				150	3	24.45	1.21	3.44	95.33					0, 0, 0	0	
				150	3	17.95	1.15	3.97	88					0, 0, 1	0.033	
				150	3	27.97	1.31	7.58	98					0, 0, 0	0	0.08 ± 0.08
				75	3	36.97	1.97	5.95	100					0, 1, 0	0.033	
				75	3	42.83	2.06	4.73	100					3, 0, 2	0.167	
				75	3	25.17*	1.80	1.89	96	8	28.47 ± 8.28	95.69 ± 3.93	29.50 ± 7.71	3, 2, 2	0.233	
			100	75	3	27.25	1.50	3.92	100					0, 0, 0	0	
				75	3	42.55	1.94	2.37	100					0, 0, 0	0	0
				125	2	30.51	1.23	—	100	3	33.46 ± 8.06	100	33.46 ± 8.06	0, 0, —	0	
			250	75	3	31.79	1.76	5.24	100					8, 13, 8	0.967	
				75	3	36.0	2.03	6.12	100	2	33.9 ± 2.98	100	33.9 ± 2.98	2, 0, 1	0.1	0.53
			500	125	3	41.63[+]	2.07	7.9	94					3, 2, 2	0.233	
				75	3	43.25	2.72	12.11	100					17, 21, 13	1.7	0.62 ± 0.720
				75	3	34.79	1.63	5.39	98.67					4, 1, 3	0.267	
				75	3	49.99	4.44	4.78	100	4	42.42 ± 6.24	98.17 ± 2.85	43.19 ± 6.08	2, 2, 4	0.267	
4AAF	AP	36	2000	100	3	− 2.46*	0.56	1.49	9					31, 26, 33	3.0	
				150	3	− 1.92*	0.46	0.75	10					74, 110, 127	10.4	10.3 ± 5.24
				150	3	− 3.52[a]	0.68	2.98	12.67					169, 182, 182	17.8	
				150	3	− 2.84[a]	0.6	0.87	11.33					88, 112, 102	10.1	
				150	3	− 4.51[a]	0.69	1.07	8.66	5	− 3.05 ± 1.00	10.33 ± 1.67	8.86 ± 1.93	118, 71, 113	10.1	

Net grain (NG) values are derived by substracting the cytoplasmic (C) grain count from nuclear (N) count.
Cells in repair are those with an NG value of 5 or greater. Some animals were also assessed for micronuclei in their bone marrow; * = 0/1000, + = 1/1000 and a = 3/1000.

UDS in the liver. A marked increase in the percentage of hepatocytes in S-phase was observed after 4AAF treatment. In Table 2, the UDS data and percentage of hepatocytes in S-phase are summarized. Individual animal data are shown in Table 1. As mentioned above, some of the animals were used for both liver UDS and the bone marrow micronucleus test, and results from these two genetic endpoints in the same animal are shown in Table 3. No effect of 2AAF on the incidence of bone marrow micronuclei was observed in these rats. However, with an increased number of animals, there was a weak but statistically significant induction of micronuclei after exposure to 500 mg 2AAF/kg body weight. The details regarding the bone marrow micronucleus test are given in the IPCS report "Micronucleus Test in Bone Marrow of Rats Exposed to 2AAF and 4AAF" (Beije and Ashby, this volume).

4 Discussion

In the present study, only quite high dose levels of the test compounds were used. For 2AAF, the lowest dose was 50 mg/kg, at which level the induction of liver UDS seems to have reached a maximum. When the dose level was further increased, only minor changes in effect were observed. Another set of experiments reported separately by Ashby, Burlinson, and Lefevre (this volume), was done concurrently in the same laboratory. In those experiments, lower dose levels of 2AAF (0.1–5 mg/kg) were given to the same rat strain, indicating an induction of liver UDS at 5 mg/kg. Similar results have been demonstrated by Mirsalis and Butterworth (1980).

No effect of 4AAF on liver UDS activity was observed in the present study, although the cell turnover was dramatically increased. The effect of 4AAF at lower dose levels and earlier time points is discussed in the

Table 2. *Summary of liver UDS data in rats and percentage of hepatocytes in S-phase after in vivo treatment with 2AAF or 4AAF*

Compound	Exposure (hs)	Dose (mg/kg)	No. of animals	(NG) ± SD	% in repair ± S.D.[a]	No. of animals	% cells in S-phase ± S.D.
Corn oil	12	10 ml/kg	3	− 2.85 ± 0.35	1.67 ± 0.88	1	0.05
	24	10 ml/kg	2	− 2.64	2.0	2	0.13
	36	10 ml/kg	4	− 2.79 ± 1.24	2.83 ± 2.9	4	0.05 ± 0.02
2AAF	5	100	2	9.25	67.67	—	—
	12	50	9	42.02 ± 4.97	97.78 ± 2.0	—	—
	12	100	5	52.07 ± 13.48	100	3	0.12 ± 0.08
	24	50	5	19.27 ± 2.27	92.93 ± 2.48	5	0.06 ± 0.03
	24	500	2	21.43	92.43	2	0.02
	36	50	8	28.47 ± 8.28	95.69 ± 3.93	8	0.08 ± 0.08
	36	100	3	33.46 ± 8.06	100	3	0
	36	250	2	33.9 ± 2.98	100	2	0.53
	36	500	4	42.42 ± 6.24	98.17 ± 2.85	4	0.62 ± 0.72
4AAF	36	2000	5	− 3.05 ± 1.0	10.33 ± 1.67	5	10.3 ± 5.24

[a] Percentage of hepatocytes with ⩾ 5 NG.

Table 3. *Liver UDS and bone marrow micronucleus data obtained from the same animal at the times and dose levels shown*

Compound	Exposure (h)	Dose (mg/kg)	Liver UDS (NG)	Bone marrow (MPE/1000PEs)[a]
Corn oil	36	10 ml/kg	− 1.8	0
2AAF	36	50	25.17	0
	36	500	41.36	1
4AAF	36	2000	− 2.46	0
	36	2000	− 1.92	0
	36	2000	− 3.52	3
	36	2000	− 2.84	3
	36	2000	− 4.51	3
Cumulative Controls (mean ± SD)				
Corn oil n = 4	36	10 ml/kg	− 2.79 ± 1.24	—
Water n = 6	36	10 ml/kg	—	0.83 ± 0.41
Cyclophosphamide n = 6	36	15	—	15.5 ± 6.12

[a] Micronucleated polychromatic erythrocytes/1000 polychromatic erythrocytes.

report by Ashby, Burlinson, and Lefevre (this volume).

5 Acknowledgement

The financial support to B. Beije from the Swedish Work Environment Protection Fund and the technical assistance from Brian Burlinson are acknowledged.

6 References

Ashby, J., P.A. Lefevre, B. Burlinson and M.G. Penman (1985) An assessment of the *in vivo* hepatocyte DNA repair assay, *Mutation Res.*, **156**, 1–18.

Ashby, J., B. Burlinson and P.A. Lefevre (1985) Activity of 2AAF and 4AAF in an *in vitro* and an *in vivo/in vitro* assay for UDS in the rat liver. *This volume.*

Beije B. and J. Ashby (1985) Use of the *in vivo/in vitro* rat liver DNA repair assay to predict the relative rodent hepatocarcinogenic potency of 3 new azo mutagens, *Carcinogenesis*, **6(4)**, 611–615.

Beije, B. and J. Ashby (1985) Micronucleus test in bone marrow of rats exposed to 2AAF and 4AAF. *This volume.*

Mirsalis, J.C. and B.E. Butterworth (1980) Detection of unscheduled DNA synthesis in hepatocytes isolated from rats treated with genotoxic agents: an *in vivo* assay for potential carcinogens and mutagens, *Carcinogenesis*, **1**, 621–625.

Charles B. Salocks and Gary M. Decad
Department of Materials Toxicology, IBM Corporation, San Jose, California, USA

60

Induction of unscheduled DNA synthesis and DNA replication in rat hepatocytes following in vivo administration of 2-acetylaminofluorene and 4-acetylaminofluorene

1 Summary

The genotoxicity of 2-acetylaminofluorene (2AAF) and 4-acetylaminofluorene (4AAF) was investigated in the *in vivo/in vitro* hepatocyte DNA repair assay. Adult male Fischer-344 rats were administered the compounds by gavage at 2, 12, or 24 h prior to hepatocyte isolation. The cells were cultured with ^{3}H-thymidine, and induction of unscheduled DNA synthesis (UDS) was measured by quantitative autoradiography. 2AAF, at doses of 10 and 50 mg/kg, produced a clear dose-related increase in UDS, with maximal response occurring 12 h posttreatment. The fraction of cells undergoing DNA replication was not altered by 2AAF treatment. 4AAF, at doses of 200 and 500 mg/kg, produced a slight increase in UDS at 12 h posttreatment and a marked stimulation of DNA replication at 24 h posttreatment. A separate *in vivo* experiment was carried out to determine if 4AAF induces DNA replication indirectly by virtue of being hepatotoxic. Serum concentrations of bilirubin, cholesterol, lactate dehydrogenase, sorbitol dehydrogenase, and glutamic-pyruvic transaminase were measured 24 h after 4AAF administration. No evidence of hepatotoxicity was detected, suggesting that 4AAF induces DNA replication by a direct mechanism.

2 Introduction

The measurement of DNA repair as unscheduled DNA synthesis (UDS) following exposure to DNA-damaging agents has proved to be a useful method for assessing the potential genotoxicity of chemicals. The assay is based on the capacity of genotoxic chemicals to produce structural alterations in DNA. These alterations are recognized by repair enzymes, which excise the damaged portion and synthesize a new polynucleotide strand utilising exogenously added ^{3}H-thymidine. The amount of nuclear ^{3}H-thymidine incorporation is quantified by autoradiography.

There are several advantages in using primary cultures of hepatocytes as an *in vitro* model for quantifying UDS. Primary hepatocytes have been shown to metabolise a variety of chemicals in a manner similar to *in vivo* metabolism. For this reason, they are particularly useful for evaluating compounds that require metabolic activation to exert their genotoxic effect. Thus, activation and interaction with DNA occur within the same cell, and there is no need for an exogenous enzyme activation system. Furthermore, hepatocytes in primary culture undergo an extremely low rate of DNA replication, so potential confounding effects of a high background rate of ^{3}H-thymidine incorporation are minimised.

As originally developed, the primary hepatocyte DNA repair assay entailed *in vitro* exposure of cell cultures to the compounds of interest. Although responsive to a variety of genotoxins, the *in vitro* assay does not always reflect the true *in vivo* genotoxic potential of some agents. For example, chemicals requiring metabolism by nitroreductase or azoreductase are generally not detected in this assay, because these enzymes are inhibited by oxygen and are not present in appreciable amounts in hepatocytes. In a recent modification of this assay, it was demonstrated that such chemicals do in fact elicit a increase in UDS if they are administered *in vivo*, and the hepatocytes, isolated at appropriate times thereafter, are cultured *in vitro* with ^{3}H-thymidine. Chemicals that are metabolised to proximate carcinogens by gut flora, such as nitroaromatics and benzidine-derived azo dyes, are not detected in the *in vitro* assay, but are clearly positive with the *in vivo/in vitro* protocol (Mirsalis and Butterworth, 1982; Joachim and Decad, 1984).

This report describes the results of studies utilising the *in vivo/in vitro* hepatocyte DNA repair assay. The structural isomers 2AAF and 4AAF were tested. The former is a potent animal carcinogen and is positive in most short-term mutagenesis assays. 4-AAF was considered noncarcinogenic based on limited negative data from two rat bioassays. However, as a result of positive responses in a number of *in vitro* assays for genotoxicity, its status as a noncarcinogen is uncertain at present (Ashby, 1981).

3 Methods

Animals and treatments

Male Fischer-344 rats were purchased from Simonsen Laboratories, Inc. (Gilory, CA) and were housed in groups of four in plastic cages with wood shavings for bedding. They were kept in a controlled-environment room with filtered air and a 12-h light-dark cycle. Animals had free access to water and Purina No. 5001 Laboratory Rodent Chow (Ralston Purina Co., St. Louis, MO) at all times. When hepatocyte isolations were carried out, animals weighed 200–260 g.

2AAF and 4AAF were administered by gavage at 2, 12, or 24 h before hepatocyte isolations. Doses of 2AAF were 10 and 50 mg/kg; doses of 4AAF were 200 and 500 mg/kg. Controls received corn oil only. 2AAF was dissolved in corn oil by sonication in a bath sonicator for approximately 1 h. During this time, the solution warmed to 50 °C. 4AAF was not soluble in corn oil at the concentrations tested, but the 4AAF solutions were sonicated for 15 min to produce a fine, uniform suspension. Both doses of each chemical were administered to three animals at each of the three time intervals between dosing and sacrifice. Concentrations of the test chemicals were adjusted so that dosing volumes were equivalent to 1.0 ml/100 g body weight. Once dosed, animals were housed separately but still had free access to food and water.

Chemicals

2AAF and 4AAF were provided by ICI Limited (Cheshire, England). Williams' medium E (without L-glutamine) was obtained from Grand Island Biological Co. (Grand Island, NY). Fetal bovine serum was purchased from Flow Laboratories (McLean, VA). (methyl-^{3}H) Thymidine (60–70 Ci/m mole, in a 7:3 solution of ethanol:water) was obtained from ICN Pharmaceuticals (Irvine, CA). Collagenase (type I, 200 units/mg), EGTA (ethylene glycol-bis (β-aminothyl ether) N, N′-tetraacetic acid), HEPES (N-2-hydroxyethylpiperazine-N′-2-ethanesulfonic acid, free acid), gentamicin sulfate, and all other salts and biochemicals were purchased from Sigma Chemical Co. (St. Louis, MO).

Perfusates and culture medium

EGTA perfusate and collagenase perfusate, both of which are similar in composition to HBSS, were prepared as described previously (Dougherty *et al.*, 1980; Salocks *et al.*, 1981). However, the concentration of $NaHCO_3$ was increased to 1.29 g/liter, and was added to each solution after adjusting the pH to 7.4. The solution for dispersing and washing the cells was identical to EGTA perfusate but lacked EGTA. All three solutions were buffered with 0.05 M HEPES and sterilised by filtration.

Culture medium was prepared by adding 5 ml L-glutamine (200 mM), 0.84 ml gentamicin sulfate (50 mg/ml), and 25 μl insulin (40 IU/ml) to 500 ml Williams' medium E (WME). During hepatocyte attachment (the first 90 min of culture), the medium was supplemented with 10% heat-inactivated fetal bovine serum. Heat inactivation was achieved by heating at 56 °C for 30 min.

Isolation and culture of hepatocytes

Hepatocytes were isolated by perfusion of liver biopsies as described by Reese and Byard (1981). The isolation procedure and all culture manipulations were carried out in a containment hood under sterile conditions.

Rats were anaesthetised by intraperitoneal (ip) injection of sodium pentobarbital (Abbott Laboratories, North Chicago, IL) at a dose of 10 mg/100 g body weight. The distal one-half to two-thirds of the largest lobe of the liver was removed. The biopsy was placed in a sterile beaker containing 25 ml EGTA perfusate, then transferred to a petri dish having a small hole in the center for drainage. The tissue was perfused by cannulation of a single vein on the cut surface of the biopsy. It was usually possible to find one vein which perfused at least 80% of the tissue. EGTA perfusate was pumped through the tissue without recirculation at a rate of 10 ml/min. At this time 45 mg of collagenase was dissolved in 100 ml of collagenase perfusate. Perfusion with collagenase was initiated immediately after 100 ml of EGTA perfusate had been pumped through the tissue. Once perfusion with collagenase began, the biopsy was transferred to a recirculating perfusion apparatus. Flow rate was often reduced slightly during collagenase perfusion to minimise swelling of the tissue. Digestion usually required 15–18 min.

Both EGTA and collagenase perfusates were equilibrated with 95% O_2–5% CO_2 prior to use. Perfusion temperature was maintained at 30 °C throughout the entire procedure.

Following digestion, the tissue was transferred to a crystallizing dish containing 50–100 ml of ice-cold washing buffer. Several incisions were made in the capsule, and the tissue was swirled gently through the buffer to release individual cells. The suspension was filtered through two layers of nylon cloth, the upper layer being 263 μm mesh and the lower layer 64 μm mesh. The filtrate was allowed to stand for 1–2 min, then centrifuged once at 30 g for 2 min. The supernatant was discarded, and the cell pellet resuspended in 10 ml ice-cold WME. Viability was determined after incubating 0.2 ml of the cell suspension with 4.6 ml of culture medium and 0.2 ml of 0.4% trypan blue for 3–4 min. Viable (clear) and nonviable (blue) cells were counted using a hemocytometer.

The cells were diluted in WME containing 10% heat-inactivated fetal bovine serum to a concentration of 0.625×10^6 viable hepatocytes/ml. They were seeded onto collagen-coated plastic tissue culture dishes at a density of 2.5×10^6 viable cells/60 mm dish and placed in a 37 °C incubator having a humidified 5% CO_2-95% air atmosphere.

Incubation with Labelled and Unlabelled Thymidine
After 90 min, the serum-supplemented medium was aspirated, and the cultures were washed once with 3 ml WME. The cells were then incubated with 2 ml of medium containing 10 μCi/ml ^{3}H-thymidine and buffered with 10 mM HEPES. The stock solution of ^{3}H-thymidine was made up in 70% ethanol, and final solvent concentration was 0.7%.

Following a 4 h incubation, medium containing ^{3}H-thymidine was aspirated, and each culture was washed once with 3 ml of WME. Three ml of medium containing 0.25 mM unlabelled thymidine was added, and the cultures were incubated overnight (14–16 h).

Fixation and autoradiography
The following morning, each culture was washed twice with 3 ml of WME. The cells were fixed in methanol:acetic acid (3:1) for 10 min, and allowed to dry for 10 min. The fixation-drying sequence was carried out three times. The cultures were stored in a cool, dry cabinet for up to one month before being processed for autoradiography.

It was found that autoradiograms of cultures processed by the above procedures were difficult to quantify. In particular, nuclear grains on cells from animals exposed to 2AAF were often contiguous, to the extent that it was not possible to count them accurately. Therefore additional steps, as described by Mitchell and Mirsalis (1984), were taken to reduce nonspecific ^{3}H-thymidine incorporation. Fixative was added to the cultures once more, again for 10 min. The fixative was aspirated, and the cultures were washed six times with 3–4 ml of distilled water. The additional washing resulted in autoradiograms with significantly fewer cytoplasmic and nuclear grains, improving the accuracy of grain quantification.

The fixed and washed cultures were coated with Kodak NTB-2 photographic emulsion (Eastman Kodak Company, Rochester, NY), which had been diluted 1:1 with distilled water. They were placed in light-tight boxes and stored for 10 days in a −20 °C freezer.

Cultures were allowed to thaw for several hours before developing. They were developed in Kodak D-19 developer for 3 min, rinsed with distilled water, and fixed for 8 min. Developer, rinse water and fixative were all cooled to 15 °C. Development at room temperature (22–24 °C) produced larger grains, aggravating the problem of excessive nuclear grain density described above. After fixation, the cultures were rinsed in deionized water for 10 min, and stained with a solution of 3% Giemsa in 13.3 mM phosphate buffer, pH 6.8, for 30–40 sec. Excess stain was washed off with distilled water, and the cultures were allowed to dry.

Counting autoradiographic grains
An Artek model 880 colony counter, interfaced to a Nikon Optiphot microscope via an external Artek television camera, was used to quantify autoradiographic grains. Grain counts were fed to a Texas Instruments TI 59 calculator via an Artek Compu-Print 700 interface for calculation of net nuclear grains. Software for data processing and recording was written specifically for this procedure.

Culture dishes were divided into three wedge-shaped sections, and 25 cells were scored in each section for a total of 75 cells scored from each culture. Scoring was carried out under 1000X magnification using oil immersion. All counts were made with the "mode" switch of the Artek counter set on "count." (The "obj. area" counting mode was not available on our instrument.) Microscope illumination, as well as the "sensitivity" and "size" adjustments on the Artek counter, were adjusted to maximise the number of grains flagged without producing spurious flags over areas of uneven staining intensity. Readjustments were made as necessary during the course of counting.

Criteria for determining whether or not a cell should be counted were based exclusively on morphological characteristics of the cell. Because the aperture of the Artek counter was round, it was necessary that the cell nucleus be approximately round also. A cell was not counted if it did not have at least two nuclear-size areas of cytoplasm clearly associated with the nucleus under consideration. This latter requirement meant that, in general, binucleate hepatocytes were not counted. Cells with pyknotic nuclei, debris, or unusual staining artifacts were not counted. Areas with unusually uneven grain intensity (perhaps caused by a bubble in the emulsion, or an area adjacent to a replicating cell) were also avoided.

If a cell met the above criteria, the counting apereture of the Artek counter was centered directly over the nucleus, and the diameter of the aperture was adjusted to match that of the nucleus. Flagging accuracy was checked, and the number of nuclear grains was recorded. Using the same aperture and sensitivity settings, three cytoplasmic areas immediately adjacent to the nucleus were counted. Cytoplasmic areas occasionally overlapped. The average of counts from the three cytoplasmic areas was subtracted from the nuclear count to give the net grains/nucleus (N.G.). A cell was determined to be undergoing UDS (or "in repair") if the net grain count was $\geqslant 5$.

Determination of induction of replication

The percentage of cells undergoing replication was also determined using the Artek counter. Magnification was reduced to 100X. At this magnification, a cell in "S" phase was clearly distinguished by the presence of totally confluent grains over the nucleus and was indicated by a single flag. Individual grains were not flagged. The number of cells in S phase was determined for 10 randomly selected areas on each culture.

The total number of cells in an area of identical diameter was determined in matching replicate cultures that had not been processed for autoradiography. These cultures were stained for 10 min with 3% Giemsa, producing nuclei that were dark enough to be flagged by the Artek counter. Cytoplasmic staining was not sufficiently intense to produce a significant number of non-nuclear flags. The total number of cells was also determined in ten different areas of each culture. The percentage of cells undergoing replication was calculated from the average number of cells in S phase divided by the average total number of cells in an equivalent area.

Determination of hepatotoxicity of 4AAF

Male Fischer-344 rats weighing 220–240 g were gavaged with corn oil or 4AAF using the same protocol as the one described for the UDS experiments. Four control animals received 1.0 ml/100 g corn oil. Groups of three animals each received 200 mg/kg or 500 mg/kg 4AAF in corn oil, and the dosing volume was also 1.0 ml/100 g. Blood samples were withdrawn by cardiac puncture 24 h after dosing. Serum was separated from whole blood and analysed for bilirubin, cholesterol, lactate dehydrogenase (LDH), sorbitol dehydrogenase (SDH), and glutamic-pyruvic transaminase (SGPT). All analyses were standard clinical spectrophotometric assays. Dosing, blood sampling, and serum analyses were carried out by the Biomedical Research Laboratory of SRI International (Menlo Park, CA).

4 Results

2AAF produced a dose-related increase in UDS at all three posttreatment intervals (Table 1). Maximal response for both doses occurred 12 h posttreatment,

Table 1. *Induction of unscheduled DNA synthesis and DNA replication in hepatocytes isolated from Fischer-344 rats following in vivo treatment with corn oil, 2AAF, or 4AAF*

Treatment	Dose	Time (h)[a]	Net grains[b]	% ≥ 5 N.G.[c]	% in S phase
Corn Oil	10 ml/kg	12–24	− 3.09 ± 0.43	0	< 0.1
2AAF	50 mg/kg	2	13.35 ± 1.49	94.6 ± 4.0	0.13 ± 0.05
		12	14.72 ± 1.41	97.4 ± 2.3	<0.1
		24	6.68 ± 4.06	56.6 ± 38.9	< 0.1
	10 mg/kg	2	4.60 ± 2.33	40.9 ± 24.0	0.14 ± 0.03
		12	10.49 ± 2.64	87.2 ± 12.5	0.21 ± 0.10
		24	4.42 ± 1.25	41.5 ± 14.3	< 0.1
4AAF	500 mg/kg	2	− 1.55 ± 0.09	1.8 ± 3.1	0.13 ± 0.05
		12	2.24 ± 0.45	19.0 ± 1.9	0.22 ± 0.18
		24	(UDS not scorable)		34.3 ± 12.6
	200 mg/kg	2	− 1.17 ± 0.41	2.2 ± 2.0	0.12 ± 0.03
		12	1.19 ± 1.00	13.0 ± 8.3	0.18 ± 0.09
		24	(UDS not scorable)		8.51 ± 3.91

[a] Time elapsed between dosing and hepatocyte isolation. Corn oil controls were dosed 12 h (one rat) or 24 h (two rats) prior to hepatocyte isolations.
[b] Magnitude of UDS is expressed as net nuclear grains (N.G.; nuclear grains minus grains in an equivalent area of cytoplasm). All values represent the mean ± standard deviation (variation between animals) of data obtained from three rats. UDS was not scored in cells isolated 24 h after 4AAF treatment due to very heavy grain density.
[c] Per cent of cells having 5 or more N.G.

Table 2. *Results of serum chemistry analysis following administration of corn oil or 4AAF to male Fischer-344 rats*[a]

	Control (4)[b]	200 mg/kg (3)	500 mg/kg (3)
Bilirubin (mg%)	0.60 ± 0.12	1.40 ± 0.36	1.73 ± 0.40
Cholesterol (mg%)	26.3 ± 2.8	30.0 ± 3.6	22.0 ± 4.6
LDH (IU/L)	1027 ± 270	829 ± 120	588 ± 12
SDH (IU/L)	11.5 ± 2.4	14.0 ± 3.5	13.7 ± 0.6
SGPT (IU/L)	52.3 ± 13.0	56.0 ± 12.3	77.3 ± 15.5

[a] Animals were dosed with corn oil or 4AAF 24 h before blood samples were taken. All values represent the mean ± the standard deviation (variation between animals). LDH: lactate dehydrogenase; SDH: sorbitol dehydrogenase; SGPT: glutamic-pyruvic transaminase.
[b] Numbers in parentheses indicate the number of rats tested at each dose.

although the higher dose produced nearly equivalent responses at 2 and 12 h.

4AAF produced only a slight increase in UDS, but the response at 12 h posttreatment is considered positive. This conclusion is based on both the N.G. counts and the percentage of cells with $\geqslant 5$ N.G. The 12 h response was also dose-related, although the difference between the two doses may not be significant. The 2 h response was greater than that of the corn oil controls, but was not dose-related and was not considered significant. Since the highest dose of 4AAF was 10-fold greater than the highest dose of 2AAF, the magnitude of the response produced by the two chemicals at 12 h indicates that 4AAF is substantially less potent than 2AAF in this assay.

4AAF also produced a marked increase in DNA replication at 24 h posttreatment (Table 1). The increase in scheduled DNA synthesis was dose-related: at 200 and 500 mg/kg, an average of 8.5 and 34.3% of the cells were in S phase, respectively. This contrasts with the percentage usually observed in control cultures, less than 0.1%. The density of cytoplasmic grains was also greater in these cultures and was evident even in cells that were not undergoing DNA replication. Excessive grain density precluded accurate scoring of these cultures for UDS.

There was no indication that 4AAF is hepatotoxic *in vivo* (Table 2). Apparent dose-related differences in two of the indices, bilirubin and LDH, are probably artifacts caused by interference with the analytical methods (A. Thurber, personal communication). In the case of bilirubin, a likely source of interference is lipemia, which was observed in all serum samples. In support of this conclusion, control bilirubin concentrations were two to three times greater than those of historical controls. Differences in LDH concentrations may have resulted from hemolysis of the blood samples. Since LDH in red cells is more than 100 times greater than serum levels, small differences in degree of hemolysis can produce large differences in apparent serum LDH concentrations. The key enzyme that is diagnostic of hepatotoxicity, SDH, was not elevated in any of the animals dosed with 4AAF.

5 Discussion

The results of this study are consistent with those of the other laboratories running the *in vivo/in vitro* UDS assay. 2AAF was clearly positive at both doses and at all three posttreatment times (Table 1). No increase in DNA replication following treatment with 2AAF was observed, although the latest posttreatment interval we tested was 24 h. Laboratories reporting an increase in DNA replication observed this effect at later times, 36 and 48 h after dosing. This suggests that the increase in S phase cells following 2AAF treatment is an indirect result of hepatotoxicity.

4AAF was weakly positive in the UDS assay. Dose-related increases in both the mean N.G. count and the percentage of cells having $\geqslant 5$ N.G. were observed 12 h posttreatment. This result is consistent with data from the ^{32}P-postlabelling assay (presented elsewhere in this volume), demonstrating that 4AAF generates a low level of DNA adducts in rat liver following ip dosing. Nevertheless, one may judge instead that the magnitude of the UDS response is so low that no definite conclusions regarding 4AAF can be drawn.

Clearly, the primary effect of 4AAF is stimulation of DNA replication. This observation has two implications. First, stimulation of DNA replication has been linked to the process of tumour promotion, suggesting that 4AAF may indeed be an animal carcinogen even though it is only a weak genotoxin. Second, it is possible that the slight UDS response occurring at 12 h is a nonspecific result of alterations in cellular and/or nucleic acid metabolism preceding replication. Both of these hypotheses are speculative and indicate a need for further research into chemical induction of DNA replication.

Several procedures utilised in this study are at variance with generally accepted protocols but did not appear to alter results significantly. These differences included isolation of hypatocytes by biopsy perfusion rather than in situ perfusion, and cell culture on plastic tissue culture dishes rather than plastic cover slips. Culture with ^{3}H-thymidine in the presence of 0.7% ethanol also did not affect results. The latter finding is consistent with a report by Mitchell and Mirsalis (1984), demonstrating that 1.0% ethanol did not increase N.G. counts in the *in vitro* assay. The practice of subtracting the mean, rather than the largest, of three cytoplasmic counts may explain why control N.G. counts in this study are slightly higher than values reported by others.

From a technical standpoint, grain counting was greatly facilitated if fixed cells were washed with water to remove unbound ^{3}H-thymidine and if the emulsion was developed at 15 °C rather than room temperature. These findings are in accord with the recommendations of Mitchell and Mirsalis (1984). Furthermore, it is clear that grain counting is sufficiently subjective that blind scoring is advisable. Finally, although the *in vivo/in vitro* UDS assay takes advantage of *in vivo* pharmacokinetics, it is critical that solvent vehicle, concentration of the test chemical, and the means of solubilization be carefully considered. Variations in any of these parameters may alter the pharmacokinetics of the test chemical and thereby affect the final determination as to whether or not a chemical is positive in this assay.

6 Conclusions

When tested in the *in vivo/in vitro* hepatocyte UDS assay, 2AAF was strongly positive and 4AAF was weakly positive. The primary effect of 4AAF was induction of DNA replication. *In vivo*, 4AAF was not hepatotoxic, suggesting that induction of replication occurs by a direct mechanism. 2AAF did not induce DNA replication in this study, most likely because it was not tested at times later than 24 h after dosing.

7 Acknowledgments

The authors gratefully acknowledge the excellent technical assistance of Bruce Irvine and Keith Esch. We also wish to thank Andrea Thurber of SRI International for advice regarding the interpretation of data from the serum chemistry assays.

8 References

Ashby, J. (1981) Overview of the study and test chemical activities. In F.J. de Serres and J. Ashby (Eds),

Evaluation of Short-Term Tests for Carcinogens, Elsevier/North-Holland, New York, pp. 112–171.
Dougherty, K.K., S.D. Spilman, C.E. Green, A.R. Steward and J.L. Byard (1980) Primary cultures of adult rat and mouse hepatocytes for studying the metabolism of foreign chemicals, *Biochem. Pharmacol.*, **29**, 2117–2124.
Joachim, F. and G.M. Decad (1984) Induction of unscheduled DNA synthesis in primary rat hepatocytes by benzidine-congener-derived azo dyes in the *in vitro* and *in vivo/in vitro* assays, *Mutation Res.*, **136**, 147–152.
Mirsalis, J.C. and B.E. Butterworth (1982) Induction of unscheduled DNA synthesis in rat hepatocytes following *in vivo* treatment with dinitrotoluene, *Carcinogenesis*, **3**, 241–245.
Mitchell, A.D., and J.C. Mirsalis (1984) Unscheduled DNA synthesis as an indicator of genotoxic exposure. In A.A. Ansari and F.J. de Serres (Eds), *Single-Cell Mutation Monitoring Systems*, Plenum Press, New York, pp. 165–216.
Reese, J.A. and J.L. Byard (1981) Isolation and culture of adult hepatocytes from liver biopsies, *In Vitro*, **17**, 935–940.
Salocks, C.B., D.P.H. Hsieh and J.L. Byard (1981) Butylated hydroxytoluene pretreatment protects against cytotoxicity and reduces covalent binding of aflatoxin B_1 in primary hepatocyte cultures, *Toxicol. Appl. Pharmacol.*, **59**, 331–345.
Thurber, A.B., Biomedical Research Laboratory, SRI International, Menlo Park, CA. Personal communication.

Douglas Kornbrust and Dennis Dietz[1]
Merck Sharp & Dohme Research Laboratories Department of Safety Assessment, West Point, Pennsylvania, USA
[1]Research Triangle Institute Research Triangle Park, North Carolina, USA

61

Testing of 2-acetylaminofluorene and 4-acetylaminofluorene in the in vivo/in vitro rat hepatocyte/DNA repair assay and investigation of lactational exposure of neonatal rats to 2-acetylaminofluorene

1 Summary

2-Acetylaminofluorene (2AAF) and 4-acetylaminofluorene (4AAF) were tested in the *in vivo/in vitro* rat hepatocyte/DNA repair assay. Each chemical was tested at two dose levels and at two different time points, using three male Fischer-344 (F344) rats per dose and time point. 2AAF was clearly positive at the higher dose and weakly positive at the lower dose; a greater response was observed at 12 than at 24 h postadministration. 4AAF did not induce DNA repair at either dose or time point, but produced a dose-dependent increase in the frequency of S phase cells, which is indicative of cell division. A reduction in the magnitude of the DNA repair response to the *in vitro* positive control, methyl methanesulfonate, was observed in hepatocytes derived from 4AAF-treated animals, which suggested that DNA repair processes were partially inhibited by 4AAF administration.

Preliminary findings on the secretion of 2AAF into the milk of lactating rats and induction of DNA repair in hepatocytes of neonatal rats were also obtained.

2 Introduction

The measurement of DNA repair as unscheduled DNA synthesis (UDS) in rat hepatocytes has proven to be a reliable indicator of the potential geno-toxicity of a wide variety of chemicals (Williams *et al.*, 1982; Probst *et al.*, 1981). The original *in vitro* version of the rat hepatocyte/DNA repair assay, developed by Williams (1977), has demonstrated considerable utility as a screening test for genotoxic carcinogens (Mitchell *et al.*, 1983). The main advantage of this short-term test is that the hepatocytes serve as both the source of metabolic activation and as the target cell for the genotoxic response.

A modification of this assay in which test chemicals are administered to the whole animal prior to isolation of hepatocytes and measurement of DNA repair (i.e. the *in vivo/in vitro* rat hepatocyte/DNA repair assay) was introduced by Mirsalis and Butterworth (1980). This assay system provides the additional capability to assess the impact of *in vivo* parameters of uptake, distribution, extrahepatic metabolism, and excretion on the ability of test chemicals to elicit hepatocyte DNA repair. Although a rigorous validation has not yet been performed, the assay appears to be capable of identifying genotoxic hepatocarcinogens (Mirsalis *et al.*, 1982).

In the present study, 2AAF and its structural congener, 4AAF, were tested in the *in vivo/in vitro* rat hepatocyte/DNA repair assay. 2AAF previously has been shown to elicit a positive response in this assay (Mirsalis *et al.*, 1982), whereas 4AAF has not been tested previously. Both aromatic amines were positive in a variety of bacterial mutation assays (Ashby, 1981). 4AAF elicited fewer positive responses than did 2AAF in *in vitro* eukaryotic test systems and was negative in a limited number of *in vivo* tests, in contrast to the positive responses elicited by 2AAF *in vivo* (Ashby, 1981). Since the carcinogenicity of 4AAF is undecided (Purchase *et al.*, 1981), an evaluation of its potential *in vivo* genotoxicity is critical to evaluating the status of this chemical.

In a recent series of investigations in this laboratory, the *in vivo/in vitro* rat hepatocyte/DNA repair assay was employed to characterize the ability of chemical carcinogens to be secreted into the milk of lactating rats and to subsequently induce DNA repair in the hepatocytes of neonatal animals exposed via suckling of contaminated milk (Kornbrust and Dietz, manuscript submitted). The results obtained with administration of 2AAF to lactating rats are herein reported.

3 Materials and methods

Materials

Williams Medium E (WE) was obtained from Gibco Laboratories, Grand Island, NY, in powdered form and dissolved in sterile deionised distilled water several days prior to use. At this time, 10 ml of 200 mM L-glutamine (Flow Laboratories, Rockville, MD) and 1 ml of 50 mg/ml gentamycin sulfate (Sigma Chemical Co., St. Louis, MO) were added per liter of media. This mixture constituted the incomplete media (WEI). A portion of this was mixed with 10% (v/v) heat-inactivated foetal bovine serum (Flow Laboratories) to yield the complete media (WEC).

[Methyl-^{3}H] thymidine, 70–80 Ci/m mol, was obtained from ICN Radiochemicals, Irvine, CA. NTB-2 Nuclear Track Emulsion was purchased from Kodak, Rochester, NY. Dimethylnitrosamine (DMN), DNAase (Type III) and collagenase (Type 1A) were obtained from Sigma Chemical Co., while methyl methanesulphonate (MMS) was purchased from Aldrich Chemical Co., Metuchen, NJ. N-[9-^{14}C]-2-AAF was a product of New England Nuclear, Boston, MA.

Animals

For the standard assay protocol, male F344 rats, weighing 180–200 g, were used. Prior to use, the animals were housed in humidity- and temperature-controlled rooms with 12 h light/dark cycles and allowed access to food and water ad libitum. For the neonatal DNA repair experiments, female rats were mated at approximately 70 days of age. Two days after parturition, litters were normalized to 10 pups each (5 of each sex).

Administration of test chemicals

The male rats were fasted for 4 h prior to the administration of the test chemicals. Corn oil suspensions of the chemicals were prepared at different concentrations such that the administered volume was always 1 ml/kg body weight. The doses were delivered orally via 1.5-inch, 18-g curved stainless steel dosing tubes. The dose levels of 2AAF were 10 and 50 mg/kg body weight, while the dose levels of 4AAF were 50 and 200 mg/kg body weight. The time intervals between administration of chemical and liver perfusion were 12 and 24 h. The doses and time points were selected on the basis of existing knowledge of the responses to 2AAF (Mirsalis and Butterworth, 1982) and personal communication with Dr. Byron Butterworth. Three rats were used per dose-time point. Control rats received 1 ml of corn oil/kg body weight.

2AAF was similarly administered to lactating female rats at a dose level of 200 mg/kg body weight when the pups were 12–13 days of age. At the time of dosing, the litters were culled from 10 to 6 pups (3 of each sex). Hepatocytes were isolated from neonatal livers 15 h postadministration of 2AAF to the dams.

Isolation of hepatocytes

Hepatocytes were isolated from adult male rats by an in situ liver perfusion method similar to that of Williams (1977). The rats were first anaesthetised with sodium pentobarbital and the portal vein was isolated and catheterised. Perfusion was begun with a 37 °C solution of 0.5 mM EGTA at a rate of 8 ml/min, at which time the inferior vena cava was cut below the right renal branch. After approximately 2 min, the vena cava was ligated above the right renal branch and the flow of EGTA was increased to 20 ml/min for 2–3 min. The flow was then switched to a 37 °C collagenase solution (100 units/ml of WEI) for 12 minutes. The liver was covered and kept warm with a lamp.

After perfusion, the liver was removed, placed in a sterile dish containing 30–40 ml fresh collagenase, the capsule was perforated, and loose cells were combed away from the fibrous and connective tissue. The cells were collected and poured through layers of sterile gauze into large plastic tubes, and centrifuged at 50 X g for 3 min. The supernatant was discarded, and the cell pellet was resuspended in 20–30 ml of WEC. Hepatocyte viability, as determined by trypan blue exclusion, was $81.6 \pm 9.2\%$ ($n = 28$). Treatment with 2AAF or 4AAF had no effect on the per cent viability in the cell populations isolated.

Hepatocytes were isolated from neonatal rats by a procedure similar to Devirgilis *et al.* (1981). The livers from the 6 pups in each litter were excised, pooled in a beaker containing 5–10 ml EGTA solution, and thoroughly minced until the tissue pieces were approximately 1 mm^3. The mince was transferred to a 125-ml round-bottom flask containing 30 ml of EGTA and incubated with gentle shaking for 10 min at 39 °C. The EGTA, which contained red blood cells and cellular debris was removed by aspiration and a second 10 min incubation with 30 ml EGTA was performed. The EGTA again was discarded and the tissue mince was incubated with 15 ml of a solution of collagenase (300 units/ml) and DNAase (75 units/ml) in WEI for 15 min at 37 °C. The cells liberated by this first collagenase digestion were also discarded, since the contamination of this fraction with red blood cells was appreciable. Three successive 15 min incubations of the liver mince with 15 ml of fresh collagenease/DNAase solution were subsequently performed, and the cells liberated by each incubation were collected and pooled. The cells were washed by centrifuging the cells at 50 X g for 3 min and resuspending the pellet in 10–20 ml of WEC. The viability of neonatal hepatocytes was $88.4 \pm 6.5\%$ ($n = 8$).

Determination of unscheduled DNA synthesis
Approximately $3–6 \times 10^5$ viable cells were seeded into Costar 6-well cluster dishes (Bellco Glass, Inc., Vineland, NJ) containing 25 mm Thermanox[R] plastic coverslips (Flow Laboratories) and 4 ml of WEC. The cells were allowed to attach to the coverslips by incubating for 2 h in a 37 °C/95% air/5% CO_2 incubator, following which the WEC was removed and replaced with 2 ml of WEI containing 10 uCi/ml of ^{3}H-thymidine. For the positive control incubations, DMN (dissolved in water) or MMS (dissoved in dimethylsulphoxide, DMSO) were mixed with the ^{3}H-thymidine solution to yield the desired concentrations. The concentration of DMSO in the MMS incubations was 1%, which had no effect on DNA repair. Six wells were used for each animal or positive control incubation. Hepatocyte cultures were incubated for 4 h with ^{3}H-thymidine, following which the cultures were washed with phosphate-buffered saline and incubated overnight with 4 ml of WEI. (Note: unlabeled thymidine was not added during the overnight incubation period because thymidine has been shown to influence the extent of DNA repair secondary to an inhibition of ADP-ribosyltransferase (Althaus *et al.*, 1982b).

The following morning the cultures were washed, incubated for 10 min with 1% citrate to swell the nuclei, fixed by 3 successive incubations with 3:1 methanol/acetic acid, and washed 4 times with distilled H_2O. The coverslips containing attached hepatocytes were then removed, air dried and mounted with Permount (Fisher Scientific) on glass slides.

Three of the six slides per animal or incubation condition were chosen for autoradiography, while the remaining 3 slides served as backup in the event of problems with autoradiography. In the absence of light, the slides were dipped in undiluted NTB-2 emulsion, stored for 13 days at – 20 °C, developed with Kodak D-19 developer and fixer, and stained with methyl-green pyronin Y.

Grain counting was performed with a Nikon Labophot microscope fitted with an Artek high-resolution video camera interfaced with an Artek grain counter. Net nuclear grains (NNG) were determined by counting the number of grains in the nuclei and subtracting the average number of grains present in 3 equal size adjacent cytoplasmic areas. A strict procedure for random selection of viable cells was used (similar to Mirsalis and

Table 1. *Results of testing 2AAF and 4AAF in the in vivo/in vitro rat hepatocyte/DNA repair assay*

Chemical tested	Dose (mg/kg)	Time point (h)	Average[a] NNG (± SD)	% of Cells ≥ NNG[b]	% S[c]	Response evaluation[d] Individual	Collective
Corn oil control			– 1.08 (3.45)	7.8	0.0	—	—
			– 1.75 (3.79)	4.2	1.1		
2AAF	10	12	8.51 (8.38)	66.7	0.0	P	P
			4.46 (4.03)	38.9	2.2	WP	
			7.17 (7.07)	62.2	2.2	P	
		24	3.94 (4.16)	42.2	1.1	WP	
			6.34 (4.73)	68.9	1.1	P	
			4.50 (4.03)	54.4	2.2	WP	
	50	12	13.60 (9.55)	83.3	1.1	P	
			14.14 (8.33)	93.3	0.0	P	
			10.92 (8.86)	80.0	0.0	P	
		24	7.34 (5.63)	68.9	1.1	P	
			6.78 (4.04)	74.4	2.2	P	
			10.56 (10.43)	76.7	0.0	P	
4AAF	50	12	– 0.98 (1.97)	1.1	0.0	N	N
			– 1.57 (2.46)	5.6	1.1	N	
			– 1.32 (2.80)	3.3	1.1	N	
		24	– 1.31 (3.26)	4.4	4.3	N	
			– 1.12 (3.68)	8.9	7.2	N	
			– 2.32 (4.19)	8.9	7.2	N	
	200	12	– 1.90 (2.69)	3.3	0.0	N	
			– 0.94 (2.51)	1.1	1.1	N	
			– 1.06 (2.65)	7.8	0.0	N	
		24	– 0.86 (3.45)	7.8	20.4	N	
			– 1.79 (3.44)	6.7	28.6	N	
			– 1.48 (3.55)	6.7	35.7	N	

[a] Mean net nuclear grains (± the standard deviation) from 90 cells.
[b] Percentage of cells with ≥ 5 net nuclear grains.
[c] Percentage of cells undergoing replicative DNA synthesis (S phase).
[d] P-positive, WP-weak positive, N-negative.

Butterworth, 1980). Thirty cells per slide were counted (not including S-phase cells) and the average NNG (±SD) for all 3 slides (i.e. 90 cells total) was calculated. The percentage of cells with greater than or equal to 5 NNG was also determined for each animal or incubation condition. Cells with nuclei of extremely high grain density were considered to be undergoing replicative DNA synthesis (S phase). When these cells appeared in the microscope field, they were noted and their incidence as a percentage of the total number of cells examined was recorded.

Response evaluation criteria

Consistent with the criteria employed by other investigators (Williams, 1977; Probst *et al.*, 1981; Mirsalis and Butterworth, 1980), average NNG counts of 5 or greater were assumed to constitute a positive response, since these differed from the control NNG counts by greater than 2 standard deviations. On a historical basis, including the present study, hepatocytes from untreated animals invariably yield an average value for NNG of less than zero. Therefore, average NNG from treated animals of less than zero were considered negative responses. Responses between zero and 5 average NNG were considered equivocal unless at least 25% of the cells examined contained greater than or equal to 5 NNG, in which case the response was referred to as a weak positive.

Secretion of 2AAF into the milk of lactating rats

Lactating rats with litters of six 12 to 13-day-old pups were administered by gavage a corn oil suspension of ^{14}C-labeled 2AAF mixed with unlabeled 2AAF, which yielded a dose of 50 mg of 2AAF/kg body weight and approximately 5 uCi/rat (specific radioactivity = 6.6×10^5 dpm/mg). The dams and pups were not separated except for a few minutes at the time of dosing. At 1, 3, 8 and 24 h post-administration, the dams were anaesthetized with sodium pentobarbital and milk was collected from the mammary glands by manual expression following an ip injection of 1 unit of oxytoxin. The dams were then sacrificed with an overdose of sodium pentobarbital, and blood was removed from the inferior vena cava prior to death. Whole milk and serum aliquots (400 ul) were mixed with 12 ml of ACS (Amersham) liquid scintillation cocktail in glass vials and counted in a Beckman LS 5800 liquid scintillation counter.

To determine the amount of lipid-soluble radioactivity in the samples, 600 ul aliquots of milk and serum were transferred to glass screw-cap tubes along with 1400 ul of 0.2 *M* sodium acetate buffer (pH 6.0) and 8 ml of diethyl ether. The mixture was vortexed for 30 sec and centrifuged briefly to break phases. The resulting either layer was removed and set aside, while the aqueous phase was reextracted with 8 ml of ether. The ether layers were pooled in scintillation vials, taken to dryness under a stream of N_2, and the residue was redissolved in liquid scintillation cocktail and counted in the Beckman instrument.

4 Results of discussion

2AAF and 4AAF tested in adult male rats

2AAF produced a clearly positive response in the *in vivo/in vitro* rat hepatocyte/DNA repair assay, using adult male F344 rats (Table 1). The 50 mg/kg body weight dose produced only about twice as much DNA repair in terms of the average NNG as the 10 mg/kg body weight dose. Thus, the dose response was not proportional to the administered dose. Approximately a 2-fold greater response (i.e., NNG counts) for both doses was observed at 12 h than at 24 h postadministration. The percentage of cells with 5 or more NNG was also greater at the higher dose and earlier timepoint, but the differences were even less than for the average NNG counts.

These findings are similar to the time course of 2AAF-induced DNA repair reported by Mirsalis *et al.* (1982). However, the magnitude of the response to a dose of 50 mg/kg body weight was considerably less than that observed by Mirsalis *et al.* (1982). The reasons for the lower potency of 2AAF in the present study are not clear, but one possibility may be that unlabeled thymidine was not added to the overnight incubation medium. Althaus *et al.* (1982a) have shown that inhibition of the activity of ADP-ribosyltransferase results in large increases in the amount of DNA repair produced in hepatocytes by various genotoxic agents and that thymidine has been identified as a potent inhibitor of the enzyme (Althaus *et al.*, 1982b). Furthermore, the DNA repair response elicited *in vitro* by 2AAF was shown to be increased approximately 4-fold under conditions in which ADP-ribosyltransferase was inhibited 76% by addition of 10 mM isonicotinamide to the culture medium (Althaus *et al.*, 1982a, b). Since ADP-ribosyltransferase is similarly inhibited by as little as 50 uM thymidine (Althaus *et al.*,

Table 2. *In vitro positive control compounds*

		Experiment 1		Experiment 2	
Chemical	Concentration (molar)	Average NNG (±SD)[a]	% of Cells ≥5 NNG[b]	NNG (±SD)[a]	% of Cells ≥5 NNG[b]
DMN	2×10^{-3}	9.2 ± 6.8	94	12.4 ± 7.1	98
	0.5×10^{-4}	3.5 ± 4.7	42	3.2 ± 5.8	48
MMS	5×10^{-4}	17.1 ± 11.2	99	16.5 ± 12.1	95
	1×10^{-4}	5.2 ± 6.1	65	8.4 ± 7.2	78

[a] Mean net nuclear grains (± the standard deviation) from 90 cells.
[b] Percentage of cells with ≥ 5 net nuclear grains.

1982b), it is possible that the additon by Mirsalis *et al.* (1982) of 250 uM thymidine to the medium used for overnight incubation of hepatocytes from 2AAF-treated rats resulted in enhancement of the DNA repair response. This effect would not have occured in the present study since thymidine was not added to the overnight medium.

The activity of the *in vitro* positive control compounds, DMN and MMS, in hepatocytes from untreated F344 rats was somewhat lower than that previously observed (Kornbrust and Bartknect, 1984) with hepatocytes from Sprague-Dawley rats (Table 2). Although the variabilty in responses between rat strains has not been evaluated for either the *in vitro* or *in vivo/in vitro* assays, the occurrence of a relatively large variability between experiments and even between animals in the same experiment is not uncommon with this type of assay. Therefore, the data obtained for the positive control compounds serve as a gross indicator of the proper performance of the assay, but they do not provide a sufficient basis for determining whether a general reduction in response magnitude occurred in the present experiment.

4AAF did not elicit any DNA repair at either dose or time point (Table 1). A pronounced increase in the percentage of cells in S phase (i.e., undergoing replicative DNA synthesis) was observed at 24 h postadministration, particularly with the higher (200 mg/kg) dose. These results suggest that 4AAF induces hepatocellular replication *in vivo*, either as a regenerative response to cytotoxicity, as has been observed for the hepatotoxins carbon tetrachloride (Mirsalis *et al.*, 1982) and dinitrotoluene (Mirsalis and Butterworth, 1982), or via a direct mitogenic stimulus to the hepatocytes.

An additional finding that may pertain to the question of 4AAF toxicity is that the concentration of cytoplasmic (i.e., background) grains appeared to be decreased by 4AAF treatment in hepatocytes isolated 12 h postadministration. For hepatocytes from the untreated as well as most treated animals in the present study, the average number of grains in a nuclear size area of cytoplasm was approximately 10–15, which is a typical background grain count. Equivalent size cytoplasmic areas of hepatocytes from rats treated with 4AAF 12 h prior to isolation contained approximately 3–6 grains on the average. The nature of cytoplasmic background grains has not been completely resolved, but a large component may be derived from the catabolism of ^{3}H-thymidine and subsequent metabolic incorporation into cellular macromolecules (Yager and Miller, 1978). Hence, a lower concentration of cytoplasmic grains may indicate metabolic disturbances in the cells arising from some form of compound-induced toxicity. Unfortunately, a method for characterizing subtle toxic effects of test compounds in the *in vivo/in vitro* assay has not been developed.

In contrast to the present results, Althaus *et al.* (1982a) obtained a positive response for 4AAF in an *in vitro* rat hepatocyte/DNA repair assay, although the potency was considerably less than for 2AAF. 4AAF was also active in a number of other short-term *in vitro* tests for genotoxicity, including the Ames Salmonella assay (Ashby, 1981). Two *in vivo* tests of 4AAF yielded negative responses; 2AAF was found to be positive in both cases (Ashby, 1981). Thus, the present findings are consistent with the previous results obtained for both of these compounds in *in vivo* test systems. Considering the *in vitro* genotoxic potential of 4AAF, uncertainty persists as to whether the apparent lack of activity of 4AAF *in vivo* is due to insufficient sensitivity of the assays or to some type of interference with response detection produced by 4AAF *in vivo*.

With regard to the latter possibilty, a question was raised as to whether the apparent hepatic toxicity associated with high doses of 4AAF might perturb hepatocellular DNA repair processes and, thereby, preclude detection of a response in the *in vivo/in vitro* DNA repair assay. As a preliminary attempt to address this hypothesis, the *in vitro* DNA repair response to the direct-acting alkylating agent MMS was compared in hepatocytes from control rats and rats treated with 200 mg of 4AAF/kg body weight 12 h prior to cell isolation. The amount of repair elicited by MMS was reduced over 50% in hepatocytes from 4AAF-treated rats (Table 3). It appears that exposure to a high dose of 4-AAF may partially suppress the hepatocellular capacity for DNA repair. However, it is unlikely that this degree of inhibition would be sufficient to completely abolish a DNA repair response in the event that 4AAF had exerted a DNA-damaging effect, particularly since many other agents, including N-acetoxy-2-acetylaminofluorene and MMS, are known to interfere with DNA repair processes (Brown *et al.*, 1979; Freeman and Larcom, 1983) but still produce strong responses in the rat hepatocyte/DNA repair assay. Most chemicals that inhibit DNA repair usually produce a greater inhibition of semiconservative

Table 3. *Effect of in vivo treatment with 4AAF on the in vitro DNA repair response to methyl methanesulfonate*

In vivo treatment[a]	*In vitro* chemical	Concentration (molar)	Average NNG (± SD)	Per cent of Cells ≥ 5 NNG
None	None		− 1.91 (± 2.72)	3.3
None	MMS	5×10^{-4}	20.45 (± 9.17)	97
		1×10^{-4}	8.32 (± 4.46)	78
4AAF[b]	None		− 1.13 (± 2.03)	2.2
4AAF[b]	MMS	5×10^{-4}	8.55 (± 6.25)	75
		1×10^{-4}	4.20 (± 4.14)	60

[a] Hepatocytes were pooled from two control and three 4AAF-treated rats.
[b] 200 mg/kg body weight, 12 h prior to liver perfusion.

(replicative) DNA synthesis (Bernheim and Falk, 1983). Therefore, the increased incidence of S phase in the absence of any evidence of DNA repair at 24 h postadministration suggests that inhibition of repair was not manifested at this time. The negative responses obtained with 4AAF are probably a reliable indicator of its lack of genotoxicity in this assay system.

Secretion of 2-AAF into milk and induction of DNA repair in neonatal hepatocytes

In a previous investigation, the *in vivo/in vitro* hepatocyte/DNA repair assay was used to detect DNA repair in hepatocytes from neonatal rats that were exposed to the genotoxic hepatocarcinogen DMN via secretion into milk after administration of the chemical to the maternal animals (Kornbrust and Dietz, manuscript submitted). A clearly positive response was obtained in neonatal hepatocytes at 4 and 10 h following administration of 20 mg of DMN/kg body weight to the lactating rats. Measurement of the DMN concentration in milk samples confirmed the presence of DMN at a level sufficient to constitute a neonatal exposure consistent with the observed DNA repair response. Metabolic activation of DMN in neonatal liver microsomes was found to be comparable to the adult capacity and was increased by pretreatments that induced DMN metabolism.

In the present study, 2AAF was tested in a similar manner. Table 4 shows that a weak DNA repair response was observed in hepatocytes from 12 to 13-day-old pups 15 h following administration to the dams of 200 mg of 2AAF/kg body weight, whereas no repair was detected with a 4-fold lower dose. An estimate of the amount of 2AAF secreted into milk was obtained by monitoring the appearance of radioactivity in milk and serum of the dams after administration of 50 mg/kg of ^{14}C-labeled 2AAF (Figure 1). Between 1 and 8 h postadministration, the levels of total 2AAF-derived ^{14}C in milk were 56–66% of the levels in serum. The lipid-soluble fraction of the radioactivity in milk declined from 53% of the total at 1 h to 7% by 24 h following administration. This shift of radioactivity towards the aqueous fraction over time probably was due to the metabolism of 2AAF to water-soluble glucuronide or glutathione conjugates.

Assuming that the lipid-soluble radioactivity represents parent material, and using a previously determined estimate of the rate of milk consumption for 12 to 13-day-old neonates, it was calculated that, over the 10 h period following administration, each pup may have received a "dose" of 2AAF of approximately 2.5 mg/kg body weight, which is 20-fold less than the maternal dose on a body

Figure 1. Concentration of [^{14}C] in maternal serum, milk and lipid soluble fractions of serum and milk collected at various times following administration of [^{14}C]-2AAF (50 mg/kg body weight, 5uCi/rat) to lactating rats ($n = 3$).

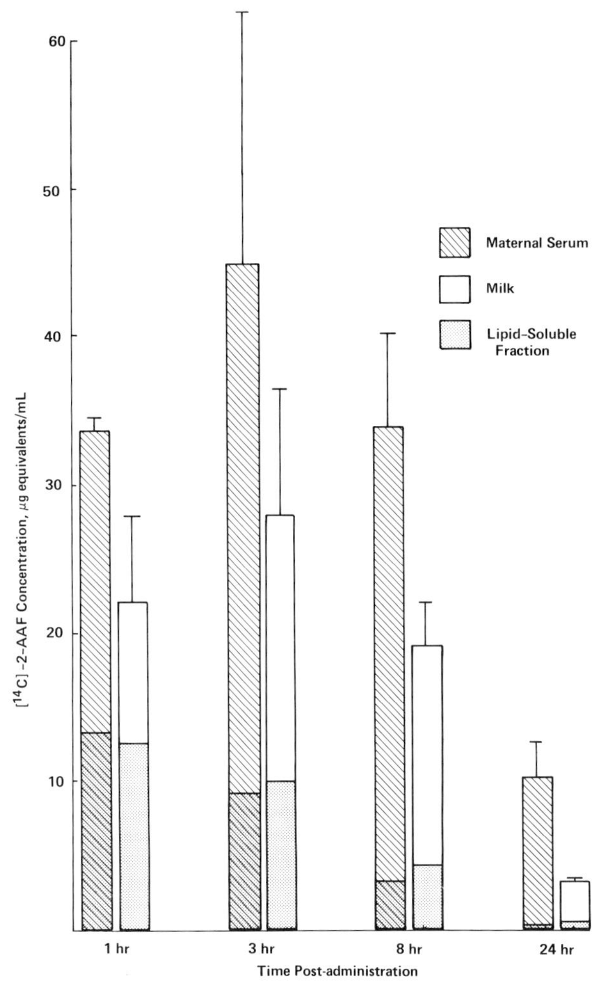

Table 4. *DNA repair responses in neonatal hepatocytes after administration of 2AAF to lactating rats*[a]

Treatment/Dose	Average NNG (±SD)	Per cent of cells ⩾5 NNG	Cells in S Phase (%)	Response evaluation
Control	−2.0 (±2.6)	6.7	12.6	
2AAF, 200 mg/kg	2.3 (±4.8)[2]	35.6	10.9	WP
	0.6 (±4.7)	22.2	15.1	
	1.7 (±4.2)	28.8	13.5	
2AAF, 50 mg/kg	−1.1 (3.4)	10.0	15.9	N
	0.2 (±3.1)	7.7	17.6	
	−0.5 (±2.9)	4.4	15.1	

[a] 200 mg of 2AAF/kg body weight was administered to lactating rats with litters of six 12 to 13-day-old pups. Hepatocytes were isolated from the neonatal livers 15 h following treatment of dams.

[b] Data from separate litters. Hepatocytes from 6 pups per litter were pooled (see Methods).

weight basis. Extrapolation of this ratio to the 200 mg/kg dose given to the dams in Table 4 yields an estimate, albeit crude, of 10 mg/kg body weight per pup. Direct comparison of the neonatal DNA repair response with the response of adult male rats to 10 mg of 2AAF/kg body weight (Table 1) is not possible since the neonatal exposure was highly protracted relative to the gavage dosing the adult animals. However, experiments have shown that *in vitro* the response of neonatal hepatocytes to 2AAF is considerably less than that of adult hepatocytes (Kornbrust and Dietz, manuscript submitted), probably due to the immaturity of metabolic activation pathways in the neonatal rat. Therefore, the weak DNA repair response observed under the lactation exposure conditions is consistent with the low concentration of 2AAF measured in maternal milk samples, as well as with a possible lower sensitivity of the neonate.

5 Acknowledgements

The authors are grateful to Dr. Byron Butterworth and Dr. Jerry Reel for their support and guidance and to Thelma Demetrius for preparation of the manuscript.

6 References

Althaus, F.R., S.D. Lawrence, G.L. Sattler, D.G. Longfellow and H.C. Pitot (1982a) Chemical quantification of unscheduled DNA synthesis in cultured hepatocytes as an assay for the rapid screening of potential chemical carcinogens, *Cancer Res.*, **42**, 3010–3015.

Althaus, F.R., S.D. Lawrence, G.L. Sattler and H.C. Pitot (1982b) ADP-ribosyltransferase activity in cultured hepatocytes. Interactions with DNA repair, *J. Biol. Chem.*, **257**, 5528–5535.

Ashby, J. (1981) Overview of study and test chemical activities. In F.J. de Serres and J. Ashby (Eds), *Elsevier*, New York, pp. 113–172.

Bernheim, N.J. and H. Falk (1983) Chemical, physical and genetic factors interfering with DNA repair – a review, *J. Am. College Toxicol.*, **2**, 23–54.

Brown, A.J., T.H. Fickel, J.E. Cleaver, P.H.M. Lohman, M.H. Wade and R. Waters (1979) Overlapping pathways for repair of damage from ultraviolet light and chemical carcinogens in human fibroblasts, *Cancer Res.*, **39**, 2522–2527.

Devirgiliis, L.C., L. Dini, A. DiPierro, S. Leoni, S. Spagnuolo and S. Stefanini (1981) An improved non-perfusion method for the isolation and purification of rat foetal and neonatal hepatocytes, *Cell. Molec. Biol.*, **27**, 687–694.

Freeman, S.E. and L.L. Larcom (1983) Inhibition of unscheduled DNA synthesis in human lymphocytes by chemical carcinogens, *Chem. Biol. Interact.*, **46**, 85–99.

Kornbrust, D.J. and T.R. Bartknecht (1984) Comparison of rat and hamster hepatocyte primary culture/DNA repair assays, *Environ. Mutagen.*, **6**, 1–11.

Mirsalis, J.C. and B.E. Butterworth (1982) Induction of unscheduled DNA synthesis in rat hepatocytes following *in vivo* treatment with dinitrotoluene, *Carcinogenesis*, **3**, 241–245.

Mirsalis, J.C. and B.E. Butterworth (1980) Detection of unscheduled DNA synthesis in hepatocytes isolated from rats treated with genotoxic agents: An *in vivo–in vitro* assay for potential carcinogens and mutagens, *Carcinogenesis*, **1**, 621–625.

Mirsalis, J.C., C.K. Tyson and B.E. Butterworth (1982) Detection of genotoxic carcinogens in the *in vivo–in vitro* hepatocyte DNA repair assay, *Environ. Mutagen.*, **4**, 553–562.

Mitchell, A.D., D.A. Casciano, M.L. Meltz, D.E. Robinson, R.H. San, G.M. Williams and E.S. Von Halle (1983) Unscheduled DNA synthesis tests. A report of the U.S. Environmental Protection Agency Gene-Tox Program, *Mutation Res.*, **123**, 363–410.

Probst, G.S., R.E. McMahon, L.E. Hill, C.Z. Thompson, J.K. Epp and S.B. Neal (1981) Chemically induced unscheduled DNA synthesis in primary rat hepatocyte cultures: A comparison with bacterial mutagenicity using 218 compounds, *Environ. Mutagen.*, **3**, 11–32.

Purchase, I.F.H., D.B. Clayson, R. Preussmann and L. Tomatis (1981) Activity of 42 compounds in animal carcinogenicity studies. In F.J. de Serres and J. Ashby, (Eds), *Evaluation of Short Term Tests for Carcinogens Elsevier*, New York, pp. 21–32.

Williams, G.M. (1977) Detection of chemical carcinogens by unscheduled DNA synthesis in rat liver primary cell cultures, *Cancer Res.*, **37**, 1845–1851.

Williams, G.M., M.F. Laspia and V.C. Dunkel (1982) Reliability of the hepatocyte primary culture/DNA repair test in testing of coded carcinogens and noncarcinogens, *Mutation Res.*, **97**, 359–370.

Yager, J.D. and J.A. Miller (1978) DNA repair in primary cultures of rat hepatocytes, *Cancer Res.*, **38**, 4385–4394.

David J. Doolittle, Gregory Muller, Janice L. Sames, and Harvey E. Scribner
Rohm and Haas Company, Toxicology Department, Spring House, Pennsylvania, USA

62

In vivo–in vitro assay for quantifying hepatic DNA repair and S phase in F-344 rats

1 Summary

The potential of a single oral dose of 2-acetylaminofluorene (2AAF) or 4-acetylaminofluorene (4AAF) to induce hepatic DNA repair and/or increase the percentage of hepatocytes in S phase has been evaluated in F-344 rats. The autoradiographic approach used in this assay allows assessment of both DNA repair and S phase in the same cell population. These two endpoints are easily distinguished by the degree of nuclear labelling. The capability to measure both events in the same animal considerably enhances the utility of the assay. The results indicate that, under the conditions of this study, 2AAF induces DNA repair but does not increase the percentage of hepatocytes in S phase whereas 4AAF increases the percentage of hepatocytes in S phase but does not induce DNA repair.

2 Introduction

To obtain the most relevant and meaningful information for assessing the genotoxic/carcinogenic potential of chemical compounds, it is important to use assays that incorporate the target organ of interest. When using short-term tests to evaluate the potential of a chemical compound to induce liver tumours in rats, the use of an *in vivo* assay system that utilizes the cells at risk (i.e., rat hepatocytes) as target cells is desirable. This approach is used in the *in vivo–in vitro* rat hepatocyte assay developed by Mirsalis and Butterworth (1980) and inherently accounts for factors such as absorption, distribution, metabolism, and excretion. In the *in vivo–in vitro* rat hepatocyte assay, the test compound is administered *in vivo*, hepatocytes are isolated, and DNA repair as well as the percentage of hepatocytes in the DNA synthesis phase of the cell cycle (S phase) are quantified. The induction of DNA repair indicates that the compound has caused DNA damage. An increase in the number of hepatocytes in S phase indicates that the compound has produced nonspecific hepatotoxicity (i.e., regenerative hyperplasia) or alternatively, that the compound has exerted a direct mitogenic effect on the hepatocytes.

In the present study two structurally related compounds, 2AAF and 4AAF, have been evaluated in this assay.

3 Materials and methods

Animals care and treatment

Male CDF (F-344)/CrIBR rats weighing between 150–

200 g were obtained from Charles River Breeding Laboratories (Kingston, NY). Upon arrival all animals were housed one per cage in hanging, stainless-steel mesh cages and allowed free access to food (Purina Rodent Lab Chow Checkers, Richmond, IN) and water. Lights were controlled by an automatic timer set for 12 h of light and 12 h of darkness. Animals were acclimated to the animal facility for 2 weeks prior to use and were fasted 3–4 h prior to dosing. 2AAF and 4AAF were supplied by the ICI (UK) and stored at 4 °C until use. Just prior to dose administration, homogeneous dispersions of the compound were made in corn oil and administered by gavage (1.0 ml/100 g body weight). Control animals received corn oil.

Isolation, culturing, fixation and autoradiographic processing of hepatocytes

At either 12 or 24 h after dosing, the livers of the animals were perfused. The animals were anaesthetized with 0.2 cc/100 gm body weight Nembutal (50 mg/ml pentobarbitol, Abbott Labs, Chicago, IL). The portal vein was cannulated with a 20-gauge, 1.25-inch Angiocath (Deseret Co., Sandy, UT). EGTA solution (0.5 mM EGTA, Sigma Chemical Co.) at 37 °C, dissolved in Hanks Balanced Salt Solution (GIBCO Labs, Grand Island, N.Y.), was perfused at a flow rate of 8 ml/min. The vena cava was cut just below the suture and the liver allowed to clear of blood for 1.5 min. The suture around the vena cava was tightened and the flow increased to 20 ml/min for 2 minutes. EGTA solution was perfused for a total of 4 minutes and then the flow changed to collagenase solution [100 units Sigma type 1 collagenase per ml of WEI (William's Medium E, GIBCO), supplemented with 2 mM L-glutamine (GIBCO), and 50 mg/ml gentamycin sulfate (Sigma)] at 37 °C for 12 minutes at a flow rate of 20 ml/min.

The liver was removed by carefully cutting away the membranes connecting it to the stomach, oesophagus, and diaphragm. The liver was placed in 40 ml of 37 °C collagenase solution and several small incisions made in the capsule of each lobe. The hepatocytes were gently combed out with a sterile metal dog comb. When only fibrous connective tissue remained in the liver, the cell suspension was transferred to a sterile centrifuge tube, placed on ice, and the cells allowed to settle for approximately 10 minutes. The supernatant was removed, the cells were suspended in 50 ml of ice cold WEI, pipetted though sterile gauze to remove large clumps of hepatocytes, and centrifuged (50X g, 5 min). The supernatant was removed and the cells gently suspended in 20–50 ml of ice cold WEI. Viability was determined by trypan blue exclusion, and 400,000 viable cells were seeded into each well of a Linbro 6 well culture dish (Flow Laboratories, McLean, VA). Each well contained 4 ml of WEC (WEI with 10% foetal calf serum, GIBCO) and a Thermanox round plastic coverslip (Flow Labs). Cultures were placed in a 37 °C incubator with 5% CO_2 in air and >90% relative humidity. Following a 90 minute incubation to permit attachment of the hepatocytes to the coverslips, cultures were gently washed once with 4 ml of 37 °C WEI per well. Two ml of WEC solution (37 °C) containing 10 uCi/ml of methyl-^{3}H-thymidine (specific activity 40–60 Ci/mmol, Amersham Corp., Arlington Heights, IL) were added to each well.

Following a 4 h incubation, the ^{3}H-thymidine containing media was removed, the cultures were washed once with 4 ml WEI (37 °C) per well, and 3 ml of WEC (37 °C) were added to each well. Cultures were incubated overnight (approximately 16 h). The cultures were washed once with 4 ml WEI (37 °C) per well; the nuclei were swollen with 1% sodium citrate for 10 minutes and fixed with 3 consecutive 10 minute exposures to an acetic acid:absolute ethanol (1:3) solution. The coverslips were washed 3 times with deionized distilled water, air dried, and mounted on coded slides. These slides were dipped in photographic emulsion (Kodak NTB-2) at 42 °C in total darkness, allowed to dry for 3–5 h, and stored at −20 °C in total darkness for 14 days. Following the 2-week exposure, the slides were developed with Kodak D-19 developer and fixed with Kodak fixer. The cells were stained with haematoxylin and eosin, and coverslips were applied with a thin layer of permount. All slides evaluated in the study were coded and scored blind.

Grain counting for quantitation of DNA repair

The silver grains that were formed in the emulsion by radiation from the ^{3}H-thymidine were counted microscopically using a Biotran III colony counter (New Brunswick Scientific Co., Edison, NJ), which was interfaced to a microscope with a TV camera and monitor. Data were fed directly into a Hewlett Packard computer. A patch of cells was randomly selected for a starting point and 50 cells were scored under a 100X oil immersion objective by moving the X axis of the microscope stage in a straight line from the starting point. If 50 scorable cells were not observed before coming to the edge of the slide, the stage was moved 2–3 fields on the Y axis and counting resumed in the opposite direction, parallel to the first line. The following criteria were used to determine if a cell was scoreable: (1) morphologically altered cells were not counted (i.e., cells with lysed nuclei or cytoplasm); (2) cells in S phase were not scored for DNA repair; (3) cells with pycnotic nuclei were not scored. All other cells encountered were scored.

The scoring of an individual hepatocyte was accomplished by centering the image of a hepatocyte on the TV monitor and placing an electronically produced aperture over the image of the nucleus on the TV monitor, while adjusting the size to occlude the nucleus. The number of grains within the nuclear area was automatically counted by the Biotran III. The aperture was moved (without changing size) to the area of cytoplasm adjacent to the nucleus that contained the greatest number of grains; this area was counted.

Data analysis consisted of determining the average number of nuclear grains, the average number of cytoplasmic grains, and the average net nuclear grains (average number of nuclear grains − average number of cytoplasmic grains). The percentage of the cell population with >6 net nuclear grains (NNG) was also determined. A positive average NNG value indicated preferential localization of ^{3}H-thymidine in the nucleus (i.e., DNA repair). A negative average NNG value indicated that the cytoplasm contained more grains than the nucleus (i.e., absence of DNA repair).

Determination of the percentage of cells in S phase
Scoring was performed on a research quality light microscope under low power magnification (100–200X). Hepatocyte nuclei in S phase appeared as dense black spots under low power microscopic evaluation. Hepatocytes in each of 5 separate areas of the slide were scored (200 cells per area, 1000 cells per slide). Within each area, the number of S-phase nuclei and the number of healthy nuclei not in S phase was counted. When 2 contiguous nuclei, both in S phase appeared, these were scored as one S-phase cell (some rat hepatocytes are binucleate). Similarly, binucleate cells not in S phase were scored as a single cell. Results were expressed as the percentage of the cells in S phase.

4 Results

The DNA repair response induced by each treatment was evaluated as: (1) average NNG and (2) percentage of hepatocytes with greater than 6 NNG. A positive DNA repair response was indicated by a NNG value equal to or greater than 1 when at least 10% of the hepatocytes scored contained more than 6 NNG. The S-phase response induced by each treatment was expressed as the percentage of the cell population in S phase; a positive response was indicated by an S-phase value of >2%.

2AAF induced DNA damage and repair at all dose levels examined (10, 50, and 200 mg/kg), and the magnitude of the observed response was dependent on both dose level and time of evaluation (Table 1). At 12 h postgavage, the 10 mg/kg dose produced a positive repair response (2.5 NNG; 26.9% >6 NNG), while the 50 mg/kg dose produced a stronger positive response (8.8 NNG; 58.7% >6 NNG). At 24 h postgavage, the 50 mg/kg dose produced a positive response of 3.1 NNG (27.1% >6 NNG), and the 200 mg/kg dose also produced a positive response of 1.7 NNG (17.3% >6 NNG). Therefore, for 2AAF, a dose-dependent DNA repair response occurred at the 12 h time point, and a DNA repair response that was not dose-dependent occurred at the 24 h time point.

2AAF did not increase replicative DNA synthesis as measured by the percentage of hepatocytes in S phase (Table 1). The S-phase values for corn oil controls at 12 h and 24 h were 0.3 and 0.6%, respectively, and all values obtained with 2AAF were within this range.

4AAF did not induce DNA repair at any dose level or any time point evaluated (Table 1). NNG values ranged from −6.2 to −3.0, and the percentage of hepatocyte nuclei containing greater than 6 NNG was low, ranging from 0.0 to 0.4%.

Although 4AAF failed to induce DNA repair, it did increase replicative DNA synthesis as measured by the percentage of hepatocytes in S phase (Table 1). Twenty-four hours after 4AAF administration a dose-dependent increase in the percentage of hepatocytes in S phase was observed. A 50 mg/kg dose of 4AAF increased the percentage of hepatocytes in S phase to 2.4%, and a 200 mg/kg dose increased the percentage to 22.0%. The S-phase values for corn oil controls at 12 and 24 h were 0.3 and 0.6%, respectively. Therefore, 4AAF was ineffective at inducing DNA damage and repair, yet effective at increasing the percentage of hepatocytes in S phase. The ability of 4AAF to stimulate replicative DNA synthesis was observed only at the 24 h time point.

5 Discussion

The measurement of chemically induced DNA repair is a valuable tool for assessing the organ-specific genotoxic potential of a chemical (Mirsalis *et al.*, 1982; Doolittle *et al.*, 1984, Doolittle and Butterworth, 1984; Working and Butterworth, 1984), and genotoxic potential has been correlated with carcinogenic potential (McCann *et al.*, 1975, 1976). The induction of hepatic DNA repair by 2AAF provides evidence that this compound is producing its hepatocarcinogenic response by a genotoxic mechanism.

Table 1. *Assessment of DNA repair and S-phase following administration of corn oil, 2AAF, or 4AAF*

Compound	Dose (mg/kg)	Time (h)[a]	DNA repair[b]	% with NNG >6	% in S phase
corn oil	10 ml/kg	12	−5.2 ± 0.6[c]	0.2	0.3 ± 0.1[d]
		24	−3.1 ± 0.5	0.0	0.6 ± 0.3
2AAF	10	12	+2.5 ± 1.4	26.9	0.3 ± 0.1
	50	12	+8.8 ± 2.9	58.7	0.4 ± 0.1
		24	+3.1 ± 0.8	27.1	0.4 ± 0.1
	200	24	+1.7 ± 0.3	17.3	0.4 ± 0.0
4AAF	50	12	−3.9 ± 0.8	0.2	0.7 ± 0.3
		24	−4.2 ± 1.2	0.0	2.4 ± 0.7
	200	24	−6.2 ± 0.4	0.4	22.0 ± 1.2
	500	12	−3.0 ± 0.8	0.2	0.6 ± 0.1

[a]Time elapsed from dosing to hepatocyte isolation.
[b]Expressed as net nuclear grains (nuclear grains − cytoplasmic grains). A negative number indicates that there were more grains in the cytoplasm than in the nucleus.
[c]Values represent the mean ± S.E. of the values obtained from 3 animals; 3 slides were prepared per animal and 50 cells were scored per slide (450 cells scored per point).
[d]Values represent the mean ± S.E. of the values obtained from 3 animals; 3 slides were prepared per animal and 1000 cells were scored per slide (9000 cells scored per point).

The mechanism(s) of action of nongenotoxic carcinogens is(are) unknown. There are two lines of evidence indicating that exposure to an agent that increases cell proliferation in the target organ may produce or enhance a carcinogenic response: (1) Initiation of chemical carcinogenesis has been shown to require cell proliferation (Cayama *et al.*, 1978; Columbano *et al.*, 1981). (2) The promotional phase of carcinogenesis, while extremely complex and multifactorial (Slaga *et al.*, 1980a, 1980b) has been shown to be intimately involved with the induction of hyperplasia (Furstenberger *et al.*, 1985; Argyris, 1980; Argyris and Slaga, 1981; Kinzel *et al.*, 1984). It is possible that agents that induce hyperplasia in a particular target organ may be involved in both the initiation and promotional phases of carcinogenesis. Consistent with this are observations that tissue-specific nonmutagenic carcinogens generally induce a hyperplastic response in the target organ. This has been demonstrated in the rat colon (Bull *et al.*, 1984; Jacobs, 1984) in rat urinary bladder (Fukushima and Cohen 1980), in mouse skin (Furstenberger *et al.*, 1985; Yuspa *et al.*, 1976), in mouse liver (Doolittle *et al.*, 1985), and in rat liver (Schulte-Hermann *et al.*, 1981; Butterworth *et al.*, 1984). The observation that 4AAF increases the percentage of hepatocytes in S phase indicates that this compound may induce or enhance hepatic carcinogenesis by increasing the rate of hepatocyte DNA replication.

In conclusion, under the conditions of this study, 2AAF induces DNA repair but does not increase the percentage of hepatocytes in S phase, whereas, 4AAF increases the percentage of hepatocytes in S phase but does not induce DNA repair. The *in vivo–in vitro* rat hepatocyte assay for assessing DNA repair and S phase induction has utility as a short-term screening test for detecting potential rat hepatocarcinogens and may also provide information on the mode of action of known hepatocarcinogens.

6 Acknowledgement

The authors wish to thank V.L. Pruitt for excellent secretarial assistance.

7 References

Argyris, T.S. (1980) Tumor promotion by abrasion induced epidermal hyperplasia in the skin of mice, *J. Invest. Dermatol.*, **75**, 360–362.

Argyris, T.S. and T.J. Slaga (1981) Promotion of carcinomas by repeated abrasion in initiated skin of mice, *Cancer Res.*, **41**, 5193–5195.

Bull, A.W., N.D. Nigro, W.A. Golembieski, J.D. Crissman and L.J. Marnett (1984) *In vivo* stimulation of DNA synthesis and induction of ornithine decarboxylase in rat colon by fatty acid hydroperoxides, autoxidation products of unsaturated fatty acids, *Cancer Res.*, **44**, 4924–4928.

Butterworth, B.E., E. Bermudez, T. Smith-Oliver, L. Earle, R. Cattley, J. Martin, J.A. Popp, S. Strom, R. Jirtle and G. Michalopoulos (1984) Lack of genotoxic activity of di(2-ethylhexyl)phthalate (DEHP) in rat and human hepatocytes, *Carcinogenesis*, **5**, 1329–1335.

Cayama, E., H. Tsuda, D.S.R. Sarma and E. Farber (1978) Initiation of chemical carcinogenesis requires cell proliferation, *Nature* (London) **275**, 60–62.

Columbano, A., S. Rajalakshmi and D.S.R. Sarma (1981) Requirement of cell proliferation for the initiation of liver carcinogenesis as assayed by three different procedures, *Cancer Res.*, **41**, 2079–2083.

Doolittle, D.J. and B.E. Butterworth (1984) Assessment of chemically-induced DNA repair in rat tracheal epithelial cells, *Carcinogenesis*, **5**, 773–779.

Doolittle, D.J., E. Bermudez, P.K. Working and B.E. Butterworth (1984) Measurement of genotoxic activity in multiple tissues following inhalation exposure to dimethylnitrosamine, *Mutation Res.*, **141**, 123–127.

Doolittle, D.J., G. Muller, J.M. Smith and H.E. Scribner (1985) A short-term *in vivo* test for detecting hepatocarcinogens in the CD-1 mouse, *Environ. Mutagen.*, **7 Supp. 3**, 69–70 (abstract).

Fukushima, S. and S.M. Cohen (1980) Saccharin-induced hyperplasia of the rat urinary bladder, *Cancer Res.*, **40**, 734–736.

Furstenberger, G., J. Schweizer and F. Marks (1985) Development of phorbol ester responsiveness in neonatal mouse epidermis: correlation between hyperplastic response and sensitivity to first-stage tumor promotion, *Carcinogenesis*, **6**, 289–294.

Jacobs, L.R. (1984) Stimulation of rat colonic crypt cell proliferative activity by wheat bran consumption during the stage of 1,2-dimethylhydrazine administration, *Cancer Res.*, **44**, 2458–2463.

Kinzel, V., H. Loehrke, K. Goerttler, G. Furstenberger and F. Marks (1984) Suppression of the first stage of phorbol 12-tetradecanoate 13-acetate-effected tumor promotion in mouse skin by nontoxic inhibition of DNA synthesis, *Proc. Natl. Acad. Sci. USA*, **81**, 5858–5862.

McCann, J., E. Choi, E. Yamasaki and B.N. Ames (1975) Detection of carcinogens as mutagens in the Salmonella/microsome test: Assay of 300 chemicals, *Proc. Natl. Acad. Sci. USA*, **72**, 5135–5139.

McCann, J. and B.N. Ames (1976) Detection of carcinogens as mutagens in the Salmonella/microsome test: Assay of 300 chemicals: Discussion, *Proc. Natl. Acad. Sci. USA*, **73**, 950–954.

Mirsalis, J.C. and B.E. Butterworth (1980) Detection of unscheduled DNA synthesis in hepatocytes isolated from rats treated with genotoxic agents: an *in vivo–in vitro* assay for potential carcinogens and mutagens, *Carcinogenesis*, **1**, 621–625.

Mirsalis, J.C., C.K. Tyson and B.E. Butterworth (1982) Detection of genotoxic carcinogens in the *in vivo–in vitro* hepatocyte DNA repair assay, *Environ. Mutagen.*, **4**, 553–562.

Schulte-Herman, R., G. Ohde, J. Schuppler and I. Timmermann-Trosiener (1981) Enhanced proliferation of putative preneoplastic cells in rat liver following treatment with the tumor promoters phenobarbital, hexachlorocyclohexane, steroid compounds, and nafenopin, *Cancer Res.*, **41**, 2556–2562.

Slaga, T.J., A.J.P. Klein-Szanto, S.M. Fischer, C.E. Weeks, K. Nelson and S. Major (1980a) Studies on mechanism of action of anti-tumor-promoting agents: Their specificity in two-stage promotion, *Proc. Natl. Acad. Sci. USA*, **77**, 2251–2254.

Slaga, T.J., S.M. Fischer, K. Nelson and G.L. Gleason (1980b) Studies on the mechanism of skin tumor promotion: Evidence for several stages in promotion, *Proc. Natl. Acad. Sci. USA*, **77**, 3659–3663.

Working, P.K. and B.E. Butterworth (1984) An assay to detect chemically induced DNA repair in rat spermatocytes, *Environ. Mutagen.*, **6**, 273–286.

Yuspa, S.H., U. Lichti, T. Ben, E. Patterson, H. Hennings T.J. Slaga, N. Colburn and W. Kelsey (1976) Phorbol esters stimulate DNA synthesis and ornithine decarboxylase activity in mouse epidermal cell cultures, Nature (London) 262, 402–404.

Brian Burlinson and John Ashby
Central Toxicology Laboratory, Imperial Chemical Industries PLC, Alderley Park, Cheshire, UK

63

Inactivity of benzo(a)pyrene in a weanling rat liver, unscheduled DNA synthesis assay

1 Summary

3,4-Benzo(a)pyrene (BP) is known to elicit a genotoxic response in rat liver if dosing is followed by partial hepatectomy (i.e. in a regenerating liver). This fact, together with the general inactivity of BP in the liver assays included in this study led to its limited evaluation for UDS induction in weanling rats (i.e. in growing and dividing liver).

No activity was evident in what must be emphasised is a limited study.

2 Methods

The methods are those described in Ashby *et al.* (1985) except that the superior vena cava was not cannulated (i.e. the collagenase buffer was not recirculated) and that 10-day old, weanling Alderley Park rats were used.

The number of silver grains in the nucleus and corresponding area of cytoplasm were measured automatically using an Olympus BH2 microscope (× 100, oil immersion) linked to an Analytical Measuring System Ltd., 40–10 Image Analyser. The grain area was converted to grain number by means of a calibration factor gen-

Table 1. *Incidence of S-phase cells in the livers of the treated animals shown*

Compound	No S phase cells per slide	Total	S phase (%)	SD
Control	20	1025		
	24	1063	1.95	0.31
	17	1038		
6BT* (40 mg/kg)	36	1057		
	64	1063	4.59	1.32
	46	1061		
BP (100 mg/kg)	22	1020		
	41	1001	3.45	1.12
	42	1023		
	33	1022		
	29	1070	3.15	0.42
	36	1017		

*6BT (6-dimethylaminophenylazobenzthiazole) is a potent hepatocarcinogen used by us as a positive control chemical for *in vivo* UDS assays. Its preparation has been described earlier (Ashby *et al.*, 1985).

erated by measuring the area of discrete grains in 10 separate locations for each slide.

The test rats were dosed orally with BP in corn oil at 100 mg/kg; negative control rats were given corn oil at 10 ml/kg. The positive control rats were given the hepatocarcinogen 6BT (Ashby *et al.*, 1985) at 40 mg/kg. The animals were killed and their livers perfused 4 h after dosing.

To confirm the generative state of the liver, the incidence of S-phase cells was also assessed.

3 Results

The animal livers were in a generative phase (Table 1), and the animals given BP or 6BT had an increased percentage of S-phase cells when compared with the control.

The positive control chemical 6BT gave a positive response, but the 2 rats given BP showed no sign of UDS. The negative controls gave the expected result (Table 2).

4 Conclusion

In this very limited study, BP did not initiate UDS in weanling rat liver.

5 Reference

Ashby J, P.A. Lefevre, B. Burlinson and M.P. Penman (1985), An assessment of the *in vivo* liver UDS assay, *Mutation Res.*, **156**, 1–18.

Table 2. *Levels of UDS in the liver of the treated animals*

					Individual animal data		Cumulative data			
Cmpd	Length of exp.	Dose (mg/kg)	No. of cells	No. of slides	NG (a)	% repair	No. of animals	(NG) ± SD	% repair ± SD	NG for cells in repair (N-C ≥ 5) ± SD (b)
Corn Oil	4	10 ml/kg	75	3	−0.23 ± 0.4	5	1	−0.23	5	7.8
6BT	4	40	75	3	4.7 ± 0.7	47	1	4.7	47	9.1
BP	4	100	100	3	−0.95 ± 0.3	2				
			25	1	−0.08 ± 0.4	0	2	−0.5 ± 0.6	1	6
Historical Controls (Adult Rats)			1225	25			40	−2.12 ± 1.09	3.0 ± 3.7	6.3 ± 1.8

[a] NG represents the net grain count.
[b] The NG count is derived from the nuclear count (N-C). When this number is 5 or greater the cell is considered to be undertaking DNA repair. The extent of this repair is shown by this column.

Anthony L. Meyer and Ad D. Tates
(Workgroup Co-Chairmen)[1]
[1] *Department of Radiation Genetics and Chemical Mutagenesis, State University of Leiden, Leiden, The Netherlands*

64

Summary report on the performance of cytogenetic assays in rat hepatic cells

1 Cytogenetic analysis of hepatic cells from rats exposed to BP, PYR, 2AAF, or 4AAF

A. In vitro chromosome studies with primary hepatic cells prepared from Wistar rats exposed to test chemicals (in vivo/in vitro)

A novel approach to assess the genotoxic activity of test chemicals has been developed. This assay system is designed to monitor chromosome damage and SCE induction in primary cultures of hepatic cells prepared from rats previously exposed to the test compound. In this way, 2 firmly established assay endpoints are employed to monitor the effects of chemicals subjected to the modifying influences of whole body metabolism.

The method of primary liver cell culture preparation involves trypsin treatment of finely minced liver tissue from treated rats. The resulting cell suspension, when transferred to glass microscope slides in culture medium, gives rise to a steadily dividing population of predominantly epithelioid cells. The precise nature of the cells analysed in this study needs clarification, as it is generally accepted that hepatocytes are unable to divide in culture. Nevertheless, using this technique, it is possible to achieve an adequate number of harlequin-stained cells for SCE analyses, indicating at least 2 rounds of cell division in culture.

This type of *in vivo/in vitro* assay has been carried out by only one investigator in the present IPCS study. Therefore, direct comparisons of results are not possible. The data show that no statistically significant clastogenic responses were observed with any of the compounds in the chromosome assay when cells were analysed for chromosome damage after 48 h in culture. However, there were small, but consistent, increases in chromosome aberrations in cultures exposed to BP and 2AAF compared with cultures exposed to PYR and 4AAF. Pooled data of 2 experiments with BP indicate a statistically significant increase in the frequency of SCE. A marginal increase in SCE induction was seen with 2AAF after 68 h in culture. Both PYR and 4AAF failed to induce SCE. The fact that BP gave positive, albeit weak, responses in both chromosome and SCE tests is of interest, because a substantial number of other liver-specific assays employing different endpoints failed to register BP as positive and suggests that these assays may represent more sensitive endpoints. There was a degree of variation in response among animals that could be ascribed to factors such as genetic differences which affect metabolic activation and rate of detoxification.

Although there are certain aspects of this system needing further investigation (such as definition of the target cells), the data generated in this study indicate that *in vivo* dosing followed by the *in vitro* detection of genetic damage represents a promising short-term test procedure. A wide range of carcinogens and noncarcinogens need to be assayed before the sensitivity and the discriminatory power of the system can be adequately assessed.

B. Micronucleus test with rat hepatocytes to detect induction of clastogenic damage

This test has been developed to measure clastogenic effects of genotoxic agents directly in cells of the most active organ for the majority of drug-metabolizing systems (Tates, this volume). The system is suitable for the detection of indirect mutagens, but direct mutagens can be assessed as well. In particular, it is useful for liver carcinogens and for indirect mutagens that form reactive intermediates that are too short-lived to reach or are produced in such small quantities that they do not reach the classical target cells for cytogenetic studies (e.g., bone marrow cells and peripheral blood lymphocytes).

The liver of adult mammals does not seem to be a suitable organ for cytogenetic studies because of its very low mitotic activity. This activity can be enhanced considerably by surgical removal of 2/3 of the liver or partial hepatectomy (PH). About 1 to 4 days after PH, many dividing cells are present and hepatocytes can be isolated via liver perfusion. Cytogenetic damage in hepatocytes is then assessed by determining the frequency of micronuclei in 1000 hepatocytes per animal. Analysis of chromosomal aberrations can also be performed, but this is much more laborious and difficult because many cells are tetraploid and even octaploid.

When micronuclei are screened, it is essential that the cell population is also analysed for signs of mitotic activity to avoid false negatives. As indices for proliferative activity, use is made of the frequencies of dividing and binucleated cells in 1000 hepatocytes (with good proliferative activity, the frequency of binucleated cells rapidly decreases with time after PH).

The micronucleus method for hepatocytes appears to be a cumbersome method, but trained personnel can perform about 8 hepatectomies and 4 perfusions per hour, and 1000 hepatocytes can be screened for micronuclei in about 20 minutes.

Depending on the objectives of the investigator, the test chemical can be administered before (procedure I) or after PH (procedure II). Procedure I is followed when optimal use is to be made of the capacity of the liver to activate indirect mutagens. By dosing the animals at different times before PH, information can be obtained on the capacity of liver cells to repair preclastogenic lesions (Tates, this volume). With procedure I, the following chemicals were positive: N, N-diethylnitrosamine (DEN), N, N-dimethylnitrosamine (DMN), N-ethyl-N-nitrosourea (ENU), mitomycin C (MMC), 2AAF. Negative results were found for: 4AAF, PYR, BP (positive in a small experiment and negative in a large one), methylmethanesulphonate (MMS) and ethylmethanesulphonate (EMS). The latter 2 chemicals (and also DEN, DMN and ENU) were strongly positive with procedure II (i.e., when the chemicals were given during the peak of DNA synthesis activity at 17 h after PH). Apparently, preclastogenic lesions produced by MMS and EMS are effectively removed when the chemicals are given before PH.

For direct mutagens, procedure II is indicated, but for indirect mutagens this procedure has the disadvantage that the drug metabolising activity of hepatocytes decreases rapidly after PH.

2 Summary of cytogenetic tests with liver cells

There are a number of established animal carcinogens that yield negative or ambiguous results with conventional short-term *in vitro* mutation tests. However, it has been demonstrated that certain short-term *in vivo* tests with liver tissue have the capacity to detect genotoxicants that are difficult to identify in other systems. Two distinct assay systems employing cytogenetic endpoints in liver tissues are presented here. Although the *in vivo/in vitro* technique showed promise in distinguishing between the carcinogenic/noncarcinogen chemical pairs, the positive responses with BP and 2AAF were weak. It was only possible to identify 2AAF as a fairly strong clastogen in hepatocyte micronuclei studies.

There is hope for the further development and validation of both assays. In the absence of more extensive data, the relative merits of the assays are impossible to evaluate at present.

M. Gillian Clare, Anthony L. Meyer,
Alison V. Ruck and Brian J. Dean
Shell Research Ltd., Sittingbourne Research Centre, Kent, UK

65

Chromosome studies in hepatic cells from rats exposed to genotoxic chemicals

1 Summary

Four compounds were tested in a rat liver *in vivo/in vitro* assay for their ability to induce structural chromosome aberrations or sister chromatid exchanges (SCE).

Young male Wistar rats were given single oral doses of benzo(a)pyrene (BP), pyrene (PYR), 2-acetylaminofluorene (2AAF) or 4-acetylaminofluorene (4AAF). Cyclophosphamide (CPM) was administered either orally or intraperitoneally (ip) as a positive control compound. Four hours after dosing, primary cultures of liver cells from each animal were prepared on microscope slides and, after an appropriate period in culture, these were processed and stained for analysis.

Under the experimental conditions described, BP induced a small number of chromatid aberrations and 2AAF caused a very slight increase in the frequency of chromatid aberrations and chromosome breaks, whilst PYR and 4AAF did not do so. BP significantly increased the frequency of SCE, 2AAF gave equivocal results, and 4AAF and PYR were not considered to affect the frequency of SCE.

2 Introduction

There are a number of established animal carcinogens that yield negative or ambiguous results in conventional short-term *in vitro* mutation tests (Ishidate and Odashima, 1977; Purchase *et al.*, 1981). Some of these carcinogens are thought to induce neoplasia by routes other than the direct modification of nuclear DNA (Rubin, 1980). The remainder may not be reactive in the *in vitro* assays because the nature of metabolic activation is inappropriate to the particular chemical being studied.

Measurement of chromosome damage in rodent bone marrow cells has sucessfully discriminated between carcinogenic and noncarcinogenic analogues (Ashby, 1981). Therefore, it may be possible to clarify some of the equivocal or conflicting results obtained in short-term *in vitro* tests by using short-term *in vivo* exposure to the chemical, followed by demonstration of genotoxic effects in primary cultures of cells from the treated animals. The rat liver *in vivo/in vitro* assay appears to offer significant advantages over the bone marrow procedure, as the target cells for chromosome damage are situated in the organ where the major biotransformation occurs. Thus, the need for transport of electrophilic reactants from the liver to marrow cells is avoided (Clare and Dean, 1985).

3 Materials and methods

Methodology

Young male Wistar rats were exposed to the test compound for 4 h and sacrificed. The livers were removed and primary cultures of epithelial-type cells were established on microscope slides. These were incubated in darkness in tissue culture medium containing 15 μM bromodeoxyuridine for a sufficient time for the cells to divide once or twice. The cultures were harvested and the slides stained to reveal sister chromatid exchanges (SCE). Cells at the first metaphase in culture were examined for chromosome damage, and those at the second division were examined for SCE.

Experimental details relating to the initiation of cultures of principally epithelial-like cells and the incubation, processing, and staining procedures have been described by Clare and Dean (1985) and will not be referred to in the following procedure.

Procedure

Chemicals The chemicals included in this study were BP, PYR, 2AAF, 4AAF and CPM. With the exception of CPM, which was dissolved in sterile distilled water for ip injection, all compounds were formulated in corn oil shortly before use. Since all of the test compounds formed a precipitate in corn oil, the vials were shaken to ensure an even distribution immediately before each animal was dosed. The same batches of chemicals were used throughout the experiments.

Dosing The highest concentrations of the test substances were based on information provided in the study guidelines. If there was an equivocal result, slightly higher concentrations were also used.

For each experiment, 14 young (approximately 10-day old) male Wistar rats were used. Groups of 3 or 4 rats were administered single oral doses of either corn oil, BP or PYR at 20 or 100 mg kg^{-1}, or 2AAF or 4AAF at 10, 50, or 100 mg kg^{-1}. Also, 2 to 4 animals were injected ip, and on some occasions 2 additional animals were dosed by gavage with CPM at 25 mg kg^{-1} as a positive control compound. Animals were exposed to the compounds for 4 h.

Analysis The stained slides were randomised and the diamond-etched information was obscured by a paper label. For each animal, up to 100 cells at the first metaphase in culture were analysed microscopically for chromosome abnormalities; at the second division about 30 cells with clearly differentially stained chromosomes were chosen and SCE counted.

Five hundred cells from cultures from animals exposed to corn oil or to the highest concentration of each of the test compounds were counted as being at mitosis or interphase, and the percentage of mitotic cells was calculated.

Statistical analysis For each experiment, the frequency of all structural chromatid and chromosome aberrations and also the incidence of SCE per chromosome for individual animals were analysed. In studies where CPM had been administered by 2 different routes, data from animals dosed by ip injection only were analysed, since ip injection was used in every experiment. The data for each experiment were assessed for departures from the assumptions of normality and constant error variance. Provided that departures from these assumptions were judged not to have a deleterious effect upon the analysis, the data were analysed by a one-way analysis of variance, and Dunnett's test (Dunnett, 1964) was used to assess the significance of the differences between the treated group mean and the untreated control mean.

To improve the power of the test, in certain cases a combined analysis was performed on 2 experiments with identical designs, using a one-way analysis of covariance in which the effect of experiments was accounted for by dummy covariates taking the values 0 and 1. A Dunnett's test was used to test for significant differences between the combined treated groups and combined control group means.

All significance tests were two-sided and made at the 0.05 and 0.01 levels.

4 Results

The results of the analysis of structural chromosome damage and polyploidy are shown in Tables 1–4 and those of the SCE assays in Table 5. The chromosome aberrations observed were usually acentric fragments. The mitotic index results are presented in the footnotes to the Tables. There was variation in the mitotic index between animals within a treatment and also between experiments, but the highest concentrations of BP and 2AAF tended to reduce the proportion of cells at metaphase compared with control values, whilst PYR and 4AAF generally did not do so.

The findings are summarised in Table 6. No concentration-dependent change in the incidence of polyploidy was observed in any experiment.

BP induced a small, dose-related increase in the frequency of chromatid gaps. At dose levels of 20 and 100 mg kg^{-1}, there was a small but consistent increase in the number of breaks and exchanges compared with the controls.

BP raised the frequency of SCE in 2 experiments, and, when the results were combined, a concentration of 100 mg kg^{-1} BP induced a statistically significant increase in the frequency of SCE compared with the control ($P \leqslant 0.01$), although there was variation between the 2 experiments.

PYR did not significantly raise the frequency of chromosome damage or SCE.

In the first experiment with 2AAF, cells from animals given 50 mg kg^{-1} 2AAF contained slightly more chromatid aberrations compared with those from control animals. This was confirmed when 2AAF appeared to induce a small increase in the percentage of cells showing chromatid gaps, breaks and exchanges, and chromosome breaks (second experiment). A concentration of 50 mg kg^{-1} was slightly more effective in causing aberrations than 10 mg kg^{-1}.

There was a very small elevation in the frequency of SCE in cultures from 2AAF-treated animals which did not reach statistical significance when compared with controls in any of the 3 experiments and was similar,

Table 1. *Structural chromosome aberrations and polyploidy in liver cells of rats treated with BP*

				% Cells showing							Frequency per cell of					
Compound	Conc. (mg kg^{-1})	No. of animals	No. of Cells analysed	Poly-ploidy[a]	Chroma-tid gaps[b]	Chromatid aberra-tions[c]	Chromosome aberra-tions[d]	Multiple aberra-tions[e]	All aberra-tions excl. gaps[f]	All aberra-tions incl. gaps[g]	Chroma-tid gaps[b]	Chroma-tid breaks[h]	Chroma-tid exchanges	Chromosome aberra-tions[d]	All aberra-tions excl. gaps[f]	All aberra-tions incl. gaps[g]
Corn oil	10	2	131	0	1.5	0	0	0	0	1.5	0.015	0	0	0	0	0.015
BP	20	1	156	0	2.6	2.6	0	0	2.6	5.1	0.032	0.019	0.006	0	0.026	0.058
BP	100	3	190	0	4.2	2.1	0	0	2.1	6.3	0.047	0.005	0.016	0	0.021	0.068
CPM(gav)i	25	2	151	0	6.0	2.6	0	0	2.6	8.6	0.066	0.013	0.013	0	0.026	0.093
CPM(ip)	25	2	171	0	5.8	1.2	0	0	1.2	7.0	0.070	0	0.018	0	0.018	0.088
CPM(both)	25	4	322	0	5.9	1.9	0	0	1.9	7.8	0.068	0.006	0.015	0	0.022	0.090

Cells at the first division analysed from cultures harvested after about 48 h and 68.5 h; data combined from two experiments.
[a] Polyploidy or endoreduplication.
[b] Gaps or iso-gaps or both.
[c] Breaks or single fragments or exchange figures or single rings.
[d] Acentric fragments or dicentrics or translocation or rings or minutes.
[e] Any combination of identifiable chromatid or chromosome aberrations.
[f] All types of chromatid and chromosome-type aberrations excluding chromatid gaps and polyploidy.
[g] All types of chromatid and chromosome-type aberrations including chromatid gaps, but excluding polyploidy.
[h] Breaks or single fragments or both.
[i] gav. = administered orally; ip = administered intraperitoneally.
The mitotic index of control cultures was 1.7%, whilst that of cultures from animals exposed to 100 mg kg^{-1} BP was 1.2% on average.

Table 2. *Structural chromosome aberrations and polyploidy in liver cells of rats treated with PYR*

				% Cells showing							Frequency per cell of					
Compound	Conc. (mg kg^{-1})	No. of animals	No. of cells analysed	Poly-ploidy[a]	Chroma-tid gaps[b]	Chromatid aberra-tions[c]	Chromosome aberra-tions[d]	Multiple aberra-tions[e]	All aberra-tions excl. gaps[f]	All aberra-tions incl. gaps[g]	Chroma-tid gaps[b]	Chroma-tid breaks[h]	Chroma-tid exchanges	Chromosome aberra-tions[d]	All aberra-tions excl. gaps[f]	All aberra-tions incl. gaps[g]
Corn oil	10	3	300	0	1.0	0	0	0	0	1.0	0.010	0	0	0	0	0.010
PYR	20	3	226	0.4	0.4	0	0	0	0	0.4	0.004	0	0	0	0	0.004
PYR	100	3	300	0	0.7	0.7	0	0	0.7	1.3	0.010	0.007	0	0	0.007	0.017
CPM(gav)i	25	2	200	0	2.0	1.5	0	0	1.5	3.0	0.020	0.015	0	0	0.015	0.035
CPM(i.p)	25	2	200	0	1.5	1.0	1.5	0	2.5	4.0	0.015	0	0.010	0.015	0.025	0.040
CPM(both)	25	4	400	0	1.8	1.3	0.8	0	2.0	3.8	0.018	0.008	0.005	0.008	0.020	0.038

Cells at the first division analysed from cultures harvested after about 48 h.
a–i as for Table 1.
The mitotic index of control cultures was 2.7%, whilst that of cultures from animals exposed to 100 mg kg^{-1} PYR was 4.5% on average.
Data not subjected to statistical analysis.

Table 3. *Structural chromosome aberrations and polyploidy in liver cells of rats treated with 2AAF*

				%Cells showing							Frequency per cell of					
Compound	Conc. (mg kg^{-1})	No. of animals	No. of cells analysed	Polyploidy[a]	Chromatid gaps[b]	Chromatid aberrations[c]	Chromosome aberrations[d]	Multiple aberrations[e]	All aberrations excl. gaps[f]	All aberrations incl. gaps[g]	Chromatid gaps[b]	Chromatid breaks[h]	Chromatid exchanges	Chromosome aberrations[d]	All aberrations excl. gaps[f]	All aberrations incl. gaps[g]
Corn oil	10	3	300	0	2.0	0.3	0	0	0.3	2.3	0.020	0.003	0	0	0.003	0.023
2AAF	10	3	300	0	1.3	0.7	0	0	0.7	2.0	0.017	0.007	0	0	0.007	0.023
2AAF	50	3	300	0	2.3	1.0	0	0	1.0	3.3	0.023	0.007	0.003	0	0.010	0.033
CPM(gav)	25	2	200	0.5	4.0	0.5	0.5	0	1.0	5.0	0.050	0	0.005	0.005	0.010	0.060
CPM(ip)	25	2	200	0	2.5	5.5	0.5	0	6.0	8.0	0.025	0.015	0.050	0.005	0.070	0.095(+)
CPM(both)	25	4	400	0.3	3.3	3.0	0.5	0	3.5	6.5	0.038	0.008	0.028	0.005	0.040	0.078
Corn oil	10	3	300	0	1.0	0.7	0	0	0.7	1.7	0.013	0.010	0	0	0.010	0.023
2AAF	10	4	375	0	1.6	1.6	0.8	0	2.4	4.0	0.016	0.008	0.008	0.008	0.024	0.040
2AAF	50	4	323	0.6	1.9	2.2	1.2	0	3.4	5.0	0.019	0.015	0.006	0.012	0.034	0.053
CPM(ip)	25	3	186	0.5	4.8	2.7	2.2	0	4.8	9.7	0.048	0.011	0.016	0.022	0.048	0.097(+ +)

Cells at the first division analysed from cultures harvested after about 48 h.
(+) $P < 0.05$ is significance of difference between control and treated means using Dunnett's test.
(+ +) $P < 0.01$ is significance of difference between control and treated means using Dunnett's test.
a–i as for Table 1.
The mitotic index of control cultures was 4.3%, whilst that of cultures from animals exposed to 50 mg kg^{-1} 2AAF was 3.8% on average in the first experiment.
The mitotic index of control cultures was 2.1%, whilst that of cultures from animals exposed to 50 mg kg^{-1} 2AAF was 1.0% on average in the second experiment.

Table 4. *Structural chromosome aberrations and polyploidy in liver cells of rats treated with 4AAF*

				% Cells showing							Frequency per cell of					
Compound	Conc. (mg kg^{-1})	No. of animals	No. of cells analysed	Polyploidy[a]	Chromatid gaps[b]	Chromatid aberrations[c]	Chromosome aberrations[d]	Multiple aberrations[e]	All aberrations excl. gaps[f]	All aberrations incl. gaps[g]	Chromatid gaps[b]	Chromatid breaks[h]	Chromatid exchanges	Chromosome aberrations[d]	All aberrations excl. gaps[f]	All aberrations incl. gaps[g]
Corn oil	10	3	191	0	2.1	0.5	0	0	0.5	2.6	0.021	0.005	0	0	0.005	0.026
4AAF	10	3	146	0.7	3.4	0	0	0	0	3.4	0.034	0	0	0	0	0.034
4AAF	50	3	239	0	2.5	0.8	0	0	0.8	3.3	0.025	0.008	0	0	0.008	0.033
CPM(gav)	25	2	55	0	14.5	5.4	1.8	0	7.3	21.8	0.145	0.036	0.073	0.018	0.127	0.273
CPM(ip)	25	1	21	0	14.3	0	0	4.8	4.8	19.0	0.143	0	0.095	0.048	0.143	0.286
CPM(both)	25	3	76	0	14.5	3.9	1.3	1.3	6.6	21.1	0.145	0.026	0.079	0.026	0.132	0.276

Cells at the first division analysed from cultures harvested after about 48 h and 68.5; data combined from two experiments.
a–i as for Table 1.
The mitotic index of control cultures was 1.2%, whilst that of cultures from animals exposed to 50 mg kg^{-1} 4AAF was 1.9% on average.
Data not subjected to statistical analysis.

Table 5. *Sister chromatid exchanges in rat liver cells*

Compound	Conc. (mg kg^{-1})	No. of animals	Total cells scored	No. of chromosomes	No. of SCE	SCE per chromosome	SCE per cell
Corn oil[a]	10 ml kg^{-1}	2	54	2126	711	0.33	14
BP	20	1	21	836	305	0.37	15
BP	100	3	34	1335	733	0.55	23
CPM(gav)	25	2	35	1414	582	0.41	17
CPM(ip)	25	2	4	165	70	0.42	18
Corn oil[b]	10 ml kg^{-1}	3	90	3623	1494	0.41	17
BP	20	4	120	4904	2391	0.49	21
BP	100	4	120	4581	2550	0.56	23
CPM(ip)	25	3	26	1082	741	0.68**	29
Corn oil[c]	10 ml kg^{-1}	3	90	3475	1293	0.37	16
PYR	20	3	71	2867	897	0.31	13
PYR	100	3	91	3412	1301	0.38	16
CPM(gav)	25	2	60	2351	926	0.39	17
CPM(ip)	25	2	37	1394	806	0.58	24
Corn oil[d]	10 ml kg^{-1}	3	67	2818	1286	0.46	19
PYR	20	4	120	4772	2309	0.48	20
PYR	100	4	120	4712	2147	0.46	19
CPM(ip)	25	3	90	3607	2759	0.76**	32
Corn oil[e]	10 ml kg^{-1}	3	68	2424	520	0.21	9
2AAF	10	3	90	3244	948	0.29	12
2AAF	50	3	69	2495	652	0.26	11
CPM(gav)	25	2	40	1432	460	0.32	13
CPM(ip)	25	2	40	1509	421	0.28	12
Corn oil[f]	10 ml kg^{-1}	3	90	3624	1545	0.43	18
2AAF	10	4	120	4858	2263	0.47	20
2AAF	50	4	120	4660	2188	0.47	20
CPM(ip)	25	3	90	3475	2472	0.71**	30
Corn oil[g]	10 ml kg^{-1}	6	180	6821	2958	0.43	18
2AAF	100	6	180	6870	3241	0.47	20
CPM(ip)	25	4	120	4730	2720	0.58**	24
Corn oil[h]	10 ml kg^{-1}	3	70	2869	1056	0.37	15
4AAF	10	3	28	1106	448	0.41	17
4AAF	50	3	76	3017	1299	0.43	18
CPM(gav)	25	2	38	1499	672	0.45	19
CPM(ip)	25	1	4	157	92	0.59**	25
Corn oil[i]	10 mg kg^{-1}	3	90	3539	1595	0.45	19
4AAF	10	4	120	4932	2099	0.43	18
4AAF	50	4	114	4613	2281	0.49	21
CPM(ip)	25	3	64	2610	1803	0.69**	29

**$P < 0.01$ (Significance of difference between controls and treated means using Dunnett's test).
[a] M.I.* controls = 2.4%, M.I. BP at 100 mg kg^{-1} = 1.7%.
[b] M.I. controls = 2.9%, M.I. BP at 100 mg kg^{-1} = 1.9%.
[c] M.I. controls = 4.2%, M.I. PYR at 100 mg kg^{-1} = 3.4%.
[d] M.I. controls = 0.7%, M.I. PYR at 100 mg kg^{-1} = 1.7%.
[e] M.I. controls = 5.1%, M.I. 2AAF at 50 mg kg^{-1} = 4.5%.
[f] M.I. controls = 2.5%, M.I. 2AAF at 50 mg kg^{-1} = 2.1%.
[g] M.I. controls = 2.8%, M.I. 2AAF at 100 mg kg^{-1} = 2.9%.
[h] M.I. controls = 1.5%, M.I. 4AAF at 50 mg kg^{-1} = 2.5%.
[i] M.I. controls = 1.3%, M.I. 4AAF at 50 mg kg^{-1} = 0.8%.
* M.I. = Mitotic index.

Table 6. *Summary of results*

Compound	Chromosome aberrations	SCE
BP	+/–[a]	+
PYR	–	–
2AAF	+/–[a]	+/–[a]
4AAF	–	–

[a] These experiments all yielded a marginal positive response, which must be regarded as equivocal in the absence of statistical significance.

irrespective of whether animals were given 2AAF at 10, 50, or 100 mg kg^{-1}. These data are regarded as equivocal.

In view of the lack of a consistent, compound-related increase in the frequency of chromosome damage or SCE in individual animals and the small amounts of damage or SCE involved, 4AAF was judged to be negative in these assays.

CPM, the control compound, markedly decreased the proportion of cells at metaphase so that in some experiments, there were few cells available for analysis. CPM increased the percentage of cells showing chromatid gaps, breaks and exchanges, and chromosome breaks; also the frequency of SCE increased, although there was some variation in response from one experiment to the next. CPM appeared to be more toxic to cells when dosed ip rather than by gavage, as shown by the number of second division metaphases available for analysis of SCE in the experiments with BP and 4AAF.

5 Discussion

Under the experimental conditions described, BP induced a small number of chromatid aberrations and 2AAF caused a slight increase in the frequency of chromatid aberrations and chromosome breaks, whilst PYR and 4AAF were negative in the chromosome assay. The system described shows promise in identifying qualitative differences in clastogenic potential between the polycyclic hydrocarbon and aromatic amine analogues. However, more experimentation is required for a comprehensive assessment.

In the SCE assay, BP was positive and PYR was clearly negative, and it was possible to discriminate between these chemically similar compounds. 2AAF reproducibly increased the frequency of SCE very slightly, which suggests a marginal positive response, whilst 4AAF did not. The increase observed in animals exposed to 4AAF compared with the increase of those exposed to corn oil was confined to a small number of animals within each treatment group. On the basis of these equivocal data, it was not possible to discriminate between 2AAF and 4AAF.

In this study, the resolving power of the chromosome assay appeared to be higher than that of the SCE assay in contrast to the generally accepted perception of the high sensitivity of the SCE test. However, excluding CPM, none of the test compounds increased the frequency of chromosome damage to a statistically significant level ($P < 0.05$) and only BP raised the number of SCE significantly ($P < 0.01$), when the results of 2 experiments were combined. This absence of statistically significant effect may be attributed to a number of factors. Small numbers of structural chromosome aberrations and SCE were observed, even in animals exposed to the highest concentrations of the 4 test compounds. Also small numbers of animals were used, particularly in the positive control group owing to the cytotoxicity of CPM; although the experiment in which 6 males were dosed with corn oil and 6 with 2AAF at 100 mg kg^{-1} did not improve the detection of the ability of 2AAF to induce SCE. Finally, there was a variation in response between animals, which may be partly explained by the precipitation of the chemicals in corn oil, although the vials were shaken to ensure an even distribution before dosing each animal. The variation in response between animals is not exclusive to this study and has been ascribed to genetic differences that affect factors such as the metabolic activation of BP, (Galloway, *et al.*, 1980; Schreck and Latt, 1980; Raineri *et al.*, 1983) and rates of detoxification of 2AAF (Takehisa and Wolff, 1978).

2AAF did not induce a strong positive dose-dependent increase in the frequency of SCE in the bone marrow cells of male mice (Schreck *et al.*, 1979; Parodi *et al.*, 1983) or rabbits (Takehisa and Wolff, 1978), even when concentrations of up to 255 mg kg^{-1} were used (Parodi *et al.*, 1983). The results are in general agreement with those observed in this study. In the present study, 2AAF was not markedly cytotoxic as measured by the depression in mitotic index (apart from in the second chromosome aberration assay), and concentrations of 10, 50, or 100 mg kg^{-1} all affected the SCE frequency to the same limited extent. It is possible that the lack of a definitive positive response was a consequence of an insufficient length of exposure, since 2AAF has a complex activation pathway (Miller, 1978; Abe *et al.*, 1983) and it has been demonstrated that the biological activity of 2AAF was enhanced by a longer activation period *in vivo* (Schreck *et al.*, 1979). Alternatively, it has been suggested that SCE induced by aromatic amines are correlated with DNA damage and not with carcinogenic potency (Parodi *et al.*, 1983).

6 Conclusions

Under the experimental conditions described, BP induced a small number of chromatid aberrations and 2AAF caused a very slight increase in the frequency of chromatid aberrations and chromosome breaks, whilst PYR and 4AAF did not do so. BP significantly increased the frequency of SCE, 2AAF gave equivocal results, and 4AAF and PYR were not considered to affect the frequency of SCE.

7 Acknowledgements

The authors are indebted to Ms. R. Eagle for her skilful technical assistance in conducting these studies, to Dr. D.E. Wiggins and Mr. D.D. Lad for providing formulation chemistry support, to Mr. D. Daniel for advice on statistical analysis, to Dr. A.S. Wright for reviewing, and to Julie Smith for typing the manuscript.

8 References

Abe, S., N. Nemoto and M. Sasaki (1983) Sister-chromatid exchange induction by indirect mutagens/carcinogens, aryl hydrocarbon hydroxylase activity and benzo(a)pyrene metabolism in cultured human hepatoma cells, *Mutation Res.*, **109**, 83–90.

Ashby, J. (1981) Overview of study and test chemical activities. In F.J. de Serres and J. Ashby (Eds), *Progress in Mutation Research*, Vol. 1. Evaluation of Short-term Tests for Carcinogens, Elsevier/North Holland, pp. 112–171.

Clare, M.G. and B.J. Dean (1985) Chromosome studies in hepatic cells from rats exposed to genotoxic chemicals. In C. Arlett and J.M. Parry (Eds), *Comparative Genetic Toxicology: the Second UKEMS Collaborative Study of Short-Term Tests for Mutagenicity and Carcinogenicity*, McMillans Scientific Publishers/London, pp. 441–451.

Dunnett, C.W. (1964) New table for multiple comparisons with a control, *Biometrics*, **20**, 482–491.

Galloway, S.M., P.E. Perry, J. Meneses, D.W. Nebert and R.A. Pedersen (1980) Cultured mouse embryos metabolize benzo(a)pyrene during early gestation: Genetic differences detectable by sister chromatid exchange, *Proc. Natl. Acad. Sci. USA*, **77**, 3524–3528.

Ishidate, M. and S. Odashima (1977) Chromosome tests with 134 compounds on Chinese hamster cells *in vitro* – a screening for chemical carcinogens, *Mutation Res.*, **48**, 337–354.

Miller, E.C. (1978) Some current perspectives on chemical carcinogenesis in humans and experimental animals: Presidential address, *Cancer Res.*, **38**, 1479–1496.

Parodi, S., A. Zunino, L. Ottaggio, M. De Ferrari and L. Santi (1983) Lack of correlation between the capability of inducing sister-chromatid exchanges *in vivo* and carcinogenic potency, for 16 aromatic amines and azo derivatives, *Mutation Res.*, **108**, 225–238.

Purchase, I.F.H., V. Ray, J. Heddle, B. Kirkhart, S.A. Latt, M. Salamone, J. Topham, T. Tsuchimoto, R. Valencia, E.M. Vogel, F.E. Wurgler and A. Wyrobek (1981) Summary report on the performance of *in vivo* assays. In F.J. de Serres and J. Ashby (Eds), *Progress in Mutation Research*, Vol. 1. Evaluation of Short-term Tests for Carcinogens Elsevier/North Holland, pp. 86–95.

Raineri, R., T. Hillesund, M.K. Ernst, D.M. Cavanaugh and J.A. Poiley (1983) Metabolic activation, detoxification and transformation studies with benzo(a)pyrene using rat and hamster organ cells *in vitro*, *J. Appl. Toxicol.*, **3**, 196–202.

Rubin, H. (1980) Is somatic mutation the major mechanism of malignant transformation? *J. Natl. Cancer Inst.*, **64**, 995–1000.

Schreck, R.R. and S.A. Latt (1980) Comparison of benzo(a)pyrene metabolism and sister chromatid exchange induction in mice, *Nature* (London), **288**, 407–408.

Schreck, R.R., I.J. Paika and S.A. Latt (1979) *In vivo* induction of sister chromatid exchanges in liver and marrow cells by drugs requiring metabolic activation, *Mutation Res.*, **64**, 315–328.

Takehisa, S. and S. Wolff (1978) Sister-chromatid exchanges induced in rabbitlymphocytes by 2-aminofluorene and 2-acetylaminofluorene after *in vitro* and *in vivo* metabolic activation, *Mutation Res.*, **58**, 321–329.

Ad D. Tates and Isaac Neuteboom
Department of Radiation Genetics and Chemical Mutagenesis, Sylvius Laboratories, State University of Leiden, Leiden, The Netherlands

66

Micronucleus test with rat hepatocytes to detect induction of clastogenic damage

1 Summary

The clastogenic potential of 2AAF, 4AAF, BP, and PYR in hepatocytes of adult male Wistar rats was assessed after oral administration of these chemicals before and/or after partial hepatectomy. Frequencies of micronuclei (indicative of chromosome breakage events) were significantly enhanced when 2AAF (6.9 and 13.8 mg/kg) was given one day before partial hepatectomy. At higher doses (27.6–223 mg/kg), 2AAF was negative due to inhibition of proliferative activity after partial hepatectomy. As a consequence, preclastogenic damage induced by 2AAF could not be expressed in the form of micronuclei. Results obtained with the other three chemicals were not statistically significant.

2 Introduction

In 1980, a micronucleus technique for hepatocytes of rats was developed in our laboratory to measure clastogenic effects of genotoxic agents directly in cells of the most active organ for the majority of drug-metabolising systems (Tates *et al.*, 1980). Using this test system, it was possible to demonstrate clastogenic effects of indirect mutagens (e.g., N, N-diethylnitrosamine (DEN) and N, N-dimethylnitrosamine (DMN)), which are difficult to detect in other organ systems such as bone marrow and peripheral blood (den Engelse *et al.*, 1983; Menkveld *et al.*, 1983; Tates *et al.*, 1980, 1983; Tates and den Engelse, 1981, 1982). In view of these findings it was considered of importance to further evaluate the hepatocyte micronucleus test by using it for detection of possible clastogenic effects of the 4 test compounds: 2AAF, 4AAF, BP and PYR. Although a considerable number of experiments have been carried out, particularly for 2AAF, this report should be considered as a status report rather than as a final report on an exhaustive series of experiments.

3 Materials and methods

Animals

Male Wistar rats were obtained from the specific pathogen-free colony of the Department of Pharmacology, State University of Leiden. Animals were kept under standard laboratory conditions. Sterilized rodent pellets (Diet SRM 12/10, Hope Farms, The Netherlands) and water were given *ad libitum*. Weight of the animals varied from 200–250 g.

Chemical compounds

The 4 test compounds were received through the IPCS organization and stored at 4 °C. Olive oil (Ph. Ned.) was used as a solvent. Test compounds were administered orally in 2.2–2.5 ml quantities (1% of body weight). 2AAF (0.2–224 mg/kg) and 4AAF (6.9–892 mg/kg) remained in suspension. BP (32–250 mg/kg) could be dissolved in oil (with the help of sonication) without problems. PYR dissolved easily at the dose of 202 mg/kg. At 404 mg/kg, some small crystals remained visible and at 808 mg/kg, PYR remained in suspension despite sonication for 5 min. Positive and negative controls were not run for every experiment, because these were abundantly available in the context of concurrent liver studies with mitomycin C and phosphate-buffered saline (PBS) that will be published elsewhere.

Broad outline of the experimental design

Two procedures can be followed to measure clastogenicity of genotoxants in hepatocytes.

Procedure I

Injection of test compound	Partial hepatectomy (PH)	Hepatocyte isolation and analysis
← hours or days →	← 2, 3, or 4 days →	

This procedure is followed when optimal use is to be made of the capacity of liver to activate indirect mutagens. A prerequisite for detection of micronuclei is a mitotically active population of liver cells. In normal adult livers, the mitotic activity is very low, but it can be artificially induced by partial hepatectomy (PH). Using a perfusion technique (Tates *et al.*, 1980), hepatocytes are isolated at 2, 3, or more days after PH. Cells are then fixed, stained, and screened for the presence of micronuclei.

Procedure II

Partial hepatectomy	Injection of test compound	Hepatocyte isolation and analysis
← 3–17 h →	← about 2, 3, or 4 days →	

Past experience has shown that the well-known clastogen methylmethanesulfonate (MMS) registered negative with test procedure I. Apparently, the preclastogenic lesions induced by MMS in G_0 hepatocytes are effectively removed before they become irreversibly fixed during the peak of the S-phase at about 17 h after PH. However, MMS was clearly positive when administered at about 17 h after PH. Under these circumstances, there is little or no time for repair of DNA damage, so that preclastogenic lesions are converted into chromosomal breakage events, partially manifesting themselves as micronuclei in the cytoplasm of hepatocytes.

A disadvantage of procedure II is that the drug-metabolising activity of hepatocytes rapidly decreases after PH, so that the conditions for activation of indirect mutagens are decidedly less optimal than before PH.

In the present report, we mainly have applied procedure I. Procedure II was used on a limited scale for 2AAF and BP. For experiments with procedure I, rats were deprived of food pellets at 3 h before oral administration of the test compounds. They were given a 5% glucose solution during 1 day after PH. With procedure II, food pellets were removed at 3 h before PH and the glucose solution was given during 1 day after PH.

A more detailed description of the methodology for partial hepatectomy and for isolation, fixation, staining, and analysis of hepatocytes has been presented elsewhere (Tates *et al.*, 1980).

4 Results

Results are presented in Tables 1–3. The third table contains primary data for statistical analysis of some of the data obtained for 2AAF.

Effects of 2AAF

Prior to the present liver studies we found that doses of 2AAF ranging from 55.8–223 mg/kg significantly increased the frequencies of micronuclei (MN) in bone marrow of mice (Tates *et al.*, this volume). In the present study, these doses were ineffective in raising the frequencies of MN in hepatocytes above control level (Table 1, expts. 3 and 6). At these doses, the negative result clearly is due to inhibition of cell proliferation after PH. This is indicated by a mitotic index of zero (expt. 3) or low (expt. 6) compared with that found in the olive oil control and a high frequency of binucleated hepatocytes in 2AAF-treated rats compared with the low frequency in the control. In the absence of liver proliferation after PH it is not surprising to find no MN. As can be seen in expt. 2, a lowering of the concentration of 2AAF resulted in a positive effect at 6.9 (48 and 72 h after PH) and 13.8 mg/kg (72 h after PH). In line with earlier findings with other chemicals, the effect was more pronounced at 72 h after PH than at 48 h. Experience with other chemicals makes it seem probable that the 2AAF effect would have been even stronger when hepatocytes had also been sampled at 96 h after PH. Since a further lowering of the 2AAF dose (expt. 2) gave rise to lower frequencies of MN, it looks as if the greatest effect of 2AAF is detectable between 6.9–13.8 mg/kg. Inspection of the primary data in Table 3 indicates that some rats exposed to 13.8 mg/kg and sampled at 72 h after PH reacted positively (as evidenced by a high frequency of MN, a high mitotic index, and a low frequency of binucleated hepatocytes), whereas, others showed a negative response.

Thus far, a positive effect of 2AAF could be demonstrated only by administering the chemical before PH (procedure I). When 2AAF was given after PH (procedure II), there was no induction of MN. As liver cell proliferation in the latter tests was quite normal at most of the doses tested, it seems that insufficient metabolic activation of 2AAF must be responsible for the negative results with procedure II.

Results with 4AAF

4AAF has only been tested in one experiment but over a wide range of doses. Using procedure I, it was found that 4AAF did not induce clastogenic damage. 4AAF was much less toxic than 2AAF. At all doses tested, 4AAF had little effect on cellular proliferation after PH, so that the negative result for 4AAF cannot be explained by this factor but rather by an absence of clastogenic potential.

Results with BP

BP was administered before and after PH (Table 2) and appeared to have little effect on the proliferative response of hepatocytes after PH. In a small pilot experiment (expt. 1), it looked as if there was a positive effect of BP at 250 mg/kg. In a second and larger experiment, this effect could not be repeated. Compared with the negative control (olive oil), the results obtained with BP do not indicate a statistically significant positive effect. With considerable expansion of the number of treated animals per group, a weak positive effect of BP may eventually be detectable.

Results with PYR

At the doses tested, PYR had little or no adverse effects on cellular proliferation after PH. Therefore, the absence of clearly enhanced frequencies of MN is attributable to the absence of clear-cut clastogenic activity of PYR. At best, future experiments with more rats per group may result in the demonstration of a weak clastogenic effect.

Table 1. *Induction of chromosomal damage in hepatocytes of rats exposed to 2AAF and 4AAF*

Expt. No.	Chemical: dose (mg/kg)	Procedure I or II	Time between injection and hepatectomy (hours)	Time between hepatectomy and injection (hours)	Time between hepatectomy and liver perfusion (hours)	Frequency $\times 10^{-3}$ of cells with one or more micronuclei ± S.D.	Mitotic index in hepatocytes ± S.D.	Frequency $\times 10^{-3}$ of binucleated hepatocytes ± S.D.
	Olive oil (control of expt. 5)	I	24		48	1.2 ± 1.3 (5)[a]	28 ± 13	42 ± 13
					72	0.4 ± 0.5 (5)	12 ± 4	20 ± 8
1	2AAF 0.55	I	12		72	3.3 ± 2 (3)	25 ± 13	57 ± 22
					96	3.3 ± 1.5 (3)	13 ± 7	36 ± 6
	5.5				72	4.0 ± 4 (3)	22 ± 2	50 ± 6
					96	4.0 ± 1 (3)	32 ± 14	64 ± 11
2	6.9	I	24		48	7.8 ± 2.5 (5)[b]	21 ± 5	83 ± 13
					72	13.2 ± 5.6 (5)[b]	24 ± 10	36 ± 9
	13.8				48	3.6 ± 2.5 (5)	22 ± 14	125 ± 50
					72	10.0 ± 8.3 (5)[b]	13 ± 10	96 ± 33
	27.6				48	0.6 ± 0.6 (5)	1 ± 0.4	151 ± 28
					72	0.8 ± 1.0 (4)	2 ± 2	130 ± 34
3	56	I	24		48	0.0 ± 0.0 (5)	0	143 ± 49
					72	0.3 ± 0.5 (5)	0	159 ± 23
	112				48	0.8 ± 1.3 (5)	0	119 ± 15
					72	0.2 ± 0.5 (5)	0	153 ± 49
	223				48	0.3 ± 0.5 (4)	0	126 ± 34
					72	0.0 (5)	0	174 ± 19
4	0.2	II		6	78	1.0 ± 1.4 (2)	22 ± 17	25 ± 4
					102	0.5 ± 0.7 (2)	7 ± 4	21 ± 4
	1				78	0.5 ± 0.7 (2)	21 ± 11	14 ± 10
					102	0.5 ± 0.7 (2)	6 ± 4	10 ± 14
	5				78	0.5 ± 0.7 (2)	19 ± 6	47 ± 31
					102	3.0 ± 0.0 (2)	11 ± 6	27 ± 18
	0.2			17	89	0.5 ± 0.7 (2)	12 ± 5	17 ± 3
	1					0.0 ± 0.0 (2)	12 ± 6	15 ± 3
	5					2.0 ± 1.4 (2)	9 ± 1	18 ± 3
5	6.9	II		17	89	4.5 ± 3.5 (2)	4 ± 5	124 ± 13
	13.8					2.5 ± 0.7 (2)	5 ± 4	119 ± 7
	27.6					2.0 ± 1.4 (2)	1 ± 1	134 ± 0
6	223	II		17	89	1.0 ± 0.0 (4)	8 ± 8	99 ± 37
7	6.9	II		3	51	0.4 ± 0.5 (5)	5 ± 3	123 ± 19
					75	3.2 ± 2.7 (5)	12 ± 9	125 ± 26
					99	1.8 ± 2.2 (4)	3 ± 3	95 ± 36
	13.8				51	0.2 ± 0.5 (5)	1 ± 1	155 ± 34
					75	1.0 ± 1.7 (5)	6 ± 5	120 ± 32
					99	0.8 ± 1.5 (5)	7 ± 6	159 ± 7
1	4AAF 6.9	I	24		48	1.0 ± 1.4 (2)	9 ± 4	21 ± 1
					72	2.0 ± 2.9 (2)	5 ± 5	10 ± 6
	56				48	3.5 ± 2.1 (2)	16 ± 8	25 ± 11
					72	2.0 (1)	4	9
	223				48	2.5 ± 0.7 (2)	14 ± 3	32 ± 6
					72	3.0 (2)	5 ± 1	21 ± 8
	892				48	3.5 ± 0.7 (2)	9 ± 9	20 ± 5
					72	1.5 ± 2.1 (2)	5 ± 5	10 ± 8

[a] Number of rats between brackets. In each rat, 1000 hepatocytes were analyzed for micronuclei, mitotic activity, and frequency of binucleated cells.
[b] $P < 0.01$ (one-way analysis of variance followed by the Scheffé method for contrasts between experimental groups).

5 Discussion

Of the 4 compounds tested, only 2AAF was found to be clastogenic in rat hepatocytes. The positive effect was detected when the chemical was administered orally 1 day before PH at doses of 6, 9 and 13 mg/kg. The frequencies of MN observed are similar or about 50% lower than those found after ip administration (with the same experimental design) of DEN (50 mg/kg), DMN (10 mg/kg) or ethylnitrosamine (EMS 140 mg/kg). With the same experimental procedures, EMS (200 mg/kg) and MMS (40 mg/kg) proved to be negative. However, the latter 2 chemicals clearly registered positive when given about 17 h after PH (procedure II). From these findings (Tates *et al.*, 1980, 1983; Tates and den Engelse, 1981, 1982, and unpublished data), it appears that the hepatocarcinogen 2AAF is a fairly potent clastogen in hepatocytes.

After completion of the AAF experiments we found data indicating sex differences in the 50% lethal dose (LD_{50}) of N-hydroxy-2-acetylaminofluorene (N-OH-AAF) and in hepatic N-hydroxy-2-acetylaminofluorenesulfotransferase activity. For The strain of rats used in the present study LD_{50}s for males and females were 52 and 116 mg/kg, respectively, whereas, sulfotransferase activity in males was 1.7 times greater than in females (Irving *et al.*, 1975). Consequently, it appears that females are less susceptible to toxic effects of N-OH-AAF than males. In addition, there are also indications for sex differences in DNA adduct formation. These findings taken together suggest that the response of females to the clastogenic effect of 2AAF in rat liver cells may be different from that observed in the present study with male Wistar rats. These sex differences may be even more pronounced in Sprague-Dawley rats, because in this

Table 2. *Induction of chromosomal damage in hepatocytes of rats exposed to BP and PYR*

Expt. No.	Chemical:	dose (mg/kg)	Procedure I or II	Time between injection and hepatectomy (hours)	Time between hepatectomy and injection (hours)	Time between hepatectomy and liver perfusion (hours)	Frequency × 10^{-3} of cells with one or more micronuclei ± S.D.	Mitotic index in hepatocytes ± S.D.	Frequency × 10^{-3} of binucleated hepatocytes ± S.D.
1	BP	63	I	24		72	2.5 ± 0.7 (2)	15 ± 3	25 ± 1
		125					4.0 ± 1.4 (2)	23 ± 1	26 ± 6
		250					6.5 ± 0.7 (2)	9 ± 1	30 ± 1
2		63	I	24		48	1.2 ± 1.1 (5)	27 ± 14	61 ± 15
						72	2.4 ± 1.1 (5)	30 ± 6	25 ± 10
		125				48	2.3 ± 1.0 (4)	26 ± 10	40 ± 18
						72	1.6 ± 1.1 (5)	24 ± 12	24 ± 9
		250				48	1.5 ± 1.0 (4)	53 ± 22	50 ± 15
						72	1.2 ± 0.9 (5)	13 ± 6	33 ± 16
3		250	II		17	89	3.0 ± 1.0 (3)	12 ± 4	18 ± 8
4		32	II		17	89	4.0 ± 0.0 (2)	9 ± 6	19 ± 1
		63					2.0 (1)	21	21
		126					3.0 ± 1.4 (2)	17 ± 1	28 ± 6
		126	II		6	72	0.0 ± 0.0 (2)	15 ± 7	22 ± 9
						96	1.5 ± 0.7 (2)	10 ± 1	21 ± 8
1	PYR	202	I	24		48	3.2 ± 1.9 (5)	15 ± 5	32 ± 14
						72	4.4 ± 1.8 (5)	22 ± 7	47 ± 9
		404				48	1.8 ± 1.1 (5)	15 ± 4	49 ± 22
						72	3.6 ± 1.8 (5)	9 ± 4	49 ± 13
		808				48	2.4 ± 2.2 (5)	12 ± 5	47 ± 6
						72	4.4 ± 0.9 (5)	12 ± 5	42 ± 11

Table 3. *Primary data 2AAF experiment No. 2 plus accompanying control*

Chemical	Time between hepatectomy and perfusion	Frequency × 10^{-3} of cells with one or more micronuclei	Mitotic index	Frequency × 10^{-3} of binucleated hepatocytes
Oil	48	3[a], 0[b], 2, 0, 1	6[a], 35[b], 33, 38, 30	27[a], 51[b], 46, 30, 57
	72	1, 1, 0, 0, 0	9, 13, 19, 8, 11	19, 31, 13, 23, 13
2AAF 6.9 mg/kg	48	8, 6, 12, 6, 7	28, 17, 21, 21, 17	77, 95, 96, 66, 82
	72	12, 18, 19, 12, 5	12, 21, 36, 32, 19	31, 22, 37, 46, 42
2AAF 13.8 mg/kg	48	2, 8, 3, 3, 2	40, 34, 14, 9, 11	100, 66, 201, 132, 125
	72	22, 3, 3, 15, 7	15, 3, 8, 29, 8	61, 119, 120, 60, 122
2AAF 27.6 mg/kg	48	0, 1, 0, 1, 1	1, 1, 1, 1, 2	172, 135, 115, 148, 186
	72	0, 1, 0, 2	0, 5, 2, 0	179, 126, 103, 113

[a] these 3 parameters have been measured in the same-male.
[b] these 3 parameters have been measured in the same-male.
The same holds for all other values presented in the table.

strain, sex differences in LD_{50} and sulfotransferase activity are much more pronounced than in Wistar rats (Irving *et al.*, 1975).

6 Conclusions

On the basis of the present results, it is concluded that 2AAF is a fairly strong clastogen in hepatocytes of male rats. With respect to the results with 4AAF, BP, and PYR, the conclusion is reached that these 3 compounds are not clastogenic under the conditions of the test. It is believed at present that more exhaustive testing of these compounds with our liver test system will not significantly alter the results obtained. In some experimental series, we noticed a slight increase of the frequency of MN, which may become statistically when the numbers of animals per experimental group are considerably increased. In view of the amount of work involved, this is not a realistic procedure for routine tests.

7 Acknowledgement

This work was financially supported by "Ministerie Volkshuisvesting, Ruimtelijke Ordening and Milieuhygiëne".

8 References

Engelse, L. den, B.G.J.Floot, R.-J. de Brij and A.D. Tates (1983) The induction of chromosomal damage in rat hepatocytes and lymphocytes. II. Alkylation damage and repair of rat-liver DNA after diethylnitrosamine, dimethylnitrosamine and ethylmethanesulphonate in relation to clastogenic effects, *Mutation Res.*, **107**, 153–166.

Irving, C.C. (1975) Comparative toxicity of N-hydroxy-2-acetylaminofluorene in several strains of rats, *Cancer Res.*, **35**, 2959–2961.

Menkveld, G.J., A.D. Tates and L. den Engelse (1983) Persistence of chromosomal damage induced by (m)ethylating agents in rat hepatocytes in relation to alkylating pattern of liver DNA (Abstract), *Eur. J. Cancer Clin. Oncol.*, **19**, 1301.

Tates, A.D., I. Neuteboom, M. Hofker and L. den Engelse (1980) A micronucleus technique for detecting clastogenic effects of mutagens/carcinogens (DEN, DMN) in hepatocytes of rat liver *in vivo, Mutation Res.*, **74**, 11–20.

Tates, A.D. and L. den Engelse (1981) Time-dependent repair of chemically induced DNA damage in rat hepatocytes in relation to changes in the frequencies of chromosomal damage (Abstract), *Mutation Res.*, **85**, 277–278.

Tates, A.D. and L. den Engelse (1982) Time-dependent induction of chromosome damage in rat hepatocytes in relation to alkylation damage of DNA. In A.T. Natarajan, G. Obe and H. Altmann (Eds), *Progress in Mutation Research*, Vol. 4, Elsevier Biomedical Press, Amsterdam, pp. 267–278.

Tates, A.D., I. Neuteboom, N. de Vogel and L. den Engelse (1983) The induction of chromosomal damage in rat hepatocytes and lymphocytes. I. Time-dependent changes of the clastogenic effects of diethylnitrosamine, dimethylnitrosamine and ethylmethanesulfonate, *Mutation Res.*, **107**, 131–151.

Tates *et al.*, this volume.

PART IV
In vivo assays: liver

C
DNA Strand breaks

Edilberto Bermudez (Workgroup Chairman)[1]
Chemical Industry Institute of Toxicology, Research Triangle Park, North Carolina, USA

67

Summary report of the measurement of DNA strand breaks

1 Summary

Four chemicals were administered to rats mice and the DNA of the hepatocytes was examined for the presence of DNA strand breaks (SB). Of 3 laboratories that treated rats, 2 found no induction of SB by benzo(a)pyrene (BP), pyrene (PYR), 2-acetylaminofluorene (2AAF) or 4-acetylaminofluorene (4AAF), when the DNA was examined using the alkaline elution technique. In contrast, 1 laboratory using the same technique, found that all 4 chemicals tested had an effect on DNA (except for PYR in male F344 rats). Treatment of mice with BP, PYR, and 4AAF did not result in the induction of SB, however a weak response was observed following treatment with 2AAF. It was concluded that, although the alkaline elution assay is not presently suited for the general screening of potential carcinogens administered *in vivo*, a further understanding of the factors operational in such assays will result in systems with an endpoint very relevant to the detection of these chemicals.

2 Introduction

In the present studies 2 techniques were utilized to determine the induction of SB following the *in vivo* administration of 4 compounds. Three laboratories used the alkaline elution assay and 1 used the fluorimetric analysis of DNA unwinding (FADU). The major parameters of the assays, as they were performed, are shown in Table 1. Presumably, both assays can detect frank SB, DNA breaks due to incision sites generated by repair mechanisms, and alkali labile sites (i.e., apurinic sites, alkylated bases, and phosphotriesters).

3 Results

The results obtained by each laboratory and for each compound are shown in Table 2.

BP and PYR

The livers of rats and mice were examined for SB after i.g. administration of the 2 test compounds in corn oil. All 4 laboratories were in agreement in finding that BP gave no detectable SB, even though other laboratories have shown that DNA adducts are formed. Presumably the BP adducts are alkali stable and do not lead directly or indirectly to strand scission. The highest dose of BP tested

[1] Workgroup participants: Hyman C. Birnboim, Barry M. Elliot, and Patricia Russo

was 250 mg/kg in the rat and 100 mg/kg in the mouse; animals were examined at times ranging from 3 to 48 h after administration of BP. Despite these negative results, one laboratory reported that BP had some effects on liver DNA. In the case of rat liver, 52 mg/kg BP gave detectable DNA-protein crosslinking (DPC), as indicated by the alkaline elution technique. Operationally, DPC were defined as lesions that inhibit the increase in DNA elution normally produced by a standard dose of dimethylsulfate. The presence of BP-induced SB was detected in mouse liver, using the FADU technique, where adenine aribinofuranoside (ara A, a presumed inhibitor of DNA repair) was administered in combination with BP (100 mg/kg).

The same profile was observed for SB with PYR as for BP, except that a weak, but statistically significant positive was observed in 1 laboratory in Sprague Dawley rats (Table 2). Combined treatment of PYR and ara A did not result in the detection of SB in the mouse. No DPC effect was observed following treatment with PYR at concentrations up to 136 mg/kg.

We tentatively conclude that BP produces adducts in DNA that are relatively alkali-stable and, therefore, do not lead to SB. Strand breaks may however be generated in the course of the repair of these lesions and may be accumulated by the inhibition of the repair process. Furthermore, there was an indication that some of DNA crosslinks can be produced by BP. In contrast, PYR appears to be inactive under the conditions utilized.

2AAF and 4AAF

In vivo administration of 2AAF to rats did not result in the induction of SB in 2 laboratories performing the alkaline elution assay. In these 2 laboratories, doses ranging from 50 to 700 mg/kg from 2 to 24 h after treatment were examined. Positive results were found by the third laboratory at a minimum of 20 mg/kg and 4 h after administration of 2AAF. Similarly, negative results were obtained when 4AAF was administered at 200 to 700 mg/kg and 12 to 24 h by 2 laboratories, but positive in the other laboratory following 20 mg/kg and 4 h after treatment. Although procedural differences between laboratories were found (Table 1), none could be pinpointed as the plausible reason(s) for the observed discrepancies.

In vivo administration of 50 mg/kg 2AAF to mice resulted in the induction of SB. Furthermore, an effect could be demonstrated at 10 mg/kg when treatment with 2AAF was followed by treatment with ara A to block DNA repair. Presumably this reflects an accumulation of SB by the DNA repair mechanism(s). In contrast, the administration of 4AAF to mice did not result in the detection of SB with or without the use of ara A.

Table 1. *Major parameters in the assays used to detect DNA strand breaks*

	Alkaline elution		
	Russo	Elliot	Bermudez
Rat strain	Sprague Dawley; F344	Alderley Park	F344
Sex	Male	Male	Male
Cell isolation	Mince, obtain Cells and nuclei	Mince, obtain cells and nuclei	Perfuse and obtain cells
Cell lysis	EDTA + Sarcosyl ± Prot. K, 20 min	EDTA + Sarcosyl − Prot. K, 6 min	EDTA + Sarcosyl + Prot. K, 30 min
Elution	TEAH[a] − 1, 20 h pH 12.3	TEAH − 40 min pH 12.3	TPAH[b] − 10 h pH12.2
Criteria for a positive	Dose response, statistically significant increase over control	Dose response, statistically significant increase over control	Dose response, statistically significant increase over control
	DNA unwinding		
	Birnboim		
Mouse	CD-1		
Sex	Female		
Cell isolation	Core samples and obtain cells and nuclei		
Cell lysis	urea + detergent		
Unwinding	1 h alkali pH 12.8 (± ara A *in vivo*)		
Criteria for a positive	Dose response, statistically significant increase over control		

[a] tetraethyl ammonium hydroxide.
[b] tetrapropyl ammonium hydroxide.

Based on these results, no joint conclusion could be drawn as to the induction of SB by 2AAF or 4AAF.

4 Recommendations

The most notable aspect of the data obtained in this study was the inability of the investigators using 2AAF and 4AAF to achieve comparable responses with the alkaline elution assay. It is recognized that there were many detailed differences in the protocols employed, however, these were considered in turn and it was not possible to associate the differences in response observed to any one parameter or simple combination of parameters. Moreover, it has recently been reported that the elution of DNA from filters depends not only on the length of the DNA but on the molecular conformation of the DNA molecule. If this is correct, then the endpoint of the alkaline elution technique and, therefore, those parameters important for its correct execution, require a critical reappraisal. In this context, it is interesting to note that the same investigators who reported 2AAF and 4AAF as positive also obtained positive data for the noncarcinogens caprolactam and benzoin. In view of the negative responses obtained for these latter chemicals in several other assays measuring DNA effects, it does appear that under certain conditions, the elution of DNA from filters in alkali is dependent on factors other than SB.

Table 2. *The induction of DNA strand breaks in the assay systems*

	Russo	Elliot	Bermudez	Birnboim
2AAF	+ [a]	–	–	+ [b]
4AAF	+	–	–	–
BP	– [a]	–	– [c]	– [d]
PYR	+ [e]	–	–	–
	– [f]	–	–	–

[a] DNA-protein crosslinks found.
[b] Weak response.
[c] Marginal negative found at 12 h, 200 mg/kg.
[d] DNA strand breaks found when ara A treatment was included.
[e] Small, but statistically significant increase over control.
[f] In F344 rat.

The results obtained in this study and the information in the literature alluded to above, leads us to conclude that the alkaline elution assay in its present form is unsuitable for the general screening of potential carcinogens for the induction of SB in liver DNA *in vivo*. The use of SB as an endpoint has obvious relevance to the potential induction of cancer and we consider it of importance to encourage further studies aimed at understanding the underlying factors governing the elution of DNA as described herein. Such studies would not only clarify the potential of this assay as a screening tool but would also result in advancing the understanding of chemical-DNA interactions. In addition, the other available assays measuring SB (such as DNA unwinding, as reported here) should also receive further consideration.

Finally, inhibitors of DNA repair may prove useful increasing the sensitivity of the assay, where the chemical being tested does not result in the formation of alkali-labile sites.

Edilberto Bermudez and Loretta Lyon Delehanty
Department of Genetic Toxicology, Chemical Industry Institute of Toxicology Research Traingle Park, North Carolina, USA

68

DNA strand break induction in Fischer-344 rat hepatocytes as assessed by the alkaline elution assay

1 Summary

Four chemicals were administered to Fischer 344 (F344) rats by gavage and the DNA of the hepatocytes was examined for the presence of single-strand breaks (SSB) using the alkaline elution assay. There were no statistically significant increases of SSB found following treatment of rats with benzo(a)pyrene (BP), pyrene (PYR), 2-acetylaminofluorene (2AAF) or 4-acetylaminofluorene (4AAF).

2 Introduction

The alkaline elution assay was first described by Kohn and Ewig (1973) and is a sensitive means of detecting the presence of strand breaks in the DNA of cells (Petzold and Swenberg, 1978; Parodi *et al.*, 1978, Brambilla *et al.*, 1978; Erickson *et al.*, 1980; Bermudez *et al.*, 1982b; Sina *et al.*, 1983). The basis of the assay is the differential elution of single-stranded DNA from a filter support by an alkaline solution, with the shortest lengths of DNA eluting first. The assay detects DNA breaks from several sources: directly due to the treatment (as with lonizing radiation); generated by repair mechanisms; or as a result of alkali labile sites (i.e., apurinic sites, alkylated bases, phosphotriesters). With some modification of the assay, DNA crosslinks may also be detected.

In the present study 4 chemicals, 2 with known carcinogenic potential (2AAF and BP) and 2 remaining unclassified as to their carcinogenicity (PYR and 4AAF), were examined for their capacity to induce SSB in the DNA of rat hepatocytes. The presence of DNA-protein crosslinking (DPC) can diminish the capacity to detect SSB by increasing the apparent retention of DNA. In this study, proteinase K was included in the lysis step to remove DPC, thereby limiting the assay to the evaluation of SSB occurrence.

3 Experimental protocol

Male F344 rats [CDF(G-344)/CrlBr] weighing 200–280 grams were obtained from Charles River Breeding Laboratories (Kingston, NY) and housed in hanging, stainless steel mesh cages (Hazleton Systems, Vienna, VA) in a mass air displacement room (Bioclean, Hazleton Systems, Vienna, VA) with controlled temperature (22 ± 1 °C), relative humidity (50 ± 10%), and light cycles (12 h). The animals were provided food (NIH-07 open formula) and water ad libitum. Randomly selected animals were tested and found free of known pathogens

(Standard Screening Necropsy, Microbiological Associates, Bethesda, MD).

Rats were gavaged with PYR, BP, 2AAF or 4AAF (Imperial Chemical Industries, UK) at various doses using corn oil as the vehicle. Control animals received only vehicle. At various times after treatment the animals were anaesthetised and the livers perfused with collagenase to obtain single cell suspensions of hepatocytes as described by Williams *et al.* (1977) and Bermudez *et al.* (1979). The cell suspensions were filtered through gauze, centrifuged at 50 × g for 4 min and resuspended in cold phosphate (5 mM) buffered saline, pH 7.2 (PBS). At this time the cell density and viability, as determined by trypan blue exclusion, were assessed and the cell density adjusted to 5×10^5 cells/ml. *In vitro* exposure of control hepatocytes to methylmethanesulfonate (MMS; 5 mM, 20 min) was utilized as the positive control. In some experiments, control cells were also treated with 2AAF *in vitro*.

The alkaline elution assay (Kohn, 1979) was performed as described by Bermudez *et al.* (1982b). Control and treated cells (5×10^5) suspended in PBS were deposited onto PVC filters (2 μm, BSWP25, Millipore), lysed (0.02 M Na_3 EDTA, 2 M NaCl, 0.3% Sarkosyl, 0.5 mg/ml proteinase K, pH 10.00) for 30 min and eluted for 10 h with 0.02 M H_4 EDTA (brought to pH 12.20 with 10% tetrapropylammonium hydroxide) at a flow rate of 0.05 ml/min. Two sets of triplicate filters were prepared for each sample, one set to be eluted and one to serve as a reference for the quantity of DNA initially deposited on the filters. The relative amounts of DNA present on the eluted and reference filters were determined by quantitation of the fluorescent product obtained after reacting DNA with 3,5-diaminobenzoic acid hydrochloride (0.3 g/ml, Aldrich, Milwaukee, WI) at 60 °C for 90 min. Fluorescense was measured at an excitation wavelength of 414 nm and emission wavelength of 508 nm. The data is expressed as the percentage of the DNA retained (percentage retention) and is determined by dividing the adjusted mean relative fluorescense of the eluted filters by the mean relative fluorescense of the reference filters.

Significance at the $p < 0.05$ level was determined by comparing the percentage retention of the treated samples to the percentage retention of the pooled control samples [84.7 ± 1.3 (SEM)], using the t test.

4 Results

No statistically significant increases in the induction of SSB in hepatocyte DNA were observed when rats were treated with various doses and examined at various times after treatment with: PYR (Table 1, 5); 2AAF (Table 4, 5); 4AAF (Table 3, 5); or BP (Table 2, 5). In the case of BP there was a marginal response (significance level $0.1 > p > 0.05$) observed 12 h after the administration of 200 mg/kg. *In vitro* treatment of control cells with 2AAF (Figure 1) or with MMS (Table 5) resulted in the induction of SSB.

5 Discussion

The present study shows no statistically significant ($p < 0.05$) induction of SSB by BP, however, a weak response was observed 12 h after the administration of 200 mg/kg. BP has been shown to be a carcinogen for

Table 1. *Retentions for all animals treated with PYR*

	H	MG/KG	Retention[a] (%) Test	Control
PYR	12	100	101	91
	12	100	75	76
	12	100	105	92
	12	200	76	91
	12	200	80	80
	12	200	83	80
	12	200	80	84
	12	200	92	84
	12	500	96	84
	12	500	87	84
	24	100	97	91
	24	100	94	76
	24	100	84	92
	24	200	83	91
	24	200	87	80
	24	200	89	80
	24	500	84	87
	24	500	78	87
	48	100	121	91
	48	100	94	76
	48	100	83	92
	48	200	77	91
	48	200	83	80
	48	500	92	81
	48	500	98	81
	48	500	84	81

$$^{a}\text{Retention} = \frac{\text{Mean of Eluted Filters}}{\text{Mean of Reference Filters}} \times 100$$

Table 2. *Retentions for all animals treated with BP*

	H	MG/KG	Retention[a] (%) Test	Control
BP	12	100	125	103
	12	100	65	73
	12	100	78	92
	12	100	100	87
	12	100	78	88
	12	200	59	76
	12	200	80	89
	12	200	57	89
	12	200	89	89
	24	100	105	103
	24	100	92	73
	24	100	71	92
	24	100	87	87
	24	200	74	76
	24	200	82	89
	24	200	79	89
	24	200	90	90
	48	100	66	73
	48	100	97	73
	48	100	100	92
	48	100	88	87
	48	200	76	76
	48	200	93	89
	48	200	85	89
	48	200	96	90

$$^{a}\text{Retention} = \frac{\text{Mean of Eluted Filters}}{\text{Mean of Reference Filters}} \times 100$$

several species including the rat (IARC, 1973) and to induce various genotoxic effects *in vivo* and *in vitro* (Brookes and de Serres, 1981). Previous experiments using the alkaline elution assay to determine DNA damaging capacity have shown that BP induces SSB in hepatocytes (Sina, 1983), V79 cells (Swenberg *et al.*, 1976) and both SSB and DPC in hamster tracheal epithelial cells (Eastman *et al.*, 1981; Ireland *et al.*, 1982). Hepatocytes repair BP induced DNA damage (Williams, 1977) and metabolise BP to mutagenic metabolites (Bermudez *et al.*, 1982a) *in vitro*. *In vivo* studies to examine DNA damage (Parodi *et al.*, 1981) and DNA repair (Mirsalis *et al.*, 1982) following the administration of BP, concluded that no induction was apparent, consistent with our results. The lack of an *in vivo* response in the alkaline elution assay may be due to several factors including: the production of few adducts in hepatocyte DNA (Parodi *et al.*, 1981); formation of ad-

Table 3. *Retentions for all animals treated with 2AAF*

	H	MG/KG	Retention[a] Test	Control
2AAF	2	50	77	95
	2	50	89	84
	2	50	87	81
	12	50	80	95
	12	50	90	84
	12	50	92	80
	12	50	85	93
	12	50	79	80
	12	50	79	80
	12	50	86	79
	12	50	87	79
	12	50	95	83
	12	50	97	83
	12	50	80	83
	12	100	77	80
	12	100	76	80
	12	200	90	96
	12	400	65	96
	12	400	97	77
	12	400	94	81
	12	400	87	81
	12	700	103	77
	12	700	72	81
	12	700	102	81
	24	50	90	95
	24	50	78	84

[a] $\text{Retention} = \dfrac{\text{Mean of Eluted Filters}}{\text{Mean of Reference Filters}} \times 100$

Table 4. *Retentions for all animals treated with 4AAF*

	H	MG/KG	Retention[a] Test	Control
4AAF	12	200	81	95
	12	200	100	79
	12	200	84	87
	12	200	83	87
	12	200	80	85
	12	700	93	85
	12	700	81	85
	12	700	70	85
	24	200	77	95
	24	200	88	79
	24	200	84	79

[a] $\text{Retention} = \dfrac{\text{Mean of Eluted Filters}}{\text{Mean of Reference Filters}} \times 100$

Table 5. *Summary of the conclusions for each chemical at each time and dose tested*

	H	MG/KG	Number of animals	Conclusion[a]
PYR	12	100	3	NEG
	12	200	5	NEG
	12	500	2	NEG
	24	100	3	NEG
	24	200	3	NEG
	24	500	2	NEG
	48	100	3	NEG
	48	200	2	NEG
	48	500	3	NEG
BP	12	100	5	NEG
	12	200	4	NEG[b]
	24	100	4	NEG
	24	200	4	NEG
	48	100	4	NEG
	48	200	4	NEG
4AAF	12	200	5	NEG
	12	700	3	NEG
	24	200	3	NEG
2AAF	2	50	3	NEG
	12	50	11	NEG
	12	100	2	NEG
	12	400	4	NEG
	12	700	3	NEG
	24	50	2	NEG
MMS	0.33	5 mM	7	POS

[a] NEG = NEG = Where no significant ($P < 0.05$) difference from controls was observed. The mean retention for all controls ($n = 34$) was 84.7 ± 1.3 (S.E.M.). Comparisons were made using the T test.
[b] Significance level = 0.096.

Table 6. *Pooled control data compared with the test data to assess statistical significance*

No.	Retention[a] (%)	No.	Retention (%)
1	103	19	81
2	91	19	86
4	73	20	86
5	76	23	96
6	95	23	77
7	72	28	83
8	91	29	90
10	95	30	88
11	65	30	86
12	84	32	83
13	89	32	88
14	80	33	85
15	92	33	88
16	80	34	81
16	93	35	84
17	81	36	81
18	76	36	81

[a] $\text{Retention} = \dfrac{\text{Mean of Eluted Filters}}{\text{Mean of Reference Filters}} \times 100$

ducts that do not induce true DNA breaks and are not alkali labile; absorption from the gut as a limiting step; or organ specific factors.

Treatment of rats with PYR did not result in the induction of SSB. Although PYR has been examined in bioassays for carcinogenicity the resultant data have been determined to be inconclusive (Purchase *et al.*, 1981). Various *in vitro* assays for genotoxicity have shown PYR to be positive (Ashby, 1981). In contrast, no activity was observed in *in vivo* studies (Ashby, 1981). The results in the present study are consistent with PYR not inducing DNA damage after *in vivo* administration and are supported by the lack of induction of UDS in rat hepatocytes treated *in vivo* with PYR (Butterworth and Smith Oliver, this volume).

Treatment, by gavage, of F344 rats with 2AAF did not result in the induction of SSB at any of the doses or times examined. The carcinogenicity of 2AAF in the rodent has been well established (Purchase *et al.*, 1981). Most *in vitro* assays in which this compound has been tested have shown it to be positive for the various endpoints examined (Ashby, 1981). SSB induced by 2AAF are detectable by alkaline elution after *in vitro* treatment of V79 cells (Swenberg *et al.*, 1978) and hepatocytes (Sina *et al.*, 1983; Bermudez, *et al.*, 1982b). *In vivo* exposure has resulted in the induction of SSB after i.p. administration in some studies (Parodi *et al.*, 1978; Petzold and Swenberg, 1978) but not in others (Parodi *et al.*, 1981; Schwarz *et al.*, 1979). The induction of SSB after administration by gavage has also been reported (Bermudez *et al.*, 1982b). As this laboratory had previously reported a positive response with 2AAF *in vivo* it was of interest to investigate the possible reasons for the apparent discrepancy. The material provided for this study was examined by HPLC and found to be > 99.99% 2AAF whereas, the commercially produced materials are in the range of 95 to 97% pure. In addition, the purified material was tested for the induction of SSB in hepatocyte DNA after *in vitro* exposure and found to induce SSB. Although higher concentrations were necessary to induce SSB *in vitro* than previously observed in this laboratory, they were within the range observed by other investigators (Swenberg *et al.*, 1976; Sina *et al.*, 1983). Simultaneous examination by alkaline elution and autoradiography of hepatocytes from rats treated with purified 2AAF for the induction of SSB and UDS showed that UDS was induced (Butterworth and Smith-Oliver, this volume) but SSB were not detected, indicating that the compound had reached the liver. The adducts found in DNA as a result of treatment with 2AAF are bulky, probably repaired by a long-patch repair mechanism (Teebor and Frenkel, 1983) and are not alkali labile. It is to be expected that the sensitivity of a UDS assay is likely to be much greater for this compound than is the alkaline elution assay since only excision sites might be detected by the latter. The detectable response in studies with impure 2AAF may be due to the induction of additional adducts, which are alkali labile, or perhaps to a synergistic effect between components.

Figure 1. The induction of single-strand breaks in rat hepatocytes treated with 2AAF *in vitro* for 2 h.

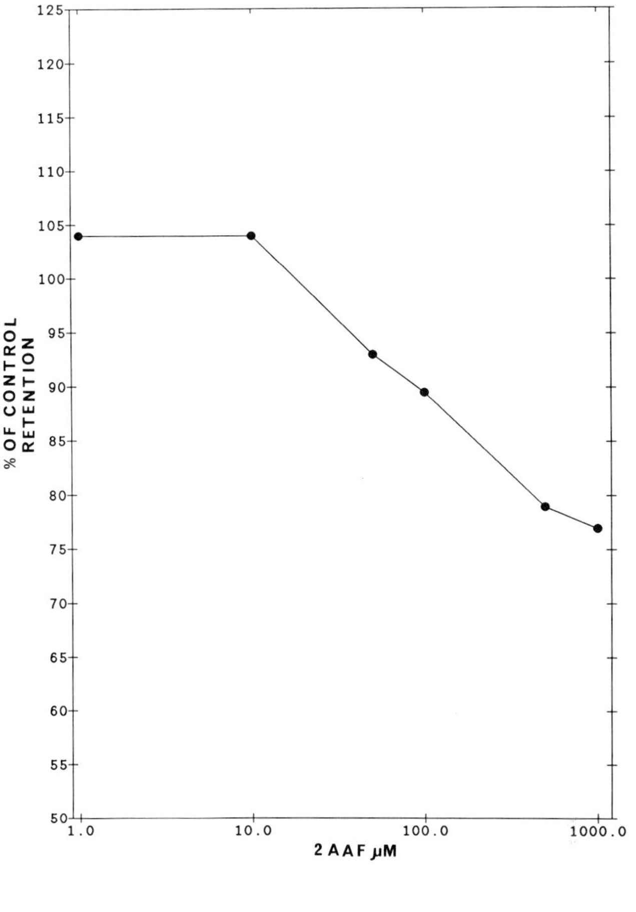

4AAF did not induce SSB at doses up to 700 mg/kg. Evidence for carcinogenicity has been determined as equivocal and therefore 4AAF is unclassified (Purchase *et al.*, 1981). Short-term *in vitro* assays for genotoxicity are also equivocal in that there were mixed positive and negative results (Ashby, 1981). Exposure of rats to 4AAF *in vivo* resulted in an increase in scheduled but not in unscheduled DNA synthesis in hepatocytes (Butterworth and Smith-Oliver, this volume) indicating that the compound reaches the liver. The results in the present study are consistent with the lack of SSB production in rat hepatocytes by 4AAF.

In conclusion, no induction of SSB was observed with BP, PYR, 2AAF, or 4AAF when these compounds were examined at various times and doses.

6 References

Ashby, J. (1981) Overview of Study and Test Chemical Activities. In F.J. de Serres and J. Ashby, (Eds), *Evaluation of Short-Term Tests for Carcinogens, Progress in Mutation Research*, **1**, pp. 112–171.

Bermudez, E., D. Tillery and B.E. Butterworth (1979) The Effect of 2,4-Diaminotoluene and Isomers of Dinitrotoluene on Unscheduled DNA Synthesis in Primary Rat Hepatocytes, *Environ. Mutagen.*, **1**, 391–398.

Bermudez, E., D.B. Couch and D. Tillery (1982a) The Use of Primary Rat Hepatocytes to Achieve Metabolic Activation of Promutagens in the Chinese Hamster Ovary/Hypoxanthine-Guanine Phosphoribosyl Transferase Mutational Assay, *Environ. Mutagen.*, **4**, 55–64.

Bermudez, E., J.C. Mirsalis and H.C. Eales (1982b) Detection of DNA Damage in Primary Cultures of Rat Hepatocytes Following *In Vivo* and *In Vitro* Exposure to Genotoxic Agents, *Environ. Mutagen.*, **4**, 667–679.

Brambilla, G., M. Cavanna, S. Parodi, L. Sciaba, A. Pino and L. Robiano (1978) DNA Damage in Liver, Colon, Stomach, Lung and Kidney of BALB/c Mice Treated

With 1, 2-Dimethylhydrazine, *Int. J. Cancer*, **22**, 174–180.

Brookes, P. and F.J. de Serres (1981) Overview of Assay System Performance. In F.J. de Serres and J. Ashby, (Eds), *Evaluation of Short-Term Tests for Carcinogens, Progress in Mutation Research*, **1**, pp. 96–111.

Eastman, A., B.T. Moosman and E. Bresnick (1981) Formation and Removal of Benzo(a)pyrene Adducts of DNA in Hamster Tracheal Epithelial Cells, *Cancer Res.*, **41**, 2605–2610.

Erickson, L.C., R. Osieka, N.A. Sharkey, K.W. Kohn, (1980) Measurement of DNA Damage in Unlabeled Mammalian Cells Analyzed by Alkaline Elution and a Fluorometric DNA, *Assay. Anal. Biochem.*, **106**, 169–174.

IARC Monographs on the Evaluation of Carcinogenic Risk of Chemicals to Man (1973) *Certain Polycyclic Aromatic Hydrocarbons and Heterocyclic Compounds*, **3**, 91–136.

Ireland, C.M., A. Eastman and E. Bresnick (1980) DNA-Protein Crosslinks Induced in a Hamster Tracheal Epithelial Cell Line by Benzo(a)pyrene, *Biochem. Biophys. Res. Comm.*, **109**, 1291–1296.

Kohn, K.W. and R.A.G. Ewig (1973) Alkaline Elution Analysis, a New Approach to the Study of DNA Single-Strand Interruptions in Cells, *Cancer. Res.*, **33**, 1849–1853.

Kohn, K.W. (1979) DNA as a Target in Cancer Chemotherapy: A New Approach to the Study of Macromolecular Damage Produced in Mammalian Cells by Anticancer Agents and Carcinogens. In *Methods of Cancer Research*, **16**, (Cancer Drug Dev., Part A), pp. 291–345.

Mirsalis, J., C.K. Tyson and B.E. Butterworth (1982) Detection of Genotoxic Carcinogens in the *In Vivo/In Vitro* Hepatocyte DNA Repair Assay. *Environ. Mutagen.*, **4**, 553–562.

Parodi, S., M. Taningher, L. Santi, M. Cavanna, L. Sciaba, A. Maura, and G. Brambilla (1978) A Practical Procedure for Testing DNA Damage *In Vivo* Proposed for a Prescreening of Chemical Carcinogens. *Mutation Res.*, **53**, 39–46.

Parodi, S., M. Taningher, M. Pala and L. Santi (1981) Alkaline DNA Fragmentation *In Vivo*: Borderline or Negative Results Obtained Respectively With 7, 12-Dimethylbenz(a)anthracene and Benzo(a)pyrene, *Tumori*, **67**, 87–93.

Petzold, G.L. and J.A. Swenberg (1978) Detection of DNA Damage Induced *In Vivo* Following Exposure of Rats to Carcinogens, *Cancer Res.*, **38**, 1589–1594.

Purchase, I.F. H., D.B. Clayson, R. Preussmann and L. Tomatis (1981) Activity of 42 Compounds in Animal Carcinogenicity Studies. In F.J. de Serres and J. Ashby, (Eds), Evaluation of Short-Term Tests for Carcinogens, *Progress in Mutation Research*, **1**, pp. 21–32.

Schwarz, M., J. Hummel, K.E. Appel, R. Rickart and W. Kunz (1979) DNA Damage Induced *In Vivo* Evaluated With a Non-Radioactive Alkaline Elution Technique, *Cancer Lett.*, **6**, 221–226.

Sina, J.F., C.L. Bean, G.R. Dysart, V.I. Taylor and M.O. Bradley (1983) Evaluation of the Alkaline Elution/Rat Hepatocyte Assay as a Predictor of Carcinogenic/Mutagenic Potential, *Mutation Res.*, **113**, 357–391.

Swenberg, J.A., G.L. Pedzold and P.R. Harbach (1976) *In Vitro* DNA Damage/Alkaline Elution Assay for Predicting Carcinogenic Potential, *Biochem. Biophys. Res. Comm.*, **72**, 732–738.

Teebor, G.W. and K. Frenkel (1983) The Initiation of DNA Excision-Repair, *Advances in Cancer Research*, **38**, 23–59.

Williams, G.M., E. Bermudez, and D. Scaramuzzino (1977) Rat Hepatocyte Primary Cultures. III. Improved Dissociation and Attachment Techniques and the Enhancement of Survival by Culture Medium, *In Vitro*, **13**, 809–817.

Williams, G.M. (1977) Detection of Chemical Carcinogens by Unscheduled DNA Synthesis in Rat Liver Primary Cultures, *Cancer Res.*, **37**, 1845–1851.

Barry M. Elliott and Elizabeth Riley
Genetic Toxicology Section, Imperial Chemical Industries PLC
Central Toxicology Laboratory, Alderley Park, Cheshire, UK

69

DNA strand breakage monitored by alkaline elution in the livers of male Wistar-derived rats administered carcinogenic/noncarcinogenic analogues

1 Summary

The carcinogens 2-acetylaminofluorene (2AAF) and benzo(a)pyrene (BP) and their noncarcinogenic analogues 4-acetylaminofluorene (4AAF) and pyrene (PYR) were administered orally to male Alderley Park rats and their livers were examined for DNA strand breaks using the technique of alkaline elution. Neither 2AAF (examined at 4, 12 and 24 h after dosing) nor BP (examined at 24 h after dosing) nor either of the noncarcinogen analogues produced an increased rate of DNA elution from the filters compared with that seen with control animals. DNA from the livers of animals treated with dimethylnitrosamine showed a marked increase in elution rate compared with controls. The results obtained are discussed with reference to the number and nature of the DNA adducts likely to have been produced, and it is concluded that the DNA alkaline elution technique may not be sensitive in the detection of DNA single-strand breaks for all classes of carcinogens.

2 Introduction

The rate at which DNA elutes from a filter under alkaline conditions has been widely used to examine for DNA damage induced by mutagenic/carcinogenic chemicals (Bradley and Sina, 1984; Parodi *et al.*, 1981c; Swenberg *et al.*, 1976; Petzold and Swenberg, 1978; Brambilla *et al.*, 1981; Bermudez *et al.*, 1982). The presence of DNA strand breaks causes the single-stranded DNA to elute more rapidly from the filter compared with control DNA. The assay was originally established using *in vitro* cell cultures and monitoring the DNA by prelabelling the cells with radioactive DNA precursors (Kohn *et al.*, 1976; Swenberg *et al.*, 1976). More recently, sensitive fluorimetric techniques have allowed the direct measurement of the DNA eluted from the filter and have thus enabled tissues from animals treated with chemicals to be examined directly for DNA damage (Parodi *et al.*, 1978; Stout and Becker, 1982; Erickson *et al.*, 1980). The assay has taken on the role of a rapid *in vivo* screening test for the potential carcinogenicity of chemicals in animals (Parodi *et al.*, 1978; Bradley and Sina, 1984).

As part of an international study to evaluate the responses of several *in vivo* screening tests, we have evaluated the DNA from the livers of rats administered 2AAF, 4AAF, BP, PYR, by alkaline elution. The method used is basically that described by Parodi *et al.* (1981c) but using a direct fluorescent assay with Hoechst 33258 to determine the DNA eluted and retained on the filter (Stout and Becker, 1982).

This elution assay has been used previously to examine many carcinogenic chemicals for their ability to produce DNA damage following the *in vivo* administration of the chemical (Barbin *et al.*, 1983; Parodi *et al.*, 1981, a, c; Brambilla *et al.*, 1981; Petzold and Swenberg, 1978; Pearson and Edwards, 1982).

Both 2AAF and BP are established carcinogens in rodents (IARC, 1973; Conney, 1982, Clayson and Garner, 1976); only 2AAF is reported as being carcinogenic to the liver in adult rats or mice. BP has instead been found to produce tumours locally, for example, of the skin, following dermal administration, and of the forestomach, following oral administration. It is only reported to have produced a significant increase of liver tumours when tested in the newborn mouse (IARC, 1973; Conney 1982). The profile of DNA binding of BP reflects this site specificity, with much greater amounts of material being bound to DNA in the tissue susceptible to tumour formation than in the liver (Borchert and Wattenberg, 1976; Lutz *et al.*, 1978). 2AAF binds to liver DNA to a greater extent than does BP [by comparing values for the Covalent Binding Index (CBI)], and has the liver as the primary target organ for its carcinogenic response (Lutz, 1979; Clayson and Garner, 1976). It is an interesting challenge to *in vivo* screening assays to compare results obtained for these two carcinogens, as well as for the reportedly noncarcinogenic analogues 4AAF and PYR. Dimethylnitrosamine (DMN), also a potent rodent hepatocarcinogen, has been examined as a positive control for the alkaline elution assay.

3 Methods

Administration of chemicals
Male Alderley Park (Alpk/Ap, Wistar-derived) rats, 180–220 g bodyweight, were administered a single dose of the test chemicals by gavage (1 cm^3/100 g bodyweight in corn oil) or of DMN (0.5 cm^3/100 g bodyweight in physiological saline) intraperitoneally. The DMN was obtained from Aldrich Chemical Co., Gillingham, Dorset, UK, of Gold Label grade, and the dosing solution was prepared just prior to use. The other test chemicals were obtained from Imperial Chemical Industries PLC, Macclesfield, UK, from the batches designated for this international study and were ball-milled overnight in corn oil before dosing. Control animals were administered an equivalent volume of vehicle alone.

Preparation of hepatocyte nuclei
At times up to 24 h after dosing, groups of test and control animals were sacrificed by inhalation of ether vapour and the livers placed on ice. A 33% liver homogenate was prepared by homogenising in 2 volumes of ice-cold Merchants solution (0.14 M NaCl, 1.47 mM KH_2PO_4, 2.7 mM KCl, 8.1 mM Na_2HPO_4, 0.53 mM Na_2EDTA, pH 7.4 at 20 °C) using 5 complete strokes of a Potter-Elvejhem type homogeniser with a total clearance of 0.508 mm. The homogenate was filtered through 150 μ nylon mesh and the cell/nuclei sedimented at 50 g for 4 minutes. The loose pellet was resuspended to the original volume in cold Merchants solution and diluted 10-fold in the same.

Alkaline elution assay
Cells/nuclei (approx. 1×10^6) were applied to a 5.0 μ, 25 mm diameter millipore filter (mixed esters of cellulose, SMWP02500) held in a Swinnex filter assembly fitted with a 20 cm^3 syringe barrel as reservoir and washed with approximately 5 cm^3 of cold Merchants solution. The cells were then lysed with 4.5 cm^3 of a solution of 0.2% sodium lauryl sarcosinate, 2 M NaCl, 20 mM Na_2EDTA (pH 10.0) and washed with 5.0 cm^3 of 20 mM Na_2EDTA (pH 10.0). The above loading and washing steps were done using a flow rate of 0.8 cm^3/min. Single-stranded DNA was eluted from the filter with 60 mM $(C_2H_5)_4NOH$, 10 mM Na_2EDTA (pH 12.3), by collecting 8 cm^3 at a flow rate of 0.2 cm^3/min. The DNA remaining on the filter was recovered by breaking up and extracting the filter in 3.0 cm^3 of eluting medium.

DNA assay
The DNA eluted and that remaining on the filter was assayed fluorimetrically using Hoechst 33258 as described by Stout and Becker (1982). Hoechst 33258 (4.2×10^{-7} M final concentration) was added to aliquots of filter and eluant (100–400 μl) contained in 1.4 cm^3 of 1 M phosphate buffer (pH 7.0) and the fluorescence read using an excitation wavelength of 360 nm and an emission wavelength of 450 nm. A standard curve was obtained using calf thymus DNA similarly prepared in eluant.

Results of the alkaline elution assay are expressed as an elution rate constant *K*, using the formula

$$K = \frac{-\ln \text{(fraction of DNA retained on filter)}}{V}$$

where *K* has units of cm^{-3} and *V* is the eluted volume in cm^3 (Parodi *et al.*, 1981c).

4 Results and discussion

The average elution rate constants observed for each of the chemicals examined are shown in Table 1. For DMN, a significant increase in the elution rate constant over that for control animals was observed, and this is considered a positive response. For all 4 test chemicals (2AAF, 4AAF, BP, and PYR), no increase in the elution rate constant over that for control animals was observed, and it is concluded that none of the chemicals produced detectable levels of DNA damage (DNA strand breaks) in the livers of the treated animals. The liver carcinogen 2AAF was examined at 3 time periods over 24 h to increase the chance of detecting DNA damage arising from a single dose. Furthermore, parallel studies in this laboratory have shown that 12 h appears to be the optimal time to detect unscheduled DNA synthesis (UDS) in the livers of Alpk/AP rats treated similarly with 2AAF, confirming that the chemical has interacted with the hepatic DNA within the timescale examined.

The alkaline elution technique is generally considered to measure DNA single-strand breaks (SSB) (Kohn and Ewig, 1973; Kohn *et al.*, 1976, Bradley and Sina, 1984). Such SSBs can be produced in DNA by (1) direct scission, (2) enzymic scission, and (3) hydrolysis of the sugar-phosphate backbone (Bradley and Sina, 1984). Direct scission of the sugar phosphate backbone is produced by radical damage (e.g., X irradiation); most organic chemicals do not induce this mechanism. The number of detectable SSBs induced by chemicals that covalently bind to DNA is governed rather by the number and nature of DNA adducts formed. Chemicals that are simple alkylating agents (e.g., DMN and dimethylhydrazine) have been the most widely studied by alkaline elution techniques and have been found to give potent responses (Stewart and Farber, 1973; Parodi *et al.*, 1981a; 1982; Skare and Schrotel, 1984; Brambilla *et al.*, 1981, Barbin *et al.*, 1981, Bermudez *et al.*, 1982, Petzold and Swenberg, 1978). This activity is understandable since the CBI values for such chemicals (Lutz, 1979) are generally high (approximately 500–1000+) and the main lesion produced (N-7 of guanine), (O'Connor *et al.*, 1973) renders the DNA backbone alkali labile. Such damage, producing apurinic sites on DNA, leads to repair by the "short-patch" repair process. Alkylating agents may also lead to the production of phosphotriesters on DNA, lesions which are again important in the formation of DNA strand breaks, especially under alkaline conditions (Walker and Ewart, 1973; O'Connor *et al.*, 1975; Dogliotti *et al.*, 1984).

Chemicals such as BP and 2AAF which have been found to be negative in the present study, produce a different profile with their lesion on DNA. The adducts formed are predominantly at the C-8 and/or N^2 positions of guanine (Kriek, 1972, Meehan *et al.*, 1977, Straub *et al.*, 1977) and do not render the DNA backbone alkali labile in the same way as the alkylating agents. The CBI values (Lutz, 1979) for such chemicals (approximately 10–100) reveal a considerably smaller number of adducts also. Any SSBs arising from such DNA adducts presumably will come primarily from enzymic attack of DNA during the repair process. Such DNA repair is again different from that for the simple alkylating agents, with "long-patch" repair (Kornberg, 1974) being involved in the removal of the bulky adducts found with 2AAF and BP. This is usually an efficient, coupled process however, and at nontoxic doses of chemical, no SSBs would be expected to accumulate. The inhibition of the repair process with agents such as cytosine arabinoside and 2′, 3′-dideoxythymidine does cause the accumulation of DNA strand breaks and has been used *in vitro* to detect DNA damage caused by BP (Ireland *et al.*, 1984). It is interesting to consider that the nature of the "long-patch" repair process makes it intrinsically more efficient at being recognised by the UDS assays than that operative for the simple alkylating agents.

From the above discussion, it is evident that normal alkaline elution techniques will be less efficient at detecting DNA strand breaks induced by chemicals such as 2AAF and BP with their profile of DNA adduct, compared with simple alkylating agents such as DMN. This is well reflected in the literature with both positive (Bermudez *et al.*, 1982; Petzold and Swenberg, 1978; Vu *et al.*, 1985) and negative (Parodi *et al.*, 1981b, c; 1982; Brambilla *et al.*, 1983) responses being reported for chemicals like 2AAF and BP but with a consistent positive profile for chemicals like DMN (Stewart and Farber, 1973; Parodi *et al.*, 1981a; 1982; Skare and Schrotel, 1984; Brambilla *et al.*, 1981, Barbin *et al.*, 1981, Bermudez *et al.*, 1982, Petzold and Swenberg, 1978). On a cautionary note, with studies where high doses of chemicals are employed to elicit a positive response, it is important to consider that a more general toxic response involving cell damage and release of lysosomal nuclease enzymes may contribute to any SSBs observed.

In conclusion, it is considered that the detection of SSBs by the alkaline elution assay may not be sensitive for all classes of carcinogens. For this assay, it is the kinetics of the production and repair of DNA strand breaks and the degree of coupling of the enzyme processes involved that determine whether a positive response is observed. Such parameters will be determined largely by the nature of the DNA adduct/damage produced.

Finally, it has recently been suggested that the alkaline elution technique detects not only SSBs but also changes in the conformation of the DNA molecule (Parodi *et al.*, 1983; Nicolini *et al.*, 1983). It will be exciting

Table 1. *Alkaline elution data for DNA from the livers of rats administered 2AAF, 4AAF, BP, PYR*

Chemical (dose)[a]	10^2 times elution rate constant, K,[b] at various times after dosing: 4 h	12 h	24 h
Control	2.49 ± 1.09 (3)	3.05 ± 0.42 (3)	3.77 ± 0.82 (4)
2AAF (500 mg/kg)	1.89 ± 1.45 (3)	3.90 ± 0.92 (3)	3.52 ± 0.50 (7)
Control	nd[c]	4.46 ± 0.54 (3)	nd
4AAF (500 mg/kg)	nd	3.13 ± 1.48 (3)	nd
Control	nd	nd	2.82 ± 0.49 (3)
BP (250 mg/kg)	nd	nd	3.11 ± 0.24 (3)
Control	nd	nd	3.34 ± 0.60 (3)
PYR (250 mg/kg)	nd	nd	3.28 ± 0.38 (3)
Control	3.95 ± 0.52 (3)	nd	nd
DMN (40 mg/kg)	14.40 ± 2.92 (3)	nd	nd

[a] Chemicals were administered orally except DMN which was given intraperitoneally.
[b] Average elution rate constant K is given as the mean ± SD, with the number of animals used in parenthesis.
[c] Not determined.

to watch the investigation of the molecular mechanisms underlying the alkaline elution technique that must surely follow such observations.

5 Acknowledgements

We thank D. Hart, S. Beck and Miss L. Pollitt for technical assistance and Mrs. Jane Steele for secretarial assistance.

6 References

Barbin, A., J-C. Bereziat and H. Bartsch (1983) Evaluation of DNA damage by the alkaline elution technique in liver, kidney and lungs of rats and hamsters treated with N-nitrosodialkylamines, *Carcinogenesis*, **4**, 541–545.

Bermudez, E., J.C. Mirsalis and H.C. Eales (1982) Detection of DNA damage in primary cultures of rat hepatocytes following *in vivo* and *in vitro* exposure to genotoxic agents, *Environ. Mutagen.*, **4**, 667–679.

Borchert, P. and L.W. Wattenberg (1976) Inhibition of macromolecular binding of benzo(a)pyrene and inhibition of neoplasia by disulfiram in the mouse forestomach, *J. Natl. Cancer Inst.*, **57**, 173–179.

Bradley, M.O. and J.F. Sina (1984) Methods for detecting carcinogens and mutagens with the alkaline elution/rat hepatocyte assay, In B.J. Kilbey, M. Legator, W. Nichols and C. Ramel (Eds), *Handbook of mutagenicity test procedures*, pp. 71–82, Elsevier, Amsterdam.

Brambilla G., P. Carlo, R. Finollo, F.A. Bignone, A. Ledda and E. Cajelli (1983) Viscometric detection of liver DNA fragmentation in rats treated with minimal doses of chemical carcinogens, *Cancer Res.*, **43**, 202–209.

Brambilla G., M. Cavanna, A. Pino and L. Robbiano (1981) Quantitative correlation among DNA damaging potency of six N-nitroso compounds and their potency in inducing tumour growth and bacterial mutations, *Carcinogenesis*, **2**, 425–429.

Clayson, D.B, and R.C. Garner (1976) Carcinogenic aromatic amines and related compounds. In C.E. Searle (Ed.), *Chemical carcinogens*, American Chemical Society, Washington, DC, pp. 366–461.

Conney, A.H. (1982) Induction of microsomal enzymes by foreign chemicals and carcinogenesis by polycyclic aromatic hydrocarbons, *Cancer Res.*, **42**, 4875–4917.

Dogliotti, E., T. Lakhanisky, G.P. van der Schans and P.H.M. Lohman (1984) Molecular dosimetry of DNA damage caused by alkylation. II The induction and repair of different classes of single-strand breaks in cultured mammalian cells treated with ethylating agents, *Mutation Res.*, **132**, 41–49.

Erickson, L.C., R. Osieka, N.A. Sharkey and K.W. Kohn (1980) Measurement of DNA damage in unlabelled mammalian cells analysed by alkaline elution and a fluorimetric DNA assay, *Anal. Biochem.*, **106**, 169–174.

IARC (1983) *IARC monographs on the evaluation of carcinogenic risk of chemicals to man*, Vol. 3, Certain polycyclic aromatic hydrocarbons and heterocyclic compounds, Lyon.

Ireland, C.M., A. Eastman and E. Bresnick (1984) The influence of inhibitors on the repair of benzo(a)pyrene damaged DNA in hamster tracheal epithelial cells, *Carcinogenesis*, **5**, 187–191.

Kohn, K.W., L.C. Erickson, R.A.G. Ewig and C.A. Friedman (1976) Fractionation of DNA from mammalian cells by alkaline elution, *Biochemistry*, **15**, 4629–4637.

Kohn, K.W. and R.A.G. Ewig (1973) Alkaline elution analysis, a new approach to the study of DNA single strand interruptions in cells, *Cancer Res.*, **33**, 1849–1853.

Kornberg, A. (1974) DNA replication, W.H. Freeman and Company, San Francisco, pp. 609–622.

Kriek, E. (1972) Persistent binding of a new reaction product of the carcinogen N-hydroxy-N-2 acetylaminofluorene with guanine in rat liver *in vivo*, *Cancer Res.*, **32**, 2042–2048.

Lutz, W.K. (1979) *In vivo* covalent binding of organic chemicals to DNA as a quantitative indicator in the process of chemical carcinogenesis, *Mutation Res.*, **65**, 289–356.

Lutz, W.K., A. Viviani and C. Schlatter (1978) Nonlinear dose–response relationship for the binding of the carcinogen benzo(a)pyrene to rat liver DNA *in vivo*, *Cancer Res.*, **38**, 575–578.

Meehan, T., K. Straub and M. Calvin (1977) Benzo(a)pyrene diol epoxide covalently binds to deoxyguanosine and deoxyadenosine in DNA, *Nature (London)*, **269**, 725–727.

Nicolini, C., A. Belmont, S. Zietz, A. Maura, A. Pino, L. Robbiano and G. Brambilla (1983) Physico-chemical model for DNA alkaline elution: New experimental evidence and differential role of DNA length, chain flexibility and superpacking, *J. Theor. Biol.*, **100**, 341–357.

O'Connor, P.J., M.J. Capps and A.W. Craig (1973) Comparative studies of the hepatocarcinogen N-N-Dimethylnitrosamine *in vivo*: reaction sites in rat liver DNA and the significance of their relative stabilities, *Br. J. Cancer*, **27**, 153–166.

O'Connor, P.J., G.P. Margison and A.W. Craig (1975) Phosphotriesters in rat liver deoxyribonucleic acid after the administration of the carcinogen N,N-dimethylnitrosamine *in vivo, Biochem. J.*, **145**, 475–482.

Parodi, S., C. Balbi, M.L. Abelmoschi, M. Pala, P. Russo and L. Santi (1983) Studies on DNA damage: discordant responses of rate of DNA disentanglement (viscosimetrically evaluated) and alkaline elution rate, obtained for several compounds. Possible explanations of discrepancies, *Cell Biophys.*, **5**, 285–300.

Parodi, S., S. DeFlora, M. Cavanna, A. Pino, L. Robbiano, C. Bennicelli and G. Brambilla (1981a) DNA-damaging activity *in vivo* and bacterial mutagenicity of sixteen hydrazine derivatives as related quantitatively to their carcinogenicity, *Cancer Res.*, **41(4)**, 1469–1482.

Parodi, S., M. Taningher, P. Boero and L. Santi (1982) Quantitative correlations amongst alkaline DNA fragmentation, DNA covalent binding, mutagenicity in the Ames test and carcinogenicity for 21 compounds, *Mutation Res.*, **93**, 1–24.

Parodi, S., M. Taningher, M. Pala and L. Santi (1981b) Alkaline DNA fragmentation *in vivo*: borderline or negative results obtained respectively with 7, 12-dimethylbenz(a)anthracene and benzo(a)pyrene, *Tumori*, **67(2)**, 87–93.

Parodi, S., M. Taningher, P. Russo, M. Pala, M. Tamaro and C. Monti-Bragadin (1981c) DNA-damaging activity *in vivo* and bacterial mutagenicity of sixteen aromatic amines and azo-derivatives, as related quantitatively to their carcinogenicity, *Carcinogenesis*, **2**, 1317–1326.

Parodi, S., M. Taningher, L. Santi, M. Cavanna, L. Sciaba, A. Maura and G. Brambilla (1978) A practical procedure for testing DNA damage *in vivo* proposed for a prescreening of chemical carcinogens, *Mutation Res.*, **54**, 39–46.

Pearson, C.M., and K. Edwards (1982) An investigation of DNA repair phenomena induced by 4-chloromethylbiphenyl (4CMB), *Mutation Res.*, **100**, 211–214.

Petzold, G.L. and J.A. Swenberg (1978) Detection of DNA damage induced *in vivo* following exposure of rats to carcinogens, *Cancer Res.*, **38**, 1589–1594.

Skare, J.A. and K.R. Schrotel (1984) Alkaline elution of rat testicular DNA: detection of DNA strand breaks after *in vivo* treatment with chemical mutagens, *Mutation Res.*, **130**, 283–294.

Stewart, B.W. and E. Farber (1973) Strand breakage in rat liver DNA and its repair following administration of

cyclic nitrosamines, *Cancer Res.*, **33**, 3209–3215.
Stout, D.L. and F.F. Becker (1982) Fluorimetric quantitation of single-stranded DNA: A method applicable to the technique of alkaline elution, *Anal. Biochem.*, **127**, 302–307.
Straub, K.M., T. Meehan, A.L. Burlingame and M. Calvin (1977) Identification of the major adducts formed by reaction of benzo(a)pyrene diol epoxide with DNA *in vitro, Proc. Natl. Acad. Sci. (USA)*, **74**, 5285–5289.
Swenberg, J.A., G.L. Petzold and P.R. Harbach (1976) *In vitro* DNA damage/alkaline elution assay for predicting carcinogenic potential, *Biochem. Biophys. Res. Commun.*, **72**, 732–738.
Vu, V.T., M.E. Moller, P.H. Grantham, P.J. Wirth and S.S. Thorgeirsson (1985) Association between DNA strand breaks and specific DNA adducts in murine hepatocytes following *in vivo* and *in vitro* exposure to N-hydroxy-2-acetylaminofluorene and N-acetoxy-2-acetylaminofluorene, *Carcinogenesis*, **6**, 45–52.
Walker, I.G., and D.F. Ewart (1973) The nature of single-strand breaks in DNA following treatment of L-cells with methylating agents, *Mutation Res.*, **19**, 331–341.

Patrizia Russo[1], Mauro Pala[1], Annalisa Zunino[1], Antonella Melchiori[1], Leonardo Santi[2] and Silvio Parodi[2]

[1]*Istituto Nazionale per la Ricerea sul Cancro,* [2]*Department of Clinical and Experimental Oncology, University of Genoa, Genoa, Italy*

70

Alkaline elution of rat liver DNA for benzo(a)pyrene, pyrene, 2-acetylaminofluorene, and 4-acetylaminofluorene

1 Summary

Benzo(a)pyrene (BP), its analogue pyrene (PYR), 2-acetylaminofluorene (2AAF), and its analogue 4-acetylaminofluorene (4AAF) were tested for their capability to induce DNA damage. The most important effect observed was the crosslinking activity of BP. PYR was essentially negative. 2AAF and 4AAF appeared active in increasing the alkaline elution rate. An even greater elution rate was observed for both compounds after treatment with proteinase K. Treatment with 2AAF *in vivo* plus dimethylsulphate *in vitro* suggested a protein-DNA type of crosslinking activity for 2AAF.

2 Introduction

Farber and coworkers showed that both BP and 2AAF are initiators in the rat liver model (Tsuda *et al.*, 1980). As a consequence, rat liver appeared as a reasonable target organ for the study of BP, 2AAF and their analogs.

Using the alkaline elution technique we have examined both the induction of single-stranded breaks and induction of DNA–DNA or DNA–protein crosslinks, both for the two carcinogenic compounds (BP and 2AAF) and the two non carcinogenic or less carcinogenic analogs (PYR and 4AAF).

The investigation was carried out in two different strains of rats: Sprague-Dawley and Fischer male adults.

3 Materials and methods

Chemicals

The tested chemicals, BP, PYR, 2AAF and 4AAF were obtained from ICI Central Toxicology Laboratory; the other chemicals were obtained from the following sources: (ethylenedinitrilo)-tetraacetic acid disodium salt (Na_2EDTA), tetraethyl ammonium hydroxide (TEA-OH), bovine serum albumin and dimethylsulphate (DMS), E. Merck, (Darmstadt, F.R.G.); 35% N-lauroyl sarcosine sodium salt (Sarcosyl), Kock-Light (Colnbrook, U.K.); 3, 5-diaminobenzoic acid dihydrochloride (DABA), Fluka (Bucks, Switzerland); proteinase K, corn oil and carboxymethyl cellulose (CMC), Sigma Chemical Company (St. Louis, MO); 2-diacetylaminofluorene (DAAF), ICN Pharmaceuticals (New York).

Animals and treatments

Sprague-Dawley and Fischer 344 male albino rats (200–250 g body weight) purchased from Charles River Italia

(Como, Italy) were used. The animals were treated intraperitoneally (ip) with single doses of 2AAF, 4AAF, and DAAF suspended in 0.9% NaCl containing 1% CMC or by gavage with single doses of 2AAF, 4AAF, BP and PYR suspended in corn oil. The animals were sacrificed 4 h after treatment. Several controls, treated with vehicle alone, were also employed in every set of experiments.

Preparation of nuclei used for evaluation of DNA damage

Control and treated animals were killed by cervical dislocation and exanguinated. Rat liver was quickly removed and soaked in cold Merchant's solution (0.14 M NaCl, 1.47 mM KH_2PO_4, 2.7 mM KCl, 0.53 mM Na_2EDTA; pH 7.67).

Liver was briefly minced and then homogenized using a loosely fitting Potter-Elvehjem homogenizer. After sedimentation of the large tissue fragments, nuclei and single cells remaining in the supernatant were pelleted at 50 g for 2 min and resuspended in a suitable volume of Merchant's solution. Cells and crude nuclei were counted in a haemocytometer while suspended in cold Merchant's solution.

DNA damage/alkaline elution assay

The alkaline elution was performed according to Kohn *et al.* (1976) with minor modifications, as previously described (Parodi *et al.*, 1982). Approximately 10^6 cells, or nuclei, were loaded on a Millipore (mixed esters of cellulose) filter (25 mm diameter, 5-μm-pore size) and washed with cold Merchant's solution. The cells or nuclei were lysed on the filter at room temperature with 4.5 ml of 0.2% Sarcosyl, 2 M NaCl, 20 mM Na_2EDTA (pH 10), and the filter was washed with 2.5 ml of 20 mM Na_2EDTA (pH 10). Lysis volume and washing volume contained 0.5% and 2%, respectively, of total DNA. Single stranded DNA was eluted in the dark with 20 ml of 60 mM TEA-OH, 10 mM Na_2EDTA (pH 12.3) at a pump speed of 0.02 ml/min, by collecting 20 ml after 20 h of elution (slow elution). In about half of the experiments, the 20 ml were collected in 10 equal fractions to obtain elution profiles. In some cases, the pump speed was 0.2 ml/min, by collecting 13 ml after 60 min of elution (fast elution). At the end of elution, the DNA remaining on the filter was recovered by breaking up the filter in 4 ml of eluting solution. Eluted DNA and filter were assayed for DNA content by the following modification of the microfluorimetric technique of Kissane and Robins (1958). To each 1 ml aliquot of the collected fractions 0.1 ml of bovine serum albumin (2 mg/mm) and 0.1 ml of 100% (w/v) trichloroacetic acid were added. The samples were then refrigerated for several hours and the precipitated DNA was pelleted by centrifuging at 1750 g for 30 min at 4 °C. Supernatants were carefully decanted and 1 ml of 80% ethanol was layered on

Table 1. *DNA damage evaluated as single-strand breaks or DNA crosslinkage, induced by BP and PYR in liver DNA of Sprague-Dawley rats*

Compounds	Dosage (mg/kg)	No. of exp.[a] No. anim.[b]	No. elut.	$K \cdot 10^2$ average (± SD)	% retained DNA average (± SD)	Statistical significance[c]
BP	52	4	7	2.46 (0.50)	61.3 (5.7)	N.S.[d]
BP + DMS[e]	52	3	6	3.43 (0.37)	50.5 (3.7)	$p < 0.002$[f]
PYR	136	4	7	3.87 (1.39)	47.6 (13.2)	$p < 0.002$[d]
PYR + DMS[e]	136	2	4	15.91 (3.61)	5.0 (3.3)	N.S.[f]
DMS[e]		3	6	13.49 (2.26)	7.3 (3.2)	$p < 0.002$[d]
Controls		5	8	1.75 (0.47)	70.8 (7.0)	

[a] Usually more than one elution was performed for each animal: Total experimental variability is partially animal dependent and partially elution experiment dependent.
[b] The animals were sacrificed 4 h after treatment.
[c] Nonparametric Mann-Whitney test (Siegel, 1956); N.S. = $p > 0.05$.
[d] Against controls.
[e] Nuclei were incubated at 22 °C for 30 min with DMS at the concentration of 600 μg/ml.
[f] Against DMS.

Compounds	Dosage (mg/kg)	Individual value $K \cdot 10^2$ (ml^{-1})	% retained DNA
BP	52	2.60, 2.08, 2.19, 2.09, 2.47, 2.31, 3.50.	59.4, 65.9, 64.5, 65.8, 61.0, 63.0, 49.6.
BP + DMS	52	3.83, 3.47, 3.19, 3.78, 2.87, 3.40.	46.4, 49.9, 52.3, 46.9, 56.3,50.6.
PYR	136	3.82, 5.49, 4.50, 3.09, 2.66, 1.92, 5.58.	46.5, 33.3, 40.5, 53.8, 58.7, 67.8, 32.7.
PYR + DMS[c]	136	11.88, 16.90, 20.37, 14.50.	9.3, 3.4, 1.7, 6.5.
DMS[a]		12.38, 12.50, 13.51, 15.97, 10.35, 16.22.	8.4, 8.2, 6.7, 4.3, 12.6, 3.9.
Controls	—	2.02, 1.84, 1.81, 1.95, 2.15, 1.21, 2.15, 0.84.	66.7, 69.2, 69.5, 67.7, 65.0, 78.5, 65.0, 84.5.

[a] Nuclei were incubated at 22 °C for 30 min with DMS at the concentration of 600 μg/ml.

each pellet. After a second centrifugation, the pellets were air dried for about 3 h at 40 °C. Thirty microliters of a 40% (w/v) aqueous solution of DABA were added to each sample, and the tubes were incubated for 30 min at 70 °C. Due to the addition of bovine serum albumin as a carrier, a good pellet was formed on the tip of the tube. The recovery and dosage of comparable amount of added DNA was always 80% or greater, even if the pellet on the tip of the tube was covered with 30 μl of DABA. After cooling, 1.5 ml of 0.6 N perchloric acid was added to each tube. The fluorescence was read at 500 nm with an excitation wavelength of 400 nm in a Perkin-Elmer model 3000 fluorimeter. The blank readings were made from tubes containing 1 ml of eluting solution subjected to the same procedure. DABA is fluorescent by itself, and the measurement is that of differences in fluorescence between DNA containing samples and corresponding blanks. For the first fractions (richer in DNA), DABA represents 25–33% of fluorescence. At our wavelengths, no other significant contribution to fluorescence could be found.

Another set of experiments was performed for evaluating the possible crosslinking effect of the 4 compounds administered *in vivo*, considering the principles described by Ewig and Kohn (1978); both DNA–DNA and DNA–protein crosslinks were investigated, the common procedure for the 2 types of investigation were the following. Rats, treated by ip injection or by gavage with a single dose of each compound, were sacrificed 4 h after treatment. Liver nuclei were prepared. Before the lysis and DNA elution process had been performed, the nuclei were incubated with DMS dissolved in Merchant's solution (600 μg/ml/1 $\cdot 10^6$ nuclei) at 22 °C for 30 min, as previously reported (Russo *et al.*, 1982). This nuclear suspension was loaded on the filter and the incubation medium was pumped out. At this stage, to investigate the DNA–protein crosslinkage, the nuclei were lysed in standard conditions (0.2% Sarcosyl, 2 M NaCl, 20 mM Na_2EDTA, pH 10). To investigate DNA–DNA crosslinkage, the nuclei were incubated for 30 min at room temperature in the standard lysing solution added with proteinase *K* (0.5 mg/ml lysing solution). In both cases, the nuclear lysates were then washed on filters with 20 mM Na_2EDTA, and the DNA was eluted from the filters and dosed as described above in the standard elution assay. When slow elution is performed, the filter holders are covered with a tin-foil cap. The initial pH measured for eight filter holders was in a range of 12.3–12.2. The final pH after 20 h was in a range of 12.2–11.9.

All the results obtained with the alkaline elution assay were expressed as K-values. K is given by the following formula:

$$K = \frac{-\ln\,(\text{fraction of DNA retained on filter})}{V}, \text{ where}$$

Table 2. *DNA damage, evaluated as single-strand breaks or DNA crosslinkage, induced by BP and PYR in liver DNA of Fischer 344 rats*

Compounds	Dosage (mg/kg)	No. of exp.[a] No. anim.[b]	No. elut.	K·10^2 average (±SD)	% retained DNA average (±SD)	Statistical significance[c]
BP	52	3	6	1.83 (0.54)	69.6 (6.99)	N.S.[d]
	200	3	6	1.64 (0.18)	72.0 (2.54)	N.S.[d]
BP + DMS[e]	52	3	6	5.47 (1.34)	33.9 (9.75)	$p < 0.002$[f]
PYR	136	3	6	1.09 (0.56)	80.7 (8.63)	N.S.[d]
	500	3	6	2.98 (0.47)	55.1 (5.15)	N.S.[d]
DMS[e]		2	6	14.13 (1.63)	6.2 (1.98)	$p < 0.002$[d]
Controls		5	9	1.61 (0.76)	73.2 (11.57)	

[a] Usually more than one elution was performed for each animal. Total experimental variability is partially animal dependent and partially elution experiment dependent.
[b] The animals were sacrificed 4 h after treatment.
[c] Nonparametric Mann-Whitney test (Siegel, 1956); N.S. = $p > 0.05$.
[d] Against controls.
[e] Nuclei were incubated at 22 °C for 30 min with DMS at the concentration of 600 μg/ml.
[f] Against DMS.

Compounds	Dosage (mg/kg)	Individual value K·10^2(ml^{-1})	% retained DNA
BP	52	1.46, 1.52, 1.57, 1.64, 1.92, 2.89.	74.7, 73.7, 73.0, 72.0, 68.1, 56.1.
	200	1.43, 1.54, 1.55, 1.60, 1.84, 1.88.	75.0, 73.4, 73.3, 72.5, 69.2, 68.6.
BP + DMS	52	3.84, 4.08, 5.29, 5.93, 6.39, 7.30.	46.4, 44.2, 34.7, 30.5, 24.6, 23.2.
PYR	136	0.51, 0.66, 0.98, 1.08, 1.22, 2.10.	90.3, 87.6, 82.2, 80.5, 78.3, 65.7.
	500	2.37, 2.59, 2.95, 3.04, 3.32, 3.66.	62.2, 59.5, 55.4, 54.4, 51.4, 48.1.
DMS[a]		16.09, 15.85, 12.88, 11.98, 13.59. 14.41.	4.0, 4.2, 7.6, 9.1, 6.6, 5.6
Controls		1.52, 1.55, 1.24, 1.26, 1.51, 2.52. 2.48, 2.31, 0.10.	73.7, 73.2, 78.0, 77.6, 73.9, 60.4. 60.9, 63.0, 97.9.

[a] Nuclei were incubated at 22 °C for 30 min with DMS at the concentration of 600 μg/ml.

K is the average elution rate constant (ml^{-1}) of DNA and V is the eluted volume in ml. As a first approximation, K is directly proportional to the number of single strand breaks. Moreover, the above formula reflects the assumption of a first order kinetics for DNA elution, always as a first approximation (Kohn *et al.*, 1976). The percentage of DNA retained on the filter was also given.

4 Results

The dosages used for BP and PYR were 1/5 of LD_{50} at 4 days, as determined by Salamone (1981) and also approximately 3.8 times that dosage.

Literature data (Fornace and Little, 1979, Ireland *et al.*, 1982) suggested that DNA fragmentation and DNA crosslinking could be studied. As shown in Table 1, 2 and in Figure 1, BP caused DNA crosslinking. Treatment schedules are specified in Materials and Methods. The breaks induced by DMS, probably through depurination of N-7 methylguanine and subsequent rapid transformation of apurinic sytes into breaks in alkali, are masked to a significant extent by BP. This observation confirms the work of Ireland *et al.* (1982). We have not investigated if this crosslinking is of the DNA–protein type, as suggested by the above authors.

In Sprague-Dawley rats, PYR had a weak effect by itself, especially considering that we used the slow, more sensitive, elution method. PYR also had no crosslinking effect. In Fischer rats, PYR was completely negative.

2AAF had been tested in a previous study with the fast, less sensitive elution method at the dosage of 500 mg/kg ip in CMC and was found essentially negative (Parodi *et al.*, 1981b). However, Petzold and Swenberg (1978) previously had reported a weak but positive result at the doses of 10–25 mg/kg ip. We reasoned that extremely high dosages were probably saturating the metabolic activation system, and to detect a possible effect it was more reasonable to increase the sensitivity of the method than to increase the dosage. Therefore, we worked at the dosage of 20 mg/kg ip, but with the slow, about 5 times more sensitive, alkaline elution method.

Fornace and Little (1979), showed that part of the breaks are masked by a DNA-protein crosslinking effect. As a consequence, we used proteinase K, which is capable of reducing this effect (Ewig and Kohn, 1978). The results in Tables 3 and 4 show that with both Sprague-Dawley and Fischer rats, 2AAF and 4AAF can be detected clearly as positive by the slow alkaline elution method; the effect is even larger in the presence of proteinase K. Without proteinase K, 2AAF seems capable of reducing significantly the DNA damage induced by DMS. This reduction seems to confirm the presence of a DNA–protein crosslinking effect as suggested by Fornace and Little (1979).

We have confirmed that little can be seen with 2AAF by the less sensitive fast elution method, without proteinase K.

Figure 2 shows the profiles of 2AAF, 4AAF, and another analogue DAAF, that we have tested in Sprague-Dawley rats. All the profiles are in the presence of proteinase K. The presence of an activity for all 3 compounds seems clear. Treatment schedules are reported in Materials and Methods. All the samples are in the presence of proteinase K (0.5 mg/ml in lysing solution for 30 min). Each profile represents the average value (± S.E.) of 4–8 experiments.

Liver sections from animals treated with BP, PYR, 2AAF and 4AAF were histologically examined at the dosage tested, 4 h after treatment, with standard haematoxylin and eosin staining. No toxic effects were detected.

5 Discussion

The data on carcinogenicity by BP, PYR, 2AAF, and 4AAF seem to suggest that only BP and 2AAF are carcinogens (N.C.I., 1972–73, 1978).

We were able to find some alkaline elution data for BP. The compound was essentially negative, as a direct inducer of breaks, *in vivo* in the liver (Parodi *et al.* 1981a),

Figure 1. Crosslinking effect of BP on liver DNA of Sprague-Dawley rats: (○) Controls; (●) BP; (■) DMS; (□) BP + DMS. Each profile represents the average value (± S.E.) of 4–8 experiments.

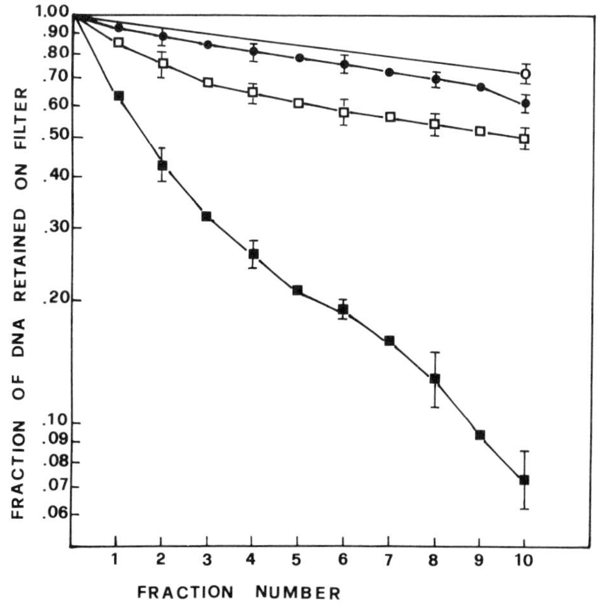

Figure 2. Elution profiles of liver DNA of Sprague-Dawley rats: (○) Controls; (□) 2AAF; (●) 4AAF; (■) DAAF. Each profile represents the average value (± S.E.) of 4–8 experiments.

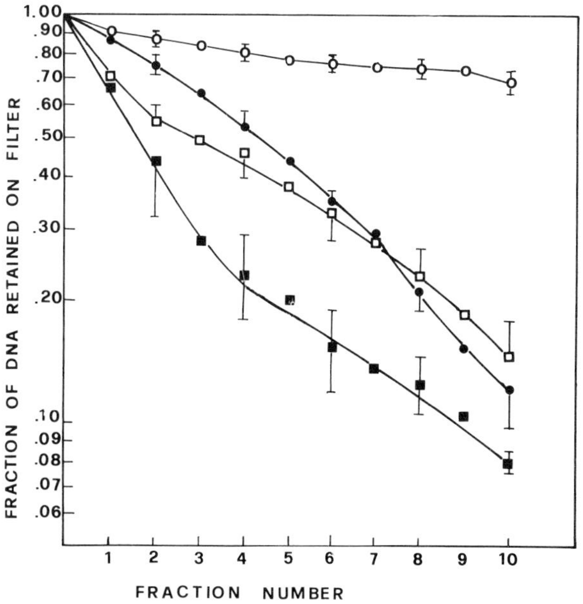

Table 3. *DNA damage evaluated as single-strand breaks or DNA-protein crosslink or DNA–DNA crosslink induced by 2AAF and 4AAF in liver DNA Sprague-Dawley rats*

Compounds		Dosage (mg/kg)	No. of exp.[a] No. anim.[b]	No. elut.	K·10² average (±SD)	% retained DNA average (±SD)	Statistical significance[c]
2AAF	+ prot.K	20	4	4	9.98(2.69)	14.85(5.98)	$p = 0.002$[d]
	− prot.K		4	4	5.86(0.79)	31.20(4.55)	$p = 0.005$[e]
DMS[f]	+ prot.K		4	4	15.14(4.18)	7.50(3.72)	$p = 0.002$[d]
	− prot.K		4	4	13.72(4.66)	8.32(5.54)	$p = 0.005$[e]
2AAF + DMS	+ prot.K	20	4	4	13.83(6.20)	10.35(10.38)	N.S.[g]
	− prot.K		4	4	7.99(1.79)	21.25(7.88)	$p = 0.014$[h]
4AAF	+ prot.K	20	6	12	11.59(3.42)	12.38(9.58)	$p = 0.002$[d]
	− prot.K		3	6	8.88(1.48)	17.60(5.54)	$p = 0.001$[e]
Controls	+ prot.K		6	8	1.88(0.50)	68.73(6.75)	
	− prot.K		5	6	1.05(0.83)	81.80 (12.65)	
2AAF[i]	+ prot.K	20	2	4	4.69(0.38)	54.40(2.69)	$p = 0.014$[j]
	− prot.K		5	7	2.29(1.17)	74.93(11.26)	N.S.[K]
Controls[i]	+ prot.K		2	4	2.19(1.13)	75.80(11.62)	
	− prot.K		2	3	1.79(1.00)	79.70(10.74)	

[a] Usually more than one elution was performed for each animal. Total experimental variability is partially animal dependent and partially elution experiment dependent.
[b] The animals were sacrificed 4 hr after treatment.
[c] According to the non parametric Mann-Whitney test (Siegel, 1956); N.S. = $p > 0.05$.
[d] Against controls + proteinase K (slow elution).
[e] Against controls − proteinase K (slow elution).
[f] Nuclei were incubated at 22 °C for 30 min with DMS at the concentration of 600 μg/ml.
[g] Against DMS + proteinase K.
[h] Against DMS − proteinase K.
[i] At pump speed of 0.2 ml/min (fast elution).
[j] Against controls + proteinase K (fast elution).
[k] Against controls − proteinase K (fast elution).

Compounds		Dosage (mg/kg)	Individual value K·10² (ml⁻¹)	% retained DNA
2AAF	+ prot.K	20	9.19, 13.98, 8.38, 8.38.	15.9, 6.1, 18.7, 18.7.
	− prot.K	20	6.99, 5.16, 5.72, 5.57.	24.7, 35.3, 31.8, 32.8.
DMS	+ prot.K		10.24, 14.97, 14.88, 20.45.	12.9, 5.0, 5.1, 4.0.
	− prot.K		20.09, 10.01, 14.32, 10.48.	1.8, 13.5, 5.7, 2.3
2AAF + DMS	+ prot.K	20	20.99, 10.64, 16.62, 7.05.	1.5, 11.9, 3.6, 4.4.
	− prot.K	20	8.68, 7.64, 9.93, 5.69.	17.5, 21.7, 13.7, 32.0.
4AAF	+ prot.K	20	11.22, 11.17, 8.98, 11.13. 5.68, 5.89, 15.85, 14.23, 14.32, 11.66, 12.95, 15.97.	10.6, 10.7, 16.6, 10.8, 32.1, 30.8, 4.2, 5.8, 5.7, 9.7, 7.5, 4.1.
	− prot.K	20	10.60, 9.19, 8.02, 6.49, 10.08, 8.85.	12.0, 15.9, 20.1, 27.3, 13.3, 17.0.
Controls	+ prot.K		1.60, 1.24, 1.47, 1.52, 2.63, 2.28, 2.37, 1.98.	72.5, 77.9, 74.4, 73.7, 59.0, 63.3, 62.2, 67.3.
	− prot.K		0.43, 0.56, 0.53, 0.54, 1.90, 2.31.	91.6, 89.4, 89.8, 88.7, 68.3, 63.0.
2AAF	+ prot.K	20 (fast el.)	4.23, 4.68, 5.16, 4.68.	57.7, 54.4, 51.1, 54.4,
	− prot.K	20 „	1.86, 0.75, 1.29, 3.71, 1.88, 2.78, 3.78.	78.5, 90.7, 84.6, 61.7, 78.3, 69.6, 61.1.
Controls	+ prot.K	„	0.60, 3.29, 2.42, 2.45.	92.4, 65.2, 73.0, 72.7.
	− prot.K	„	0.65, 2.16, 2.55.	91.9, 75.5, 71.7.

with the fast elution technique, and *in vitro* (Eastman *et al.*, 1981), using hamster tracheal epithelial cells with the slow elution technique. A very important DNA–protein crosslinking effect was observed *in vitro*, with the same type of cells mentioned before (Ireland *et al.*, 1982). Our data confirm what was reported in the literature. BP was very weak or negative as a direct inducer of breaks with the slow alkaline elution technique, but an important crosslinking effect could be demonstrated.

We were not able to find alkaline elution data for PYR in the literature. A very weak direct break inducing effect was observed only with Sprague-Dawley rats, but there was no DNA crosslinking effect.

We found 3 reports in the literature for 2AAF. Petzold and Swenberg (1978) had reported a weak effect in liver DNA *in vivo* for a dosage of 20 mg/kg ip, using the fast elution technique. With very similar conditions, but for a dosage of 500 mg/kg, we had previously reported an essentially negative result (Parodi *et al.*, 1981b). In both cases, only direct breaks without proteinase *K* were investigated. Fornace and Little (1979) had reported a DNA–protein crosslinking effect for 2-acetoxy-acetylaminofluorene.

While our results remained substantially negative using the fast elution technique as in previous experiments (Parodi *et al.*, 1981b), this time, at the dosage level of 20 mg/kg i.p., a positive result was obtained in slow elution. We have also confirmed DNA–protein crosslinking with 2AAF observed initially by Fornace and Little (1979). The addition of proteinase *K* allows the detection of a weak but positive result even with the fast alkaline elution technique.

For 4AAF, we are not aware of results reported in literature using the alkaline elution technique. Our results are clearly positive, both with and without proteinase *K*.

An element of caution should be added to the interpretation of results obtained with the alkaline elution technique. We previously reported that a different viscosimetric method for detecting the same type of damage was in clear disagreement with the alkaline elution results for a significant number of compounds (Parodi *et al.*, 1983). An example of these discrepancies is also reported by Brambilla *et al.* (1983); 2AAF showed a similar sensitivity using the fast alkaline elution method and a viscosimetric method that for many alkilating agents is about 50 times more sensitive than alkaline elution. Using the above mentioned viscosimetric method, we could confirm the almost negligible effect of 2AAF in this system (unpublished data). Moreover, we have observed that incubating the nuclei with different types of solutions not containing alkylating or arylating agents can reversibly modify the alkaline elution rate (Parodi *et al.*, 1985).

Table 4. *DNA damage, evaluated as single-strand breaks or DNA-protein crosslink, induced by 2AAF and 4AAF in liver of Fischer rats*

Compounds		Dosage (mg/kg)	No. of exp.[b] No. anim.[a]	No. elut.	$K \cdot 10^2$ average ($\pm$SD)	% retained DNA average ($\pm$SD)	Statistical significance[c]
2AAF	+ prot.K	20	3	6	11.98 (2.31)	9.95 (4.59)	$p < 0.002$[d]
	− prot.K	20	3	6	7.81 (2.60)	23.13 (10.55)	$p < 0.002$[e]
	+ prot.K	200	3	6	13.74 (4.26)	8.48 (6.61)	$p < 0.002$[d]
4AAF	+ prot.K	20	3	6	16.02 (1.57)	4.22 (1.23)	$p < 0.002$[d]
	− prot.K	20	3	6	6.10 (1.50)	30.55 (8.54)	$p < 0.002$[e]
	+ prot.K	200	3	6	11.37 (2.75)	11.45 (5.14)	$p < 0.002$[d]
Controls	+ prot.K		3	6	1.59 (0.20)	72.63 (2.91)	
	− prot.K		5	9	1.61 (0.76)	73.18 (11.57)	

[a] The animals were sacrificed 4 hr after treatment.
[b] Usually more than one elution was performed for each animal. Total experimental variability is partially animal dependent and partially elution experiment dependent.
[c] According to the non parametric Mann-Whitney test (Siegel, 1956); N.S. = $p > 0.05$.
[d] Against controls + proteinase K (slow elution).
[e] Against controls − proteinase K (slow elution).

Compounds		Dosage (mg/kg)	Individual value $K \cdot 10^2$ (ml^{-1})	% retained DNA
2AAF	+ prot.K	20	8.85, 10.04, 11.22, 13.09, 13.90, 14.78.	17.0, 13.4, 10.6, 7.3, 6.2, 5.2.
	− prot.K	20	4.51, 6.60, 7.17, 7.89, 8.35, 12.38.	40.5, 26.7, 23.8, 20.6, 18.8, 8.4.
	+ prot.K	200	8.22, 10.20, 12.56, 14.41, 18.44, 18.64.	19.3, 13.0, 8.1, 5.6, 2.5, 2.4.
4AAF	+ prot.K	20	14.41, 14.51, 15.50, 16.09, 17.21, 18.44.	5.6, 5.5, 4.5, 4.0, 3.2, 2.5.
	− prot.K	20	4.45, 4.68, 6.08, 6.22, 6.60, 8.60	41.1, 39.2, 29.6, 28.8, 26.7, 17.9
	+ prot.K	200	8.60, 9.10, 10.51, 11.56, 12.26, 16.22.	17.9, 16.2, 12.2, 9.9, 8.6, 3.9.
Controls	+ prot.K		1.43, 1.69, 1.97, 1.48, 1.46, 1.55.	75.1, 71.2, 67.4, 74.3, 74.6, 73.2.
	− prot.K		1.55, 1.24, 1.51, 1.52, 2.48, 1.26, 0.10, 2.31, 2.52.	73.2, 78.0, 73.9, 73.7, 60.9, 77.6, 97.9, 63.0, 60.4.

6 Conclusions

We have found:

1. BP is positive, but only as a crosslinking agent.
2. PYR is very weak or negative, both in terms of DNA fragmentation and crosslinking.
3. 2AAF is positive, both in terms of DNA fragmentation and DNA protein crosslinking.
4. 4AAF is positive in terms of DNA fragmentations, the crosslinking activity of 4AAF was not tested.
5. Substantial agreement was observed for the two strains of rats examined.

7 Acknowledgements

This research was supported by grants N. 84.00719.44 and N. 84.02542.56 from Consiglio Nazionale delle Ricerche.

We are grateful to Dr G. Nicolò and Dr L. Rossi for their help in histological examination of specimens from controls and treated tissues and to Miss L. Vigo for her excellent technical assistance.

8 References

Brambilla, G., P. Carlo, R. Finollo, F.A. Bignone, A. Ledda and E. Cajelli (1983) Viscosimetric detection of liver DNA fragmentation in rats treated with minimal doses of chemical carcinogens, *Cancer Res.*, **43**, 202–209.

Eastman, A., B.T. Mossman and E. Bresnick (1981) Formation and removal of Benzo(a)pyrene adducts of DNA in hamster tracheal epithelial cells, *Cancer Res.*, **41**, 2605–2610.

Ewig, R.A.G. and K.W. Kohn (1978) DNA–protein crosslinking and DNA interstrand crosslinking by haloethylnitrosourea in L1210 cells, *Cancer Res.*, **38**, 3197–3202.

Fornace, A.J. and J.B. Little (1979) DNA–protein crosslinking by chemical carcinogens in mammalian cells, *Cancer Res.*, **39**, 704–710.

Ireland, C.M., A. Eastman and E. Bresnick (1982) DNA–protein crosslinkings induced in a hamster tracheal epithelial cell line by Benzo(a)pyrene, *Biochem. Biophys. Res. Comm.*, **109**, 1291–1296.

Kissane, J.M. and E. Robins (1985) The fluorometric measurement of deoxyribonucleic acid in animal tissues with special reference to the central nervous system, *J. Biol. Chem.*, **233**, 184–188.

Kohn, K.W., L.C. Erichson, R.A.G. Ewig and C.A. Friedman (1976) Fractionation of DNA from mammalian cells by alkaline elution, *Biochemistry*, **15**, 4629–4637.

National Cancer Institute (1972–73, 1978) *Survey of compounds which have been tested for carcinogenic activity*, Government Printing Office, Washington D.C.

Parodi, S., C. Balbi, M.L. Abelmoschi, M. Pala, P. Russo and L. Santi (1983) Studies of DNA damage: discordant responses of rate of DNA disentanglement (viscosimetrically evaluated) and alkaline elution rate, obtained for several compounds; possible explanations of the discrepancies, *Cell Biophys.*, **5**, 285–300.

Parodi, S., M. Pala, P. Russo, A. Zunino, C. Balbi, A. Albini, F. Valerio, M.R. Cimberle and L. Santi (1982) DNA damage in liver, kidney, bone marrow, and spleen of rats and mice treated with commercial and purified Aniline as determined by alkaline elution assay and sister chromatid exchange induction, *Cancer Res.*, **42**, 2277–2283.

Parodi, S., P. Russo, M. Pala, G.F. Fassina, A. Melchiori, C. Balbi, R. Roner and L. Santi (1985) Different types of evidence suggesting that DNA alkaline elution rate can be increased not only as an effect of DNA damage, but also as an effect of changes in chromatin conformation, Proceeding "XVII Congresso della Società Italiana di Patologia, Firenze 11–13 Aprile", in press.

Parodi, S., M. Taningher, M. Pala and L. Santi (1981a) Alkaline DNA fragmentation *in vivo*: borderline or negative results obtained respectively with 7, 12-Dimethylbenz(a)anthracene and Benzo(a)pyrene, *Tumori*, **67**, 87–93.

Parodi, S., M. Taningher, P. Russo, M. Pala, M. Tamaro and C. Monti-Bragadin (1981b) DNA–damaging activity *in vivo* and bacterial mutagenicity of sixteen aromatic amines and azo-derivatives, as related quantitatively to their carcinogenicity, *Carcinogenesis*, **2**, 1317–1326.

Petzold, G.L. and A.J. Swenberg (1978) Detection of DNA damage induced *in vivo* following exposure of rats to carcinogens, *Cancer Res.*, **38**, 1589–1594.

Russo, P., M. Pala, L. Ottaggio, D. Vecchio, L. Santi and S. Parodi (1982) Detectability *in vivo* of stabilized intercalating agents with the alkaline elution technique. Comparison with *in vivo* sister chromatid exchanges induction, *J. Appl. Toxicol.*, **3**, 58–62.

Salamone M.F. (1981) Toxicity of 41 carcinogens and noncarcinogenic analogs. In F.J. de Serres and J. Ashby (Eds), Evaluation of Short-Term Tests for Carcinogens. Report of the International Collaborative Program. *Progress in Mutation Research*, Elsevier/North Holland Vol. 1, pp. 682–686.

Siegel, S. (1956) *Non parametric statistic for the behavioral sciences*, McGraw-Hill Book Company, inc., N.Y., pp. 116–127.

Tsuda, H., G. Lee and E. Farber (1980) Induction of resistant hepatocytes as a new principle for a possible short term *in vivo* test for carcinogens, *Cancer Res.*, **40**, 1157–1164.

Hyman C. Birnboim
Health Sciences Division, Chalk River Nuclear Laboratories, Chalk River Ontario, Canada

71

Use of the fluorometric analysis of DNA unwinding (FADU) assay to detect DNA strand breaks in the liver of CD-1 mice exposed to carcinogens and a repair inhibitor

1 Summary

The FADU technique for measuring DNA strand breaks is a procedure that employs a fluorescent dye to determine the rate of DNA unwinding on exposure of crude cell lysates to alkaline conditions. This technique is particularly useful where cellular DNA cannot be readily radiolabelled, such as in the case of nongrowing cells or cells freshly removed from an animal. We have previously applied the method to the analysis of DNA in soft organs such as spleen and thymus, from which cell suspensions can readily be prepared. In the present study, the liver was examined as a target organ for DNA damage by the carcinogens under investigation, so conditions for applying FADU to this organ were developed. As a positive control, we were able to demonstrate that DNA strand breakage in liver cells was induced by treatment of whole animals (CD-1 female mice) with a direct acting genotoxin (MMS). By contrast, only one of the test carcinogens (2AAF) at a moderately high dose (50 mg/kg) gave a clear indication of DNA strand breakage. Therefore, an alternative strategy was tested. On the assumption that base damage but not alkali-labile lesions or frank strand breaks were occurring, a DNA repair inhibitor (adenine arabinoside) was used in combination with the test carcinogens to determine if enzymatically induced breaks could be revealed. Both carcinogens (benzo(a)pyrene and 2-acetylaminofluorine) but not their noncarcinogenic counterparts, gave positive results when tested in this way.

2 Introduction

Our participation in the IPCS collaborative study on *in vivo* short-term tests for genotoxins and carcinogens has involved the measurement of DNA strand breaks in the livers of mice exposed to the 4 test compounds. The method chosen for detecting strand breaks was Fluorometric Analysis of DNA Unwinding (FADU), which had been developed earlier in our laboratory (Birnboim and Jevcak, 1981). Although previously used successfully for examining *in vivo* DNA strand-break damage in cells from soft organs such as thymus, spleen (McLean *et al.*, 1982), lymph node and bone marrow (McLean and Birnboim, unpublished), FADU had not been used to examine the integrity of DNA in a solid organ such as the liver. Initially, we encountered considerable difficulty in obtaining reasonably intact DNA from the liver of the mouse, as compared with the relative ease of obtaining intact DNA from the soft organs listed above.

The method we developed proved suitable for the present study, but simultaneous analysis of controls was necessary since day-to-day variability in preparing liver extracts was encountered.

Positive controls were readily obtained with MMS with liver DNA using the FADU method. Despite the use of several different doses, routes of administration, and times of treatment, only one of the 4 test chemicals (at a moderately high dose) gave an indication of causing DNA strand breaks. We explored the possibility of using *in vivo* administration of a repair inhibitor to enhance the number of detectable breaks on the assumption that stable base adducts were being formed and removed by an excision repair process. The inhibitor was expected to allow incision but not ligation to occur, leading to an accumulation of a strand break corresponding to each base adduct. The repair inhibitor that proved to be most effective in potentiating the number of strand breaks produced by MMS was adenine arabinoside; it also proved effective in revealing genotoxicity by the 2 carcinogenic test chemicals.

3 Materials and methods

Preparation of test chemicals

The 4 test chemicals used were supplied by the organizers of the study. The chemicals used were benzo(a)pyrene (BP), pyrene (PYR), 2-acetylaminofluorene (2AAF), and 4-acetylaminofluorene (4AAF). All were dissolved or suspended in corn oil for administration by gavage. Adenine-9-β-D-arabinofuranoside (ara-A) was supplied by Sigma Chemical Company; it was dissolved in a balanced salt solution (BSS: 137 mM NaCl, 5 mM KCl, 1.8 mM $CaCl_2$, 0.8 mM $MgCl_2$, 10 mM HEPES, pH 7.4) immediately before intraperitoneal (ip) administration.

Treatment schedule

At least 40 CD-1 female mice (approximately 25 g body weight, 4–8 weeks of age) were used to test each dose of the 4 chemicals under investigation. Eight animals were used per day and the experiment was repeated 5 or more times, usually on consecutive days. The 8 animals were divided into 4 groups of 2: C, solvent controls, received corn oil and BSS only; A, ara-A controls, received corn oil and ara-A; T, test animals, received test compound and BSS; A + T, ara-A + test animals, received ara-A and a test compound. The animals were allowed drinking water only for 16 h prior to the experiment. At 8:00 a.m., the animals received 0.2 mL of corn oil (with or without the test chemical) intragastrically with a stainless steel tube. Thirty minutes later, they received an ip injection of BSS (with or without 50 mg/kg ara-A). At 11:00 a.m. (3 h after the first treatment), they were killed by cervical dislocation and their livers were removed.

Preparation of liver homogenate for DNA damage analysis

The homogenising solution (HS: 75 mM NaCl, 24 mM EDTA, pH 7.5) and the dissecting tools were well chilled in ice before use. A special borer was constructed consisting of a sharpened piece of stainless steel tubing (2.5-mm i.d.) and set into an aluminum handle that acted as a heat sink, helping to keep the core sample cold. Animals were killed by cervical dislocation and immediately the peritoneal cavity was opened. A small wooden tongue depressor covered with a few thicknesses of Parafilm was used to support the liver while 4 core samples were rapidly taken with the borer from different parts of the liver. These samples were immediately ejected from the borer to a siliconised 15-mL conical glass centrifuge tube containing 4 mL of ice-cold HS, and 4 more samples were taken and combined. The following manipulations were carried out in a cold room at 0–4 °C. The samples were gently homogenised by passing up and down 5 times in a large bore plastic pipet (5 mL tip of a Gilson Pipetman). Large fragments were allowed to settle out for 5 min and about 80% of the supernate was removed with a # 16 (37 mm) needle and syringe to a fresh centrifuge tube. The suspension was centrifuged at 600 rpm (approximately 100 × g) for 5 min at 0 °C, the supernate discarded, and the pellet suspended in 3.0 mL of HS with the aid of a 5 mL plastic tip. This suspension, containing small cell clumps, single cells and nuclei, was used directly for the FADU procedure.

DNA damage analysis by FADU

The technique was carried out as described previously (Birnboim and Jevcak, 1981), except that HS was used in place of solution B (inositol solution). In brief, the cell/nuclei suspension was distributed equally to 12 tubes and lysed with a urea/detergent solution. An alkaline solution was added and DNA strand unwinding was allowed to occur for a 60-min period at 15 °C. Following this treatment, the samples were neutralized and the percentage of DNA remaining double-strand (D) was estimated, using a fluorescent dye, ethidium bromide. 'D' is a measure of the rate of DNA unwinding. Empirically, it has been found that log(D) is a linear function of strand breakage induced by treating cells with gamma radiation. Damage induced by a given agent is seen as a difference between the log(D) values of untreated control and treated samples, expressed as $Q_d = 100(\log(D_c) - \log(D))$. Q_d is correlated with the number of strand breaks per cell by assuming that 1 Gy of gamma rays produces 1200 strand breaks per cell. A more complete description of the derivation of this relationship is given in McWilliams *et al.* (1983). One Q_d unit corresponds to the induction of about 100 strand breaks per cell. Single-strand breaks, double-strand breaks and alkali-labile lesions are detected but not distinguishable by this method.

Statistical analysis of the results

Each day's experiment involved 8 animals, grouped into C, A, T, and A + T pairs (see treatment schedule). The '100·log (D)' values for the duplicates were averaged and the Q_d values calculated. A block of 5–7 experiments was carried out with each chemical at each of 2 doses. Comparisons between blocks of experiments were carried out using a nonpaired T test.

4 Results and discussion

Earlier studies with FADU indicated that *in vivo* DNA damage could readily be detected in organs such as spleen and thymus following irradiation of mice (McLean et al., 1982). Preliminary results with MMS indicated that this direct-acting alkylating agent was able to induce dose-dependent damage to the DNA of mouse thymus, spleen, and bone marrow as detectable by the FADU technique (data not shown). On the other hand, neither BP nor 2AAF could produce such damage, so we concentrated on the liver, a possible target organ for the latter agents after oral administration.

The technique developed to prepare mouse liver homogenates for analysis of DNA damage is outlined above. Compared with the soft organs, initially it proved difficult to work with the liver and to prepare homogenates containing 'intact' DNA. Attention to details such as low temperature and speed of sampling after killing of the animal seemed to be most important. With this technique, DNA damage in liver could readily be detected

Table 1. *Study of DNA strand breakage in liver of mice treated with pyrene (PYR) and benzo(A)pyrene (BP), with or without adenine arabinoside (A)*

	Comparison of [a]	$\bar{X}$	SEM	DF	Significance level
1.	C_{PYR} vs C_{BP} (low)	136.5[b] 138.8[b]	7.4 3.8	8	$p > 0.10$
2.	C_{PYR} vs C_{BP} (high)	145.5[b] 149.1[b]	6.9 4.4	12	$p > 0.10$
3.	A_{PYR} vs A_{BP} (low)	17.1 17.4	5.5 8.3	8	$p > 0.10$
4.	A_{PYR} vs A_{BP} (high)	8.8 11.7	2.6 2.9	12	$p > 0.10$
5.	C vs A	143.3[b] 130.1[b]	2.8 4.0	46	$p < 0.005$**
6.	C_{PYR} vs PYR (low)	136.5[b] 111.4[b,c]	7.4 31.3	8	$p > 0.10$
7.	C_{PYR} vs PYR (high)	145.5[b] 143.4[b]	6.9 6.2	12	$p > 0.10$
8.	C_{BP} vs BP (low)	138.8[b] 137.6[b]	3.9 3.9	8	$p > 0.10$
9.	C_{BP} vs BP (high)	149.1[b] 146.1[b]	4.4 4.6	12	$p > 0.10$
10.	A_{PYR} vs A + PYR (low)	17.1 19.2	5.5 7.9	8	$p > 0.10$
11.	A_{PYR} vs A + PYR (high)	8.8 15.6	2.6 3.8	12	$p < 0.10$
12.	A_{BP} vs A + BP (low)	17.4 30.5	8.3 22.2	8	$p > 0.10$
13.	A_{BP} vs A + BP (high)	11.7 35.7	2.9 7.9	12	$p < 0.005$**
14.	PYR vs BP (low)	25.2[c] 1.2	25.0 5.9	8	$p > 0.10$
15.	PYR vs BP (high)	2.1 3.1	2.9 3.7	12	$p > 0.10$
16.	A + PYR vs A + BP (low)	19.2 30.5	7.9 22.2	8	$p > 0.10$
17.	A + PYR vs A + BP (high)	15.6 35.7	3.8 7.9	12	$p < 0.025$*

[a] Abbreviations used: C-control animals; A, PYR and BP – animals who received ara-A, pyrene or benzo(a) pyrene (singly or in combination), respectively. Low-dose of 20 mg/kg; high-dose of 100 mg/kg. C_{PYR}, A_{PYR}, etc., refer to the untreated or ara-A treated animals used as controls for animals treated with pyrene or ara-A + pyrene, etc.

[b] Average of '100·log(D)' values. Other entries in this column are averages of 'Q_d' values. These units are defined in Materials and Methods.

[c] Large difference and large SEM attributable to high level of DNA damage in a single animal.

at 1/2, 1, and 2 h following ip injection of MMS at 100 mg/kg (data not shown). This 'positive control' indicated that the FADU technique could be applied to liver and we proceeded to examine the test chemicals.

The treatment schedule is given above. Animals were fasted for 16 h (water only) prior to administration of the test chemicals in corn oil via a stomach tube. Thirty min later, the animals received an ip injection of ara-A (50 mg/kg). Each day's experiment consisted of 4 sets (in duplicate): C received solvent only; A received ara-A only; T received a chemical at a particular dose to be tested; A + T received ara-A plus the test chemical. The duplicate results from each day's experiment were averaged. The data from each block of experiments (5–7 days) were grouped and compared by a nonpaired T test. These comparisons are shown in Table 1 and 2. Statistically significant differences are indicated by one or two asterisks.

Table 2. *Study of DNA strand breakage in liver of mice treated with 4-acetylaminofluorene (4) and 2-acetylaminofluorene (2), with or without adenine arabinoside (A)*

Comparison of [a]	$\bar{X}$	SEM	DF	Significance level
1. C_4 vs C_2 (low dose)	146.8[b] 143.8[b]	4.7 5.0	10	$p > 0.10$
2. C_4 vs C_2 (high dose)	150.2[b] 146.9[b]	3.1 2.5	10	$p > 0.10$
3. A_4 vs A_2 (low)	10.2 19.7	7.9 8.3	10	$p > 0.10$
4. A_4 vs A_2 (high)	5.4 10.2	2.0 3.4	10	$p > 0.10$
5. C vs A	147.0[b] 135.5[b]	1.8 3.5	46	$p < 0.005$**
6. C_4 vs 4 (low)	143.9[b] 140.4[b]	5.0 5.8	10	$p > 0.10$
7. C_4 vs 4 (high)	146.9[b] 147.0[b]	2.5 3.6	10	$p > 0.10$
8. C_2 vs 2 (low)	146.8[b] 138.8[b]	4.7 8.8	10	$p > 0.10$
9. C_2 vs 2 (high)	150.2[b] 141.3[b]	3.1 3.6	10	$p < 0.05$*
10. A_4 vs A + 4 (low)	10.2 10.5	7.9 3.8	10	$p > 0.10$
11. A_4 vs A + 4 (high)	5.4 6.8	2.1 1.9	10	$p > 0.10$
12. A_2 vs A + 2 (low)	19.7 34.3	8.3 5.1	10	$p < 0.10$
13. A_2 vs A + 2 (high)	10.2 26.5	3.7 5.4	10	$p < 0.01$**
14. 4 vs 2 (low)	3.3 8.1	3.0 4.4	10	$p > 0.10$
15. 4 vs 2 (high)	−0.2 8.9	2.1 3.5	10	$p < 0.025$*
16. A + 4 vs A + 2 (low)	10.5 34.3	3.8 5.1	10	$p < 0.005$**
17. A + 4 vs A + 2 (high)	6.8 26.5	1.9 5.4	10	$p < 0.005$**

[a] Abbreviations used: C-control animals; A, 4 and 2 – animals who received ara-A 4AAF or 2AAF (singly or in combination), respectively, as described in section 2.b. Low-dose of 10 mg/kg; high dose of 50 mg/kg. C_4, A_4, etc., refer to the untreated or ara A treated animals used as controls for animals treated with 4AAF or ara-A + 4AAF, etc.

[b] Average of '100˙log(D)' values. Other entries in this column are averages of 'Q_d' values. These units are defined in Materials and Methods.

The PYR and BP comparisons are shown in Table 1. As expected, neither the untreated control animals (lines 1 and 2) nor the ara-A treated control animals (lines 3 and 4) used in the PYR and BP experiments showed any statistically significant differences. However, a significant amount of strand breakage was caused by the ara-A alone (line 5 in Table 1). No difference from untreated controls was seen with either PYR or BP, at either low or high dose (lines 6–9). A marginally significant difference with a high dose of PYR was seen in ara-A treated animals (line 11). A highly significant amount of DNA damage was seen only with a high dose of BP in the presence of ara-A, as compared with animals receiving ara-A only (line 13) or ara-A + P (line 17). The conclusion from this set of experiments is that the genotoxic effect of BP can be distinguished from that of PYR, but only at the high dose (100 mg/kg) in the presence of a repair inhibitor, ara-A.

The 4AAF and 2AAF comparisons are shown in Table 2. Again, neither the untreated control animals (lines 1 and 2) nor the ara-A treated control animals (lines 3 and 4) used for the 4AAF and 2AAF experiments showed any statistically significant differences, but again DNA strand breakage was seen by ara-A treatment alone (line 5 in Table 2). A difference from untreated controls and from 4AAF was seen with 2AAF, but only at a high dose (lines 9 and 15). This difference was greater and statistically more highly significant when ara-A was used (lines 13 and 17) and, furthermore, a significant effect at the low dose of 2AAF was seen in one comparison (line 16), and a marginally significant effect was seen in the comparison of line 12. The conclusion from this set of experiments is that 2AAF can be distinguished from 4AAF, but only at the high dose (50 mg/kg); in the presence of the repair inhibitor, ara-A, the difference is accentuated, permitting detection of the effects of a low dose (10 mg/kg) of 2AAF.

The *in vivo* use of a repair inhibitor to enhance the sensitivity of a strand-break assay such as FADU or alkaline elution should have wide application. We did not determine if strand breaks could be detected in organs other than liver. Our presumption is that the effectiveness of ara-A is due to interference with the closing of strand breaks introduced by the incision step of a repair enzyme, although other mechanisms are possible. Thus FADU, which is a sensitive assay for alkali-labile bonds and frank breaks, can also be used to detect base damage that can be recognized by cellular repair enzymes having an incision step. Our present results indicate that a hard organ such as liver, as well as soft organs such as thymus, spleen, lymph node and bone marrow, can be used for analysis of *in vivo* DNA damage. However, liver initially showed greater variability in the number of DNA strand breaks seen in control animals, as compared with the soft organs; this improved as we developed more experience in preparing liver homogenates. Intact hepatocytes prepared by a collagenase method could be a useful alternative to explore, since the FADU method has already been used successfully with such a preparation (Althaus *et al.*, 1982). In conclusion, with ara-A as a repair inhibitor, this short-term test (3 h animal treatment plus 4 h analysis) shows promise as a system for screening of genotoxins since our results correlated well with the number of base adducts in mouse liver DNA as measured by Randerath (this volume).

5 Conclusions

Of the 4 chemicals tested, only 2AAF (at 50 mg/kg) produced a detectable amount of genotoxicity when administered to animals. However, administration of ara-A (50 mg/kg), a repair inhibitor, allowed detection of genotoxicity by BP (at 100 mg/kg) and 2AAF (at both 10 and 50 mg/kg). Thus, the carcinogenic chemicals could be distinguished from their presumed noncarcinogenic counterparts in this short-term *in vivo* test by their ability to produce genotoxicity.

6 Acknowledgements

I thank Mrs. W.C. Wolfgram for her skillful technical assistance in the execution of these experiments.

7 References

Althaus, F.R., S.D. Lawrence, G.L. Sattler, D.G. Longfellow and H.C. Pitot (1982) Chemical quantification of unscheduled DNA synthesis in cultured hepatocytes as an assay for the rapid screening of potential chemical carcinogens, *Cancer Res.*, **42**, 3010–3015.

Birnboim, H.C. and J.J. Jevcak (1981) Fluorometric method for rapid detection of DNA strand breaks in human white blood cells produced by low doses of radiation, *Cancer Res.*, **41**, 1889–1892.

McLean, J.R., R.S. McWilliams, J.G. Kaplan and H.C. Birnboim (1982) Rapid detection of DNA strand breaks in human peripheral white blood cells and animal organs following treatment with physical and chemical agents. In K.C. Bora, G.R. Douglas, and E.R. Nestmann (Eds), *Progress in Mutations Research*, Vol. 3, Elsevier, Amsterdam, pp. 137–141.

McWilliams, R.S., W.G. Cross, J.G., Kaplan and H.C. Birnboim (1983) Rapid rejoining of DNA strand breaks in resting human lymphocytes after irradiation by low doses of Co-60 gamma rays or 14.6 MeV neutrons, *Radiation Res.*, **94**, 499–507.

Author Index

Subject Index